Intermediate
Algebra

SIXTH EDITION

Charles P. McKeague

Cuesta College

SAUNDERS COLLEGE PUBLISHING

A DIVISION OF HARCOURT COLLEGE PUBLISHERS

FORT WORTH   PHILADELPHIA   SAN DIEGO   NEW YORK   AUSTIN   ORLANDO   SAN ANTONIO

TORONTO   MONTREAL   LONDON   SYDNEY   TOKYO

Publisher: Emily Barrosse
Executive Editor: Angus McDonald
Product Manager: Julia Downs
Developmental Editor: Carol Loyd
Project Editor: Robin C. Bonner
Production Manager: Alicia Jackson
Art Director: Lisa Adamitis
Text Designer: Lisa Adamitis

Cover Credit: Vienna, Austria (©Murat Ayranci/SuperStock)

INTERMEDIATE ALGEBRA 6/e

ISBN: 0-03-026286-0
Library of Congress Catalog Card Number: 99-067247

Address for domestic orders:
Saunders College Publishing, 6277 Sea Harbor Drive, Orlando, FL 32887-6777
1-800-782-4479
email: collegesales@harcourt.com

Address for international orders:
International Customer Service, Harcourt Brace & Company
6277 Sea Harbor Drive, Orlando, FL 32887-6777
(407) 345-3800
Fax (407) 345-4060
email: hbintl@harcourtbrace.com

Address for editorial correspondence:
Saunders College Publishing, Public Ledger Building, Suite 1250,
150 S. Independence Mall West, Philadelphia, PA 19106-3412

Web Site Address
http://www.harcourtcollege.com

Printed in the United States of America

9012345678 032 10987654321

SAUNDERS
COLLEGE
PUBLISHING

soon to become

A Harcourt Higher Learning Company

Soon you will find Saunders College Publishing's distinguished innovation, leadership, and support under a different name . . . a new brand that continues our unsurpassed quality, service, and commitment to education.

We are combining the strengths of our college imprints into one worldwide brand: ◀Harcourt Our mission is to make learning accessible to anyone, anywhere, anytime—reinforcing our commitment to lifelong learning.

We'll soon be Harcourt College Publishers. Ask for us by name.

One Company
"Where Learning Comes to Life."

CONTENTS

Rational Expressions *348*

Rational Exponents and Roots *426*

Quadratic Functions *494*

Exponential and Logarithmic Functions *566*

Sequences and Series   *622*

Conic Sections   *666*

Appendix A
Synthetic Division   *A.1*

PREFACE TO THE INSTRUCTOR

This sixth edition of Intermediate Algebra retains the same basic format and style as the fifth edition. The book is intended to be used in a lecture format class. Each section of the book can be discussed in a forty-five- to fifty-minute class session.

FEATURES OF THE BOOK

Early Coverage of Graphing   The material on graphing equations in two variables has been moved from Chapter 5 to Chapter 3 and combined with an introduction to functions and function notation.

Early Coverage of Functions   The material on functions and graphs has been rewritten for this edition. It starts in Section 3.1 with coverage of paired data, bar charts, and line graphs. Functions are introduced in Section 3.5 and function notation in Section 3.6. The concept of a function is then carried through to the remaining chapters in the book.

Using Technology   Scattered throughout the book is material that shows how graphing calculators, spreadsheet programs, and computer graphing programs can be used to enhance the topics being covered. This material is easy to find because it appears in boxes under the heading "Using Technology."

Blueprint for Problem Solving   The Blueprint for Problem Solving is a detailed outline of the steps needed to successfully attempt application problems. Intended as a guide to problem solving in general, the Blueprint overlays the solution process to all the application problems in the first few chapters of the book. As students become more familiar with problem solving, the steps in the Blueprint are streamlined.

Increased Visualization of Topics   This edition contains many more diagrams, charts, and graphs than the previous edition. The purpose is to give the students additional information, in visual form, to help them understand the topics we cover.

Facts from Geometry   Many important facts from geometry are listed under this heading. In most cases, an example or two accompanies each of the facts, to give students a chance to see how topics from geometry are related to the algebra they are learning.

Conditional Statements   An introductory look at the type of reasoning that is the foundation of mathematics is contained in Section 1.5. This section gives students experience with conditional statements and their associated forms: inverse, converse, and contrapositive.

Number Sequences and Inductive Reasoning   An introductory coverage of number sequences and inductive reasoning is contained in Section 1.6. This material is carried into Chapter 3 and other parts of the book. It is expanded upon in Chapter 10. I find that there are many interesting topics I can cover if the students have some experience with number sequences. It is also the easiest way to demonstrate inductive reasoning.

Unit Analysis   Chapter 6 contains problems requiring students to convert from one unit of measure to another. The method used to accomplish the conversions is the method they will use if they take a chemistry class. Since this method is the method we use to multiply rational expressions, unit analysis is covered in Section 6.7 as an application of multiplication and division of rational expressions.

Getting Ready for Class   Just before each problem set is a list of four problems under the heading *Getting Ready for Class*. These problems require written responses from students and can be answered by reading the preceding section. They are to be done before the students come to class.

Chapter Openings   Each chapter opens with an introduction in which a real-world application is used to stimulate interest in the chapter. Most of them are explained using the rule of four: in words, numerically, graphically, and algebraically.

Study Skills   Found in the first six chapters, after the opening application, is a list of study skills intended to help students become organized and efficient with their time. The study skills point students in the direction of success.

Organization of the Problem Sets   Six main ideas are incorporated into the problem sets.

1. **Drill**   There are enough problems in each set to ensure student proficiency in the material.

2. **Progressive Difficulty**   The problems increase in difficulty as the problem set progresses.

3. **Odd-Even Similarities**   Each pair of consecutive problems is similar. Since the answers to the odd-numbered problems are listed in the back of the book, the similarity of the odd-even pairs allows your students to check their work on an odd-numbered problem and then to try the similar even-numbered problem.

4. **Applying-the-Concepts Problems**   Students are always curious about how the algebra they are learning can be applied, but at the same time many of them are apprehensive about attempting application problems. I have found that they are more likely to put some time and effort into trying application problems if they do not have to work an overwhelming number of them at one time and if they work on them every day. For these reasons, I have placed a few application problems in almost every problem set in the book.

5. **Review Problems**   Each problem set, beginning with Chapter 2, contains a few review problems. Where appropriate, the review problems cover material that will be needed in the next section. Otherwise, they cover material from the pre-

vious chapter. That is, the review problems in Chapter 5 cover the important material from Chapter 4. Likewise, the review problems in Chapter 6 review the important material from Chapter 5. If you give tests on two chapters at a time, you will find this to be a timesaving feature. Your students will review one chapter as they study the next.

6. **Extending-the-Concepts Problems**   Many of the problem sets end with a few problems under this heading. These problems are more challenging than those in the problem sets, or they are problems that extend some of the topics covered in the section.

End-of-Chapter Retrospective   Each chapter ends with the following items, which together give a comprehensive reexamination of the chapter and some of the important problems from the previous chapters.

Chapter Summary   This lists all main points from the chapter. In the margin, next to each topic being reviewed, is an example that illustrates that type of problem associated with the topics being reviewed.

Chapter Review   The Chapter Review is a set of review problems covering the essential ideas of the chapter. Numbers in brackets refer to the section(s) in the text where similar problems can be found.

Chapter Projects   The Projects include one group project, for students to work on in class, and one research project, for students to do outside of class.

Chapter Test   The Chapter Test is a set of problems representative of all the main points of the chapter.

Cumulative Review   Starting in Chapter 2, each chapter ends with a set of problems that reviews material from all preceding chapters. The cumulative review keeps students current with past topics and helps them retain the information they study.

CHANGES IN THE SIXTH EDITION

Chapter 1   Section 1.7 has been deleted and the material there has been distributed among Sections 1.6, 2.4, and 3.1.

Chapter 2   The material in Sections 2.4 and 2.5 is now contained in a single section, Section 2.4. Section 2.8 has been eliminated and the material there integrated into Chapter 3.

Chapter 3   Graphing in two dimensions has been moved from Chapter 5 to Chapter 3 and now includes an introduction to functions and function notation.

Chapter 4   Systems of equations have been moved from Chapter 8 to Chapter 4. The section on matrix solutions to systems of equations has been moved to the appendix.

Chapter 5   There are two fewer sections in this chapter than in the previous edition. The properties of exponents are covered in one section at the beginning of Chapter 5, instead of two sections, and special factoring and the general review of factoring have been combined into a single section.

Chapter 8   The section on solving quadratic inequalities has been moved to the end of the chapter, so that technology can be used in the solution process. The section at the end of the chapter on graphing parabolas that opened left and right has been removed from this edition.

Chapter 9   Exponential functions and inverse functions have been combined with the material on logarithms. Together they make up Chapter 9 of this edition.

Chapter 11   The material on conic sections has been moved to Chapter 11, the last chapter of the book.

SUPPLEMENTS TO THE TEXTBOOK

This sixth edition of *Intermediate Algebra* is accompanied by a number of useful supplements.

For the Instructor

- ***Printed Test Bank and Prepared Tests***   The test bank consists of multiple-choice and short-answer test items organized by chapter, section, and difficulty level. The prepared tests comprise 12 sets of ready-to-copy tests, one set for each chapter and one set for the entire book. Each set comprises 2 multiple-choice and 4 show-your-work tests. Answers for every test item are provided. Also included are 15 sample problems for each section of the text. These problems range in difficulty from easy to hard.

- ***ExaMaster+***™ ***Computerized Test Bank***   A flexible, powerful, computerized testing system, *Examaster+*™ offers teachers a wide range of integrated testing options and features. Available in Macintosh or Windows format, it offers teachers the ability to select, edit, and create not only test items but algorithms for test items as well. Teachers can tailor tests according to a variety of criteria, scramble the order of test items, and administer tests on-line. *Examaster+*™ also includes full-function gradebook and graphing features.

For the Student

- ***Videotape Package***   Free to adopters, the videotape package consists of 11 VHS videotapes, one for each chapter of the book. Each chapter tape is an hour to an hour and a half in length and is divided into lessons that correspond to each section of the chapter. In the videotapes, I work out selected problems from the text.

- ***Core Concepts Video***   This single videotape is over 4 hours in length and contains more than 50 problem-solving sessions, covering most sections of the text. Tailor-made as a take-home tutorial for students with a demanding schedule, this video can be used as a preview of what is to be covered in class, as an aid to completing homework assignments, or as a tool to aid in review for a test.

- **Student Solutions Manual** This manual contains complete annotated solutions to every other odd problem in problem sets and to all chapter review and chapter test problems.

- **MathCue Tutorial** This computer software package of tutorials has problems that correspond to every section in the text. The software presents problems to solve and tutors students by displaying annotated, step-by-step solutions. Students may view partial solutions to get started on a problem, see continuous record of progress, and back up to review missed problems. Students' scores can also be printed. Available for Windows and Macintosh.

- **MathCue Solution Finder** This software allows students to enter their own problems into the computer and obtain annotated, step-by-step solutions in return. This unique program simulates working with a tutor, tracks students' progress, refers students to specific sections in the text when appropriate, and prints students' scores. Available for Windows and Macintosh.

- **MathCue Practice** This algorithm-based software allows students to generate large numbers of practice problems keyed to problem types from each section of the book. *Practice* scores students' performance and saves students' scores session to session. Available for Windows and Macintosh.

- **MathCue F/C Graph** For use with *Intermediate Algebra* 6th edition, this computer program allows students to graph and analyze any polynomial, logarithmic, exponential, or trigonometric function or conic equation they choose. *F/C Graph* can zoom, trace, display function values or coordinates of selected points, graph up to four functions simultaneously, and save and retrieve set-ups. Students can use *F/C Graph* to relate algebraic and visual forms of functions and conic sections and to explore how changing parameters affect the graph. The set of computer lab exercises accompanying the software directs student investigations of a variety of topics. Available for Windows and Macintosh.

- **Intermediate Algebra and the Graphing Calculator: A Learning Resource** This workbook helps both instructors and students understand how to use graphing calculators and presents exercises and exploratory investigations into graphing calculators, linear equations and inequalities, polynomials and factoring, lines and inequalities, quadratic equations, and systems of linear equations.

Saunders College Publishing, a division of Harcourt College Publishers, may provide complimentary instructional aids and supplements or supplement packages to those adopters qualified under our adoption policy. Please contact your sales representative for more information. If as an adopter or potential user you receive supplements you do not need, please return them to your sales representative or send them to:

Attn: Returns Department
Troy Warehouse
465 South Lincoln Drive
Troy, MO 63379

ACKNOWLEDGEMENTS

A project of this size cannot be completed without help from many people. Angus McDonald, my editor at Saunders College Publishing, has been very helpful and encouraging in all parts of the revision process. Carol Loyd, my developmental editor, worked especially hard to get the book ready for production. Anne Scanlan-Rohrer and Robin Bonner handled the production part of this revision. They each did an outstanding job, and I am very lucky to have these two on our team. Judy Barclay and Denise Chellsen, my colleagues at Cuesta College, suggested many new exercises for the problem sets. Warren Hawley, the Latin School of Chicago and DePaul University, Mary Lee Seitz, Erie Community College, Mike Rosenborg, and Héctor Muñoz of Santiago, Chile, submitted many new application problems that strengthened this revision. My son, Patrick, assisted me with this revision, and his influence on the use of technology and new application problems has made this a better book. Ann Ostberg and Sandra Beken did an excellent job of accuracy checking and proofreading. I also received a number of excellent suggestions from Mark Turner, Pat Hughes, Steve Herbekian, and Rebecca Otteson, my colleagues at Cuesta College. My thanks to all these people; this book would not have been possible without them.

Thanks also to Diane McKeague and Amy Jacobs for their encouragement with all my writing endeavors.

Finally, I am grateful to the following instructors for their suggestions and comments on this revision. Some reviewed the entire manuscript, while others were asked to evaluate the development of specific topics or the overall sequence of topics. My thanks go to the following:

Susan Akers, Northeast State Technical Community College
Gerald Allen, Angelo State University
Jon Becker, Indiana University Northwest
Sandra Beken, Horry-Georgetown Technical College
Sandra Belcher, Midwestern State University
Karen Sue Cain, Eastern Kentucky University
Pat Roux, Delgado Community College
Margaret Donaldson, East Tennessee State University
Dorothy Gotway, University of Missouri — St. Louis
Cathy Hayes, University of Mobile
Josephine Johansen, Rutgers University
Harriet Kiser, Floyd College
Gene Majors, Fullerton College
Aimee Martin, Amarillo College
Annette Noble, University of Maryland — Eastern Shore
Richard Riggs, New Jersey City University
Margaret Schmid, Black Hawk College
Mark Sigfrids, Kalamazoo Valley Community College
Mark Serebransky, Camden County College
Heidi Staebler, Texas A&M University — Commerce
Katalin Szucs, East Carolina University
Radu Teodorescu, Western Michigan University
Lucy Thrower, Francis Marion University
Pete Witt, Glendale Community College
Alice Wong, Miami-Dade Community College

Charles P. McKeague — October, 1999

PREFACE TO THE STUDENT

Many of my algebra students are apprehensive at first because they are worried they will not understand the topics we cover. When I present a new topic that they do not grasp completely, they think something is wrong with them for not understanding it. On the other hand, some students are excited about the course from the beginning. They are not worried about understanding algebra, and, in fact, expect to find some topics difficult.

What is the difference between these two types of students? Those who are excited about the course know from experience (as you do) that a certain amount of confusion is associated with most new topics in mathematics. They don't worry about it because they also know that the confusion gives way to understanding in the process of reading the textbook, working the problems, and getting questions answered. If they find a topic they are having difficulty with, they work as many problems as necessary to grasp the subject. They don't wait for the understanding to come to them; they go out and get it by working lots of problems. In contrast, the students who lack confidence tend to give up when they become confused. Instead of working more problems, they sometimes stop working problems altogether, and that, of course, guarantees that they will remain confused.

If you are worried about this course because you lack confidence in your ability to understand algebra, and you want to change the way you feel about mathematics, then look forward to the first topic that causes confusion. As soon as that topic comes along, make it your goal to master it, in spite of your apprehension. You will see that each and every topic covered in this course is one you can eventually master, even if your initial introduction to it is accompanied by some confusion. As long as you have passed a college-level beginning algebra course (or its equivalent), you are ready to take this course. If you have decided to do well in algebra, the following list will be important to you:

HOW TO BE SUCCESSFUL IN ALGEBRA

1. **Attend all class sessions on time.** You cannot know exactly what goes on in class unless you are there. Missing class and then expecting to find out what went on from someone else is not the same as being there yourself.

2. **Read the book.** It is the best to read the section that will be covered in class beforehand. Reading in advance, even if you do not understand everything you read, is still better than going to class with no idea of what will be discussed.

3. **Work problems every day, and check your answers.** The key to success in mathematics is working problems. The more problems you work, the better you will become at working them. The answers to the odd-numbered problems are given in the back of the book. When you have finished an assignment, be sure to compare your answers with those in the book. If you have made a mistake, find out what it is, and correct it.

4. **Do it on your own.**   Don't be misled into thinking someone else's work is your own. Having someone else show you how to work a problem is not the same as working the same problem yourself. It is okay to get help when you are stuck. As a matter of fact, it is a good idea. Just be sure you do the work yourself.

5. **Review every day.**   After you have finished the problems your instructor has assigned, take another fifteen minutes and review a section you have already completed. The more you review, the longer you will retain the material you have learned.

6. **Don't expect to understand every topic the first time you see it.**   Sometimes you will understand everything you are doing, and sometimes you won't. That's just the way things are in mathematics. Expecting to understand each new topic the first time you see it can lead to disappointment and frustration. The process of understanding algebra takes time. It requires that you read the book, work problems, and get your questions answered.

7. **Spend as much time as it takes for you to master the material.**   No set formula exists for the exact amount of time you need to spend on algebra to master it. You will find out as you go along what is or isn't enough time for you. If you end up spending two or more hours on each section in order to master the material there, then that's how much time it takes; trying to get by with less will not work.

8. **Relax.**   It's probably not as difficult as you think.

Charles P. McKeague — October, 1999

FEATURES OF THE NEW EDITION

New Design and Art Program

The revamped look of this edition is more open and inviting. Featuring numerous new mathematical and situational figures, the enhanced art program will assist students in visualizing the concepts.

Equations and Inequalities in Two Variables

(© SuperStock)

INTRODUCTION

A student is heating water in a chemistry lab. As the water heats, she records the temperature readings from two thermometers, one giving temperature in degrees Fahrenheit and the other in degrees Celsius. Table 1 shows some of the data she collects. Figure 1 is a scatter diagram that gives a visual representation of the data in Table 1.

TABLE 1	Corresponding Temperatures
In Degrees Fahrenheit	In Degrees Celsius
77	25
95	35
167	75
	100

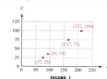

FIGURE 1

...tionship between the Fahrenheit and Celsius temperature scales is ...rmula

$$C = \frac{5}{9}(F - 32)$$

... ways to describe the relationship between the two temperature ... a graph, and an equation. But, most important to us, we don't need ...formula on faith. In Problem Set 3.3, you will derive the formula ... Table 1.

Linear Inequalities in Two Variables

A small movie theater holds 100 people. The owner charges more for adults than for children, so it is important to know the different combinations of adults and children that can be seated at one time. The shaded region in Figure 1 contains all the seating combinations. The line $x + y = 100$ shows the combinations for a full theater: The y-intercept corresponds to a theater full of adults, and the x-intercept corresponds to a theater full of children. In the shaded region below the line $x + y = 100$ are the combinations that occur if the theater is not full.

Shaded regions like the one shown in Figure 1 are produced by linear inequalities in two variables, which is the topic of this section.

FIGURE 1

A *linear inequality in two variables* is any expression that can be put in the form

$$ax + by < c$$

where a, b, and c are real numbers (a and b not both 0). The inequality symbol can be any one of the following four: $<, \leq, >, \geq$.

Some examples of linear inequalities are

$$2x + 3y < 6 \qquad y \geq 2x + 1 \qquad x - y \leq 0$$

Although not all of these examples have the form $ax + by < c$, each one can be put in that form.

The solution set for a linear inequality is a *section of the coordinate plane*. The *boundary* for the section is found by replacing the inequality symbol with an equal sign and graphing the resulting equation. The boundary is included in the solution set (and is represented with a *solid line*) if the inequality symbol used originally is $\leq$ or $\geq$. The boundary is not included (and is represented with a *broken line*) if the original symbol is $<$ or $>$.

EXAMPLE 1 Graph the solution set for $x + y \leq 4$.

Solution The boundary for the graph is the graph of $x + y = 4$. The boundary is included in the solution set because the inequality symbol is $\leq$.

Figure 2 is the graph of the boundary:

173

Chapter Openers Each chapter begins with a real-world application, helping students to make the connection immediately between mathematics and their everyday lives. Similar problems on a variety of subjects also appear in the problem sets and end-of-chapter retrospectives.

Basic Properties and Definitions

(Bruce Ayers/Tony Stone Images)

INTRODUCTION

The following table and diagram show how the concentration of a popular antidepressant changes over time, once the patient stops taking it. In this particular case, the concentration in the patient's system is 80 ng/ml (nanograms per milliliter) when the patient stops taking the antidepressant, and the half-life of the antidepressant is 5 days.

TABLE 1	Concentration of an Antidepressant
Days Since Discontinuing	**Concentration (ng/ml)**
0	80
5	40
10	20
15	10
20	5

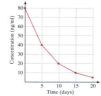

FIGURE 1 Concentration of antidepressant over time

The half-life of a medication tells how quickly the medication is eliminated from a person's system: Medications with a long half-life are eliminated slowly, whereas those with a short half-life are more quickly eliminated. Half-life is the key to constructing the preceding table and graph. When you are finished with Section 1.6 you ... the half-life of a medication to construct the table and graph.

STUDY SKILLS

At the beginning of each of the first few chapters of this book, you will find a section like this, in which we list the skills necessary for success in algebra. If you have just completed an introductory algebra class successfully, you have acquired most of these skills. If it has been some time since you last took a math class, you must pay attention to the sections on study skills.

Here is a list of things you can do to begin to develop effective study skills.

1. **Put Yourself on a Schedule** The general rule is that you spend 2 hours on homework for every hour you are in class. Make a schedule for yourself in which you set aside 2 hours each day to work on algebra. Once you make the schedule, stick to it. Don't just complete your assignments and stop. Use all the time you have set aside. If you complete an assignment and have time left over, read the next section in the book and work more problems. As the course progresses, you may find that 2 hours a day is not enough time for you to master the material in this course. If it takes you longer than 2 hours a day to reach your goals for this course, then that's how much time it takes. Trying to get by with less will not work.

2. **Find Your Mistakes and Correct Them** There is more to studying algebra than just working problems. You must always check your answers with the answers in the back of the book. When you make a mistake, find out what it is and correct it. Making mistakes is part of the process of learning mathematics. One of the number sequences we will study in this chapter is known as the *Fibonacci sequence*, after the Italian mathematician Leonardo Fibonacci (c. 1170–c. 1250). Fibonacci published a book titled *The Book of Squares* in 1225. In the Prologue he has this to say about his writing:

   I have come to request indulgence if in any place it contains something more or less than right or necessary; for to remember everything and be mistaken in nothing is divine rather than human . . .

   Fibonacci knew, as you know, that human beings make mistakes. You cannot learn algebra without making mistakes. The key to discovering what you do not understand can be found by correcting your mistakes.

3. **Imitate Success** Your work should look like the work you see in this book and the work your instructor shows. The steps shown in solving problems in this book were written by someone who has been successful in mathematics. The same is true of your instructor. Your work should imitate the work of people who have been successful in mathematics.

Study Skills List of skills necessary for success in algebra; teaches students to become organized and efficient with their time. The Study Skills appear in Chapters 1 through 6.

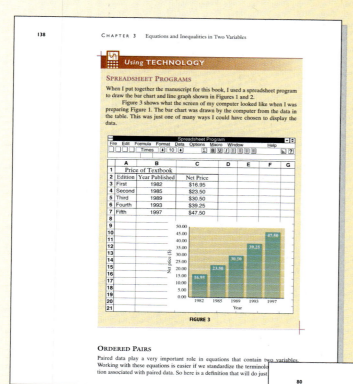

Using Technology This material illustrates the ways in which graphing calculators, spreadsheet programs, and computer graphing programs can be used to enhance the topics covered. An icon makes the information easy to locate.

Facts from Geometry Important facts and associated examples are included under this heading, illustrating to students how geometry relates to the algebra they are learning.

---

80   CHAPTER 2   Equations and Inequalities in One Variable

Solution Substituting 486.7 for S, 5 for r, and 3.14 for π into the formula $S = \pi r^2 + 2\pi rh$, we have

$$486.7 = (3.14)(5^2) + 2(3.14)(5)h$$

$$486.7 = 78.5 + 31.4h$$

$$408.2 = 31.4h \qquad \text{Add } -78.5 \text{ to each side.}$$

$$13 = h \qquad \text{Divide each side by 31.4.}$$

The height of each can is 13 centimeters.

Surface area = 486.7 cm²

FACTS FROM Geometry

Formulas for Area and Perimeter

To review, here are the formulas for the area and perimeter of some common geometric objects.

A Square

Perimeter = $4s$
Area = s^2

A Rectangle

Perimeter = $2l + 2w$
Area = lw

A Triangle

Perimeter = $a + b + c$
Area = $\frac{1}{2}bh$

The formula for perimeter gives us the distance around the outside of the object along its sides, while the formula for area gives us a measure of the amount of surface the object covers.

Note: The pink line labeled h in the triangle is its height, or altitude. It extends from the top of the triangle down to the base, meeting the base at an angle of 90°. The altitude of a triangle is always perpendicular to the base. The small square shown where the altitude meets the base is used to indicate that the angle formed is 90°.

EXAMPLE 4 Given the formula $P = 2w + 2l$, solve for w.

Solution To solve for w, we must isolate it on one side of the equation. We can accomplish this if we delete the $2l$ term and the coefficient 2 from the right side of the equation.

Getting Ready for Class
New element designed to reinforce the concepts students have learned by reading the discourse, prior to attending class. Over 250 new writing exercises appear throughout the book under this heading. Students are asked to write on various topics using their own words, thus enhancing their understanding and helping to alleviate math anxiety.

EXAMPLES   Let $A = \{1, 3, 5\}$, $B = \{0, 2, 4\}$, and $C = \{1, 2, 3, \ldots\}$. Then

12. $A \cup B = \{0, 1, 2, 3, 4, 5\}$
13. $A \cap B = \varnothing$ (*A* and *B* have no elements in common)
14. $A \cap C = \{1, 3, 5\} = A$
15. $B \cup C = \{0, 1, 2, 3, \ldots\}$

Another notation we can use to describe sets is called *set builder* notation. Here is how we write our definition for the union of two sets A and B using set builder notation:

$$A \cup B = \{x \mid x \in A \text{ or } x \in B\}$$

The right side of this statement is read "the set of all *x* such that *x* is a member of *A* or *x* is a member of *B*." As you can see, the vertical line after the first *x* is read "such that."

EXAMPLE 16   If $A = \{1, 2, 3, 4, 5, 6\}$, find $C = \{x \mid x \in A \text{ and } x \geq 4\}$.

Solution   We are looking for all the elements of *A* that are also greater than or equal to 4. They are 4, 5, and 6. Using set notation, we have

$$C = \{4, 5, 6\}$$

 Getting Ready for Class

Each section of the book will end with some problems and questions like the ones that follow. They are for you to answer after you have read through the section, but before you go to class. All of them require that you give written responses, in complete sentences. Writing about mathematics is a valuable exercise. If you write with the intention of explaining and communicating what you know to someone else, you will find that you understand the topic you are writing about even better than you did before you started writing. As with all problems in this course, you want to approach these writing exercises with a positive point of view. You will get better at giving written responses to questions as you progress through the course. Even if you never feel comfortable writing about mathematics, just the process of attempting to do so will increase your understanding and ability in mathematics.

After reading through the preceding section, respond in your own words and in complete sentences.

A. What is the meaning of the expression 2^3?
B. Why do we have a rule for the order of operations?
C. What is the intersection of two sets of numbers?
D. Explain the operations that are associated with the words *sum, difference, product,* and *quotient.*

PROBLEM SET 3.7

For the following problems, *y* varies directly with *x*.

1. If *y* is 10 when *x* is 2, find *y* when *x* is 6.
2. If *y* is 20 when *x* is 5, find *y* when *x* is 3.
3. If *y* is -32 when *x* is 4, find *x* when *y* is -40.
4. If *y* is -50 when *x* is 5, find *x* when *y* is -70.

For the following problems, *r* is inversely proportional to *s*.

5. If *r* is -3 when *s* is 4, find *r* when *s* is 2.
6. If *r* is -10 when *s* is 6, find *r* when *s* is -5.
7. If *r* is 8 when *s* is 3, find *s* when *r* is 48.
8. If *r* is 12 when *s* is 5, find *s* when *r* is 30.

For the following problems, *d* varies directly with the square of *r*.

9. If $d = 10$ when $r = 5$, find *d* when $r = 10$.
10. If $d = 12$ when $r = 6$, find *d* when $r = 9$.
11. If $d = 100$ when $r = 2$, find *d* when $r = 3$.
12. If $d = 50$ when $r = 5$, find *d* when $r = 7$.

For the following problems, *y* varies inversely with the square of *x*.

13. If $y = 45$ when $x = 3$, find *y* when x is 5.
14. If $y = 12$ when $x = 2$, find *y* when *x* is 6.
15. If $y = 18$ when $x = 3$, find *y* when *x* is 2.
16. If $y = 45$ when $x = 4$, find *y* when *x* is 5.

For the following problems, *z* varies jointly with *x* and the square of *y*.

17. If *z* is 54 when *x* and *y* are 3, find *z* when $x = 2$ and $y = 4$.
18. If *z* is 80 when *x* is 5 and *y* is 2, find *z* when $x = 2$ and $y = 5$.
19. If *z* is 64 when $x = 1$ and $y = 4$, find *x* when $z = 32$ and $y = 1$.
20. If *z* is 27 when $x = 6$ and $y = 3$, find *x* when $z = 50$ and $y = 4$.

Applying the Concepts

21. **Length of a Spring**   The length a spring stretches is directly proportional to the force applied. If a force of 5 pounds stretches a spring 3 inches, how much force is necessary to stretch the same spring 10 inches?

22. **Weight and Surface Area**   The weight of a certain material varies directly with the surface area of that material. If 8 square feet weighs half a pound, how much will 10 square feet weigh?

23. **Pressure and Temperature**   The temperature of a gas varies directly with its pressure. A temperature of 200 K produces a pressure of 50 pounds per square inch.
    (a) Find the equation that relates pressure and temperature.
    (b) Graph the equation from part (a) in the first quadrant only.
    (c) What pressure will the gas have at 280 K?

24. **Circumference and Diameter**   The circumference of a wheel is directly proportional to its diameter. A wheel has a circumference of 8.5 feet and a diameter of 2.7 feet.
    (a) Find the equation that relates circumference and diameter.
    (b) Graph the equation from part (a) in the first quadrant only.
    (c) What is the circumference of a wheel that has a diameter of 11.3 feet?

25. **Volume and Pressure**   The volume of a gas is inversely proportional to the pressure. If a pressure of 36 pounds per square inch corresponds to a volume of 25 cubic feet, what pressure is needed to produce a volume of 75 cubic feet?

26. **Wave Frequency**   The frequency of an electromagnetic wave varies inversely with the wavelength. If a wavelength of 200 meters has a frequency of 800 kilocycles per second, what frequency will be associated with a wavelength of 500 meters?

27. **f-Stop and Aperture Diameter**   The relative aperture, or f-stop, for a camera lens is inversely proportional to the diameter of the aperture. An f-stop of 2 corresponds to an aperture diameter of 40 millimeters for the lens on an automatic camera.
    (a) Find the equation that relates f-stop and diameter.
    (b) Graph the equation from part (a) in the first quadrant only.

Problem Sets
Just the right number of drill problems is included in each problem set to ensure student proficiency in the material. The problems are progressive in level of difficulty, and each pair of consecutive problems is similar.

Section 3.3   Problem Set   **171**

33.

34.

35. Give the slope and y-intercept of $y = -2$. Sketch the graph.

36. For the line $x = -3$, sketch the graph, give the slope, and name any intercepts.

37. Find the equation of the line parallel to the graph of $3x - y = 5$ that contains the point $(-1, 4)$.

38. Find the equation of the line parallel to the graph of $2x - 4y = 5$ that contains the point $(0, 3)$.

39. Line l is perpendicular to the graph of the equation $2x - 5y = 10$ and contains the point $(-4, -3)$. Find the equation for l.

40. Line l is perpendicular to the graph of the equation $-3x - 5y = 2$ and contains the point $(2, -6)$. Find the equation for l.

41. Give the equation of the line perpendicular to the graph of $y = -4x + 2$ that has an x-intercept of -1.

42. Write the equation of the line parallel to the graph of $7x - 2y = 14$ that has an x-intercept of 5.

43. Find the equation of the line with x-intercept 3 and y-intercept 2.

44. Find the equation of the line with x-intercept 2 and y-intercept 3.

Applying the Concepts

45. Deriving the Temperature Equation The table below resembles the table from the introduction to this section. The rows of the table give us ordered pairs (C, F).

Degrees Celsius	Degrees Fahrenheit
C	F
0	32
25	77
50	122
75	167
100	212

(a) Use any two of the ordered pairs from the table to derive the equation $F = \frac{9}{5}C + 32$.

(b) Use the equation from part (a) to find the Fahrenheit temperature that corresponds to a Celsius temperature of 30°.

46. Maximum Heart Rate The table gives the maximum heart rate for adults 30, 40, 50, and 60 years old. Each row of the table gives us an ordered pair (A, M).

Age (years)	Maximum Heart Rate (beats per minute)
A	M
30	190
40	180
50	170
60	160

(a) Use any two of the ordered pairs from the table to derive the equation $M = 220 - A$, which gives the maximum heart rate M for an adult whose age is A.

Applying the Concepts These problems are designed to show students how the algebra they are learning can be applied to real-life situations. Over 400 new application problems have been added, and approximately 100 of these emphasize reading and interpreting graphical and tabular data, in keeping with AMATYC standards.

CHAPTER 5   Exponents and Polynomials

Factor the right side of this equation, and then find h when t is 6 seconds and when t is 3 seconds. [Find $h(6)$ and $h(3)$.]

78. Height of an Arrow An arrow is shot into the air with an upward velocity of 16 feet per second from a hill 32 feet high. The equation that gives the height of the arrow at any time t is

$$h(t) = 32 + 16t - 16t^2$$

Factor the right side of this equation, and then find h when t is 2 seconds and when t is 1 second. [Find $h(2)$ and $h(1)$.]

Review Problems

The following problems review material we covered in Section 5.3. Reviewing these problems will help you with the next section.

Multiply.

79. $(2x - 3)(2x + 3)$ **80.** $(4 - 5x)(4 + 5x)$
81. $(2x - 3)^2$ **82.** $(4 - 5x)^2$
83. $(2x - 3)(4x^2 + 6x + 9)$
84. $(2x + 3)(4x^2 - 6x + 9)$

Extending the Concepts

Factor completely.

85. $8x^6 + 26x^3y^2 + 15y^4$ **86.** $24x^4 + 6x^2y^3 - 45y^6$
87. $3x^2 + 295x - 500$ **88.** $3x^2 + 594x - 1,200$
89. $\frac{1}{8}x^2 + x + 2$ **90.** $\frac{1}{9}x^2 + x + 2$
91. $2x^2 + 1.5x + 0.25$ **92.** $6x^2 + 2x + 0.16$

93. Factoring and Area The following area model gives us a way to visualize the factorization of the trinomial $x^2 + 5x + 6$. Construct a similar diagram that will allow you to visualize the factorization of $x^2 + 5x + 4$.

94. Factoring and Area Refer to the area model for factoring trinomials mentioned in the preceding exercise. Write the factoring problem represented by each of the following diagrams.

(a)

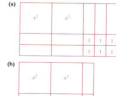

(b)

Extending the Concepts Many problem sets end with a group of these challenging exercises. Unlike any of the previous problems students have encountered in the chapter, these questions prompt them to synthesize ideas, thus extending their understanding of the topics.

Examples
The margins of the chapter summaries will be used for brief examples of the topics being reviewed,
whenever it is convenient.

1. $2^5 = 2 \cdot 2 \cdot 2 \cdot 2 \cdot 2 = 32$
   $5^2 = 5 \cdot 5 = 25$
   $10^3 = 10 \cdot 10 \cdot 10 = 1{,}000$
   $1^4 = 1 \cdot 1 \cdot 1 \cdot 1 = 1$

The numbers in brackets refer to the section(s) in which the topic can
be found.

Exponents [1.1]
Exponents represent notation used to indicate repeated multiplication.
In the expression 3^4, 3 is the *base* and 4 is the *exponent*.

$$3^4 = 3 \cdot 3 \cdot 3 \cdot 3 = 81$$

The expression 3^4 is said to be in *exponential form*, while the expression $3 \cdot 3 \cdot 3 \cdot 3$ is in *expanded form*.

2. $10 + (2 \cdot 3^2 - 4 \cdot 2)$
   $= 10 + (2 \cdot 9 - 4 \cdot 2)$
   $= 10 + (18 - 8)$
   $= 10 + 10$
   $= 20$

Order of Operations [1.1]
When evaluating a mathematical expression, we will perform the operations in the following order, beginning with the expression in the innermost parentheses or brackets and working our way out.

1. Simplify all numbers with exponents, working from left to right if
   more than one of these numbers is present.
2. Then, do all multiplications and divisions left to right.
3. Finally, perform all additions and subtractions left to right.

3. If $A = \{0, 1, 2\}$ and
   $B = \{2, 3\}$, then
   $A \cup B = \{0, 1, 2, 3\}$ and
   $A \cap B = \{2\}$.

Sets [1.1]
A *set* is a collection of objects or things.
   The *union* of two sets A and B, written $A \cup B$, is all the elements
that are in A *or* are in B.
   The *intersection* of two sets A and B, written $A \cap B$, is the set con-
sisting of all elements common to both A *and* B.
   Set A is a *subset* of set B, written $A \subset B$, if all elements in set A are
also in set B.

4. 5 is a counting number, a whole
   number, an integer, a rational number, and a real number.
   $\frac{3}{4}$ is a rational number and a
   real number.
   $\sqrt{2}$ is an irrational number and
   a real number.

Special Sets [1.2]
   Counting numbers $= \{1, 2, 3, \ldots\}$
   Whole numbers $= \{0, 1, 2, 3, \ldots\}$
   Integers $= \{\ldots, -3, -2, -1, 0, 1, 2, 3, \ldots\}$
   Rational numbers $= \{\frac{a}{b} \mid a \text{ and } b \text{ are integers}, b \neq 0\}$
   Irrational numbers $= \{x \mid x \text{ is real, but not rational}\}$
   Real numbers $= \{x \mid x \text{ is rational or } x \text{ is irrational}\}$
   Prime numbers $= \{2, 3, 5, 7, 11, \ldots\} = \{x \mid x \text{ is a positive inte-}$
   ger greater than 1 whose only positive divisors are itself and 1$\}$

59

Chapter Summary Highlights key
points of the chapter and illustrates the
concepts with example problems.
Contains a list of new properties and definitions.

---

5. The perimeter of a rectangle is
   32 inches. If the length is 3 times
   the width, find the dimensions.
   Step 1: This step is done mentally.
   Step 2: Let $x =$ the width.
   Then the length is $3x$.
   Step 3: The perimeter is 32;
   therefore,
   $$2x + 2(3x) = 32$$
   Step 4: $8x = 32$
   $$x = 4$$
   Step 5: The width is 4 inches.
   The length is
   $3(4) = 12$ inches.
   Step 6: The perimeter is
   $2(4) + 2(12)$, which is
   32. The length is 3
   times the width.

Blueprint for Problem Solving [2.3]

Step 1: **Read** the problem, and then mentally **list** the items that are
known and the items that are unknown.

Step 2: **Assign a variable** to one of the unknown items. (In most cases
this will amount to letting $x =$ the item that is asked for in the
problem.) Then **translate** the other **information** in the problem
to expressions involving the variable.

Step 3: **Reread** the problem, and then **write an equation**, using the
items and variables listed in steps 1 and 2, that describes the
situation.

Step 4: **Solve the equation** found in step 3.

Step 5: **Write** your **answer** using a complete sentence.

Step 6: **Reread** the problem, and **check** your solution with the original
words in the problem.

6. Adding 5 to both sides of the inequality $x - 5 < -2$ gives
   $$x - 5 + 5 < -2 + 5$$
   $$x < 3$$

   [number line marked 0 and 3]

Addition Property for Inequalities [2.4]
For expressions A, B, and C,

   if $A < B$

   then $A + C < B + C$

Adding the same quantity to both sides of an inequality never changes
the solution set.

7. Multiplying both sides of
   $-2x \geq 6$ by $-\frac{1}{2}$ gives
   $$-2x \geq 6$$
   $$-\frac{1}{2}(-2x) \leq -\frac{1}{2}(6)$$
   $$x \leq -3$$

   [number line marked -3 and 0]

Multiplication Property for Inequalities [2.4]
For expressions A, B, and C,

   if $A < B$

   then $AC < BC$ if $C > 0$ (C is positive)

   or $AC > BC$ if $C < 0$ (C is negative)

We can multiply both sides of an inequality by the same nonzero number without changing the solution set as long as each time we multiply
by a negative number we also reverse the direction of the inequality
symbol.

Blueprint for Problem Solving
A detailed outline of the steps needed to
successfully attempt application problems. Intended as a guide to problem
solving in general, the Blueprint overlays
the solution process to all the application
problems in the first few chapters of the
book. As students become more familiar
with problem solving, the steps in the
Blueprint are streamlined.

Common Mistakes Located at the end of each Chapter Summary, this section alerts students to problems they are likely to encounter and provides helpful hints for avoiding these pitfalls.

Chapter Review These exercises are representative of all the different problem types within a given chapter. The Chapter Reviews are longer and more extensive than the Chapter Tests.

---

16. The following sequence is a geometric sequence since each term is obtained from the previous term by multiplying by 3 each time.

$$4, 12, 36, 108, \ldots$$

Geometric Sequence [1.6]

A *geometric sequence* is a sequence of numbers in which each number (after the first number) comes from the number before it by multiplying by the same amount each time. The amount by which we multiply each term to obtain the next term is called the *common ratio*.

COMMON MISTAKES

1. Interpreting absolute value as changing the sign of the number inside the absolute value symbols. That is, $|-5| = +5, |+5| = -5$. To avoid this mistake, remember, absolute value is defined as a distance and distance is always measured in positive units.

2. Confusing $-(-5)$ with $-|-5|$. The first answer is $+5$, but the second answer is -5.

CHAPTER 1 REVIEW

The numbers in brackets refer to the section(s) in the text where similar problems can be found.

Translate each expression into symbols. [1.1]

1. The sum of x and 2

2. The difference of x and 2

3. The quotient of x and 2

4. The product of 2 and x

5. Twice the sum of x and y

6. The sum of twice x and y

Expand and multiply. [1.1]

7. 3^3 **8.** 5^3 **9.** 8^2

10. 1^8 **11.** 2^5 **12.** 3^4

Simplify each expression. [1.1]

13. $2 + 3 \cdot 5$ **14.** $10 - 2 \cdot 3$

15. $20 \div 2 + 3$ **16.** $30 \div 6 + 4 \div 2$

17. $3 + 2(5 - 2)$ **18.** $(10 - 2)(7 - 3)$

19. $3 \cdot 4^2 - 2 \cdot 3^2$ **20.** $3 + 5(2 \cdot 3^2 - 10)$

Let $A = \{1, 3, 5\}$, $B = \{2, 4, 6\}$, and $C = \{0, 1, 2, 3, 4\}$, and find each of the following. [1.1]

21. $A \cup B$ **22.** $A \cap C$

23. $\{x \mid x \in A \text{ and } x \notin C\}$

24. $\{x \mid x \in B \text{ and } x > 4\}$

25. Locate the numbers $-4, -2.5, -1, 0, 1.5, 3.1, 4.75$ on the number line. [1.2]

26. Locate the numbers $-4.75, -3, -1, 0, 1.5, 2.3, 3$ on the number line. [1.2]

Give the opposite and reciprocal of each number. [1.3]

27. 2   **28.** $-\dfrac{2}{5}$

Write each expression without absolute value symbols, then simplify. [1.3]

29. $|-3|$ **30.** $-|-5|$

31. $|-4|$ **32.** $|10 - 16|$

For the set $\{-7, -4.2, -\sqrt{3}, 0, \frac{3}{4}, \pi, 5\}$ list all the elements that are in the following sets. [1.2]

33. Integers   **34.** Rational numbers

35. Irrational numbers

36. Factor 4,356 into the product of prime factors. [1.2]

37. Reduce $\dfrac{4,356}{5,148}$ to lowest terms. [1.2]

Group Projects
Optional collaborative activities are included at the end of each chapter. These projects, intended to be completed in class, promote a discovery-based approach to learning.

---

64   CHAPTER I   Basic Properties and Definitions

For each conditional statement, assume that the statement is true, and then write another conditional statement that must also be true. [1.5]

99. If $x = 3$, then $x^4 = 81$.

100. If Maria goes to the beach, then she does not go to school.

Find the next number in each sequence. Identify any sequences that are arithmetic and any that are geometric. [1.6]

101. $11, 8, 5, 2, \ldots$

102. $10, -30, 90, -270, \ldots$

103. $1, 1, 2, 3, 5, \ldots$ **104.** $1, 16, 81, 256, \ldots$

105. $1, \dfrac{1}{2}, 0, -\dfrac{1}{2}, \ldots$ **106.** $1, \dfrac{1}{2}, \dfrac{1}{4}, \dfrac{1}{8}, \ldots$

⊕ CHAPTER 1   **PROJECTS**

BASIC PROPERTIES AND DEFINITIONS

Students and Instructors: The end of each chapter in this book will contain a page like this one containing two projects. The group project is intended to be done in class. The research projects are to be completed outside of class. They can be done in groups or individually. In my classes, I use the research projects for extra credit. I require all research projects to be done on a word processor and to be free of spelling errors.

 GROUP PROJECT

GAUSS'S METHOD FOR ADDING CONSECUTIVE INTEGERS

Number of People: 3

Time Needed: 15–20 minutes

Equipment: Paper and pencil

Background: There is a popular story about famous mathematician Karl Friedrich Gauss (1777–1855). As the story goes, a 9-year-old Gauss, and the rest of his class, was given what the teacher had hoped was busy work. They were told to find the sum of all the whole numbers from 1 to 100. While the rest of the class labored under the assignment, Gauss found the answer within a few moments. He may have set up the problem like this:

1	2	3	4	⋯	98	99	100
100	99	98	97	⋯	3	2	1

Karl Gauss, 1777–1855

Procedure: Copy the above illustration to get started.

1. Add the numbers in the column and write your answers below the line.
2. How many numbers do you have below the line?

Chapter I   Test   **65**

3. Suppose you add all the numbers below the line; how will your result be related to the sum of all the whole numbers from 1 to 100?

4. Use multiplication to add all the numbers below the line.

5. What is the sum of all the whole numbers from 1 to 100?

6. Use Gauss's method to add all the whole numbers from 1 to 10. (The answer is 55.)

7. Explain, in your own words, Gauss's method for finding this sum so quickly.

RESEARCH PROJECT   **BERTRAND RUSSELL**

To see the connection between the topics we studied in Section 1.5 and mathematics in general, here is a quote taken from the beginning of the first sentence in the book *Principles of Mathematics* by the British philosopher and mathematician, Bertrand Russell.

> Pure Mathematics is the class of all propositions of the form "*p* implies *q*," where *p* and *q* are propositions containing one or more variables . . .

He is using the phrase "*p* implies *q*" in the same way we use the phrase "If *A*, then *B*." Our work with conditional statements is actually an introduction to the foundations upon which all of mathematics is built.

Write an essay on the life of Bertrand Russell. In the essay, indicate what purpose he had for writing and publishing his book *Principles of Mathematics*. Write in complete sentences and organize your work just as you would if you were writing a paper for an English class.

Bertrand Russell *(Corbis/Bettmann-UPI)*

⊕ CHAPTER 1   **TEST**

The numbers in brackets indicate the section(s) to which the problems correspond.

Write each of the following in symbols. [1.1]

1. Twice the sum of $3x$ and $4y$
2. The difference of $2a$ and $3b$ is less than their sum

Simplify each expression using the rule for order of operations. [1.1]

3. $3 \cdot 2^2 + 5 \cdot 3^2$ 4. $6 + 2(4 \cdot 3 - 10)$

5. $12 - 8 \div 4 + 2 \cdot 3$

6. $20 - 4[3^2 - 2(2^3 - 6)]$

Research Projects
Designed to be completed outside of the classroom, these activities encourage students to read and write critically about mathematical topics. The Research Projects help to illustrate the connections between the concepts in a given chapter and their broader application.

CHAPTER 3   TEST

For each of the following straight lines, identify the x-intercept, y-intercept, and slope, and sketch the graph. [3.1–3.3]

1. $2x + y = 6$
2. $y = -2x - 3$
3. $y = \frac{3}{2}x + 4$
4. $x = -2$

Find the equation for each line. [3.3]

5. Give the equation of the line through $(-1, 3)$ that has slope $m = 2$.
6. Give the equation of the line through $(-3, 2)$ and $(4, -1)$.
7. Line l contains the point $(5, -3)$ and has a graph parallel to the graph of $2x - 5y = 10$. Find the equation for l.
8. Line l contains the point $(-1, -2)$ and has a graph perpendicular to the graph of $y = 3x - 1$. Find the equation for l.
9. Give the equation of the vertical line through $(4, -7)$.

Graph the following linear inequalities. [3.4]

10. $3x - 4y < 12$
11. $y \le -x + 2$

State the domain and range for the following relations, and indicate which relations are also functions. [3.5]

12. $\{(-2, 0), (-3, 0), (-2, 1)\}$
13. $y = x^2 - 9$

Let $f(x) = x - 2$, $g(x) = 3x + 4$, and $h(x) = 3x^2 - 2x - 8$, and find the following. [3.6]

14. $f(3) + g(2)$
15. $h(0) + g(0)$
16. $f[g(2)]$
17. $g[f(2)]$

Constructing a Box A piece of cardboard is in the shape of a square with each side 8 inches long. You are going to make a box out of the piece of cardboard by cutting four equal squares from each corner and then folding up the sides, as shown in Figure 1. Use this information in Problems 18–20. [3.5, 3.6]

FIGURE 1

18. If the length of each side of the squares you are cutting from the corners is represented by the variable x, state the restrictions on x.
19. Use function notation to write the volume of the resulting box $V(x)$ as a function of x.
20. Find $V(2)$, and explain what it represents.

Solve the following variation problems. [3.7]

21. **Direct Variation** Quantity y varies directly with the square of x. If y is 50 when x is 5, find y when x is 3.
22. **Joint Variation** Quantity z varies jointly with x and the cube of y. If z is 15 when x is 5 and y is 2, find z when x is 2 and y is 3.
23. **Maximum Load** The maximum load (L) a horizontal beam can safely hold varies jointly with the width (w) and the square of the depth (d) and inversely with the length (l). If a 10-foot beam with width 3 feet and depth 4 feet will safely hold up to 800 pounds, how many pounds will a 12-foot beam with width 3 feet and depth 4 feet hold?

Chapter Test This set of problems tests the student's understanding of all key points in the chapter.

CHAPTERS 1–3   CUMULATIVE REVIEW

Simplify.

1. $9 - 6 \div 2 + 2 \cdot 3$
2. $7(2x + 3) + 2(3x - 1)$
3. $12 - 7 - 2 - (-3)$
4. $-5(2x + 3) + 4x$
5. $5(3 - 7)^2 - 2(4 - 6)^3$
6. $\frac{3}{4} - \frac{2}{3}$

Solve.

7. $\frac{2}{3}a - 4 = 6$
8. $-6x - 7 = 3x + 2$
9. $-4 + 3(3x + 2) = 7$
10. $\frac{3}{4}(8x - 3) + \frac{3}{4} = 2$
11. $-4x + 9 = -3(-2x + 1) - 2$
12. $|-5x + 2| - 3 = -5$
13. $|2x - 3| - 7 = 1$
14. $|2y + 3| = |2y - 7|$
15. $|3x - 1| - 1 \le 2$
16. Solve for x: $ax + 4 = bx - 9$

Solve and graph the solution.

17. $3y - 6 \ge 3$ or $3y - 6 \le -3$
18. $|3x - 2| \le 7$
19. $|3x + 4| - 7 > 3$
20. $5 \le \frac{1}{4}x + 3 \le 8$

Graph each equation on a rectangular coordinate system.

21. $6x - 5y = 30$
22. $x = 2$
23. $y = \frac{1}{2}x - 1$
24. $2x - y \le -3$
25. Find the slope of the line through $(-3, -2)$ and $(-5, 3)$.
26. Find y if the line through $(-1, 2)$ and $(-3, y)$ has a slope of -2.
27. Find the slope of the line $y = 3$.
28. Give the equation of a line with slope $-\frac{3}{4}$ and y-intercept $= 2$.
29. Find the slope and y-intercept of $4x - 5y = 20$.

30. Find the equation of the line with x-intercept -2 and y-intercept -1.
31. Find the equation of the line that is parallel to $5x + 2y = -1$ that contains the points $(-4, -2)$.
32. Write the equation of the line in slope-intercept form given the slope is $\frac{2}{3}$ and if $(6, 8)$ is on the line.

If $f(x) = x^2 - 3x$, $g(x) = x - 1$, and $h(x) = 3x - 1$, find the solutions to Problems 33–35.

33. $h(0)$
34. $f(-1) + g(4)$
35. $f[g(-2)]$
36. Add $-\frac{3}{4}$ to the product of -3 and $\frac{5}{12}$.
37. Subtract $\frac{2}{5}$ from the product of -3 and $\frac{6}{15}$.
38. State the properties that justify $2 + (a + b) = (2 + b) + a$.
39. Reduce to lowest terms: $\frac{721}{927}$
40. Write in symbols: The sum of $3a$ and $4b$ is less than their difference.
41. If $A = \{0, 3, 6, 7\}$ and $B = \{1, 2, 3, 6\}$, find $A \cup B$.
42. Give the opposite and reciprocal of $-\frac{3}{4}$.
43. Identify hypothesis and conclusion: If a quadrangle is a square, then all its sides are equal.
44. Give the next number of the sequence and state whether it is arithmetic or geometric: -4, -1, 2, . . .
45. 12 is 4% of what number?
46. Write the first five terms of the sequence $a_n = (-3)^n$.
47. Specify the domain and range for the relation $\{(-1, 3), (2, -1), (3, 3)\}$. Is the relation also a function?
48. y varies inversely with the square of x. If $y = 4$ when $x = 4$, find y when $x = 6$.
49. **Geometry** The length of a rectangle is 3 feet more than twice the width. The perimeter is 48 feet. Find the dimensions.
50. **Geometry** Find all three angles in a triangle if the smallest angle is one-fifth the largest angle and the remaining angle is 12 degrees more than the smallest angle.

Cumulative Review Beginning with Chapter 2, this new element follows the Chapter Review and includes material from all preceding chapters, to help students retain information as they progress.

Basic Properties and Definitions

(Bruce Ayers/Tony Stone Images)

INTRODUCTION

The following table and diagram show how the concentration of a popular antidepressant changes over time, once the patient stops taking it. In this particular case, the concentration in the patient's system is 80 ng/ml (nanograms per milliliter) when the patient stops taking the antidepressant, and the half-life of the antidepressant is 5 days.

TABLE 1	Concentration of an Antidepressant
Days Since Discontinuing	**Concentration (ng/ml)**
0	80
5	40
10	20
15	10
20	5

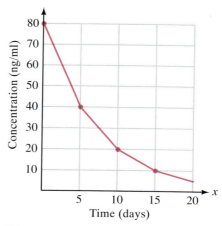

FIGURE 1 Concentration of antidepressant over time

The half-life of a medication tells how quickly the medication is eliminated from a person's system: Medications with a long half-life are eliminated slowly, whereas those with a short half-life are more quickly eliminated. Half-life is the key to constructing the preceding table and graph. When you are finished with Section 1.6 you will be able to use the half-life of a medication to construct the table and graph.

STUDY SKILLS

At the beginning of each of the first few chapters of this book, you will find a section like this, in which we list the skills necessary for success in algebra. If you have just completed an introductory algebra class successfully, you have acquired most of these skills. If it has been some time since you last took a math class, you must pay attention to the sections on study skills.

Here is a list of things you can do to begin to develop effective study skills.

1. **Put Yourself on a Schedule** The general rule is that you spend 2 hours on homework for every hour you are in class. Make a schedule for yourself in which you set aside 2 hours each day to work on algebra. Once you make the schedule, stick to it. Don't just complete your assignments and stop. Use all the time you have set aside. If you complete an assignment and have time left over, read the next section in the book and work more problems. As the course progresses, you may find that 2 hours a day is not enough time for you to master the material in this course. If it takes you longer than 2 hours a day to reach your goals for this course, then that's how much time it takes. Trying to get by with less will not work.

2. **Find Your Mistakes and Correct Them** There is more to studying algebra than just working problems. You must always check your answers with the answers in the back of the book. When you make a mistake, find out what it is and correct it. Making mistakes is part of the process of learning mathematics. One of the number sequences we will study in this chapter is known as the *Fibonacci sequence,* after the Italian mathematician Leonardo Fibonacci (c. 1170– c. 1250). Fibonacci published a book titled *The Book of Squares* in 1225. In the Prologue he has this to say about his writing:

   > I have come to request indulgence if in any place it contains something more or less than right or necessary; for to remember everything and be mistaken in nothing is divine rather than human . . .

   Fibonacci knew, as you know, that human beings make mistakes. You cannot learn algebra without making mistakes. The key to discovering what you do not understand can be found by correcting your mistakes.

3. **Imitate Success** Your work should look like the work you see in this book and the work your instructor shows. The steps shown in solving problems in this book were written by someone who has been successful in mathematics. The same is true of your instructor. Your work should imitate the work of people who have been successful in mathematics.

Fundamental Definitions and Notation

The diagram shown at the top of page 4 is called a *bar chart*. This one shows the net price of a popular intermediate algebra textbook. (The net price is the price the bookstore pays for the book.)

From the chart, we can find many relationships between numbers. We may notice that the price of the third edition was less than the price of the fourth edition. In mathematics we use symbols to represent relationships between quantities. If we let P represent the price of the book, then the relationship just mentioned, between the price of the third edition and the price of the fourth edition, can be written this way:

$$P(3) < P(4)$$

This section is, for the most part, simply a list of many of the basic symbols and definitions we will be using throughout the book.

COMPARISON SYMBOLS

In Symbols	*In Words*
$a = b$	a is equal to b
$a \neq b$	a is not equal to b
$a < b$	a is less than b
$a \leq b$	a is less than or equal to b
$a \not< b$	a is not less than b
$a > b$	a is greater than b
$a \geq b$	a is greater than or equal to b
$a \not> b$	a is not greater than b
$a \Leftrightarrow b$	a is equivalent to b

OPERATION SYMBOLS

Operation	*In Symbols*	*In Words*
Addition	$a + b$	The sum of a and b
Subtraction	$a - b$	The difference of a and b
Multiplication	ab, $a \cdot b$, $a(b)$, $(a)b$, $(a)(b)$, or $(a \times b)$	The product of a and b
Division	$a \div b$, a/b, or $\frac{a}{b}$	The quotient of a and b

The key words are *sum, difference, product,* and *quotient.* They are used frequently in mathematics. For instance, we may say the product of 3 and 4 is 12. We mean both the statements $3 \cdot 4$ and 12 are called the products of 3 and 4. The important idea here is that the word *product* implies multiplication, regardless of whether it is written $3 \cdot 4$, 12, 3(4), (3)4, or (3×4).

The following example shows how we translate sentences written in English into expressions written in symbols.

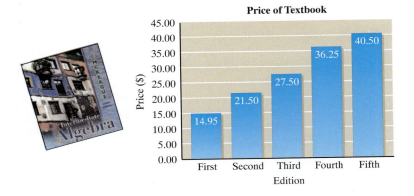

Price of Textbook

EXAMPLE I

In English	*In Symbols*
The sum of x and 5 is less than 2.	$x + 5 < 2$
The product of 3 and x is 21.	$3x = 21$
The quotient of y and 6 is 4.	$\frac{y}{6} = 4$
Twice the difference of b and 7 is greater than 5.	$2(b - 7) > 5$
The difference of twice b and 7 is greater than 5.	$2b - 7 > 5$

EXPONENTS

Consider the expression 3^4. The 3 is called the *base,* and the 4 is called the *exponent.* The exponent 4 tells us the number of times the base appears in the product. That is,

$$3^4 = 3 \cdot 3 \cdot 3 \cdot 3 = 81$$

The expression 3^4 is said to be in exponential form, while $3 \cdot 3 \cdot 3 \cdot 3$ is said to be in expanded form.

EXAMPLES   Expand and multiply.

2. $5^2 = 5 \cdot 5 = 25$ Base 5, exponent 2

3. $2^5 = 2 \cdot 2 \cdot 2 \cdot 2 \cdot 2 = 32$ Base 2, exponent 5

4. $4^3 = 4 \cdot 4 \cdot 4 = 64$ Base 4, exponent 3

ORDER OF OPERATIONS

It is important when evaluating arithmetic expressions in mathematics that each expression have only one answer in reduced form. Consider the expression

$$3 \cdot 7 + 2$$

If we find the product of 3 and 7 first, then add 2, the answer is 23. On the other hand, if we first combine the 7 and 2, then multiply by 3, we have 27. The problem seems to have two distinct answers depending on whether we multiply first or add first. To avoid this situation, we will decide that multiplication in a situation like this will always be done before addition. In this case, only the first answer, 23, is correct.

Here is the complete set of rules for evaluating expressions.

RULE (ORDER OF OPERATIONS)

When evaluating a mathematical expression, we will perform the operations in the following order, beginning with the expression in the innermost parentheses or brackets and working our way out.

1. Simplify all numbers with exponents, working from left to right if more than one of these expressions is present.
2. Then, do all multiplications and divisions left to right.
3. Perform all additions and subtractions left to right.

EXAMPLES   Simplify each expression using the rule for order of operations.

5. $5 + 3(2 + 4) = 5 + 3(6)$ Simplify inside parentheses.
$= 5 + 18$ Then, multiply.
$= 23$ Add.

6. $5 \cdot 2^3 - 4 \cdot 3^2 = 5 \cdot 8 - 4 \cdot 9$ Simplify exponents left to right.
$= 40 - 36$ Multiply left to right.
$= 4$ Subtract.

7. $20 - (2 \cdot 5^2 - 30) = 20 - (2 \cdot 25 - 30)$ Simplify inside parentheses,
$= 20 - (50 - 30)$ evaluating exponents first,
$= 20 - (20)$ then multiply, and finally
$= 0$ subtract.

8. $40 - 20 \div 5 + 8 = 40 - 4 + 8$ Divide first.
$= 36 + 8$ Then, add and subtract left to right.
$= 44$

9. $2 + 4[5 + (3 \cdot 2 - 2)] = 2 + 4[5 + (6 - 2)]$ Simplify inside
$= 2 + 4(5 + 4)$ innermost parentheses.
$= 2 + 4(9)$
$= 2 + 36$ Then, multiply.
$= 38$ Add.

The next concept we will cover, that of a set, can be considered the starting point for all the branches of mathematics.

SETS

DEFINITION   A **set** is a collection of objects or things. The objects in the set are called *elements*, or *members*, of the set.

Sets are usually denoted by capital letters and elements of sets by lowercase letters. We use braces, { }, to enclose the elements of a set.

To show that an element is contained in a set we use the symbol $\in$. That is,

$$x \in A \text{ is read "}x \text{ is an element (member) of set } A\text{"}$$

For example, if A is the set $\{1, 2, 3\}$, then $2 \in A$. On the other hand, $5 \notin A$ means 5 is not an element of set A.

> **DEFINITION** Set A is a **subset** of set B, written $A \subset B$, if every element in A is also an element of B. That is,
>
> $$A \subset B \qquad \text{if and only if} \qquad A \text{ is contained in } B$$

EXAMPLES

10. The set of numbers used to count things is $\{1, 2, 3, . . .\}$. The dots mean the set continues indefinitely in the same manner. This is an example of an *infinite* set.

11. The set of all numbers represented by the dots on the faces of a regular die is $\{1, 2, 3, 4, 5, 6\}$. This set is a subset of the set in Example 10. It is an example of a *finite* set, since it has a limited number of elements.

> **DEFINITION** The set with no members is called the **empty**, or **null**, **set**. It is denoted by the symbol $\varnothing$.
>
> The empty set is considered a subset of every set.

The diagrams shown here are called *Venn diagrams* after John Venn (1834–1923). They can be used to visualize operations with sets. The region inside the circle labeled A is set A; the region inside the circle labeled B is set B.

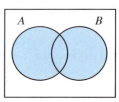

$A \cup B$

FIGURE 1 The *union* of two sets

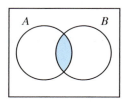

$A \cap B$

FIGURE 2 The *intersection* of two sets

OPERATIONS WITH SETS

Two basic operations are used to combine sets: union and intersection.

> **DEFINITION** The **union** of two sets A and B, written $A \cup B$, is the set of all elements that are either in A or in B, or in both A and B. The key word here is *or*. For an element to be in $A \cup B$ it must be in A or B. In symbols, the definition looks like this:
>
> $$x \in A \cup B \qquad \text{if and only if} \qquad x \in A \text{ or } x \in B$$

> **DEFINITION** The **intersection** of two sets A and B, written $A \cap B$, is the set of elements in both A and B. The key word in this definition is the word *and*. For an element to be in $A \cap B$ it must be in both A and B, or
>
> $$x \in A \cap B \qquad \text{if and only if} \qquad x \in A \text{ and } x \in B$$

EXAMPLES   Let $A = \{1, 3, 5\}$, $B = \{0, 2, 4\}$, and $C = \{1, 2, 3, \ldots\}$. Then

12. $A \cup B = \{0, 1, 2, 3, 4, 5\}$
13. $A \cap B = \varnothing$ (A and B have no elements in common)
14. $A \cap C = \{1, 3, 5\} = A$
15. $B \cup C = \{0, 1, 2, 3, \ldots\}$

Another notation we can use to describe sets is called *set builder* notation. Here is how we write our definition for the union of two sets A and B using set builder notation:

$$A \cup B = \{x \mid x \in A \text{ or } x \in B\}$$

The right side of this statement is read "the set of all x such that x is a member of A or x is a member of B." As you can see, the vertical line after the first x is read "such that."

EXAMPLE 16   If $A = \{1, 2, 3, 4, 5, 6\}$, find $C = \{x \mid x \in A \text{ and } x \geq 4\}$.

Solution   We are looking for all the elements of A that are also greater than or equal to 4. They are 4, 5, and 6. Using set notation, we have

$$C = \{4, 5, 6\}$$

Getting Ready for Class

Each section of the book will end with some problems and questions like the ones that follow. They are for you to answer after you have read through the section, but before you go to class. All of them require that you give written responses, in complete sentences. Writing about mathematics is a valuable exercise. If you write with the intention of explaining and communicating what you know to someone else, you will find that you understand the topic you are writing about even better than you did before you started writing. As with all problems in this course, you want to approach these writing exercises with a positive point of view. You will get better at giving written responses to questions as you progress through the course. Even if you never feel comfortable writing about mathematics, just the process of attempting to do so will increase your understanding and ability in mathematics.

After reading through the preceding section, respond in your own words and in complete sentences.

A. What is the meaning of the expression 2^3?
B. Why do we have a rule for the order of operations?
C. What is the intersection of two sets of numbers?
D. Explain the operations that are associated with the words *sum, difference, product,* and *quotient.*

PROBLEM SET 1.1

Translate each of the following statements into symbols.

1. The sum of x and 5 is 2.

2. The sum of y and -3 is 9.

3. The difference of 6 and x is y.

4. The difference of x and 6 is $-y$.

5. The product of t and 2 is less than y.

6. The product of $5x$ and y is equal to z.

7. The sum of x and y is less than the difference of x and y.

8. Twice the sum of a and b is 15.

9. Three times the difference of x and 5 is more than y.

10. The product of x and y is greater than or equal to the quotient of x and y.

Expand and multiply.

11. 6^2 **12.** 8^2 **13.** 10^2

14. 10^3 **15.** 2^3 **16.** 5^3

17. 2^4 **18.** 1^4 **19.** 10^4

20. 4^3 **21.** 11^2 **22.** 10^5

Simplify each expression using the rule for order of operations.

23. $3 \cdot 5 + 4$

24. $3 \cdot 7 - 6$

25. $3(5 + 4)$

26. $3(7 - 6)$

27. $2 + 8 \cdot 5$

28. $12 - 3 \cdot 3$

29. $(2 + 8)5$

30. $(12 - 3)3$

31. $6 + 3 \cdot 4 - 2$

32. $8 + 2 \cdot 7 - 3$

33. $6 + 3(4 - 2)$

34. $8 + 2(7 - 3)$

35. $(6 + 3)(4 - 2)$

36. $(8 + 2)(7 - 3)$

37. $(7 - 4)(7 + 4)$

38. $(8 - 5)(8 + 5)$

39. $7^2 - 4^2$

40. $8^2 - 5^2$

41. $2 + 3 \cdot 2^2 + 3^2$

42. $3 + 4 \cdot 4^2 + 5^2$

43. $2 + 3(2^2 + 3^2)$

44. $3 + 4(4^2 + 5^2)$

45. $(2 + 3)(2^2 + 3^2)$

46. $(3 + 4)(4^2 + 5^2)$

47. $40 - 10 \div 5 + 1$

48. $20 - 10 \div 2 + 3$

49. $(40 - 10) \div 5 + 1$

50. $(20 - 10) \div 2 + 3$

51. $(40 - 10) \div (5 + 1)$

52. $(20 - 10) \div (2 + 3)$

53. $24 \div 4 + 8 \div 2$

54. $36 \div 3 + 9 \div 3$

55. $24 \div (4 + 8) \div 2$

56. $36 \div (3 + 9) \div 3$

57. $5 \cdot 10^3 + 4 \cdot 10^2 + 3 \cdot 10 + 1$

58. $6 \cdot 10^3 + 5 \cdot 10^2 + 4 \cdot 10 + 3$

59. $40 - [10 - (4 - 2)]$

60. $50 - [17 - (8 - 3)]$

61. $40 - 10 - 4 - 2$

62. $50 - 17 - 8 - 3$

63. $3 + 2(2 \cdot 3^2 + 1)$

64. $4 + 5(3 \cdot 2^2 - 5)$

65. $(3 + 2)(2 \cdot 3^2 + 1)$

66. $(4 + 5)(3 \cdot 2^2 - 5)$

67. $3[2 + 4(5 + 2 \cdot 3)]$

68. $2[4 + 2(6 + 3 \cdot 5)]$

69. $6[3 + 2(5 \cdot 3 - 10)]$

70. $8[7 + 2(6 \cdot 9 - 14)]$

71. $5(7 \cdot 4 - 3 \cdot 4) + 8(5 \cdot 9 - 4 \cdot 9)$

72. $4(3 \cdot 9 - 2 \cdot 9) + 5(6 \cdot 8 - 5 \cdot 8)$

73. $5^3 + 4^3 \div 2^4 - 3^4$

74. $6^3 + 2^6 \div 4^2 - 3^4$

For the following problems, let $A = \{0, 2, 4, 6\}$, $B = \{1, 2, 3, 4, 5\}$, and $C = \{1, 3, 5, 7\}$.

75. $A \cup B$ **76.** $A \cap B$

77. $A \cap C$ **78.** $B \cup C$

79. $A \cup (B \cap C)$ **80.** $C \cup (A \cap B)$

81. $\{x \mid x \in A \text{ and } x < 4\}$

82. $\{x \mid x \in B \text{ and } x > 3\}$

83. $\{x \mid x \in A \text{ and } x \notin B\}$

84. $\{x \mid x \in B \text{ and } x \notin C\}$

85. $\{x \mid x \in A \text{ or } x \in C\}$

86. $\{x \mid x \in A \text{ or } x \in B\}$

Applying the Concepts

Median Incomes   The bar chart gives the top five median incomes for women age 30 and older in the United States, according to the field of study in which they earned a bachelor's degree. Use the chart to answer questions 87 and 88.

The following bar chart gives the top five median incomes for men age 30 and older in the United States, according to the field of study in which they earned a bachelor's degree. Use the chart to answer questions 89–92.

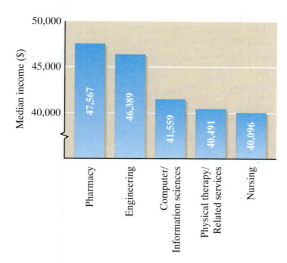

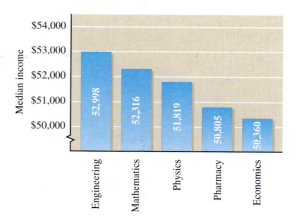

89. What is the difference, in dollars, between the highest median wage and the lowest median wage?

90. Find the difference in median wage between men studying engineering and men studying physics.

91. Using the information in both of the median income charts, find the difference between the median incomes for men studying engineering and women studying engineering.

87. What is the difference, in dollars, between the highest median wage and the lowest median wage?

92. What is the difference, in dollars, between the median incomes of men studying pharmacy and women studying pharmacy?

88. Find the difference in median wage between women studying engineering and women studying nursing.

The following bar chart provides information about the average per-student debt from federal student loans from community colleges, public four-year institutions, and private four-year institutions. The average was determined among those students who borrowed money from student loans and graduated in 1993 or 1996. Use the bar chart to solve Problems 93 to 96.

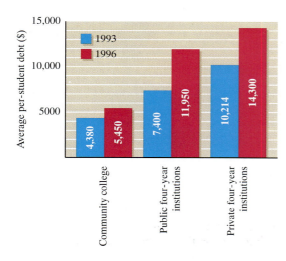

93. Determine the increase in average per-student debt from 1993 to 1996 at private institutions.

94. Determine the increase in average per-student debt from 1993 to 1996 at community colleges.

95. Determine the difference between the average per-student debt at public four-year institutions and community colleges in 1996.

96. Why is there such a large difference in average debt for students borrowing money at private four-year institutions compared with students at community colleges in both 1993 and 1996?

97. Ahmes Papyrus In approximately 1650 B.C. a mathematical document was written in ancient Egypt called the *Ahmes Papyrus*. An "exercise" in this document asked the reader to find "a quantity such that when it is added to one-fourth of itself results in 15." Verify that this quantity must be 12.

98. Ambiguous Statements In mathematics an ambiguous statement is a statement that has more than one meaning. Explain why the statement "three times a number decreased by x" is am-

biguous, but the statement "three times a number, decreased by x" is not ambiguous.

99. Order of Operations Place any of the symbols $+$, $-$, $\div$, $\cdot$ between the following numbers, so that the resulting statement is true. Also, you may use parentheses wherever you like.

a. 16　4　8　2 = 0　**e.** 16　4　8　2 = 5
b. 16　4　8　2 = 1　**f.** 16　4　8　2 = 8
c. 16　4　8　2 = 2　**g.** 16　4　8　2 = 10
d. 16　4　8　2 = 3　**h.** 16　4　8　2 = 64

100. Venn Diagrams The Swiss mathematican *Leonhard Euler* (1707–1783), one of the greatest mathematicians of all time, did much of his greatest work after he became totally blind. It was Euler who first used circles to represent sets and show the relationships between sets. In 1894, the English scholar *John Venn* improved on Euler's circles by drawing a rectangle around them. This rectangle represents the universal set, U, which is a large set that contains all of the sets under consideration. If $A = \{1, 2, 3, 4, 5\}$ and $B = \{2, 4, 6, 8, 10\}$, then Venn would have shown the relationship between these sets as follows:

Let $C = \{1, 5, 10, 15\}$, $D = \{10, 15, 20, 30\}$, and $E = (10, 100, 1{,}000\}$, and draw a Venn diagram for these sets.

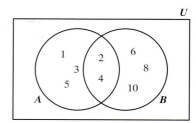

Extending the Concepts

Many of the problem sets in this book end with a few problems like the ones that follow. These problems challenge you to extend your knowledge of the material in the problem set. In most cases, there are no examples in the text similar to these problems. You should approach these problems with a positive point of view, because even though you may not complete them correctly, just the process of attempting them will increase your knowledge and ability in algebra.

The notation $n(A)$ is used to denote the *number* of elements in set A. For example, if $A = \{a, b, c\}$, then $n(A) = 3$.

101. If A and B are sets such that $n(A) = 4$, $n(B) = 5$, and $A \cap B = \varnothing$, find $n(A \cup B)$. (How many elements are in the union of A and B?)

102. In general, if $A \cap B = \varnothing$, what is $n(A \cap B)$?

103. If $n(A) = 7$, $n(B) = 8$, and $n(A \cap B) = 2$, find $n(A \cup B)$.

104. If $n(A) = 4$, $n(B) = 10$, and $n(A \cap B) = 3$, find $n(A \cup B)$.

1.2   The Real Numbers

THE REAL NUMBER LINE

The real number line is constructed by drawing a straight line and labeling a convenient point with the number 0. Positive numbers are in increasing order to the right of 0; negative numbers are in decreasing order to the left of 0. The point on the line corresponding to 0 is called the *origin*.

The numbers on the number line increase in size as we move to the right. When we compare the size of two numbers on the number line, the one on the left is always the smaller number.

The numbers associated with the points on the line are called *coordinates* of those points. Every point on the line has a number associated with it. The set of all these numbers makes up the set of real numbers.

> **DEFINITION**   A **real number** is any number that is the coordinate of a point on the real number line.

EXAMPLE 1   Locate the numbers -4.5, -0.75, $\frac{1}{2}$, $\sqrt{2}$, π, and 4.1 on the real number line.

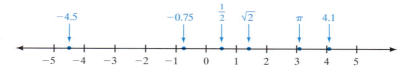

Note: In this book we will refer to real numbers as being on the real number line. Actually, real numbers are *not* on the line; only the points representing them are on the line. We can save some writing, however, if we simply refer to real numbers as being on the number line.

We can use the real number line to give a visual representation to inequality statements.

EXAMPLE 2   Graph $\{x \mid x \le 3\}$.

Solution We want to graph all the real numbers less than or equal to 3 — that is, all the real numbers below 3 and including 3. We label 0 on the number line for reference as well as 3, since the latter is what we call the endpoint. The graph is as follows:

We use a solid circle at 3, since 3 is included in the graph.

EXAMPLE 3   Graph $\{x \mid x < 3\}$.

Solution The graph is identical to the graph in Example 2 except at the endpoint 3. In this case we use an open circle, since 3 is not included in the graph.

In Section 1.1 we defined the **union** of two sets A and B to be the set of all elements that are in either A or B. The word *or* is the key word in the definition. The **intersection** of two sets A and B is the set of all elements contained in both A and B, the key word here being *and*. We can put the words *and* and *or* together with our methods of graphing inequalities to graph some **compound inequalities.**

EXAMPLE 4   Graph $\{x \mid x \le -2 \text{ or } x > 3\}$.

Solution The two inequalities connected by the word *or* are referred to as a **compound inequality.** We begin by graphing each inequality separately.

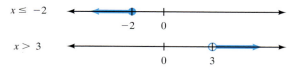

Since the two inequalities are connected by the word *or,* we graph their union. That is, we graph all points on either graph.

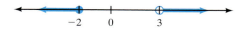

EXAMPLE 5   Graph $\{x \mid x > -1 \text{ and } x < 2\}$.

Solution We first graph each inequality separately.

Since the two inequalities are connected by the word *and,* we graph their intersection — the part they have in common.

....

Notation   Sometimes compound inequalities that use the word *and* as the connecting word can be written in a shorter form. For example, the compound inequality $-3 \leq x$ and $x \leq 4$ can be written $-3 \leq x \leq 4$. The word *and* does not appear when an inequality is written in this form. It is implied. Inequalities of the form $-3 \leq x \leq 4$ are called *continued inequalities.* This new notation is useful because it takes fewer symbols to write it. The graph of $-3 \leq x \leq 4$ is

EXAMPLE   6   Graph $\{x \mid 1 \leq x < 2\}$.

Solution   The word *and* is implied in the continued inequality $1 \leq x < 2$. That is, the continued inequality $1 \leq x < 2$ is equivalent to $1 \leq x$ and $x < 2$. Therefore, we graph all the numbers between 1 and 2 on the number line, including 1 but not including 2.

EXAMPLE   7   Graph $\{x \mid x < -2 \text{ or } 2 < x < 6\}$.

Solution   Here we have a combination of compound and continued inequalities. We want to graph all real numbers that are either less than -2 or between 2 and 6.

In addition to the phrases that translate directly into inequality statements, we have the following translations:

In Words	*In Symbols*
x is at least 40	$x \geq 40$
x is at most 30	$x \leq 30$
x is no more than 20	$x \leq 20$
x is no less than 10	$x \geq 10$
x is between 4 and 5	$4 < x < 5$

In the last case, we can include the endpoints 4 and 5 by saying "x is between 4 and 5, inclusive," which translates to $4 \leq x \leq 5$.

EXAMPLE   8   Suppose you have a part-time job that requires you to work at least 10 hours, but no more than 20 hours, each week. Use the letter t to write an inequality that shows the number of hours you work per week.

Solution   If t is at least 10 but no more than 20, then $10 \leq t$ and $t \leq 20$, or equivalently, $10 \leq t \leq 20$. Note that the word *but,* as used here, has the same meaning as the word *and.*

EXAMPLE 9   If the highest temperature on Tuesday was 76°F and the lowest temperature was 55°F, write an inequality using the letter x that gives the range of temperatures on Tuesday.

Solution   Since the smallest value of x is 55 and the largest value of x is 76, then $55 \leq x \leq 76$. We could say that the temperature on Tuesday was between 55°F and 76°F, inclusive.

SUBSETS OF THE REAL NUMBERS

Next, we consider some of the more important subsets of the real numbers. Each set listed here is a subset of the real numbers:

Counting (or **natural**) **numbers** = $\{1, 2, 3, \ldots\}$
Whole numbers = $\{0, 1, 2, 3, \ldots\}$
Integers = $\{\ldots, -3, -2, -1, 0, 1, 2, 3, \ldots\}$
Rational numbers = $\left\{\dfrac{a}{b} \,\middle|\, a \text{ and } b \text{ are integers, } b \neq 0\right\}$

Remember, the notation used to write the rational numbers is read "the set of numbers a/b, such that a and b are integers and b is not equal to 0." Any number that can be written in the form

$$\frac{\text{Integer}}{\text{Integer}}$$

is a rational number. Rational numbers are numbers that can be written as the ratio of two integers. Each of the following is a rational number:

$\dfrac{3}{4}$	Because it is the ratio of the integers 3 and 4
-8	Because it can be written as the ratio of -8 to 1
0.75	Because it is the ratio of 75 to 100 (or 3 to 4 if you reduce to lowest terms)
$0.333\ldots$	Because it can be written as the ratio of 1 to 3

Still other numbers on the number line are not members of the subsets we have listed so far. They are real numbers, but they cannot be written as the ratio of two integers. That is, they are not rational numbers. For that reason, we call them irrational numbers.

$$\text{Irrational numbers} = \{x \,|\, x \text{ is real, but not rational}\}$$

The following are irrational numbers:

$$\sqrt{2}, \qquad -\sqrt{3}, \qquad 4 + 2\sqrt{3}, \qquad \pi, \qquad \pi + 5\sqrt{6}$$

Note: We can find decimal approximations to some irrational numbers by using a calculator or a table. For example, on an eight-digit calculator

$$\sqrt{2} = 1.4142136$$

This is not exactly $\sqrt{2}$ but simply an approximation to it. There is no decimal that gives $\sqrt{2}$ exactly.

EXAMPLE 10   For the set $\{-5, -3.5, 0, \frac{3}{4}, \sqrt{3}, \sqrt{5}, 9\}$, list the numbers that are: (a) whole numbers, (b) integers, (c) rational numbers, (d) irrational numbers, and (e) real numbers.

Solution

(**a**) Whole numbers: 0, 9

(**b**) Integers: $-5, 0, 9$

(**c**) Rational numbers: $-5, -3.5, 0, \frac{3}{4}, 9$

(**d**) Irrational numbers: $\sqrt{3}, \sqrt{5}$

(**e**) They are all real numbers.

The following diagram gives a visual representation of the relationships among subsets of the real numbers.

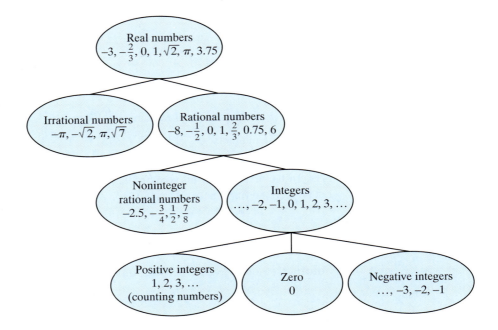

Prime Numbers and Factoring

The following diagram shows the relationship between multiplication and factoring:

Multiplication

$$\text{Factors} \longrightarrow 3 \cdot 4 = 12 \longleftarrow \text{Product}$$

Factoring

When we read the problem from left to right, we say the product of 3 and 4 is 12. Or we multiply 3 and 4 to get 12. When we read the problem in the other direction, from right to left, we say we have *factored* 12 into 3 times 4, or 3 and 4 are *factors* of 12.

The number 12 can be factored still further:

$$12 = 4 \cdot 3$$
$$= 2 \cdot 2 \cdot 3$$
$$= 2^2 \cdot 3$$

The numbers 2 and 3 are called *prime* factors of 12 because neither can be factored any further.

> **DEFINITION**   If a and b represent integers, then a is said to be a **factor** (or divisor) of b if a divides b evenly—that is, if a divides b with no remainder.

> **DEFINITION**
> A **prime** number is any positive integer larger than 1 whose only positive factors (divisors) are itself and 1. An integer greater than 1 that is not prime is said to be **composite**.

Here is a list of the first few prime numbers:

Prime numbers = {2, 3, 5, 7, 11, 13, 17, 19, 23, 29, 31, 37, 41, . . .}

When a number is not prime, we can factor it into the product of prime numbers. To factor a number into the product of primes, we simply factor it until it cannot be factored further.

EXAMPLE 11   Factor 525 into the product of primes.

Solution   Since 525 ends in 5, it is divisible by 5:

$$525 = 5 \cdot 105$$
$$= 5 \cdot 5 \cdot 21$$
$$= 5 \cdot 5 \cdot 3 \cdot 7$$
$$= 3 \cdot 5^2 \cdot 7$$

EXAMPLE 12   Reduce $\frac{210}{231}$ to lowest terms.

Solution   First we factor 210 and 231 into the product of prime factors. Then we reduce to lowest terms by dividing the numerator and denominator by any factors they have in common.

$$\frac{210}{231} = \frac{2 \cdot 3 \cdot 5 \cdot 7}{3 \cdot 7 \cdot 11} \qquad \text{Factor the numerator and denominator completely.}$$

$$= \frac{2 \cdot \cancel{3} \cdot 5 \cdot \cancel{7}}{\cancel{3} \cdot \cancel{7} \cdot 11} \qquad \text{Divide the numerator and denominator by } 3 \cdot 7.$$

$$= \frac{2 \cdot 5}{11}$$

$$= \frac{10}{11}$$

The small lines we have drawn through the factors that are common to the numerator and denominator are used to indicate that we have divided the numerator and denominator by those factors.

Getting Ready for Class

After reading through the preceding section, respond in your own words and in complete sentences.

A. Explain why some, but not all, rational numbers are also integers.

B. Explain the difference between a prime number and a composite number.

C. Give a written description of the set $\{x \mid -2 \le x < 3\}$.

D. What is an irrational number?

PROBLEM SET 1.2

Graph the following on the real number line.

1. $\{x \mid x < 1\}$
2. $\{x \mid x > -2\}$
3. $\{x \mid x \le 1\}$
4. $\{x \mid x \ge -2\}$
5. $\{x \mid x \ge 4\}$
6. $\{x \mid x \le -3\}$
7. $\{x \mid x > 4\}$
8. $\{x \mid x < -3\}$
9. $\{x \mid x > 0\}$
10. $\{x \mid x < 0\}$
11. $\{x \mid 3 \ge x\}$
12. $\{x \mid 4 \le x\}$

Graph the following compound inequalities.

13. $\{x \mid x \le -3 \text{ or } x \ge 1\}$
14. $\{x \mid x < 1 \text{ or } x > 4\}$
15. $\{x \mid -3 \le x \text{ and } x \le 1\}$
16. $\{x \mid 1 < x \text{ and } x < 4\}$
17. $\{x \mid -3 < x \text{ and } x < 1\}$
18. $\{x \mid 1 \le x \text{ and } x \le 4\}$

19. $\{x \mid x < -1 \text{ or } x \geq 3\}$

20. $\{x \mid x < 0 \text{ or } x \geq 3\}$

21. $\{x \mid x \leq -1 \text{ and } x \geq 3\}$

22. $\{x \mid x \leq 0 \text{ and } x \geq 3\}$

23. $\{x \mid x > -4 \text{ and } x < 2\}$

24. $\{x \mid x > -3 \text{ and } x < 0\}$

Graph the following continued inequalities.

25. $\{x \mid -1 \leq x \leq 2\}$ **26.** $\{x \mid -2 \leq x \leq 1\}$

27. $\{x \mid -1 < x < 2\}$ **28.** $\{x \mid -2 < x < 1\}$

29. $\{x \mid -4 < x \leq 1\}$ **30.** $\{x \mid -1 < x \leq 5\}$

Graph each of the following.

31. $\{x \mid x < -3 \text{ or } 2 < x < 4\}$

32. $\{x \mid -4 \leq x \leq -2 \text{ or } x \geq 3\}$

33. $\{x \mid x \leq -5 \text{ or } 0 \leq x \leq 3\}$

34. $\{x \mid -3 < x < 0 \text{ or } x > 5\}$

35. $\{x \mid -5 < x < -2 \text{ or } 2 < x < 5\}$

36. $\{x \mid -3 \leq x \leq -1 \text{ or } 1 \leq x \leq 3\}$

Translate each of the following phrases into an equivalent inequality.

37. x is at least 5

38. x is at least -2

39. x is no more than -3

40. x is no more than 8

41. x is at most 4

42. x is at most -5

43. x is between -4 and 4

44. x is between -3 and 3

45. x is between -4 and 4, inclusive

46. x is between -3 and 3, inclusive

For $\{-6, -5.2, -\sqrt{7}, -\pi, 0, 1, 2, 2.3, \frac{9}{2}, \sqrt{17}\}$, list all the elements of the set that are named in each of the following problems.

47. Counting numbers   **48.** Whole numbers

49. Rational numbers   **50.** Integers

51. Irrational numbers   **52.** Real numbers

53. Nonnegative integers   **54.** Positive integers

Factor each number into the product of prime factors.

55. 60   **56.** 154

57. 266   **58.** 385

59. 111   **60.** 735

61. 369   **62.** 1,155

Reduce each fraction to lowest terms.

63. $\frac{165}{385}$ **64.** $\frac{550}{735}$

65. $\frac{385}{735}$ **66.** $\frac{266}{285}$

67. $\frac{111}{185}$ **68.** $\frac{279}{310}$

69. $\frac{525}{630}$ **70.** $\frac{205}{369}$

Applying the Concepts

71. Name two numbers that are 5 units from 2 on the number line.

72. Name two numbers that are 6 units from -3 on the number line.

73. Suppose you have a job that requires that you work at least 20 hours but less than 40 hours per week. Write an inequality, using t, that gives the number of hours you work each week.

74. Suppose you study at least 2 hours but no more than 5 hours each night. Write an inequality, using t, that indicates the number of hours you study each night.

75. Write an inequality that gives all numbers that are less than 5 units from 8 on the number line.

76. Write an inequality that gives all numbers that are less than 3 units from 5 on the number line.

77. Write an inequality that gives all numbers that are more than 5 units from 8 on the number line.

78. Write an inequality that gives all numbers that are more than 3 units from 5 on the number line.

79. Goldbach's Conjecture   Goldbach's conjecture states that any even number (except 2) can be expressed as the sum of two prime numbers. (This conjecture has never been proven — that's why it is called a conjecture.) For example, $4 = 2 + 2$, $12 = 5 + 7$. Test Goldbach's conjecture by seeing if you are able to write the following even numbers as the sum of two primes:

(a) 8   (b) 16   (c) 24   (d) 36

80. Twin Primes   Any pair of prime numbers with a difference of 2 is called a twin prime. For example, 11 and 13 are twin primes, and so are 59 and 61. Mathematicians are still not sure if there are infinitely many pairs of twin primes. Find another pair of twin primes besides those already mentioned.

81. Perfect Numbers   More than 2,200 years ago, the Greek mathematician Euclid wrote: "A perfect number is a whole number that is equal to the sum of all its divisors, except itself." The first perfect number is 6 because $6 = 1 + 2 + 3$. Find the next perfect number.

Euclid *(Brown Brothers)*

82. Abundant Numbers A number is abundant if the sum of its divisors (except itself) is greater than the number. The number 12 is abundant because $12 < 1 + 2 + 3 + 4 + 6$. Find two more abundant numbers.

Extending the Concepts

The expression $n!$ is read "n factorial" and is the product of all the consecutive integers from n down to 1. For example,

$$1! = 1$$

$$2! = 2 \cdot 1 = 2$$

$$3! = 3 \cdot 2 \cdot 1 = 6$$

$$4! = 4 \cdot 3 \cdot 2 \cdot 1 = 24$$

83. Calculate 5!

84. Calculate 6!

85. Show that this statement is true: $6! = 6 \cdot 5!$

86. Show that the following statement is false: $(2 + 3)! = 2! + 3!$

1.3   ## Properties of Real Numbers

The area of the large rectangle shown here can be found in two ways: We can multiply its length a by its width $b + c$, or we can find the areas of the two smaller rectangles and add those areas to find the total area.

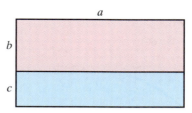

Area of large rectangle: $a(b + c)$
Sum of the areas of two smaller rectangles: $ab + ac$

Since the area of the large rectangle is the sum of the areas of the two smaller rectangles, we can write

$$a(b + c) = ab + ac$$

This expression is called the *distributive property*. It is one of the properties we will be discussing in this section. Before we arrive at the distributive property, we need to review some basic definitions and vocabulary.

OPPOSITES AND RECIPROCALS

> **DEFINITION**   Any two real numbers the same distance from 0, but in opposite directions from 0 on the number line, are called **opposites**, or *additive inverses*.

EXAMPLE 1   The numbers -3 and 3 are opposites. So are π and $-\pi$, $\frac{3}{4}$ and $-\frac{3}{4}$, and $\sqrt{2}$ and $-\sqrt{2}$.

The negative sign in front of a number can be read in a number of different ways. It can be read as "negative" or "the opposite of." We say -4 is the opposite of 4, or negative 4. The one we use will depend on the situation. For instance, the expression $-(-3)$ is best read "the opposite of negative 3." Since the opposite of -3 is 3, we have $-(-3) = 3$. In general, if a is any positive real number, then

$$-(-a) = a \qquad \text{(The opposite of a negative is a positive.)}$$

REVIEW OF MULTIPLICATION WITH FRACTIONS

Before we go further with our study of the number line, we need to review multiplication with fractions. Recall that for the fraction $\frac{a}{b}$, a is called the numerator and b is called the denominator. To multiply two fractions we simply multiply numerators and multiply denominators.

EXAMPLES   Multiply.

2. $\dfrac{3}{5} \cdot \dfrac{7}{8} = \dfrac{3 \cdot 7}{5 \cdot 8} = \dfrac{21}{40}$

3. $8 \cdot \dfrac{1}{5} = \dfrac{8}{1} \cdot \dfrac{1}{5} = \dfrac{8 \cdot 1}{1 \cdot 5} = \dfrac{8}{5}$

4. $\left(\dfrac{2}{3}\right)^4 = \dfrac{2}{3} \cdot \dfrac{2}{3} \cdot \dfrac{2}{3} \cdot \dfrac{2}{3} = \dfrac{16}{81}$

Note: In past math classes you may have written fractions like $\frac{8}{5}$ (improper fractions) as mixed numbers, such as $1\frac{3}{5}$. In algebra, it is usually better to leave them as improper fractions.

The idea of multiplication of fractions is useful in understanding the concept of the reciprocal of a number. Here is the definition.

> **DEFINITION**   Any two real numbers whose product is 1 are called **reciprocals**, or *multiplicative inverses*.

EXAMPLES   Give the reciprocal of each number.

	Number	*Reciprocal*	
5.	3	$\frac{1}{3}$	Because $3 \cdot \frac{1}{3} = \frac{3}{1} \cdot \frac{1}{3} = \frac{3}{3} = 1$
6.	$\frac{1}{6}$	6	Because $\frac{1}{6} \cdot 6 = \frac{1}{6} \cdot \frac{6}{1} = \frac{6}{6} = 1$
7.	$\frac{4}{5}$	$\frac{5}{4}$	Because $\frac{4}{5} \cdot \frac{5}{4} = \frac{20}{20} = 1$
8.	a	$\frac{1}{a}$	Because $a \cdot \frac{1}{a} = \frac{a}{1} \cdot \frac{1}{a} = \frac{a}{a} = 1$ $(a \neq 0)$

Although we will not develop multiplication with negative numbers until later in this chapter, you should know that the reciprocal of a negative number is also a negative number. For example, the reciprocal of -5 is $-\frac{1}{5}$.

THE ABSOLUTE VALUE OF A REAL NUMBER

> **DEFINITION**   The **absolute value** of a number (also called its *magnitude*) is the distance the number is from 0 on the number line. If x represents a real number, then the absolute value of x is written $|x|$.

This definition of absolute value is geometric in form since it defines absolute value in terms of the number line. Here is an alternative definition of absolute value that is algebraic in form since it involves only symbols.

> **ALTERNATIVE DEFINITION**   If x represents a real number, then the **absolute value** of x is written $|x|$, and is given by
> $$|x| = \begin{cases} x & \text{if } x \geq 0 \\ -x & \text{if } x < 0 \end{cases}$$

If the original number is positive or 0, then its absolute value is the number itself. If the number is negative, its absolute value is its opposite (which must be positive).

Note: It is important to recognize that if x is a real number, $-x$ is not necessarily negative. For example, if x is 5, then $-x$ is -5. On the other hand, if x were -5, then $-x$ would be $-(-5)$, which is 5.

EXAMPLES   Write each expression without absolute value symbols.

9. $|5| = 5$ **10.** $|-2| = 2$

11. $\left|-\frac{1}{2}\right| = \frac{1}{2}$ **12.** $-|-3| = -3$

13. $-|5| = -5$ **14.** $-|-\sqrt{2}| = -\sqrt{2}$

PROPERTIES OF REAL NUMBERS

We know that adding 3 and 7 gives the same answer as adding 7 and 3. The order of two numbers in an addition problem can be changed without changing the result. This fact about numbers and addition is called the **commutative property of addition.**

For all the properties listed in this section, *a*, *b*, and *c* represent real numbers.

COMMUTATIVE PROPERTY OF ADDITION

In symbols: $a + b = b + a$

In words: The *order* of the numbers in a sum does not affect the result.

COMMUTATIVE PROPERTY OF MULTIPLICATION

In symbols: $a \cdot b = b \cdot a$

In words: The *order* of the numbers in a product does not affect the result.

EXAMPLES

15. The statement $3 + 7 = 7 + 3$ is an example of the commutative property of addition.

16. The statement $3 \cdot x = x \cdot 3$ is an example of the commutative property of multiplication.

The other two basic operations (subtraction and division) are not commutative. If we change the order in which we are subtracting or dividing two numbers, we change the result.

Another property of numbers you have used many times has to do with grouping. When adding $3 + 5 + 7$, we can add the 3 and 5 first and then the 7, or we can add the 5 and 7 first and then the 3. Mathematically, it looks like this: $(3 + 5) + 7 = 3 + (5 + 7)$. Operations that behave in this manner are called *associative* operations.

ASSOCIATIVE PROPERTY OF ADDITION

In symbols: $a + (b + c) = (a + b) + c$

In words: The *grouping* of the numbers in a sum does not affect the result.

ASSOCIATIVE PROPERTY OF MULTIPLICATION

In symbols: $a(bc) = (ab)c$

In words: The *grouping* of the numbers in a product does not affect the result.

The following examples illustrate how the associative properties can be used to simplify expressions that involve both numbers and variables.

EXAMPLES   Simplify by using the associative property.

17. $2 + (3 + y) = (2 + 3) + y$ Associative property

$\qquad\qquad = 5 + y$ Addition

18. $5(4x) = (5 \cdot 4)x$ Associative property

$\qquad\quad = 20x$ Multiplication

19. $\dfrac{1}{4}(4a) = \left(\dfrac{1}{4} \cdot 4\right)a$ Associative property

$\qquad\qquad = 1a$ Multiplication

$\qquad\qquad = a$

Our next property involves both addition and multiplication. It is called the *distributive property* and is stated as follows.

DISTRIBUTIVE PROPERTY

In symbols: $a(b + c) = ab + ac$

In words: Multiplication *distributes* over addition.

You will see as we progress through the book that the distributive property is used very frequently in algebra. To see that the distributive property works, compare the following:

$$
\begin{array}{ll}
3(4 + 5) & 3(4) + 3(5) \\
3(9) & 12 + 15 \\
27 & 27
\end{array}
$$

In both cases the result is 27. Since the results are the same, the original two expressions must be equal; or, $3(4 + 5) = 3(4) + 3(5)$.

EXAMPLES   Apply the distributive property to each expression and then simplify the result.

20. $5(4x + 3) = 5(4x) + 5(3)$ Distributive property

$\qquad\qquad\;\; = 20x + 15$ Multiplication

21. $6(3x + 2y) = 6(3x) + 6(2y)$ Distributive property

$\qquad\qquad\;\; = 18x + 12y$ Multiplication

22. $\dfrac{1}{2}(3x + 6) = \dfrac{1}{2}(3x) + \dfrac{1}{2}(6)$ Distributive property

$\qquad\qquad\;\; = \dfrac{3}{2}x + 3$ Multiplication

23. $2(3y + 4) + 2 = 2(3y) + 2(4) + 2$ Distributive property
$\qquad\qquad\quad = 6y + 8 + 2$ Multiplication
$\qquad\qquad\quad = 6y + 10$ Addition

Note: Although the properties we are listing are stated for only two or three real numbers, they hold for as many numbers as needed. For example, the distributive property holds for expressions like $3(x + y + z + 2)$. That is,

$$3(x + y + z + 2) = 3x + 3y + 3z + 6$$

COMBINING SIMILAR TERMS

The distributive property can also be used to combine similar terms. (For now, a term is the product of a number with one or more variables.) Similar terms are terms with the same variable part. The terms $3x$ and $5x$ are similar, as are $2y$, $7y$, and $-3y$, because the variable parts are the same.

EXAMPLES   Use the distributive property to combine similar terms.

24. $3x + 5x = (3 + 5)x$ Distributive property
$\qquad\quad = 8x$ Addition

25. $3y + y = (3 + 1)y$ Distributive property
$\qquad\quad = 4y$ Addition

REVIEW OF ADDITION WITH FRACTIONS

To add fractions, each fraction must have the same denominator.

> **DEFINITION**   The **least common denominator** (LCD) for a set of denominators is the smallest number divisible by *all* the denominators.

The first step in adding fractions is to find a common denominator for all the denominators. We then rewrite each fraction (if necessary) as an equivalent fraction with the common denominator. Finally, we add the numerators and reduce to lowest terms if necessary.

EXAMPLE 26   Add $\frac{5}{12} + \frac{7}{18}$.

Solution   The least common denominator for the denominators 12 and 18 must be the smallest number divisible by both 12 and 18. We can factor 12 and 18 completely and then build the LCD from these factors.

$$12 = 2 \cdot 2 \cdot 3 \Big\}$$
$$18 = 2 \cdot 3 \cdot 3 \Big\}$$

12 divides the LCD

$$LCD = 2 \cdot 2 \cdot 3 \cdot 3 = 36$$

18 divides the LCD

Next, we rewrite our original fractions as equivalent fractions with denominators of 36. To do so, we multiply each original fraction by an appropriate form of the number 1:

$$\frac{5}{12} + \frac{7}{18} = \frac{5}{12} \cdot \frac{3}{3} + \frac{7}{18} \cdot \frac{2}{2}$$

$$= \frac{15}{36} + \frac{14}{36}$$

Finally, we add numerators and place the result over the common denominator, 36. (Remember, this is an application of the distributive property.)

$$\frac{15}{36} + \frac{14}{36} = \frac{15 + 14}{36} = \frac{29}{36}$$

SIMPLIFYING EXPRESSIONS

We can use the commutative, associative, and distributive properties together to simplify expressions.

EXAMPLE 27   Simplify $7x + 4 + 6x + 3$.

Solution   We begin by applying the commutative and associative properties to group similar terms:

$$7x + 4 + 6x + 3 = (7x + 6x) + (4 + 3) \qquad \text{Commutative and associative properties}$$

$$= (7 + 6)x + (4 + 3) \qquad \text{Distributive property}$$

$$= 13x + 7 \qquad \text{Addition}$$

EXAMPLE 28   Simplify $4 + 3(2y + 5) + 8y$.

Solution   Since our rule for order of operations indicates that we are to multiply before adding, we must distribute the 3 across $2y + 5$ first:

$$4 + 3(2y + 5) + 8y = 4 + 6y + 15 + 8y \qquad \text{Distributive property}$$

$$= (6y + 8y) + (4 + 15) \qquad \text{Commutative and associative properties}$$

$$= (6 + 8)y + (4 + 15) \qquad \text{Distributive property}$$
$$= 14y + 19 \qquad \text{Addition}$$

The remaining properties of real numbers have to do with the numbers 0 and 1.

ADDITIVE IDENTITY PROPERTY

There exists a unique number 0 such that
In symbols: $a + 0 = a$ and $0 + a = a$

MULTIPLICATIVE IDENTITY PROPERTY

There exists a unique number 1 such that
In symbols: $a(1) = a$ and $1(a) = a$

Note: 0 and 1 are called the *additive identity* and *multiplicative identity,* respectively. Combining 0 with a number, under addition, does not change the identity of the number. Likewise, combining 1 with a number, under multiplication, does not alter the identity of the number. We see that 0 is to addition what 1 is to multiplication.

ADDITIVE INVERSE PROPERTY

For each real number a, there exists a unique real number $-a$ such that
In symbols: $a + (-a) = 0$
In words: Opposites add to 0.

MULTIPLICATIVE INVERSE PROPERTY

For every real number a, except 0, there exists a unique real number $\frac{1}{a}$ such that
In symbols: $a(\frac{1}{a}) = 1$
In words: Reciprocals multiply to 1.

EXAMPLES

29. $7(1) = 7$ Multiplicative identity property
30. $4 + (-4) = 0$ Additive inverse property
31. $6\left(\dfrac{1}{6}\right) = 1$ Multiplicative inverse property
32. $(5 + 0) + 2 = 5 + 2$ Additive identity property

Getting Ready for Class

After reading through the preceding section, respond in your own words and in complete sentences.

A. Describe the commutative property of multiplication.

B. Give definitions for each of the following:

   1. The opposite of a number

   2. The absolute value of a number

C. Explain why subtraction and division are not commutative operations.

D. Explain why zero doesn't have a multiplicative inverse.

PROBLEM SET I.3

Complete the following table.

	Number	Opposite	Reciprocal
1.	4		
2.	-3		
3.	$-\frac{1}{2}$		
4.	$\frac{5}{6}$		
5.		-5	
6.		7	
7.		$-\frac{3}{8}$	
8.		$\frac{1}{2}$	
9.			-6
10.			-3
11.			$\frac{1}{3}$
12.			$-\frac{1}{4}$

13. Name two numbers that are their own reciprocals.

14. Give the number that has no reciprocal.

15. Name the number that is its own opposite.

16. The reciprocal of a negative number is negative — true or false?

Write each of the following without absolute value symbols.

17. $|-2|$

18. $|-7|$

19. $\left|-\frac{3}{4}\right|$

20. $\left|\frac{5}{6}\right|$

21. $|\pi|$

22. $|-\sqrt{2}|$

23. $-|4|$

24. $-|5|$

25. $-|-2|$

26. $-|-10|$

27. $-\left|-\frac{3}{4}\right|$

28. $-\left|\frac{7}{8}\right|$

Multiply the following.

29. $\frac{3}{5} \cdot \frac{7}{8}$

30. $\frac{6}{7} \cdot \frac{9}{5}$

31. $\frac{1}{3} \cdot 6$

32. $\frac{1}{4} \cdot 8$

33. $\left(\frac{2}{3}\right)^3$

34. $\left(\frac{4}{5}\right)^2$

35. $\left(\frac{1}{10}\right)^4$

36. $\left(\frac{1}{2}\right)^5$

37. $\frac{3}{5} \cdot \frac{4}{7} \cdot \frac{6}{11}$

38. $\frac{4}{5} \cdot \frac{6}{7} \cdot \frac{3}{11}$

39. $\frac{4}{3} \cdot \frac{3}{4}$

40. $\frac{5}{8} \cdot \frac{8}{5}$

Use the associative property to rewrite each of the following expressions and then simplify the result.

41. $4 + (2 + x)$

42. $6 + (5 + 3x)$

43. $(a + 3) + 5$

44. $(4a + 5) + 7$

45. $5(3y)$

46. $7(4y)$

47. $\dfrac{1}{3}(3x)$

48. $\dfrac{1}{5}(5x)$

49. $4\left(\dfrac{1}{4}a\right)$

50. $7\left(\dfrac{1}{7}a\right)$

51. $\dfrac{2}{3}\left(\dfrac{3}{2}x\right)$

52. $\dfrac{4}{3}\left(\dfrac{3}{4}x\right)$

Apply the distributive property to each expression. Simplify when possible.

53. $3(x + 6)$

54. $5(x + 9)$

55. $2(6x + 4)$

56. $3(7x + 8)$

57. $5(3a + 2b)$

58. $7(2a + 3b)$

59. $\dfrac{1}{3}(4x + 6)$

60. $\dfrac{1}{2}(3x + 8)$

61. $\dfrac{1}{5}(10 + 5y)$

62. $\dfrac{1}{6}(12 + 6y)$

63. $(5t + 1)8$

64. $(3t + 2)5$

Apply the distributive property to each expression. Simplify when possible.

65. $3(5x + 2) + 4$

66. $4(3x + 2) + 5$

67. $4(2y + 6) + 8$

68. $6(2y + 3) + 2$

69. $5(1 + 3t) + 4$

70. $2(1 + 5t) + 6$

71. $3 + (2 + 7x)4$

72. $4 + (1 + 3x)5$

Add the following fractions.

73. $\dfrac{2}{5} + \dfrac{1}{15}$

74. $\dfrac{5}{8} + \dfrac{1}{4}$

75. $\dfrac{17}{30} + \dfrac{11}{42}$

76. $\dfrac{19}{42} + \dfrac{13}{70}$

77. $\dfrac{9}{48} + \dfrac{3}{54}$

78. $\dfrac{6}{28} + \dfrac{5}{42}$

79. $\dfrac{25}{84} + \dfrac{41}{90}$

80. $\dfrac{23}{70} + \dfrac{29}{84}$

Use the commutative, associative, and distributive properties to simplify the following.

81. $5a + 7 + 8a + a$

82. $6a + 4 + a + 4a$

83. $3y + y + 5 + 2y + 1$

84. $4y + 2y + 3 + y + 7$

85. $2(5x + 1) + 2x$

86. $3(4x + 1) + 9x$

87. $7 + 2(4y + 2)$

88. $6 + 3(5y + 2)$

89. $3 + 4(5a + 3) + 4a$

90. $8 + 2(4a + 2) + 5a$

91. $5x + 2(3x + 8) + 4$

92. $7x + 3(4x + 1) + 7$

Identify the property of real numbers that justifies each of the following.

93. $3 + 2 = 2 + 3$

94. $3(ab) = (3a)b$

95. $5x = x5$

96. $2 + 0 = 2$

97. $4 + (-4) = 0$

98. $1(6) = 6$

99. $x + (y + 2) = (y + 2) + x$

100. $(a + 3) + 4 = a + (3 + 4)$

101. $4(5 \cdot 7) = 5(4 \cdot 7)$

102. $6(xy) = (xy)6$

103. $4 + (x + y) = (4 + y) + x$

104. $(r + 7) + s = (r + s) + 7$

105. $3(4x + 2) = 12x + 6$

106. $5\left(\dfrac{1}{5}\right) = 1$

Applying the Concepts

Area and the Distributive Property   Find the area of each of the following rectangles in two ways: first, by multiplying length and width, and then by adding the areas of the two smaller rectangles together.

107.

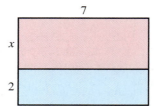

108.

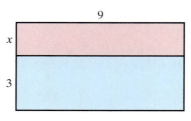

109.

 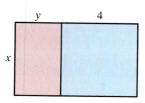

110.

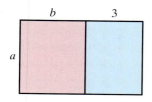

111. Distributive Property In 1996, Massachusetts raised 1.212 million pounds of tobacco, Connecticut raised 3.795 million pounds, and Virginia raised 103.543 million pounds. Show how the distributive property is used to determine the total number of tons of tobacco that were raised in these three states. (2,000 pounds = 1 ton.)

112. Mental Arithmetic The commutative and associative properties of addition are useful for doing mental arithmetic. Consider the calculation $26 + 140 + 4$. We can rearrange the order and grouping of the numbers mentally to make the calculation easier to do in our heads. Here is how we think about the problem mentally: $(26 + 4) + 140 = 30 + 140 = 170$. Use mental arithmetic to find the following sums.

(a) $125 + 210 + 175$

(b) $150 + 140 + 350 + 160$

(c) $96 + 125 + 90 + 4 + 175 + 210$

(d) $107 + 90 + 3 + 6 + 74 + 25 + 75$

113. Clock Arithmetic In a normal clock with 12 hours on its face, 12 is the additive identity, because adding 12 hours to any time on the clock will not change the hands of the clock. Also, if we think of the hour hand of a clock, the problem $10 + 4$ can be taken to mean: The hour hand is pointing at 10; if we add 4 more hours, it will be pointing at what number? Reasoning this way, we see that in clock arithmetic $10 + 4 = 2$ and $9 + 6 = 3$. Find the following in clock arithmetic:

(a) $6 + 7$

(b) $3 + 11 + 8$

(c) $3 + 11 + 8 + 12$

114. More Clock Arithmetic The clock arithmetic shown in the preceding problem is an example of a more general type of arithmetic called **modular arithmetic.** When we add numbers in clock arithmetic we are doing "addition mod 12" in modular arithmetic. That is $(10 + 4) \bmod 12 = 2$. Arithmetic mod 8 would be like clock arithmetic if there were 8 hours on the face of a clock instead of 12. Therefore, $(4 + 7) \bmod 8 = 3$, and $(3 + 7) \bmod 8 = 2$. Find the following:

(a) $(3 + 8) \bmod 9$

(b) $(4 + 6) \bmod 8$

(c) $(2 + 5) \bmod 7$

The temperature at the airport is 70°F. A plane takes off and reaches its cruising altitude of 28,000 feet, where the temperature is −40°F. Find the difference in the temperatures at takeoff and at cruising altitude.

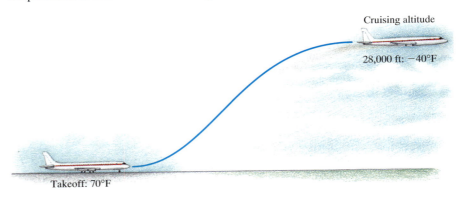

Cruising altitude

28,000 ft: −40°F

Takeoff: 70°F

We know intuitively that the difference in temperature is 110°F. If we write this problem using symbols, we have

$$70 - (-40) = 110$$

In this section we review the rules for arithmetic with real numbers, which will include problems such as this one.

ADDING REAL NUMBERS

The purpose of this section is to review the rules for arithmetic with real numbers and the justification for those rules. We can justify the rules for addition of real numbers geometrically by use of the real number line. Consider the sum of −5 and 3:

$$-5 + 3$$

We can interpret this expression as meaning "start at the origin and move 5 units in the negative direction and then 3 units in the positive direction." With the aid of a number line we can visualize the process.

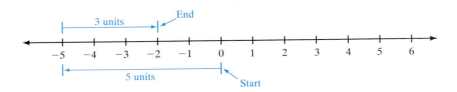

Since the process ends at −2, we say the sum of −5 and 3 is −2:

$$-5 + 3 = -2$$

We can use the real number line in this way to add any combination of positive and negative numbers.

The sum of -4 and -2, $-4 + (-2)$, can be interpreted as starting at the origin, moving 4 units in the negative direction, and then 2 more units in the negative direction:

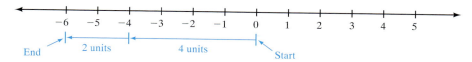

Since the process ends at -6, we say the sum of -4 and -2 is -6:

$$-4 + (-2) = -6$$

We can eliminate actually drawing a number line by simply visualizing it mentally. The following example gives the results of all possible sums of positive and negative 5 and 7.

EXAMPLE 1   Add all combinations of positive and negative 5 and 7.

Solution

$$5 + 7 = 12$$
$$-5 + 7 = 2$$
$$5 + (-7) = -2$$
$$-5 + (-7) = -12$$

Looking closely at the relationships in Example 1 (and trying other similar examples if necessary), we can arrive at the following rule for adding two real numbers.

TO ADD TWO REAL NUMBERS

With the *same* sign:

Step 1: Add their absolute values.
Step 2: Attach their common sign. If both numbers are positive, their sum is positive; if both numbers are negative, their sum is negative.

With *opposite* signs:

Step 1: Subtract the smaller absolute value from the larger.
Step 2: Attach the sign of the number whose absolute value is larger.

SUBTRACTING REAL NUMBERS

In order to have as few rules as possible, we will not attempt to list new rules for the *difference* of two real numbers. We will instead define it in terms of addition and apply the rule for addition.

> **DEFINITION**   If a and b are any two real numbers, then the **difference** of a and b, written $a - b$, is given by
>
> $$\underbrace{a - b}_{\text{To \textbf{subtract} } b,} = \underbrace{a + (-b)}_{\text{add the opposite of } b.}$$

We define the process of subtracting b from a as being equivalent to adding the opposite of b to a. In short, we say, "subtraction is addition of the opposite."

EXAMPLES   Subtract.

2. $5 - 3 = 5 + (-3)$ Subtracting 3 is equivalent to adding -3.
$\quad\quad\quad = 2$

3. $-7 - 6 = -7 + (-6)$ Subtracting 6 is equivalent to adding -6.
$\quad\quad\quad\quad = -13$

4. $9 - (-2) = 9 + 2$ Subtracting -2 is equivalent to adding 2.
$\quad\quad\quad\quad = 11$

5. $-6 - (-5) = -6 + 5$ Subtracting -5 is equivalent to adding 5.
$\quad\quad\quad\quad\quad = -1$

EXAMPLE 6   Subtract -3 from -9.

Solution   Since subtraction is not commutative, we must be sure to write the numbers in the correct order. Because we are subtracting -3, the problem looks like this when translated into symbols:

$$-9 - (-3) = -9 + 3 \quad\quad \text{Change to addition of the opposite.}$$
$$= -6 \quad\quad \text{Add.}$$

EXAMPLE 7   Add -4 to the difference of -2 and 5.

Solution   The difference of -2 and 5 is written $-2 - 5$. Adding -4 to that difference gives us

$$(-2 - 5) + (-4) = -7 + (-4) \quad\quad \text{Simplify inside parentheses.}$$
$$= -11 \quad\quad \text{Add.}$$

MULTIPLYING REAL NUMBERS

Multiplication with whole numbers is simply a shorthand way of writing repeated addition.

For example, $3(-2)$ can be evaluated as follows:

$$3(-2) = -2 + (-2) + (-2) = -6$$

We can evaluate the product $-3(2)$ in a similar manner if we first apply the commutative property of multiplication:

$$-3(2) = 2(-3) = -3 + (-3)$$

From these results it seems reasonable to say that the product of a positive and a negative is a negative number.

The last case we must consider is the product of two negative numbers, such as $-3(-2)$. To evaluate this product we will look at the expression $-3[2 + (-2)]$ in two different ways. First, since $2 + (-2) = 0$, we know the expression $-3[2 + (-2)]$ is equal to 0. On the other hand, we can apply the distributive property to get

$$-3[2 + (-2)] = -3(2) + (-3)(-2) = -6 + ?$$

Since we know the expression is equal to 0, it must be true that our ? is 6, since 6 is the only number we can add to -6 to get 0. Therefore, we have

$$-3(-2) = 6$$

Here is a summary of what we have so far:

Original Numbers Have		The Answer Is
The same sign	$3(2) = 6$	Positive
Different signs	$3(-2) = -6$	Negative
Different signs	$-3(2) = -6$	Negative
The same sign	$-3(-2) = 6$	Positive

TO MULTIPLY TWO REAL NUMBERS

To multiply two real numbers, simply

Step 1: Multiply their absolute values.

Step 2: If the two numbers have the *same* sign, the product is positive. If the two numbers have *opposite* signs, the product is negative.

EXAMPLE 8   Multiply all combinations of positive and negative 7 and 3.

Solution

$$7(3) = 21$$

$$7(-3) = -21$$

$$-7(3) = -21$$

$$-7(-3) = 21$$

DIVIDING REAL NUMBERS

> **DEFINITION**   If a and b are any two real numbers, where $b \neq 0$, then the
> **quotient** of a and b, written $\dfrac{a}{b}$, is given by
>
> $$\frac{a}{b} = a \cdot \left(\frac{1}{b}\right)$$

Dividing a by b is equivalent to multiplying a by the reciprocal of b. In short, we say, "division is multiplication by the reciprocal."

Since division is defined in terms of multiplication, the same rules hold for assigning the correct sign to a quotient as held for assigning the correct sign to a product. That is, *the quotient of two numbers with like signs is positive, while the quotient of two numbers with unlike signs is negative.*

EXAMPLES   Divide.

9. $\dfrac{6}{3} = 6 \cdot \left(\dfrac{1}{3}\right) = 2$ Notice these examples indicate that if a and b are positive real numbers then

10. $\dfrac{6}{-3} = 6 \cdot \left(-\dfrac{1}{3}\right) = -2$ $\dfrac{-a}{b} = \dfrac{a}{-b} = -\dfrac{a}{b}$

11. $\dfrac{-6}{3} = -6 \cdot \left(\dfrac{1}{3}\right) = -2$ and

12. $\dfrac{-6}{-3} = -6 \cdot \left(-\dfrac{1}{3}\right) = 2$ $\dfrac{-a}{-b} = \dfrac{a}{b}$

The second step in the preceding examples is written only to show that each quotient can be written as a product. It is not actually necessary to show this step when working problems.

In the examples that follow, we find a combination of operations. In each case we use the rule for order of operations.

EXAMPLES   Simplify each expression as much as possible.

13. $(-2 - 3)(5 - 9) = (-5)(-4)$ Simplify inside parentheses.
$\qquad\qquad\qquad\quad = 20$ Multiply.

14. $2 - 5(7 - 4) - 6 = 2 - 5(3) - 6$ Simplify inside parentheses.
$\qquad\qquad\qquad\quad = 2 - 15 - 6$ Then, multiply.
$\qquad\qquad\qquad\quad = -19$ Finally, subtract, left to right.

15. $2(4 - 7)^3 + 3(-2 - 3)^2 = 2(-3)^3 + 3(-5)^2$ Simplify inside parentheses.

$$= 2(-27) + 3(25) \qquad \text{Evaluate numbers with exponents.}$$

$$= -54 + 75 \qquad \text{Multiply.}$$

$$= 21 \qquad \text{Add.}$$

EXAMPLES   Simplify as much as possible.

16. $\dfrac{-5(-4) + 2(-3)}{2(-1) - 5} = \dfrac{20 - 6}{-2 - 5}$

$$= \dfrac{14}{-7}$$

$$= -2$$

17. $\dfrac{2^3 + 3^3}{2^2 - 3^2} = \dfrac{8 + 27}{4 - 9}$

$$= \dfrac{35}{-5}$$

$$= -7$$

Remember, since subtraction is defined in terms of addition, we can restate the distributive property in terms of subtraction. That is, if a, b, and c are real numbers, then $a(b - c) = ab - ac$.

EXAMPLE 18   Simplify $3(2y - 1) + y$.

Solution   We begin by multiplying the 3 and $2y - 1$. Then, we combine similar terms:

$$3(2y - 1) + y = 6y - 3 + y \qquad \text{Distributive property}$$

$$= 7y - 3 \qquad \text{Combine similar terms.}$$

EXAMPLE 19   Simplify $8 - 3(4x - 2) + 5x$.

Solution   First, we distribute the -3 across the $4x - 2$.

$$8 - 3(4x - 2) + 5x = 8 - 12x + 6 + 5x$$

$$= -7x + 14$$

EXAMPLE 20   Simplify $5(2a + 3) - (6a - 4)$.

Solution   We begin by applying the distributive property to remove the parentheses. The expression $-(6a - 4)$ can be thought of as $-1(6a - 4)$. Thinking of it in this way allows us to apply the distributive property.

$$-1(6a - 4) = -1(6a) - (-1)(4) = -6a + 4$$

Here is the complete problem:

$$5(2a + 3) - (6a - 4) = 10a + 15 - 6a + 4 \qquad \text{Distributive property}$$
$$= 4a + 19 \qquad \text{Combine similar terms.}$$

DIVIDING FRACTIONS

We end this section by reviewing division with fractions and division with the number 0.

EXAMPLES   Divide and reduce to lowest terms.

21. $\dfrac{3}{4} \div \dfrac{6}{11} = \dfrac{3}{4} \cdot \dfrac{11}{6}$ Definition of division

$\qquad\qquad = \dfrac{33}{24}$ Multiply numerators; multiply denominators.

$\qquad\qquad = \dfrac{11}{8}$ Divide numerator and denominator by 3.

22. $10 \div \dfrac{5}{6} = \dfrac{10}{1} \cdot \dfrac{6}{5}$ Definition of division

$\qquad\qquad = \dfrac{60}{5}$ Multiply numerators; multiply denominators.

$\qquad\qquad = 12$ Divide.

23. $-\dfrac{3}{8} \div 6 = -\dfrac{3}{8} \cdot \dfrac{1}{6}$ Definition of division

$\qquad\qquad = -\dfrac{3}{48}$ Multiply numerators; multiply denominators.

$\qquad\qquad = -\dfrac{1}{16}$ Divide numerator and denominator by 3.

DIVISION WITH THE NUMBER 0

For every division problem an associated multiplication problem involving the same numbers exists. For example, the following two problems say the same thing about the numbers 2, 3, and 6:

Division   *Multiplication*

$\dfrac{6}{3} = 2$ $6 = 2(3)$

We can use this relationship between division and multiplication to clarify division involving the number 0.

First of all, dividing 0 by a number other than 0 is allowed and always results in 0. To see this, consider dividing 0 by 5. We know the answer is 0 because of the relationship between multiplication and division. This is how we write it:

$\dfrac{0}{5} = 0$ because $0 = 0(5)$

On the other hand, dividing a nonzero number by 0 is not allowed in the real numbers. Suppose we were attempting to divide 5 by 0. We don't know whether there is an answer to this problem, but if there is, let's say the answer is a number that we can represent with the letter n. If 5 divided by 0 is a number n, then

$\dfrac{5}{0} = n$ and $5 = n(0)$

But this is impossible; because no matter what number n is, when we multiply it by 0 the answer must be 0. It can never be 5. In algebra, we say expressions like $\frac{5}{0}$ are undefined, because there is no answer to them. That is, division by 0 is not allowed in the real numbers.

Getting Ready for Class

After reading through the preceding section, respond in your own words and in complete sentences.

A. For each of the following expressions, give an example of an everyday situation that is modeled by the expression.
   1. $\$35 - \$12 = \$23$
   2. $-\$35 - \$12 = -\$47$

B. For each of the following expressions, give an example of an everyday situation that is modeled by the expression.
   1. $3(-\$25) = -\$75$
   2. $(-\$100) \div 5 = -\$20$

C. Why is division by 0 not allowed?

D. Why isn't the statement "two negatives make a positive" true?

PROBLEM SET 1.4

Find each of the following sums.

1. $6 + (-2)$ **2.** $11 - 5$

3. $-6 + 2$ **4.** $-11 + 5$

Find each of the following differences.

5. $-7 - 3$ **6.** $-6 - 9$

7. $-7 - (-3)$ **8.** $-6 - (-9)$

9. $\dfrac{3}{4} - \left(-\dfrac{5}{6}\right)$ **10.** $\dfrac{2}{3} - \left(-\dfrac{7}{5}\right)$

11. $\dfrac{11}{42} - \dfrac{17}{30}$ **12.** $\dfrac{13}{70} - \dfrac{19}{42}$

13. Subtract 5 from -3.

14. Subtract -3 from 5.

15. Find the difference of -4 and 8.

16. Find the difference of 8 and -4.

17. Subtract $4x$ from $-3x$.

18. Subtract $-5x$ from $7x$.

19. What number do you subtract from 5 to get -8?

20. What number do you subtract from -3 to get 9?

21. Add -7 to the difference of 2 and 9.

22. Add -3 to the difference of 9 and 2.

23. Subtract $3a$ from the sum of $8a$ and a.

24. Subtract $-3a$ from the sum of $3a$ and $5a$.

Find the following products.

25. $3(-5)$ **26.** $-3(5)$

27. $-3(-5)$ **28.** $4(-6)$

29. $2(-3)(4)$ **30.** $-2(3)(-4)$

31. $-2(5x)$ **32.** $-5(4x)$

33. $-\dfrac{1}{3}(-3x)$ **34.** $-\dfrac{1}{6}(-6x)$

35. $-\dfrac{2}{3}\left(-\dfrac{3}{2}y\right)$ **36.** $-\dfrac{2}{5}\left(-\dfrac{5}{2}y\right)$

37. $-2(4x - 3)$ **38.** $-2(-5t + 6)$

39. $-\dfrac{1}{2}(6a - 8)$ **40.** $-\dfrac{1}{3}(6a - 9)$

Simplify each expression as much as possible.

41. $3(-4) - 2$ **42.** $-3(-4) - 2$

43. $4(-3) - 6(-5)$ **44.** $-6(-3) - 5(-7)$

45. $2 - 5(-4) - 6$

46. $3 - 8(-1) - 7$

47. $4 - 3(7 - 1) - 5$

48. $8 - 5(6 - 3) - 7$

49. $2(-3)^2 - 4(-2)^3$

50. $5(-2)^2 - 2(-3)^3$

51. $(2 - 8)^2 - (3 - 7)^2$

52. $(5 - 8)^2 - (4 - 8)^2$

53. $7(3 - 5)^3 - 2(4 - 7)^3$

54. $3(-7 + 9)^3 - 5(-2 + 4)^3$

55. $-3(2 - 9) - 4(6 - 1)$

56. $-5(5 - 6) - 7(2 - 8)$

57. $2 - 4[3 - 5(-1)]$

58. $6 - 5[2 - 4(-8)]$

59. $(8 - 7)[4 - 7(-2)]$

60. $(6 - 9)[15 - 3(-4)]$

61. $-3 + 4[6 - 8(-3 - 5)]$

62. $-2 + 7[2 - 6(-3 - 4)]$

63. $5 - 6[-3(2 - 9) - 4(8 - 6)]$

64. $9 - 4[-2(4 - 8) - 5(3 - 1)]$

Simplify each expression.

65. $3(5x + 4) - x$

66. $4(7x + 3) - x$

67. $6 - 7(3 - m)$

68. $3 - 5(5 - m)$

69. $7 - 2(3x - 1) + 4x$

70. $8 - 5(2x - 3) + 4x$

71. $5(3y + 1) - (8y - 5)$

72. $4(6y + 3) - (6y - 6)$

73. $4(2 - 6x) - (3 - 4x)$

74. $7(1 - 2x) - (4 - 10x)$

75. $10 - 4(2x + 1) - (3x - 4)$

76. $7 - 2(3x + 5) - (2x - 3)$

Use the definition of division to write each division problem as a multiplication problem, then simplify.

77. $\dfrac{8}{-4}$ **78.** $\dfrac{-8}{4}$

79. $\dfrac{4}{0}$ **80.** $\dfrac{-7}{0}$

81. $\dfrac{0}{-3}$ **82.** $\dfrac{0}{5}$

83. $-\dfrac{3}{4} \div \dfrac{9}{8}$ **84.** $-\dfrac{2}{3} \div \dfrac{4}{9}$

85. $-8 \div \left(-\dfrac{1}{4}\right)$ **86.** $-12 \div \left(-\dfrac{2}{3}\right)$

87. $-40 \div \left(-\dfrac{5}{8}\right)$ **88.** $-30 \div \left(-\dfrac{5}{6}\right)$

89. $\dfrac{4}{9} \div (-8)$ **90.** $\dfrac{3}{7} \div (-6)$

Simplify as much as possible.

91. $\dfrac{3(-1) - 4(-2)}{8 - 5}$ **92.** $\dfrac{6(-4) - 5(-2)}{7 - 6}$

93. $8 - (-6)\left[\dfrac{2(-3) - 5(4)}{-8(6) - 4}\right]$

94. $-9 - 5\left[\dfrac{11(-1) - 9}{4(-3) + 2(5)}\right]$

95. $6 - (-3)\left[\dfrac{2 - 4(3 - 8)}{1 - 5(1 - 3)}\right]$

96. $8 - (-7)\left[\dfrac{6 - 1(6 - 10)}{4 - 3(5 - 7)}\right]$

97. Subtract -5 from the product of 12 and $-\frac{2}{3}$.

98. Subtract -3 from the product of -12 and $\frac{3}{4}$.

99. Add $8x$ to the product of -2 and $3x$.

100. Add $7x$ to the product of -5 and $-2x$.

Applying the Concepts

Baseball Major league baseball has various player awards at the end of the year. One of them is the Ro-laids Relief Man of the Year. To compute the Relief Man standings, points are gained or taken away based on the number of wins, losses, saves, and blown saves a relief pitcher has at the end of the year. The pitcher with the most Rolaids Points is the Rolaids Relief Man of the Year. Given the following scoring formula, com-plete the tables in Problems 101 and 102 by filling in the number of Rolaids points for each pitcher:

Scoring formula: Saves = 3 PTS, relief wins = 2 PTS, relief losses = −2 PTS, blown saves = −2 PTS

101. National League

Pitcher, Team	W	L	Saves	Blown Saves	Rolaids Points
Trevor Hoffman, San Diego	4	2	53	1	
Rod Beck, Chicago	3	4	51	7	
Jeff Shaw, Los Angeles	3	8	48	9	
Robb Nen, San Francisco	7	7	40	5	
Ugueth Urbina, Montreal	6	3	34	4	
Kerry Ligtenberg, Atlanta	3	2	30	4	

102. American League

Pitcher, Team	W	L	Saves	Blown Saves	Rolaids Points
Tom Gordon, Boston	7	4	46	1	
John Wetteland, Texas	3	1	42	5	
Mike Jackson, Cleveland	1	1	40	5	
Troy Percival, Anaheim	2	7	42	6	
Mariano Rivera, New York	3	0	36	5	
Jeff Montgomery, Kansas City	2	5	36	5	

103. Time Zones The continental United States is di-vided into four time zones. When it is 4:00 in the Pacific zone, it is 5:00 in the Mountain zone, 6:00 in the Central zone, and 7:00 in the Eastern zone.

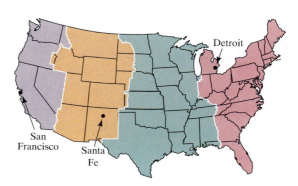

You board a plane at 6:55 P.M. in California (Pacific zone), and take 2 hours and 15 minutes to arrive in Santa Fe, New Mexico (Mountain zone), and another 3 hours and 20 minutes to arrive at your final destination, Detroit, Michigan (Eastern zone). At what local times did you arrive at Santa Fe and Detroit?

104. Temperature   The coldest recorded temperature in the United States was $-80°F$ at Prospect Creek Camp, Alaska, on January 23, 1971. The warmest recorded temperature was 134°F at Greenland Ranch, California, on July 10, 1913. What is the difference of these two temperatures?

105. Oceans and Mountains   The deepest ocean depth is 35,840 feet, found in the Pacific Ocean's Mariana Trench. The tallest mountain is Mount Everest, with a height of 29,028 feet.

Mount Everest *(Galen Rowell/FPG International)*

(a) The deepest gorge in the United States is 7,900 feet, located in Hell's Canyon, Snake River, Oregon. How many times deeper is the Mariana Trench than Hell's Canyon?

(b) What is the difference in the lowest and highest points on Earth?

106. Income and Expenditures   The following table shows approximate expenditures and income of the federal government for 6 randomly selected years prior to World War I. Amounts are in millions of dollars.

(a) Write an expression which, when simplified, represents the total amount of money that the government "overspent" or "saved" these 6 years.

(b) Simplify your expression from part a. What is the significance of the final sign?

Year	Expenditure	Income
1912	$720.2	$710.2
1867	377.6	447.3
1790	5.8	5.7
1897	457.5	434.9
1832	24.5	30.5
1802	9.1	13.1

1.5   Conditional Statements

The day after my daughter Amy turned 17, she and her friend Jenny had the following conversation with me.

AMY: Dad, now that I'm 17, can my curfew be extended?

ME: To how late?

AMY: One o'clock.

ME: No.

AMY: Dad, everyone else gets to stay out as long as they want. It's not fair that I have to be in by midnight.

> **ME:**   There is nothing to do here after midnight, except get in trouble.
> **AMY:**   Okay then, can I spend the night at Jenny's?

There are messages within this conversation, some explicit and some implied, that mathematics, and the forms of reasoning on which mathematics is built, can clarify. For instance, one fact begins to emerge from this conversation:

> If I don't give Amy permission to stay out late, then she will anyway.

In this section we cover the kind of reasoning I would use to understand the conversation with my daughter. You may be surprised to find that it is the same kind of reasoning that accompanies most topics in mathematics.

Consider the two statements below:

Statement 1: If I study, then I will get good grades.
Statement 2: If x is a negative number, then $-x$ is a positive number.

In Statement 1, if we let A represent the phrase "I study" and B represent the phrase "I will get good grades," then Statement 1 has the form

$$\text{If } A, \text{ then } B$$

Likewise, in Statement 2, if A is the phrase "x is a negative number" and B is "$-x$ is a positive number," then Statement 2 has the form

$$\text{If } A, \text{ then } B$$

Each statement has the same form: If A, then B. We call this the "if/then" form, and any statement that has this form is called a *conditional statement.* For every conditional statement, the first phrase, A, is called the *hypothesis,* and the second phrase, B, is called the *conclusion.* All conditional statements can be written in the form

$$\text{If } hypothesis, \text{ then } conclusion$$

NOTATION

A shorthand way to write an "if/then" statement is with the implies symbol

$$A \Rightarrow B$$

This statement is read "A implies B." It is equivalent to saying "If A, then B."

EXAMPLE I   Identify the hypothesis and conclusion in each statement.

(a) If a and b are positive numbers, then $-a(-b) = ab$.
(b) If it is raining, then the streets are wet.
(c) $C = 90° \Rightarrow c^2 = a^2 + b^2$.

Solution

(a) *Hypothesis:* a and b are positive numbers
   Conclusion: $-a(-b) = ab$

(b) *Hypothesis:* It is raining.
   Conclusion: The streets are wet.
(c) *Hypothesis:* $C = 90°$
   Conclusion: $c^2 = a^2 + b^2$

For each conditional statement $A \Rightarrow B$, we can find three related statements that may or may not be true depending on whether or not the original conditional statement is true.

DEFINITION   For every conditional statement $A \Rightarrow B$, there exist the following associated statements:

The converse:	$B \Rightarrow A$	If B, then A
The inverse:	not $A \Rightarrow$ not B	If not A, then not B
The contrapositive:	not $B \Rightarrow$ not A	If not B, then not A

EXAMPLE 2   For the statement below, write the converse, the inverse, and the contrapositive.

   If it is raining, then the streets are wet.

Solution   It is sometimes easier to work a problem like this if we write out the phrases A, B, not A, and not B.

   Let $A =$ it is raining; then not $A =$ it is not raining.
   Let $B =$ the streets are wet; then not $B =$ the streets are not wet.

Here are the three associated statements:

   The converse: (If B, then A.) If the streets are wet, then it is raining.
   The inverse: (If not A, then not B.) If it is not raining, then the streets are not wet.
   The contrapositive: (If not B, then not A.) If the streets are not wet, then it is not raining.

TRUE OR FALSE?

Next, we want to answer this question: "If a conditional statement is true, which, if any, of the associated statements are true also?" Consider the statement below.

   If it is a square, then it has four sides.

We know from our experience with squares that this is a true statement. Now, is the converse necessarily true? Here is the converse:

   If it has four sides, then it is a square.

Obviously the converse is *not* true, because there are many four-sided figures that are not squares — rectangles, parallelograms, and trapezoids, to mention a few. So, the converse of a true conditional statement is not necessarily true itself.

Next, we consider the inverse of our original statement.

> If it is not a square, then it does not have four sides.

Again, the inverse is not true since there are many nonsquare figures that do have four sides. For example, a 3-inch by 5-inch rectangle fits that description.

Finally, we consider the contrapositive:

> If it does not have four sides, then it is not a square.

As you can see, the contrapositive is true. That is, if something doesn't have four sides, it can't possibly be a square.

The preceding discussion leads us to the following theorem.

THEOREM

If a conditional statement is true, then so is its contrapositive. That is, the two statements

> If *A*, then *B*   and   If not *B*, then not *A*

are equivalent; one can't be true without the other being true also. That is, they are either both true or both false.

The theorem doesn't mention the inverse and the converse because they are true or false independent of the original statement. That is, knowing that a conditional statement is true tells us that the contrapositive is also true — but the truth of the inverse and the converse does not follow from the truth of the original statement.

The next two examples are intended to clarify the preceding discussion and our theorem. As you read through them, be careful not to let your intuition, experience, or opinion get in the way.

EXAMPLE   3   If the statement "If you are guilty, then you will be convicted" is true, give another statement that must also be true.

Solution   From our theorem, and the discussion preceding it, we know that the contrapositive of a true conditional statement is also true. Here is the contrapositive of our original statement:

> If you are not convicted, then you are not guilty.

Remember, we are not asking for your opinion; we are simply asking for another conditional statement that must be true if the original statement is true. The answer is *always* the contrapositive. Now, you may be wondering about the converse:

> If you are convicted, then you must be guilty.

It may be that the converse is actually true. But if it is, it is not because of the original conditional statement. That is, the truth of the converse *does not follow* from the truth of the original statement.

EXAMPLE 4　If the following statement is true, what other conditional statement must also be true?

$$\text{If } a = b, \text{ then } a^2 = b^2.$$

Solution　Again, every true conditional statement has a true contrapositive. Therefore, the statement below is also true:

$$\text{If } a^2 \neq b^2, \text{ then } a \neq b.$$

In this case, we know from experience that the original statement is true; that is, if two numbers are equal, then so are their squares. We also know from experience that the contrapositive is true; if the squares of two numbers are not equal, then the numbers themselves can't be equal. Do you think the inverse and converse are true also? Here is the converse:

$$\text{If } a^2 = b^2, \text{ then } a = b.$$

The converse is not true. If a is -3 and b is 3, then a^2 and b^2 are equal, but a and b are not. This same kind of reasoning will show that the inverse is not necessarily true. This example, then gives further evidence that our theorem is true: A true conditional statement has a true contrapositive. No conclusion can be drawn about the inverse or the converse.

EVERYDAY LANGUAGE

In everyday life, we don't always use the "if/then" form exactly as we have illustrated it here. Many times we use shortened, reversed, or otherwise altered forms of "if/then" statements. For instance, each of the following statements is a variation of the "if/then" form, and each carries the same meaning.

If it is raining, then the streets are wet.

If it is raining, the streets are wet.

When it rains, the streets get wet.

The streets are wet if it is raining.

The streets are wet because it is raining.

Rain will make the streets wet.

EXAMPLE 5　Write the following statement in "if/then" form.

Romeo loves Juliet.

Solution　We must be careful that we do not change the meaning of the statement when we write it in "if/then" form. Here is an "if/then" form that has the same meaning as the original statement:

> If he is Romeo, then he loves Juliet.

We can see that it would be incorrect to rewrite the original statement as

> If she is Juliet, then she loves Romeo

because the original statement is Romeo loves Juliet, not Juliet loves Romeo. It would also be incorrect to rewrite our statement as either

> If he is not Romeo, then he does not love Juliet

or

> If he loves Juliet, then he is Romeo

because people other than Romeo may also love Juliet. (The preceding statements are actually the inverse and converse, respectively, of the original statement.) Finally, another statement that has the same meaning as our original statement is

> If he does not love Juliet, then he is not Romeo

because this is the contrapositive of our original statement, and we know that the contrapositive is always true when the original statement is true.

Getting Ready for Class

After reading through the preceding section, respond in your own words and in complete sentences.

A. What is a conditional statement?

B. State the hypothesis and conclusion in the following statement: "Staying up all night will make you tired."

C. Determine if the following statement is true or false and explain your answer: "If a conditional statement $A \Rightarrow B$ is true, then its converse must also be true."

D. Give an example of a conditional statement and its converse from your own life. State whether the converse is true or false.

PROBLEM SET 1.5

For each conditional statement, state the hypothesis and the conclusion.

1. If you argue for your limitations, then they are yours.

2. If you think you can, then you can.

3. If x is an even number, then x is divisible by 2.

4. If x is an odd number, then x is not divisible by 2.

5. If a triangle is equilateral, then all of its angles are equal.

6. If a triangle is isosceles, then two of its angles are equal.

7. If $x + 5 = -2$, then $x = -7$.

8. If $x - 5 = -2$, then $x = 3$.

For each of the following conditional statements, give the converse, the inverse, and the contrapositive.

9. If $a = 8$, then $a^2 = 64$.

10. If $x = y$, then $x^2 = y^2$.

11. If $\frac{a}{b} = 1$, then $a = b$.

12. If $a + b = 0$, then $a = -b$.

13. If it is a square, then it is a rectangle.

14. If you live in a glass house, then you shouldn't throw stones.

15. If better is possible, then good is not enough.

16. If a and b are positive, then ab is positive.

For each statement below, write an equivalent statement in "if/then" form.

17. $E \Rightarrow F$

18. $a^3 = b^3 \Rightarrow a = b$

19. Misery loves company.

20. Rollerblading is not a crime.

21. The squeaky wheel gets the grease.

22. The girl who can't dance says the band can't play.

In each of the following problems a conditional statement is given. If the conditional statement is true, which of the three statements that follow it must also be true?

23. If you heard it first, then you heard it on Eyewitness News.
   (a) If you heard it on Eyewitness News, then you heard it first.
   (b) If you didn't hear it first, then you didn't hear it on Eyewitness News.
   (c) If you didn't hear it on Eyewitness News, then you didn't hear it first.

24. If it is raining, then the streets are wet.
   (a) If the streets are wet, then it is raining.
   (b) If the streets are not wet, then it is not raining.
   (c) If it is not raining, then the streets are not wet.

25. If $C = 90°$, then $c^2 = a^2 + b^2$.
   (a) If $c^2 \neq a^2 + b^2$, then $C \neq 90°$.
   (b) If $c^2 = a^2 + b^2$, then $C = 90°$.
   (c) If $C \neq 90°$, then $c^2 \neq a^2 + b^2$.

26. If you graduate from college, then you will get a good job.
   (a) If you get a good job, then you graduated from college.
   (b) If you do not get a good job, then you did not graduate from college.
   (c) If you do not graduate from college, then you will not get a good job.

27. If you get a B average, then your car insurance will cost less.
   (a) If your car insurance costs less, then you got a B average.
   (b) If you do not get a B average, then your car insurance will not cost less.
   (c) If your car insurance does not cost less, then you did not get a B average.

28. If you go out without a sweater, then you will get sick.
   (a) If you do not get sick, then you went out with a sweater.
   (b) If you go out with a sweater, then you will not get sick.
   (c) If you get sick, then you went out without a sweater.

29. If a and b are both negative, then $a + b$ is negative.
   (a) If a and b are not both negative, then $a + b$ is not negative.
   (b) If $a + b$ is not negative, then a and b are not both negative.
   (c) If $a + b$ is negative, then a and b are both negative.

30. If it is a square, then all its sides have the same length.
   (a) If all its sides are not the same length, then it is not a square.
   (b) If all its sides have the same length, then it is a square.
   (c) If it is not a square, then its sides are not all the same length.

Extending the Concepts

31. Contrapositive of the Contrapositive You have learned that the opposite of a negative number is a positive number. In the same way, the contrapositive of a statement's contrapositive is the statement itself. Illustrate this idea with the following statement: "If you are sleeping, then your eyes are closed."

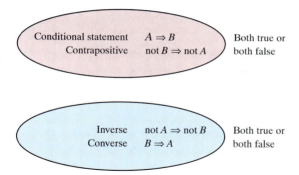

Conditional statement $A \Rightarrow B$ Both true or
Contrapositive $not\ B \Rightarrow not\ A$ both false

Inverse $not\ A \Rightarrow not\ B$ Both true or
Converse $B \Rightarrow A$ both false

32. True Statements Come in Pairs   The theorem in this section states that a conditional statement and its contrapositive are either both true or both false. However, it is also true that any conditional statement's inverse and converse are both true or both false. Explain why this second relationship is also true.

33. Truth Table   Using the information from the preceding exercise, fill in the blank spaces in the following truth table with true or false. (One part of this exercise contains a contradiction and is impossible.)

	Statement	Inverse	Converse	Contrapositive
(a)	True	True	?	?
(b)	?	True	?	False
(c)	?	?	False	True
(d)	False	?	True	?
(e)	?	False	True	?

34. Logic   Modern mathematicians who use the rules of logic developed in this section are called logicians, and include such notables as *George Boole, Alfred North Whitehead,* and *Bertrand Russell.* Logicians divide conditional statements into two types, deductive and inductive. The conclusion of a deductive statement must follow from its hypothesis, while the conclusion of an inductive statement will probably follow from the hypothesis. The statement "If $x = 2$, then $x + 7 = 9$" is a deductive statement because the conclusion, $x +$

$7 = 9$, must follow from the hypothesis, $x = 2$. On the other hand, the statement "If I study, then I will earn an A" is an inductive statement because the conclusion (earning an A) will probably (but not necessarily) follow from studying.

Write three conditional statements that are also deductive statements, and three conditional statements that are also inductive statements.

35. More About Amy   Suppose after reading the introduction to this section you decide that the following statement is true:

> If I don't extend her curfew,
> Amy will stay out late anyway.

Write another statement that must also be true if this statement is true. Now, if the preceding statement is true, and I want Amy to be home at her regular time, what should I tell her when she asks if she can have her curfew extended?

36. Critical Thinking and Subway Tokens   Pat and Tom visit New York City for a week. At the beginning of the trip, they each buy ten subway tokens. Even though they each pay the same amount for their tokens, Tom can ride the subway for half price because he is a senior. Here is how it happens: Each time Tom uses a token to ride the subway, he shows his driver's license to prove he is a senior and then he is given a pass he can use instead of a token for his next subway ride. So, one token gets Tom a ride on the subway and a pass for another ride. After 2 days of riding the subway together, Pat has four tokens left, and Tom has seven tokens left. Has Tom lost a token, or is everything the way it should be? Explain your answer.

Much of what we do in mathematics is concerned with recognizing patterns and classifying together groups of numbers that share a common characteristic. For instance, suppose you were asked to give the next number in this sequence:

$$3, 5, 7, \ldots$$

Looking for a pattern, you may observe that each number is 2 more than the number preceding it. That being the case, the next number in the sequence will be 9 because 9 is 2 more than 7. Reasoning in this manner is called *inductive reasoning.* In mathematics, we use inductive reasoning when we notice a pattern to a sequence of numbers and then extend the sequence using the pattern.

EXAMPLE 1 Use inductive reasoning to find the next term in each sequence.

(a) 5, 8, 11, 14, . . .
(b) $\triangle, \triangleright, \triangledown, \triangleleft, \ldots$
(c) 1, 4, 9, 16, . . .

Solution In each case we use the pattern we observe in the first few terms to write the next term.

(a) Each term comes from the previous term by adding 3. Therefore, the next term would be 17.
(b) The triangles rotate a quarter turn to the right each time. The next term would be a triangle that points up, $\triangle$.
(c) This looks like the sequence of squares, $1^2, 2^2, 3^2, 4^2, \ldots$. The next term is $5^2 = 25$.

Now that we have an intuitive idea of inductive reasoning, here is a formal definition.

DEFINITION Inductive reasoning is reasoning in which a conclusion is drawn based on evidence and observations that support that conclusion. In mathematics this usually involves noticing that a few items in a group have a trait or characteristic in common, and then concluding that all items in the group have that same trait.

ARITHMETIC SEQUENCES

We can extend our work with sequences by classifying together sequences that share a common characteristic. Our first classification is for sequences that are constructed by adding the same number each time.

> **DEFINITION**   An **arithmetic sequence** is a sequence of numbers in which each number (after the first number) comes from adding the same amount to the number before it.

The sequence

$$4, 7, 10, 13, \ldots$$

is an example of an arithmetic sequence, since each term is obtained from the preceding term by adding 3 each time. The number we add each time — in this case, 3 — is the *common difference,* since it can be obtained by subtraction.

EXAMPLE 2   Each sequence shown here is an arithmetic sequence. Find the next two numbers in each sequence.

(a) $10, 16, 22, \ldots$

(b) $\frac{1}{2}, 1, \frac{3}{2}, \ldots$

(c) $5, 0, -5, \ldots$

Solution   Since we know that each sequence is arithmetic, we know how to look for the number that is added to each term to produce the next consecutive term.

(a) $10, 16, 22, \ldots$: Each term is found by adding 6 to the term before it. Therefore, the next two terms will be 28 and 34.

(b) $\frac{1}{2}, 1, \frac{3}{2}, :$ Each term comes from the term before it by adding $\frac{1}{2}$. The fourth term will be $\frac{3}{2} + \frac{1}{2} = 2$, while the fifth term will be $2 + \frac{1}{2} = \frac{5}{2}$.

(c) $5, 0, -5, \ldots$: Each term comes from adding -5 to the term before it. Therefore, the next two terms will be $-5 + (-5) = -10$, and $-10 + (-5) = -15$.

GEOMETRIC SEQUENCES

Our second classification of sequences with a common characteristic involves sequences that are constructed using multiplication.

> **DEFINITION**   A **geometric sequence** is a sequence of numbers in which each number (after the first number) comes from the number before it by multiplying the same amount each time.

The sequence

$$4, 12, 36, 108, \ldots$$

is a geometric sequence. Each term is obtained from the previous term by multiplying by 3. The amount by which we multiply each term to obtain the next term — in this case, 3 — is called the *common ratio.*

EXAMPLE 3 Each sequence shown here is a geometric sequence. Find the next number in each sequence.

(a) 2, 10, 50, . . .

(b) 3, − 15, 75, . . .

(c) $\frac{1}{8}, \frac{1}{4}, \frac{1}{2},$ . . .

Solution Since each sequence is a geometric sequence, we know that each term is obtained from the previous term by multiplying by the same number each time.

(a) 2, 10, 50, . . . : Starting with 2, each number is obtained from the previous number by multiplying by 5 each time. The next number will be $50 \cdot 5 = 250$.

(b) 3, − 15, 75, . . . : The sequence starts with 3. After that, each number is obtained by multiplying by − 5 each time. The next number will be $75(-5) = -375$.

(c) $\frac{1}{8}, \frac{1}{4}, \frac{1}{2},$ . . . : This sequence starts with $\frac{1}{8}$. Multiplying each number in the sequence by 2 produces the next number in the sequence. To extend the sequence we multiply $\frac{1}{2}$ by 2:

$$\frac{1}{2} \cdot 2 = 1$$

The next number in the sequence is 1.

THE FIBONACCI SEQUENCE

In the introduction to this chapter we quoted the mathematician Fibonacci. There is a special sequence in mathematics named for Fibonacci.

Fibonacci sequence: 1, 1, 2, 3, 5, 8, . . .

To construct the Fibonacci sequence we start with two 1's. The rest of the numbers in the sequence are found by adding the two previous terms. Adding the first two terms, 1 and 1, we have 2. Then, adding 1 and 2 we have 3. In general, adding any two consecutive terms of the Fibonacci sequence gives us the next term.

A MATHEMATICAL MODEL

One of the reasons we study number sequences is because they can be used to model some of the patterns and events we see in the world around us. The discussion that follows shows how the Fibonacci sequence can be used to predict the number of bees in each generation of the family tree of a male honeybee. It is based on an example from Chapter 2 of the book *Mathematics: A Human Endeavor*, by Harold Jacobs. If you find that you enjoy discovering patterns in mathematics, Mr. Jacobs's book has many interesting examples and problems involving patterns in mathematics.

A male honeybee has one parent, its mother, while a female honeybee has two parents, a mother and a father. (A male honeybee comes from an unfertilized egg; a female honeybee comes from a fertilized egg.) Using these facts, we construct the

Generation   Number of bees

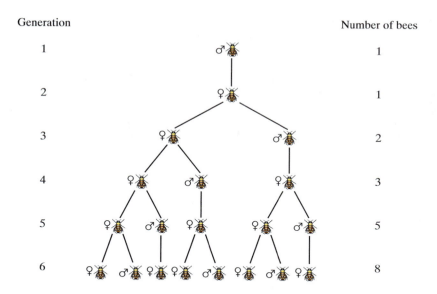

family tree of a male honeybee using ♂ to represent a male honeybee and ♀ to represent a female honeybee.

Looking at the numbers in the right column in our diagram, the sequence that gives us the number of bees in each generation of the family tree of a male honeybee is

$$1 \qquad 1 \qquad 2 \qquad 3 \qquad 5 \qquad 8$$

As you can see, this is the Fibonacci sequence. We have taken our original diagram (the family tree of the male honeybee) and reduced it to a mathematical model (the Fibonacci sequence). The model can be used in place of the diagram to find the number of bees in any generation back from our first bee.

EXAMPLE 4   Find the number of bees in the tenth generation of the family tree of a male honeybee.

Solution   We can continue the previous diagram and simply count the number of bees in the tenth generation, or we can use inductive reasoning to conclude that the number of bees in the tenth generation will be the tenth term of the Fibonacci sequence. Let's make it easy on ourselves and find the first ten terms of the Fibonacci sequence.

Generation:	1	2	3	4	5	6	7	8	9	10
Number of bees:	1	1	2	3	5	8	13	21	34	55

As you can see, the number of bees in the tenth generation is 55.

CONNECTING TWO SEQUENCES: PAIRED DATA

We can use the discussion that opened this chapter to illustrate how we connect sequences. Suppose an antidepressant has a half-life of 5 days, and its concentration in a patient's system is 80 ng/ml (nanograms/milliliter) when the patient stops taking the antidepressant. To find the concentration after a half-life passes, we multiply the previous concentration by $\frac{1}{2}$. Here are two sequences that together summarize this information.

Days: 0, 5, 10, 15, . . .

Concentration: 80, 40, 20, 10, . . .

The sequence of days is an arithmetic sequence in which we start with 0 and add 5 each time. The sequence of concentrations is a geometric sequence in which we start with 80, and multiply by $\frac{1}{2}$ each time. Here is the same information displayed in a table.

TABLE 1	
Days Since Discontinuing	**Concentration (ng/ml)**
0	80
5	40
10	20
15	10
20	5

The information in Table 1 is shown visually in Figures 1 and 2. The diagram in Figure 1 is called a **scatter diagram.** If we connect the points in the scatter dia-

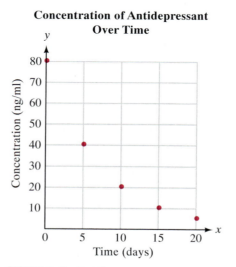

FIGURE 1 Scatter diagram

Concentration of Antidepressant Over Time

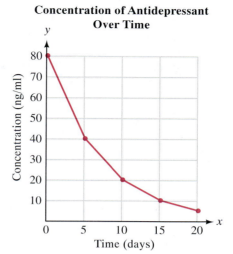

FIGURE 2 Line graph

gram with straight lines, we have the diagram in Figure 2, which is called a **line graph.** In both figures, the horizontal line that shows days is called the **horizontal axis,** while the vertical line that shows concentration is called the **vertical axis.**

The data in Table 1 is called **paired data** because each number in the days column is paired with a specific number in the concentration column. For instance, after 15 days, the concentration is 10 ng/ml. Each pair of numbers from Table 1 is associated with one of the dots in the scatter diagram and line graph.

Table 1, Figure 1, and Figure 2 each describe the same relationship. The figures are visual descriptions, while the table is a numerical description. The discussion that precedes Table 1 is a written description of that relationship. This gives us three ways to describe the relationship: written, numerical, and visual. As we proceed further into our study of algebra, you will see this same relationship described a fourth way when we give the equation that relates time and concentration: $C = 80 \cdot 2^{-t/5}$.

 ## Getting Ready for Class

After reading through the preceding section, respond in your own words and in complete sentences.

A. What is inductive reasoning?

B. If you were to describe the Fibonacci sequence in words, you would start this way: "The first two numbers are ones. After that, each number is found by . . . ." Finish the sentence so that someone reading it will know how to find the numbers in the Fibonacci sequence.

C. Create an arithmetic sequence and explain how it was formed.

D. Create a geometric sequence and explain how it was formed.

PROBLEM SET 1.6

Here are some sequences that we will be referring to throughout the book. Find the next number in each sequence.

1. 1, 2, 3, 4, . . .   (The sequence of counting numbers)

2. 0, 1, 2, 3, . . .   (The sequence of whole numbers)

3. 2, 4, 6, 8, . . .   (The sequence of even numbers)

4. 1, 3, 5, 7, . . .   (The sequence of odd numbers)

5. 1, 4, 9, 16, . . .   (The sequence of squares.)

6. 1, 8, 27, 64, . . .   (The sequence of cubes.)

Find the next number in each sequence.

7. 1, 8, 15, 22, . . .   **8.** 1, 8, 64, 512, . . .

9. 1, 8, 27, 64, . . .   **10.** 1, 8, 16, 25, . . .

Give one possibility for the next term in each sequence.

11. △, ◁, ▽, ▷, . . .   **12.** →, ↓, ←, ↑, . . .

13. △, □, ○, △, ⊡, . . .

14. □, ⊏⊐, ⊞, ⊞, ⊏⊐⊐, . . .

Each sequence shown here is an arithmetic sequence. In each case, find the next two numbers in the sequence.

15. 1, 5, 9, 13, . . .   **16.** 10, 16, 22, 28, . . .

17. 1, 0, − 1, . . .   **18.** 6, 0, − 6, . . .

19. 5, 2, − 1, . . .   **20.** 8, 4, 0, . . .

21. $\dfrac{1}{4}$, 0, $-\dfrac{1}{4}$, . . .   **22.** $\dfrac{2}{5}$, 0, $-\dfrac{2}{5}$, . . .

23. 1, $\dfrac{3}{2}$, 2, . . .   **24.** $\dfrac{1}{3}$, 1, $\dfrac{5}{3}$, . . .

Each sequence shown here is a geometric sequence. In each case, find the next number in the sequence.

25. 1, 3, 9, . . .   **26.** 1, 7, 49, . . .

27. 10, − 30, 90, . . .   **28.** 10, − 20, 40, . . .

29. 1, $\dfrac{1}{2}$, $\dfrac{1}{4}$, . . .   **30.** 1, $\dfrac{1}{3}$, $\dfrac{1}{9}$, . . .

31. 20, 10, 5, . . .   **32.** 8, 4, 2, . . .

33. 5, − 25, 125, . . .   **34.** − 4, 16, − 64, . . .

35. 1, $-\dfrac{1}{5}$, $\dfrac{1}{25}$, . . .   **36.** 1, $-\dfrac{1}{2}$, $\dfrac{1}{4}$, . . .

37. Find the next number in the sequence 4, 8, . . . if the sequence is
(a) An arithmetic sequence
(b) A geometric sequence

38. Find the next number in the sequence 1, − 4, . . . if the sequence is
(a) An arithmetic sequence
(b) A geometric sequence

39. Find the 12th term of the Fibonacci sequence.

40. Find the 13th term of the Fibonacci sequence.

Any number in the Fibonacci sequence is a *Fibonacci number.*

41. Name three Fibonacci numbers that are prime numbers.

42. Name three Fibonacci numbers that are composite numbers.

43. In the first ten terms of the Fibonacci sequence, which ones are even numbers?

44. In the first ten terms of the Fibonacci sequence, which ones are odd numbers?

Applying the Concepts

45. Climbing Snail A snail is climbing straight up a brick wall at a constant rate. It takes the snail 8 hours to climb up 6 feet. After climbing for 8 hours, the snail rests for 4 hours, during which time it slides 2 feet down the wall. The snail starts at the bottom of the wall and repeats this climbing up/sliding down process continually. Write a sequence of numbers that gives the snail's height above the ground every 4 hours, starting when the snail is at the bottom of the wall and ending 36 hours later.

46. **Climbing Snail**   If the snail from Problem 45 is climbing a 10-foot wall, how long will it take the snail to reach the top of the wall?

47. **Temperature and Altitude**   A pilot checks the weather conditions before flying and finds that the air temperature drops 3.5°F every 1,000 feet above the surface of the Earth. (The higher up he flies, the colder the air.) If the air temperature is 41°F when the plane reaches 10,000 feet, write a sequence of numbers that gives the air temperature every 1,000 feet as the plane climbs from 10,000 feet to 15,000 feet. Is this sequence an arithmetic sequence?

48. **Temperature and Altitude**   For the plane mentioned in Problem 47, at what altitude will the air temperature be 20°F?

49. **Temperature and Altitude**   The weather conditions on a certain day are such that the air temperature drops 4.5°F every 1,000 feet above the surface of the Earth. If the air temperature is 41°F at 10,000 feet, write a sequence of numbers that gives the air temperature every 1,000 feet starting at 10,000 feet and ending at 5,000 feet. Is this sequence an arithmetic sequence?

50. **Value of a Painting**   Suppose you own a painting that doubles in value every 5 years. If you bought the painting for $125 in 1990, write a sequence of numbers that gives the value of the painting every 5 years from the time you purchased it until the year 2010. Is this sequence a geometric sequence?

Half-Life   Reread the introduction to this chapter and the discussion of paired data at the end of this section, then work the following problems.

51. The half-lives of two antidepressants are given in the following table. A person on antidepressant 1

tells his doctor that he begins to feel sick if he misses his morning dose, while a patient taking antidepressant 2 tells his doctor that he doesn't notice a difference if he misses a day of taking the medication. Explain these situations in terms of the half-life of the medication.

Half-Life	
Antidepressant 1	*Antidepressant 2*
11 hours	5 days

52. Two patients are taking the antidepressants mentioned in Problem 51. Both decide to stop their current medications and take another medication. The physician tells the patient on antidepressant 2 to simply stop taking it, but instructs the patient on antidepressant 1 to take half the normal dose for 3 days, then one-fourth the normal dose for another 3 days before stopping the medication altogether. Use half-life to explain why the doctor uses two different methods of taking the patients off their medications.

53. **Half-Life**   The half-life of a drug is 4 hours. A patient has been taking the drug on a regular basis for a few months and then discontinues taking it. The concentration of the drug in a patient's system is 60 ng/ml when the patient stops taking the medication. Complete the table, and then use the results to construct a line graph of that data (a template is provided in Figure 3).

Hours Since Discontinuing	Concentration (ng/ml)
0	60
4	
8	
12	
16	

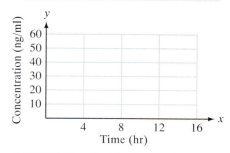

FIGURE 3 Template for line graph

54. Half-Life The half-life of a drug is 8 hours. A patient has been taking the drug on a regular basis for a few months and then discontinues taking it. The concentration of the drug in a patient's system is 120 ng/ml when the patient stops taking the medication. Complete the table, and then use the results to construct a line graph of that data (a template is provided in Figure 4).

Hours Since Discontinuing	Concentration (ng/ml)
0	120
8	
16	
24	
32	

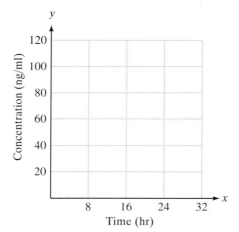

FIGURE 4 Template for line graph

55. Boiling Point The boiling point of water at sea level is 212°F. The boiling point of water drops 1.8°F every 1,000 feet above sea level, and it rises 1.8°F every 1,000 feet below sea level. Complete the following table to write the sequence that gives the boiling points of water from 2,000 feet below sea level to 3,000 fet above sea level.

Elevation (feet)	Boiling Point (°F)
−2,000	
−1,000	
0	
1,000	
2,000	
3,000	

56. Chlorine Level A swimming pool needs 1 ppm (part per million) of chlorine to keep bacteria growth to a minimum. A pool contains 3 ppm chlorine and loses 20% of that amount each day if no more chlorine is added. Fill in the following table, assuming that no additional chlorine is added to a pool containing 3 ppm chlorine, until the chlorine level drops below 1 ppm.

Day	1	2	3	4	5	6	7	8
Chlorine Level	3							

57. Doubling Time Currently, the world's population is increasing 2% each year, which means that the population will double in 35 years. Using a 1997 world population of 5.852 billion, write a sequence of numbers that will give the world's population every 35 years from 1997 until the year 2137.

58. Harmonic Sequence The sequence $1, \frac{1}{2}, \frac{1}{3}, \frac{1}{4}, \frac{1}{5}, \ldots$ is a harmonic sequence and can be mathematically associated with a vibrating musical string.
(a) Find the next term in this sequence.
(b) Find the 78th term in this sequence.
(c) Write an expression for term number n of the sequence.

Extending the Concepts

59. Reading Graphs A patient is taking a prescribed dose of a medication every 4 hours during the day to relieve the symptoms of a cold. Figure 5 shows how the concentration of that medication in the patient's system changes over time. The 0 on the horizontal axis corresponds to the time the patient takes the first dose of the medication.

(a) Explain what the steep vertical line segments show with regard to the patient and his medication.
(b) What has happened to make the graph fall off on the right?
(c) What is the maximum concentration of the medication in the patient's system during the time period shown in Figure 5?
(d) Find the values of A, B, and C.

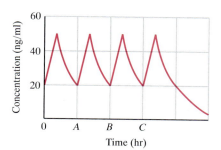

FIGURE 5

60. **Reading Graphs** Figure 6 shows the number of people in line at a theater box office to buy tickets for a movie that starts at 7:30. The box office opens at 6:45.
(a) How many people are in line at 6:30?
(b) How many people are in line when the box office opens?
(c) How many people are in line when the show starts?
(d) At what time are there 60 people in line?
(e) How long after the show starts is there no one left in line?

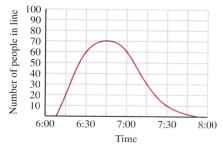

FIGURE 6

61. **Constructing a Line Graph** The diagram on the milk carton (Figure 7) shows the relationship between the temperature at which the milk is stored and the length of time before it spoils and cannot be consumed. Use the information on the carton to complete Table 2. Then use the information in the table to construct a line graph (Figure 8).

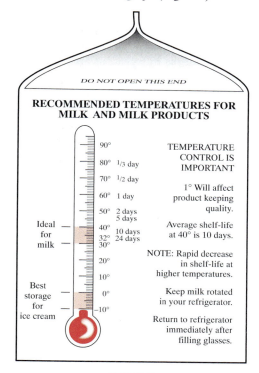

FIGURE 7

| TABLE 2 | Shelf-Life of Milk |
Temperature (Fahrenheit)	Shelf-Life (days)
32°	24
40°	
50°	
60°	
70°	

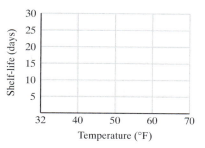

FIGURE 8 Template for line graph

Another Mathematical Model

Suppose you are going to toss a coin three times and you want to know how many different orders are possible for the sequence of heads and tails that results. To find out, you can draw what is called a *tree diagram*. Each branch of the tree gives one of the possible outcomes of the coin toss. The tree diagram is similar to the pattern we used to find the number of bees in the family tree of the male honeybee.

First toss — 2 possibilities

Second toss — 4 possibilities

Third toss — 8 possibilities

The branches that run down the tree correspond to the possible outcomes of tossing a coin three times. For example, the first branch on the left is H — H — H, which corresponds to the coin coming up heads on all three tosses. The third branch from the left is H — T — H, which corresponds to a head on the first toss, a tail on the second toss, and a head on the third toss.

62. Copy the diagram and continue it one more row. How many branches does the tree have to the fourth row?
63. Write out the sequence of numbers that corresponds to the number of branches for each toss of the coin. (This is your mathematical model of the situation.)
64. How many branches will the tree have if the coin is tossed five times?
65. How many branches will the tree have if the coin is tossed ten times?

Examples

The margins of the chapter summaries will be used for brief examples of the topics being reviewed, whenever it is convenient.

1. $2^5 = 2 \cdot 2 \cdot 2 \cdot 2 \cdot 2 = 32$
$5^2 = 5 \cdot 5 = 25$
$10^3 = 10 \cdot 10 \cdot 10 = 1,000$
$1^4 = 1 \cdot 1 \cdot 1 \cdot 1 = 1$

The numbers in brackets refer to the section(s) in which the topic can be found.

Exponents [1.1]

Exponents represent notation used to indicate repeated multiplication. In the expression 3^4, 3 is the *base* and 4 is the *exponent*.

$$3^4 = 3 \cdot 3 \cdot 3 \cdot 3 = 81$$

The expression 3^4 is said to be in *exponential form,* while the expression $3 \cdot 3 \cdot 3 \cdot 3$ is in *expanded form.*

2. $10 + (2 \cdot 3^2 - 4 \cdot 2)$
$= 10 + (2 \cdot 9 - 4 \cdot 2)$
$= 10 + (18 - 8)$
$= 10 + 10$
$= 20$

Order of Operations [1.1]

When evaluating a mathematical expression, we will perform the operations in the following order, beginning with the expression in the innermost parentheses or brackets and working our way out.

1. Simplify all numbers with exponents, working from left to right if more than one of these numbers is present.

2. Then, do all multiplications and divisions left to right.

3. Finally, perform all additions and subtractions left to right.

3. If $A = \{0, 1, 2\}$ and $B = \{2, 3\}$, then
$A \cup B = \{0, 1, 2, 3\}$ and
$A \cap B = \{2\}$.

Sets [1.1]

A *set* is a collection of objects or things.

The *union* of two sets A and B, written $A \cup B$, is all the elements that are in A *or* are in B.

The *intersection* of two sets A and B, written $A \cap B$, is the set consisting of all elements common to both A *and* B.

Set A is a *subset* of set B, written $A \subset B$, if all elements in set A are also in set B.

4. 5 is a counting number, a whole number, an integer, a rational number, and a real number.

$\frac{3}{4}$ is a rational number and a real number.

$\sqrt{2}$ is an irrational number and a real number.

Special Sets [1.2]

Counting numbers $= \{1, 2, 3, \ldots\}$

Whole numbers $= \{0, 1, 2, 3, \ldots\}$

Integers $= \{\ldots, -3, -2, -1, 0, 1, 2, 3, \ldots\}$

Rational numbers $= \{\frac{a}{b} \mid a \text{ and } b \text{ are integers}, b \neq 0\}$

Irrational numbers $= \{x \mid x \text{ is real, but not rational}\}$

Real numbers $= \{x \mid x \text{ is rational or } x \text{ is irrational}\}$

Prime numbers $= \{2, 3, 5, 7, 11, \ldots\} = \{x \mid x \text{ is a positive integer greater than 1 whose only positive divisors are itself and 1}\}$

5. Graph each inequality mentioned at the right.

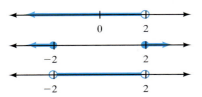

Inequalities [1.2]
The set $\{x \mid x < 2\}$ is the set of all real numbers that are less than 2. To graph this set we place an open circle at 2 on the real number line and then draw an arrow that starts at 2 and points to the left.
     The set $\{x \mid x \leq -2 \text{ or } x \geq 2\}$ is the set of all real numbers that are either less than or equal to -2 or greater than or equal to 2.
     The set $\{x \mid -2 < x < 2\}$ is the set of all real numbers that are between -2 and 2, that is, the real numbers that are greater than -2 and less than 2.

6. The numbers 5 and -5 are opposites; their sum is 0.

$$5 + (-5) = 0$$

Opposites [1.3, 1.4]
Any two real numbers the same distance from 0 on the number line, but in opposite directions from 0, are called *opposites,* or *additive inverses.* Opposites always add to 0.

7. The numbers 3 and $\frac{1}{3}$ are reciprocals; their product is 1.

$$3\left(\frac{1}{3}\right) = 1$$

Reciprocals [1.3, 1.4]
Any two real numbers whose product is 1 are called *reciprocals,* or *multiplicative inverses.* Every real number has a reciprocal except 0.

8. $|5| = 5$
$|-5| = 5$

Absolute Value [1.3]
The *absolute value* of a real number is its distance from 0 on the number line. If $|x|$ represents the absolute value of x, then

$$|x| = \begin{cases} x & \text{if } x \geq 0 \\ -x & \text{if } x < 0 \end{cases}$$

The absolute value of a real number is never negative.

Properties of Real Numbers [1.3]

	For Addition	*For Multiplication*
Commutative	$a + b = b + a$	$ab = ba$
Associative	$a + (b + c) = (a + b) + c$	$a(bc) = (ab)c$
Identity	$a + 0 = a$	$a \cdot 1 = a$
Inverse	$a + (-a) = 0$	$a\left(\frac{1}{a}\right) = 1$
Distributive	$a(b + c) = ab + ac$	

9. $5 + 3 = 8$
$5 + (-3) = 2$
$-5 + 3 = -2$
$-5 + (-3) = -8$

Addition [1.4]
To add two real numbers with

1. *The same sign:* Simply add absolute values and use the common sign.

2. *Different signs:* Subtract the smaller absolute value from the larger absolute value. The answer has the same sign as the number with the larger absolute value.

10. $6 - 2 = 6 + (-2) = 4$
$6 - (-2) = 6 + 2 = 8$

Subtraction [1.4]

If a and b are real numbers,

$$a - b = a + (-b)$$

To subtract b, add the opposite of b.

11. $5(4) = 20$
$5(-4) = -20$
$-5(4) = -20$
$-5(-4) = 20$

Multiplication [1.4]

To multiply two real numbers, simply multiply their absolute values. Like signs give a positive answer. Unlike signs give a negative answer.

12. $\frac{12}{-3} = -4$
$\frac{-12}{-3} = 4$

Division [1.4]

If a and b are real numbers and $b \neq 0$, then

$$\frac{a}{b} = a \cdot \left(\frac{1}{b}\right)$$

To divide by b, multiply by the reciprocal of b.

13. The following statement is a conditional statement.

If $x = 3$, then $x^2 = 9$.

The hypothesis is $x = 3$, while the conclusion is $x^2 = 9$.

For the preceding conditional statement, we have the following three associated statements:

The converse:
If $x^2 = 9$, then $x = 3$.
The inverse:
If $x \neq 3$, then $x^2 \neq 9$.
The contrapositive:
If $x^2 \neq 9$, then $x \neq 3$.

Conditional Statements [1.5]

A *conditional statement* is any statement that can be written in the form "If A, then B." The first phrase, A, is called the *hypothesis,* and the second phrase, B, is called the *conclusion.* A shorthand way to write an "if/then" statement is with the implies symbol:

$$A \Rightarrow B$$

For every conditional statement "If A, then B," there exist the following associated statements:

The converse: If B, then A.
The inverse: If not A, then not B.
The contrapositive: If not B, then not A.

If the original conditional statement is true, then the contrapositive is also true.

14. We use inductive reasoning when we conclude that the next number in the following sequence is 25.

$1, 4, 9, 16, \ldots$

Inductive Reasoning [1.6]

Inductive reasoning is reasoning in which a conclusion is drawn based on evidence and observations that support that conclusion. In mathematics this usually involves noticing that a few items in a group have a trait or characteristic in common, and then concluding that all items in the group have that same trait.

15. The following sequence is an arithmetic sequence since each term is obtained from the preceding term by adding 3 each time.

$4, 7, 10, 13, \ldots$

Arithmetic Sequence [1.6]

An *arithmetic sequence* is a sequence of numbers in which each number (after the first number) comes from adding the same amount to the number before it. The number we add to each term to obtain the next term is called the *common difference.*

16. The following sequence is a geometric sequence since each term is obtained from the previous term by multiplying by 3 each time.

$$4, 12, 36, 108, \ldots$$

Geometric Sequence [1.6]

A *geometric sequence* is a sequence of numbers in which each number (after the first number) comes from the number before it by multiplying by the same amount each time. The amount by which we multiply each term to obtain the next term is called the *common ratio*.

COMMON MISTAKES

1. Interpreting absolute value as changing the sign of the number inside the absolute value symbols. That is, $|-5| = +5, |+5| = -5$. To avoid this mistake, remember, absolute value is defined as a distance and distance is always measured in positive units.

2. Confusing $-(-5)$ with $-|-5|$. The first answer is $+5$, but the second answer is -5.

CHAPTER 1   REVIEW

The numbers in brackets refer to the section(s) in the text where similar problems can be found.

Translate each expression into symbols. [1.1]

1. The sum of x and 2
2. The difference of x and 2
3. The quotient of x and 2
4. The product of 2 and x
5. Twice the sum of x and y
6. The sum of twice x and y

Expand and multiply. [1.1]

7. 3^3 8. 5^3 9. 8^2
10. 1^8 11. 2^5 12. 3^4

Simplify each expression. [1.1]

13. $2 + 3 \cdot 5$ 14. $10 - 2 \cdot 3$
15. $20 \div 2 + 3$ 16. $30 \div 6 + 4 \div 2$
17. $3 + 2(5 - 2)$ 18. $(10 - 2)(7 - 3)$
19. $3 \cdot 4^2 - 2 \cdot 3^2$ 20. $3 + 5(2 \cdot 3^2 - 10)$

Let $A = \{1, 3, 5\}$, $B = \{2, 4, 6\}$, and $C = \{0, 1, 2, 3, 4\}$, and find each of the following. [1.1]

21. $A \cup B$ 22. $A \cap C$
23. $\{x | x \in A \text{ and } x \notin C\}$
24. $\{x | x \in B \text{ and } x > 4\}$
25. Locate the numbers $-4, -2.5, -1, 0, 1.5,$ $3.1, 4.75$ on the number line. [1.2]
26. Locate the numbers $-4.75, -3, -1, 0, 1.5,$ $2.3, 3$ on the number line. [1.2]

Give the opposite and reciprocal of each number. [1.3]

27. 2   28. $-\dfrac{2}{5}$

Write each expression without absolute value symbols, then simplify. [1.3]

29. $|-3|$ 30. $-|-5|$
31. $|-4|$ 32. $|10 - 16|$

For the set $\{-7, -4.2, -\sqrt{3}, 0, \frac{3}{4}, \pi, 5\}$ list all the elements that are in the following sets. [1.2]

33. Integers   34. Rational numbers
35. Irrational numbers
36. Factor 4,356 into the product of prime factors. [1.2]
37. Reduce $\dfrac{4,356}{5,148}$ to lowest terms. [1.2]

PEMDAS

Multiply. [1.3]

38. $\dfrac{3}{4} \cdot \dfrac{8}{5} \cdot \dfrac{5}{6}$

39. $\left(\dfrac{3}{4}\right)^3$

40. $\dfrac{1}{4} \cdot 8$

Graph each inequality. [1.2]

41. $\{x \mid x < -2 \text{ or } x > 3\}$

42. $\{x \mid x > 2 \text{ and } x < 5\}$

43. $\{x \mid -3 \le x \le 4\}$

44. $\{x \mid 0 \le x \le 5 \text{ or } x > 10\}$

Translate each statement into an equivalent inequality. [1.2]

45. x is at least 4.   **46.** x is no more than 5.

47. x is between 0 and 8.

48. x is between 0 and 8, inclusive.

Combine similar terms. [1.4]

49. $-2y + 4y$ **50.** $-3x - x + 7x$

51. $3x - 2 + 5x + 7$ **52.** $2y + 4 - y - 2$

Match each numbered expression with the letter of the appropriate property (or properties) below. [1.3]

53. $x + 3 = 3 + x$

54. $(x + 2) + 3 = x + (2 + 3)$

55. $3(x + 4) = 3(4 + x)$

56. $(5x)y = x(5y)$

57. $(x + 2) + y = (x + y) + 2$

58. $3(1) = 3$

59. $5 + 0 = 5$

60. $5 + (-5) = 0$

   (a) Commutative property of addition
   (b) Commutative property of multiplication
   (c) Associative property of addition
   (d) Associative property of multiplication
   (e) Additive identity
   (f) Multiplicative identity
   (g) Additive inverse
   (h) Multiplicative inverse

Find the following sums and differences. [1.4]

61. $5 - 3$ **62.** $-5 - (-3)$

63. $7 + (-2) - 4$ **64.** $6 - (-3) + 8$

65. $|-4| - |-3| + |-2|$

66. $|7 - 9| - |-3 - 5|$

67. $6 - (-3) - 2 - 5$ **68.** $2 \cdot 3^2 - 4 \cdot 2^3 + 5 \cdot 4^2$

69. $-\dfrac{1}{12} - \dfrac{1}{6} - \dfrac{1}{4} - \dfrac{1}{3}$

70. $-\dfrac{1}{3} - \dfrac{1}{4} - \dfrac{1}{6} - \dfrac{1}{12}$

Find the following products. [1.4]

71. $6(-7)$ **72.** $-3(5)(-2)$

73. $7(3x)$ **74.** $-3(2x)$

Apply the distributive property. [1.4]

75. $-2(3x - 5)$ **76.** $-3(2x - 7)$

77. $-\dfrac{1}{2}(2x - 6)$ **78.** $-3(5x - 1)$

Divide. [1.4]

79. $-\dfrac{5}{8} \div \dfrac{3}{4}$ **80.** $-12 \div \dfrac{1}{3}$

81. $\dfrac{3}{5} \div 6$ **82.** $\dfrac{4}{7} \div (-2)$

Simplify each expression as much as possible. [1.4]

83. $2(-5) - 3$ **84.** $3(-4) - 5$

85. $6 + 3(-2)$ **86.** $7 + 2(-4)$

87. $-3(2) - 5(6)$ **88.** $-4(3)^2 - 2(-1)^3$

89. $8 - 2(6 - 10)$ **90.** $(8 - 2)(6 - 10)$

91. $\dfrac{3(-4) - 8}{-5 - 5}$ **92.** $\dfrac{9(-1)^3 - 3(-6)^2}{6 - 9}$

93. $4 - (-2)\left[\dfrac{6 - 3(-4)}{1 + 5(-2)}\right]$

Simplify. [1.3, 1.4]

94. $7 - 2(3y - 1) + 4y$

95. $4(3x - 1) - 5(6x + 2)$

96. $4(2a - 5) - (3a + 2)$

For each conditional statement, write the converse, the inverse, and the contrapositive. [1.5]

97. If $x = -7$, then $|x| = 7$.

98. If Therese lives in Amarillo, then she lives in Texas.

For each conditional statement, assume that the statement is true, and then write another conditional statement that must also be true. [1.5]

99. If $x = 3$, then $x^4 = 81$.

100. If Maria goes to the beach, then she does not go to school.

Find the next number in each sequence. Identify any sequences that are arithmetic and any that are geometric. [1.6]

101. $11, 8, 5, 2, \ldots$

102. $10, -30, 90, -270, \ldots$

103. $1, 1, 2, 3, 5, \ldots$ **104.** $1, 16, 81, 256, \ldots$

105. $1, \dfrac{1}{2}, 0, -\dfrac{1}{2}, \ldots$ **106.** $1, -\dfrac{1}{2}, \dfrac{1}{4}, -\dfrac{1}{8}, \ldots$

CHAPTER 1 PROJECTS

BASIC PROPERTIES AND DEFINITIONS

Students and Instructors: The end of each chapter in this book will contain a page like this one containing two projects. The group project is intended to be done in class. The research projects are to be completed outside of class. They can be done in groups or individually. In my classes, I use the research projects for extra credit. I require all research projects to be done on a word processor and to be free of spelling errors.

GROUP PROJECT

GAUSS'S METHOD FOR ADDING CONSECUTIVE INTEGERS

Karl Gauss, 1777–1855
(Corbis/Bettmann)

Number of People: 3

Time Needed: 15–20 minutes

Equipment: Paper and pencil

Background: There is a popular story about famous mathematician Karl Friedrich Gauss (1777–1855). As the story goes, a 9-year-old Gauss, and the rest of his class, was given what the teacher had hoped was busy work. They were told to find the sum of all the whole numbers from 1 to 100. While the rest of the class labored under the assignment, Gauss found the answer within a few moments. He may have set up the problem like this:

1	2	3	4	$\cdots$	98	99	100
100	99	98	97	$\cdots$	3	2	1

Procedure: Copy the above illustration to get started.

1. Add the numbers in the column and write your answers below the line.

2. How many numbers do you have below the line?

3. Suppose you add all the numbers below the line; how will your result be related to the sum of all the whole numbers from 1 to 100?

4. Use multiplication to add all the numbers below the line.

5. What is the sum of all the whole numbers from 1 to 100?

6. Use Gauss's method to add all the whole numbers from 1 to 10. (The answer is 55.)

7. Explain, in your own words, Gauss's method for finding this sum so quickly.

RESEARCH PROJECT

BERTRAND RUSSELL

To see the connection between the topics we studied in Section 1.5 and mathematics in general, here is a quote taken from the beginning of the first sentence in the book *Principles of Mathematics* by the British philosopher and mathematician, Bertrand Russell.

> Pure Mathematics is the class of all propositions of the form "*p* implies *q*," where *p* and *q* are propositions containing one or more variables . . .

He is using the phrase "*p* implies *q*" in the same way we use the phrase "If *A*, then *B*." Our work with conditional statements is actually an introduction to the foundations upon which all of mathematics is built.

Write an essay on the life of Bertrand Russell. In the essay, indicate what purpose he had for writing and publishing his book *Principles of Mathematics*. Write in complete sentences and organize your work just as you would if you were writing a paper for an English class.

Bertrand Russell *(Corbis/ Bettmann-UPI)*

CHAPTER 1 TEST

The numbers in brackets indicate the section(s) to which the problems correspond.

Write each of the following in symbols. [1.1]

1. Twice the sum of $3x$ and $4y$

2. The difference of $2a$ and $3b$ is less than their sum

Simplify each expression using the rule for order of operations. [1.1]

3. $3 \cdot 2^2 + 5 \cdot 3^2$

4. $6 + 2(4 \cdot 3 - 10)$

5. $12 - 8 \div 4 + 2 \cdot 3$

6. $20 - 4[3^2 - 2(2^3 - 6)]$

If $A = \{1, 2, 3, 4\}$, $B = \{2, 4, 6\}$, and $C = \{1, 3, 5\}$, find: [1.1]

7. $A \cap B$ **8.** $\{x \mid x \in B \text{ and } x \in C\}$

For the set $\{-5, -4.1, -3.75, -\frac{5}{6}, -\sqrt{2}, 0, \sqrt{3}, 1, 1.8, 4\}$, list all the elements belonging to the following sets. [1.2]

9. Integers   **10.** Rational numbers

11. Irrational numbers

12. Factor 585 into the product of prime factors. [1.2]

Give the opposite and reciprocal of each of the following. [1.3]

13. -3 **14.** $\dfrac{4}{3}$

Simplify each of the following. [1.3]

15. $-(-3)$ **16.** $-|-2|$

Graph each of the following. [1.2]

17. $\{x \mid x \le -1 \text{ or } x > 5\}$
18. $\{x \mid -2 \le x \le 4\}$

State the property or properties that justify each of the following. [1.3]

19. $4 + x = x + 4$ **20.** $5(1) = 5$
21. $3(x \cdot y) = (3y) \cdot x$
22. $(a + 1) + b = (a + b) + 1$

Simplify each of the following as much as possible. [1.4]

23. $5(-4) + 1$ **24.** $-4(-3) + 2$
25. $3(2 - 4)^3 - 5(2 - 7)^2$

26. $-2 + 5[7 - 3(-4 - 8)]$

27. $\dfrac{-4(-1) - (-10)}{5 - (-2)}$

28. $3 - 2\left[\dfrac{8(-1) - 7}{-3(2) - 4}\right]$

29. $-\dfrac{3}{8} + \dfrac{5}{12} - \left(-\dfrac{7}{9}\right)$

30. $-\dfrac{1}{2}(8x)$

31. $-4(3x + 2) + 7x$
32. $5(2y - 3) - (6y - 5)$
33. $3 + 4(2x - 5) - 5x$
34. $2 + 5a + 3(2a - 4)$
35. Add $-\frac{2}{3}$ to the product of -2 and $\frac{5}{6}$.
36. Subtract $\frac{3}{4}$ from the product of -4 and $\frac{7}{16}$.
37. For the following statement, write the converse, the inverse, and the contrapositive. [1.5]

    If Emily goes out at night, then she does not study.

38. Assume the following conditional statement is true, then write another conditional statement that must also be true. [1.5]

    If $x = -3$, then $|x| = 3$.

Find the next number in each sequence. Identify any sequences that are arithmetic and any that are geometric. [1.6]

39. $5, -20, 80, \ldots$ **40.** $1, 0, -1, \ldots$

41. $7, 8, 10, 13, \ldots$ **42.** $1, \dfrac{1}{5}, \dfrac{1}{25}, \ldots$

Equations and Inequalities in One Variable

INTRODUCTION

A recent newspaper article gave the following guideline for college students taking out loans to finance their education: The maximum monthly payment on the amount borrowed should not exceed 8% of their monthly starting salary. In this situation, the maximum monthly payment can be described mathematically with the formula

$$L(x) = 0.08x$$

(Telegraph Colour Library/FPG International)

Using this formula, we can construct a table and a line graph that show the maximum student loan payments that can be made for a variety of starting salaries.

TABLE 1	Maximum Student Loan Payments
Monthly Starting Salary	**Maximum Loan Payment**
$2,000	$160
$2,500	$200
$3,000	$240
$3,500	$280
$4,000	$320
$4,500	$360
$5,000	$400

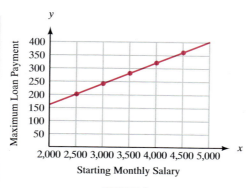

FIGURE 1

In this chapter we begin our work with connecting tables and line graphs with algebraic formulas.

STUDY SKILLS

If you have successfully completed Chapter 1, then you have made a good start at developing the study skills necessary to succeed in all math classes. Some of the study skills for this chapter are a continuation of the skills from Chapter 1, while others are new to this chapter.

1. **Continue to Set and Keep a Schedule** Sometimes I find students do well in Chapter 1 and then become overconfident. They will begin to put in less time with their homework. Don't do it. Keep to the same schedule.

2. **Increase Effectiveness** You want to become more and more effective with the time you spend on your homework. Increase those activities that are the most beneficial and decrease those that have not given you the results you want.

3. **List Difficult Problems** Begin to make lists of problems that give you the most difficulty. These are the problems in which you are repeatedly making mistakes.

4. **Begin to Develop Confidence With Word Problems** It seems that the main difference between people who are good at working word problems and those who are not is confidence. People with confidence know that no matter how long it takes them, they will eventually be able to solve the problem. Those without confidence begin by saying to themselves, "I'll never be able to work this problem." If you are in this second category, then instead of telling yourself that you can't do word problems, decide to do whatever it takes to master them. The more word problems you work, the better you will become at them.

Many of my students keep a notebook that contains everything that they need for the course: class notes, homework, quizzes, tests, and research projects. A three-ring binder with tabs is ideal. Organize your notebook so that you can easily get to any item you wish to look at.

Linear Equations in One Variable

A **linear equation in one variable** is any equation that can be put in the form

$$ax + b = c$$

where a, b, and c are constants and $a \neq 0$. For example, each of the equations

$$5x + 3 = 2 \qquad 2x = 7 \qquad 2x + 5 = 0$$

are linear because they can be put in the form $ax + b = c$. In the first equation, $5x$, 3, and 2 are called **terms** of the equation — $5x$ is a variable term; 3 and 2 are constant terms.

> **DEFINITION** The **solution set** for an equation is the set of all numbers that, when used in place of the variable, make the equation a true statement.

EXAMPLE 1 The solution set for $2x - 3 = 9$ is $\{6\}$, since replacing x with 6 makes the equation a true statement.

$$
\begin{aligned}
\text{If} \qquad\qquad x &= 6 \\
\text{then} \qquad\qquad 2x - 3 &= 9 \\
\text{becomes} \qquad 2(6) - 3 &= 9 \\
12 - 3 &= 9 \\
9 &= 9 \qquad \text{A true statement}
\end{aligned}
$$

> **DEFINITION** Two or more equations with the same solution set are called **equivalent equations.**

EXAMPLE 2 The equations $2x - 5 = 9$, $x - 1 = 6$, and $x = 7$ are all equivalent equations since the solution set for each is $\{7\}$.

PROPERTIES OF EQUALITY

The first property states that adding the same quantity to both sides of an equation preserves equality. Or, more importantly, adding the same amount to both sides of an equation *never changes* the solution set. This property is called the **addition property of equality** and is stated in symbols as follows.

ADDITION PROPERTY OF EQUALITY

For any three algebraic expressions A, B, and C,

$$\text{if} \qquad A = B$$

$$\text{then} \qquad A + C = B + C$$

In words: Adding the same quantity to both sides of an equation will not change the solution.

Our second new property is called the **multiplication property of equality** and is stated like this.

MULTIPLICATION PROPERTY OF EQUALITY

For any three algebraic expressions A, B, and C, where $C \neq 0$,

$$\text{if} \qquad A = B$$

$$\text{then} \qquad AC = BC$$

In words: Multiplying both sides of an equation by the same nonzero quantity will not change the solution.

Note: Since subtraction is defined in terms of addition, and division is defined in terms of multiplication, we do not need to introduce separate properties for subtraction and division. The solution set for an equation will never be changed by subtracting the same amount from both sides or by dividing both sides by the same nonzero quantity.

The following examples illustrate how we use the properties from Chapter 1 along with the addition property of equality and the multiplication property of equality to solve linear equations.

EXAMPLE 3   Solve $\dfrac{3}{4}x + 5 = -4$.

Solution   We begin by adding -5 to both sides of the equation. Once this has been done, we multiply both sides by the reciprocal of $\frac{3}{4}$, which is $\frac{4}{3}$.

$$\frac{3}{4}x + 5 = -4$$

$$\frac{3}{4}x + 5 + (\mathbf{-5}) = -4 + (\mathbf{-5}) \qquad \text{Add } \mathbf{-5} \text{ to both sides.}$$

$$\frac{3}{4}x = -9$$

$$\frac{\mathbf{4}}{\mathbf{3}}\left(\frac{3}{4}x\right) = \frac{\mathbf{4}}{\mathbf{3}}(-9) \qquad \text{Multiply both sides by } \frac{\mathbf{4}}{\mathbf{3}}.$$

$$x = -12 \qquad\qquad \frac{4}{3}(-9) = \frac{4}{3}\left(-\frac{9}{1}\right) = -\frac{36}{3} = -12$$

Our next example involves solving an equation that has variable terms on both sides of the equal sign.

EXAMPLE   4   Find the solution set for $3a - 5 = -6a + 1$.

Solution   To solve for a we must isolate it on one side of the equation. Let's decide to isolate a on the left side by adding $6a$ to both sides of the equation.

$$3a - 5 = -6a + 1$$

$$3a + \mathbf{6a} - 5 = -6a + \mathbf{6a} + 1 \qquad \text{Add } \mathbf{6a} \text{ to both sides.}$$

$$9a - 5 = 1$$

$$9a - 5 + \mathbf{5} = 1 + \mathbf{5} \qquad \text{Add } \mathbf{5} \text{ to both sides.}$$

$$9a = 6$$

$$\frac{\mathbf{1}}{\mathbf{9}}(9a) = \frac{\mathbf{1}}{\mathbf{9}}(6) \qquad \text{Multiply both sides by } \frac{\mathbf{1}}{\mathbf{9}}.$$

$$a = \frac{2}{3} \qquad\qquad \frac{1}{9}(6) = \frac{6}{9} = \frac{2}{3}$$

Note: From Chapter 1 we know that multiplication by a number and division by its reciprocal always produce the same result. Because of this fact, instead of multiplying each side of our equation by $\frac{1}{9}$, we could just as easily divide each side by 9. If we did so, the last two lines in our solution would look like this:

$$\frac{9a}{9} = \frac{6}{9}$$

$$a = \frac{2}{3}$$

There will be times when we solve equations and end up with a negative sign in front of the variable. The next example shows how to handle this situation.

EXAMPLE   5   Solve each equation.

(a) $-x = 4$ **(b)** $-y = -8$

Solution   Neither equation can be considered solved because of the negative sign in front of the variable. To eliminate the negative signs we simply multiply each side of the equations by -1.

(a) $-x = 4$ **(b)** $-y = -8$

$\quad -1(-x) = -1(4)$ $\quad -1(-y) = -1(-8)$ Multiply each side

$\quad\quad x = -4$ $\quad\quad y = 8$ by -1.

The next example involves fractions. The least common denominator, which is the smallest expression that is divisible by each of the denominators, can be used with the multiplication property of equality to simplify equations containing fractions.

E X A M P L E   6   Solve $\dfrac{2}{3}x + \dfrac{1}{2} = -\dfrac{3}{8}$.

Solution   We can solve this equation by applying our properties and working with fractions, or we can begin by eliminating the fractions. Let's use both methods.

Method 1   Working with the fractions.

$$\frac{2}{3}x + \frac{1}{2} + \left(-\frac{1}{2}\right) = -\frac{3}{8} + \left(-\frac{1}{2}\right) \qquad \text{Add } -\frac{1}{2} \text{ to each side.}$$

$$\frac{2}{3}x = -\frac{7}{8} \qquad\qquad -\frac{3}{8} + \left(-\frac{1}{2}\right) = -\frac{3}{8} + \left(-\frac{4}{8}\right)$$

$$\frac{3}{2}\left(\frac{2}{3}x\right) = \frac{3}{2}\left(-\frac{7}{8}\right) \qquad \text{Multiply each side by } \frac{3}{2}.$$

$$x = -\frac{21}{16}$$

Method 2   Eliminating the fractions in the beginning.

Our original equation has denominators of 3, 2, and 8. The least common denominator, abbreviated LCD, for these three denominators is 24, and it has the property that all three denominators will divide it evenly. Therefore, if we multiply both sides of our equation by 24, each denominator will divide into 24, and we will be left with an equation that does not contain any denominators other than 1.

$$24\left(\frac{2}{3}x + \frac{1}{2}\right) = 24\left(-\frac{3}{8}\right) \qquad \begin{array}{l}\text{Multiply each side}\\ \text{by the LCD } \mathbf{24.}\end{array}$$

$$24\left(\frac{2}{3}x\right) + 24\left(\frac{1}{2}\right) = 24\left(-\frac{3}{8}\right) \qquad \begin{array}{l}\text{Distributive property}\\ \text{on the left side}\end{array}$$

$$16x + 12 = -9 \qquad \text{Multiply.}$$

$$16x = -21 \qquad \text{Add } -12 \text{ to each side.}$$

$$x = -\frac{21}{16} \qquad \text{Multiply each side by } \frac{1}{16}.$$

As the third line above indicated, multiplying each side of the equation by the LCD eliminates all the fractions from the equation. Both methods yield the same solution. To check our solution, we substitute $x = -21/16$ back into our original equation to obtain

$$\frac{2}{3}\left(-\frac{21}{16}\right) + \frac{1}{2} \overset{?}{=} -\frac{3}{8}$$

$$-\frac{7}{8} + \frac{1}{2} \overset{?}{=} -\frac{3}{8}$$

$$-\frac{7}{8} + \frac{4}{8} \overset{?}{=} -\frac{3}{8}$$

$$-\frac{3}{8} = -\frac{3}{8}$$

As we can see, our solution checks.

Note: We are placing question marks over the equal signs because we don't know yet if the expressions on the left will be equal to the expressions on the right.

EXAMPLE 7   Solve the equation $0.06x + 0.05(10{,}000 - x) = 560$.

Solution   We can solve the equation in its original form by working with the decimals, or we can eliminate the decimals first by using the multiplication property of equality and solve the resulting equation. Here are both methods.

Method 1   Working with the decimals.

$0.06x + 0.05(10{,}000 - x) = 560$	Original equation
$0.06x + 0.05(10{,}000) - 0.05x = 560$	Distributive property
$0.01x + 500 = 560$	Simplify the left side.
$0.01x + 500 + (\mathbf{-500}) = 560 + (\mathbf{-500})$	Add $\mathbf{-500}$ to each side.
$0.01x = 60$	
$\dfrac{0.01x}{\mathbf{0.01}} = \dfrac{60}{\mathbf{0.01}}$	Divide each side by $\mathbf{0.01}$.
$x = 6{,}000$	

Method 2   Eliminating the decimals in the beginning.

To move the decimal point two places to the right in $0.06x$ and 0.05, we multiply each side of the equation by 100.

$0.06x + 0.05(10{,}000 - x) = 560$	Original equation
$0.06x + 500 - 0.05x = 560$	Distributive property
$\mathbf{100}(0.06x) + \mathbf{100}(500) - \mathbf{100}(0.05x) = \mathbf{100}(560)$	Multiply each side by $\mathbf{100}$.
$6x + 50{,}000 - 5x = 56{,}000$	Multiply.
$x + 50{,}000 = 56{,}000$	Simplify the left side.
$x = 6{,}000$	Add $-50{,}000$ to each side.

Using either method, the solution to our equation is 6,000. We check our work (to be sure we have not made a mistake in applying the properties or an arithmetic mistake) by substituting 6,000 into our original equation and simplifying each side of the result separately.

Check: Substituting 6,000 for x in the original equation, we have

$$0.06(6,000) + 0.05(10,000 - 6,000) \overset{?}{=} 560$$

$$0.06(6,000) + 0.05(4,000) \overset{?}{=} 560$$

$$360 + 200 \overset{?}{=} 560$$

$$560 = 560 \qquad \text{A true statement}$$

Here is a list of steps to use as a guideline for solving linear equations in one variable.

STRATEGY FOR SOLVING LINEAR EQUATIONS IN ONE VARIABLE

Step 1a: Use the distributive property to separate terms, if necessary.

1b: If fractions are present, consider multiplying both sides by the LCD to eliminate the fractions. If decimals are present, consider multiplying both sides by a power of 10 to clear the equation of decimals.

1c: Combine similar terms on each side of the equation.

Step 2: Use the addition property of equality to get all variable terms on one side of the equation and all constant terms on the other side. A variable term is a term that contains the variable (e.g., $5x$). A constant term is a term that does not contain the variable (the number 3, for example).

Step 3: Use the multiplication property of equality to get x (i.e., $1x$) by itself on one side of the equation.

Step 4: Check your solution in the original equation to be sure that you have not made a mistake in the solution process.

As you will see as you work through the problems in the problem set, it is not always necessary to use all four steps when solving equations. The number of steps used depends on the equation. In Example 8 there are no fractions or decimals in the original equation, so step 1b will not be used.

EXAMPLE 8   Solve $3(2y - 1) + y = 5y + 3$.

Solution   Applying the steps outlined in the strategy, we have

Step 1a: $\begin{cases} 3(2y - 1) + y = 5y + 3 \\ \quad \downarrow \quad \downarrow \\ 6y - 3 + y = 5y + 3 \end{cases}$ Distributive property

Step 1c: $7y - 3 = 5y + 3$ Simplify.

Step 2: $\begin{cases} 7y + (-5y) - 3 = 5y + (-5y) + 3 & \text{Add } -5y \text{ to both sides.} \\ 2y - 3 = 3 \\ 2y - 3 + 3 = 3 + 3 & \text{Add } +3 \text{ to both sides.} \\ 2y = 6 \end{cases}$

Step 3: $\begin{cases} \dfrac{1}{2}(2y) = \dfrac{1}{2}(6) & \text{Multiply by } \dfrac{1}{2}. \\ y = 3 \end{cases}$

Check:

Step 4: $\begin{cases} \text{When} & y = 3 \\ \text{the equation} & 3(2y - 1) + y = 5y + 3 \\ \text{becomes} & 3(2 \cdot 3 - 1) + 3 \overset{?}{=} 5 \cdot 3 + 3 \\ & 3(5) + 3 \overset{?}{=} 15 + 3 \\ & 18 = 18 \quad\quad \text{A true statement} \end{cases}$

Our solution checks in the original equation.

EXAMPLE 9   Solve the equation $8 - 3(4x - 2) + 5x = 35$.

Solution   We must begin distributing the -3 across the quantity $4x - 2$. (It would be a mistake to subtract 3 from 8 first, since the rule for order of operations indicates we are to do multiplication before subtraction.) After we have simplified the left side of our equation, we apply the addition property and the multiplication property. In this example, we will show only the result:

Step 1a: $\begin{cases} 8 - 3(4x - 2) + 5x = 35 & \text{Original equation} \\ \quad\quad \downarrow \quad\quad \downarrow \\ 8 - 12x + 6 + 5x = 35 & \text{Distributive property} \end{cases}$

Step 1c: $-7x + 14 = 35$ Simplify.

Step 2: $-7x = 21$ Add -14 to each side.

Step 3: $x = -3$ Multiply by $-\dfrac{1}{7}$.

Step 4: When x is replaced by -3 in the original equation, a true statement results. Therefore, -3 is the solution to our equation.

IDENTITIES AND EQUATIONS WITH NO SOLUTION

There are two special cases associated with solving linear equations in one variable, which are illustrated in the following examples.

EXAMPLE 10   Solve for x: $2(3x - 4) = 3 + 6x$

Solution   Applying the distributive property to the left side gives us

$$6x - 8 = 3 + 6x \qquad \text{Distributive property}$$

Now, if we add $-6x$ to each side, we are left with

$$-8 = 3$$

which is a false statement. This means that there is no solution to our equation. Any number we substitute for x in the original equation will lead to a similar false statement.

EXAMPLE 11   Solve for x: $-15 + 3x = 3(x - 5)$

Solution   We start by applying the distributive property to the right side.

$$-15 + 3x = 3x - 15 \qquad \text{Distributive property}$$

If we add $-3x$ to each side, we are left with the true statement

$$-15 = -15$$

In this case, our result tells us that any number we use in place of x in the original equation will lead to a true statement. Therefore, all real numbers are solutions to our equation. We say the original equation is an **identity,** because the left side is always identically equal to the right side.

 Getting Ready for Class

After reading through the preceding section, respond in your own words and in complete sentences.

A. What is a solution to an equation?

B. What are equivalent equations?

C. Describe how to eliminate fractions in an equation.

D. Suppose when solving an equation your result is the statement "$3 = -3$." What would you conclude about the solution to the equation?

PROBLEM SET 2.1

Solve each of the following equations.

1. $x - 5 = 3$

2. $x + 2 = 7$

3. $2x - 4 = 6$

4. $3x - 5 = 4$

5. $7 = 4a - 1$

6. $10 = 3a - 5$

7. $3 - y = 10$

8. $5 - 2y = 11$

9. $-3 - 4x = 15$

10. $-8 - 5x = -6$

11. $-3 = 5 + 2x$

12. $-12 = 6 + 9x$

13. $-300y + 100 = 500$　**14.** $-20y + 80 = 30$

15. $160 = -50x - 40$　**16.** $110 = -60x - 50$

17. $3 - 4a = -11$　**18.** $8 - 2a = -13$

19. $9 + 5a = -2$　**20.** $3 + 7a = -7$

21. $\frac{2}{3}x = 8$　**22.** $\frac{3}{2}x = 9$

23. $-\frac{3}{5}a + 2 = 8$　**24.** $-\frac{5}{3}a + 3 = 23$

25. $8 = 6 + \frac{2}{7}y$　**26.** $1 = 4 + \frac{3}{7}y$

27. $9 - \frac{3}{4}t = 12$　**28.** $3 - \frac{2}{3}t = 1$

29. $-x = 2$　**30.** $-x = \frac{1}{2}$

31. $-a = -\frac{3}{4}$　**32.** $-a = -5$

33. $7y - 4 = 2y + 11$　**34.** $8y - 2 = 6y - 10$

35. $5 - 2x = 3x + 1$　**36.** $7 - 3x = 8x - 4$

37. $5y - 2 + 4y = 2y + 12$

38. $7y - 3 + 2y = 7y - 9$

39. $11x - 5 + 4x - 2 = 8x$

40. $2x + 7 - 3x + 4 = -2x$

41. $k = 2(3k - 5)$

42. $9k = 3(4k - 1)$

43. $-5(2x + 1) + 5 = 3x$

44. $-3(5x + 7) - 4 = -10x$

45. $5(y + 2) - 4(y + 1) = 3$

46. $6(y - 3) - 5(y + 2) = 8$

47. $6 - 7(m - 3) = -1$

48. $3 - 5(2m - 5) = -2$

49. $4(a - 3) + 5 = 7(3a - 1)$

50. $6(a - 4) + 6 = 2(5a + 2)$

51. $7 + 3(x + 2) = 4(x - 1)$

52. $5 + 2(4x - 4) = 3(2x - 1)$

53. $5 = 7 - 2(3x - 1) + 4x$

54. $20 = 8 - 5(2x - 3) + 4x$

55. $10 - 4(2x + 1) - (3x - 4) = -9x + 4 - 4x$

56. $7 - 2(3x + 5) - (2x - 3) = -5x + 3 - 2x$

57. $\frac{1}{2}x + \frac{1}{4} = \frac{1}{3}x + \frac{5}{4}$　**58.** $\frac{2}{3}x - \frac{3}{4} = \frac{1}{6}x + \frac{21}{4}$

59. $-\frac{2}{5}x + \frac{2}{15} = \frac{2}{3}$　**60.** $-\frac{1}{6}x + \frac{2}{3} = \frac{1}{4}$

61. $\frac{1}{2}y - \frac{2}{7} = \frac{1}{7}y + \frac{11}{14}$　**62.** $\frac{1}{2}y - \frac{1}{8} = \frac{3}{8}y - \frac{5}{8}$

63. $\frac{3}{4}(8x - 4) = \frac{2}{3}(6x - 9)$

64. $\frac{3}{5}(5x + 10) = \frac{5}{6}(12x - 18)$

65. $\frac{1}{4}(12a + 1) - \frac{1}{4} = 5$　**66.** $\frac{2}{3}(6x - 1) + \frac{2}{3} = 4$

67. $\frac{1}{2}x + \frac{1}{3}x + \frac{1}{4}x = \frac{13}{12}$　**68.** $\frac{1}{3}x + \frac{1}{4}x + \frac{1}{5}x = \frac{47}{60}$

69. $0.08x + 0.09(9{,}000 - x) = 750$

70. $0.08x + 0.09(9{,}000 - x) = 500$

71. $0.12x + 0.10(15{,}000 - x) = 1{,}600$

72. $0.09x + 0.11(11{,}000 - x) = 1{,}150$

73. $0.35x - 0.2 = 0.15x + 0.1$

74. $0.25x - 0.05 = 0.2x + 0.15$

75. $0.42 - 0.18x = 0.48x - 0.24$

76. $0.3 - 0.12x = 0.18x + 0.06$

77. There is no solution to the linear equation $6x - 2(x - 5) = 4x + 3$. That is, there is not a real number to use in place of x that will turn the equation into a true statement. What happens when you try to solve the equation?

78. The equation $5(x - 2) - 2x = 3x + 7$ has no solution. What happens when you try to solve the equation?

79. The equation $4x - 8 = 2(2x - 4)$ is called an *identity* because every real number is a solution. That is, replacing x with any real number will result in a true statement. Try to solve the equation.

80. Every real number is a solution to the equation $7(x + 2) - 4x = 3x + 14$. What happens when you try to solve the equation?

Solve each equation, if possible.

81. $3x - 6 = 3(x + 4)$

82. $7x - 14 = 7(x - 2)$

83. $4y + 2 - 3y + 5 = 3 + y + 4$

84. $7y + 5 - 2y - 3 = 6 + 5y - 4$

85. $2(4t - 1) + 3 = 5t + 4 + 3t$

86. $5(2t - 1) + 1 = 2t - 4 + 8t$

Applying the Concepts

87. Cost of a Taxi Ride　The taximeter was invented in 1891 by Wilhelm Bruhn. The city of Chicago charges \$1.80 plus \$0.40 per mile for a taxi ride.

(a) A woman paid a fare of \$6.60. Write an equation that connects the fare the woman paid, the miles she traveled, n, and the charges the taximeter computes.

(b) Solve the equation from part (a) to determine how many miles the woman traveled.

88. Coughs and Earaches　In 1992, twice as many people visited their doctor because of a cough than an earache. The total number of doctor's visits for these two ailments was reported to be 45 million.

(a) Let x represent the number of earaches reported in 1992, then write an expression using x for the number of coughs reported in 1992.

(b) Write an equation that relates 45 million to the variable x.

(c) Solve the equation from part (b) to determine the number of people who visited their doctor in 1992 to complain of an earache.

89. Population Density In the 1990 census, the population of Puerto Rico was given as 3,522,037, with a population density (people per square mile) of 1,025.

(a) Let A represent the area of Puerto Rico, and write an equation using the fact that the population is equal to the product of the area and the population density.

(b) Solve the equation from part (a) for the area of Puerto Rico.

90. Solving Equations by Trial and Error Sometimes equations can be solved most easily by trial and error. Solve the following equations by trial and error.

(a) Find x and y if $x \cdot y + 1 = 36$, and both x and y are prime.

(b) Find w, t, and z if $w + t + z + 10 = 52$, and w, t, and z are consecutive terms of a Fibonacci sequence.

(c) Find x and y if $x \neq y$ and $x^y = y^x$.

Review Problems

From now on, each problem set will include a series of review problems. In mathematics it is very important to review. The more you review, the better you will understand the topics we cover and the longer you will remember them. Also, there are times when material that seemed confusing earlier will be less confusing the second time around. The following problems review some of the material we covered in Section 1.3.

Identify the property (or properties) that justifies each of the following statements.

91. $ax = xa$ **92.** $5(\frac{1}{5}) = 1$

93. $3 + (x + y) = (3 + x) + y$

94. $3 + (x + y) = (x + y) + 3$

95. $3 + (x + y) = (3 + y) + x$

96. $7(3x - 5) = 21x - 35$

97. $5(1) = 5$ **98.** $5 + 0 = 5$

99. $4(xy) = 4(yx)$ **100.** $4(xy) = (4y)x$

101. $2 + 0 = 2$ **102.** $2 + (-2) = 0$

2.2   Formulas

A formula in mathematics is an equation that contains more than one variable. Some formulas are probably already familiar to you — for example, the formula for the area (A) of a rectangle with length l and width w is $A = lw$.

To begin our work with formulas, we will consider some examples in which we are given numerical replacements for all but one of the variables.

EXAMPLE 1 Find y when x is 4 in the formula $3x - 4y = 2$.

Solution We substitute 4 for x in the formula and then solve for y:

When $x = 4$

the formula $3x - 4y = 2$

becomes $3(4) - 4y = 2$

$$12 - 4y = 2 \qquad \text{Multiply 3 and 4.}$$

$$-4y = -10 \qquad \text{Add } -12 \text{ to each side.}$$

$$y = \frac{5}{2} \qquad \text{Divide each side by } -4.$$

Note that, in the last line of Example 1, we divided each side of the equation by -4. Remember that this is equivalent to multiplying each side of the equation by $-\frac{1}{4}$. For the rest of the examples in this section, it will be more convenient to think in terms of division rather than multiplication.

EXAMPLE 2   A store selling art supplies finds that they can sell x sketch pads each week at a price of p dollars each, according to the formula $x = 900 - 300p$. What price should they charge for each sketch pad if they want to sell 525 pads each week?

Solution   Here we are given a formula, $x = 900 - 300p$, and asked to find the value of p if x is 525. To do so, we simply substitute 525 for x and solve for p:

When $x = 525$

the formula $x = 900 - 300p$

becomes $525 = 900 - 300p$

$$-375 = -300p \qquad \text{Add } -900 \text{ to each side.}$$

$$1.25 = p \qquad \text{Divide each side by } -300.$$

In order to sell 525 sketch pads, the store should charge \$1.25 for each pad.

Our next example involves a formula that contains the number π. In this book, you can use either 3.14 or $\frac{22}{7}$ as an approximation for π. Generally speaking, if the problem contains decimals, use 3.14 to approximate π.

EXAMPLE 3   A company manufacturing coffee cans uses 486.7 square centimeters of material to make each can. Because of the way the cans are to be placed in boxes for shipping, the radius of each can must be 5 centimeters. If the formula for the surface area of a can is $S = \pi r^2 + 2\pi rh$ (the cans are open at the top), find the height of each can. (Use 3.14 as an approximation for π.)

Surface area = 486.7 cm²

Solution   Substituting 486.7 for S, 5 for r, and 3.14 for π into the formula $S = \pi r^2 + 2\pi rh$, we have

$$486.7 = (3.14)(5^2) + 2(3.14)(5)h$$

$$486.7 = 78.5 + 31.4h$$

$$408.2 = 31.4h \qquad\qquad \text{Add } -78.5 \text{ to each side.}$$

$$13 = h \qquad\qquad\qquad \text{Divide each side by 31.4.}$$

The height of each can is 13 centimeters.

FACTS FROM
Geometry

Formulas for Area and Perimeter

To review, here are the formulas for the area and perimeter of some common geometric objects.

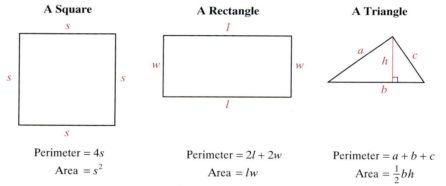

A Square
Perimeter = 4s
Area = s^2

A Rectangle
Perimeter = $2l + 2w$
Area = lw

A Triangle
Perimeter = $a + b + c$
Area = $\frac{1}{2}bh$

The formula for perimeter gives us the distance around the outside of the object along its sides, while the formula for area gives us a measure of the amount of surface the object covers.

Note: The pink line labeled h in the triangle is its height, or altitude. It extends from the top of the triangle down to the base, meeting the base at an angle of 90°. The altitude of a triangle is always perpendicular to the base. The small square shown where the altitude meets the base is used to indicate that the angle formed is 90°.

E X A M P L E   4   Given the formula $P = 2w + 2l$, solve for w.

Solution   To solve for w, we must isolate it on one side of the equation. We can accomplish this if we delete the $2l$ term and the coefficient 2 from the right side of the equation.

To begin, we add $-2l$ to both sides:

$$P + (-2l) = 2w + 2l + (-2l)$$

$$P - 2l = 2w$$

To delete the 2 from the right side, we can multiply both sides by $\frac{1}{2}$:

$$\frac{1}{2}(P - 2l) = \frac{1}{2}(2w)$$

$$\frac{P - 2l}{2} = w$$

The two formulas

$$P = 2w + 2l \qquad \text{and} \qquad w = \frac{P - 2l}{2}$$

give the relationship between P, l, and w. They look different, but they both say the same thing about P, l, and w. The first formula gives P in terms of l and w, and the second formula gives w in terms of P and l.

Note: We know we are finished solving a formula for a specified variable when that variable appears alone on one side of the equal sign and not on the other.

E X A M P L E   5   Solve the formula $S = 2\pi rh + 2\pi r^2$ for h.

Solution   This is the formula for the surface area of a right circular cylinder, with radius r and height h, that is closed at both ends. To isolate h, we first add $-2\pi r^2$ to both sides:

Right circular cylinder with radius r and height h

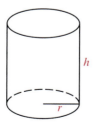

Surface area if open at one end:
  $S = \pi r^2 + 2\pi rh$
Surface area if closed at both ends:
  $S = 2\pi r^2 + 2\pi rh$
Volume: $V = \pi r^2 h$

$$S + (-2\pi r^2) = 2\pi rh + 2\pi r^2 + (-2\pi r^2)$$

$$S - 2\pi r^2 = 2\pi rh$$

Next, we divide each side by $2\pi r$.

$$\frac{S - 2\pi r^2}{2\pi r} = h$$

Exchanging the two sides of our answer, we have

$$h = \frac{S - 2\pi r^2}{2\pi r}$$

E X A M P L E   6   Solve for x: $ax - 3 = bx + 5$

Solution   In this example, we must begin by collecting all the variable terms on the left side of the equation and all the constant terms on the other side (just like we did when we were solving linear equations in Section 2.1):

$$ax - 3 = bx + 5$$

$$ax - bx - 3 = 5 \qquad \text{Add } -bx \text{ to each side.}$$

$$ax - bx = 8 \qquad \text{Add 3 to each side.}$$

At this point we need to apply the distributive property to write the left side as $(a - b)x$. After that, we divide each side by $a - b$:

$$(a - b)x = 8 \qquad \text{Distributive property}$$

$$x = \frac{8}{a - b} \qquad \text{Divide each side by } a - b.$$

BASIC PERCENT PROBLEMS

The next examples in this section show how basic percent problems can be translated directly into equations. To understand these examples, we must recall that percent means "per hundred." That is, 75% is the same as 75/100, 0.75, and, in reduced fraction form, $\frac{3}{4}$. Likewise, the decimal 0.25 is equivalent to 25%. To change a decimal to a percent, we move the decimal point two places to the right and write the % symbol. To change from a percent to a decimal, we drop the % symbol and move the decimal point two places to the left. The table that follows gives some of the most commonly used fractions and decimals and their equivalent percents.

Fraction	Decimal	Percent
$\frac{1}{2}$	0.5	50%
$\frac{1}{4}$	0.25	25%
$\frac{3}{4}$	0.75	75%
$\frac{1}{3}$	$0.\overline{3}$	$33\frac{1}{3}\%$
$\frac{2}{3}$	$0.\overline{6}$	$66\frac{2}{3}\%$
$\frac{1}{5}$	0.2	20%
$\frac{2}{5}$	0.4	40%
$\frac{3}{5}$	0.6	60%
$\frac{4}{5}$	0.8	80%

EXAMPLE 7   What number is 15% of 63?

Solution   To solve a problem like this, we let x = the number in question and then translate the sentence directly into an equation. Here is how it is done:

$$\underbrace{\text{What number}}_{x} \overset{\downarrow\ \downarrow\ \ \downarrow\ \downarrow}{\text{is 15\% of 63?}}$$

$$x \qquad\quad = 0.15 \ \cdot \ 63$$

$$= 9.45$$

The number 9.45 is 15% of 63.

Note: We write 0.15 instead of 15% when we translate the sentence into an equation because we cannot do calculations with the % symbol. Since percent (%) means per hundred, we think of 15% as 15 hundredths, or 0.15.

E X A M P L E   8   What percent of 42 is 21?

Solution   We translate the sentence as follows:

$$\underbrace{\text{What percent}}_{x} \overset{\downarrow\ \downarrow\ \downarrow\ \downarrow}{\text{of 42 is 21?}}$$

$$x \qquad \cdot \ \ 42 = 21$$

Next, we divide each side by 42:

$$x = \frac{21}{42}$$

$$= 0.50, \text{ or } 50\%$$

E X A M P L E   9   25 is 40% of what number?

Solution   Again, we translate the sentence directly:

$$\overset{\downarrow\ \downarrow\ \downarrow\ \ \downarrow}{\text{25 is 40\% of }} \underbrace{\text{what number?}}$$

$$25 = 0.40 \ \cdot \qquad x$$

We solve the equation by dividing both sides by 0.40:

$$\frac{25}{0.40} = \frac{0.40 \cdot x}{0.40}$$

$$62.5 = x$$

25 is 40% of 62.5.

FORMULAS FOR NUMBER SEQUENCES

In Chapter 1 we used inductive reasoning to find the next number in a sequence of numbers. What we are going to do now is extend our work with number sequences so that we can find any number in the sequence. To begin, we introduce notation that allows us to associate each number in a number sequence with its position in the sequence. To illustrate, consider the sequence of even numbers:

Even numbers: 2, 4, 6, 8, 10, . . .

The number 8 is the fourth number in the sequence. Its *position* in the sequence is 4, and its *value* is 8. Here is the sequence of even numbers again, this time written so that the position of each term is noted.

Position: 1, 2, 3, 4, 5, . . .

Value: 2, 4, 6, 8, 10, . . .

The notation we use to indicate the position of a number in a number sequence involves *subscripts*. Subscripts are numbers written to the right and just below another number or variable. In the expression a_4, the 4 is a subscript. We use subscripts to designate the terms of a sequence this way:

The *first term* of a sequence is designated by a_1.
The *second term* of a sequence is designated by a_2.
The *third term* of a sequence is designated by a_3.
The *fourth term* of a sequence is designated by a_4.

Using this notation with the sequence of even numbers, we write

Sequence of Even Numbers

First term $= a_1 = 2$

Second term $= a_2 = 4$

Third term $= a_3 = 6$

Fourth term $= a_4 = 8$

If you look closely at the subscripts and the terms in the preceding list, you will see that each number in the sequence of even numbers is twice the term number (subscript). In general, if n is a positive integer, then

$a_n = 2n$ for any term in the sequence of even numbers.

The expression a_n is called the *n*th term, or the *general term,* of the sequence. The general term is used to define the other terms of the sequence. That is, if we are given the formula for the general term a_n, we can find any other term in the sequence. The following examples illustrate the concept.

EXAMPLE 10   Find the first four terms of the sequence whose general term is given by $a_n = 2n - 1$.

Solution   To find the first, second, third, and fourth terms of this sequence, we simply substitute 1, 2, 3, and 4 for n in the formula $2n - 1$:

If   The general term is $a_n = 2n - 1$.

then   The first term is $a_1 = 2(1) - 1 = 1$.

The second term is $a_2 = 2(2) - 1 = 3$.

The third term is $a_3 = 2(3) - 1 = 5$.

The fourth term is $a_4 = 2(4) - 1 = 7$.

The first four terms of this sequence are the odd numbers 1, 3, 5, and 7. The whole sequence can be written as

$$1, 3, 5, \ldots, 2n - 1, \ldots$$

Since each term in this sequence is larger than the preceding term, we say the sequence is an *increasing sequence*.

EXAMPLE 11   Write the first four terms of the sequence defined by

$$a_n = \frac{1}{n + 1}$$

Solution   Replacing n with 1, 2, 3, and 4, we have, respectively, the first four terms:

$$\text{First term} = a_1 = \frac{1}{1 + 1} = \frac{1}{2}$$

$$\text{Second term} = a_2 = \frac{1}{2 + 1} = \frac{1}{3}$$

$$\text{Third term} = a_3 = \frac{1}{3 + 1} = \frac{1}{4}$$

$$\text{Fourth term} = a_4 = \frac{1}{4 + 1} = \frac{1}{5}$$

The sequence defined by

$$a_n = \frac{1}{n + 1}$$

can be written as

$$\frac{1}{2}, \frac{1}{3}, \frac{1}{4}, \ldots, \frac{1}{n + 1}, \ldots$$

Since each term in the sequence is smaller than the term preceding it, the sequence is said to be a *decreasing sequence*.

Using TECHNOLOGY

GRAPHING CALCULATORS

Entering, Recalling, and Editing Formulas

Graphing calculators can be useful in evaluating formulas. For example, the formula for the volume of a right circular cylinder is

$$V = \pi r^2 h$$

To find V when $r = 3$ and $h = 4$, we must do three things: First we use the $\boxed{\text{STO} \rightarrow}$ key to store 3 in the variable r. Using the same key we store 4 in the variable h. Then we enter the formula V. All three of these things can be done at one time by using the $\boxed{:}$ key. Here are the characters that will appear on your calculator screen when you have done the preceding three tasks:

Store 3 in R   Store 4 in H   Enter formula for V

$$3 \rightarrow R : 4 \rightarrow H : \pi R \wedge 2\,H$$

When you press the $\boxed{\text{ENTER}}$ key, the calculator will display a little more than 113.0973355, which is the volume of a right circular cylinder for which $r = 3$ and $h = 4$, accurate to 10 digits.

    If we want to evaluate the same formula for other values of r or h, we can recall it to the screen by pressing $\boxed{\text{2nd}}$ $\boxed{\text{ENTER}}$. (If we have done some other calculations in the meantime, we can repeatedly press the sequence $\boxed{\text{2nd}}$ $\boxed{\text{ENTER}}$ until the formula appears on the screen. The Texas Instruments TI-83 will recall as many previous entries as can be stored in its memory.) Once the entry has been recalled, we can use the arrow keys on the calculator to move around the formula, in order to change any part of it that we choose to change. Try it yourself. Recall the formula, and then change the number stored in R to 5, and the number stored in H to 13 (you will have to use the $\boxed{\text{INS}}$ key to do your editing). When you press $\boxed{\text{ENTER}}$, you should see a number that matches the volume of the cylinder given in Example 3.

Getting Ready for Class

After reading through the preceding section, respond in your own words and in complete sentences.

A. What is a formula in mathematics?

B. Create a formula for a sequence and explain how to obtain the first four terms of your sequence.

C. Write a percent problem that can be solved by the equation $30 = 0.25x$.

D. Explain in words the formula for the perimeter of a rectangle.

PROBLEM SET 2.2

Use the formula $3x - 4y = 12$ to find y if

1. x is 0

2. x is -2

3. x is 4

4. x is -4

Use the formula $y = 2x - 3$ to find x when

5. y is 0

6. y is -3

7. y is 5

8. y is -5

Solve each of the following formulas for the indicated variable.

9. $A = lw$ for l

10. $A = \dfrac{1}{2}bh$ for b

11. $I = prt$ for t

12. $I = prt$ for r

13. $PV = nRT$ for T

14. $PV = nRT$ for R

15. $y = mx + b$ for x

16. $A = P + Prt$ for t

17. $C = \dfrac{5}{9}(F - 32)$ for F

18. $F = \dfrac{9}{5}C + 32$ for C

19. $h = vt + 16t^2$ for v

20. $h = vt - 16t^2$ for v

21. $A = a + (n - 1)d$ for d

22. $A = a + (n - 1)d$ for n

23. $2x + 3y = 6$ for y

24. $2x - 3y = 6$ for y

25. $-3x + 5y = 15$ for y

26. $-2x - 7y = 14$ for y

27. $2x - 6y + 12 = 0$ for y

28. $7x - 2y - 6 = 0$ for y

29. $ax + 4 = bx + 9$ for x

30. $ax - 5 = cx - 2$ for x

31. $A = P + Prt$ for P

32. $ax + b = cx + d$ for x

Solve for y.

33. $\dfrac{x}{8} + \dfrac{y}{2} = 1$

34. $\dfrac{x}{7} + \dfrac{y}{9} = 1$

35. $\dfrac{x}{5} + \dfrac{y}{-3} = 1$

36. $\dfrac{x}{16} + \dfrac{y}{-2} = 1$

Translate each of the following into a linear equation and then solve the equation.

37. What number is 54% of 38?

38. What number is 11% of 67?

39. What percent of 36 is 9?

40. What percent of 50 is 5?

41. 37 is 4% of what number?

42. 8 is 2% of what number?

Write the first five terms of the sequences with the following general terms.

43. $a_n = 3n + 1$

44. $a_n = 4n - 1$

45. $a_n = n^2 + 3$

46. $a_n = n^3 + 1$

47. $a_n = \dfrac{n}{n + 3}$

48. $a_n = \dfrac{n}{n + 2}$

49. $a_n = \dfrac{1}{n^2}$

50. $a_n = \dfrac{1}{n^3}$

51. $a_n = 2^n$

52. $a_n = 3^n$

53. $a_n = 1 + \dfrac{1}{n}$

54. $a_n = 1 - \dfrac{1}{n}$

Applying the Concepts

Pricing   A company that manufactures typewriter ribbons finds that they can sell x ribbons each week at a price of p dollars each, according to the formula $x = 1,300 - 100p$. What price should they charge for each ribbon if they want to sell

55. 800 ribbons each week?

56. 400 ribbons each week?

57. 300 ribbons each week?

58. 900 ribbons each week?

Volume and Height of a Cylinder   The volume of a cylinder is given by the formula $V = \pi r^2 h$, where r is the radius and h is the height. Find the height if

59. the volume is 308 cubic centimeters and the radius is 7 centimeters. (Use $\frac{22}{7}$ for π.)

60. the volume is 308 cubic centimeters and the radius is $\frac{7}{2}$ centimeters.

61. the volume is 628 cubic inches and the radius is 10 inches. (Use 3.14 for π.)

62. the volume is 12.56 cubic inches and the radius is 5 inches.

Surface Area and Height of a Cylinder   The surface area of a cylinder that is closed at the top and bottom is given by the formula $S = 2\pi r^2 + 2\pi rh$. Find the height if

63. the surface area is 942 square feet and the radius is 10 feet. (Use 3.14 for π.)

64. the surface area is 471 square feet and the radius is 5 feet.

Baseball   In Section 1.4 we worked a problem in which we found Rolaids points earned in 1998 for some relief pitchers. Now we are going to do a similar problem, but this time using a formula. The formula $P = 3s + 2w - 2l - 2b$ gives the number of Rolaids points a pitcher earns, where s = saves, w = wins, l = losses, and b = blown saves. Use this formula to complete the following tables.

65. National League

Pitcher, Team	W	L	Saves	Blown Saves	Rolaids Points
John Franco, New York	0	8	38	8	
Billy Wagner, Houston	4	3	30	5	
Gregg Olson, Arizona	3	4	30	4	
Bob Wickman, Milwaukee	6	9	25	7	

66. American League

Pitcher, Team	W	L	Saves	Blown Saves	Rolaids Points
Rick Aguilera, Minnesota	4	9	38	11	
Billy Taylor, Oakland	4	9	33	4	
Randy Myers, Toronto	3	4	28	5	
Todd Jones, Detroit	1	4	28	4	

Estimating Vehicle Weight

If you can measure the area that the tires on your car contact the ground, and you know the air pressure in the tires, then you can estimate the weight of your car with the following formula:

$$W = \frac{APN}{2,000}$$

where W is the vehicle's weight in tons, A is the average tire contact area with a hard surface in square inches, P is the air pressure in the tires in pounds per square inch (psi, or lb/in.^{2}), and N is the number of tires.

Tire pressure P

Contact area A

67. What is the approximate weight of a car if the average tire contact area is a rectangle 6 inches by 5 inches, and if the air pressure in the tires is 30 psi?

68. What is the approximate weight of a car if the average tire contact area is a rectangle 5 inches by 4 inches, and the tire pressure is 30 psi?

Ordinary and Exact Interest   When paying simple interest, a bank computes interest on the original principal during the whole time the money is invested. There is no compounding. There are two types of simple interest: ordinary interest and exact interest. If R is the interest rate and D is the number of days that P dollars have been invested, then the two types of interest are calculated as follows:

$$\text{Ordinary interest} = PR\,\frac{D}{360}$$

$$\text{Exact interest} = PR\,\frac{D}{365}$$

69. Find the ordinary and exact interest on $4,000 invested at 7% for 87 days.

70. Explain why ordinary interest will always be greater than exact interest.

Digital Video   The biggest video download of all time was a *Star Wars* movie trailer. The video was compressed so it would be small enough for people to download over the Internet. A formula for estimating the size, in kilobytes, of a compressed video is

$$S = \frac{height \cdot width \cdot fps \cdot time}{35{,}000}$$

where *height* and *width* are in pixels, *fps* is the number of frames per second the video is to play (television plays at 30 fps), and time is given in seconds.

(Courtesy of Lucasfilm Ltd. **Star Wars: Episode I — The Phantom Menace** *© Lucasfilm Ltd. & ™. All rights reserved. Used under authorization)*

71. Estimate the size in kilobytes of the *Star Wars* trailer that has a height of 480 pixels, a width of 216 pixels, plays at 30 fps, and runs for 150 seconds.

72. Estimate the size in kilobytes of the *Star Wars* trailer that has a height of 320 pixels, a width of 144 pixels, plays at 15 fps, and runs for 150 seconds.

Compound Interest   The following formula gives the amount of money A in an account in which P dollars has been invested for t years at an interest rate of r compounded n times a year:

$$A = P\left(1 + \frac{r}{n}\right)^{nt}$$

Use the recall formula function on your calculator to answer the following questions.

73. How much money is in an account in which $500 has been invested at 5%, compounded quarterly, for 6 years (i.e., find A when $P = 500$, $r = 0.05$, $n = 4$, and $t = 6$)?

74. How much money is in an account in which $500 has been invested at 5%, compounded monthly, for 6 years?

75. How much money is in an account in which $500 has been invested at 5%, compounded daily, for 6 years?

76. How much money is in an account in which $500 has been invested at 6%, compounded monthly, for 6 years?

77. How much money is in an account in which $500 has been invested at 7%, compounded monthly, for 6 years?

78. In general, is it more desirable to increase the interest rate or the number of compounding periods to increase the amount of money in the account?

Interest   If interest in an investment is compounded just once a year, then the formula given before Problem 73 is simply $A = P(1 + r)^t$.

79. When Benjamin Franklin died in 1790, he left the city of Boston $5,000. If the city had invested this money at 4% compounded yearly, how much would Franklin's gift have been worth in 2000?

80. How much more would Franklin's gift be worth in 2000 if his gift had been compounded quarterly?

Review Problems

The following problems review some of the material we covered in Section 1.1. Reviewing these problems will help you with the next section.

Translate each of the following into symbols.

81. Twice the sum of x and 3

82. The sum of twice x and 3

83. Twice the sum of x and 3 is 16.

84. The sum of twice x and 3 is 16.

85. Five times the difference of x and 3

86. Five times the difference of x and 3 is 10.

87. The sum of $3x$ and 2 is equal to the difference of x and 4.

88. The sum of x and $x + 2$ is 12 more than their difference.

Extending the Concepts

89. Solve for x: $\dfrac{x}{a} + \dfrac{y}{b} = 1$

90. Solve for y: $\dfrac{x}{a} + \dfrac{y}{b} = 1$

91. Solve for a: $\dfrac{1}{a} + \dfrac{1}{b} = \dfrac{1}{c}$

92. Solve for b: $\dfrac{1}{a} + \dfrac{1}{b} = \dfrac{1}{c}$

93. Solve for R: $\dfrac{1}{R} = \dfrac{1}{a} + \dfrac{1}{b} + \dfrac{1}{c}$

94. Solve for a: $\dfrac{1}{R} = \dfrac{1}{a} + \dfrac{1}{b} + \dfrac{1}{c}$

Exercise Physiology   In exercise physiology, a person's maximum heart rate, in beats per minute, is found by subtracting his age, in years, from 220. So, if A represents your age in years, then your maximum heart rate is

$$M = 220 - A$$

A person's training heart rate, in beats per minute, is her resting heart rate plus 60% of the difference between her maximum heart rate and her resting heart rate. If resting heart rate is R and maximum heart rate is M, then the formula that gives training heart rate is

$$T = R + 0.6(M - R)$$

95. Training Heart Rate   Shar is 46 years old. Her daughter, Sara, is 26 years old. If they both have a resting heart rate of 60 beats per minute, find the training heart rate for each.

96. Training Heart Rate   Shane is 30 years old and has a resting heart rate of 68 beats per minute. His mother, Carol, is 52 years old and has the same resting heart rate. Find the training heart rate for Shane and for Carol.

2.3   Applications

In this section we use the skills we have developed for solving equations to solve problems written in words. You may find that some of the examples and problems are more realistic than others. Since we are just beginning our work with application problems, even the ones that seem unrealistic are good practice. What is important in this section is the *method* we use to solve application problems, not the applications themselves. The method, or strategy, that we use to solve application problems is called the *Blueprint for Problem Solving*. It is an outline that will overlay the solution process we use on all application problems.

Blueprint for Problem Solving

> **Step 1:** **Read** the problem, and then mentally **list** the items that are known and the items that are unknown.
>
> **Step 2:** **Assign a variable** to one of the unknown items. (In most cases this will amount to letting x = the item that is asked for in the problem.)

Then ***translate*** the other ***information*** in the problem to expressions involving the variable.

Step 3: ***Reread*** the problem, and then ***write an equation,*** using the items and variables listed in steps 1 and 2, that describes the situation.

Step 4: ***Solve the equation*** found in step 3.

Step 5: ***Write your answer*** using a complete sentence.

Step 6: ***Reread*** the problem, and ***check*** your solution with the original words in the problem.

A number of substeps occur within each of the steps in our blueprint. For instance, with steps 1 and 2 it is always a good idea to draw a diagram or picture, if it helps you visualize the relationship between the items in the problem.

EXAMPLE 1   The length of a rectangle is 3 inches less than twice the width. The perimeter is 45 inches. Find the length and width.

Solution   When working problems that involve geometric figures, a sketch of the figure helps organize and visualize the problem.

Step 1: ***Read and list.***

Known items: The figure is a rectangle. The length is 3 inches less than twice the width. The perimeter is 45 inches.

Unknown items: The length and the width

Step 2: ***Assign a variable and translate information.***

Since the length is given in terms of the width (the length is 3 less than twice the width), we let $x =$ the width of the rectangle. The length is 3 less than twice the width, so it must be $2x - 3$. The diagram in Figure 1 is a visual description of the relationships we have listed so far.

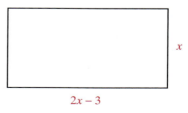

$2x - 3$

FIGURE 1

Step 3: ***Reread and write an equation.***

The equation that describes the situation is

Twice the length + twice the width is the perimeter.

$$2(2x - 3) \quad + \quad 2x \quad = \quad 45$$

Step 4: Solve the equation.

$$2(2x - 3) + 2x = 45$$
$$4x - 6 + 2x = 45$$
$$6x - 6 = 45$$
$$6x = 51$$
$$x = 8.5$$

Step 5: Write the answer.
The width is 8.5 inches. The length is $2x - 3 = 2(8.5) - 3 =$ 14 inches.

Step 6: Reread and check.
If the length is 14 inches and the width is 8.5 inches, then the perimeter must be $2(14) + 2(8.5) = 28 + 17 = 45$ inches. Also, the length, 14, is 3 less than twice the width.

Remember as you read through the steps in the solutions to the examples in this section that step 1 is done mentally. Read the problem and then *mentally* list the items that you know and the items that you don't know. The purpose of step 1 is to give you direction as you begin to work application problems. Finding the solution to an application problem is a process; it doesn't happen all at once. The first step is to read the problem with a purpose in mind. That purpose is to mentally note the items that are known and the items that are unknown.

E X A M P L E   2   In April 1998, Pat bought a new Ford Mustang with a 5.0-liter engine. The total price, which includes the price of the car plus sales tax, was $17,481.75. If the sales tax rate is 7.25%, what was the price of the car?

Solution

Step 1: Read and list.
Known items: The total price is $17,481.75. The sales tax rate is 7.25%, which is 0.0725 in decimal form.
Unknown item: The price of the car

Step 2: Assign a variable and translate information.
If we let $x =$ the price of the car, then to calculate the sales tax, we multiply the price of the car x by the sales tax rate:

$$\text{Sales tax} = (\text{price of the car})(\text{sales tax rate})$$
$$= 0.0725x$$

Step 3: Reread and write an equation.

$$\text{Car price} + \text{sales tax} = \text{total price}$$
$$x + 0.0725x = 17,481.75$$

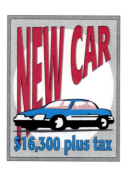
$16,300 plus tax

Step 4: Solve the equation.

$$x + 0.0725x = 17{,}481.75$$

$$1.0725x = 17{,}481.75$$

$$x = \frac{17{,}481.75}{1.0725}$$

$$= 16{,}300.00$$

Step 5: Write the answer.

The price of the car is $16,300.00.

Step 6: Reread and check.

The price of the car is $16,300.00. The tax is $0.0725(16{,}300) =$ $1,181.75. Adding the retail price and the sales tax we have a total bill of $17,481.75.

FACTS FROM
Geometry

Angles in General

An angle is formed by two rays with the same endpoint. The common endpoint is called the *vertex* of the angle, and the rays are called the *sides* of the angle.

In Figure 2, angle θ (theta) is formed by the two rays *OA* and *OB*. The vertex of θ is *O*. Angle θ is also denoted as angle *AOB*, where the letter associated with the vertex is always the middle letter in the three letters used to denote the angle.

Degree Measure

The angle formed by rotating a ray through one complete revolution about its endpoint (Figure 3) has a measure of 360 degrees, which we write as 360°.

One complete revolution = 360°

FIGURE 2   **FIGURE 3**

One degree of angle measure, written 1°, is $\frac{1}{360}$ of a complete rotation of a ray about its endpoint; there are 360° in one full rotation. (The number 360 was decided upon by early civilizations because it was believed that Earth was at the center of the universe and the sun would rotate once around Earth every 360 days.)

Similarly, 180° is half of a complete rotation, and 90° is a quarter of a full rotation. Angles that measure 90° are called *right angles,* and angles that measure 180° are called *straight angles.* If an angle measures between 0° and 90° it is called an *acute angle,* and an angle that measures between 90° and 180° is an *obtuse angle.* Figure 4 illustrates further.

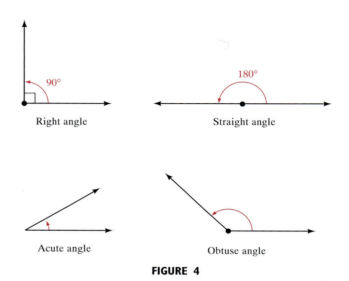

FIGURE 4

Complementary Angles and Supplementary Angles

If two angles add up to 90°, we call them *complementary angles,* and each is called the *complement* of the other. If two angles have a sum of 180°, we call them *supplementary angles,* and each is called the *supplement* of the other. Figure 5 illustrates the relationship between angles that are complementary and angles that are supplementary.

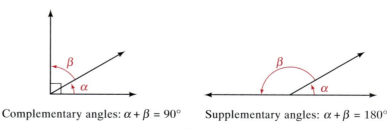

Complementary angles: $\alpha + \beta = 90°$ Supplementary angles: $\alpha + \beta = 180°$

FIGURE 5

Special Triangles

The two special triangles we will discuss next are important triangles. If you go on to take a trigonometry class, you will see them often.

An *equilateral triangle* (Figure 6) is a triangle with three sides of equal length. If all three sides in a triangle have the same length, then the three interior angles in the triangle must also be equal. Since the sum of the interior angles in a triangle is always 180°, the three interior angles in any equilateral triangle must be 60°.

An *isosceles triangle* (Figure 7) is a triangle with two sides of equal length. Angles *A* and *B* in the isosceles triangle in Figure 7 are called the *base angles;* they are the angles opposite the two equal sides. In every isosceles triangle, the base angles are equal.

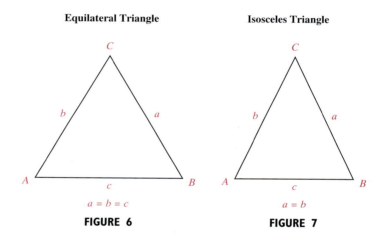

FIGURE 6   **FIGURE 7**

Note: As you can see from Figures 6 and 7, one way to label the important parts of a triangle is to label the vertices with capital letters and the sides with small letters: side *a* is opposite vertex *A*, side *b* is opposite vertex *B*, and side *c* is opposite vertex *C*.

Also, since each vertex is the vertex of one of the angles of the triangle, we refer to the three interior angles as *A*, *B*, and *C*.

Finally, in any triangle, the sum of the interior angles is 180°. For the triangles shown in Figures 6 and 7, the relationship is written

$$A + B + C = 180°$$

E X A M P L E   3   Two complementary angles are such that one is twice as large as the other. Find the two angles.

Solution   Applying the Blueprint for Problem Solving, we have:

Step 1: *Read and list.*

Known items: Two complementary angles. One is twice as large as the other.

Unknown items: The size of the angles

Step 2: *Assign a variable and translate information.*
Let $x =$ the smaller angle. The larger angle is twice the smaller, so we represent the larger angle with $2x$.

Step 3: *Reread and write an equation.*
Since the two angles are complementary, their sum is 90. Therefore,

$$x + 2x = 90$$

Step 4: *Solve the equation.*

$$x + 2x = 90$$

$$3x = 90$$

$$x = 30$$

Step 5: *Write the answer.*
The smaller angle is $30°$, and the larger angle is $2 \cdot 30 = 60°$.

Step 6: *Reread and check.*
The larger angle is twice the smaller angle, and their sum is $90°$.

Suppose we know that the sum of two numbers is 50. If we let x represent one of the two numbers, how can we represent the other? Let's suppose for a moment that x turns out to be 30. Then the other number will be 20, because their sum is 50. That is, if two numbers add up to 50, and one of them is 30, then the other must be $50 - 30 = 20$. Generalizing this to any number x, we see that if two numbers have a sum of 50, and one of the numbers is x, then the other must be $50 - x$. The following table shows some additional examples:

If Two Numbers Have a Sum of	And One of Them Is	Then the Other Must Be
50	x	$50 - x$
10	y	$10 - y$
12	n	$12 - n$

INTEREST PROBLEM

EXAMPLE 4 Suppose a person invests a total of $10,000 in two accounts. One account earns 5% annually, and the other earns 6% annually. If the total interest earned from both accounts in a year is $560, how much is invested in each account?

Solution

Step 1: *Read and list.*
Known items: Two accounts. One pays interest of 5%, and the other pays 6%. The total dollars invested is $10,000.

Unknown items: The number of dollars invested in each individual account.

Step 2: ***Assign a variable and translate information.***

If we let x equal the amount invested at 6%, then $10,000 - x$ is the amount invested at 5%. The total interest earned from both accounts is $560. The amount of interest earned on x dollars at 6% is $0.06x$, and interest earned on $10,000 - x$ dollars at 5% is $0.05(10,000 - x)$.

	Dollars at 6%	Dollars at 5%	Total
Number of	x	$10,000 - x$	10,000
Interest on	$0.06x$	$0.05(10,000 - x)$	560

Step 3: ***Reread and write an equation.***

The last line gives us the equation we are after:

$$0.06x + 0.05(10,000 - x) = 560$$

Step 4: ***Solve the equation.***

To make this equation a little easier to solve, we begin by multiplying both sides by 100 to move the decimal point two places to the right:

$$6x + 5(10,000 - x) = 56,000$$

$$6x + 50,000 - 5x = 56,000$$

$$x + 50,000 = 56,000$$

$$x = 6,000$$

Step 5: ***Write the answer.***

The amount of money invested at 6% is $6,000. The amount of money invested at 5% is $10,000 - $6,000 = $4,000.

Step 6: ***Reread and check.***

To check our results, we find the total interest from the two accounts:

The interest on $6,000 at 6% is $0.06(6,000) = 360$

The interest on $4,000 at 5% is $0.05(4,000) = 200$

The total interest $= \$560$

INTEREST PROBLEM

E X A M P L E   5   The Gateway 2000 Duo Line MasterCard charges customers two different rates: 11.9% for products purchased from Gateway 2000 (computers) and 12.9% for all other purchases. Last year, Ron spent $3,750 using his new MasterCard and was charged $462.75 in interest. How much did Ron spend on Gateway 2000 products? How much on other products?

Solution   We solve this problem using our six-step Blueprint for Problem Solving:

Step 1: *Read and list.*
Known items: Two rates. One charges 11.9% and the other 12.9%. Ron spent $3,750. He was charged $462.75 in interest.
Unknown items: The amount spent at each rate.

Step 2: *Assign a variable and translate information.*
If we let $x =$ Ron's Gateway 2000 purchases (amount charged at 11.9%), then $3,750 - x$ is the amount charged at 12.9%. The total interest charged is $462.75. The amount charged on x dollars at 11.9% is $0.119x$, and that charged on $3,750 - x$ dollars at 12.9% is $0.129(3,750 - x)$.

	Dollars at 11.9%	Dollars at 12.9%	Total
Number of	x	$3,750 - x$	3,750
Interest on	$0.119x$	$0.129(3,750 - x)$	462.75

Step 3: *Reread and write an equation.*
The last line gives us the equation we are after:

$$0.119x + 0.129(3,750 - x) = 462.75$$

Step 4: *Solve the equation.*
To make the equation a little easier to solve, we begin by multiplying both sides by 1,000 to move the decimal point three places to the right.

$$119x + 129(3,750 - x) = 462,750$$
$$119x + 483,750 - 129x = 462,750$$
$$-10x + 483,750 = 462,750$$
$$-10x = -21,000$$
$$x = 2,100$$

Step 5: *Write the answer.*
Ron spent $2,100 at Gateway 2000 (11.9%). The amount of his other purchases (12.9%) is $3,750 - 2,100 = \$1,650$.

Step 6: *Reread and check.*
To check our results, we find the total interest from the two accounts:

The interest charged on $2,100 at 11.9% is $0.119(2,100) = 249.90$

The interest charged on $1,650 at 12.9% is $0.129(1,650) = 212.85$

The total interest $= \$462.75$

TABLE BUILDING

We can use our knowledge of formulas from Section 2.2 to build tables of paired data. As you will see, equations or formulas that contain exactly two variables produce pairs of numbers that can be used to construct tables.

EXAMPLE 6 A piece of string 12 inches long is to be formed into a rectangle. Build a table that gives the length of the rectangle if the width is 1, 2, 3, 4, or 5 inches. Then find the area of each of the rectangles formed.

12 inches

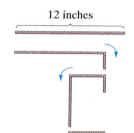

Solution Since the formula for the perimeter of a rectangle is $P = 2l + 2w$, and our piece of string is 12 inches long, then the formula we will use to find the lengths for the given widths is $12 = 2l + 2w$. To solve this formula for l, we divide each side by 2 and then subtract w. The result is $l = 6 - w$. Table 1 organizes our work so that the formula we use to find l for a given value of w is shown, and we have added a last column to give us the areas of the rectangles formed. The units for the first three columns are inches, and the units for the numbers in the last column are square inches.

TABLE 1	**Length, Width, and Area**		
Width (in.)	**Length (in.)**		**Area (in.²)**
w	$l = 6 - w$	l	$A = lw$
1	$l = 6 - 1$	5	5
2	$l = 6 - 2$	4	8
3	$l = 6 - 3$	3	9
4	$l = 6 - 4$	2	8
5	$l = 6 - 5$	1	5

Figures 8 and 9 show two bar charts constructed from the information in Table 1.

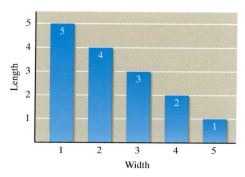

FIGURE 8 Length and width of rectangles with perimeters fixed at 12 inches

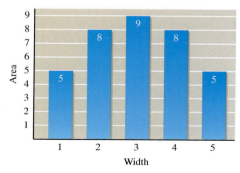

FIGURE 9 Area and width of rectangles with perimeters fixed at 12 inches

Using TECHNOLOGY

GRAPHING CALCULATORS AND SPREADSHEET PROGRAMS

Graphing Calculators

A number of graphing calculators have table-building capabilities. We can let the calculator variable X represent the widths of the rectangles in Example 6. To find the lengths, we set variable Y_1 equal to $6 - X$. The area of each rectangle can be found by setting variable Y_2 equal to $X * Y_1$. To have the calculator produce the table automatically, we use a table minimum of 0 and a table increment of 1. Here is a summary of how the graphing calculator is set up:

	Table Setup		*Y Variables Setup*
	Table minimum $= 0$		$Y_1 = 6 - X$
	Table increment $= 1$		$Y_2 = X * Y_1$
	Independent variable: Auto		
	Dependent variable: Auto		

The table will look like this:

X	Y_1	Y_2
0	6	0
1	5	5
2	4	8
3	3	9
4	2	8
5	1	5
6	0	0

Getting Ready for Class

After reading through the preceding section, respond in your own words and in complete sentences.

A. What is the first step in solving an application problem?

B. What is the biggest obstacle between you and success in solving application problems?

C. Write an application problem for which the solution depends on solving the equation $2x + 2 \cdot 3 = 18$.

D. What is the last step in solving an application problem? Why is this step important?

PROBLEM SET 2.3

Solve each application problem. Be sure to follow the steps in the Blueprint for Problem Solving.

Geometry Problems

1. **Rectangle**   A rectangle is twice as long as it is wide. The perimeter is 60 feet. Find the dimensions.

2. **Rectangle**   The length of a rectangle is 5 times the width. The perimeter is 48 inches. Find the dimensions.

3. **Square**   A square has a perimeter of 28 feet. Find the length of each side.

4. **Square**   A square has a perimeter of 36 centimeters. Find the length of each side.

5. **Triangle**   A triangle has a perimeter of 23 inches. The medium side is 3 inches more than the shortest side, and the longest side is twice the shortest side. Find the shortest side.

6. **Triangle**   The longest side of a triangle is 2 times the shortest side, while the medium side is 3 meters more than the shortest side. The perimeter is 27 meters. Find the dimensions.

7. **Rectangle**   The length of a rectangle is 3 meters less than twice the width. The perimeter is 18 meters. Find the width.

8. **Rectangle**   The length of a rectangle is 1 foot more than twice the width. The perimeter is 20 feet. Find the dimensions.

9. **Livestock Pen**   A livestock pen is built in the shape of a rectangle that is twice as long as it is wide. The perimeter is 48 feet. If the material used to build the pen is $1.75 per foot for the longer sides and $2.25 per foot for the shorter sides (the shorter sides have gates, which increase the cost per foot), find the cost to build the pen.

10. **Garden**   A garden is in the shape of a square with a perimeter of 42 feet. The garden is surrounded by two fences. One fence is around the perimeter of the garden, while the second fence is 3 feet from the first fence, as Figure 10 indicates. If the material used to build the two fences is $1.28 per foot, what was the total cost of the fences?

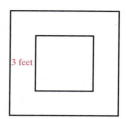

3 feet

FIGURE 10

Percent Problems

11. **Truck Price**   Suppose the total price, including sales tax, of a new Ford F-150 pickup truck is $10,039.43. If the sales tax rate is 7.5%, what was the price of the truck? Round your answer to the nearest cent.

12. **Money**   Shane returned from a trip to Las Vegas with $300.00, which was 50% more money than he had at the beginning of the trip. How much money did Shane have at the beginning of his trip?

13. **Items Sold**   Every item in the Just a Dollar store is priced at $1.00. When Mary Jo opens the store, there is $125.50 in the cash register. When she counts the money in the cash register at the end of the day, the total is $1,058.60. If the sales tax rate is 8.5%, how many items were sold that day?

14. **TV Show Ratings**   During one ratings period in 1993, it was estimated that 5.4 million people watched *The Tonight Show* with Jay Leno. If the number of people watching Jay Leno was 1.8% more than the number of viewers watching *Nightline* with Ted Koppel, approximately how many viewers did Ted Koppel have? Round your answer to the nearest tenth of a million viewers.

15. **Workforce**   On Monday, August 16, 1993, Continental Airlines announced it was laying off 2,500 workers, or 6% of its workforce, by December 31 of that year. How many workers did Continental Airlines have on Friday, August 13, 1993?

16. **Textbook Price**   Suppose a college bookstore buys a textbook from a publishing company and then marks up the price they paid for the book 33% and sells it to a student at the marked-up price. If the student pays $45.00 for the textbook, what did

the bookstore pay for it? Round your answer to the nearest cent.

17. Monthly Salary An accountant earns $3,440 per month after receiving a 5.5% raise. What was the accountant's monthly income before the raise? Round your answer to the nearest cent.

18. Hourly Wage A sheet metal worker earns $26.80 per hour after receiving a 4.5% raise. What was the sheet metal worker's hourly pay before the raise? Round your answer to the nearest cent.

19. Movies *Batman Forever* grossed $52.8 million on its opening weekend and had one of the most successful movie launches in history. If *Batman Forever* accounted for approximately 53% of all box office receipts that weekend, what were the total box office receipts?

20. Movies The two top money-making films from 1990–1995 were *Jurassic Park* and *Lion King*. If together these two films grossed $651 million and *Lion King* grossed $25 million less than *Jurassic Park*, what were the gross earnings for *Jurassic Park*?

21. Loans Upon graduation in 1996, 52% of bachelor degree recipients from public institutions and 25% of associate degree recipients from community colleges had federal loan debt. This represents a significant increase from 1993 in the percent of students incurring debt. In addition, the average per-student loan total upon graduation significantly increased from 1993 to 1996.

(a) The average per-student debt at graduation from community colleges increased from $4,380 to $5,450. Find the percent of increase.

(b) In 1993, the average per-student debt for private school students was $10,214. By 1996, this average had risen by 40%. Find the 1996 average debt per student to the nearest dollar.

(c) The per-student debt averaged $11,950 at public institutions in 1996, which represented a 61.48% increase over the average in 1993. What was the average debt per student (rounded to the nearest dollar) in 1993?

22. Loans In 1997, 5.4 million students obtained federal loans. Currently, the default rate on student loans has been reduced to approximately 11%. If this rate holds, how many of the students who borrowed in 1997 will default on their loan (with serious consequences)?

23. Fat Content in Milk I was reading the information on the milk carton in Figure 11 at breakfast

one morning when I was working on this book. According to the carton, this milk contains 70% less fat than whole milk. The nutrition label on the other side of the carton states that one serving of this milk contains 2.5 grams of fat. How many grams of fat are in an equivalent serving of whole milk?

FIGURE 11

24. Calories in Milk The information on the milk carton in Figure 11 states that it contains 20% fewer calories than whole milk. If one serving of this milk contains 120 calories, how many calories are in an equivalent serving of whole milk?

More Geometry Problems

25. Angles Two supplementary angles are such that one is 8 times larger than the other. Find the two angles.

26. Angles Two complementary angles are such that one is 5 times larger than the other. Find the two angles.

27. Angles One angle is 12° less than 4 times another. Find the measure of each angle if
(a) they are complements of each other.
(b) they are supplements of each other.

28. Angles One angle is 4° more than 3 times another. Find the measure of each angle if
(a) they are complements of each other.
(b) they are supplements of each other.

29. Triangles A triangle is such that the largest angle is 3 times the smallest angle. The third angle is 9°

less than the largest angle. Find the measure of each angle.

30. Triangles   The smallest angle in a triangle is half of the largest angle. The third angle is 15° less than the largest angle. Find the measure of all three angles.

31. Triangles   The smallest angle in a triangle is one-third of the largest angle. The third angle is 10° more than the smallest angle. Find the measure of all three angles.

32. Triangles   The third angle in an isosceles triangle is half as large as each of the two base angles. Find the measure of each angle.

33. Isosceles Triangles   The third angle in an isosceles triangle is 8° more than twice as large as each of the two base angles. Find the measure of each angle.

34. Isosceles Triangles   The third angle in an isosceles triangle is 4° more than one-fifth of each of the two base angles. Find the measure of each angle.

Interest Problems

35. Investing   A woman has a total of $9,000 to invest. She invests part of the money in an account that pays 8% per year and the rest in an account that pays 9% per year. If the interest earned in the first year is $750, how much did she invest in each account?

	Dollars at 8%	Dollars at 9%	Total
Number of			
Interest on			

36. Investing   A man invests $12,000 in two accounts. If one account pays 10% per year and the other pays 7% per year, how much was invested in each account if the total interest earned in the first year was $960?

	Dollars at 10%	Dollars at 7%	Total
Number of			
Interest on			

37. Investing   A total of $15,000 is invested in two accounts. One of the accounts earns 12% per year, and the other earns 10% per year. If the total interest earned in the first year is $1,600, how much was invested in each account?

38. Investing   A total of $11,000 is invested in two accounts. One of the two accounts pays 9% per year, and the other account pays 11% per year. If the total interest paid in the first year is $1,150, how much was invested in each account?

39. Investing   Stacey has a total of $6,000 in two accounts. The total amount of interest she earns from both accounts in the first year is $500. If one of the accounts earns 8% interest per year and the other earns 9% interest per year, how much did she invest in each account?

40. Investing   Travis has a total of $6,000 invested in two accounts. The total amount of interest he earns from the accounts in the first year is $410. If one account pays 6% per year and the other pays 8% per year, how much did he invest in each account?

41. Credit Cards   The Shell MasterCard lets customers earn free gasoline based on purchases made with the card. On purchases of Shell products, including gasoline, 3% of the purchase price goes toward free gasoline. On all other purchases, 2% of the purchase price goes toward the free gasoline. Last year Melissa spent $2,850 with her Shell MasterCard and accumulated $64.00 worth of free gasoline. How much were her Shell purchases? How much were her other purchases?

42. Credit Cards   Angela owns the same type of Shell MasterCard as Melissa in Problem 41. Last year Angela spent $5,625 with her Shell Master Card and accumulated $134.15 worth of free gasoline. How much were her Shell purchases? How much were her other purchases?

43. Credit Cards   A credit card lets users accumulate cash refunds on purchases made with the card. On purchases from Company A, 2.65% of all purchases goes toward a cash refund. On purchases from all other companies, 1.65% of the purchase price goes toward the cash refund. Last year Chad spent $6,680 with his credit card and accumulated a $152.22 cash refund. How much were his purchases from Company A? How much were his other purchases?

44. Credit Cards   A certain credit card charges people two different rates: one rate, 9.85%, for products purchased from Store A and another rate, 13.85%, for all other purchases. Last year Alex spent $11,185.00 using his new credit card and was charged $1,404.12 in interest. How much did Alex spend on Store A products? How much on other products?

Miscellaneous Problems

45. Tickets   Tickets for the father-and-son breakfast were $2.00 for fathers and $1.50 for sons. If a total of 75 tickets were sold for $127.50, how many fathers and how many sons attended the breakfast?

46. Tickets   A Girl Scout troop sells 62 tickets to their mother-and-daughter dinner, for a total of $216. If the tickets cost $4.00 for mothers and $3.00 for daughters, how many of each ticket did they sell?

47. Sales Tax   A woman owns a small, cash-only business in a state that requires her to charge a 6% sales tax on each item she sells. At the beginning of the day she has $250 in the cash register. At the end of the day she has $1,204 in the register. How much money should she send to the state government for the sales tax she collected?

48. Sales Tax   A store is located in a state that requires a 6% tax on all items sold. If the store brings in a total of $3,392 in one day, how much of that total was sales tax?

49. Long-Distance Phone Calls   Patrick goes away to college. The first week he is away from home he calls his girlfriend, using his parents' telephone credit card, and talks for a long time. The telephone company charges 40 cents for the first minute and 30 cents for each additional minute, and then adds on a 50-cent service charge for using the credit card. If his parents receive a bill for $13.80 for Patrick's call, how long did he talk?

50. Long-Distance Phone Calls   A person makes a long distance person-to-person call to Santa Barbara, California. The telephone company charges 41 cents for the first minute and 32 cents for each additional minute. Because the call is person-to-person, there is also a service charge of $3.00. If the cost of the call is $6.29, how many minutes did the person talk?

Problems 51–54 may be solved using a graphing calculator.

Table Building

51. Livestock Pen   A farmer buys 48 feet of fencing material to build a rectangular livestock pen. Fill in the second column of the table to find the length of the pen if the width is 2, 4, 6, 8, 10, or 12 feet. Then find the area of each of the pens formed.

w	l	A
2		
4		
6		
8		
10		
12		

52. Art Supplies   A store selling art supplies finds that they can sell x sketch pads each week at a price of p dollars each according to the formula $x = 900 - 300p$. Use this formula to build a table that gives the number of pads sold each week if the price per pad is $0.50, $1.00, $1.50, $2.00, or $2.50.

53. Model Rocket   A small rocket is projected straight up into the air with a velocity of 128 feet per second. The formula that gives the height h of the rocket t seconds after it is launched is

$$h = -16t^2 + 128t$$

Use this formula to find the height of the rocket after 1, 2, 3, 4, 5, and 6 seconds.

Time (seconds)	Height (feet)
1	
2	
3	
4	
5	
6	

54. Cellular Phone   If you were to buy a cellular phone in California and access it through GTE Wireless, you would have a choice of rates. One rate card lists a flat rate of $19.95 per month plus $0.38 for each minute you use the phone. Write an equation in two variables that will let you calculate the monthly charge for talking x minutes. Then build a table that shows the cost for talking 10, 15, 20, 25, or 30 minutes in one month.

Maximum Heart Rate In exercise physiology, a person's maximum heart rate, in beats per minute, is found by subtracting his age in years from 220. So, if A represents your age in years, then your maximum heart rate is

$$M = 220 - A$$

Use this formula to complete the following tables.

55.

Age (years)	Maximum Heart Rate (beats per minute)
18	
19	
20	
21	
22	
23	

56.

Age (years)	Maximum Heart Rate (beats per minute)
15	
20	
25	
30	
35	
40	

Training Heart Rate A person's training heart rate, in beats per minute, is his resting heart rate plus 60% of the difference between his maximum heart rate and his resting heart rate. If resting heart rate is R and maximum heart rate is M, then the formula that gives training heart rate is

$$T = R + 0.6(M - R)$$

Use this formula along with the results of Problems 55 and 56 to fill in the following two tables.

57. For a 20-year-old person

Resting Heart Rate (beats per minute)	Training Heart Rate (beats per minute)
60	
62	
64	
68	
70	
72	

58. For a 40-year-old person

Resting Heart Rate (beats per minute)	Training Heart Rate (beats per minute)
60	
62	
64	
68	
70	
72	

59. Length of String Use a graphing calculator or a spreadsheet program to produce Table 1 from this section (page 99). Start with a width of 0, but increase the width by 0.5 inch each time. What is the largest area produced by this table?

60. Length of String Use a graphing calculator or a spreadsheet program to reproduce Table 1 from this section (page 99). This time, start with a width of 0 and increase the width by 0.25 inch each time. What is the largest area produced by this table?

61. Investing If you invest $100 in an account that earns 6% annual interest, compounded monthly, the amount of money in that account t months later is given by the formula $A = 100(1.005)^t$. Use a graphing calculator or a spreadsheet program to produce a table that gives the amount of money in the account each month for the first 2 years it is on deposit. (On your graphing calculator, let X represent t, and Y_1 represent A. The minimum X is 0 and the table increment is 1.)

62. Volume of a Balloon A spherical balloon is being inflated at a constant rate so that the radius increases by 1 centimeter every second. Since the balloon is a sphere, its volume when the radius is r is given by the formula $V = \frac{4}{3}\pi r^3$. Construct a table that gives the volume of the balloon every second, starting when the radius is 0 centimeters and ending when the radius is 10 centimeters. (Round to the nearest whole number.)

Review Problems

The following problems review material we covered in Section 1.2. Reviewing these problems will help you with the next section.

Graph each inequality.

63. $\{x \mid x > -5\}$
64. $\{x \mid x \leq 4\}$
65. $\{x \mid x \leq -2 \text{ or } x > 5\}$
66. $\{x \mid x < 3 \text{ or } x \geq 5\}$

67. $\{x \mid x > -4 \text{ and } x < 0\}$
68. $\{x \mid x \geq 0 \text{ and } x \leq 2\}$
69. $\{x \mid 1 \leq x \leq 4\}$
70. $\{x \mid -4 < x < -2\}$

2.4 Linear Inequalities in One Variable

A linear inequality in one variable is any inequality that can be put in the form

$$ax + b < c \qquad (a, b, \text{ and } c \text{ constants}, a \neq 0)$$

where the inequality symbol ($<$) can be replaced with any of the other three inequality symbols ($\leq$, $>$, or $\geq$).

Some examples of linear inequalities are

$$3x - 2 \geq 7 \qquad -5y < 25 \qquad 3(x - 4) > 2x$$

Our first property for inequalities is similar to the addition property we used when solving equations.

ADDITION PROPERTY FOR INEQUALITIES

For any algebraic expressions, A, B, and C,

$$\text{if} \qquad A < B$$

$$\text{then} \qquad A + C < B + C$$

In words: Adding the same quantity to both sides of an inequality will not change the solution set.

Note: Since subtraction is defined as addition of the opposite, our new property holds for subtraction as well as addition. That is, we can subtract the same quantity from each side of an inequality and always be sure that we have not changed the solution.

EXAMPLE 1 Solve $3x + 3 < 2x - 1$, and graph the solution.

Solution We use the addition property for inequalities to write all the variable terms on one side and all constant terms on the other side:

$$3x + 3 < 2x - 1$$

$$3x + (-2x) + 3 < 2x + (-2x) - 1 \qquad \text{Add } -2x \text{ to each side.}$$

$$x + 3 < -1$$

$$x + 3 + (-3) < -1 + (-3) \qquad \text{Add } -3 \text{ to each side.}$$

$$x < -4$$

The solution set is all real numbers that are less than -4. To show this we can use set notation and write

$$\{x \mid x < -4\}$$

Or we can graph the solution set on the number line using an open circle at -4 to show that -4 is not part of the solution set:

This graph gives rise to the following notation, called *interval notation,* that is an alternative way to write the solution set.

$$(-\infty, -4)$$

The preceding expression indicates that the solution set is all real numbers from negative infinity up to, but not including, -4.

We have three equivalent representations for the solution set to our original inequality. Here are all three together.

The English mathematician John Wallis (1616–1703) was the first person to use the ∞ symbol to represent infinity. When we encounter the interval $(3, \infty)$ we read it as "the interval from 3 to infinity," and we mean the set of real numbers that are greater than 3. Likewise, the interval $(-\infty, -4)$ is read "the interval from negative infinity to -4," which is all real numbers less than -4.

Set Notation	Line Graph	Interval Notation
$\{x \mid x < -4\}$	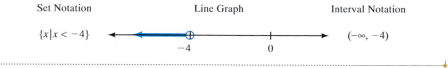	$(-\infty, -4)$

Before we state the multiplication property for inequalities, we will take a look at what happens to an inequality statement when we multiply both sides by a positive number and what happens when we multiply by a negative number.

We begin by writing three true inequality statements:

$$3 < 5 \qquad -3 < 5 \qquad -5 < -3$$

We multiply both sides of each inequality by a positive number, say, 4:

$$4(3) < 4(5) \qquad 4(-3) < 4(5) \qquad 4(-5) < 4(-3)$$

$$12 < 20 \qquad -12 < 20 \qquad -20 < -12$$

Notice in each case that the resulting inequality symbol points in the same direction as the original inequality symbol. Multiplying both sides of an inequality by a positive number preserves the *sense* of the inequality.

Let's take the same three original inequalities and multiply both sides by -4:

$$3 < 5 \qquad\qquad -3 < 5 \qquad\qquad -5 < -3$$

$$\downarrow \qquad\qquad\qquad \downarrow \qquad\qquad\qquad \downarrow$$

$$-4(3) > -4(5) \qquad -4(-3) > -4(5) \qquad -4(-5) > -4(-3)$$

$$-12 > -20 \qquad\qquad 12 > -20 \qquad\qquad 20 > 12$$

Notice in this case that the resulting inequality symbol always points in the opposite direction from the original one. Multiplying both sides of an inequality by a negative number *reverses* the sense of the inequality. Keeping this in mind, we will now state the multiplication property for inequalities.

MULTIPLICATION PROPERTY FOR INEQUALITIES

Let A, B, and C represent algebraic expressions.

$$\text{If} \qquad A < B$$

$$\text{then} \qquad AC < BC \qquad \text{if} \qquad C \text{ is positive } (C > 0)$$

$$\text{or} \qquad AC > BC \qquad \text{if} \qquad C \text{ is negative } (C < 0)$$

In words: Multiplying both sides of an inequality by a positive number always produces an equivalent inequality. Multiplying both sides of an inequality by a negative number reverses the sense of the inequality.

Note: Since division is defined as multiplication by the reciprocal, we can apply our new property to division as well as to multiplication. We can divide both sides of an inequality by any nonzero number as long as we reverse the direction of the inequality when the number we are dividing by is negative.

The multiplication property for inequalities does not limit what we can do with inequalities. We are still free to multiply both sides of an inequality by any nonzero number we choose. If the number we multiply by happens to be *negative,* then we *must also reverse* the direction of the inequality.

EXAMPLE 2 Find the solution set for $-2y - 3 \le 7$.

Solution   We begin by adding 3 to each side of the inequality:

$$-2y - 3 \le 7$$

$$-2y \le 10 \qquad\qquad \text{Add } +3 \text{ to both sides.}$$

$$\downarrow$$

$$-\frac{1}{2}(-2y) \ge -\frac{1}{2}(10) \qquad\qquad \text{Multiply by } -\frac{1}{2} \text{ and reverse the}$$
$$\text{direction of the inequality symbol.}$$

$$y \ge -5$$

The solution set is all real numbers that are greater than or equal to -5. The following are three equivalent ways to represent this solution set.

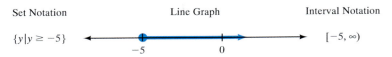

Set Notation	Line Graph	Interval Notation
$\{y \mid y \geq -5\}$		$[-5, \infty)$

Notice how a bracket is used with interval notation to show that 5 is part of the solution set.

When our inequalities become more complicated, we use the same basic steps we used in Section 2.1 when we were solving equations. That is, we simplify each side of the inequality before we apply the addition property or multiplication property. When we have solved the inequality, we graph the solution on a number line.

EXAMPLE 3   Solve $3(2x - 4) - 7x \leq -3x$.

Solution   We begin by using the distributive property to separate terms. Next, simplify both sides.

$$3(2x - 4) - 7x \leq -3x \qquad \text{Original inequality}$$

$$6x - 12 - 7x \leq -3x \qquad \text{Distributive property}$$

$$-x - 12 \leq -3x \qquad 6x - 7x = (6 - 7)x = -x$$

$$-12 \leq -2x \qquad \text{Add } x \text{ to both sides.}$$

$$-\frac{1}{2}(-12) \geq -\frac{1}{2}(-2x) \qquad \begin{array}{l}\text{Multiply both sides by } -\frac{1}{2} \text{ and reverse}\\ \text{the direction of the inequality symbol.}\end{array}$$

$$6 \geq x$$

This last line is equivalent to $x \leq 6$. The solution set can be represented with any of the three following items.

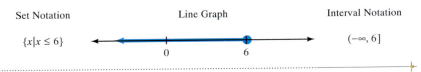

Set Notation	Line Graph	Interval Notation
$\{x \mid x \leq 6\}$		$(-\infty, 6]$

Note: In Examples 2 and 3, notice that each time we multiplied both sides of the inequality by a negative number we also reversed the direction of the inequality symbol. Failure to do so would cause our graph to lie on the wrong side of the endpoint.

EXAMPLE 4   Solve and graph $-3 \le 2x - 5 \le 3$.

Solution   We can extend our properties for addition and multiplication to cover this situation. If we add a number to the middle expression, we must add the same number to the outside expressions. If we multiply the center expression by a number, we must do the same to the outside expressions, remembering to reverse the direction of the inequality symbols if we multiply by a negative number. We begin by adding 5 to all three parts of the inequality:

$$-3 \le 2x - 5 \le 3$$

$$2 \le \quad 2x \quad \le 8 \qquad \text{Add 5 to all three members.}$$

$$1 \le \quad x \quad \le 4 \qquad \text{Multiply through by } \tfrac{1}{2}.$$

Here are three ways to write this solution set:

Set Notation	Line Graph	Interval Notation

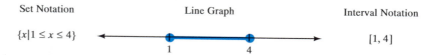

Set Notation: $\{x \mid 1 \le x \le 4\}$ Interval Notation: $[1, 4]$

EXAMPLE 5   Solve the compound inequality.

$$3t + 7 \le -4 \qquad \text{or} \qquad 3t + 7 \ge 4$$

Solution   We solve each half of the compound inequality separately, then we graph the solution set:

$$3t + 7 \le -4 \qquad \text{or} \qquad 3t + 7 \ge 4$$

$$3t \le -11 \qquad \text{or} \qquad 3t \ge -3 \qquad \text{Add } -7.$$

$$t \le -\frac{11}{3} \qquad \text{or} \qquad t \ge -1 \qquad \text{Multiply by } \tfrac{1}{3}.$$

The solution set can be written in any of the following ways:

Set Notation	Line Graph	Interval Notation

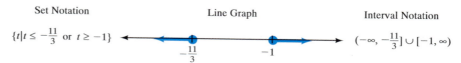

Set Notation: $\{t \mid t \le -\frac{11}{3} \text{ or } t \ge -1\}$ Interval Notation: $(-\infty, -\frac{11}{3}] \cup [-1, \infty)$

EXAMPLE 6   A company that manufactures typewriter ribbons finds that they can sell x ribbons each week at a price of p dollars each, according to the formula $x = 1,300 - 100p$. What price should they charge for each ribbon if they want to sell at least 300 ribbons a week?

Solution   Since x is the number of ribbons they sell each week, an inequality that corresponds to selling at least 300 ribbons a week is

$$x \ge 300$$

Substituting $1{,}300 - 100p$ for x gives us an inequality in the variable p.

$$1{,}300 - 100p \geq 300$$

$$-100p \geq -1{,}000 \qquad \text{Add } -1{,}300 \text{ to each side.}$$

$$p \leq 10 \qquad \text{Divide each side by } -100, \text{ and reverse} \\ \text{the direction of the inequality symbol.}$$

In order to sell at least 300 ribbons each week, the price per ribbon should be no more than $10. That is, selling the ribbons for $10 or less will produce weekly sales of 300 or more ribbons.

EXAMPLE 7   The formula $F = \frac{9}{5}C + 32$ gives the relationship between the Celsius and Fahrenheit temperature scales. If the temperature range on a certain day is 86° to 104° Fahrenheit, what is the temperature range in degrees Celsius?

Solution   From the given information we can write $86 \leq F \leq 104$. But, since F is equal to $\frac{9}{5}C + 32$, we can also write

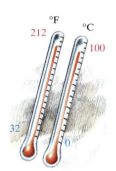

$$86 \leq \frac{9}{5}C + 32 \leq 104$$

$$54 \leq \frac{9}{5}C \leq 72 \qquad \text{Add } -32 \text{ to each number.}$$

$$\frac{5}{9}(54) \leq \frac{5}{9}\left(\frac{9}{5}C\right) \leq \frac{5}{9}(72) \qquad \text{Multiply each number by } \frac{5}{9}.$$

$$30 \leq C \leq 40$$

A temperature range of 86° to 104° Fahrenheit corresponds to a temperature range of 30° to 40° Celsius.

Getting Ready for Class

After reading through the preceding section, respond in your own words and in complete sentences.

A. What is the addition property for inequalities?

B. When we use interval notation to denote a section of the real number line, when do we use parentheses, (and), and when do we use brackets, [and]?

C. Explain the difference between the multiplication property of equality and the multiplication property for inequalities.

D. When solving an inequality, when do we change the direction of the inequality symbol?

PROBLEM SET 2.4

Solve each of the following inequalities, and graph each solution.

1. $2x \le 3$

2. $5x \ge -115$

3. $\dfrac{1}{2}x > 2$

4. $\dfrac{1}{3}x > 4$

5. $-5x \le 25$

6. $-7x \ge 35$

7. $-\dfrac{3}{2}x > -6$

8. $-\dfrac{2}{3}x < -8$

9. $-12 \le 2x$

10. $-20 \ge 4x$

11. $-1 \ge -\dfrac{1}{4}x$

12. $-1 \le -\dfrac{1}{5}x$

Solve each of the following inequalities, and graph each solution.

13. $-3x + 1 > 10$

14. $-2x - 5 \le 15$

15. $\dfrac{1}{2} - \dfrac{m}{12} \le \dfrac{7}{12}$

16. $\dfrac{1}{2} - \dfrac{m}{10} > -\dfrac{1}{5}$

17. $\dfrac{1}{2} \ge -\dfrac{1}{6} - \dfrac{2}{9}x$

18. $\dfrac{9}{5} > -\dfrac{1}{5} - \dfrac{1}{2}x$

19. $-40 \le 30 - 20y$

20. $-20 > 50 - 30y$

21. $\dfrac{2}{3}x - 3 < 1$

22. $\dfrac{3}{4}x - 2 > 7$

23. $10 - \dfrac{1}{2}y \le 36$

24. $8 - \dfrac{1}{3}y \ge 20$

Simplify each side first, then solve the following inequalities. Write your answers with interval notation.

25. $2(3y + 1) \le -10$

26. $3(2y - 4) > 0$

27. $-(a + 1) - 4a \le 2a - 8$

28. $-(a - 2) - 5a \le 3a + 7$

29. $\dfrac{1}{3}t - \dfrac{1}{2}(5 - t) < 0$

30. $\dfrac{1}{4}t - \dfrac{1}{3}(2t - 5) < 0$

31. $-2 \le 5 - 7(2a + 3)$

32. $1 < 3 - 4(3a - 1)$

33. $-\dfrac{1}{3}(x + 5) \le -\dfrac{2}{9}(x - 1)$

34. $-\dfrac{1}{2}(2x + 1) \le -\dfrac{3}{8}(x + 2)$

35. $5(y + 3) + 4 < 6y - 1 - 5y$

36. $4(y - 1) + 2 \ge 3y + 8 - 2y$

Solve the following continued inequalities. Use both a line graph and interval notation to write each solution set.

37. $-2 \le m - 5 \le 2$

38. $-3 \le m + 1 \le 3$

39. $-60 < 20a + 20 < 60$

40. $-60 < 50a - 40 < 60$

41. $0.5 \le 0.3a - 0.7 \le 1.1$

42. $0.1 \le 0.4a + 0.1 \le 0.3$

43. $3 < \dfrac{1}{2}x + 5 < 6$

44. $5 < \dfrac{1}{4}x + 1 < 9$

45. $4 < 6 + \dfrac{2}{3}x < 8$

46. $3 < 7 + \dfrac{4}{5}x < 15$

Graph the solution sets for the following compound inequalities. Then write each solution set using interval notation.

47. $x + 5 \le -2$ or $x + 5 \ge 2$

48. $3x + 2 < -3$ or $3x + 2 > 3$

49. $5y + 1 \le -4$ or $5y + 1 \ge 4$

50. $7y - 5 \le -2$ or $7y - 5 \ge 2$

51. $2x + 5 < 3x - 1$ or $x - 4 > 2x + 6$

52. $3x - 1 > 2x + 4$ or $5x - 2 < 3x + 4$

Applying the Concepts

Art Supplies A store selling art supplies finds that they can sell x sketch pads each week at a price of p dollars each, according to the formula $x = 900 - 300p$. What price should they charge if they want to sell

53. At least 300 pads each week?

54. More than 600 pads each week?

55. Less than 525 pads each week?

56. At most 375 pads each week?

Temperature Range   Each of the following tempera-ture ranges is in degrees Fahrenheit. Use the formula $F = \frac{9}{5}C + 32$ to find the corresponding temperature range in degrees Celsius.

57. 95° to 113°

58. 68° to 86°

59. −13° to 14°

60. −4° to 23°

61. Polling Error   The October 24, 1998, issue of the *Chicago Tribune* reported the results of a telephone poll of 1,099 registered Illinois voters for the office of governor. This poll indicated that 49% would vote for the Republican, George Ryan, and only 34% would vote for his Democratic opponent, Glen Poshard. The margin of error of this poll was given as ±3%.

(a) Based on the results of this poll, write an in-equality for the minimum and maximum per-centage of the vote, R, that George Ryan can expect to receive, and an inequality for the min-imum and maximum percentage of the vote, P, that Glen Poshard can expect.

(b) Based on the results of this poll, what is the minimum winning margin that George Ryan can expect?

(c) Is George Ryan "guaranteed" to win this elec-tion? Explain.

62. Photocopying Money   Federal law states that it is legal to photocopy United States currency if the photocopy is not color, and the copy is smaller than three-fourths of the actual size, or greater than one and one-half times the actual size. A $10 bill mea-sures approximately 6 inches by $2\frac{1}{2}$ inches. What are the maximum and minimum size photo repro-ductions of this bill that may be legally made from a black and white photocopy machine? Express your answer as an inequality.

63. Sound Frequencies   Nearly every animal can emit and hear sound waves. A dog can emit sound waves with between 452 and 1,080 vibrations per second, but hear from 15 to 50,000 vibrations per second. Humans can emit sound waves from 85 to 1,100 vi-brations per second and hear between 20 and 20,000 vibrations per second.

(a) Write as an inequality the number of vibrations per second, D, that a dog can hear but a human cannot.

(b) Write as an inequality the number of vibrations per second, H, that a human can emit but a dog cannot.

(c) Refer to your answer for part (a). Explain what this means in terms of the ability to hear sound for a dog versus a human.

64. Cell Phone Cost   Suppose you have a cellular phone in which you have committed to spend less than $30 on each monthly bill. Your rate plan costs $20 a month and includes 30 free minutes. Each ad-ditional minute costs $0.40. Set up and solve an in-equality to determine how many minutes you may use each month to keep your bill under $30.

65. Student Loan   When considering how much debt to incur in student loans, you learn that it is wise to keep your student loan payment to 8% or less of your starting monthly income. Suppose you antici-pate a starting annual salary of $24,000. Set up and solve an inequality that represents the amount of monthly debt for student loans that would be con-sidered manageable.

66. Student Loan   Suppose you have already bor-rowed $30,000 in student loans, which you will pay back at $368 a month for 10 years. Set up and solve an inequality that determines how much you should make per month as your starting salary to keep your debt at 8% or less of your monthly income.

Review Problems

The following problems review material we covered in Section 1.3. Reviewing the problems will help you with the next section.

Simplify.

67. $|-3|$

68. $|3|$

69. $-|-3|$

70. $-(-3)$

71. Give a definition for the absolute value of x that in-volves the number line. (This is the geometric defi-nition.)

72. Give a definition of the absolute value of x that does not involve the number line. (This is the alge-braic definition.)

Extending the Concepts

Assume a, b, and c are positive, and solve each formula for x.

73. $ax + b < c$

74. $\dfrac{x}{a} + \dfrac{y}{b} < 1$

75. $-c < ax + b < c$

76. $-1 < \dfrac{ax + b}{c} < 1$

Equations With Absolute Value

In Chapter 1 we defined the absolute value of x, $|x|$, to be the distance between x and 0 on the number line. The absolute value of a number measures its distance from 0.

EXAMPLE 1 Solve for x: $|x| = 5$

Solution Using the definition of absolute value, we can read the equation as, "The distance between x and 0 on the number line is 5." If x is 5 units from 0, then x can be 5 or -5:

$$\text{If } |x| = 5 \quad \text{then } x = 5 \quad \text{or} \quad x = -5$$

In general, then, we can see that any equation of the form $|a| = b$ is equivalent to the equations $a = b$ or $a = -b$, as long as $b > 0$.

EXAMPLE 2 Solve $|2a - 1| = 7$.

Solution We can read this question as "$2a - 1$ is 7 units from 0 on the number line." The quantity $2a - 1$ must be equal to 7 or -7:

$$|2a - 1| = 7$$
$$2a - 1 = 7 \quad \text{or} \quad 2a - 1 = -7$$

We have transformed our absolute value equation into two equations that do not involve absolute value. We can solve each equation using the method in Section 2.1:

$$
\begin{array}{llll}
2a - 1 = 7 & \text{or} & 2a - 1 = -7 & \\
2a = 8 & \text{or} & 2a = -6 & \text{Add } +1 \text{ to both sides.} \\
a = 4 & \text{or} & a = -3 & \text{Multiply by } \tfrac{1}{2}.
\end{array}
$$

Our solution set is $\{4, -3\}$.

To check our solutions, we put them into the original absolute value equation:

When	$a = 4$	When	$a = -3$
the equation	$|2a - 1| = 7$	the equation	$|2a - 1| = 7$
becomes	$|2(4) - 1| = 7$	becomes	$|2(-3) - 1| = 7$
	$|7| = 7$		$|-7| = 7$
	$7 = 7$		$7 = 7$

EXAMPLE 3   Solve $\left|\frac{2}{3}x - 3\right| + 5 = 12$.

Solution   In order to use the definition of absolute value to solve this equation, we must isolate the absolute value on the left side of the equal sign. To do so, we add -5 to both sides of the equation to obtain

$$\left|\frac{2}{3}x - 3\right| = 7$$

Now that the equation is in the correct form, we can write

$$\frac{2}{3}x - 3 = 7 \qquad \text{or} \qquad \frac{2}{3}x - 3 = -7$$

$$\frac{2}{3}x = 10 \qquad \text{or} \qquad \frac{2}{3}x = -4 \qquad \text{Add } +3 \text{ to both sides.}$$

$$x = 15 \qquad \text{or} \qquad x = -6 \qquad \text{Multiply by } \frac{3}{2}.$$

The solution set is $\{15, -6\}$.

EXAMPLE 4   Solve $|3a - 6| = -4$.

Solution   The solution set is $\varnothing$ because the left side cannot be negative and the right side is negative. No matter what we try to substitute for the variable a, the quantity $|3a - 6|$ will always be positive or zero. It can never be -4.

Consider the statement $|a| = |b|$. What can we say about a and b? We know they are equal in absolute value. By the definition of absolute value, they are the same distance from 0 on the number line. They must be equal to each other or opposites of each other. In symbols, we write

$$|a| = |b| \Leftrightarrow a = b \qquad \text{or} \qquad a = -b$$

$$\uparrow \qquad\qquad \uparrow \qquad\qquad\qquad \uparrow$$

Equal in   Equals   or   Opposites
absolute value

EXAMPLE 5   Solve $|3a + 2| = |2a + 3|$.

Solution   The quantities $3a + 2$ and $2a + 3$ have equal absolute values. They are, therefore, the same distance from 0 on the number line. They must be equals or opposites:

$$|3a + 2| = |2a + 3|$$

Equals		*Opposites*
$3a + 2 = 2a + 3$	or	$3a + 2 = -(2a + 3)$
$a + 2 = 3$		$3a + 2 = -2a - 3$
$a = 1$		$5a + 2 = -3$
		$5a = -5$
		$a = -1$

The solution set is $\{1, -1\}$.

It makes no difference in the outcome of the problem if we take the opposite of the first or second expression. It is very important, once we have decided which one to take the opposite of, that we take the opposite of both its terms and not just the first term. That is, the opposite of $2a + 3$ is $-(2a + 3)$, which we can think of as $-1(2a + 3)$. Distributing the -1 across *both* terms, we have

$$-1(2a + 3) = -2a - 3$$

EXAMPLE 6   Solve $|x - 5| = |x - 7|$.

Solution   As was the case in Example 5, the quantities $x - 5$ and $x - 7$ must be equal or they must be opposites, because their absolute values are equal:

Equals		*Opposites*
$x - 5 = x - 7$	or	$x - 5 = -(x - 7)$
$-5 = -7$		$x - 5 = -x + 7$
No solution here		$2x - 5 = 7$
		$2x = 12$
		$x = 6$

Since the first equation leads to a false statement, it will not give us a solution. (If either of the two equations were to reduce to a true statement, it would mean all real numbers would satisfy the original equation.) In this case, our only solution is $x = 6$.

Getting Ready for Class

After reading through the preceding section, respond in your own words and in complete sentences.

A. Why do some of the equations in this section have two solutions instead of one?

B. Translate $|x| = 6$ into words using the definition of absolute value.

C. Explain in words what the equation $|x - 3| = 4$ means with respect to distance on the number line.

D. When is the statement $|x| = x$ true?

PROBLEM SET 2.5

Use the definition of absolute value to solve each of the following problems.

1. $|x| = 4$

2. $|x| = 7$

3. $2 = |a|$

4. $5 = |a|$

5. $|x| = -3$

6. $|x| = -4$

7. $|a| + 2 = 3$

8. $|a| - 5 = 2$

9. $|y| + 4 = 3$

10. $|y| + 3 = 1$

11. $4 = |x| - 2$

12. $3 = |x| - 5$

13. $|x - 2| = 5$

14. $|x + 1| = 2$

15. $|a - 4| = \dfrac{5}{3}$

16. $|a + 2| = \dfrac{7}{5}$

17. $1 = |3 - x|$

18. $2 = |4 - x|$

19. $\left|\dfrac{3}{5}a + \dfrac{1}{2}\right| = 1$

20. $\left|\dfrac{2}{7}a + \dfrac{3}{4}\right| = 1$

21. $60 = |20x - 40|$

22. $800 = |400x - 200|$

23. $|2x + 1| = -3$

24. $|2x - 5| = -7$

25. $\left|\dfrac{3}{4}x - 6\right| = 9$

26. $\left|\dfrac{4}{5}x - 5\right| = 15$

27. $\left|1 - \dfrac{1}{2}a\right| = 3$

28. $\left|2 - \dfrac{1}{3}a\right| = 10$

Solve each equation.

29. $|3x + 4| + 1 = 7$

30. $|5x - 3| - 4 = 3$

31. $|3 - 2y| + 4 = 3$

32. $|8 - 7y| + 9 = 1$

33. $3 + |4t - 1| = 8$

34. $2 + |2t - 6| = 10$

35. $\left|9 - \dfrac{3}{5}x\right| + 6 = 12$

36. $\left|4 - \dfrac{2}{7}x\right| + 2 = 14$

37. $5 = \left|\dfrac{2x}{7} + \dfrac{4}{7}\right| - 3$

38. $7 = \left|\dfrac{3x}{5} + \dfrac{1}{5}\right| + 2$

39. $2 = -8 + \left|4 - \dfrac{1}{2}y\right|$

40. $1 = -3 + \left|2 - \dfrac{1}{4}y\right|$

Solve the following equations.

41. $|3a + 1| = |2a - 4|$

42. $|5a + 2| = |4a + 7|$

43. $\left|x - \dfrac{1}{3}\right| = \left|\dfrac{1}{2}x + \dfrac{1}{6}\right|$

44. $\left|\dfrac{1}{10}x - \dfrac{1}{2}\right| = \left|\dfrac{1}{5}x + \dfrac{1}{10}\right|$

45. $|y - 2| = |y + 3|$

46. $|y - 5| = |y - 4|$

47. $|3x - 1| = |3x + 1|$

48. $|5x - 8| = |5x + 8|$

Solve the following equations.

49. $|3 - m| = |m + 4|$

50. $|5 - m| = |m + 8|$

51. $|0.03 - 0.01x| = |0.04 + 0.05x|$

52. $|0.07 - 0.01x| = |0.08 - 0.02x|$

53. $|x - 2| = |2 - x|$

54. $|x - 4| = |4 - x|$

55. $\left|\dfrac{x}{5} - 1\right| = \left|1 - \dfrac{x}{5}\right|$

56. $\left|\dfrac{x}{3} - 1\right| = \left|1 - \dfrac{x}{3}\right|$

Applying the Concepts

57. Triangle Inequality Five important properties are associated with absolute value. They are the following:

(i) $|a| \geq 0$

(ii) $|a| = |-a|$

(iii) $|a \cdot b| = |a| \cdot |b|$

(iv) $|a - b| = |b - a|$

(v) $|a + b| \leq |a| + |b|$

Property (v) is often called the triangle inequality. Verify that this triangle inequality is true by choosing instances in which a and b are equal; a and b are not equal, but both positive; a and b are not equal, but both negative; and a and b have different signs.

58. Absolute Error Given two numbers, n_1 and n_2, then $|n_1 - n_2|$ is defined as the **absolute error** of n_1 from n_2. Most polls that are used to predict election results have an absolute error of the actual vote percentage from the predicted vote percentage of 3%. If P = predicted vote percentage and A = actual vote percentage, write an absolute value equation that relates P, A, and the absolute error of 3%.

Table Building

To obtain a visual representation for expressions that contain absolute value, we can use number sequences. For each of the following sequences, use the formula to construct a table that gives the first five terms in the sequence by substituting 1, 2, 3, 4, and 5 for n in the formula. (Label the first column in the table n, and the second column a_n.) Then use the paired data from the table to construct a scatter diagram.

59. $a_n = |n - 3|$

60. $a_n = |3 - n|$

61. $a_n = |2n - 6|$

62. $a_n = |6 - 2n|$

Review Problems

The problems below review material we covered in Sections 1.2 and 2.4. Reviewing these problems will help you with the next section.

Graph each inequality. [1.2]

63. $x < -2$ or $x > 8$

64. $1 < x < 4$

65. $-2 \leq x \leq 1$

66. $x \leq -\dfrac{3}{2}$ or $x \geq 3$

Solve each inequality. [2.4]

67. $4t - 3 \leq -9$

68. $-3 < 2a - 5 < 3$

69. $-3x > 15$

70. $-2x \leq 10$

71. $\dfrac{1}{2} < \dfrac{3}{4}a < \dfrac{3}{5}$

72. $\dfrac{3}{7}a + 2 < \dfrac{1}{4}$

Extending the Concepts

Solve each formula for x. (Assume a, b, and c are positive.)

73. $|x - a| = b$

74. $|x + a| - b = 0$

75. $|ax + b| = c$

76. $|ax - b| - c = 0$

77. $\left| \dfrac{x}{a} + \dfrac{y}{b} \right| = 1$

78. $\left| \dfrac{x}{a} + \dfrac{y}{b} \right| = c$

2.6   Inequalities Involving Absolute Value

In this section we will again apply the definition of absolute value to solve inequalities involving absolute value. Again, the absolute value of x, which is $|x|$, represents the distance that x is from 0 on the number line. We will begin by considering three absolute value expressions and their verbal translations:

Expression	In Words
$|x| = 7$	x is exactly 7 units from 0 on the number line.
$|a| < 5$	a is less than 5 units from 0 on the number line.
$|y| \geq 4$	y is greater than or equal to 4 units from 0 on the number line.

Once we have translated the expression into words, we can use the translation to graph the original equation or inequality. The graph is then used to write a final equation or inequality that does not involve absolute value.

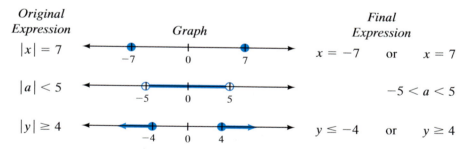

Although we will not always write out the verbal translation of an absolute value inequality, it is important that we understand the translation. Our second expression, $|a| < 5$, means a is within 5 units of 0 on the number line. The graph of this relationship is

which can be written with the following continued inequality:

$$-5 < a < 5$$

We can follow this same kind of reasoning to solve more complicated absolute value inequalities.

EXAMPLE 1   Graph the solution set: $|2x - 5| < 3$

Solution   The absolute value of $2x - 5$ is the distance that $2x - 5$ is from 0 on the number line. We can translate the inequality as, "$2x - 5$ is less than 3 units from 0 on the number line." That is, $2x - 5$ must appear between -3 and 3 on the number line.
    A picture of this relationship is

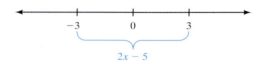

Using the picture, we can write an inequality without absolute value that describes the situation:

$$-3 < 2x - 5 < 3$$

Next, we solve the continued inequality by first adding $+5$ to all three members and then multiplying all three by $\frac{1}{2}$.

$$-3 < 2x - 5 < 3$$

$$2 < \quad 2x \quad < 8 \qquad \text{Add} + 5 \text{ to all three expressions.}$$

$$1 < \quad x \quad < 4 \qquad \text{Multiply each expression by } \tfrac{1}{2}.$$

The graph of the solution set is

    We can see from the solution that for the absolute value of $2x - 5$ to be within 3 units of 0 on the number line, x must be between 1 and 4.

EXAMPLE 2   Solve and graph $|3a + 7| \leq 4$.

Solution   We can read the inequality as, "The distance between $3a + 7$ and 0 is less than or equal to 4." Or, "$3a + 7$ is within 4 units of 0 on the number line." This relationship can be written without absolute value as

$$-4 \leq 3a + 7 \leq 4$$

Solving as usual, we have

$$-4 \leq 3a + 7 \leq 4$$

$$-11 \leq \quad 3a \quad \leq -3 \qquad \text{Add} -7 \text{ to all three members.}$$

$$-\frac{11}{3} \leq \quad a \quad \leq -1 \qquad \text{Multiply each expression by } \tfrac{1}{3}.$$

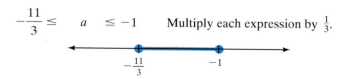

We can see from Examples 1 and 2 that to solve an inequality involving absolute value, we must be able to write an equivalent expression that does not involve absolute value.

EXAMPLE 3 Solve $|x - 3| > 5$, and graph the solution.

Solution We interpret the absolute value inequality to mean that $x - 3$ is more than 5 units from 0 on the number line. The quantity $x - 3$ must be either above $+5$ or below -5. Here is a picture of the relationship:

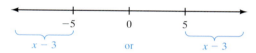

An inequality without absolute value that also describes this situation is

$$x - 3 < -5 \qquad \text{or} \qquad x - 3 > 5$$

Adding $+3$ to both sides of each inequality we have

$$x < -2 \qquad \text{or} \qquad x > 8$$

the graph of which is

EXAMPLE 4 Graph the solution set: $|4t - 3| \geq 9$.

Solution The quantity $4t - 3$ is greater than or equal to 9 units from 0. It must be either above $+9$ or below -9.

$$4t - 3 \leq -9 \qquad \text{or} \qquad 4t - 3 \geq 9$$

$$4t \leq -6 \qquad \text{or} \qquad 4t \geq 12 \qquad \text{Add } +3.$$

$$t \leq -\frac{6}{4} \qquad \text{or} \qquad t \geq \frac{12}{4} \qquad \text{Multiply by } \tfrac{1}{4}.$$

$$t \leq -\frac{3}{2} \qquad \text{or} \qquad t \geq 3$$

We can use the results of our first few examples and the material in the previous section to summarize the information we have related to absolute value equations and inequalities.

REWRITING ABSOLUTE VALUE EQUATIONS AND INEQUALITIES

If c is a positive real number, then each of the following statements on the left
is equivalent to the corresponding statement on the right.

With Absolute Value   *Without Absolute Value*

$$|x| = c \qquad\qquad x = -c \quad \text{or} \quad x = c$$
$$|x| < c \qquad\qquad -c < x < c$$
$$|x| > c \qquad\qquad x < -c \quad \text{or} \quad x > c$$
$$|ax + b| = c \qquad ax + b = -c \quad \text{or} \quad ax + b = c$$
$$|ax + b| < c \qquad\qquad -c < ax + b < c$$
$$|ax + b| > c \qquad ax + b < -c \quad \text{or} \quad ax + b > c$$

EXAMPLE 5   Solve and graph $|2x + 3| + 4 < 9$.

Solution   Before we can apply the method of solution we used in the previous examples, we must isolate the absolute value on one side of the inequality. To do so,
we add -4 to each side.

$$|2x + 3| + 4 < 9$$
$$|2x + 3| + 4 + (-4) < 9 + (-4)$$
$$|2x + 3| < 5$$

From this last line we know that $2x + 3$ must be between -5 and $+5$.

$$-5 < 2x + 3 < 5$$
$$-8 < \quad 2x \quad < 2 \qquad \text{Add} - 3 \text{ to each expression.}$$
$$-4 < \quad x \quad < 1 \qquad \text{Multiply each expression by } \tfrac{1}{2}.$$

The graph is

EXAMPLE 6   Solve and graph $|4 - 2t| > 2$.

Solution   The inequality indicates that $4 - 2t$ is less than -2 or greater than $+2$.
Writing this without absolute value symbols, we have

$$4 - 2t < -2 \qquad \text{or} \qquad 4 - 2t > 2$$

To solve these inequalities we begin by adding -4 to each side.

$$4 + (-4) - 2t < -2 + (-4) \quad \text{or} \quad 4 + (-4) - 2t > 2 + (-4)$$

$$-2t < -6 \quad \text{or} \quad -2t > -2$$

Next we must multiply both sides of each inequality by $-\frac{1}{2}$. When we do so, we must also reverse the direction of each inequality symbol.

$$-2t < -6 \quad \text{or} \quad -2t > -2$$

$$-\frac{1}{2}(-2t) > -\frac{1}{2}(-6) \quad \text{or} \quad -\frac{1}{2}(-2t) < -\frac{1}{2}(-2)$$

$$t > 3 \quad \text{or} \quad t < 1$$

Although in situations like this we are used to seeing the "less than" symbol written first, the meaning of the solution is clear. We want to graph all real numbers that are either greater than 3 or less than 1. Here is the graph.

Since absolute value always results in a nonnegative quantity, we sometimes come across special solution sets when a negative number appears on the right side of an absolute value inequality.

EXAMPLE 7 Solve $|7y - 1| < -2$.

Solution The *left* side is never negative because it is an absolute value. The *right* side is negative. We have a positive quantity less than a negative quantity, which is impossible. The solution set is the empty set, $\varnothing$. There is no real number to substitute for y to make this inequality a true statement.

EXAMPLE 8 Solve $|6x + 2| > -5$.

Solution This is the opposite case from that in Example 7. No matter what real number we use for x on the *left* side, the result will always be positive, or zero. The *right* side is negative. We have a positive quantity greater than a negative quantity. Every real number we choose for x gives us a true statement. The solution set is the set of all real numbers.

Getting Ready for Class

After reading through the preceding section, respond in your own words and in complete sentences.

A. Write an inequality containing absolute value, the solution to which is all the numbers between -5 and 5 on the number line.

B. Translate $|x| \geq 3$ into words using the definition of absolute value.

C. Explain in words what the inequality $|x - 5| < 2$ means with respect to distance on the number line.

D. Why is there no solution to the inequality $|2x - 3| < 0$?

PROBLEM SET 2.6

Solve each of the following inequalities using the definition of absolute value. Graph the solution set in each case.

1. $|x| < 3$

2. $|x| \leq 7$

3. $|x| \geq 2$

4. $|x| > 4$

5. $|x| + 2 < 5$

6. $|x| - 3 < -1$

7. $|t| - 3 > 4$

8. $|t| + 5 > 8$

9. $|y| < -5$

10. $|y| > -3$

11. $|x| \geq -2$

12. $|x| \leq -4$

13. $|x - 3| < 7$

14. $|x + 4| < 2$

15. $|a + 5| \geq 4$

16. $|a - 6| \geq 3$

Solve each inequality and graph the solution set.

17. $|a - 1| < -3$

18. $|a + 2| \geq -5$

19. $|2x - 4| < 6$

20. $|2x + 6| < 2$

21. $|3y + 9| \geq 6$

22. $|5y - 1| \geq 4$

23. $|2k + 3| \geq 7$

24. $|2k - 5| \geq 3$

25. $|x - 3| + 2 < 6$

26. $|x + 4| - 3 < -1$

27. $|2a + 1| + 4 \geq 7$

28. $|2a - 6| - 1 \geq 2$

29. $|3x + 5| - 8 < 5$

30. $|6x - 1| - 4 \leq 2$

Solve each inequality and graph the solution set. Keep in mind that if you multiply or divide both sides of an inequality by a negative number you must reverse the sense of the inequality.

31. $|5 - x| > 3$

32. $|7 - x| > 2$

33. $\left| 3 - \dfrac{2}{3}x \right| \geq 5$

34. $\left| 3 - \dfrac{3}{4}x \right| \geq 9$

35. $\left| 2 - \dfrac{1}{2}x \right| > 1$

36. $\left| 3 - \dfrac{1}{3}x \right| > 1$

Solve each inequality.

37. $|x - 1| < 0.01$

38. $|x + 1| < 0.01$

39. $|2x + 1| \geq \dfrac{1}{5}$

40. $|2x - 1| \geq \dfrac{1}{8}$

41. $\left| \dfrac{3x - 2}{5} \right| \leq \dfrac{1}{2}$

42. $\left| \dfrac{4x - 3}{2} \right| \leq \dfrac{1}{3}$

43. $\left| 2x - \dfrac{1}{5} \right| < 0.3$

44. $\left| 3x - \dfrac{3}{5} \right| < 0.2$

45. Write the continued inequality $-4 \leq x \leq 4$ as a single inequality involving absolute value.

46. Write the continued inequality $-8 \leq x \leq 8$ as a single inequality involving absolute value.

47. Write $-1 \leq x - 5 \leq 1$ as a single inequality involving absolute value.

48. Write $-3 \leq x + 2 \leq 3$ as a single inequality involving absolute value.

Applying the Concepts

49. Television Channel Capacity Channel capacity varies significantly on televisions throughout the United States. A recent study found that 30% of the televisions in the United States are equipped to receive c channels, where c satisfies $|c - 24| \leq 12$. What is the minimum and maximum number of channels that these television sets are capable of receiving?

50. Speed Limits The interstate speed limit for cars is 75 miles per hour in Nebraska, Nevada, New Mexico, Oklahoma, South Dakota, Utah, and Wyoming, and is the highest in the nation. To discourage passing, minimum speeds are also posted, so that the difference between the fastest and slowest moving traffic is no more than 20 miles per hour. Write an absolute value inequality that describes the relationship between the minimum allowable speed and a maximum speed of 75 miles per hour.

51. Speed Limits Suppose the speed limit on Highway 101 in California is 65 miles per hour, and slower cars cannot travel less than 20 miles per hour slower than the fastest cars. Write an absolute value inequality that describes the relationship between the minimum allowable speed and a maximum speed of 65 miles per hour.

52. **Wavelengths of Light** When white light from the sun passes through a prism, it is broken down into bands of light that form colors. The wavelength of each color is different, but may be expressed as an inequality. The wavelength, v, (in nanometers) of some common colors are:

Blue: $424 < v < 491$
Green: $491 < v < 575$
Yellow: $575 < v < 585$
Orange: $585 < v < 647$
Red: $647 < v < 700$

When a fireworks display made of copper is burned, it lets out light with wavelengths, v, that satisfy the relationship $|v - 455| < 23$. Write this inequality without absolute values, find the range of possible values for v, and then using the preceding list of wavelengths, determine the color of that copper fireworks display.

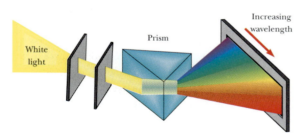

White light

Prism

Increasing wavelength

Review Problems

The following problems review material we covered in Section 1.4.

Simplify each expression as much as possible.

53. $-9 \div \dfrac{3}{2}$ 54. $-\dfrac{4}{5} \div (-4)$

55. $3 - 7(-6 - 3)$ 56. $(3 - 7)(-6 - 3)$

57. $-4(-2)^3 - 5(-3)^2$

58. $4(2 - 5)^3 - 3(4 - 5)^5$

59. $-2(-3 + 8) - 7(-9 + 6)$

60. $-3 - 6[5 - 2(-3 - 1)]$

61. $\dfrac{2(-3) - 5(-6)}{-1 - 2 - 3}$ 62. $\dfrac{4 - 8(3 - 5)}{2 - 4(3 - 5)}$

Extending the Concepts

Assume a, b, and c, are positive, and solve each formula for x.

63. $|x - a| < b$ 64. $|x - a| > b$

65. $|ax - b| > c$ 66. $|ax - b| < c$

Examples

1. We can solve

$$x + 3 = 5$$

by adding -3 to both sides:

$$x + 3 + (-3) = 5 + (-3)$$

$$x = 2$$

2. We can solve $3x = 12$ by multiplying both sides by $\frac{1}{3}$:

$$3x = 12$$

$$\tfrac{1}{3}(3x) = \tfrac{1}{3}(12)$$

$$x = 4$$

3. Solve: $3(2x - 1) = 9$

$$3(2x - 1) = 9$$

$$6x - 3 = 9$$

$$6x - 3 + 3 = 9 + 3$$

$$6x = 12$$

$$\tfrac{1}{6}(6x) = \tfrac{1}{6}(12)$$

$$x = 2$$

4. Solve for w:

$$P = 2l + 2w$$

$$P - 2l = 2w$$

$$\frac{P - 2l}{2} = w$$

Addition Property of Equality [2.1]

For algebraic expressions A, B, and C,

$$\text{if} \qquad A = B$$

$$\text{then} \qquad A + C = B + C$$

This property states that we can add the same quantity to both sides of an equation without changing the solution set.

Multiplication Property of Equality [2.1]

For algebraic expressions A, B, and C,

$$\text{if} \qquad A = B$$

$$\text{then} \qquad AC = BC \qquad C \neq 0$$

Multiplying both sides of an equation by the same nonzero quantity never changes the solution set.

Strategy for Solving Linear Equations in One Variable [2.1]

Step 1a: Use the distributive property to separate terms, if necessary.

1b: If fractions are present, consider multiplying both sides by the LCD to eliminate the fractions. If decimals are present, consider multiplying both sides by a power of 10 to clear the equation of decimals.

1c: Combine similar terms on each side of the equation.

Step 2: Use the addition property of equality to get all variable terms on one side of the equation and all constant terms on the other side. A variable term is a term that contains the variable (e.g., $5x$). A constant term is a term that does not contain the variable (e.g., the number 3).

Step 3: Use the multiplication property of equality to get x (i.e., $1x$) by itself on one side of the equation.

Step 4: Check your solution in the original equation to be sure that you have not made a mistake in the solution process.

Formulas [2.2]

A **formula** in algebra is an equation involving more than one variable. To solve a formula for one of its variables, simply isolate that variable on one side of the equation.

5. The perimeter of a rectangle is 32 inches. If the length is 3 times the width, find the dimensions.

   Step 1: This step is done mentally.
   Step 2: Let x = the width. Then the length is $3x$.
   Step 3: The perimeter is 32; therefore,

$$2x + 2(3x) = 32$$

   Step 4: $$8x = 32$$
$$x = 4$$

   Step 5: The width is 4 inches. The length is $3(4) = 12$ inches.
   Step 6: The perimeter is $2(4) + 2(12)$, which is 32. The length is 3 times the width.

Blueprint for Problem Solving [2.3]

Step 1: **Read** the problem, and then mentally **list** the items that are known and the items that are unknown.

Step 2: **Assign a variable** to one of the unknown items. (In most cases this will amount to letting x = the item that is asked for in the problem.) Then **translate** the other **information** in the problem to expressions involving the variable.

Step 3: **Reread** the problem, and then **write an equation,** using the items and variables listed in steps 1 and 2, that describes the situation.

Step 4: **Solve the equation** found in step 3.

Step 5: **Write** your **answer** using a complete sentence.

Step 6: **Reread** the problem, and **check** your solution with the original words in the problem.

6. Adding 5 to both sides of the inequality $x - 5 < -2$ gives

$$x - 5 + 5 < -2 + 5$$
$$x < 3$$

Addition Property for Inequalities [2.4]
For expressions A, B, and C,

$$\text{if} \qquad A < B$$
$$\text{then} \qquad A + C < B + C$$

Adding the same quantity to both sides of an inequality never changes the solution set.

7. Multiplying both sides of $-2x \geq 6$ by $-\frac{1}{2}$ gives

$$-2x \geq 6$$
$$\downarrow$$
$$-\tfrac{1}{2}(-2x) \leq -\tfrac{1}{2}(6)$$
$$x \leq -3$$

Multiplication Property for Inequalities [2.4]
For expressions A, B, and C,

$$\text{if} \qquad A < B$$
$$\text{then} \qquad AC < BC \quad \text{if} \quad C > 0 \ (C \text{ is positive})$$
$$\text{or} \qquad AC > BC \quad \text{if} \quad C < 0 \ (C \text{ is negative})$$

We can multiply both sides of an inequality by the same nonzero number without changing the solution set as long as each time we multiply by a negative number we also reverse the direction of the inequality symbol.

8. To solve

$$|2x - 1| + 2 = 7$$

we first isolate the absolute value on the left side by adding -2 to each side to obtain

$$|2x - 1| = 5$$

$$2x - 1 = 5 \quad \text{or} \quad 2x - 1 = -5$$

$$2x = 6 \quad \text{or} \quad 2x = -4$$

$$x = 3 \quad \text{or} \quad x = -2$$

9. To solve

$$|x - 3| + 2 < 6$$

we first add -2 to both sides to obtain

$$|x - 3| < 4$$

which is equivalent to

$$-4 < x - 3 < 4$$

$$-1 < \quad x \quad < 7$$

Absolute Value Equations [2.5]

To solve an equation that involves absolute value, we isolate the absolute value on one side of the equation and then rewrite the absolute value equation as two separate equations that do not involve absolute value. In general, if b is a positive real number, then

$$|a| = b \quad \text{is equivalent to} \quad a = b \quad \text{or} \quad a = -b$$

Absolute Value Inequalities [2.6]

To solve an inequality that involves absolute value, we first isolate the absolute value on the left side of the inequality symbol. Then we rewrite the absolute value inequality as an equivalent continued or compound inequality that does not contain absolute value symbols. In general, if b is a positive real number, then

$$|a| < b \quad \text{is equivalent to} \quad -b < a < b$$

and

$$|a| > b \quad \text{is equivalent to} \quad a < -b \quad \text{or} \quad a > b$$

COMMON MISTAKES

A very common mistake in solving inequalities is to forget to reverse the direction of the inequality symbol when multiplying both sides by a negative number. When this mistake occurs, the graph of the solution set is always drawn on the wrong side of the endpoint.

CHAPTER 2 REVIEW

Solve each equation. [2.1]

1. $x - 3 = 7$

2. $5x - 2 = 8$

3. $400 - 100a = 200$

4. $5 - \frac{2}{3}a = 7$

5. $4x - 2 = 7x + 7$

6. $\frac{3}{2}x - \frac{1}{6} = -\frac{7}{6}x - \frac{1}{6}$

7. $7y - 5 - 2y = 2y - 3$

8. $\frac{3y}{4} - \frac{1}{2} + \frac{3y}{2} = 2 - y$

9. $3(2x + 1) = 18$

10. $-\frac{1}{2}(4x - 2) = -x$

11. $8 - 3(2t + 1) = 5(t + 2)$

12. $8 + 4(1 - 3t) = -3(t - 4) + 2$

Substitute the given values in each formula and then solve for the variable that does not have a numerical replacement. [2.2]

13. $P = 2b + 2h$: $P = 40$, $b = 3$

14. $A = P + Prt$: $A = 2,000$, $P = 1,000$, $r = 0.05$

Solve each formula for the indicated variable. [2.2]

15. $I = prt$ for p

16. $y = mx + b$ for x

17. $4x - 3y = 12$ for y

18. $d = vt + 16t^2$ for v

For Problems 19–22, the general term of a sequence is given. Use the general term to find the first four terms of each sequence. Identify sequences that are increasing, decreasing, or alternating. [2.2]

19. $a_n = 3n$

20. $a_n = n + 3$

21. $a_n = (-2)^n$

22. $a_n = \dfrac{1}{n}$

Solve each application. In each case, be sure to show the equation that describes the situation. [2.3]

23. Geometry   The length of a rectangle is 3 times the width. The perimeter is 32 feet. Find the length and width.

24. Geometry   The three sides of a triangle are given by three consecutive integers. If the perimeter is 12 meters, find the length of each side.

25. Brick Laying   The formula $N = 7 \cdot L \cdot H$ gives the number of standard bricks needed in a wall of L feet long and H feet high, and is called the *bricklayer's formula*.
(a) How many bricks would be required in a wall 45 feet long by 12 feet high?
(b) How long could an 8-foot-high wall be built from a load of 35,000 bricks?

26. Salary   A teacher has a salary of $25,920 for her second year on the job. If this is 4.2% more than her first-year salary, how much did she earn her first year?

Solve each inequality. Write your answer using interval notation. [2.4]

27. $-8a > -4$

28. $6 - a \geq -2$

29. $\frac{3}{4}x + 1 \leq 10$

30. $800 - 200x < 1,000$

31. $\frac{1}{3} \leq \frac{1}{6}x \leq 1$

32. $-0.01 \leq 0.02x - 0.01 \leq 0.01$

33. $5t + 1 \leq 3t - 2$ or $-7t \leq -21$

34. $3(x + 1) < 2(x + 2)$ or $2(x - 1) \geq x + 2$

Solve each equation. [2.5]

35. $|x| = 2$

36. $|a| - 3 = 1$

37. $|x - 3| = 1$

38. $|2y - 3| = 5$

39. $|4x - 3| + 2 = 11$

40. $\left|\dfrac{7}{3} - \dfrac{x}{3}\right| + \dfrac{4}{3} = 2$

41. $|5t - 3| = |3t - 5|$

42. $\left|\frac{1}{2} - x\right| = \left|x + \frac{1}{2}\right|$

Solve each inequality and graph the solution set. [2.6]

43. $|x| < 5$

44. $|0.01a| \geq 5$

45. $|x| < 0$

46. $|2t + 1| - 3 < 2$

Solve each equation or inequality, if possible. [2.1, 2.5, 2.6]

47. $2x - 3 = 2(x - 3)$

48. $3(5x - \frac{1}{2}) = 15x + 2$

49. $|4y + 8| = -1$

50. $|x| > 0$

51. $|5 - 8t| + 4 \leq 1$

52. $|2x + 1| \geq -4$

CHAPTER 2   **PROJECTS**

LINEAR EQUATIONS AND INEQUALITIES IN ONE VARIABLE

GROUP   PROJECT

FINDING THE MAXIMUM HEIGHT OF A
MODEL ROCKET

Number of People: 3

Time Needed: 20 minutes

Equipment: Paper and pencil

Background: In this chapter we used formulas to do some table building. Once we have a table, it is sometimes possible to use just the table information to extend what we know about the situation described by the table. In this project we take some basic information from a table and then look for patterns among the table entries. Once we have established the patterns we continue them and, in so doing, solve a realistic application problem.

Procedure: A model rocket is launched into the air. Table 1 gives the height of the rocket every second after takeoff, for the first 5 seconds. Figure 1 is a graphical representation of the information in Table 1.

TABLE 1	Height of a Model Rocket
Time (seconds)	**Height (feet)**
0	0
1	176
2	320
3	432
4	512
5	560

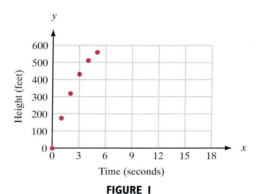

FIGURE 1

1. Table 1 is shown again as follows with two new columns. Fill in the first five entries in the First Differences column by finding the difference of consecutive heights. For example, the second entry in the First Differences column will be the difference of 320 and 176, which is 144.

TABLE 1 Height of a Model Rocket			
Time (seconds)	Height (feet)	First Differences	Second Differences
0	0		
1	176		
2	320		
3	432		
4	512		
5	560		

2. Start filling in the Second Differences column by finding the differences of the First Differences.

3. Once you see the pattern in the Second Differences table, fill in the rest of the entries.

4. Now, using the results in the Second Differences table, go back and complete the First Differences table.

5. Now, using the results in the First Differences table, go back and complete the Heights column in the original table.

6. Plot the rest of the points from Table 1 on the graph in Figure 1.

7. What is the maximum height of the rocket?

8. How long was the rocket in the air?

THE EQUAL SIGN

We have been using the equal sign, $=$, for some time now. It is interesting to note that the first published use of the symbol was in 1557, with the publication of *The Whetstone of Witte* by the English mathematician and physician Robert Recorde. Research the first use of the symbols we use for addition, subtraction, multiplication, and division and then write an essay on the subject from your results.

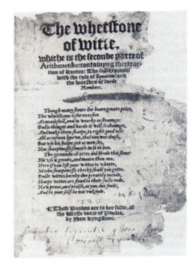

The Whetstone of Witte, 1557
(*Source: Smithsonian Institution Libraries, Photo No. 92-338*)

CHAPTER 2   TEST

Solve the following equations. [2.1]

1. $x - 5 = 7$ **2.** $3y = -4$

3. $5 - \frac{4}{7}a = -11$

4. $\frac{1}{5}x - \frac{1}{2} - \frac{1}{10}x + \frac{2}{5} = \frac{3}{10}x + \frac{1}{2}$

5. $5(x - 1) - 2(2x + 3) = 5x - 4$

6. $0.07 - 0.02(3x + 1) = -0.04x + 0.01$

Solve for the indicated variable. [2.2]

7. $P = 2l + 2w$ for w **8.** $A = \frac{1}{2}h(b + B)$ for B

For Problems 9–10, the general term of a sequence is given. Use the general term to find the first four terms of each sequence. Identify any sequences that are increasing, decreasing, or alternating. [2.2]

9. $a_n = 2n - 3$ **10.** $a_n = (-3)^n$

Solve each of the following. [2.3]

11. Geometry A rectangle is twice as long as it is wide. The perimeter is 36 inches. Find the dimensions.

12. Baseball The winning percentage, P, of a team with W wins and L losses is given by the formula

$$P = \frac{W}{W + L}$$

In 1998 postseason play, the New York Yankees won three games and lost none to the Texas Rangers, won four games and lost two to the Cleveland Indians, and won four games and lost none to the San Diego Padres. What was the postseason winning percentage of the Yankees?

13. Sales Tax At the beginning of the day, the cash register at a coffee shop contains $75. At the end of the day, it contains $881.25. If the sales tax rate is 7.5%, how much of the total is sales tax?

14. Geometry Two angles are supplementary. If the larger angle is 15° more than twice the smaller angle, find the measure of each angle.

Solve the following inequalities. Write the solution set using interval notation, then graph the solution set. [2.4]

15. $-5t \leq 30$

16. $5 - \frac{3}{2}x > -1$

17. $1.6x - 2 < 0.8x + 2.8$

18. $3(2y + 4) \geq 5(y - 8)$

Solve the following equations. [2.5]

19. $\left| \frac{1}{4}x - 1 \right| = \frac{1}{2}$

20. $\left| \frac{2}{3}a + 4 \right| = 6$

21. $\left| 3 - 2x \right| + 5 = 2$

22. $5 = \left| 3y + 6 \right| - 4$

Solve the following inequalities and graph the solutions. [2.6]

23. $\left| 6x - 1 \right| > 7$

24. $\left| 3x - 5 \right| - 4 \leq 3$

25. $\left| 5 - 4x \right| \geq -7$

26. $\left| 4t - 1 \right| < -3$

Simplify.

1. 7^2

2. $|-3|$

3. $-5 + 3$

4. $-(-5)$

5. $-5 - 6$

6. $-8(4)$

7. $3 + 2(7 + 4)$

8. $35 - 15 \div 3 + 6$

9. $-(6 - 4) - (3 - 8)$

10. $3(-5)^2 - 6(-2)^3$

11. $\dfrac{5(-3) + 6}{8(-3) + 7(3)}$

12. $8\left(\dfrac{1}{8}x\right)$

13. $5(2x + 4) + 6$

14. $\dfrac{9}{64} + \dfrac{5}{56}$

15. $2x + 6 + 3x + 5$

Solve.

16. Find the difference of -3 and 6.

17. Add -4 to the product of 5 and -6.

18. Subtract 8 from the quotient of 28 and 2.

Let $A = \{3, 6, 9\}$; $B = \{4, 5, 6, 7\}$; $C = \{2, 4, 6, 8\}$. Find the following.

19. $A \cup C$

20. $A \cap B$

21. $\{x \mid x \in B \text{ and } x \leq 5\}$

For the set $\{-13, -6.7, -\sqrt{5}, 0, \frac{1}{2}, 2, \frac{5}{2}, \pi, \sqrt{13}\}$ list all the numbers that are in each of the following sets.

22. Whole numbers

23. Rational numbers

24. Integers

25. Irrational numbers

26. Reduce $\dfrac{231}{616}$ to its lowest terms.

27. Find the next number in the sequence 2, -6, 18, . . .

28. State the converse of the following statement.

$$\text{If } x = -3, \text{ then } x^2 = 9.$$

Solve each equation.

29. $3 - \frac{4}{5}a = -5$

30. $8x - 3 = 2x - 6$

31. $\dfrac{y}{4} - 1 + \dfrac{3y}{8} = \dfrac{3}{4} - y$

32. $6(4x - 2) = 12$

33. $7 - 2(8t - 3) = 4(t - 2)$

Substitute the given values in each formula, and then solve for the variable that does not have a numerical replacement.

34. $A = P + Prt$: $A = 1{,}000, P = 500, r = 0.1$

35. $A = a + (n - 1)d$: $A = 40, a = 4, d = 9$

Solve each formula for the indicated variable.

36. $4x - 3y = 12$ for x

37. $F = \frac{9}{5}C + 32$ for C

For Problems 38–39, the general term of a sequence is given. Use the general term to find the first four terms of each sequence. Identify sequences that are increasing, decreasing, or alternating.

38. $a_n = \frac{1}{3}n$

39. $a_n = \dfrac{1}{n^2}$

Solve each application. In each case, be sure to show the equation that describes the situation.

40. Basketball The formula for the volume, V, of the thickness of a spherical object is

$$V = \frac{4}{3}\pi(R^3 - r^3)$$

in which R is the outer radius and r is the inner radius of the object. Use this formula to find the volume of material that makes up a basketball if the outer radius is 4.5 inches, and the thickness of the material that makes up the basketball is .05 inch. (*Note:* This question is not asking for the volume of the basketball.)

41. Geometry Two angles are complementary. If the larger angle is 15° more than twice the smaller angle, find the measure of each angle.

Solve each inequality. Write your answer using interval notation.

42. $600 - 300x < 900$ **43.** $-\frac{1}{2} \leq \frac{1}{6}x \leq \frac{1}{3}$

44. $6t - 3 \leq t + 1$ or $-8t \leq -16$

Solve each equation.

45. $|a| - 2 = 3$ **46.** $|3y - 2| = 7$

47. $|6x - 2| + 4 = 16$

Solve each inequality, and graph the solution set.

48. $|y - 2| < 3$ **49.** $|5x - 1| > 3$

50. $|2t + 1| - 1 < 5$

Equations and Inequalities
in Two Variables

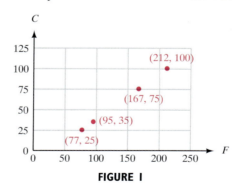

INTRODUCTION

A student is heating water in a chemistry lab. As the water heats, she records the temperature readings from two thermometers, one giving temperature in degrees Fahrenheit and the other in degrees Celsius. Table 1 shows some of the data she collects. Figure 1 is a scatter diagram that gives a visual representation of the data in Table 1.

TABLE 1 Corresponding Temperatures	
In Degrees Fahrenheit	**In Degrees Celsius**
77	25
95	35
167	75
212	100

FIGURE 1

The exact relationship between the Fahrenheit and Celsius temperature scales is given by the formula

$$C = \frac{5}{9}(F - 32)$$

We have three ways to describe the relationship between the two temperature scales: a table, a graph, and an equation. But, most important to us, we don't need to accept this formula on faith. In Problem Set 3.3, you will derive the formula from the data in Table 1.

STUDY SKILLS

The study skills for this chapter are about attitude. They are points of view that point toward success.

1. **Be Focused, Not Distracted**   I have students who begin their assignments by asking themselves "Why am I taking this class?" If you are asking yourself similar questions, you are distracting yourself from doing the things that will produce the results you want in this course. Don't dwell on questions and evaluations of the class that can be used as excuses for not doing well. If you want to succeed in this course, focus your energy and efforts toward success, rather than distracting yourself from your goals.
2. **Be Resilient**   Don't let setbacks keep you from your goals. You want to put yourself on the road to becoming a person who can succeed in this class, or any class in college. Failing a test or quiz, or having a difficult time on some topics, is normal. No one goes through college without some setbacks. Don't let a temporary disappointment keep you from succeeding in this course. A low grade on a test or quiz is simply a signal that you need to reevaluate your study habits.
3. **Intend to Succeed**   I have a few students who simply go through the motions of studying without intending to master the material. It is more important to them to look like they are studying than to actually study. You need to study with the intention of being successful in the course. Intend to master the material, no matter what it takes.

Paired Data and the Rectangular Coordinate System

In this section we begin our work with the visual component of algebra. We are setting the stage for two important topics: *functions* and *graphs.* Both topics have a wide variety of applications and are found in all the math classes that follow this one.

Table 1 gives the net price of a popular intermediate algebra text at the beginning of each year in which a new edition was published. (The net price is the price the bookstore pays for the book, not the price you pay for it.)

TABLE 1 Price of Textbook		
Edition	**Year Published**	**Net Price ($)**
First	1982	16.95
Second	1985	23.50
Third	1989	30.50
Fourth	1993	39.25
Fifth	1997	47.50

The information in Table 1 is shown visually in Figures 1 and 2. The diagram in Figure 1 is called a *bar chart.* The diagram in Figure 2 is called a *line graph.* In both figures, the horizontal line that shows years is called the *horizontal axis,* and the vertical line that shows prices is called the *vertical axis.*

Recall from Chapter 1, that the data in Table 1 are called *paired data* because each number in the years column is paired with a specific number in the price column. Each of these pairs of numbers from Table 1 is associated with one of the solid bars in the bar chart (Figure 1) and one of the dots in the line graph (Figure 2).

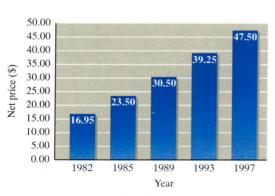

FIGURE 1

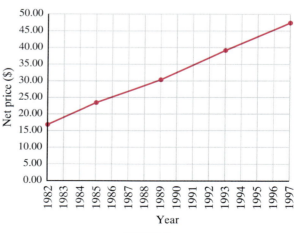

FIGURE 2

Using TECHNOLOGY

SPREADSHEET PROGRAMS

When I put together the manuscript for this book, I used a spreadsheet program to draw the bar chart and line graph shown in Figures 1 and 2.

Figure 3 shows what the screen of my computer looked like when I was preparing Figure 1. The bar chart was drawn by the computer from the data in the table. This was just one of many ways I could have chosen to display the data.

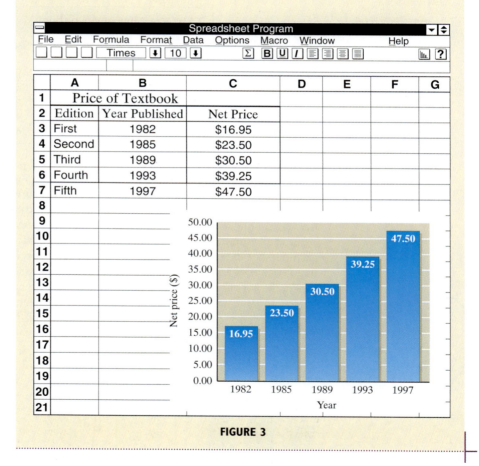

FIGURE 3

ORDERED PAIRS

Paired data play a very important role in equations that contain two variables. Working with these equations is easier if we standardize the terminology and notation associated with paired data. So here is a definition that will do just that.

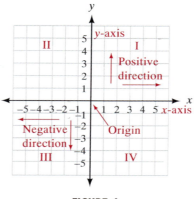

FIGURE 4

> **DEFINITION**   A pair of numbers enclosed in parentheses and separated by a comma, such as $(-2, 1)$, is called an **ordered pair** of numbers. The first number in the pair is called the ***x*-coordinate** of the ordered pair; the second number is called the ***y*-coordinate**. For the ordered pair $(-2, 1)$, the *x*-coordinate is -2 and the *y*-coordinate is 1.

In order to standardize the way in which we display paired data visually, we use a rectangular coordinate system.

A *rectangular coordinate system* is made by drawing two real number lines at right angles to each other. The two number lines, called *axes*, cross each other at 0. This point is called the *origin*. Positive directions are to the right and up. Negative directions are down and to the left. The rectangular coordinate system is shown in Figure 4.

The horizontal number line is called the *x-axis*, and the vertical number line is called the *y-axis*. The two number lines divide the coordinate system into four quadrants, which we number I through IV in a counterclockwise direction. Points on the axes are not considered as being in any quadrant.

To graph the ordered pair (a, b) on a rectangular coordinate system, we start at the origin and move *a* units right or left (right if *a* is positive, left if *a* is negative). Then we move *b* units up or down (up if *b* is positive and down if *b* is negative). The point where we end up is the graph of the ordered pair (a, b).

Note: A rectangular coordinate system allows us to connect algebra and geometry by associating geometric shapes (the curves shown in the diagrams) with algebraic equations. The French philosopher and mathematician René Descartes (1596–1650) is usually credited with the invention of the rectangular coordinate system, which is often referred to as the *Cartesian coordinate system* in his honor. As a philosopher, Descartes is responsible for the statement "I think, therefore, I am." Until Descartes invented his coordinate system in 1637, algebra and geometry were treated as separate subjects.

EXAMPLE   I   Plot (graph) the ordered pairs $(2, 5)$, $(-2, 5)$, $(-2, -5)$, and $(2, -5)$.

Solution   To graph the ordered pair (2, 5), we start at the origin and move 2 units to the right, then 5 units up. We are now at the point whose coordinates are (2, 5). We graph the other three ordered pairs in a similar manner (see Figure 5).

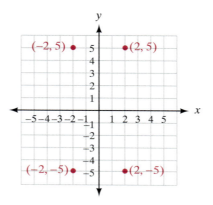

FIGURE 5

Note: From Example 1 we see that any point in quadrant I has both its x- and y-coordinates positive $(+, +)$. Points in quadrant II have negative x-coordinates and positive y-coordinates $(-, +)$. In quadrant III both coordinates are negative $(-, -)$. In quadrant IV the form is $(+, -)$.

EXAMPLE 2

Graph the ordered pairs $(1, -3)$, $(\frac{1}{2}, 2)$, $(3, 0)$, $(0, -2)$, $(-1, 0)$, and $(0, 5)$.

Solution

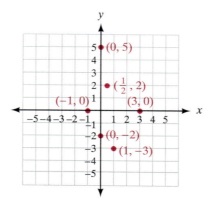

FIGURE 6

From Figure 6 we see that any point on the x-axis has a y-coordinate of 0 (it has no vertical displacement), and any point on the y-axis has an x-coordinate of 0 (no horizontal displacement).

LINEAR EQUATIONS

We can plot a single point from an ordered pair, but to draw a line, we need two points or an equation in two variables.

> **DEFINITION**   Any equation that can be put in the form $ax + by = c$, where a, b, and c are real numbers and a and b are not both 0, is called a **linear equation in two variables**. The graph of any equation of this form is a straight line (that is why these equations are called "linear"). The form $ax + by = c$ is called **standard form**.

To graph a linear equation in two variables, we simply graph its solution set. That is, we draw a line through all the points whose coordinates satisfy the equation.

EXAMPLE 3   Graph the equation $y = -\frac{1}{3}x + 2$.

Solution   We need to find three ordered pairs that satisfy the equation. To do so, we can let x equal any numbers we choose and find corresponding values of y. But, since every value of x we substitute into the equation is going to be multiplied by $-\frac{1}{3}$, let's use numbers for x that are divisible by 3, like -3, 0, and 3. That way, when we multiply them by $-\frac{1}{3}$, the result will be an integer.

$$\text{Let } x = -3; \qquad y = -\frac{1}{3}(-3) + 2$$

$$y = 1 + 2$$

$$y = 3$$

The ordered pair $(-3, 3)$ is one solution.

In table form

x	y
-3	3
0	2
3	1

$$\text{Let } x = 0; \qquad y = -\frac{1}{3}(0) + 2$$

$$y = 0 + 2$$

$$y = 2$$

The ordered pair $(0, 2)$ is a second solution.

$$\text{Let } x = 3; \qquad y = -\frac{1}{3}(3) + 2$$

$$y = -1 + 2$$

$$y = 1$$

The ordered pair $(3, 1)$ is a third solution.

Plotting the ordered pairs $(-3, 3)$, $(0, 2)$, and $(3, 1)$ and drawing a straight line through their graphs, we have the graph of the equation $y = -\frac{1}{3}x + 2$, as shown in Figure 7.

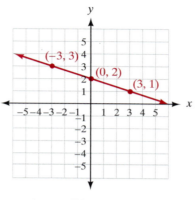

FIGURE 7

Note: It takes only two points to determine a straight line. We have included a third point for "insurance." If all three points do not line up in a straight line, we have made a mistake.

Example 3 illustrates again the connection between algebra and geometry that we mentioned in the introduction to this chapter. Descartes's rectangular coordinate system allows us to associate the equation $y = -\frac{1}{3}x + 2$ (an algebraic concept) with a specific straight line (a geometric concept). The study of the relationship between equations in algebra and their associated geometric figures is called *analytic geometry.*

INTERCEPTS

Two important points on the graph of a straight line, if they exist, are the points where the graph crosses the axes.

> **DEFINITION**   The **x-intercept** of the graph of an equation is the *x*-coordinate of the point where the graph crosses the *x*-axis. The **y-intercept** is defined similarly.

Since any point on the *x*-axis has a *y*-coordinate of 0, we can find the *x*-intercept by letting $y = 0$ and solving the equation for *x*. We find the *y*-intercept by letting $x = 0$ and solving for *y*.

EXAMPLE   4   Find the *x*- and *y*-intercepts for $2x + 3y = 6$; then graph the solution set.

Solution   To find the y-intercept we let $x = 0$.

When $\qquad\qquad\qquad\qquad x = 0$

we have $\qquad\qquad 2(0) + 3y = 6$

$$3y = 6$$

$$y = 2$$

The y-intercept is 2, and the graph crosses the y-axis at the point (0, 2).

When $\qquad\qquad\qquad\qquad y = 0$

we have $\qquad\qquad 2x + 3(0) = 6$

$$2x = 6$$

$$x = 3$$

The x-intercept is 3, so the graph crosses the x-axis at the point (3, 0). We use these results to graph the solution set for $2x + 3y = 6$. The graph is shown in Figure 8.

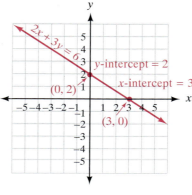

FIGURE 8

Note: Graphing straight lines by finding the intercepts works best when the coefficients of x and y are factors of the constant term.

E X A M P L E   5   Graph the line $x = 3$ and the line $y = -2$.

Solution   The line $x = 3$ is the set of all points whose x-coordinate is 3. The variable y does not appear in the equation, so the y-coordinate can be any number.

The line $y = -2$ is the set of all points whose y-coordinate is -2. The x-coordinate can be any number. The graphs are shown in Figure 9.

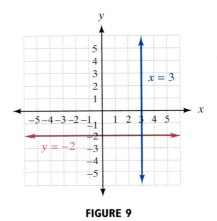

FIGURE 9

Using TECHNOLOGY

GRAPHING CALCULATORS AND COMPUTER GRAPHING PROGRAMS

A variety of computer programs and graphing calculators are currently available to help us graph equations and then obtain information from those graphs much faster than we could with paper and pencil. As we mentioned earlier, we will not give instructions for all the available calculators. Most of the instructions we give are generic in form. You will have to use the manual that came with your calculator to find the specific instructions for your calculator.

Graphing with Trace and Zoom

All graphing calculators have the ability to graph a function and then trace over the points on the graph, giving their coordinates. Furthermore, all graphing calculators can zoom in and out on a graph that has been drawn. To graph a linear equation on a graphing calculator, we first set the graph window. Most calculators call the smallest value of x Xmin and the largest value of x Xmax. The counterpart values of y are Ymin and Ymax. We will use the notation

$$\text{Window: X from } -5 \text{ to } 4, \text{Y from } -3 \text{ to } 2$$

to stand for a window in which

$$\text{Xmin} = -5 \qquad \text{Ymin} = -3$$
$$\text{Xmax} = 4 \qquad \text{Ymax} = 2$$

Set your calculator with the following window:

$$\text{Window: X from } -10 \text{ to } 10, \text{Y from } -10 \text{ to } 10$$

(continued)

Graph the equation $Y = -X + 8$. On the TI-82/83, you use the $\boxed{Y=}$ key to enter the equation; you enter a negative sign with the $\boxed{(-)}$ key, and a subtraction sign with the $\boxed{-}$ key. The graph will be similar to the one shown in Figure 10.

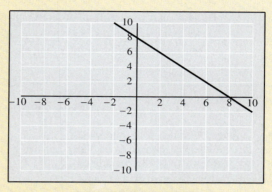

FIGURE 10

Use the Trace feature of your calculator to name three points on the graph. Next, use the Zoom feature of your calculator to zoom out so your window is twice as large.

Solving for *y* First

To graph the equation from Example 4, $2x + 3y = 6$, on a graphing calculator, you must first solve it for *y*. When you do so, you will get $y = -\frac{2}{3}x + 2$, which you enter into your calculator as $Y = -(2/3)X + 2$. Graph this equation in the window described here, and compare your results with the graph in Figure 8.

Window: X from -6 to 6, Y from -6 to 6

Hint on Tracing

If you are going to use the Trace feature and you want the *x*-coordinates to be exact numbers, set your window so that the range of X inputs is a multiple of the number of horizontal pixels on your calculator screen. On the TI-82/83, the screen has 94 pixels across. Here are a few convenient trace windows:

X from -4.7 to 4.7	To trace to the nearest tenth
X from -47 to 47	To trace to the nearest integer
X from 0 to 9.4	To trace to the nearest tenth
X from 0 to 94	To trace to the nearest integer
X from -94 to 94	To trace to the nearest even integer

Getting Ready for Class

After reading through the preceding section, respond in your own words and in complete sentences.

A. Explain how you would construct a rectangular coordinate system from two real number lines.

B. Explain in words how you would graph the ordered pair $(2, -3)$.

C. How can you tell if an ordered pair is a solution to the equation $y = 2x - 5$?

D. If you were looking for solutions to the equation $y = \frac{1}{3}x + 5$, why would it be easier to substitute 6 for x than to substitute 5 for x?

PROBLEM SET 3.1

1. Hourly Wages Suppose you have a job that pays $7.50 per hour, and you work anywhere from 0 to 40 hours per week. Table 2 gives the amount of money you will earn in 1 week for working various hours. Construct a line graph from the information in Table 2. You can copy the grid in Figure 11 to use as a guide.

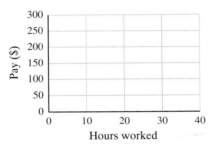

FIGURE 11 Template for constructing a line graph

TABLE 2 Weekly Wages	
Hours Worked	**Pay ($)**
0	0
10	75
20	150
30	225
40	300

2. Softball Toss Chaudra is tossing a softball into the air with an underhand motion. It takes exactly 2 seconds for the ball to come back to her. Table 3

TABLE 3 Tossing a Softball Into the Air	
Time (sec)	**Distance (ft)**
0	0
0.25	7
0.5	12
0.75	15
1	16
1.25	15
1.5	12
1.75	7
2	0

shows the distance the ball is above her hand at quarter-second intervals. Construct a line graph from the information in the table. Use the grid in Figure 12 as a guide.

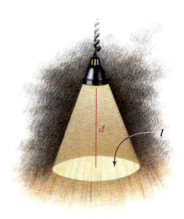

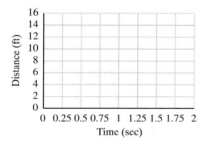

FIGURE 12 Template for constructing a line graph

TABLE 4 Light Intensity From a 100-Watt Light Bulb	
Distance Above Surface (ft)	Intensity (lumens/ft^2)
1	120.0
2	30.0
3	13.3
4	7.5
5	4.8
6	3.3

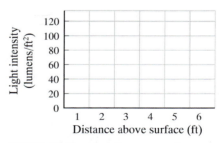

FIGURE 13 Template for constructing a bar graph

3. **Intensity of Light**   Table 4 gives the intensity of light (expressed in lumens per square foot) that falls on a surface at various distances from a 100-watt light bulb. Use the template shown in Figure 13 to construct a bar chart from the information in Table 4.

4. **Value of a Painting**   A piece of art was purchased in 1990 for $125. Table 5 shows the value of the painting at various times, assuming that it doubles in value every 5 years. Construct a bar chart from

the information in the table, using the template in Figure 14.

(KactusFoto, Santiago, Chile/ SuperStock)

TABLE 5	Value of a Painting
Year	**Value ($)**
1990	125
1995	250
2000	500
2005	1,000
2010	2,000

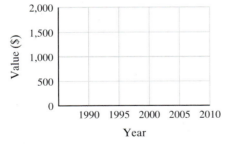

FIGURE 14 Template for constructing a bar graph

Graph each of the following ordered pairs on a rectangular coordinate system.

5. $(1, 2)$ **6.** $(-1, 2)$

7. $(-1, -2)$ **8.** $(1, -2)$

9. $(5, 0)$ **10.** $(0, -3)$

11. $(0, 2)$ **12.** $(4, 0)$

13. $(-5, -5)$ **14.** $(-4, -1)$

15. $(\frac{1}{2}, 2)$ **16.** $(3, \frac{1}{4})$

17–28. Give the coordinates of each of the points shown in the following rectangular coordinate system.

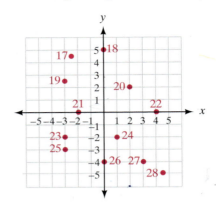

Graph each of the following linear equations by first finding the intercepts.

29. $2x - 3y = 6$ **30.** $y - 2x = 4$

31. $4x - 5y = 20$ **32.** $5x - 3y - 15 = 0$

33. $y = 2x + 3$ **34.** $y = 3x - 2$

35. Which of the following tables could be produced from the equation $y = 2x - 6$?

(a)   (b)   (c)

x	y
0	6
1	4
2	2
3	0

x	y
0	-6
1	-4
2	-2
3	0

x	y
0	-6
1	-5
2	-4
3	-3

36. Which of the following tables could be produced from the equation $3x - 5y = 15$?

(a)   (b)   (c)

x	y
0	5
-3	0
10	3

x	y
0	-3
5	0
10	3

x	y
0	-3
-5	0
10	-3

Graph each of the following straight lines.

37. $y = \frac{1}{3}x$ **38.** $y = \frac{1}{2}x$

39. $-2x + y = -3$ **40.** $-3x + y = -2$

41. $y = -\frac{2}{3}x + 1$

42. $y = -\frac{2}{3}x - 1$

43. $\frac{x}{3} + \frac{y}{4} = 1$

44. $\frac{x}{-2} + \frac{y}{3} = 1$

45. The graph shown here is the graph of which of the following equations?

   (a) $3x - 2y = 6$

   (b) $2x - 3y = 6$

   (c) $2x + 3y = 6$

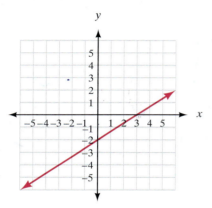

46. The graph shown here is the graph of which of the following equations?

   (a) $3x - 2y = 8$

   (b) $2x - 3y = 8$

   (c) $2x + 3y = 8$

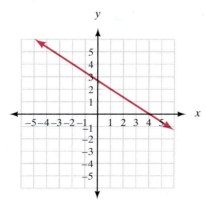

47. Graph the straight line $0.02x + 0.03y = 0.06$.

48. Graph the straight line $0.05x - 0.03y = 0.15$.

49. Graph the lines $x = -3$ and $y = 5$ on the same coordinate system.

50. Graph the lines $x = 4$ and $y = -3$ on the same coordinate system.

51. The ordered pairs that satisfy the equation $y = 3x$ all have the form $(x, 3x)$, because y is always 3 times x. Graph all ordered pairs of the form $(x, 3x)$.

52. Graph all ordered pairs of the form $(x, -3x)$.

53. Replace each question mark with the symbol $>$ or $<$.

   (a) In the first quadrant, x ? 0, and y ? 0.

   (b) In the second quadrant, x ? 0, and y ? 0.

   (c) In the third quadrant, x ? 0, and y ? 0.

   (d) In the fourth quadrant, x ? 0, and y ? 0.

54. Replace each question mark with the symbol $>$ or $<$.

   (a) In the first quadrant, the product $x \cdot y$? 0.

   (b) In the second quadrant, the product $x \cdot y$? 0.

   (c) In the third quadrant, the product $x \cdot y$? 0.

   (d) In the fourth quadrant, the product $x \cdot y$? 0.

Applying the Concepts

For each of the following applied problems, first build a table of values. Make sure each input is accompanied by an output.

55. Cost of a Phone Call If the cost of a long-distance phone call is 50¢ for the first minute and 25¢ for each additional minute, then the total cost y (in cents) of a call that goes x minutes past the first minute is $y = 25x + 50$. Let 1 unit on the x-axis equal 1 minute and 1 unit on the y-axis equal 25¢, and graph this equation.

56. Cost of a Taxi Ride If the cost of a taxi ride in Las Vegas is $1.50 for the first mile and $0.50 for each tenth of a mile after the first mile, then the total cost y (in cents) of a ride that goes x tenths of a mile after the first mile is $y = 50x + 150$. Let each unit on the x-axis equal one-tenth of a mile and each unit on the y-axis equal $0.50, and graph this equation.

57. Demand Equation A company that manufactures typewriter ribbons knows that the number of ribbons they can sell each week, x, is related to the price p of each ribbon by the equation $x + 100p = 1{,}200$. Note any restrictions on the variables; then graph the relationship between x and p, with the values of x along the horizontal axis and the values of p on the vertical axis.

58. Demand Equation   A company that manufactures diskettes for home computers finds that they can sell x diskettes each day at p dollars per diskette according to the equation $x + 100p = 800$. Note any restrictions on the variables; then graph the relationship between x and p.

59. A Snail's Pace   A snail is climbing straight up a brick wall at a constant rate. It takes the snail 4 hours to climb up 6 feet. After climbing for 4 hours, the snail rests for 4 hours, during which time it slides 2 feet down the wall. If the snail starts at the bottom of the wall and repeats this climbing up/sliding down process continuously, construct a table that gives the snail's height above the ground every 4 hours, starting at 0 and ending at 24 hours. Then construct a line graph from the information in the table.

60. Air Temperature   A pilot checks the weather conditions before flying and finds that the air temperature drops 3.5°F every 1,000 feet. If the air temperature is 41°F when the plane reaches 10,000 feet, construct a table that gives the air temperature every 2,000 feet, starting at 10,000 feet and ending at 20,000 feet. Then construct a line graph from the information in the table.

61. Between 1930 and the middle 1990s, the number of accident-related deaths at the workplace in the United States was decreasing in a linear fashion. If D represents the annual number of accident-related deaths and y represents the year, then

$$D = 325{,}870 - 159 \cdot y$$

(a) Each successive year saw a reduction of how many additional accident-related deaths?

(b) How many accident-related deaths occurred in 1930?

(c) How many accident-related deaths occurred in 1975?

(d) Can this pattern continue to the year 2050? Explain why or why not.

62. Between 1970 and 1990, inclusive, the disposable income in the United States for each person increased significantly, and may be described by the equation

$$I = 4{,}000 + 600x$$

where I is the disposable income and x is the number of years past 1970.

(a) What are the restrictions on x?

(b) What was the disposable income in 1972?

(c) By how much did the disposable income increase over any 4-year period between 1970 and 1990?

63. Price of a Textbook   The table on textbook prices shown at the beginning of this section is repeated here, along with an equation that approximates the pairs of numbers in the table. Below the table is an equation that approximates the pairs of numbers in the table.

Year of New Edition x	Net Price ($) y
1982	16.95
1985	23.50
1989	30.50
1993	39.25
1997	47.50

Approximating Equation: $y = 2.022x - 3{,}991$

(a) Use your calculator to graph the equation using the following window:

Window: X from 1981 to 1997, Y from 0 to 50

(b) Trace along the graph to find the value of y that corresponds to each of the following values of x:

From Approximating Equation	
x	y
1982	
1985	
1989	
1993	
1997	

64. Price of a Textbook   If we let the year 1980 correspond to $x = 0$, then the table on textbook prices shown at the beginning of this section can be rewritten as shown. Following the table is an equation that approximates the pairs of numbers in the table.

Year of New Edition	If $x = 0$ at 1980 x	Net Price ($) y
1982	2	16.95
1985	5	23.50
1989	9	30.50
1993	13	39.25
1997	17	47.50

Approximating Equation: $y = 2.022x + 12.94$

(a) Use your calculator to graph the equation using the following window:

Window: X from 0 to 23.5, Y from 0 to 50

(b) Trace along the graph to find the value of y that corresponds to each of the following values of x:

Year of New Edition	From Approximating Equation	
	x	y
1982	2	
1985	5	
1989	9	
1993	13	
1997	17	

Review Problems

The problems that follow review material we covered in Section 2.1.

Solve each equation.

65. $5x - 4 = -3x + 12$

66. $\dfrac{1}{2} - \dfrac{y}{5} = -\dfrac{9}{10} + \dfrac{y}{2}$

67. $\dfrac{1}{2} - \dfrac{1}{8}(3t - 4) = -\dfrac{7}{8}t$

68. $3(5t - 1) - (3 - 2t) = 5t - 8$

Extending the Concepts

Find the x- and y-intercepts for each equation. Your answers will contain the constants a, b, and c.

69. $ax + by = c$ **70.** $ax - by = c$

71. $\dfrac{x}{a} + \dfrac{y}{b} = 1$ **72.** $y = ax + b$

Spreadsheets and Drag Racing   Jim Rizzoli lives in San Luis Obispo, California. He owns and operates an alcohol-fueled dragster. The information in Table 6 was recorded by a computer in the dragster during one of his races at the 1993 Winternationals. It shows the time and speed of the dragster, along with the distance traveled past the starting line and the front axle RPM (revolutions per minute). Use a spreadsheet program to work these problems.

TABLE 6 Speed, Distance, and Front Axle RPM for a Race Car			
Time (sec)	Speed (mph)	Distance Traveled (ft)	Front Axle RPM
0	0.0	0	0
1	72.7	69	1,107
2	129.9	231	1,978
3	162.8	439	2,486
4	192.2	728	2,919
5	212.4	1,000	3,233
6	228.1	1,373	3,473

73. Construct a bar chart that shows the relationship between time and distance.

74. Construct a line graph that shows the relationship between time and front axle RPM.

75. Construct a line graph that shows the relationship between speed and distance traveled.

76. Construct a line graph that shows the relationship between speed and front axle RPM. Does your line graph suggest that the relationship between speed and front axle RPM is a linear one?

3.2   **The Slope of a Line**

A highway sign tells us we are approaching a 6% downgrade. As we drive down this hill, each 100 feet we travel horizontally is accompanied by a 6-foot drop in elevation.

In mathematics we say the slope of the highway is $-0.06 = -\dfrac{6}{100} = -\dfrac{3}{50}$. The *slope* is the ratio of the vertical change to the accompanying horizontal change.

Highway sign   Mathematical model

In defining the slope of a straight line, we are looking for a number to associate with a straight line that does two things. First, we want the slope of a line to measure the "steepness" of the line. That is, in comparing two lines, the slope of the steeper line should have the larger numerical value. Second, we want a line that *rises* going from left to right to have a *positive* slope. We want a line that *falls* going from left to right to have a *negative* slope. (A line that neither rises nor falls going from left to right must, therefore, have 0 slope.)

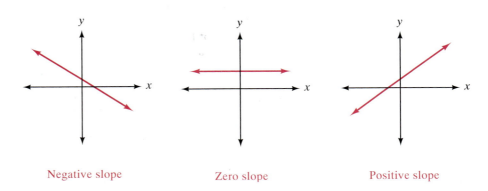

Negative slope   Zero slope   Positive slope

Geometrically, we can define the *slope* of a line as the ratio of the vertical change to the horizontal change encountered when moving from one point to another on the line. The vertical change is sometimes called the *rise*. The horizontal change is called the *run*.

EXAMPLE 1   Find the slope of the line $y = 2x - 3$.

Solution   In order to use our geometric definition, we first graph $y = 2x - 3$ (Figure 1, next page). We then pick any two convenient points and find the ratio of rise to run. By convenient points we mean points with integer coordinates. If we let $x = 2$ in the equation, then $y = 1$. Likewise if we let $x = 4$, then y is 5.

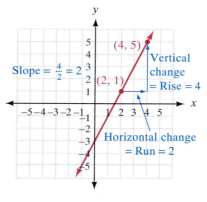

FIGURE 1

Our line has a slope of 2.

Notice that we can measure the vertical change (rise) by subtracting the y-coordinates of the two points shown in Figure 1: $5 - 1 = 4$. The horizontal change (run) is the difference of the x-coordinates: $4 - 2 = 2$. This gives us a second way of defining the slope of a line.

DEFINITION   The **slope** of the line between two points (x_1, y_1) and (x_2, y_2) is given by

$$\text{Slope} = m = \underset{\underset{\substack{\text{Geometric} \\ \text{form}}}{\uparrow}}{\frac{\text{Rise}}{\text{Run}}} = \underset{\underset{\substack{\text{Algebraic} \\ \text{form}}}{\uparrow}}{\frac{y_2 - y_1}{x_2 - x_1}}$$

EXAMPLE 2   Find the slope of the line through $(-2, -3)$ and $(-5, 1)$.

Solution

$$m = \frac{y_2 - y_1}{x_2 - x_1} = \frac{1 - (-3)}{-5 - (-2)} = \frac{4}{-3} = -\frac{4}{3}$$

Looking at the graph of the line between the two points (Figure 2, next page), we can see our geometric approach does not conflict with our algebraic approach.

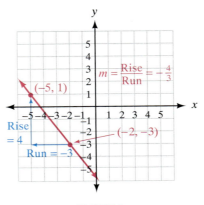

FIGURE 2

We should note here that it does not matter which ordered pair we call (x_1, y_1) and which we call (x_2, y_2). If we were to reverse the order of subtraction of both the x- and y-coordinates in the preceding example, we would have

$$m = \frac{-3 - 1}{-2 - (-5)} = \frac{-4}{3} = -\frac{4}{3}$$

which is the same as our previous result.

Note: The two most common mistakes students make when first working with the formula for the slope of a line are

1. Putting the difference of the x-coordinates over the difference of the y-coordinates.
2. Subtracting in one order in the numerator and then subtracting in the opposite order in the denominator. You would make this mistake in Example 2 if you wrote $1 - (-3)$ in the numerator and then $-2 - (-5)$ in the denominator.

EXAMPLE 3   Find the slope of the line containing $(3, -1)$ and $(3, 4)$.

Solution   Using the definition for slope, we have

$$m = \frac{-1 - 4}{3 - 3} = \frac{-5}{0}$$

The expression $\frac{-5}{0}$ is undefined. That is, there is no real number to associate with it. In this case, we say the line *has no slope*.

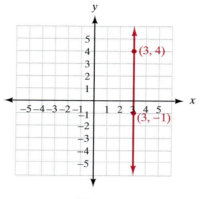

FIGURE 3

The graph of our line is shown in Figure 3. Our line with no slope is a vertical line. All vertical lines have no slope. (And all horizontal lines, as we mentioned earlier, have 0 slope.)

SLOPES OF PARALLEL AND PERPENDICULAR LINES

In geometry we call lines in the same plane that never intersect parallel. For two lines to be nonintersecting, they must rise or fall at the same rate. In other words, two lines are *parallel* if and only if they have the *same slope.*

Although it is not as obvious, it is also true that two nonvertical lines are *perpendicular* if and only if the *product of their slopes is* -1. This is the same as saying their slopes are negative reciprocals.

We can state these facts with symbols as follows: If line l_1 has slope m_1 and line l_2 has slope m_2, then

$$l_1 \text{ is parallel to } l_2 \Leftrightarrow m_1 = m_2$$

and

$$l_1 \text{ is perpendicular to } l_2 \Leftrightarrow m_1 \cdot m_2 = -1$$

$$\left(\text{or } m_1 = \frac{-1}{m_2} \right)$$

For example, if a line has a slope of $\frac{2}{3}$, then any line parallel to it has a slope of $\frac{2}{3}$. Any line perpendicular to it has a slope of $-\frac{3}{2}$ (the negative reciprocal of $\frac{2}{3}$).

Although we cannot give a formal proof of the relationship between the slopes of perpendicular lines at this level of mathematics, we can offer some justification for the relationship. Figure 4, next page, shows the graphs of two lines. One of the lines has a slope of $\frac{2}{3}$; the other has a slope of $-\frac{3}{2}$. As you can see, the lines are perpendicular.

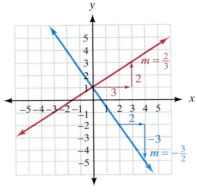

FIGURE 4

Using TECHNOLOGY

FAMILIES OF CURVES

We can use a graphing calculator to investigate the effects of the numbers a and b on the graph of $y = ax + b$. To see how the number b affects the graph, we can hold a constant and let b vary. Doing so will give us a *family* of curves. Suppose we set $a = 1$ and then let b take on integer values from -3 to 3. The equations we obtain are

$$y = x - 3$$

$$y = x - 2$$

$$y = x - 1$$

$$y = x$$

$$y = x + 1$$

$$y = x + 2$$

$$y = x + 3$$

We will give three methods of graphing this set of equations on a graphing calculator.

Method 1: Y-Variables List

To use the Y-variables list, enter each equation at one of the Y variables, set the graph window, then graph. The calculator will graph the equations in order,

(continued)

starting with Y_1 and ending with Y_7. Following is the Y-variables list, an appropriate window, and a sample of the type of graph obtained (Figure 5).

$Y_1 = X - 3$

$Y_2 = X - 2$

$Y_3 = X - 1$

$Y_4 = X$

$Y_5 = X + 1$

$Y_6 = X + 2$

$Y_7 = X + 3$

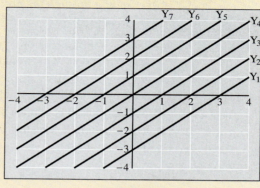

FIGURE 5

Window: X from -4 to 4, Y from -4 to 4

Method 2: Programming

The same result can be obtained by programming your calculator to graph $y = x + b$ for $b = -3, -2, -1, 0, 1, 2,$ and 3. Here is an outline of a program that will do this. Check the manual that came with your calculator to find the commands for your calculator.

Step 1:   Clear screen

Step 2:   Set window for X from -4 to 4 and Y from -4 to 4

Step 3: $-3 \rightarrow B$

Step 4:   Label 1

Step 5:   Graph $Y = X + B$

Step 6: $B + 1 \rightarrow B$

Step 7:   If $B < 4$, Goto 1

Step 8:   End

Method 3: Using Lists

On the TI-82/83 you can set Y_1 as follows

$$Y_1 = X + \{-3, -2, -1, 0, 1, 2, 3\}$$

When you press $\boxed{\text{GRAPH}}$ the calculator will graph each line from $y = x + (-3)$ to $y = x + 3$.

Each of the three methods will produce graphs similar to those in Figure 5.

Getting Ready for Class

After reading through the preceding section, respond in your own words and in complete sentences.

A. If you were looking at a graph that described the performance of a stock you had purchased, why would it be better if the slope of the line were positive, rather than negative?

B. Describe the behavior of a line with a negative slope.

C. Would you rather climb a hill with a slope of $\frac{1}{2}$ or a slope of 3? Explain why.

D. Describe how to obtain the slope of a line if you know the coordinates of two points on the line.

PROBLEM SET 3.2

Find the slope of each of the following lines from the given graph.

1.

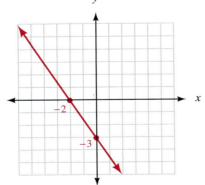

3.

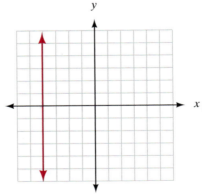

2.

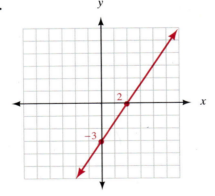

4.

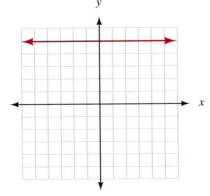

5.

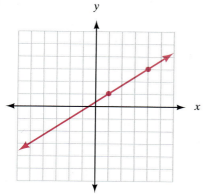

6.

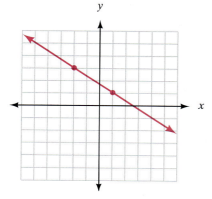

Find the slope of the line through the following pairs of points. Then, plot each pair of points, draw a line through them, and indicate the rise and run in the graph in the manner shown in Example 2.

7. $(2, 1), (4, 4)$ **8.** $(3, 1), (5, 4)$

9. $(1, 4), (5, 2)$ **10.** $(1, 3), (5, 2)$

11. $(1, -3), (4, 2)$ **12.** $(2, -3), (5, 2)$

13. $(-3, -2), (1, 3)$ **14.** $(-3, -1), (1, 4)$

15. $(-3, 2), (3, -2)$ **16.** $(-3, 3), (3, -1)$

17. $(2, -5), (3, -2)$ **18.** $(2, -4), (3, -1)$

For each of the equations in Problems 19–22, complete the table, and then use the results to find the slope of the graph of the equation.

19. $2x + 3y = 6$ **20.** $3x - 2y = 6$

x	y
0	
	0

x	y
0	
	0

21. $y = \frac{2}{3}x - 5$ **22.** $y = -\frac{3}{4}x + 2$

x	y
0	
3	

x	y
0	
4	

23. Finding Slope From Intercepts   Graph the line that has an x-intercept of 3 and a y-intercept of -2. What is the slope of this line?

24. Finding Slope From Intercepts   Graph the line that has an x-intercept of 2 and a y-intercept of -3. What is the slope of this line?

25. Finding Slope From Intercepts   Graph the line with x-intercept 4 and y-intercept 2. What is the slope of this line?

26. Finding Slope From Intercepts   Graph the line with x-intercept -4 and y-intercept -2. What is the slope of this line?

27. Parallel Lines   Find the slope of any line parallel to the line through $(2, 3)$ and $(-8, 1)$.

28. Parallel Lines   Find the slope of any line parallel to the line through $(2, 5)$ and $(5, -3)$.

29. Perpendicular Lines   Line l contains the points $(5, -6)$ and $(5, 2)$. Give the slope of any line perpendicular to l.

30. Perpendicular Lines   Line l contains the points $(3, 4)$ and $(-3, 1)$. Give the slope of any line perpendicular to l.

31. Determine if each of the following tables could represent ordered pairs from an equation of a line.

(a)

x	y
0	5
1	7
2	9
3	11

(b)

x	y
-2	-5
0	-2
2	0
4	1

32. The following lines have slope 2, $\frac{1}{2}$, 0, and -1. Match each line to its slope value.

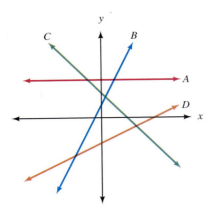

37. Find the slope of the line segment labeled A. What units would you attach to this number?

38. Find the slope of the line segment labeled C. Be sure to attach units to your answer.

39. Is the temperature changing faster during the 1st minute or the 16th minute?

40. Line segments B and D both have 0 slope. Explain what this means in terms of the melting ice.

Value of a Used Car The 1998 edition of a popular consumer car price book gives the values shown in the table for Volkswagen Jettas in good condition. The line graph was drawn from the information in the table. Use this information to solve Problems 41–44.

Applying the Concepts

33. Slope of a Sand Pile A pile of sand at a construction site is in the shape of a cone. If the slope of the side of the pile is $\frac{2}{3}$ and the pile is 8 feet high, how wide is the diameter of the base of the pile?

34. Slope of a Pyramid The slope of the sides of one of the ancient pyramids in Egypt is $\frac{13}{10}$. If the base of the pyramid is 750 feet, how tall is the pyramid?

Heating a Block of Ice A block of ice with an initial temperature of $-20°C$ is heated at a steady rate. The graph shows how the temperature changes as the ice melts to become water and the water boils to become steam and water. (Problems 35–40)

Year	Age in 1998	Value ($)
1994	4	8,525
1995	3	9,575
1996	2	11,950
1997	1	13,200
1998	0	15,250

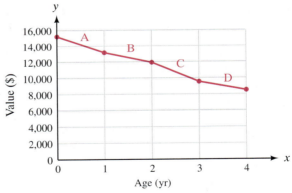

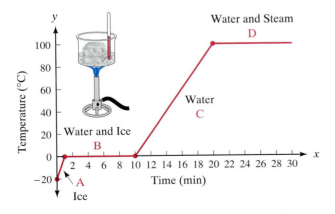

35. How long does it take all the ice to melt?

36. From the time the heat is applied to the block of ice, how long is it before the water boils?

41. Find the slope of the line segment labeled B. What units should you attach to this number?

42. Find the slope of line segment C. Be sure to include units with your answer.

43. From the graph, does the value of the car decrease more from 1 to 2 years, or from 2 to 3 years?

44. From the graph, does the value of the car decrease more from 2 to 3 years, or from 3 to 4 years?

45. Slope of a Highway A sign at the top of the Cuesta Grade, outside of San Luis Obispo, reads

"7% downgrade next 3 miles." The diagram in Figure 6 is a model of the Cuesta Grade that takes into account the information on that sign.

(a) At point B, the graph crosses the y-axis at 1,106 feet. How far is it from the origin to point A?

(b) What is the slope of the Cuesta Grade?

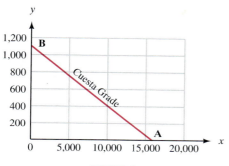

FIGURE 6

46. **Graph Reading**   A fraternity is planning a fund-raising concert at the Veterans Memorial Building, which can accommodate a maximum of 300 ticket holders. Tickets are sold at $10 each. Their expenses are $455 for the band and $125 rent for the Veterans building. If 150 or more tickets are sold, the fraternity is required to hire a security guard for $200. The following diagram (Figure 7) can be used to find the profit they will make by selling various numbers of tickets. As you can see, if no one buys a ticket, their profit will be −$580, the amount they will pay for the band and for rent.

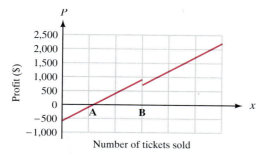

Number of tickets sold

FIGURE 7

(a) Both line segments have the same slope. What is that slope?

(b) Point A is the break-even point. It corresponds to the number of tickets they must sell so that their income is equal to their expenses. How many tickets is this?

(c) What is the value of point B?

(d) How many tickets must they sell to make a profit of $500?

47. **Baseball Salaries**   The average salary of a major league baseball player in 1967 was $25,000, and in 1975 it was $40,000. What is the slope of the line that connects these two points; that is, what is the average yearly salary increase for this period?

   The average salary of a major league baseball player in 1990 was $600,000, and in 1994 it was $1,300,000. What is the slope of the line that connects these two points; that is, what is the average yearly salary increase for this period?

48. **Blood Pressure**   The average systolic blood pressure for a 26-year-old is 113 and for a 40-year-old is 120.

(a) Using the age as the horizontal axis and the blood pressure as the vertical axis, plot this blood pressure data. Draw a line connecting them.

(b) Determine the slope of the line connecting the two points from part (a).

(c) What does the slope from part (b) represent in terms of blood pressure and age?

Review Problems

The following problems review material we covered in Section 2.3.

49. If $3x + 2y = 12$, find y when x is 4.

50. If $y = 3x − 1$, find x when y is 0.

51. Solve the formula $3x + 2y = 12$ for y.

52. Solve the formula $y = 3x − 1$ for x.

53. Solve the formula $A = P + Prt$ for t.

54. Solve the formula $S = \pi r^2 + 2\pi rh$ for h.

Extending the Concepts

55. Use your Y-variables list or write a program to graph the family of curves $Y = 2X + B$ for $B = −3, −2, −1, 0, 1, 2,$ and 3.

56. Use your Y-variables list or write a program to graph the family of curves $Y = −2X + B$ for $B = −3, −2, −1, 0, 1, 2,$ and 3.

57. Use your Y-variables list or write a program to graph the family of curves $Y = AX$ for $A = -3$, -2, -1, 0, 1, 2, and 3.

58. Use your Y-variables list or write a program to graph the family of curves $Y = AX + 2$ for $A = -3$, -2, -1, 0, 1, 2, and 3.

59. Use your Y-variables list or write a program to graph the family of curves $Y = AX$ for $A = \frac{1}{4}, \frac{1}{3}, \frac{1}{2}, 1, 2$, and 3.

60. Use your Y-variables list or write a program to graph the family of curves $Y = AX - 2$ for $A = \frac{1}{4}, \frac{1}{3}, \frac{1}{2}, 1, 2$, and 3.

3.3 **The Equation of a Line**

The table and illustrations show some corresponding temperatures on the Fahrenheit and Celsius temperature scales. For example, water freezes at 32°F and 0°C, and boils at 212°F and 100°C.

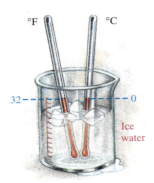

Degrees Celsius	Degrees Fahrenheit
0	32
25	77
50	122
75	167
100	212

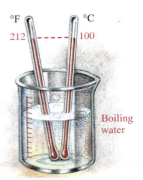

If we plot all the points in the table using the *x*-axis for temperatures on the Celsius scale and the *y*-axis for temperatures on the Fahrenheit scale, we see that they line up in a straight line (Figure 1). This means that a linear equation in two

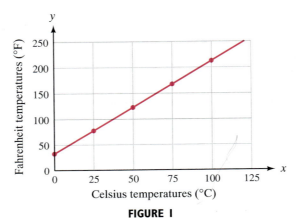

FIGURE 1

variables will give a perfect description of the relationship between the two scales. That equation is

$$F = \frac{9}{5}C + 32$$

The techniques we use to find the equation of a line from a set of points is what this section is all about.

Suppose line l has slope m and y-intercept b. What is the equation of l? Since the y-intercept is b, we know the point $(0, b)$ is on the line. If (x, y) is any other point on l, then using the definition for slope, we have

$$\frac{y - b}{x - 0} = m \qquad \text{Definition of slope}$$

$$y - b = mx \qquad \text{Multiply both sides by } x.$$

$$y = mx + b \qquad \text{Add } b \text{ to both sides.}$$

This last equation is known as the *slope-intercept form* of the equation of a straight line.

SLOPE-INTERCEPT FORM OF THE EQUATION OF A LINE

The equation of any line with slope m and y-intercept b is given by

$$y = mx + b$$

$$\nearrow \qquad \uparrow$$

Slope y-intercept

When the equation is in this form, the *slope* of the line is always the *coefficient of x,* and the *y-intercept* is always the *constant term.*

EXAMPLE I   Find the equation of the line with slope $-\frac{4}{3}$ and y-intercept 5. Then graph the line.

Solution   Substituting $m = -\frac{4}{3}$ and $b = 5$ into the equation $y = mx + b$, we have

$$y = -\frac{4}{3}x + 5$$

Finding the equation from the slope and y-intercept is just that easy. If the slope is m and the y-intercept is b, then the equation is always $y = mx + b$. Now, let's graph the line.

Since the y-intercept is 5, the graph goes through the point $(0, 5)$. To find a second point on the graph, we start at $(0, 5)$ and move 4 units down (that's a rise of -4) and 3 units to the right (a run of 3). The point we end up at is $(3, 1)$. Drawing a line that passes through $(0, 5)$ and $(3, 1)$, we have the graph of our equation. (Note that we could also let the rise = 4 and the run = -3 and obtain the same graph.) The graph is shown in Figure 2, on the next page.

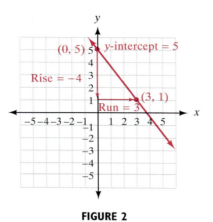

FIGURE 2

EXAMPLE 2   Give the slope and y-intercept for the line $2x - 3y = 5$.

Solution   To use the slope-intercept form, we must solve the equation for y in terms of x:

$$2x - 3y = 5$$
$$-3y = -2x + 5 \qquad \text{Add } -2x \text{ to both sides.}$$
$$y = \frac{2}{3}x - \frac{5}{3} \qquad \text{Divide by } -3.$$

The last equation has the form $y = mx + b$. The slope must be $m = \frac{2}{3}$, and the y-intercept is $b = -\frac{5}{3}$.

EXAMPLE 3   Graph the equation $2x + 3y = 6$ using the slope and y-intercept.

Solution   Although we could graph this equation using the methods developed in Section 3.1 (by finding ordered pairs that are solutions to the equation and drawing a line through their graphs), it is sometimes easier to graph a line using the slope-intercept form of the equation.

Solving the equation for y, we have

$$2x + 3y = 6$$
$$3y = -2x + 6 \qquad \text{Add } -2x \text{ to both sides.}$$
$$y = -\frac{2}{3}x + 2 \qquad \text{Divide by 3.}$$

The slope is $m = -\frac{2}{3}$ and the y-intercept is $b = 2$. Therefore, the point $(0, 2)$ is on the graph, and the ratio of rise to run going from $(0, 2)$ to any other point on the line is $-\frac{2}{3}$. If we start at $(0, 2)$ and move 2 units up (that's a rise of 2) and 3 units to the left (a run of -3), we will be at another point on the graph. (We could also go down 2 units and right 3 units and still be assured of ending up at another point on the line, since $\frac{2}{-3}$ is the same as $\frac{-2}{3}$.)

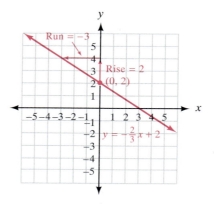

FIGURE 3

Note: As we mentioned in the introduction to this chapter, the rectangular coordinate system is the tool we use to connect algebra and geometry. Example 3 illustrates this connection, as do the many other examples in this chapter. In Example 3, Descartes's rectangular coordinate system allows us to associate the equation $2x + 3y = 6$ (an algebraic concept) with the straight line (a geometric concept) shown in Figure 3.

A second useful form of the equation of a straight line is the point-slope form.

Let line *l* contain the point (x_1, y_1) and have slope *m*. If (x, y) is any other point on *l*, then by the definition of slope we have

$$\frac{y - y_1}{x - x_1} = m$$

Multiplying both sides by $(x - x_1)$ gives us

$$(x - x_1) \cdot \frac{y - y_1}{x - x_1} = m(x - x_1)$$

$$y - y_1 = m(x - x_1)$$

This last equation is known as the *point-slope form* of the equation of a straight line.

POINT-SLOPE FORM OF THE EQUATION OF A LINE

The equation of the line through (x_1, y_1) with slope *m* is given by

$$y - y_1 = m(x - x_1)$$

This form of the equation of a straight line is used to find the equation of a line, either given one point on the line and the slope, or given two points on the line.

EXAMPLE 4   Find the equation of the line with slope -2 that contains the point $(-4, 3)$. Write the answer in slope-intercept form.

Solution

Using $(x_1, y_1) = (-4, 3)$ and $m = -2$

in $y - y_1 = m(x - x_1)$ Point-slope form

gives us $y - 3 = -2(x + 4)$ *Note*: $x - (-4) = x + 4$

$y - 3 = -2x - 8$ Multiply out right side.

$y = -2x - 5$ Add 3 to each side.

Figure 4 is the graph of the line that contains $(-4, 3)$ and has a slope of -2. Notice that the y-intercept on the graph matches that of the equation we found.

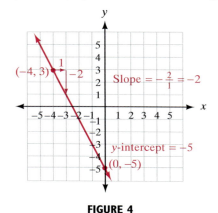

FIGURE 4

EXAMPLE 5   Find the equation of the line that passes through the points $(-3, 3)$ and $(3, -1)$.

Solution   We begin by finding the slope of the line:

$$m = \frac{3 - (-1)}{-3 - 3} = \frac{4}{-6} = -\frac{2}{3}$$

Using $(x_1, y_1) = (3, -1)$ and $m = -\frac{2}{3}$ in $y - y_1 = m(x - x_1)$ yields

$$y + 1 = -\frac{2}{3}(x - 3)$$

$$y + 1 = -\frac{2}{3}x + 2 \qquad \text{Multiply out right side.}$$

$$y = -\frac{2}{3}x + 1 \qquad \text{Add } -1 \text{ to each side.}$$

Figure 5 shows the graph of the line that passes through the points $(-3, 3)$ and $(3, -1)$. As you can see, the slope and y-intercept are $-\frac{2}{3}$ and 1, respectively.

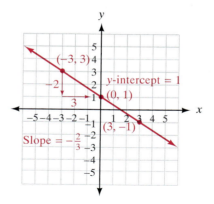

FIGURE 5

Note: We could have used the point $(-3, 3)$ instead of $(3, -1)$ and obtained the same equation. That is, using $(x_1, y_1) = (-3, 3)$ and $b = -\frac{2}{3}$ in $y - y_1 = m(x - x_1)$ gives us

$$y - 3 = -\frac{2}{3}(x + 3)$$

$$y - 3 = -\frac{2}{3}x - 2$$

$$y = -\frac{2}{3}x + 1$$

which is the same result we obtained using $(3, -1)$.

The last form of the equation of a line that we will consider in this section is called the standard form. It is used mainly to write equations in a form that is free of fractions and is easy to compare with other equations.

STANDARD FORM FOR THE EQUATION OF A LINE

If a, b, and c are integers, then the equation of a line is in standard form when it has the form

$$ax + by = c$$

If we were to write the equation

$$y = -\frac{2}{3}x + 1$$

in standard form, we would first multiply both sides by 3 to obtain

$$3y = -2x + 3$$

Then we would add $2x$ to each side, yielding

$$2x + 3y = 3$$

which is a linear equation in standard form.

EXAMPLE 6 Give the equation of the line through $(-1, 4)$ whose graph is perpendicular to the graph of $2x - y = -3$. Write the answer in standard form.

Solution To find the slope of $2x - y = -3$, we solve for y:

$$2x - y = -3$$
$$y = 2x + 3$$

The slope of this line is 2. The line we are interested in is perpendicular to the line with slope 2 and must, therefore, have a slope of $-\frac{1}{2}$.

Using $(x_1, y_1) = (-1, 4)$ and $m = -\frac{1}{2}$, we have

$$y - y_1 = m(x - x_1)$$
$$y - 4 = -\frac{1}{2}(x + 1)$$

Since we want our answer in standard form, we multiply each side by 2.

$$2y - 8 = -1(x + 1)$$
$$2y - 8 = -x - 1$$
$$x + 2y - 8 = -1$$
$$x + 2y = 7$$

The last equation is in standard form.

As a final note, we should mention again that all horizontal lines have equations of the form $y = b$ and slopes of 0. Vertical lines have no slope and have equations of the form $x = a$. These two special cases do not lend themselves well to either the slope-intercept form or the point-slope form of the equation of a line.

Using TECHNOLOGY

GRAPHING CALCULATORS

One advantage of using a graphing calculator to graph lines is that a calculator does not care whether the equation has been simplified or not. To illustrate, in Example 5 we found that the equation of the line with slope $-\frac{2}{3}$ that passes through the point $(3, -1)$ is

$$y + 1 = -\frac{2}{3}(x - 3)$$

Normally, to graph this equation we would simplify it first. With a graphing calculator we add -1 to each side and enter the equation this way:

$$Y_1 = -(2/3)(X - 3) - 1$$

(continued)

No simplification is necessary. We can graph the equation in this form, and the graph will be the same as the simplified form of the equation, which is $y = -\frac{2}{3}x + 1$. To convince yourself that this is true, graph both the simplified form for the equation and the unsimplified form in the same window. As you will see, the two graphs coincide.

Getting Ready for Class

After reading through the preceding section, respond in your own words and in complete sentences.

A. How would you graph the line $y = \frac{1}{2}x + 3$?

B. What is the slope-intercept form of the equation of a line?

C. Describe how you would find the equation of a line if you knew the slope and the y-intercept of the line.

D. If you had the graph of a line, how would you use it to find the equation of the line?

PROBLEM SET 3.3

Give the equation of the line with the following slope and y-intercept.

1. $m = 2, b = 3$

2. $m = -4, b = 2$

3. $m = 1, b = -5$

4. $m = -5, b = -3$

5. $m = \frac{1}{2}, b = \frac{3}{2}$

6. $m = \frac{2}{3}, b = \frac{5}{6}$

7. $m = 0, b = 4$

8. $m = 0, b = -2$

Give the slope and y-intercept for each of the following equations. Sketch the graph using the slope and y-intercept. Give the slope of any line perpendicular to the given line.

9. $y = 3x - 2$

10. $y = 2x + 3$

11. $2x - 3y = 12$

12. $3x - 2y = 12$

13. $4x + 5y = 20$

14. $5x - 4y = 20$

For each of the following lines, name the slope and y-intercept. Then write the equation of the line in slope-intercept form.

15.

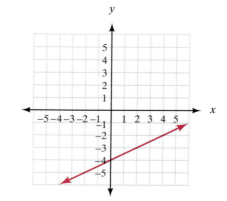

16.

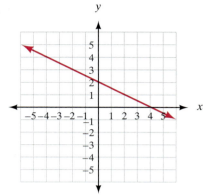

For each of the following problems, the slope and one point on a line are given. In each case, find the equation of that line. (Write the equation for each line in slope-intercept form.)

19. $(-2, -5)$; $m = 2$ **20.** $(-1, -5)$; $m = 2$

21. $(-4, 1)$; $m = -\frac{1}{2}$ **22.** $(-2, 1)$; $m = -\frac{1}{2}$

23. $(-\frac{1}{3}, 2)$; $m = -3$ **24.** $(-\frac{2}{3}, 5)$; $m = -3$

Find the equation of the line that passes through each pair of points. Write your answers in standard form.

25. $(-2, -4), (1, -1)$ **26.** $(2, 4), (-3, -1)$

27. $(-1, -5), (2, 1)$ **28.** $(-1, 6), (1, 2)$

29. $(\frac{1}{3}, -\frac{1}{5}), (-\frac{1}{3}, -1)$ **30.** $(-\frac{1}{2}, -\frac{1}{2}), (\frac{1}{2}, \frac{1}{10})$

17.

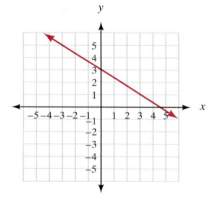

For each of the following lines, name the coordinates of any two points on the line. Then use those two points to find the equation of the line.

31.

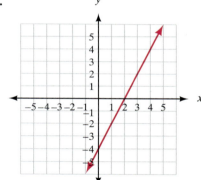

18.

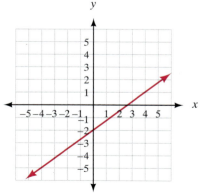

32.

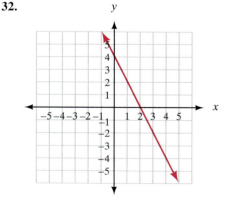

33.

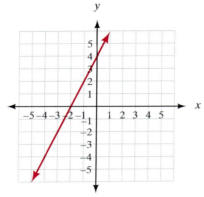

34.

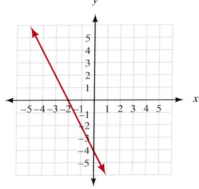

35. Give the slope and y-intercept of $y = -2$. Sketch the graph.

36. For the line $x = -3$, sketch the graph, give the slope, and name any intercepts.

37. Find the equation of the line parallel to the graph of $3x - y = 5$ that contains the point $(-1, 4)$.

38. Find the equation of the line parallel to the graph of $2x - 4y = 5$ that contains the point $(0, 3)$.

39. Line l is perpendicular to the graph of the equation $2x - 5y = 10$ and contains the point $(-4, -3)$. Find the equation for l.

40. Line l is perpendicular to the graph of the equation $-3x - 5y = 2$ and contains the point $(2, -6)$. Find the equation for l.

41. Give the equation of the line perpendicular to the graph of $y = -4x + 2$ that has an x-intercept of -1.

42. Write the equation of the line parallel to the graph of $7x - 2y = 14$ that has an x-intercept of 5.

43. Find the equation of the line with x-intercept 3 and y-intercept 2.

44. Find the equation of the line with x-intercept 2 and y-intercept 3.

Applying the Concepts

45. Deriving the Temperature Equation   The table below resembles the table from the introduction to this section. The rows of the table give us ordered pairs (C, F).

Degrees Celsius	Degrees Fahrenheit
C	F
0	32
25	77
50	122
75	167
100	212

(a) Use any two of the ordered pairs from the table to derive the equation $F = \frac{9}{5}C + 32$.

(b) Use the equation from part (a) to find the Fahrenheit temperature that corresponds to a Celsius temperature of 30°.

46. Maximum Heart Rate   The table gives the maximum heart rate for adults 30, 40, 50, and 60 years old. Each row of the table gives us an ordered pair (A, M).

Age (years)	Maximum Heart Rate (beats per minute)
A	M
30	190
40	180
50	170
60	160

(a) Use any two of the ordered pairs from the table to derive the equation $M = 220 - A$, which gives the maximum heart rate M for an adult whose age is A.

(b) Use the equation from part (a) to find the maximum heart rate for a 25-year-old adult.

47. Oxygen Consumption and Exercise A person's oxygen consumption (in liters per minute) is linearly related to that person's heart rate (in beats per minute). Suppose that at 98 beats per minute a person consumes 1 liter of oxygen per minute, and after exercising with a heart rate of 155 beats per minute consumes 1.5 liters of oxygen per minute.
(a) Find an equation that relates oxygen consumption to heart rate.
(b) What restrictions seem reasonable on values of the independent variable?

48. Exercise Heart Rate In an aerobics class, the instructor indicates that her students' exercise heart rate is 60% of their maximum heart rate, where maximum heart rate is 220 minus their age.
(a) Determine the equation that gives exercise heart rate E in terms of age A.
(b) Use the equation to find the exercise heart rate of a 22-year-old student.
(c) Sketch the graph of the equation for students from 18 to 80 years of age.

49. AIDS Cases The number of AIDS cases in the United States from 1984 through 1988 increased in a linear relationship. In 1984 there were 3,000 known cases, and in 1988 the number of cases had risen to 20,000.
(a) Write the equation of the line that relates the number of AIDS cases to the year for the years 1984 through 1988.
(b) Use the equation found in part (a) to determine how many AIDS cases were reported in 1986.

50. Cell Phone Rates Suppose you have a cellular phone with a rate plan that costs $20 per month, with no additional charge for the first 30 minutes of use, and then $0.40 for each minute after the first 30 minutes.
(a) Write an equation that gives the monthly cost C to talk for x minutes, if x is less than 30 minutes.
(b) Write an equation that gives the monthly cost C to talk for x minutes, if x is greater than 30 minutes.
(c) On the same coordinate system, graph the first equation for $x < 30$ and the second equation for $x > 30$.

Review Problems

The problems that follow review material we covered in Section 2.3.

51. The length of a rectangle is 3 inches more than 4 times the width. The perimeter is 56 inches. Find the length and width.

52. One angle is 10 degrees less than 4 times another. Find the measure of each angle if
(a) The two angles are complementary.
(b) The two angles are supplementary.

53. The cash register in a candy shop contains $66 at the beginning of the day. At the end of the day, it contains $732.50. If the sales tax rate is 7.5%, how much of the total is sales tax?

54. The third angle in an isosceles triangle is 20 degrees less than twice as large as each of the two base angles. Find the measure of each angle.

Extending the Concepts

55. Label the units on a sheet of graph paper in multiples of 10, and graph the line $2x + 5y = 100$.

56. Label the units on a sheet of graph paper in multiples of 20, and graph the line $-4x + 10y = 100$.

57. Label the y-axis in multiples of 10 and the x-axis in multiples of 1, and graph the equation $y = 20x - 50$.

58. Label the y-axis in multiples of 10 and the x-axis in multiples of 1, and graph the equation $y = -20x + 30$.

Write each equation in slope-intercept form. Then name the slope, the y-intercept, and the x-intercept.

59. $\dfrac{x}{2} + \dfrac{y}{3} = 1$

60. $\dfrac{x}{5} + \dfrac{y}{4} = 1$

61. $\dfrac{x}{-2} + \dfrac{y}{3} = 1$

62. $\dfrac{x}{2} + \dfrac{y}{-3} = 1$

63. When a linear equation is written in the form

$$\frac{x}{a} + \frac{y}{b} = 1$$

it is said to be in *two-intercept form*. Find the x-intercept, the y-intercept, and the slope of this line.

Linear Inequalities in Two Variables

A small movie theater holds 100 people. The owner charges more for adults than for children, so it is important to know the different combinations of adults and children that can be seated at one time. The shaded region in Figure 1 contains all the seating combinations. The line $x + y = 100$ shows the combinations for a full theater: The y-intercept corresponds to a theater full of adults, and the x-intercept corresponds to a theater full of children. In the shaded region below the line $x + y = 100$ are the combinations that occur if the theater is not full.

Shaded regions like the one shown in Figure 1 are produced by linear inequalities in two variables, which is the topic of this section.

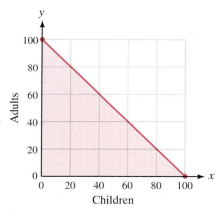

FIGURE 1

A *linear inequality in two variables* is any expression that can be put in the form

$$ax + by < c$$

where a, b, and c are real numbers (a and b not both 0). The inequality symbol can be any one of the following four: $<, \leq, >, \geq$.

Some examples of linear inequalities are

$$2x + 3y < 6 \qquad y \geq 2x + 1 \qquad x - y \leq 0$$

Although not all of these examples have the form $ax + by < c$, each one can be put in that form.

The solution set for a linear inequality is a *section of the coordinate plane*. The *boundary* for the section is found by replacing the inequality symbol with an equal sign and graphing the resulting equation. The boundary is included in the solution set (and is represented with a *solid line*) if the inequality symbol used originally is $\leq$ or $\geq$. The boundary is not included (and is represented with a *broken line*) if the original symbol is $<$ or $>$.

EXAMPLE 1   Graph the solution set for $x + y \leq 4$.

Solution   The boundary for the graph is the graph of $x + y = 4$. The boundary is included in the solution set because the inequality symbol is $\leq$.

Figure 2 is the graph of the boundary:

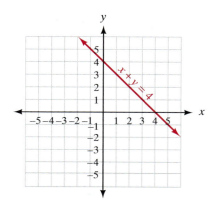

FIGURE 2

The boundary separates the coordinate plane into two regions: the region above the boundary and the region below it. The solution set for $x + y \leq 4$ is one of these two regions along with the boundary. To find the correct region, we simply choose any convenient point that is *not* on the boundary. We then substitute the coordinates of the point into the original inequality $x + y \leq 4$. If the point we choose satisfies the inequality, then it is a member of the solution set, and we can assume that all points on the same side of the boundary as the chosen point are also in the solution set. If the coordinates of our point do not satisfy the original inequality, then the solution set lies on the other side of the boundary.

In this example, a convenient point that is not on the boundary is the origin.

Substituting $(0, 0)$

into $x + y \leq 4$

gives us $0 + 0 \leq 4$

$0 \leq 4$ A true statement

Since the origin is a solution to the inequality $x + y \leq 4$, and the origin is below the boundary, all other points below the boundary are also solutions.

Figure 3 is the graph of $x + y \leq 4$.

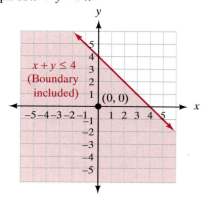

FIGURE 3

The region above the boundary is described by the inequality $x + y > 4$.

Here is a list of steps to follow when graphing the solution set for linear inequalities in two variables.

TO GRAPH A LINEAR INEQUALITY IN TWO VARIABLES

Step 1: Replace the inequality symbol with an equal sign. The resulting equation represents the boundary for the solution set.

Step 2: Graph the boundary found in step 1 using a *solid line* if the boundary is included in the solution set (i.e., if the original inequality symbol was either $\leq$ or $\geq$). Use a *broken line* to graph the boundary if it is *not* included in the solution set. (It is not included if the original inequality was either $<$ or $>$.)

Step 3: Choose any convenient point not on the boundary and substitute the coordinates into the *original* inequality. If the resulting statement is *true,* the graph lies on the *same* side of the boundary as the chosen point. If the resulting statement is *false,* the solution set lies on the *opposite* side of the boundary.

EXAMPLE   2   Graph the solution set for $y < 2x - 3$.

Solution   The boundary is the graph of $y = 2x - 3$, a line with slope 2 and y-intercept -3. The boundary is not included since the original inequality symbol is $<$. We therefore use a broken line to represent the boundary in Figure 4.

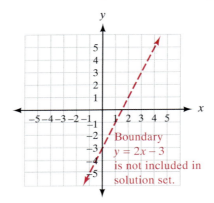

Boundary
$y = 2x - 3$
is not included in
solution set.

FIGURE 4

A convenient test point is again the origin:

Using $(0, 0)$

in $y < 2x - 3$

we have $0 < 2(0) - 3$

                $0 < -3$ A false statement

Since our test point gives us a false statement and it lies above the boundary, the solution set must lie on the other side of the boundary (Figure 5).

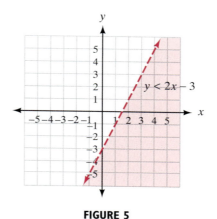

FIGURE 5

Using TECHNOLOGY

GRAPHING CALCULATORS

Most graphing calculators have a Shade command that allows a portion of a graphing screen to be shaded. With this command we can visualize the solution sets to linear inequalities in two variables. Since most graphing calculators cannot draw a dotted line, however, we are not actually "graphing" the solution set, only visualizing it.

STRATEGY FOR VISUALIZING A LINEAR INEQUALITY IN TWO VARIABLES ON A GRAPHING CALCULATOR

Step 1: Solve the inequality for y.
Step 2: Replace the inequality symbol with an equal sign. The resulting equation represents the boundary for the solution set.
Step 3: Graph the equation in an appropriate viewing window.
Step 4: Use the Shade command to indicate the solution set:
For inequalities having the $<$ or $\leq$ sign, use Shade(Xmin, Y_1).
For inequalities having the $>$ or $\geq$ sign, use Shade(Y_1, Xmax).

Note: On the TI-83, step 4 can be done by manipulating the icons in the left column in the list of Y variables.

Figures 6 and 7 show the graphing calculator screens that help us visualize the solution set to the inequality $y < 2x - 3$ that we graphed in Example 2.

Windows: X from -5 to 5, Y from -5 to 5

(continued)

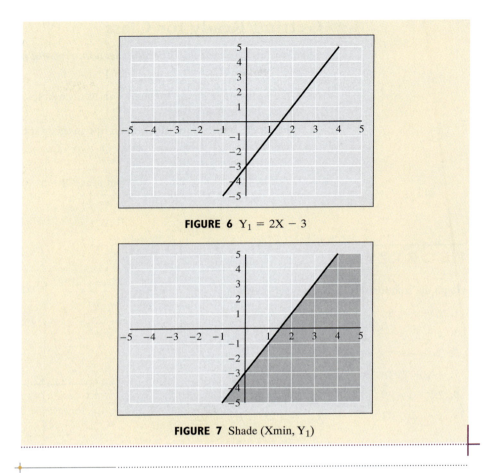

FIGURE 6 $Y_1 = 2X - 3$

FIGURE 7 Shade (Xmin, Y_1)

E X A M P L E   3   Graph the solution set for $x \leq 5$.

Solution   The boundary is $x = 5$, which is a vertical line. All points in Figure 8 to the left have x-coordinates less than 5 and all points to the right have x-coordinates greater than 5.

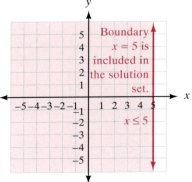

FIGURE 8

Getting Ready for Class

After reading through the preceding section, respond in your own words and in complete sentences.

A. When graphing a linear inequality in two variables, how do you find the equation of the boundary line?

B. What is the significance of a broken line in the graph of an inequality?

C. When graphing a linear inequality in two variables, how do you know which side of the boundary line to shade?

D. Describe the set of ordered pairs that are solutions to $x + y < 6$.

PROBLEM SET 3.4

Graph the solution set for each of the following.

1. $x + y < 5$

2. $x + y \le 5$

3. $x - y \ge -3$

4. $x - y > -3$

5. $2x + 3y < 6$

6. $2x - 3y > -6$

7. $-x + 2y > -4$

8. $-x - 2y < 4$

9. $2x + y < 5$

10. $2x + y < -5$

11. $y < 2x - 1$

12. $y \le 2x - 1$

14.

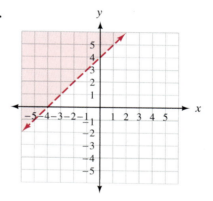

For each graph shown here, name the linear inequality in two variables that is represented by the shaded region.

13.

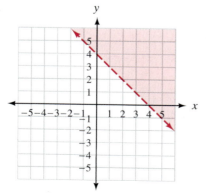

15.

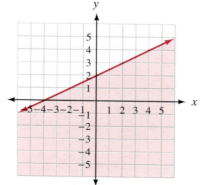

16.

Graph each inequality.

17. $x \geq 3$ **18.** $x > -2$

19. $y \leq 4$ **20.** $y > -5$

21. $y < 2x$ **22.** $y > -3x$

23. $y \geq \frac{1}{2}x$ **24.** $y \leq \frac{1}{3}x$

25. $y \geq \frac{3}{4}x - 2$ **26.** $y > -\frac{2}{3}x + 3$

27. $\frac{x}{3} + \frac{y}{2} > 1$ **28.** $\frac{x}{5} + \frac{y}{4} < 1$

Applying the Concepts

29. Number of People in a Dance Club A dance club holds a maximum of 200 people. The club charges one price for students and a higher price for nonstudents. If the number of students in the club at any time is x and the number of nonstudents is y, shade the region in the first quadrant that contains all combinations of students and nonstudents that are in the club at any time.

30. Many Perimeters Suppose you have 500 feet of fencing that you will use to build a rectangular livestock pen. Let x represent the length of the pen and y represent the width. Shade the region in the first quadrant that contains all possible values of x and y that will give you a rectangle from 500 feet of fencing. (You don't have to use all of the fencing, so the perimeter of the pen could be less than 500 feet.)

31. Gas Mileage You have two cars. The first car travels an average of 12 miles on a gallon of gasoline, and the second averages 22 miles per gallon. Suppose you can afford to buy up to 30 gallons of gasoline this month. If the first car is driven x miles this month, and the second car is driven y miles this

month, shade the region in the first quadrant that gives all the possible values of x and y that will keep you from buying more than 30 gallons of gasoline this month.

32. Number Problem The sum of two positive numbers is at most 20. If the two numbers are represented by x and y, shade the region in the first quadrant that shows all the possibilities for the two numbers.

33. Student Loan Payments When considering how much debt to incur in student loans, it is advisable to keep your student loan payment after graduation to 8% or less of your starting monthly income. Let x represent your starting monthly salary and let y represent your monthly student loan payment, and write an inequality that describes this situation. Shade the region in the first quadrant that is a solution to your inequality.

34. Student Loan Payments During your college career you anticipate borrowing money two times: once as an undergraduate student and once as a graduate student. You anticipate a starting monthly salary after graduate school of approximately $3,500. You are advised to keep your total in student loan payments to 8% or less of your starting monthly salary. If x represents the monthly payment on the first loan and y represents the monthly payment on the second loan, shade the region in the first quadrant representing all combinations of possible monthly payments on the two loans that will total 8% or less of your starting monthly salary.

Review Problems

The problems that follow review material we covered in Section 2.4.

Solve each of the following inequalities.

35. $\frac{1}{3} + \frac{y}{5} \leq \frac{26}{15}$ **36.** $-\frac{1}{3} \geq \frac{1}{6} - \frac{y}{2}$

37. $5t - 4 > 3t - 8$

38. $-3(t - 2) < 6 - 5(t + 1)$

39. $-9 < -4 + 5t < 6$

40. $-3 < 2t + 1 < 3$

Extending the Concepts

Graph each inequality.

41. $y < |x + 2|$ **42.** $y > |x - 2|$

43. $y > |x - 3|$ **44.** $y < |x + 3|$

45. The Associated Students organization holds a *Night at the Movies* fund-raiser. Students tickets are $1.00 each and nonstudent tickets are $2.00 each. The theater holds a maximum of 200 people. The club needs to collect at least $100 to make money. Shade the region in the first quadrant that contains all combinations of students and non-students that could attend the movie night so that the club makes money.

3.5 Introduction to Functions

The ad shown in the margin appeared in the help wanted section of the local newspaper the day I was writing this section of the book. We can use the information in the ad to start an informal discussion of our next topic: functions.

AN INFORMAL LOOK AT FUNCTIONS

To begin with, suppose you have a job that pays $7.50 per hour and that you work anywhere from 0 to 40 hours per week. The amount of money you make in one week depends on the number of hours you work that week. In mathematics we say that your weekly earnings are a *function* of the number of hours you work. If we let the variable x represent hours and the variable y represent the money you make, then the relationship between x and y can be written as

$$y = 7.5x \qquad \text{for} \quad 0 \leq x \leq 40$$

EXAMPLE 1 Construct a table and graph for the function

$$y = 7.5x \qquad \text{for} \quad 0 \leq x \leq 40$$

Solution Table 1 gives some of the paired data that satisfy the equation $y = 7.5x$. Figure 1 is the graph of the equation with the restriction $0 \leq x \leq 40$.

TABLE 1 Weekly Wages

Hours Worked	Rule	Pay
x	$y = 7.5x$	y
0	$y = 7.5(0)$	0
10	$y = 7.5(10)$	75
20	$y = 7.5(20)$	150
30	$y = 7.5(30)$	225
40	$y = 7.5(40)$	300

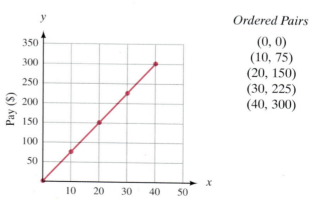

Ordered Pairs

(0, 0)
(10, 75)
(20, 150)
(30, 225)
(40, 300)

FIGURE 1 Weekly wages at $7.50 per hour

The equation $y = 7.5x$ with the restriction $0 \le x \le 40$, Table 1, and Figure 1 are three ways to describe the same relationship between the number of hours you work in one week and your gross pay for that week. In all three, we *input* values of x, and then use the function rule to *output* values of y.

DOMAIN AND RANGE OF A FUNCTION

We began this discussion by saying that the number of hours worked during the week was from 0 to 40, so these are the values that x can assume. From the line graph in Figure 1, we see that the values of y range from 0 to 300. We call the complete set of values that x can assume the *domain* of the function. The values that are assigned to y are called the *range* of the function.

EXAMPLE 2   State the domain and range for the function

$$y = 7.5x, \quad 0 \le x \le 40$$

Solution   From the previous discussion we have

$$\text{Domain} = \{x \mid 0 \le x \le 40\}$$

$$\text{Range} = \{y \mid 0 \le y \le 300\}$$

FUNCTION MAPS

Another way to visualize the relationship between x and y is with the diagram in Figure 2, which we call a *function map*:

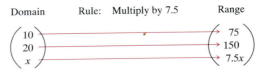

FIGURE 2 A function map

Although the diagram in Figure 2 does not show all the values that x and y can assume, it does give us a visual description of how x and y are related. It shows that values of y in the range come from values of x in the domain according to a specific rule (multiply by 7.5 each time).

A FORMAL LOOK AT FUNCTIONS

What is apparent from the preceding discussion is that we are working with paired data. The solutions to the equation $y = 7.5x$ are pairs of numbers; the points on the line graph in Figure 1 come from paired data; and the diagram in Figure 2 pairs numbers in the domain with numbers in the range. We are now ready for the formal definition of a function.

> **DEFINITION**   A **function** is a rule that pairs each element in one set, called the **domain,** with exactly one element from a second set, called the **range.**

In other words, a function is a rule for which each input is paired with exactly one output.

FUNCTIONS AS ORDERED PAIRS

The function rule $y = 7.5x$ from Example 1 produces ordered pairs of numbers (x, y). The same thing happens with all functions: The function rule produces ordered pairs of numbers. We use this result to write an alternative definition for a function.

> **ALTERNATIVE DEFINITION**   A **function** is a set of ordered pairs in which no two different ordered pairs have the same first coordinate. The set of all first coordinates is called the **domain** of the function. The set of all second co-ordinates is called the **range** of the function.

The restriction on first coordinates in the alternative definition keeps us from assigning a number in the domain to more than one number in the range.

A RELATIONSHIP THAT IS NOT A FUNCTION

You may be wondering if any sets of paired data fail to qualify as functions. The answer is yes, as the next example reveals.

EXAMPLE 3   Table 2 shows the prices of used Ford Mustangs that were listed in the local newspaper. The diagram in Figure 3 is called a *scatter diagram.* It gives a visual representation of the data in Table 2. Why is this data not a function?

TABLE 2 Used Mustang Prices	
Year	**Price ($)**
x	*y*
1997	13,925
1997	11,850
1997	9,995
1996	10,200
1996	9,600
1995	9,525
1994	8,675
1994	7,900
1993	6,975

FIGURE 3 Scatter diagram of data in Table 2

Ordered Pairs

(1997, 13,925)
(1997, 11,850)
(1997, 9,995)
(1996, 10,200)
(1996, 9,600)
(1995, 9,525)
(1994, 8,675)
(1994, 7,900)
(1993, 6,975)

Solution   In Table 2, the year 1997 is paired with three different prices: $13,925, $11,850, and $9,995. That is enough to disqualify the data from belonging to a function. For a set of paired data to be considered a function, each number in the domain must be paired with exactly one number in the range.

Still, there is a relationship between the first coordinates and second coordinates in the used-car data. It is not a function relationship, but it is a relationship. In order to classify all relationships specified by ordered pairs, whether they are functions or not, we include the following two definitions.

> **DEFINITION**   A **relation** is a rule that pairs each element in one set, called the **domain,** with **one or more elements** from a second set, called the **range.**

> **ALTERNATIVE DEFINITION**   A **relation** is a set of ordered pairs. The set of all first coordinates is the **domain** of the relation. The set of all second coordinates is the **range** of the relation.

Here are some facts that will help clarify the distinction between relations and functions.

1. Any rule that assigns numbers from one set to numbers in another set is a relation. If that rule makes the assignment so that no input has more than one output, then it is also a function.

2. Any set of ordered pairs is a relation. If none of the first coordinates of those ordered pairs is repeated, the set of ordered pairs is also a function.

3. Every function is a relation.

4. Not every relation is a function.

GRAPHING RELATIONS AND FUNCTIONS

To give ourselves a wider perspective on functions and relations, we consider some equations whose graphs are not straight lines.

EXAMPLE   4   Kendra is tossing a softball into the air with an underhand motion. The distance of the ball above her hand at any time is given by the function

$$h = 32t - 16t^2 \qquad \text{for} \quad 0 \leq t \leq 2$$

where h is the height of the ball in feet, and t is the time in seconds. Construct a table that gives the height of the ball at quarter-second intervals, starting with $t = 0$ and ending with $t = 2$. Construct a line graph from the table.

Solution   We construct Table 3 using the following values of t: $0, \frac{1}{4}, \frac{1}{2}, \frac{3}{4}, 1, \frac{5}{4}, \frac{3}{2}, \frac{7}{4},$ 2. The values of h come from substituting these values of t into the equation

$h = 32t - 16t^2$. (This equation comes from physics. If you take a physics class, you will learn how to derive this equation.) Then we construct the graph in Figure 4 from the table. The graph appears only in the first quadrant because neither t nor h can be negative.

Here is a summary of what we know about functions as it applies to this example: We input values of t and output values of h according to the function rule

$$h = 32t - 16t^2 \quad \text{for} \quad 0 \le t \le 2$$

TABLE 3	Tossing a Softball into the Air	
Time (sec)	Function Rule	Distance (ft)
t	$h = 32t - 16t^2$	h
0	$h = 32(0) - 16(0)^2 = 0 - 0 = 0$	0
$\frac{1}{4}$	$h = 32\left(\frac{1}{4}\right) - 16\left(\frac{1}{4}\right)^2 = 8 - 1 = 7$	7
$\frac{1}{2}$	$h = 32\left(\frac{1}{2}\right) - 16\left(\frac{1}{2}\right)^2 = 16 - 4 = 12$	12
$\frac{3}{4}$	$h = 32\left(\frac{3}{4}\right) - 16\left(\frac{3}{4}\right)^2 = 24 - 9 = 15$	15
1	$h = 32(1) - 16(1)^2 = 32 - 16 = 16$	16
$\frac{5}{4}$	$h = 32\left(\frac{5}{4}\right) - 16\left(\frac{5}{4}\right)^2 = 40 - 25 = 15$	15
$\frac{3}{2}$	$h = 32\left(\frac{3}{2}\right) - 16\left(\frac{3}{2}\right)^2 = 48 - 36 = 12$	12
$\frac{7}{4}$	$h = 32\left(\frac{7}{4}\right) - 16\left(\frac{7}{4}\right)^2 = 56 - 49 = 7$	7
2	$h = 32(2) - 16(2)^2 = 64 - 64 = 0$	0

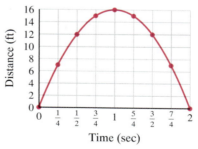

FIGURE 4

The domain is given by the inequality that follows the equation; it is

$$\text{Domain} = \{t \mid 0 \le t \le 2\}$$

The range is the set of all outputs that are possible by substituting the values of t from the domain into the equation. From our table and graph, it seems that the range is

$$\text{Range} = \{h \mid 0 \le h \le 16\}$$

Using TECHNOLOGY

MORE ABOUT EXAMPLE 4

Most graphing calculators can easily produce the information in Table 3. Simply set Y_1 equal to $32X - 16X^2$. Then set up the table so it starts at 0 and increases by an increment of 0.25 each time. (On a TI-82/83, use the TBLSET key to set up the table.)

Table Setup

Table minimum $= 0$
Table increment $= .25$
Dependent variable: Auto
Independent variable: Auto

Y Variables Setup

$Y_1 = 32X - 16X^2$

The table will look like this:

X	Y_1
0	0
.25	7
.5	12
.75	15
1	16
1.25	15
1.5	12

EXAMPLE 5   Sketch the graph of $x = y^2$.

Solution   Without going into much detail, we graph the equation $x = y^2$ by finding a number of ordered pairs that satisfy the equation, plotting these points, then drawing a smooth curve that connects them. A table of values for x and y that satisfy the equation follows, along with the graph of $x = y^2$ shown in Figure 5.

x	y
0	0
1	1
1	-1
4	2
4	-2
9	3
9	-3

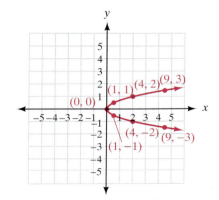

FIGURE 5

As you can see from looking at the table and the graph in Figure 5, several ordered pairs whose graphs lie on the curve have repeated first coordinates, for instance $(1, 1)$ and $(1, -1)$, $(4, 2)$ and $(4, -2)$, as well as $(9, 3)$ and $(9, -3)$. The graph is therefore not the graph of a function.

VERTICAL LINE TEST

Look back at the scatter diagram for used Mustang prices shown in Figure 3. Notice that some of the points on the diagram lie above and below each other along vertical lines. This is an indication that the data do not constitute a function. Two data points that lie on the same vertical line must have come from two ordered pairs with the same first coordinates.

Now, look at the graph shown in Figure 5. The reason this graph is the graph of a relation, but not of a function, is that some points on the graph have the same first coordinates, for example, the points $(4, 2)$ and $(4, -2)$. Furthermore, any time two points on a graph have the same first coordinates, those points must lie on a vertical line. [To convince yourself, connect the points $(4, 2)$ and $(4, -2)$ with a straight line. You will see that it must be a vertical line.] This allows us to write the following test that uses the graph to determine whether a relation is also a function.

VERTICAL LINE TEST

If a vertical line crosses the graph of a relation in more than one place, the relation cannot be a function. If no vertical line can be found that crosses a graph in more than one place, then the graph is the graph of a function.

If we look back to the graph of $h = 32t - 16t^2$ as shown in Figure 4, we see that no vertical line can be found that crosses this graph in more than one place. The graph shown in Figure 4 is therefore the graph of a function.

EXAMPLE 6   Graph $y = |x|$. Use the graph to determine whether we have the graph of a function. State the domain and range.

Solution   We let x take on values of $-4, -3, -2, -1, 0, 1, 2, 3,$ and 4. The corresponding values of y are shown in the table. The graph is shown in Figure 6.

x	y
−4	4
−3	3
−2	2
−1	1
0	0
1	1
2	2
3	3
4	4

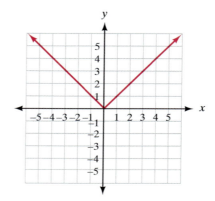

FIGURE 6

Since no vertical line can be found that crosses the graph in more than one place, $y = |x|$ is a function. The domain is all real numbers. The range is $\{y \,|\, y \geq 0\}$.

 ## Getting Ready for Class

After reading through the preceding section, respond in your own words and in complete sentences.

A. What is a function?

B. What is the vertical line test?

C. Is every line the graph of a function? Explain.

D. Which variable is usually associated with the domain of a function?

PROBLEM SET 3.5

1. Suppose you have a job that pays $8.50 per hour and you work anywhere from 10 to 40 hours per week.

(a) Write an equation, with a restriction on the variable x, that gives the amount of money, y, you will earn for working x hours in one week.

(b) Use the function rule you have written in part (a) to complete Table 4.

TABLE 4   Weekly Wages

Hours Worked	Function Rule	Gross Pay ($)
x		y
10		
20		
30		
40		

(c) Use the template in Figure 7 to construct a line graph from the information in Table 4.

FIGURE 7 Template for line graph

(d) State the domain and range of this function.
(e) What is the minimum amount you can earn in a week with this job? What is the maximum amount?

2. The ad shown here was in the local newspaper. Suppose you are hired for the job described in the ad.
   (a) If x is the number of hours you work per week and y is your weekly gross pay, write the equation for y. (Be sure to include any restrictions on the variable x that are given in the ad.)
   (b) Use the function rule you have written in part (a) to complete Table 5.

TABLE 5 **Weekly Wages**		
Hours Worked	**Function Rule**	**Gross Pay ($)**
x		y
15		
20		
25		
30		

(c) Use the template in Figure 8 to construct a line graph from the information in Table 5.

FIGURE 8 Template for line graph

(d) State the domain and range of this function.
(e) What is the minimum amount you can earn in a week with this job? What is the maximum amount?

For each of the following relations, give the domain and range, and indicate which are also functions.

3. $\{(1, 3), (2, 5), (4, 1)\}$ **4.** $\{(3, 1), (5, 7), (2, 3)\}$

5. $\{(-1, 3), (1, 3), (2, -5)\}$

6. $\{(3, -4), (-1, 5), (3, 2)\}$

7. $\{(7, -1), (3, -1), (7, 4)\}$

8. $\{(5, -2), (3, -2), (5, -1)\}$

State whether each of the following graphs represents a function.

9.

10.

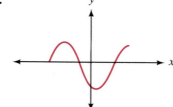

11.

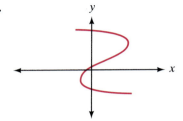

12.

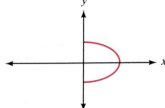

13.

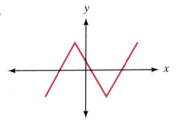

14.

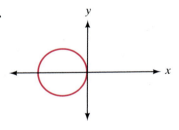

15.

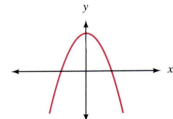

16.

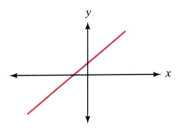

17.

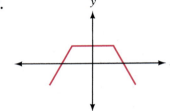

18.

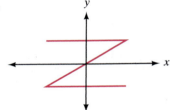

19. Tossing a Coin   Hali is tossing a quarter into the air with an underhand motion. The distance the quarter is above her hand at any time is given by the function

$$h = 16t - 16t^2 \qquad \text{for} \quad 0 \le t \le 1$$

where h is the height of the quarter in feet, and t is the time in seconds.
(a) Fill in the table.

Time (sec)	Function Rule	Distance (ft)
t	$h = 16t - 16t^2$	h
0		
0.1		
0.2		
0.3		
0.4		
0.5		
0.6		
0.7		
0.8		
0.9		
1		

(b) State the domain and range of this function.

(c) Use the data from the table to graph the function.

20. Intensity of Light The following formula gives the intensity of light that falls on a surface at various distances from a 100-watt light bulb:

$$I = \frac{120}{d^2} \qquad \text{for} \quad d > 0$$

where I is the intensity of light (in lumens per square foot), and d is the distance (in feet) from the light bulb to the surface.

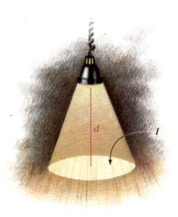

(a) Fill in the table.

Distance (ft)	Function Rule	Intensity
d		I
1		
2		
3		
4		
5		
6		

(b) Use the data from the table in part (a) to construct a graph of the function.

Determine the domain and range of the following functions. Assume the *entire* function is shown.

21.

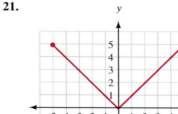
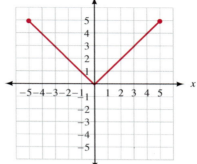

22.

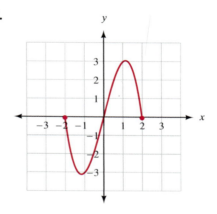

23.

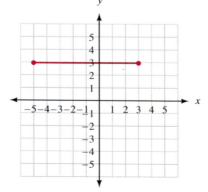

24.

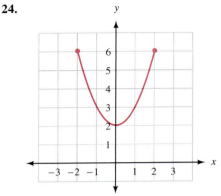

Graph each of the following relations. In each case, use the graph to find the domain and range, and indicate whether the graph is the graph of a function.

25. $y = x^2 - 1$ *if the exponent on y is greater than 1*

26. $y = x^2 + 1$

27. $y = x^2 + 4$

28. $y = x^2 - 9$

29. $x = y^2 - 1$

30. $x = y^2 + 1$

31. $x = y^2 + 4$ *than 1*

32. $x = y^2 - 9$

33. $y = |x - 2|$ *then if is not a function*

34. $y = |x| - 2$

Area of a Circle   The formula for the area A of a circle with radius r is given by $A = \pi r^2$. The formula shows that A is a function of r.

35. Graph the function $A = \pi r^2$ for $0 \le r \le 3$. (On the graph, let the horizontal axis be the r-axis, and let the vertical axis be the A-axis.)

36. State the domain and range of the function $A = \pi r^2, 0 \le r \le 3$. (Use $\pi \approx 3.14$.)

Area and Perimeter of a Rectangle   A rectangle is 2 inches longer than it is wide. Let $x =$ the width, $P =$ the perimeter, and $A =$ the area of the rectangle (Problems 37–40).

37. Write an equation that will give the perimeter P in terms of the width x of the rectangle. Are there any restrictions on the values that x can assume?

38. Graph the relationship between P and x.

39. Write an equation that will give the area A in terms of the width x of the rectangle. Are there any restrictions on the values that x can assume?

40. Graph the relationship between A and x.

41. Tossing a Ball   A ball is thrown straight up into the air from ground level. The relationship between the height h of the ball at any time t is illustrated by the following graph:

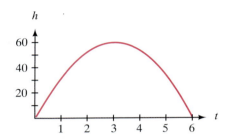

The horizontal axis represents time t, and the vertical axis represents height h.
(a) Is this graph the graph of a function?
(b) State the domain and range.
(c) At what time does the ball reach its maximum height?
(d) What is the maximum height of the ball?
(e) At what time does the ball hit the ground?

42. The following graph shows the relationship between a company's profits P and the number of items it sells x. (P is in dollars.)

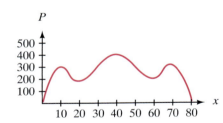

(a) Is this graph the graph of a function?
(b) State the domain and range.
(c) How many items must the company sell to make their maximum profit?
(d) What is their maximum profit?

Reading Graphs

43. Match each of the following statements to the appropriate graph indicated by Figures 9–12.
(a) Sarah works 25 hours in a week to earn $250.
(b) Justin works 35 hours in a week to earn $560.

(c) Rosemary works 30 hours in a week to earn $360.

(d) Marcus works 40 hours in a week to earn $320.

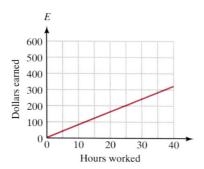

FIGURE 9

FIGURE 10

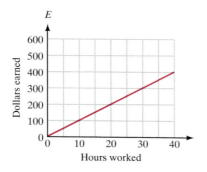

FIGURE 11

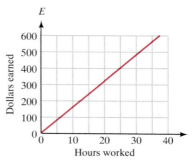

FIGURE 12

44. Find an equation for each of the functions shown in Figures 9–12. Show dollars earned, E, as a function of hours worked, t. Then, indicate the domain and range of each function.

(a) Figure 9: $E =$ ____ Domain $= \{t |$ $\}$;
Range $= \{E |$ $\}$

(b) Figure 10: $E =$ ____ Domain $= \{t |$ $\}$;
Range $= \{E |$ $\}$

(c) Figure 11: $E =$ ____ Domain $= \{t |$ $\}$;
Range $= \{E |$ $\}$

(d) Figure 12: $E =$ ____ Domain $= \{t |$ $\}$;
Range $= \{E |$ $\}$

Review Problems

The problems that follow review material we covered in Section 2.2. Reviewing these problems will help you with the next section.

For the equation $y = 3x - 2$:

45. Find y if x is 4.   **46.** Find y if x is 0.

47. Find y if x is -4.   **48.** Find y if x is -2.

For the equation $y = x^2 - 3$:

49. Find y if x is 2.   **50.** Find y if x is -2.

51. Find y if x is 0.   **52.** Find y if x is -4.

Let's return to the discussion that introduced us to functions. If a job pays $7.50 per hour for working from 0 to 40 hours a week, then the amount of money y earned in one week is a function of the number of hours worked x. The exact relationship between x and y is written

$$y = 7.5x \qquad \text{for} \quad 0 \le x \le 40$$

Since the amount of money earned y depends on the number of hours worked x, we call y the *dependent variable* and x the *independent variable*. Furthermore, if we let f represent all the ordered pairs produced by the equation, then we can write

$$f = \{(x, y) \mid y = 7.5x \text{ and } 0 \le x \le 40\}$$

Once we have named a function with a letter, we can use an alternative notation to represent the dependent variable y. The alternative notation for y is $f(x)$. It is read "f of x" and can be used instead of the variable y when working with functions. The notation y and the notation $f(x)$ are equivalent. That is,

$$y = 7.5x \Leftrightarrow f(x) = 7.5x$$

When we use the notation $f(x)$ we are using *function notation*. The benefit of using function notation is that we can write more information with fewer symbols than we can by using just the variable y. For example, asking how much money a person will make for working 20 hours is simply a matter of asking for $f(20)$. Without function notation, we would have to say "find the value of y that corresponds to a value of $x = 20$." To illustrate further, using the variable y, we can say "y is 150 when x is 20." Using the notation $f(x)$, we simply say "$f(20) = 150$." Each expression indicates that you will earn $150 for working 20 hours.

EXAMPLE I If $f(x) = 7.5x$, find $f(0)$, $f(10)$, and $f(20)$.

Solution To find $f(0)$ we substitute 0 for x in the expression $7.5x$ and simplify. We find $f(10)$ and $f(20)$ in a similar manner — by substitution.

$$\text{If} \qquad f(x) = 7.5x$$

$$\text{then} \qquad f(\mathbf{0}) = 7.5(\mathbf{0}) = 0$$

$$f(\mathbf{10}) = 7.5(\mathbf{10}) = 75$$

$$f(\mathbf{20}) = 7.5(\mathbf{20}) = 150$$

If we changed the example in the discussion that opened this section so that the hourly wage was $6.50 per hour, we would have a new equation to work with, namely,

$$y = 6.5x \qquad \text{for} \quad 0 \le x \le 40$$

Suppose we name this new function with the letter g. Then

$$g = \{(x, y) \mid y = 6.5x \text{ and } 0 \le x \le 40\}$$

Input x

Function machine

Output $f(x)$

Some students like to think of functions as machines. Values of x are put into the machine, which transforms them into values of $f(x)$, which are then output by the machine.

193

and

$$g(x) = 6.5x$$

If we want to talk about both functions in the same discussion, having two different letters, f and g, makes it easy to distinguish between them. For example, since $f(x) = 7.5x$ and $g(x) = 6.5x$, asking how much money a person makes for working 20 hours is simply a matter of asking for $f(20)$ or $g(20)$, avoiding any confusion over which hourly wage we are talking about.

The diagrams shown in Figure 1 further illustrate the similarities and differences between the two functions we have been discussing.

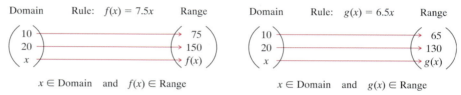

$x \in \text{Domain}$ and $f(x) \in \text{Range}$ $x \in \text{Domain}$ and $g(x) \in \text{Range}$

FIGURE 1   Function maps

FUNCTION NOTATION AND GRAPHS

We can visualize the relationship between x and $f(x)$ or $g(x)$ on the graphs of the two functions. Figure 2 shows the graph of $f(x) = 7.5x$ along with two additional line segments. The horizontal line segment corresponds to $x = 20$, and the vertical line segment corresponds to $f(20)$. Figure 3 shows the graph of $g(x) = 6.5x$ along with the horizontal line segment that corresponds to $x = 20$, and the vertical line segment that corresponds to $g(20)$. (Note that the domain in each case is restricted to $0 \leq x \leq 40$.)

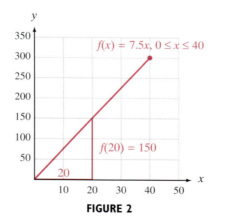

FIGURE 2

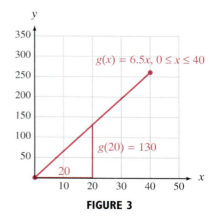

FIGURE 3

USING FUNCTION NOTATION

The remaining examples in this section show a variety of ways to use and interpret function notation.

EXAMPLE 2   If it takes Lorena t minutes to run a mile, then her average speed $s(t)$ in miles per hour, is given by the formula

$$s(t) = \frac{60}{t} \qquad \text{for} \quad t > 0$$

Find $s(10)$ and $s(8)$, and then explain what they mean.

Solution   To find $s(10)$, we substitute 10 for t in the equation and simplify:

$$s(\mathbf{10}) = \frac{60}{\mathbf{10}} = 6$$

In words: When Lorena runs a mile in 10 minutes, her average speed is 6 miles per hour.

We calculate $s(8)$ by substituting 8 for t in the equation. Doing so gives us

$$s(\mathbf{8}) = \frac{60}{\mathbf{8}} = 7.5$$

In words: Running a mile in 8 minutes is running at a rate of 7.5 miles per hour.

EXAMPLE 3   A painting is purchased as an investment for \$125. If its value increases continuously so that it doubles every 5 years, then its value is given by the function

$$V(t) = 125 \cdot 2^{t/5} \qquad \text{for} \quad t \geq 0$$

where t is the number of years since the painting was purchased, and $V(t)$ is its value (in dollars) at time t. Find $V(5)$ and $V(10)$, and explain what they mean.

Solution   The expression $V(5)$ is the value of the painting when $t = 5$ (5 years after it is purchased). We calculate $V(5)$ by substituting 5 for t in the equation $V(t) = 125 \cdot 2^{t/5}$. Here is our work:

$$V(\mathbf{5}) = 125 \cdot 2^{5/5} = 125 \cdot 2^1 = 125 \cdot 2 = 250$$

In words: After 5 years, the painting is worth \$250.

The expression $V(10)$ is the value of the painting after 10 years. To find this number, we substitute 10 for t in the equation:

$$V(\mathbf{10}) = 125 \cdot 2^{\mathbf{10}/5} = 125 \cdot 2^2 = 125 \cdot 4 = 500$$

In words: The value of the painting 10 years after it is purchased is \$500.

EXAMPLE 4   A balloon has the shape of a sphere with a radius of 3 inches. Use the following formulas to find the volume and surface area of the balloon.

$$V(r) = \frac{4}{3}\pi r^3 \qquad S(r) = 4\pi r^2$$

Solution   As you can see, we have used function notation to write the two formulas for volume and surface area, because each quantity is a function of the radius.

To find these quantities when the radius is 3 inches, we evaluate $V(3)$ and $S(3)$:

$$V(3) = \frac{4}{3}\pi 3^3 = \frac{4}{3}\pi 27 = 36\pi \text{ cubic inches, or 113 cubic inches}$$ To the nearest whole number

$$S(3) = 4\pi 3^2 = 36\pi \text{ square inches, or 113 square inches}$$ To the nearest whole number

The fact that $V(3) = 36\pi$ means that the ordered pair $(3, 36\pi)$ belongs to the function V. Likewise, the fact that $S(3) = 36\pi$ tells us that the ordered pair $(3, 36\pi)$ is a member of function S.

We can generalize the discussion at the end of Example 4 this way:

$$(a, b) \in f \qquad \text{if and only if} \qquad f(a) = b$$

Using TECHNOLOGY

MORE ABOUT EXAMPLE 4

If we look back at Example 4, we see that when the radius of a sphere is 3, the numerical values of the volume and surface area are equal. How unusual is this? Are there other values of r for which $V(r)$ and $S(r)$ are equal? We can answer this question by looking at the graphs of both V and S.

To graph the function $V(r) = \frac{4}{3}\pi r^3$, set $Y_1 = 4\pi X^3/3$. To graph $S(r) = 4\pi r^2$, set $Y_2 = 4\pi X^2$. Graph the two functions in each of the following windows:

Window 1:   X from -4 to 4, Y from -2 to 10

Window 2:   X from 0 to 4, Y from 0 to 50

Window 3:   X from 0 to 4, Y from 0 to 150

Then use the Trace and Zoom features of your calculator to locate the point in the first quadrant where the two graphs intersect. How do the coordinates of this point compare with the results in Example 4?

EXAMPLE 5   If $f(x) = 3x^2 + 2x - 1$, find $f(0)$, $f(3)$, and $f(-2)$.

Solution   Since $f(x) = 3x^2 + 2x - 1$, we have

$$f(0) = 3(0)^2 + 2(0) - 1 \qquad = 0 + 0 - 1 = -1$$

$$f(3) = 3(3)^2 + 2(3) - 1 \qquad = 27 + 6 - 1 = 32$$

$$f(-2) = 3(-2)^2 + 2(-2) - 1 = 12 - 4 - 1 = 7$$

In Example 5, the function f is defined by the equation $f(x) = 3x^2 + 2x - 1$. We could just as easily have said $y = 3x^2 + 2x - 1$. That is, $y = f(x)$. Saying $f(-2) = 7$ is exactly the same as saying y is 7 when x is -2.

EXAMPLE 6   If $f(x) = 4x - 1$ and $g(x) = x^2 + 2$, then

$$f(5) = 4(5) - 1 = 19 \qquad \text{and} \qquad g(5) = 5^2 + 2 = 27$$

$$f(-2) = 4(-2) - 1 = -9 \qquad \text{and} \qquad g(-2) = (-2)^2 + 2 = 6$$

$$f(0) = 4(0) - 1 = -1 \qquad \text{and} \qquad g(0) = 0^2 + 2 = 2$$

$$f(z) = 4z - 1 \qquad \text{and} \qquad g(z) = z^2 + 2$$

$$f(a) = 4a - 1 \qquad \text{and} \qquad g(a) = a^2 + 2$$

 Using TECHNOLOGY

MORE ABOUT EXAMPLE 6

Most graphing calculators can use tables to evaluate functions. To work Example 6 using a graphing calculator table, set Y_1 equal to $4X - 1$ and Y_2 equal to $X^2 + 2$. Then set the independent variable in the table to Ask instead of Auto. Go to your table and input 5, -2, and 0. Under Y_1 in the table, you will find $f(5)$, $f(-2)$, and $f(0)$. Under Y_2, you will find $g(5)$, $g(-2)$, and $g(0)$.

Table Setup   *Y Variables Setup*

Table minimum = 0 $Y_1 = 4X - 1$
Table increment = 1 $Y_2 = X^2 + 2$
Independent variable: Ask
Dependent variable: Ask

The table will look like this:

X	Y_1	Y_2
5	19	27
-2	-9	6
0	-1	2

Although the calculator asks us for a table increment, the increment doesn't matter since we are inputting the X values ourselves.

EXAMPLE 7   If the function f is given by

$$f = \{(-2, 0), (3, -1), (2, 4), (7, 5)\}$$

then $f(-2) = 0$, $f(3) = -1$, $f(2) = 4$, and $f(7) = 5$.

EXAMPLE 8   If $f(x) = 2x^2$ and $g(x) = 3x - 1$, find
(a) $f[g(2)]$ (b) $g[f(2)]$

Solution   The expression $f[g(2)]$ is read "f of g of 2."
(a) Since $g(2) = 3(2) - 1 = 5$,

$$f[g(2)] = f(5) = 2(5)^2 = 50$$

(b) Since $f(2) = 2(2)^2 = 8$,

$$g[f(2)] = g(8) = 3(8) - 1 = 23$$

Getting Ready for Class

After reading through the preceding section, respond in your own words and in complete sentences.

A. Explain what you are calculating when you find $f(2)$ for a given function f.

B. If $s(t) = \dfrac{60}{t}$, how do you find $s(10)$?

C. If $f(2) = 3$ for a function f, what is the relationship between the numbers 2 and 3 and the graph of f?

D. If $f(6) = 0$ for a particular function f, then you can immediately graph one of the intercepts. Explain.

PROBLEM SET 3.6

Let $f(x) = 2x - 5$ and $g(x) = x^2 + 3x + 4$. Evaluate the following.

1. $f(2)$

2. $f(3)$

3. $f(-3)$

4. $g(-2)$

5. $g(-1)$

6. $f(-4)$

7. $g(-3)$

8. $g(2)$

9. $g(4) + f(4)$

10. $f(2) - g(3)$

11. $f(3) - g(2)$

12. $g(-1) + f(-1)$

Let $f(x) = 3x^2 - 4x + 1$ and $g(x) = 2x - 1$. Evaluate the following.

13. $f(0)$

14. $g(0)$

15. $g(-4)$

16. $f(1)$

17. $f(-1)$

18. $g(-1)$

19. $g(10)$

20. $f(10)$

21. $f(3)$

22. $g(3)$

23. $g\left(\frac{1}{2}\right)$

24. $g\left(\frac{1}{4}\right)$

25. $f(a)$ **26.** $g(b)$

If $f = \{(1, 4), (-2, 0), (3, \frac{1}{2}), (\pi, 0)\}$ and $g = \{(1, 1), (-2, 2), (\frac{1}{2}, 0)\}$, find each of the following values of f and g.

27. $f(1)$ **28.** $g(1)$ **29.** $g(\frac{1}{2})$

30. $f(3)$ **31.** $g(-2)$ **32.** $f(\pi)$

Let $f(x) = 2x^2 - 8$ and $g(x) = \frac{1}{2}x + 1$. Evaluate each of the following.

33. $f(0)$ **34.** $g(0)$ **35.** $g(-4)$

36. $f(1)$ **37.** $f(a)$ **38.** $g(z)$

39. $f(b)$ **40.** $g(t)$ **41.** $f[g(2)]$

42. $g[f(2)]$ **43.** $g[f(-1)]$ **44.** $f[g(-2)]$

45. $g[f(0)]$ **46.** $f[g(0)]$

47. Graph the function $f(x) = \frac{1}{2}x + 2$. Then draw and label the line segments that represent $x = 4$ and $f(4)$.

48. Graph the function $f(x) = -\frac{1}{2}x + 6$. Then draw and label the line segments that represent $x = 4$ and $f(4)$.

49. For the function $f(x) = \frac{1}{2}x + 2$, find the value of x for which $f(x) = x$.

50. For the function $f(x) = -\frac{1}{2}x + 6$, find the value of x for which $f(x) = x$.

51. Graph the function $f(x) = x^2$. Then draw and label the line segments that represent $x = 1$ and $f(1)$, $x = 2$ and $f(2)$, and, finally, $x = 3$ and $f(3)$.

52. Graph the function $f(x) = x^2 - 2$. Then draw and label the line segments that represent $x = 2$ and $f(2)$ and the line segments corresponding to $x = 3$ and $f(3)$.

Applying the Concepts

53. Investing in Art   A painting is purchased as an investment for $150. If its value increases continuously so that it doubles every 3 years, then its value is given by the function

$$V(t) = 150 \cdot 2^{t/3} \qquad \text{for} \quad t \geq 0$$

where t is the number of years since the painting was purchased, and $V(t)$ is its value (in dollars) at time t. Find $V(3)$ and $V(6)$, and then explain what they mean.

54. Average Speed   If it takes Minke t minutes to run a mile, then her average speed $s(t)$, in miles per hour, is given by the formula

$$s(t) = \frac{60}{t} \qquad \text{for} \quad t > 0$$

Find $s(4)$ and $s(5)$, and then explain what they mean.

55. Dimensions of a Rectangle   The length of a rectangle is 3 inches more than twice the width. Let x represent the width of the rectangle and $P(x)$ represent the perimeter of the rectangle. Use function notation to write the relationship between x and $P(x)$, noting any restrictions on the variable x.

56. Dimensions of a Rectangle   The length of a rectangle is 3 inches more than twice the width. Let x represent the width of the rectangle and $A(x)$ represent the area of the rectangle. Use function notation to write the relationship between x and $A(x)$, noting any restrictions on the variable x.

Area of a Circle   The formula for the area A of a circle with radius r can be written with function notation as $A(r) = \pi r^2$.

57. Find $A(2)$, $A(5)$, and $A(10)$. (Use $\pi \approx 3.14$.)

58. Why doesn't it make sense to ask for $A(-10)$?

59. Cost of a Phone Call   Suppose a phone company charges 33¢ for the first minute and 24¢ for each additional minute to place a long-distance call between 5 P.M. and 11 P.M. If x is the number of additional minutes and $f(x)$ is the cost of the call, then $f(x) = 24x + 33$.
   (a) How much does it cost to talk for 10 minutes?
   (b) What does $f(5)$ represent in this problem?
   (c) If a call costs $1.29, how long was it?

60. Cost of a Phone Call   The same phone company mentioned in Problem 59 charges 52¢ for the first minute and 36¢ for each additional minute to place a long-distance call between 8 A.M. and 5 P.M.
   (a) Let $g(x)$ be the total cost of a long-distance call between 8 A.M. and 5 P.M., and write an equation for $g(x)$.
   (b) Find $g(5)$.
   (c) Find the difference in price between a 10-minute call made between 8 A.M. and 5 P.M. and the same call made between 5 P.M. and 11 P.M.

Straight-Line Depreciation   Straight-line depreciation is an accounting method used to help spread the cost of new equipment over a number of years. It takes into account both the cost when new and the salvage value, which is the value of the equipment at the time it gets replaced.

61. Value of a Copy Machine The function $V(t) = -3,300t + 18,000$, where V is value and t is time in years, can be used to find the value of a large copy machine during the first 5 years of use.
   (a) What is the value of the copier after 3 years and 9 months?
   (b) What is the salvage value of this copier?
   (c) State the domain of this function.
   (d) Sketch the graph of this function.

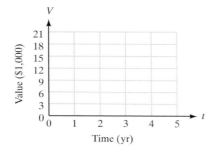

(FPG International/Telegraph Colour Library)

   (e) What is the range of this function?
   (f) After how many years will the copier be worth only $10,000?

62. Value of a Forklift The function $V(t) = -16,500t + 125,000$, where V is value and t is time in years, can be used to find the value of an electric forklift during the first 6 years of use.
   (a) What is the value of the forklift after 2 years and 3 months?

   (b) What is the salvage value of this forklift?
   (c) State the domain of this function.
   (d) Sketch the graph of this function.

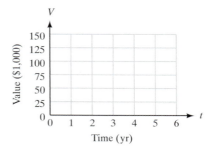

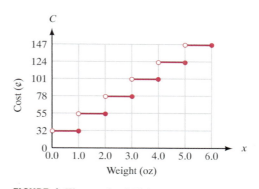

(FPG International/Telegraph Colour Library)

   (e) What is the range of this function?
   (f) After how many years will the forklift be worth only $45,000?

63. Step Function Figure 4 shows the graph of the step function C that was used to calculate the first-class postage on a letter weighing x ounces in 1997. Use this graph to answer questions (a) through (d).

FIGURE 4 The graph of $C(x)$

(a) Fill in the following table:

Weight (ounces)	0.6	1.0	1.1	2.5	3.0	4.8	5.0	5.3
Cost (cents)								

(b) If a letter cost 78 cents to mail, how much does it weigh? State your answer in words. State your answer as an inequality.

(c) If the entire function is shown in Figure 4, state the domain.

(d) State the range of the function shown in Figure 4.

64. Step Function   A taxi ride in Boston at the time I am writing this problem is $1.50 for the first $\frac{1}{4}$ mile, and then $0.25 for each additional $\frac{1}{8}$ of a mile. The following graph shows how much you will pay for a taxi ride of 1 mile or less.

(a) What is the most you will pay for this taxi ride?

(b) How much does it cost to ride the taxi for $\frac{8}{10}$ of a mile?

(c) Find points A and B on the horizontal axis.

(d) If a taxi ride costs $2.50, what distance was the ride?

(e) If the complete function is shown in Figure 5, find the domain and range of the function.

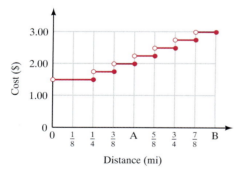

FIGURE 5

Review Problems

The problems that follow review material we covered in Section 2.5.

Solve each equation.

65. $|3x - 5| = 7$

66. $|0.04 - 0.03x| = 0.02$

67. $|4y + 2| - 8 = -2$

68. $4 = |3 - 2y| - 5$

69. $5 + |6t + 2| = 3$

70. $7 + |3 - \frac{3}{4}t| = 10$

Extending the Concepts

71. The graphs of two functions are shown in Figures 6 and 7. Use the graphs to find the following.

(a) $f(2)$

(b) $f(-4)$

(c) $g(0)$

(d) $g(3)$

(e) $g(2) - f(2)$

(f) $f(1) + g(1)$

(g) $f[g(3)]$

(h) $g[f(3)]$

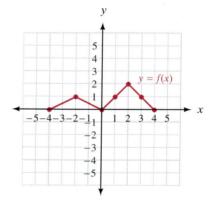

FIGURE 6

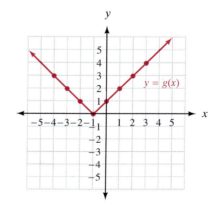

FIGURE 7

If you are a runner and you average t minutes for every mile you run during one of your workouts, then your speed s in miles per hour is given by the equation and graph shown here [the graph (Figure 1) is shown in the first quadrant only because both t and s are positive]:

$$s = \frac{60}{t}$$

Input t	Output s
4	15
6	10
8	7.5
10	6
12	5

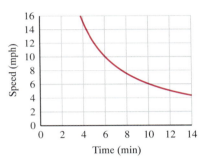

FIGURE 1

You know intuitively that as your average time per mile (t) increases, your speed (s) decreases. Likewise, lowering your time per mile will increase your speed. The equation and Figure 1 also show this to be true: Increasing t decreases s, and decreasing t increases s. Quantities that are connected in this way are said to *vary inversely* with each other. Inverse variation is one of the topics we will study in this section.

There are two main types of variation: *direct variation* and *inverse variation.* Variation problems are most common in the sciences, particularly in chemistry and physics.

DIRECT VARIATION

When we say the variable y *varies directly* with the variable x, we mean that the relationship can be written in symbols as $y = Kx$, where K is a nonzero constant called the *constant of variation* (or *proportionality constant*).

Another way of saying y varies directly with x is to say y is *directly proportional* to x.

Study the following list. It gives the mathematical equivalent of some direct variation statements.

Verbal Phrase	Algebraic Equation
y varies directly with x.	$y = Kx$
s varies directly with the square of t.	$s = Kt^2$
y is directly proportional to the cube of z.	$y = Kz^3$
u is directly proportional to the square root of v.	$u = K\sqrt{v}$

EXAMPLE   1   *y* varies directly with *x*. If *y* is 15 when *x* is 5, find *y* when *x* is 7.

Solution   The first sentence gives us the general relationship between *x* and *y*. The equation equivalent to the statement "*y* varies directly with *x*" is

$$y = Kx$$

The first part of the second sentence in our example gives us the information necessary to evaluate the constant *K*:

$$\text{When} \qquad y = 15$$
$$\text{and} \qquad x = 5$$
$$\text{the equation} \qquad y = Kx$$
$$\text{becomes} \qquad 15 = K \cdot 5$$
$$\text{or} \qquad K = 3$$

The equation can now be written specifically as

$$y = 3x$$

Letting *x* = 7, we have

$$y = 3 \cdot 7$$
$$y = 21$$

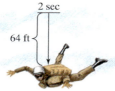

EXAMPLE   2   A skydiver jumps from a plane. Like any object that falls toward Earth, the distance the skydiver falls is directly proportional to the square of the time he has been falling, until he reaches his terminal velocity. If the skydiver falls 64 feet in the first 2 seconds of the jump, then

(a) How far will he have fallen after 3.5 seconds?   112 ft

(b) Graph the relationship between distance and time.

(c) How long will it take him to fall 256 feet?

Solution   We let *t* represent the time the skydiver has been falling, then we can let *d(t)* represent the distance he has fallen.

(a) Since *d(t)* is directly proportional to the square of *t*, we have the general function that describes this situation:

$$d(t) = Kt^2$$

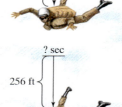

Next, we use the fact that *d*(2) = 64 to find *K*.

$$K = 16 \qquad 64 = K(2^2)$$
$$K = 16$$

The specific equation that describes this situation is

$$d(t) = 16t^2$$

3.5
×3.5
175
+105.0
122.5

19600

To find how far a skydiver will fall after 3.5 seconds, we find $d(3.5)$,

$$d(3.5) = 16(3.5^2)$$

$$d(3.5) = 196$$

A skydiver will fall 196 feet after 3.5 seconds.

(b) To graph this equation, we use a table:

Input t	Output $d(t)$
0	0
1	16
2	64
3	144
4	256
5	400

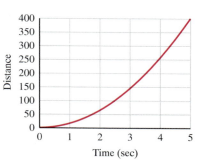

FIGURE 2

(c) From the table or the graph (Figure 2), we see that it will take 4 seconds for the skydiver to fall 256 feet.

INVERSE VARIATION

Running   From the introduction to this section, we know that the relationship between the number of minutes t it takes a person to run a mile and his or her average speed in miles per hour s can be described with the following equation and table, and with Figure 3.

$$s = \frac{60}{t}$$

Input t	Output s
4	15
6	10
8	7.5
10	6
12	5

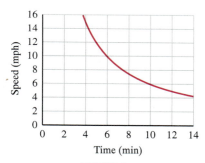

FIGURE 3

If t decreases, then s will increase, and if t increases, then s will decrease. The variable s is *inversely proportional* to the variable t. In this case, the *constant of proportionality* is 60.

Photography   If you are familiar with the terminology and mechanics associated with photography, you know that the *f*-stop for a particular lens will increase as the aperture (the maximum diameter of the opening of the lens) decreases. In mathematics we say that *f*-stop and aperture vary inversely with each other. The diagram illustrates this relationship.

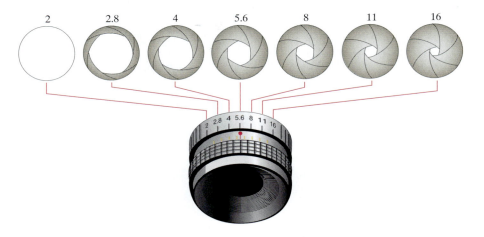

If *f* is the *f*-stop and *d* is the aperture, then their relationship can be written

$$f = \frac{K}{d}$$

In this case, K is the constant of proportionality. (Those of you familiar with photography know that K is also the focal length of the camera lens.)

In General   We generalize this discussion of inverse variation as follows: If *y* varies inversely with *x*, then

$$y = K\frac{1}{x} \qquad \text{or} \qquad y = \frac{K}{x}$$

We can also say *y* is inversely proportional to *x*. The constant K is again called the constant of variation or proportionality constant.

Verbal Phrase	Algebraic Equation
y is inversely proportional to *x*.	$y = \dfrac{K}{x}$
s varies inversely with the square of *t*.	$s = \dfrac{K}{t^2}$
y is inversely proportional to x^4.	$y = \dfrac{K}{x^4}$
z varies inversely with the cube root of *t*.	$z = \dfrac{K}{\sqrt[3]{t}}$

EXAMPLE 3   The volume of a gas is inversely proportional to the pressure of the gas on its container. If a pressure of 48 pounds per square inch corresponds to a volume of 50 cubic feet, what pressure is needed to produce a volume of 100 cubic feet?

Solution   We can represent volume with V and pressure with P:

$$V = \frac{K}{P}$$

Using $P = 48$ and $V = 50$, we have

$$50 = \frac{K}{48}$$

$$K = 50(48)$$

$$K = 2{,}400$$

Here is graph of this relationship.

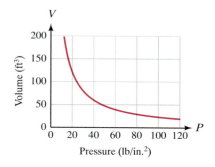

The equation that describes the relationship between P and V is

$$V = \frac{2{,}400}{P}$$

Substituting $V = 100$ into this last equation, we get

$$100 = \frac{2{,}400}{P}$$

$$100P = 2{,}400$$

$$P = \frac{2{,}400}{100}$$

$$P = 24$$

A volume of 100 cubic feet is produced by a pressure of 24 pounds per square inch.

Note: The relationship between pressure and volume as given in this example is known as Boyle's law and applies to situations such as those encountered in a piston-cylinder arrangement. It was Robert Boyle (1627–1691) who, in 1662, published the results of some of his experiments that showed, among other things, that the volume of a gas decreases as the pressure increases. This is an example of inverse variation.

JOINT VARIATION AND OTHER VARIATION COMBINATIONS

Many times relationships among different quantities are described in terms of more than two variables. If the variable y varies directly with *two* other variables, say x and z, then we say y varies *jointly* with x and z. In addition to joint variation, there are many other combinations of direct and inverse variation involving more than two variables. The following table is a list of some variation statements and their equivalent mathematical forms:

Verbal Phrase	Algebraic Equation
y varies jointly with x and z.	$y = Kxz$
z varies jointly with r and the square of s.	$z = Krs^2$
V is directly proportional to T and inversely proportional to P.	$V = \dfrac{KT}{P}$
F varies jointly with m_1 and m_2 and inversely with the square of r.	$F = \dfrac{Km_1m_2}{r^2}$

E X A M P L E   4 y varies jointly with x and the square of z. When x is 5 and z is 3, y is 180. Find y when x is 2 and z is 4.

Solution   The general equation is given by

$$y = Kxz^2$$

Substituting $x = 5$, $z = 3$, and $y = 180$, we have

$$180 = K(5)(3)^2$$

$$180 = 45K$$

$$K = 4$$

The specific equation is

$$y = 4xz^2$$

When $x = 2$ and $z = 4$, the last equation becomes

$$y = 4(2)(4)^2$$

$$y = 128$$

EXAMPLE 5 In electricity, the resistance of a cable is directly proportional to its length and inversely proportional to the square of the diameter. If a 100-foot cable 0.5 inch in diameter has a resistance of 0.2 ohm, what will be the resistance of a cable made from the same material if it is 200 feet long with a diameter of 0.25 inch?

Solution Let R = resistance, l = length, and d = diameter. The equation is

$$R = \frac{Kl}{d^2}$$

When $R = 0.2$, $l = 100$, and $d = 0.5$, the equation becomes

$$0.2 = \frac{K(100)}{(0.5)^2}$$

or

$$K = 0.0005$$

Using this value of K in our original equation, the result is

$$R = \frac{0.0005l}{d^2}$$

When $l = 200$ and $d = 0.25$, the equation becomes

$$R = \frac{0.0005(200)}{(0.25)^2}$$

$$R = 1.6 \text{ ohms}$$

 Getting Ready for Class

After reading through the preceding section, respond in your own words and in complete sentences.

A. Give an example of a direct variation statement, and then translate it into symbols.

B. Translate the equation $y = \dfrac{k}{x}$ into words.

C. For the inverse variation equation $y = \dfrac{3}{x}$, what happens to the values of y as x gets larger?

D. How are direct variation statements and linear equations in two variables related?

PROBLEM SET 3.7

For the following problems, y varies directly with x.

1. If y is 10 when x is 2, find y when x is 6.

2. If y is 20 when x is 5, find y when x is 3.

3. If y is -32 when x is 4, find x when y is -40.

4. If y is -50 when x is 5, find x when y is -70.

For the following problems, r is inversely proportional to s.

5. If r is -3 when s is 4, find r when s is 2.

6. If r is -10 when s is 6, find r when s is -5.

7. If r is 8 when s is 3, find s when r is 48.

8. If r is 12 when s is 5, find s when r is 30.

For the following problems, d varies directly with the square of r.

9. If $d = 10$ when $r = 5$, find d when $r = 10$.

10. If $d = 12$ when $r = 6$, find d when $r = 9$.

11. If $d = 100$ when $r = 2$, find d when $r = 3$.

12. If $d = 50$ when $r = 5$, find d when $r = 7$.

For the following problems, y varies inversely with the square of x.

13. If $y = 45$ when $x = 3$, find y when x is 5.

14. If $y = 12$ when $x = 2$, find y when x is 6.

15. If $y = 18$ when $x = 3$, find y when x is 2.

16. If $y = 45$ when $x = 4$, find y when x is 5.

For the following problems, z varies jointly with x and the square of y.

17. If z is 54 when x and y are 3, find z when $x = 2$ and $y = 4$.

18. If z is 80 when x is 5 and y is 2, find z when $x = 2$ and $y = 5$.

19. If z is 64 when $x = 1$ and $y = 4$, find x when $z = 32$ and $y = 1$.

20. If z is 27 when $x = 6$ and $y = 3$, find x when $z = 50$ and $y = 4$.

Applying the Concepts

21. Length of a Spring The length a spring stretches is directly proportional to the force applied. If a force of 5 pounds stretches a spring 3 inches, how much force is necessary to stretch the same spring 10 inches?

22. Weight and Surface Area The weight of a certain material varies directly with the surface area of that material. If 8 square feet weighs half a pound, how much will 10 square feet weigh?

23. Pressure and Temperature The temperature of a gas varies directly with its pressure. A temperature of 200 K produces a pressure of 50 pounds per square inch.
(a) Find the equation that relates pressure and temperature.
(b) Graph the equation from part (a) in the first quadrant only.
(c) What pressure will the gas have at 280 K?

24. Circumference and Diameter The circumference of a wheel is directly proportional to its diameter. A wheel has a circumference of 8.5 feet and a diameter of 2.7 feet.
(a) Find the equation that relates circumference and diameter.
(b) Graph the equation from part (a) in the first quadrant only.
(c) What is the circumference of a wheel that has a diameter of 11.3 feet?

25. Volume and Pressure The volume of a gas is inversely proportional to the pressure. If a pressure of 36 pounds per square inch corresponds to a volume of 25 cubic feet, what pressure is needed to produce a volume of 75 cubic feet?

26. Wave Frequency The frequency of an electromagnetic wave varies inversely with the wavelength. If a wavelength of 200 meters has a frequency of 800 kilocycles per second, what frequency will be associated with a wavelength of 500 meters?

27. f-Stop and Aperture Diameter The relative aperture, or f-stop, for a camera lens is inversely proportional to the diameter of the aperture. An f-stop of 2 corresponds to an aperture diameter of 40 millimeters for the lens on an automatic camera.
(a) Find the equation that relates f-stop and diameter.
(b) Graph the equation from part (a) in the first quadrant only.

(c) What is the *f*-stop of this camera when the aperture diameter is 10 millimeters?

28. **f-Stop and Aperture Diameter** The relative aperture, or *f*-stop, for a camera lens is inversely proportional to the diameter of the aperture. An *f*-stop of 2.8 corresponds to an aperture diameter of 75 millimeters for a certain telephoto lens.
    (a) Find the equation that relates *f*-stop and diameter.
    (b) Graph the equation from part (a) in the first quadrant only.
    (c) What aperture diameter corresponds to an *f*-stop of 5.6?

29. **Surface Area of a Cylinder** The surface area of a hollow cylinder varies jointly with the height and radius of the cylinder. If a cylinder with radius 3 inches and height 5 inches has a surface area of 94 square inches, what is the surface area of a cylinder with radius 2 inches and height 8 inches?

30. **Capacity of a Cylinder** The capacity of a cylinder varies jointly with its height and the square of its radius. If a cylinder with a radius of 3 centimeters and a height of 6 centimeters has a capacity of 169.56 cubic centimeters, what will be the capacity of a cylinder with radius 4 centimeters and height 9 centimeters?

31. **Electrical Resistance** The resistance of a wire varies directly with its length and inversely with the square of its diameter. If 100 feet of wire with diameter 0.01 inch has a resistance of 10 ohms, what is the resistance of 60 feet of the same type of wire if its diameter is 0.02 inch?

32. **Volume and Temperature** The volume of a gas varies directly with its temperature and inversely with the pressure. If the volume of a certain gas is 30 cubic feet at a temperature of 300 K and a pressure of 20 pounds per square inch, what is the volume of the same gas at 340 K when the pressure is 30 pounds per square inch?

33. **Period of a Pendulum** The time it takes for a pendulum to complete one period varies directly with the square root of the length of the pendulum. A 100-centimeter pendulum takes 2.1 seconds to complete one period.
    (a) Find the equation that relates period and pendulum length.

(b) Graph the equation from part (a) in quadrant I only.
(c) How long does it take to complete one period if the pendulum hangs 225 centimeters?

34. **Area of a Circle** The area of a circle varies directly with the square of the radius. A circle with a 5-centimeter radius has an area of 78.5 square centimeters.
    (a) Find the equation that relates area and radius.
    (b) Graph the equation from part (a) in quadrant I only.
    (c) What is the area of a circle that has an 8-centimeter radius?

35. **Music** A musical tone's pitch varies inversely with its wavelength. If one tone has a pitch of 420 vibrations each second and a wavelength of 2.2 meters, find the wavelength of a tone that has a pitch of 720 vibrations each second.

36. **Hooke's Law** Hooke's law states that the stress (force per unit area) placed on a solid object varies directly with the strain (deformation) produced.
    (a) Using the variables S_1 for stress and S_2 for strain, state this law in algebraic form.
    (b) Find the constant, K, if for one type of material $S_1 = 24$ and $S_2 = 72$.

37. **Gravity** In Book Three of his *Principia*, Isaac Newton states that there is a single force in the universe that holds everything together, called the force of universal gravity. Newton stated that the force of universal gravity, F, is directly proportional with the product of two masses, m_1 and m_2, and inversely proportional with the square of the distance d between them. Write the equation for Newton's force of universal gravity, using the symbol G as the constant of proportionality.

38. **Boyle's Law and Charles's Law** Boyle's law states that for low pressures, the pressure of an ideal gas kept at a constant temperature varies inversely with the volume of the gas. Charles's law states that for low pressures, the density of an ideal gas kept at a constant pressure varies inversely with the absolute temperature of the gas.
    (a) State Boyle's law as an equation using the symbols P, K, and V.
    (b) State Charles's law as an equation using the symbols D, K, and T.

Review Problems

The following problems review material we covered in Section 2.6.

Solve each inequality, and graph the solution set.

39. $\left| \dfrac{x}{5} + 1 \right| \geq \dfrac{4}{5}$ **40.** $|x - 6| \geq 0.01$

41. $|3 - 4t| > -5$ **42.** $|2 - 6t| < -5$

43. $-8 + |3y + 5| < 5$

44. $|6y - 1| - 4 \leq 2$

Extending the Concepts

45. Light Intensity   I found the following diagram while shopping for some track lighting for my home.

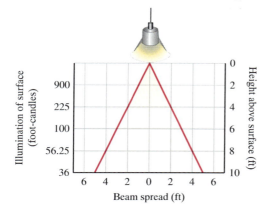

I was impressed by the diagram because it displays a lot of useful information in a very efficient manner. As the diagram indicates, the amount of light that falls on a surface depends on how far above the surface the light is placed and how much the light spreads out on the surface. Assume that this light illuminates a circle on a flat surface, and work the following problems.

(a) Fill in each table.

Height Above Surface (ft)	Illumination (foot-candles)
2	
4	
6	
8	
10	

Distance Above Surface (ft)	Area of Illuminated Region (ft²)
2	
4	
6	
8	
10	

(b) Use the templates in Figures 4 and 5 to construct line graphs from the data in the tables.

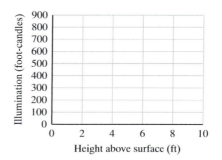

FIGURE 4

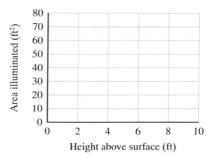

FIGURE 5

(c) Which of the relationships is direct variation, and which is inverse variation?

(d) Let F represent the number of foot-candles that fall on the surface, h the distance the light source is above the surface, and A the area of the illuminated region. Write an equation that shows the relationship between A and h, then write another equation that gives the relationship between F and h.

46. Law of Levers   Inverse variation may also be defined as occurring when the product of two variables is a constant; that is, when $x \cdot y = K$. This definition gives rise to the **Law of Levers,** or "seesaw principle." A seesaw is balanced when the variables on each side of the center, weight and distance from center, have the same inverse variation.

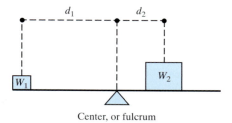

Center, or fulcrum

In the preceding figure, the weight on the left side must vary inversely with its distance from the center in the same way as the weight on the right side with its corresponding distance from the center in order for the seesaw to be balanced. This implies that $W_1 \cdot d_1 = W_2 \cdot d_2$ in order to balance. In a seesaw, an 85-pound girl is 4 feet from the center. How far from the center should her 120-pound brother place himself in order to balance the device?

CHAPTER 3 SUMMARY

Examples

1. The equation $3x + 2y = 6$ is an example of a linear equation in two variables.

Linear Equations in Two Variables [3.1, 3.3]

A *linear equation in two variables* is any equation that can be put in *standard form* $ax + by = c$, where a and b are not both zero. The graph of every linear equation is a straight line.

2. To find the x-intercept for $3x + 2y = 6$, we let $y = 0$ and get

$$3x = 6$$

$$x = 2$$

In this case the x-intercept is 2, and the graph crosses the x-axis at $(2, 0)$.

Intercepts [3.1]

The x-*intercept* of an equation is the x-*coordinate* of the point where the graph crosses the x-axis. The y-*intercept* is the y-coordinate of the point where the graph crosses the y-axis. We find the y-intercept by substituting $x = 0$ into the equation and solving for y. The x-intercept is found by letting $y = 0$ and solving for x.

3. The slope of the line through $(6, 9)$ and $(1, -1)$ is

$$m = \frac{9 - (-1)}{6 - 1} = \frac{10}{5} = 2$$

The Slope of a Line [3.2]

The *slope* of the line containing points (x_1, y_1) and (x_2, y_2) is given by

$$\text{Slope} = m = \frac{\text{Rise}}{\text{Run}} = \frac{y_2 - y_1}{x_2 - x_1}$$

Horizontal lines have 0 slope, and vertical lines have no slope.
Parallel lines have equal slopes, and perpendicular lines have slopes that are negative reciprocals.

4. The equation of the line with slope 5 and y-intercept 3 is

$$y = 5x + 3$$

The Slope-Intercept Form of a Line [3.3]

The equation of a line with slope m and y-intercept b is given by

$$y = mx + b$$

5. The equation of the line through $(3, 2)$ with slope -4 is

$$y - 2 = -4(x - 3)$$

which can be simplified to

$$y = -4x + 14$$

The Point-Slope Form of a Line [3.3]

The equation of the line through (x_1, y_1) that has slope m can be written as

$$y - y_1 = m(x - x_1)$$

6. The graph of

$$x - y \leq 3$$

is

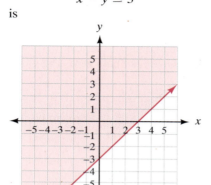

7. The relation

$$\{(8, 1), (6, 1), (-3, 0)\}$$

is also a function since no ordered pairs have the same first coordinates. The domain is $\{8, 6, -3\}$ and the range is $\{1, 0\}$.

8. The graph of $x = y^2$ shown in Figure 5 on page 185 fails the vertical line test. It is not the graph of a function.

9. If $f(x) = 5x - 3$ then

$$f(0) = 5(0) - 3$$
$$= -3$$
$$f(1) = 5(1) - 3$$
$$= 2$$
$$f(-2) = 5(-2) - 3$$
$$= -13$$
$$f(a) = 5a - 3$$

Linear Inequalities in Two Variables [3.4]

An inequality of the form $ax + by < c$ is a *linear inequality in two variables*. The equation for the boundary of the solution set is given by $ax + by = c$. (This equation is found by simply replacing the inequality symbol with an equal sign.)

To graph a linear inequality, first graph the boundary. Next, choose any point not on the boundary and substitute its coordinates into the original inequality. If the resulting statement is true, the graph lies on the same side of the boundary as the test point. A false statement indicates that the solution set lies on the other side of the boundary.

Relations and Functions [3.5]

A *function* is a rule that pairs each element in one set, called the *domain,* with exactly one element from a second set, called the *range.*

A *relation* is any set of ordered pairs. The set of all first coordinates is called the *domain* of the relation, and the set of all second coordinates is the *range* of the relation. A function is a relation in which no two different ordered pairs have the same first coordinates.

If the domain for a relation or a function is not specified, it is assumed to be all real numbers for which the relation (or function) is defined. Since we are concerned only with real number functions, a function is not defined for those values of x that give 0 in the denominator or the square root of a negative number.

Vertical Line Test [3.5]

If a vertical line crosses the graph of a relation in more than one place, the relation cannot be a function. If no vertical line can be found that crosses the graph in more than one place, the relation must be a function.

Function Notation [3.6]

The alternative notation for y is $f(x)$. It is read "f of x," and can be used instead of the variable y when working with functions. The notation y and the notation $f(x)$ are equivalent. That is, $y = f(x)$.

10. If y varies directly with x, then
$$y = Kx$$
Then if y is 18 when x is 6,
$$18 = K \cdot 6$$
or
$$K = 3$$
So the equation can be written more specifically as
$$y = 3x$$
If we want to know what y is when x is 4, we simply substitute:
$$y = 3 \cdot 4$$
$$y = 12$$

Variation [3.7]

If y varies *directly* with x (y is directly proportional to x), then we say
$$y = Kx$$

If y varies *inversely* with x (y is inversely proportional to x), then we say
$$y = \frac{K}{x}$$

If z varies *jointly* with x and y (z is directly proportional to both x and y), then we say
$$z = Kxy$$

In each case, K is called the *constant of variation.*

COMMON MISTAKES

1. When graphing ordered pairs, the most common mistake is to associate the first coordinate with the y-axis and the second with the x-axis. If you make this mistake you would graph (3, 1) by going up 3 and to the right 1, which is just the reverse of what you should do. Remember, the first coordinate is always associated with the horizontal axis, and the second coordinate is always associated with the vertical axis.

2. The two most common mistakes students make when first working with the formula for the slope of a line are
(a) Putting the difference of the x-coordinates over the difference of the y-coordinates.
(b) Subtracting in one order in the numerator and then subtracting in the opposite order in the denominator.

3. When graphing linear inequalities in two variables, remember to graph the boundary with a broken line when the inequality symbol is $<$ or $>$. The only time you use a solid line for the boundary is when the inequality symbol is $\leq$ or $\geq$.

CHAPTER 3   REVIEW

Graph each line. [3.1]

1. $3x + 2y = 6$

2. $y = -\frac{3}{2}x + 1$

3. $x = 3$

Find the slope of the line through the following pairs of points. [3.2]

4. (5, 2), (3, 6)

5. (−4, 2), (3, 2)

Find x if the line through the two given points has the given slope. [3.2]

6. $(4, x), (1, -3)$; $m = 2$

7. $(-4, 7), (2, x)$; $m = -\frac{1}{3}$

8. Find the slope of any line parallel to the line through $(3, 8)$ and $(5, -2)$. [3.2]

9. The line through $(5, 3y)$ and $(2, y)$ is parallel to a line with slope 4. What is the value of y? [3.2]

Give the equation of the line with the following slope and y-intercept. [3.3]

10. $m = 3, b = 5$ **11.** $m = -2, b = 0$

Give the slope and y-intercept of each equation. [3.3]

12. $3x - y = 6$ **13.** $2x - 3y = 9$

Find the equation of the line that contains the given point and has the given slope. [3.3]

14. $(2, 4)$, $m = 2$

15. $(-3, 1)$, $m = -\frac{1}{3}$

Find the equation of the line that contains the given pair of points. [3.3]

16. $(2, 5), (-3, -5)$

17. $(-3, 7), (4, 7)$

18. $(-5, -1), (-3, -4)$

19. Find the equation of the line that is parallel to $2x - y = 4$ and contains the point $(2, -3)$. [3.3]

20. Find the equation of the line perpendicular to $y = -3x + 1$ that has an x-intercept of 2. [3.3]

Graph each linear inequality. [3.4]

21. $y \le 2x - 3$ **22.** $x \ge -1$

State the domain and range of each relation, and then indicate which relations are also functions. [3.5]

23. $\{(2, 4), (3, 3), (4, 2)\}$

24. $\{(6, 3), (-4, 3), (-2, 0)\}$

If $f = \{(2, -1), (-3, 0), (4, \frac{1}{2}), (\pi, 2)\}$ and $g = \{(2, 2), (-1, 4), (0, 0)\}$, find the following. [3.6]

25. $f(-3)$ **26.** $f(2) + g(2)$

Let $f(x) = 2x^2 - 4x + 1$ and $g(x) = 3x + 2$, and evaluate each of the following. [3.6]

27. $f(0)$ **28.** $g(a)$

29. $f[g(0)]$ **30.** $f[g(1)]$

For the following problems, y varies directly with x. [3.7]

31. If y is 6 when x is 2, find y when x is 8.

32. If y is -3 when x is 5, find y when x is -10.

For the following problems, y varies inversely with the square of x. [3.7]

33. If y is 9 when x is 2, find y when x is 3.

34. If y is 4 when x is 5, find y when x is 2.

Solve each application problem. [3.7]

35. Tension in a Spring The tension t in a spring varies directly with the distance d the spring is stretched. If the tension is 42 pounds when the spring is stretched 2 inches, find the tension when the spring is stretched twice as far.

36. Light Intensity The intensity of a light source varies inversely with the square of the distance from the source. Four feet from the source the intensity is 9 foot-candles. What is the intensity 3 feet from the source?

CHAPTER 3 PROJECTS

EQUATIONS AND INEQUALITIES IN TWO VARIABLES

GROUP PROJECT

PHONE BILL

Number of People: 2–3

Time Needed: 20–30 minutes

Equipment: Paper, pencil, and graphing calculator

Background: Step functions, or greatest integer functions, are used in situations such as phone bills and postage rates, where the function outputs discrete numbers that occur in steps or jumps. On a graphing calculator, the notation int(X), found under the MATH menu on a TI-83, gives the greatest integer less than or equal to x. In other words, it "rounds down." For the calculator to "round up," we use the notation $-\text{int}(-X)$. Figures 1 and 2 show the graphs of $y = \text{int}(x)$ and $y = -\text{int}(-x)$. Their differences are subtle, but important. (To obtain these graphs on a TI-83, press the MODE key and then select DOT.)

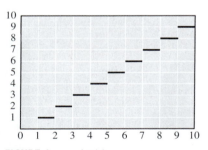

FIGURE 1 $y = \text{int}(x)$

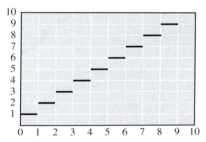

FIGURE 2 $y = -\text{int}(-x)$

Procedure: Suppose a wireless phone company charges $.50 per phone call plus $.35 per minute. Reading the fine print of your contract, you discover that you are actually paying "$.35 per minute with fractions of a minute rounded up to the next full minute."

1. Find a step function that gives the cost of a call as a function of time.

2. Sketch the graph of this function on the given grid.

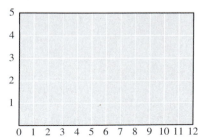

FIGURE 3

3. Fill in the following table using this step function.

Time (minutes)	0.6	1.0	1.1	2.5	6.0	6.8	7.0	8.3
Cost (dollars)								

4. If one phone call costs $5.05, how long was the call?

RESEARCH PROJECT

DESCARTES AND PASCAL

In this chapter we mentioned that René Descartes, the inventor of the rectangular coordinate system, is the person who made the statement "I think, therefore, I am." Blaise Pascal, whom we mentioned in the previous chapter, is responsible for the statement "The heart has its reasons which reason does not know." Although Pascal and Descartes were contemporaries, the philosophies of the two men differed greatly. Research the philosophy of both Descartes and Pascal, and then write an essay that gives the main points of each man's philosophy. In the essay, show how the quotations given here fit in with the philosophy of the man responsible for the quotation.

René Descartes, 1596–1650
(*David Eugene Smith Collection/Columbia University*)

Blaise Pascal, 1623–1662
(*David Eugene Smith Collection/Columbia University*)

CHAPTER 3 TEST

For each of the following straight lines, identify the x-intercept, y-intercept, and slope, and sketch the graph. [3.1–3.3]

1. $2x + y = 6$

2. $y = -2x - 3$

3. $y = \dfrac{3}{2}x + 4$

4. $x = -2$

Find the equation for each line. [3.3]

5. Give the equation of the line through $(-1, 3)$ that has slope $m = 2$.

6. Give the equation of the line through $(-3, 2)$ and $(4, -1)$.

7. Line l contains the point $(5, -3)$ and has a graph parallel to the graph of $2x - 5y = 10$. Find the equation for l.

8. Line l contains the point $(-1, -2)$ and has a graph perpendicular to the graph of $y = 3x - 1$. Find the equation for l.

9. Give the equation of the vertical line through $(4, -7)$.

Graph the following linear inequalities. [3.4]

10. $3x - 4y < 12$

11. $y \le -x + 2$

State the domain and range for the following relations, and indicate which relations are also functions. [3.5]

12. $\{(-2, 0), (-3, 0), (-2, 1)\}$

13. $y = x^2 - 9$

Let $f(x) = x - 2, g(x) = 3x + 4$, and $h(x) = 3x^2 - 2x - 8$, and find the following. [3.6]

14. $f(3) + g(2)$

15. $h(0) + g(0)$

16. $f[g(2)]$

17. $g[f(2)]$

Constructing a Box A piece of cardboard is in the shape of a square with each side 8 inches long. You are going to make a box out of the piece of cardboard by cutting four equal squares from each corner and then folding up the sides, as shown in Figure 1. Use this information in Problems 18–20. [3.5, 3.6]

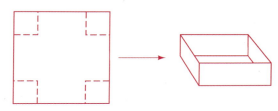

FIGURE I

18. If the length of each side of the squares you are cutting from the corners is represented by the variable x, state the restrictions on x.

19. Use function notation to write the volume of the resulting box $V(x)$ as a function of x.

20. Find $V(2)$, and explain what it represents.

Solve the following variation problems. [3.7]

21. **Direct Variation** Quantity y varies directly with the square of x. If y is 50 when x is 5, find y when x is 3.

22. **Joint Variation** Quantity z varies jointly with x and the cube of y. If z is 15 when x is 5 and y is 2, find z when x is 2 and y is 3.

23. **Maximum Load** The maximum load (L) a horizontal beam can safely hold varies jointly with the width (w) and the square of the depth (d) and inversely with the length (l). If a 10-foot beam with width 3 feet and depth 4 feet will safely hold up to 800 pounds, how many pounds will a 12-foot beam with width 3 feet and depth 4 feet hold?

Simplify.

1. $9 - 6 \div 2 + 2 \cdot 3$
2. $7(2x + 3) + 2(3x - 1)$
3. $12 - 7 - 2 - (-3)$
4. $-5(2x + 3) + 4x$
5. $5(3 - 7)^2 - 2(4 - 6)^3$
6. $\frac{3}{4} - \frac{2}{3}$

Solve.

7. $\frac{2}{3}a - 4 = 6$
8. $-6x - 7 = 3x + 2$
9. $-4 + 3(3x + 2) = 7$
10. $\frac{3}{4}(8x - 3) + \frac{3}{4} = 2$
11. $-4x + 9 = -3(-2x + 1) - 2$
12. $|-5x + 2| - 3 = -5$
13. $|2x - 3| - 7 = 1$
14. $|2y + 3| = |2y - 7|$
15. $|3x - 1| - 1 \le 2$
16. Solve for x: $ax + 4 = bx - 9$

Solve and graph the solution.

17. $3y - 6 \ge 3$ or $3y - 6 \le -3$
18. $|3x - 2| \le 7$
19. $|3x + 4| - 7 > 3$
20. $5 \le \frac{1}{4}x + 3 \le 8$

Graph each equation on a rectangular coordinate system.

21. $6x - 5y = 30$ 22. $x = 2$
23. $y = \frac{1}{2}x - 1$ 24. $2x - y \le -3$

25. Find the slope of the line through $(-3, -2)$ and $(-5, 3)$.
26. Find y if the line through $(-1, 2)$ and $(-3, y)$ has a slope of -2.
27. Find the slope of the line $y = 3$.
28. Give the equation of a line with slope $-\frac{3}{4}$ and y-intercept $= 2$.
29. Find the slope and y-intercept of $4x - 5y = 20$.

30. Find the equation of the line with x-intercept -2 and y-intercept -1.
31. Find the equation of the line that is parallel to $5x + 2y = -1$ that contains the points $(-4, -2)$.
32. Write the equation of the line in slope-intercept form given the slope is $\frac{7}{3}$ and if $(6, 8)$ is on the line.

If $f(x) = x^2 - 3x$, $g(x) = x - 1$, and $h(x) = 3x - 1$, find the solutions to Problems 33–35.

33. $h(0)$ 34. $f(-1) + g(4)$
35. $f[g(-2)]$
36. Add $-\frac{3}{4}$ to the product of -3 and $\frac{5}{12}$.
37. Subtract $\frac{2}{5}$ from the product of -3 and $\frac{6}{15}$.
38. State the properties that justify $2 + (a + b) = (2 + b) + a$.
39. Reduce to lowest terms: $\frac{721}{927}$
40. Write in symbols: The sum of $3a$ and $4b$ is less than their difference.
41. If $A = \{0, 3, 6, 7\}$ and $B = \{1, 2, 3, 6\}$, find $A \cup B$.
42. Give the opposite and reciprocal of $-\frac{3}{4}$.
43. Identify hypothesis and conclusion: If a quadrangle is a square, then all its sides are equal.
44. Give the next number of the sequence and state whether it is arithmetic or geometric: -4, -1, 2, . . .
45. 12 is 4% of what number?
46. Write the first five terms of the sequence $a_n = (-3)^n$.
47. Specify the domain and range for the relation $\{(-1, 3), (2, -1), (3, 3)\}$. Is the relation also a function?
48. y varies inversely with the square of x. If $y = 4$ when $x = 4$, find y when $x = 6$.
49. **Geometry** The length of a rectangle is 3 feet more than twice the width. The perimeter is 48 feet. Find the dimensions.
50. **Geometry** Find all three angles in a triangle if the smallest angle is one-fifth the largest angle and the remaining angle is 12 degrees more than the smallest angle.

Systems of Linear Equations

(Brandtner & Staede/Tony Stone Images)

INTRODUCTION

Suppose you decide to buy a cellular phone and are trying to decide between two rate plans. Plan A is $18.95 per month plus $.48 for each minute, or fraction of a minute, that you use the phone. Plan B is $34.95 per month plus $.36 for each minute, or fraction of a minute. The monthly cost C for each plan can be represented with a linear equation in two variables:

$$\text{Plan A:} \quad C = 0.48x + 18.95$$

$$\text{Plan B:} \quad C = 0.36x + 34.95$$

To compare the two plans, we use the table and graph shown below.

TABLE I	Monthly Cellular Phone Charges	
Number of Minutes *x*	**Monthly Cost** Plan A (dollars)	Plan B (dollars)
0	18.95	34.95
40	38.15	49.35
80	57.35	63.75
120	76.55	78.15
160	95.75	92.55
200	114.95	106.95
240	134.15	121.35

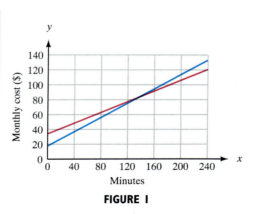

FIGURE I

The point of intersection of the two lines in Figure 1 is the point at which the monthly costs of the two plans are equal. In this chapter we will develop methods of finding that point of intersection.

STUDY SKILLS

The study skills for this chapter are concerned with getting ready to take an exam.

1. **Getting Ready to Take an Exam** Try to arrange your daily study habits so that you have very little studying to do the night before your next exam. The next two goals will help you achieve goal number 1.

2. **Review With the Exam in Mind** You should review material that will be covered on the next exam every day. Your review should consist of working problems. Preferably, the problems you work should be problems from your list of difficult problems.

3. **Continue to List Difficult Problems** This study skill was started in Chapter 2. You should continue to list and rework the problems that give you the most difficulty. It is this list that you will use to study for the next exam. Your goal is to go into the next exam knowing that you can successfully work any problem from your list of hard problems.

4. **Pay Attention to Instructions** Taking a test is different from doing homework. When you take a test, the problems will be mixed up. When you do your homework, you usually work a number of similar problems. Sometimes students who do very well on their homework become confused when they see the same problems on a test, because they have not paid attention to the instructions on their homework. For example, suppose you see the equation $y = 3x - 2$ on your next test. By itself the equation is simply a statement. There isn't anything to do unless the equation is accompanied by instructions. Each of the following is a valid instruction with respect to the equation $y = 3x - 2$, and the result of applying the instructions will be different in each case:

Find x when y is 10.	(Section 2.2)
Solve for x.	(Section 2.2)
Graph the equation.	(Section 3.1)
Find the intercepts.	(Section 3.1)
Find the slope.	(Section 3.3)

There are many things to do with the equation $y = 3x - 2$. If you train yourself to pay attention to the instructions that accompany a problem as you work through the assigned problems, you will not find yourself confused about what to do with a problem when you see it on a test.

In Chapter 3 we found the graph of an equation of the form $ax + by = c$ to be a straight line. Since the graph is a straight line, the equation is said to be a linear equation. Two linear equations considered together form a *linear system* of equations. For example,

$$3x - 2y = 6$$

$$2x + 4y = 20$$

is a linear system. The solution set to the system is the set of all ordered pairs that satisfy both equations. If we graph each equation on the same set of axes, we can see the solution set (see Figure 1).

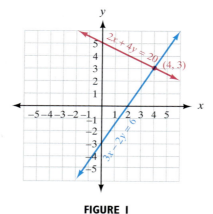

FIGURE 1

The point (4, 3) lies on both lines and therefore must satisfy both equations. It is obvious from the graph that it is the only point that does so. The solution set for the system is {(4, 3)}.

More generally, if $a_1x + b_1y = c_1$ and $a_2x + b_2y = c_2$ are linear equations, then the solution set for the system

$$a_1x + b_1y = c_1$$

$$a_2x + b_2y = c_2$$

can be illustrated through one of the graphs in Figure 2.

Case I The two lines intersect at one and only one point. The coordinates of the point give the solution to the system. This is what usually happens.

Case II The lines are parallel and therefore have no points in common. The solution set to the system is the empty set, $\varnothing$. In this case, we say the equations are *inconsistent*.

223

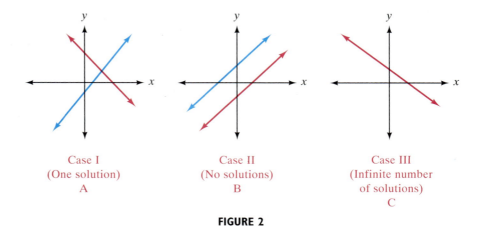

Case I
(One solution)
A

Case II
(No solutions)
B

Case III
(Infinite number
of solutions)
C

FIGURE 2

Case III   The lines coincide. That is, their graphs represent the same line. The solution set consists of all ordered pairs that satisfy either equation. In this case, the equations are said to be *dependent*.

In the beginning of this section we found the solution set for the system

$$3x - 2y = 6$$
$$2x + 4y = 20$$

by graphing each equation and then reading the solution set from the graph. Solving a system of linear equations by graphing is the least accurate method. If the coordinates of the point of intersection are not integers, it can be very difficult to read the solution set from the graph. There is another method of solving a linear system that does not depend on the graph. It is called the *addition method*.

EXAMPLE 1   Solve the system.

$$4x + 3y = 10$$
$$2x + y = 4$$

Solution   If we multiply the bottom equation by -3, the coefficients of y in the resulting equation and the top equation will be opposites:

$$4x + 3y = 10 \xrightarrow{\text{No change}} 4x + 3y = 10$$
$$2x + y = 4 \xrightarrow{\text{Multiply by } -3} -6x - 3y = -12$$

Adding the left and right sides of the resulting equations, we have

$$\begin{array}{r} 4x + 3y = 10 \\ -6x - 3y = -12 \\ \hline -2x = -2 \end{array}$$

The result is a linear equation in one variable. We have eliminated the variable y from the equations by addition. (It is for this reason we call this method of solving a linear system the *addition method*.) Solving $-2x = -2$ for x, we have

$$x = 1$$

This is the x-coordinate of the solution to our system. To find the y-coordinate, we substitute $x = 1$ into any of the equations containing both the variables x and y. Let's try the second equation in our original system:

$$2(1) + y = 4$$
$$2 + y = 4$$
$$y = 2$$

This is the y-coordinate of the solution to our system. The ordered pair $(1, 2)$ is the solution to the system.

Note: If we had put $x = 1$ into the first equation in our system, we would have obtained $y = 2$ also:

$$4(1) + 3y = 10$$
$$3y = 6$$
$$y = 2$$

EXAMPLE 2   Solve the system.

$$3x - 5y = -2$$
$$2x - 3y = 1$$

Solution   We can eliminate either variable. Let's decide to eliminate the variable x. We can do so by multiplying the top equation by 2 and the bottom equation by -3, and then adding the left and right sides of the resulting equations:

$$3x - 5y = -2 \xrightarrow{\text{Multiply by 2}} 6x - 10y = -4$$
$$2x - 3y = 1 \xrightarrow[\text{Multiply by } -3]{} \underline{-6x + 9y = -3}$$
$$-y = -7$$
$$y = 7$$

The y-coordinate of the solution to the system is 7. Substituting this value of y into any of the equations with both x- and y-variables gives $x = 11$. The solution to the system is $(11, 7)$. It is the only ordered pair that satisfies both equations.

EXAMPLE 3   Solve the system.

$$2x - 3y = 4$$
$$4x + 5y = 3$$

Solution   We can eliminate x by multiplying the top equation by -2 and adding it to the bottom equation:

$$2x - 3y = 4 \xrightarrow{\text{Multiply by } -2} -4x + 6y = -8$$
$$4x + 5y = 3 \xrightarrow[\text{No change}]{} \underline{4x + 5y = 3}$$
$$11y = -5$$
$$y = -\frac{5}{11}$$

The y-coordinate of our solution is $-\frac{5}{11}$. If we were to substitute this value of y back into either of our original equations, we would find the arithmetic necessary to solve for x cumbersome. For this reason, it is probably best to go back to the original system and solve it a second time — for x instead of y. Here is how we do that:

$$2x - 3y = 4 \xrightarrow{\text{Multiply by } 5} 10x - 15y = 20$$
$$4x + 5y = 3 \xrightarrow[\text{Multiply by } 3]{} \underline{12x + 15y = 9}$$
$$22x = 29$$
$$x = \frac{29}{22}$$

The solution to our system is $\left(\frac{29}{22}, -\frac{5}{11}\right)$.

EXAMPLE 4   Solve the system.

$$5x - 2y = 1$$
$$-10x + 4y = 3$$

Solution   We can eliminate y by multiplying the first equation by 2 and adding the result to the second equation:

$$5x - 2y = 1 \xrightarrow{\text{Multiply by } 2} 10x - 4y = 2$$
$$-10x + 4y = 3 \xrightarrow[\text{No change}]{} \underline{-10x + 4y = 3}$$
$$0 = 5$$

The result is the false statement $0 = 5$, which indicates there is no solution to the system. If we were to graph the two lines, we would find that they are parallel. In a case like this, we say the system is *inconsistent*. Whenever both variables have been eliminated and the resulting statement is false, the solution set for the system will be the empty set, $\varnothing$.

EXAMPLE 5   Solve the system.

$$4x - 3y = 2$$
$$8x - 6y = 4$$

Solution   Multiplying the top equation by -2 and adding, we can eliminate the variable x:

$$4x - 3y = 2 \xrightarrow{\text{Multiply by } -2} -8x + 6y = -4$$

$$8x - 6y = 4 \xrightarrow[\text{No change}]{} \underline{8x - 6y = 4}$$

$$0 = 0$$

Both variables have been eliminated and the resulting statement $0 = 0$ is true. In this case the lines coincide and the system is said to be *dependent*. The solution set consists of all ordered pairs that satisfy either equation. We can write the solution set as $\{(x, y)\,|\,4x - 3y = 2\}$ or $\{(x, y)\,|\,8x - 6y = 4\}$.

The previous two examples illustrate the two special cases in which the graphs of the equations in the system either coincide or are parallel. In both cases the left-hand sides of the equations were multiples of each other. In the case of the dependent equations the right-hand sides were also multiples. We can generalize these observations as follows:

The equations in the system

$$a_1x + b_1y = c_1$$
$$a_2x + b_2y = c_2$$

will be inconsistent (their graphs are parallel lines) if

$$\frac{a_1}{a_2} = \frac{b_1}{b_2} \neq \frac{c_1}{c_2}$$

and will be dependent (their graphs will coincide) if

$$\frac{a_1}{a_2} = \frac{b_1}{b_2} = \frac{c_1}{c_2}$$

EXAMPLE 6   Solve the system.

$$\frac{1}{2}x - \frac{1}{3}y = 2$$

$$\frac{1}{4}x + \frac{2}{3}y = 6$$

Solution   Although we could solve this system without clearing the equations of fractions, there is probably less chance for error if we have only integer coefficients to work with. So let's begin by multiplying both sides of the top equation by 6, and both sides of the bottom equation by 12, to clear each equation of fractions:

$$\frac{1}{2}x - \frac{1}{3}y = 2 \xrightarrow{\text{Times 6}} 3x - 2y = 12$$

$$\frac{1}{4}x + \frac{2}{3}y = 6 \xrightarrow[\text{Times 12}]{} 3x + 8y = 72$$

Now we can eliminate x by multiplying the top equation by -1 and leaving the bottom equation unchanged:

$$3x - 2y = 12 \xrightarrow{\text{Times } -1} -3x + 2y = -12$$

$$3x + 8y = 72 \xrightarrow[\text{No change}]{} \underline{\quad 3x + 8y = \quad 72 \quad}$$

$$10y = \quad 60$$

$$y = \quad 6$$

We can substitute $y = 6$ into any equation that contains both x and y. Let's use $3x - 2y = 12$.

$$3x - 2(6) = 12$$

$$3x - 12 = 12$$

$$3x = 24$$

$$x = \quad 8$$

The solution to the system is $(8, 6)$.

We end this section by considering another method of solving a linear system. The method is called the *substitution method* and is shown in the following examples.

EXAMPLE 7   Solve the system.

$$2x - 3y = -6$$

$$y = 3x - 5$$

Solution   The second equation tells us y is $3x - 5$. Substituting the expression $3x - 5$ for y in the first equation, we have

$$2x - 3(3x - 5) = -6$$

The result of the substitution is the elimination of the variable y. Solving the resulting linear equation in x as usual, we have

$$2x - 9x + 15 = -6$$

$$-7x + 15 = -6$$

$$-7x = -21$$

$$x = 3$$

Putting $x = 3$ into the second equation in the original system, we have

$$y = 3(3) - 5$$

$$= 9 - 5$$

$$= 4$$

The solution to the system is $(3, 4)$.

EXAMPLE 8   Solve by substitution.

$$2x + 3y = 5$$
$$x - 2y = 6$$

Solution   In order to use the substitution method, we must solve one of the two equations for x or y. We can solve for x in the second equation by adding $2y$ to both sides:

$$x - 2y = 6$$
$$x = 2y + 6 \qquad \text{Add } 2y \text{ to both sides.}$$

Substituting the expression $2y + 6$ for x in the first equation of our system, we have

$$2(2y + 6) + 3y = 5$$
$$4y + 12 + 3y = 5$$
$$7y + 12 = 5$$
$$7y = -7$$
$$y = -1$$

Using $y = -1$ in either equation in the original system, we find $x = 4$. The solution is $(4, -1)$.

Note: Both the substitution method and the addition method can be used to solve any system of linear equations in two variables. Systems like the one in Example 7, however, are easier to solve using the substitution method, since one of the variables is already written in terms of the other. A system like the one in Example 2 is easier to solve using the addition method, since solving for one of the variables would lead to an expression involving fractions. The system in Example 8 could be solved easily by either method, since solving the second equation for x is a one-step process.

 Using TECHNOLOGY

GRAPHING CALCULATORS

Solving Systems That Intersect in Exactly One Point

A graphing calculator can be used to solve a system of equations in two variables if the equations intersect in exactly one point. To solve the system shown in Example 3, we first solve each equation for y. Here is the result:

$$2x - 3y = 4 \qquad \text{becomes} \qquad y = \frac{4 - 2x}{-3}$$

$$4x + 5y = 3 \qquad \text{becomes} \qquad y = \frac{3 - 4x}{5}$$

Graphing these two functions on the calculator gives a diagram similar to the one in Figure 3.

(continued)

(Using Technology, continued)

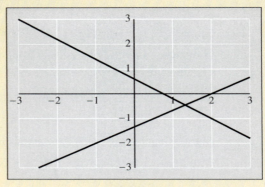

FIGURE 3

Using the Trace and Zoom features, we find that the two lines intersect at $x = 1.32$ and $y = -0.45$, which are the decimal equivalents (accurate to the nearest hundredth) of the fractions found in Example 3.

Special Cases

We cannot assume that two lines that look parallel in a calculator window are in fact parallel. If you graph the functions $y = x - 5$ and $y = 0.99x + 2$ in a window where x and y range from -10 to 10, the lines look parallel. We know this is not the case, however, since their slopes are different. As we zoom out repeatedly, the lines begin to look as if they coincide. We know this is not the case, because the two lines have different y-intercepts. To summarize: If we graph two lines on a calculator and the graphs look as if they are parallel or coincide, we should use algebraic methods, not the calculator, to determine the solution to the system.

Getting Ready for Class

After reading through the preceding section, respond in your own words and in complete sentences.

A. Two linear equations, each with the same two variables, form a system of equations. How do we define a solution to this system? That is, what form will a solution have, and what properties does a solution possess?

B. When would substitution be more efficient than the addition method in solving two linear equations?

C. Explain what an inconsistent system of linear equations looks like graphically and what would result algebraically when attempting to solve the system.

D. When might the graphing method of solving a system of equations be more desirable than the other techniques, and when might it be less desirable?

PROBLEM SET 4.1

Solve each system by graphing both equations on the same set of axes and then reading the solution from the graph.

1. $3x - 2y = 6$
$\quad\ x - \ y = 1$

2. $5x - 2y = 10$
$\quad\ x - \ y = -1$

3. $\quad\quad y = \dfrac{3}{5}x - 3$
$\quad 2x - y = -4$

4. $\quad\quad y = \dfrac{1}{2}x - 2$
$\quad 2x - y = -1$

5. $y = \dfrac{1}{2}x$
$\quad y = -\dfrac{3}{4}x + 5$

6. $y = \dfrac{2}{3}x$
$\quad y = -\dfrac{1}{3}x + 6$

7. $3x + 3y = -2$
$\quad\quad y = -x + 4$

8. $2x - 2y = 6$
$\quad\quad y = x - 3$

9. $2x - y = 5$
$\quad\ y = 2x - 5$

10. $x + 2y = 5$
$\quad\ y = -\dfrac{1}{2}x + 3$

Solve each of the following systems by the addition method.

11. $\ x + y = 5$
$\quad 3x - y = 3$

12. $\ \ x - \ y = \ 4$
$\quad -x + 2y = -3$

13. $3x + y = 4$
$\quad 4x + y = 5$

14. $6x - 2y = -10$
$\quad 6x + 3y = -15$

15. $3x - 2y = \ 6$
$\quad 6x - 4y = 12$

16. $\ \ 4x + \ 5y = -3$
$\quad -8x - 10y = \ \ 3$

17. $\ x + 2y = 0$
$\quad 2x - 6y = 5$

18. $\ x + 3y = 3$
$\quad 2x - 9y = 1$

19. $2x - 5y = 16$
$\quad 4x - 3y = 11$

20. $5x - 3y = -11$
$\quad 7x + 6y = -12$

21. $6x + 3y = -1$
$\quad 9x + 5y = \ \ 1$

22. $5x + 4y = -1$
$\quad 7x + 6y = -2$

23. $4x + 3y = 14$
$\quad 9x - 2y = 14$

24. $7x - 6y = 13$
$\quad 6x - 5y = 11$

25. $\ \ 2x - \ 5y = 3$
$\quad -4x + 10y = 3$

26. $\ \ 3x - 2y = \ 1$
$\quad -6x + 4y = -2$

27. $\ \dfrac{1}{4}x - \dfrac{1}{6}y = -2$
$\quad -\dfrac{1}{6}x + \dfrac{1}{5}y = \ \ 4$

28. $-\dfrac{1}{3}x + \dfrac{1}{4}y = 0$
$\quad \dfrac{1}{5}x - \dfrac{1}{10}y = 1$

29. $\dfrac{1}{2}x + \dfrac{1}{3}y = 13$
$\quad \dfrac{2}{5}x + \dfrac{1}{4}y = 10$

30. $\dfrac{1}{2}x + \dfrac{1}{3}y = \dfrac{2}{3}$
$\quad \dfrac{2}{3}x + \dfrac{2}{5}y = \dfrac{14}{15}$

31. $\dfrac{2}{3}x + \dfrac{2}{5}y = \ \ 4$
$\quad \dfrac{1}{3}x - \dfrac{1}{2}y = -\dfrac{1}{3}$

32. $\ \dfrac{1}{2}x - \dfrac{1}{3}y = \dfrac{5}{6}$
$\quad -\dfrac{2}{5}x + \dfrac{1}{2}y = -\dfrac{9}{10}$

Solve each of the following systems by the substitution method.

33. $7x - y = 24$
$\quad\ x = 2y + 9$

34. $3x - y = -8$
$\quad\quad y = 6x + 3$

35. $6x - y = 10$
$\quad\quad y = -\dfrac{3}{4}x - 1$

36. $2x - y = 6$
$\quad\quad y = -\dfrac{4}{3}x + 1$

37. $\ x - \ y = 4$
$\quad 2x - 3y = 6$

38. $\ x + \ y = \ 3$
$\quad 2x + 3y = -4$

39. $y = 3x - 2$
$\quad y = 4x - 4$

40. $y = \ \ 5x - 2$
$\quad y = -2x + 5$

41. $2x - \ y = \ 5$
$\quad 4x - 2y = 10$

42. $-10x + 8y = -6$
$\quad\quad\quad y = \dfrac{5}{4}x$

43. $\dfrac{1}{3}x - \dfrac{1}{2}y = 0$
$\quad\quad x = \dfrac{3}{2}y$

44. $\dfrac{2}{5}x - \dfrac{2}{3}y = 0$
$\quad\quad y = \dfrac{3}{5}x$

You may want to read Example 3 again before solving the systems that follow.

45. $4x - 7y = \ \ 3$
$\quad 5x + 2y = -3$

46. $3x - 4y = 7$
$\quad 6x - 3y = 5$

47. $9x - 8y = 4$
$\quad 2x + 3y = 6$

48. $\ \ 4x - 7y = \ \ 10$
$\quad -3x + 2y = -9$

49. $3x - 5y = 2$
$\quad 7x + 2y = 1$

50. $4x - 3y = -1$
$\quad 5x + 8y = \ \ 2$

51. Multiply both sides of the second equation in the following system by 100, and then solve as usual.

$$x + \quad\ \ y = 10,000$$
$$0.06x + 0.05y = \quad\ 560$$

52. Multiply both sides of the second equation in the following system by 10, and then solve as usual.

$$x + \quad y = 12$$

$$0.20x + 0.50y = 0.30(12)$$

53. What value of c will make the following system a dependent system (one in which the lines coincide)?

$$6x - 9y = 3$$

$$4x - 6y = c$$

54. What value of c will make the following system a dependent system?

$$5x - \quad 7y = c$$

$$-15x + 21y = 9$$

Applying the Concepts

55. Cost of a Phone Call One telephone company charges 41 cents for the first minute and 32 cents for each additional minute for a certain long-distance phone call. If the number of additional minutes after the first minute is x and the cost, in cents, for the call is y, then the equation that gives the total cost, in cents, for the call is $y = 32x + 41$.
(a) How much does it cost to make a 10-minute long-distance call under these conditions?
(b) If a second phone company charges 45 cents for the first minute and 30 cents for each additional minute, write the equation that gives the total cost y of a call in terms of the number of additional minutes x.
(c) After how many additional minutes will the two companies charge an equal amount? (What is the x-coordinate of the point of intersection of the two lines?)

56. Cost of a Taxi Ride In a certain city, a taxi ride costs 75 cents for the first $\frac{1}{7}$ of a mile and 10 cents for every additional $\frac{1}{7}$ of a mile after the first seventh. If x is the number of additional sevenths of a mile, then the total cost y of a taxi ride is

$$y = 10x + 75$$

(a) How much does it cost to ride a taxi for 10 miles in this city?

(b) Suppose a taxi ride in another city costs 50 cents for the first $\frac{1}{7}$ of a mile, and 15 cents for each additional $\frac{1}{7}$ of a mile. Write an equation that gives the total cost y, in cents, to ride x sevenths of a mile past the first seventh in this city.
(c) Solve the two equations given in part (b) simultaneously (as a system of equations), and explain in words what your solution represents.

57. Medicine From 1980 through 1992, outpatient surgery was linearly increasing in U.S. hospitals. For these years the percentage of all surgery performed that was outpatient surgery, y, may be represented by the equation

$$y = \frac{10}{3}x + 20$$

where x is the number of years past 1980, but not beyond the year 1992. Also, from 1980 through 1992, inpatient surgery was linearly decreasing in U.S. hospitals. For these years the percentage of all surgery performed that was outpatient surgery, y, may be represented by the equation

$$y = -\frac{10}{3}x + 80$$

where x is the number of years past 1980, but not beyond the year 1992.
(a) What is the range of possible values of x in these two equations?
(b) Graph these two equations.
(c) Find the solution to this system. What year does the solution represent?

58. Medicine Death from heart disease has been linearly decreasing.

Year	Deaths (Per 100,000 of the Population)
1960	286
1991	148

However, cancer deaths for females have been linearly increasing.

Year	Deaths (Per 100,000 of the Population)
1970	140
1992	180

(a) Letting the variable x represent the year beyond 1960, but not past 1991 ($1960 < x \le 1991$), write an equation relating the number of deaths from heart disease to the year.

(b) Letting the variable x represent the year beyond 1970, but not past 1992 ($1970 < x \le 1992$), write an equation relating the number of cancer deaths in females to the year.

(c) Solve the system from parts (a) and (b). What year does your solution represent?

59. Pollution   The DDT levels in fish declined in a linear manner from 1978 to 1986. In 1978 the average fish had a DDT level of 0.5 ppb (parts per billion) and in 1986 the average level was found to be only 0.18 ppb.

(a) Graph this line on a coordinate axis with the year on the horizontal axis and the DDT level on the vertical axis.

(b) Write the equation of the line found in part (a).

(c) Use the results of part (b) to find the DDT level in 1980.

(d) If the rate of DDT reduction had continued to the year 1990, what would that level have been?

60. A System With Two Solutions   Given the following system:

$$y = |x|$$

$$x - 2y = -6$$

(a) Graph this system.

(b) Solve the system by inspection.

Review Problems

The problems that follow review material we covered in Sections 3.2 and 3.3.

61. Find the slope of the line that contains $(-4, -1)$ and $(-2, 5)$. [3.2]

62. A line has a slope of $\frac{2}{3}$. Find the slope of any line [3.2]

(a) Parallel to it.

(b) Perpendicular to it.

63. Give the slope and y-intercept of the line $2x - 3y = 6$. [3.3]

64. Give the equation of the line with slope -3 and y-intercept 5. [3.3]

65. Find the equation of the line with slope $\frac{2}{3}$ that contains the point $(-6, 2)$. [3.3]

66. Find the equation of the line through $(1, 3)$ and $(-1, -5)$. [3.3]

67. Find the equation of the line with x-intercept 3 and y-intercept -2. [3.3]

68. Find the equation of the line through $(-1, 4)$ whose graph is perpendicular to the graph of $y = 2x + 3$. [3.3]

4.2   Systems of Linear Equations in Three Variables

A solution to an equation in three variables such as

$$2x + y - 3z = 6$$

is an ordered triple of numbers (x, y, z). For example, the ordered triples $(0, 0, -2)$, $(2, 2, 0)$, and $(0, 9, 1)$ are solutions to the equation $2x + y - 3z = 6$, since they produce a true statement when their coordinates are replaced for x, y, and z in the equation.

> **DEFINITION**   The **solution set** for a system of three linear equations in three variables is the set of ordered triples that satisfies all three equations.

EXAMPLE I   Solve the system.

$$x + y + z = 6 \quad (1)$$
$$2x - y + z = 3 \quad (2)$$
$$x + 2y - 3z = -4 \quad (3)$$

Solution   We want to find the ordered triple (x, y, z) that satisfies all three equations. We have numbered the equations so it will be easier to keep track of where they are and what we are doing.

There are many ways to proceed. The main idea is to take two different pairs of equations and eliminate the same variable from each pair. We begin by adding equations (1) and (2) to eliminate the y-variable. The resulting equation is numbered (4):

$$
\begin{array}{ll}
x + y + z = 6 & (1) \\
\underline{2x - y + z = 3} & (2) \\
3x \quad\quad + 2z = 9 & (4)
\end{array}
$$

Adding twice equation (2) to equation (3) will also eliminate the variable y. The resulting equation is numbered (5):

$$
\begin{array}{ll}
4x - 2y + 2z = 6 & \text{Twice (2)} \\
\underline{x + 2y - 3z = -4} & (3) \\
5x \quad\quad - z = 2 & (5)
\end{array}
$$

Equations (4) and (5) form a linear system in two variables. By multiplying equation (5) by 2 and adding the result to equation (4), we succeed in eliminating the variable z from the new pair of equations:

$$
\begin{array}{ll}
3x + 2z = 9 & (4) \\
\underline{10x - 2z = 4} & \text{Twice (5)} \\
13x \quad\quad = 13 &
\end{array}
$$

$$x = 1$$

Substituting $x = 1$ into equation (4), we have

$$3(1) + 2z = 9$$
$$2z = 6$$
$$z = 3$$

Using $x = 1$ and $z = 3$ in equation (1) gives us

$$1 + y + 3 = 6$$
$$y + 4 = 6$$
$$y = 2$$

The solution is the ordered triple $(1, 2, 3)$.

EXAMPLE 2 Solve the system.

$$2x + y - z = 3 \quad (1)$$
$$3x + 4y + z = 6 \quad (2)$$
$$2x - 3y + z = 1 \quad (3)$$

Solution It is easiest to eliminate z from the equations. The equation produced by adding (1) and (2) is

$$5x + 5y = 9 \quad (4)$$

The equation that results from adding (1) and (3) is

$$4x - 2y = 4 \quad (5)$$

Equations (4) and (5) form a linear system in two variables. We can eliminate the variable y from this system as follows:

$$5x + 5y = 9 \xrightarrow{\text{Multiply by 2}} 10x + 10y = 18$$
$$4x - 2y = 4 \xrightarrow[\text{Multiply by 5}]{} 20x - 10y = 20$$
$$\overline{30x = 38}$$

$$x = \frac{38}{30}$$
$$= \frac{19}{15}$$

Substituting $x = \frac{19}{15}$ into equation (5) or equation (4) and solving for y gives

$$y = \frac{8}{15}$$

Using $x = \frac{19}{15}$ and $y = \frac{8}{15}$ in equation (1), (2), or (3) and solving for z results in

$$z = \frac{1}{15}$$

The ordered triple that satisfies all three equations is $\left(\frac{19}{15}, \frac{8}{15}, \frac{1}{15}\right)$.

EXAMPLE 3 Solve the system.

$$2x + 3y - z = 5 \quad (1)$$
$$4x + 6y - 2z = 10 \quad (2)$$
$$x - 4y + 3z = 5 \quad (3)$$

Solution Multiplying equation (1) by -2 and adding the result to equation (2) looks like this:

$$-4x - 6y + 2z = -10 \qquad -2 \text{ times (1)}$$
$$\underline{4x + 6y - 2z = 10} \qquad (2)$$
$$0 = 0$$

All three variables have been eliminated, and we are left with a true statement. As was the case in Section 4.1, this implies that the two equations are dependent. With a system of three equations in three variables, however, a dependent system can have no solution or an infinite number of solutions. After we have concluded the examples in this section, we will discuss the geometry behind these systems. Doing so will give you some additional insight into dependent systems.

EXAMPLE 4 Solve the system.

$$x - 5y + 4z = 8 \quad (1)$$
$$3x + y - 2z = 7 \quad (2)$$
$$-9x - 3y + 6z = 5 \quad (3)$$

Solution Multiplying equation (2) by 3 and adding the result to equation (3) produces

$$\begin{array}{ll} 9x + 3y - 6z = 21 & \text{3 times (2)} \\ \underline{-9x - 3y + 6z = 5} & \text{(3)} \\ 0 = 26 & \end{array}$$

In this case all three variables have been eliminated, and we are left with a false statement. The two equations are inconsistent; there are no ordered triples that satisfy both equations. The solution set for the system is the empty set, $\varnothing$. If equations (2) and (3) have no ordered triples in common, then certainly (1), (2), and (3) do not either.

EXAMPLE 5 Solve the system.

$$x + 3y = 5 \quad (1)$$
$$6y + z = 12 \quad (2)$$
$$x - 2z = -10 \quad (3)$$

Solution It may be helpful to rewrite the system as

$$x + 3y = 5 \quad (1)$$
$$6y + z = 12 \quad (2)$$
$$x - 2z = -10 \quad (3)$$

Equation (2) does not contain the variable x. If we multiply equation (3) by -1 and add the result to equation (1), we will be left with another equation that does not contain the variable x:

$$\begin{array}{ll} x + 3y = 5 & \text{(1)} \\ \underline{-x + 2z = 10} & \text{-1 times (3)} \\ 3y + 2z = 15 & \text{(4)} \end{array}$$

Equations (2) and (4) form a linear system in two variables. Multiplying equation (2) by -2 and adding the result to equation (4) eliminates the variable z:

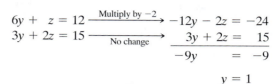

$$6y + z = 12 \xrightarrow{\text{Multiply by } -2} -12y - 2z = -24$$
$$3y + 2z = 15 \xrightarrow{\text{No change}} \underline{ 3y + 2z = 15}$$
$$-9y = -9$$
$$y = 1$$

Using $y = 1$ in equation (4) and solving for z, we have

$$z = 6$$

Substituting $y = 1$ into equation (1) gives

$$x = 2$$

The ordered triple that satisfies all three equations is $(2, 1, 6)$.

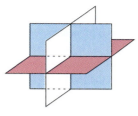

Case 1   The three planes have exactly one point in common. In this case we get one solution to our system, as in Examples 1, 2, and 5.

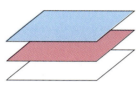

Case 2   The three planes have no points in common because they are all parallel to one another. The system they represent is an inconsistent system.

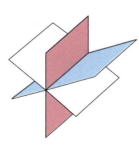

Case 3   The three planes intersect in a line. Any point on the line is a solution to the system of equations represented by the planes, so there is an infinite number of solutions to the system. This is an example of a dependent system.

The Geometry Behind Equations in Three Variables

We can graph an ordered triple on a coordinate system with three axes. The graph will be a point in space. The coordinate system is drawn in perspective; you have to imagine that the x-axis comes out of the paper and is perpendicular to both the y-axis and the z-axis. To graph the point $(3, 4, 5)$, we move 3 units in the x-direction, 4 units in the y-direction, and then 5 units in the z-direction, as shown in Figure 1.

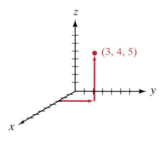

FIGURE I

Although in actual practice it is sometimes difficult to graph equations in three variables, if we were to graph a linear equation in three variables, we would find that the graph was a plane in space. A system of three equations in three variables is represented by three planes in space.

There are a number of possible ways in which these three planes can intersect, some of which are shown in the margin on this page. And there are still other possibilities that are not among those shown in the margin.

In Example 3 we found that equations (1) and (2) were dependent equations. They represent the same plane. That is, they have all their points in common. But the system of equations that they came from has either no solution or an infinite number of solutions. It all depends on the third plane. If the third plane coincides

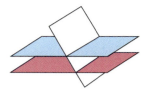

Case 4 Two of the planes are parallel; the third plane intersects each of the parallel planes. In this case the three planes have no points in common. There in no solution to the system; it is an inconsistent system.

with the first two, then the solution to the system is a plane. If the third plane is parallel to the first two, then there is no solution to the system. And, finally, if the third plane intersects the first two, but does not coincide with them, then the solution to the system is that line of intersection.

In Example 4 we found that trying to eliminate a variable from the second and third equations resulted in a false statement. This means that the two planes represented by these equations are parallel. It makes no difference where the third plane is; there is no solution to the system in Example 4. (If we were to graph the three planes from Example 4, we would obtain a diagram similar to Case 2 or Case 4 in the margin.)

If, in the process of solving a system of linear equations in three variables, we eliminate all the variables from a pair of equations and are left with a false statement, we will say the system is inconsistent. If we eliminate all the variables and are left with a true statement, then we will say the system is a dependent one.

Getting Ready for Class

After reading through the preceding section, respond in your own words and in complete sentences.

A. What is an ordered triple of numbers?

B. Explain what it means for (1, 2, 3) to be a solution to a system of linear equations in three variables.

C. Explain in a general way the procedure you would use to solve a system of three linear equations in three variables.

D. How do you know when a system of linear equations in three variables has no solution?

PROBLEM SET 4.2

Solve the following systems.

1. $x + y + z = 4$
$x - y + 2z = 1$
$x - y - 3z = -4$

2. $x - y - 2z = -1$
$x + y + z = 6$
$x + y - z = 4$

3. $x + y + z = 6$
$x - y + 2z = 7$
$2x - y - 4z = -9$

4. $x + y + z = 0$
$x + y - z = 6$
$x - y + 2z = -7$

5. $x + 2y + z = 3$
$2x - y + 2z = 6$
$3x + y - z = 5$

6. $2x + y - 3z = -14$
$x - 3y + 4z = 22$
$3x + 2y + z = 0$

7. $2x + 3y - 2z = 4$
$x + 3y - 3z = 4$
$3x - 6y + z = -3$

8. $4x + y - 2z = 0$
$2x - 3y + 3z = 9$
$-6x - 2y + z = 0$

9. $-x + 4y - 3z = 2$
$2x - 8y + 6z = 1$
$3x - y + z = 0$

10. $4x + 6y - 8z = 1$
$-6x - 9y + 12z = 0$
$x - 2y - 2z = 3$

11. $\dfrac{1}{2}x - y + z = 0$

$2x + \dfrac{1}{3}y + z = 2$

$x + y + z = -4$

12. $\dfrac{1}{3}x + \dfrac{1}{2}y + z = -1$

$x - y + \dfrac{1}{5}z = 1$

$x + y + z = 5$

13. $2x - y - 3z = 1$
$x + 2y + 4z = 3$
$4x - 2y - 6z = 2$

14. $3x + 2y + z = 3$
$x - 3y + z = 4$
$-6x - 4y - 2z = 1$

15. $2x - y + 3z = 4$
$x + 2y - z = -3$
$4x + 3y + 2z = -5$

16. $6x - 2y + z = 5$
$3x + y + 3z = 7$
$x + 4y - z = 4$

17. $x + y = 9$
$y + z = 7$
$x - z = 2$

18. $x - y = -3$
$x + z = 2$
$y - z = 7$

19. $2x + y = 2$
$y + z = 3$
$4x - z = 0$

20. $2x + y = 6$
$3y - 2z = -8$
$x + z = 5$

21. $2x - 3y = 0$
$6y - 4z = 1$
$x + 2z = 1$

22. $3x + 2y = 3$
$y + 2z = 2$
$6x - 4z = 1$

23. $\dfrac{1}{2}x + \dfrac{2}{3}y = \dfrac{5}{2}$

$\dfrac{1}{5}x - \dfrac{1}{2}z = -\dfrac{3}{10}$

$\dfrac{1}{3}y - \dfrac{1}{4}z = \dfrac{3}{4}$

24. $\dfrac{1}{2}x - \dfrac{1}{3}y = \dfrac{1}{6}$

$\dfrac{1}{3}y - \dfrac{1}{3}z = 1$

$\dfrac{1}{5}x - \dfrac{1}{2}z = -\dfrac{4}{5}$

25. $\dfrac{1}{2}x - \dfrac{1}{4}y + \dfrac{1}{2}z = -2$

$\dfrac{1}{4}x - \dfrac{1}{12}y - \dfrac{1}{3}z = \dfrac{1}{4}$

$\dfrac{1}{6}x + \dfrac{1}{3}y - \dfrac{1}{2}z = \dfrac{3}{2}$

26. $\dfrac{1}{2}x + \dfrac{1}{2}y + z = \dfrac{1}{2}$

$\dfrac{1}{2}x - \dfrac{1}{4}y - \dfrac{1}{4}z = 0$

$\dfrac{1}{4}x + \dfrac{1}{12}y + \dfrac{1}{6}z = \dfrac{1}{6}$

Applying the Concepts

27. Electric Current   In the following diagram of an electrical circuit, x, y, and z represent the amount of current (in amperes) flowing across the 5-ohm, 20-ohm, and 10-ohm resistors, respectively. (In circuit diagrams resistors are represented by ⌁ and potential differences by ⊣⊢ .)

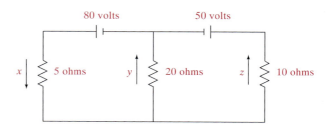

The system of equations used to find the three currents x, y, and z is

$$x - y - z = 0$$

$$5x + 20y = 80$$

$$20y - 10z = 50$$

Solve the system for all variables.

28. Cost of a Rental Car   If a car rental company charges \$10 a day and 8¢ a mile to rent one of its cars, then the cost z, in dollars, to rent a car for x days and drive y miles can be found from the equation

$$z = 10x + 0.08y$$

(a) How much does it cost to rent a car for 2 days and drive it 200 miles under these conditions?

(b) A second company charges \$12 a day and 6¢ a mile for the same car. Write an equation that gives the cost z, in dollars, to rent a car from this company for x days and drive it y miles.

(c) A car is rented from each of the companies mentioned in (a) and (b) for 2 days. To find the mileage at which the cost of renting the cars from each of the two companies will be equal, solve the following system for y:

$$z = 10x + 0.08y$$

$$z = 12x + 0.06y$$

$$x = 2$$

29. More Variables Than Equations In general, if there are more variables in a system than equations, then the system is likely to have many solutions. Given the system

$$x + y + z = 12$$

$$2x + y - z = 20$$

(a) Show that any point on the line $3x + 2y = 32$ is a solution to this system.
(b) How many solutions does the system have?

30. Fewer Variables Than Equations In general, if there are fewer variables in a system than equations, then it is likely that the system does not have a solution. Given the system

$$x + \ y = 10$$

$$x - \ y = 6$$

$$2x + 3y = 8$$

show that this system has no solution.

Review Problems

The problems that follow review material we covered in Section 3.4.

Graph each inequality.

31. $2x + 3y < 6$ **32.** $2x + y < -5$

33. $y \geq -3x - 4$ **34.** $y \geq 2x - 1$

35. $x \geq 3$ **36.** $y > -5$

Extending the Concepts

37. Graphing in Three Dimensions The graph of a linear equation in three variables is a plane in three-dimensional space. One way to graph such a plane is to find the point of intersection of that

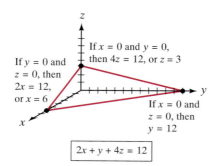

$2x + y + 4z = 12$

plane with each of the three coordinate axes. This triangular area shows where the plane cuts through the first octant. (The axes in a three-dimensional coordinate system divide space into eight octants.) The diagram shows the graph of $2x + y + 4z = 12$ as it cuts through the first octant. Sketch the graph of each of the following planes in the first octant:
(a) $x + 2y + 3z = 6$
(b) $2x + 3y + 4z = 12$
(c) $3x - 2y + z = 6$

38. Graphing by Traces Refer to the preceding exercise. The plane that is represented by the equation $2x + y + 4z = 12$ intersects each of the three coordinate planes in a line called a trace. To find the equation of a trace, substitute 0 for one of the variables in the equation of the plane.

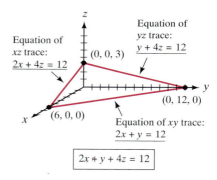

Find the equation of the xz, xy, and yz traces.
(a) $4x + 5y + 8z = 40$
(b) $x + 2y + 4z = 30$
(c) $x - 4y - 10z = 10$
(d) $4x + 5y + 10z = 8$

Solve each system for the solution (x, y, z, w).

39. $x + \ y + z + \ w = \ 10$
     $x + 2y - z + \ w = \ 6$
     $x - \ y - z + 2w = \ 4$
     $x - 2y + z - 3w = -12$

40. $x + \ y + \ z + \ w = \ 16$
     $x - \ y + 2z - \ w = \ 1$
     $x + 3y - \ z - \ w = -2$
     $x - 3y - 2z + 2w = -4$

Introduction to Determinants

In this section we will expand and evaluate *determinants*. The purpose of this section is simply to be able to find the value of a given determinant. As we will see in the next section, determinants are very useful in solving systems of linear equations. Before we apply determinants to systems of linear equations, however, we must practice calculating the value of some determinants.

DEFINITION The value of the **2 × 2** (2 by 2) determinant

$$\begin{vmatrix} a & c \\ b & d \end{vmatrix}$$

is given by

$$\begin{vmatrix} a & c \\ b & d \end{vmatrix} = ad - bc$$

From the preceding definition we see that a determinant is simply a square array of numbers with two vertical lines enclosing it. The value of a 2 × 2 determinant is found by cross-multiplying on the diagonals and then subtracting, a diagram of which looks like

$$\begin{vmatrix} a & c \\ b & d \end{vmatrix} = ad - bc$$

EXAMPLE 1   Find the value of the following 2 × 2 determinants:

(a) $\begin{vmatrix} 1 & 2 \\ 3 & 4 \end{vmatrix} = 1(4) - 3(2) = 4 - 6 = -2$

(b) $\begin{vmatrix} 3 & 5 \\ -2 & 7 \end{vmatrix} = 3(7) - (-2)5 = 21 + 10 = 31$

EXAMPLE 2   Solve for x if

$$\begin{vmatrix} x^2 & 2 \\ x & 1 \end{vmatrix} = 8$$

Solution   We expand the determinant on the left side to get

$$x^2(1) - x(2) = 8$$
$$x^2 - 2x = 8$$
$$x^2 - 2x - 8 = 0$$
$$(x - 4)(x + 2) = 0$$
$$x - 4 = 0 \quad \text{or} \quad x + 2 = 0$$
$$x = 4 \quad \text{or} \quad x = -2$$

We now turn our attention to 3×3 determinants. A 3×3 determinant is also a square array of numbers, the value of which is given by the following definition.

DEFINITION   The value of the 3×3 determinant

$$\begin{vmatrix} a_1 & b_1 & c_1 \\ a_2 & b_2 & c_2 \\ a_3 & b_3 & c_3 \end{vmatrix}$$

is given by

$$\begin{vmatrix} a_1 & b_1 & c_1 \\ a_2 & b_2 & c_2 \\ a_3 & b_3 & c_3 \end{vmatrix} = a_1b_2c_3 + a_3b_1c_2 + a_2b_3c_1 - a_3b_2c_1 - a_1b_3c_2 - a_2b_1c_3$$

At first glance, the expansion of a 3×3 determinant looks a little complicated. There are actually two different methods used to find the six products in the preceding definition that simplify matters somewhat.

Method 1   We begin by writing the determinant with the first two columns repeated on the right:

$$\begin{vmatrix} a_1 & b_1 & c_1 \\ a_2 & b_2 & c_2 \\ a_3 & b_3 & c_3 \end{vmatrix} \begin{matrix} a_1 & b_1 \\ a_2 & b_2 \\ a_3 & b_3 \end{matrix}$$

The positive products in the definition come from multiplying down the three full diagonals:

The negative products come from multiplying up the three full diagonals:

EXAMPLE 3   Find the value of

$$\begin{vmatrix} 1 & 3 & -2 \\ 2 & 0 & 1 \\ 4 & -1 & 1 \end{vmatrix}$$

Solution   Repeating the first two columns and then finding the products down the diagonals and the products up the diagonals as given in Method 1, we have

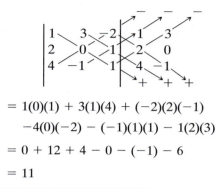

$$= 1(0)(1) + 3(1)(4) + (-2)(2)(-1)$$
$$\quad -4(0)(-2) - (-1)(1)(1) - 1(2)(3)$$
$$= 0 + 12 + 4 - 0 - (-1) - 6$$
$$= 11$$

Method 2   The second method of evaluating a 3 × 3 determinant is called *expansion by minors.*

DEFINITION   The **minor** for an element in a **3 × 3** determinant is the determinant consisting of the elements remaining when the row and column to which the element belongs are deleted. For example, in the determinant

$$\begin{vmatrix} a_1 & b_1 & c_1 \\ a_2 & b_2 & c_2 \\ a_3 & b_3 & c_3 \end{vmatrix}$$

$$\text{Minor for element } a_1 = \begin{vmatrix} b_2 & c_2 \\ b_3 & c_3 \end{vmatrix}$$

$$\text{Minor for element } b_2 = \begin{vmatrix} a_1 & c_1 \\ a_3 & c_3 \end{vmatrix}$$

$$\text{Minor for element } c_3 = \begin{vmatrix} a_1 & b_1 \\ a_2 & b_2 \end{vmatrix}$$

Before we can evaluate a 3 × 3 determinant by Method 2, we must first define what is known as the sign array for a 3 × 3 determinant.

DEFINITION   The **sign array** for a 3 × 3 determinant is a **3 × 3** array of signs in the following pattern:

$$\begin{vmatrix} + & - & + \\ - & + & - \\ + & - & + \end{vmatrix}$$

The sign array begins with a + sign in the upper left-hand corner. The signs then alternate between + and − across every row and down every column.

> ### To Evaluate a 3 × 3 Determinant by Expansion of Minors
>
> We can evaluate a 3 × 3 determinant by expanding across any row or down any column as follows:
>
> **Step 1:**   Choose a row or column to expand about.
>
> **Step 2:**   Write the product of each element in the row or column chosen in step 1 with its minor.
>
> **Step 3:**   Connect the three products in step 2 with the signs in the corresponding row or column in the sign array.

To illustrate the procedure, we will use the same determinant we used in Example 3.

EXAMPLE 4   Expand across the first row:

$$\begin{vmatrix} 1 & 3 & -2 \\ 2 & 0 & 1 \\ 4 & -1 & 1 \end{vmatrix}$$

Solution   The products of the three elements in row 1 with their minors are

$$1\begin{vmatrix} 0 & 1 \\ -1 & 1 \end{vmatrix} \qquad 3\begin{vmatrix} 2 & 1 \\ 4 & 1 \end{vmatrix} \qquad (-2)\begin{vmatrix} 2 & 0 \\ 4 & -1 \end{vmatrix}$$

Connecting these three products with the signs from the first row of the sign array, we have

$$+1\begin{vmatrix} 0 & 1 \\ -1 & 1 \end{vmatrix} - 3\begin{vmatrix} 2 & 1 \\ 4 & 1 \end{vmatrix} + (-2)\begin{vmatrix} 2 & 0 \\ 4 & -1 \end{vmatrix}$$

We complete the problem by evaluating each of the three 2 × 2 determinants and then simplifying the resulting expression:

$$+1[0 - (-1)] - 3(2 - 4) + (-2)(-2 - 0)$$
$$= 1(1) - 3(-2) + (-2)(-2)$$
$$= 1 + 6 + 4$$
$$= 11$$

The results of Examples 3 and 4 match. It makes no difference which method we use — the value of a 3 × 3 determinant is unique.

Note: This method of evaluating a determinant is actually more valuable than our first method, because it works with any size determinant from 3 × 3 to 4 × 4 to any higher order determinant. Method 1 works only on 3 × 3 determinants. It cannot be used on a 4 × 4 determinant.

EXAMPLE 5   Expand down column 2:

$$\begin{vmatrix} 2 & 3 & -2 \\ 1 & 4 & 1 \\ 1 & 5 & -1 \end{vmatrix}$$

Solution   We connect the products of elements in column 2 and their minors with the signs from the second column in the sign array:

$$\begin{vmatrix} 2 & 3 & -2 \\ 1 & 4 & 1 \\ 1 & 5 & -1 \end{vmatrix} = -3\begin{vmatrix} 1 & 1 \\ 1 & -1 \end{vmatrix} + 4\begin{vmatrix} 2 & -2 \\ 1 & -1 \end{vmatrix} - 5\begin{vmatrix} 2 & -2 \\ 1 & 1 \end{vmatrix}$$

$$= -3(-1 - 1) + 4[-2 - (-2)] - 5[2 - (-2)]$$

$$= -3(-2) + 4(0) - 5(4)$$

$$= 6 + 0 - 20$$

$$= -14$$

A Note on the History of Determinants   Determinants were originally known as resultants, a name given to them by Pierre Simon Laplace; however, the work of Gottfried Wilhelm Leibniz contains the germ of the original idea of resultants, or determinants.

 # Getting Ready for Class

After reading through the preceding section, respond in your own words and in complete sentences.

A. Describe how you evaluate a 2 × 2 determinant.

B. Name the row and column that the number 3 is in: $\begin{vmatrix} 1 & 3 \\ 5 & 7 \end{vmatrix}$

C. What is the sign array for a 3 × 3 determinant?

D. If the value of a determinant is 0, does one of the elements have to be 0? Explain.

PROBLEM SET 4.3

Find the value of the following 2 × 2 determinants.

1. $\begin{vmatrix} 1 & 0 \\ 2 & 3 \end{vmatrix}$

2. $\begin{vmatrix} 5 & 4 \\ 3 & 2 \end{vmatrix}$

3. $\begin{vmatrix} 2 & 1 \\ 3 & 4 \end{vmatrix}$

4. $\begin{vmatrix} 4 & 1 \\ 5 & 2 \end{vmatrix}$

5. $\begin{vmatrix} 0 & 1 \\ 1 & 0 \end{vmatrix}$

6. $\begin{vmatrix} 1 & 0 \\ 0 & 1 \end{vmatrix}$

7. $\begin{vmatrix} -3 & 2 \\ 6 & -4 \end{vmatrix}$

8. $\begin{vmatrix} 8 & -3 \\ -2 & 5 \end{vmatrix}$

9. $\begin{vmatrix} -3 & -1 \\ 4 & -2 \end{vmatrix}$

10. $\begin{vmatrix} 5 & 3 \\ 7 & -6 \end{vmatrix}$

Solve each of the following for x.

11. $\begin{vmatrix} 2x & 1 \\ x & 3 \end{vmatrix} = 10$ **12.** $\begin{vmatrix} 3x & -2 \\ 2x & 3 \end{vmatrix} = 26$

13. $\begin{vmatrix} 1 & 2x \\ 2 & -3x \end{vmatrix} = 21$ **14.** $\begin{vmatrix} -5 & 4x \\ 1 & -x \end{vmatrix} = 27$

15. $\begin{vmatrix} 2x & -4 \\ 2 & x \end{vmatrix} = -8x$ **16.** $\begin{vmatrix} 3x & 2 \\ 2 & x \end{vmatrix} = -11x$

17. $\begin{vmatrix} x^2 & 3 \\ x & 1 \end{vmatrix} = 10$ **18.** $\begin{vmatrix} x^2 & -2 \\ x & 1 \end{vmatrix} = 35$

Find the value of each of the following 3×3 determinants by using Method 1 of this section.

19. $\begin{vmatrix} 1 & 2 & 0 \\ 0 & 2 & 1 \\ 1 & 1 & 1 \end{vmatrix}$ **20.** $\begin{vmatrix} -1 & 0 & 2 \\ 3 & 0 & 1 \\ 0 & 1 & 3 \end{vmatrix}$

21. $\begin{vmatrix} 1 & 2 & 3 \\ 3 & 2 & 1 \\ 1 & 1 & 1 \end{vmatrix}$ **22.** $\begin{vmatrix} -1 & 2 & 0 \\ 3 & -2 & 1 \\ 0 & 5 & 4 \end{vmatrix}$

Find the value of each determinant by using Method 2 and expanding across the first row.

23. $\begin{vmatrix} 0 & 1 & 2 \\ 1 & 0 & 1 \\ -1 & 2 & 0 \end{vmatrix}$ **24.** $\begin{vmatrix} 3 & -2 & 1 \\ 0 & -1 & 0 \\ 2 & 0 & 1 \end{vmatrix}$

25. $\begin{vmatrix} 3 & 0 & 2 \\ 0 & -1 & -1 \\ 4 & 0 & 0 \end{vmatrix}$ **26.** $\begin{vmatrix} 1 & 1 & 1 \\ 1 & -1 & 1 \\ 1 & 1 & -1 \end{vmatrix}$

Find the value of each of the following determinants.

27. $\begin{vmatrix} 2 & -1 & 0 \\ 1 & 0 & -2 \\ 0 & 1 & 2 \end{vmatrix}$ **28.** $\begin{vmatrix} 5 & 0 & -4 \\ 0 & 1 & 3 \\ -1 & 2 & -1 \end{vmatrix}$

29. $\begin{vmatrix} 1 & 3 & 7 \\ -2 & 6 & 4 \\ 3 & 7 & -1 \end{vmatrix}$ **30.** $\begin{vmatrix} 2 & 1 & 5 \\ 6 & -3 & 4 \\ 8 & 9 & -2 \end{vmatrix}$

Applying the Concepts

31. Slope-Intercept Form Show that the following determinant equation is another way to write the slope-intercept form of the equation of a line.

$$\begin{vmatrix} y & x \\ m & 1 \end{vmatrix} = b$$

32. Temperature Conversions Show that the following determinant equation is another way to write the equation $F = \frac{9}{5}C + 32$.

$$\begin{vmatrix} C & F & 1 \\ 5 & 41 & 1 \\ -10 & 14 & 1 \end{vmatrix} = 0$$

33. Amusement Park Income From 1986 to 1990, the annual income of amusement parks was linearly increasing, after which time it remained fairly constant. The annual income y, in billions of dollars, may be found for one of these years by evaluating the following determinant equation, in which x represents the number of years past 1986.

$$\begin{vmatrix} x & -1.7 \\ 2 & 0.3 \end{vmatrix} = y$$

(a) Write the determinant equation in slope-intercept form.
(b) Use the equation from part (a) to find the approximate income for amusement parks in the year 1988.

34. College Enrollment From 1981 the enrollment of women in the United States armed forces was linearly increasing until 1990, after which it declined. The approximate number of women, w, enrolled in the armed forces from 1981 to 1990 may be found by evaluating the following determinant equation, in which x represents the number of years past 1981.

$$\begin{vmatrix} 6,509 & -2 \\ 85,709 & x \end{vmatrix} = w$$

Use this equation to determine the number of women enrolled in the armed forces in 1985.

Review Problems

The problems below review material we covered in Section 3.5.

For each relation below, state the domain and the range, and indicate which are also functions.

35. $\{(1, 2), (3, 4), (4, 2)\}$
36. $\{(0, 0), (1, 1), (0, 1)\}$
37. $\{(3, 1), (2, 3), (1, 2)\}$
38. $\{(-1, 1), (2, -2), (-3, -3)\}$

State whether each of the following graphs is the graph of a function.

39.

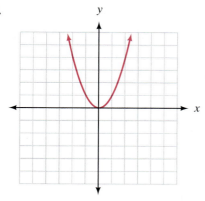

40.

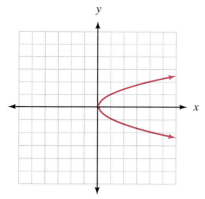

41.

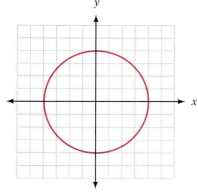

42.

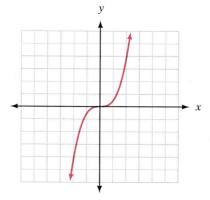

Extending the Concepts

A 4×4 determinant can be evaluated only by using Method 2, expansion by minors; Method 1 will not work. Below is a 4×4 determinant and its associated sign array.

$$\begin{vmatrix} 2 & 0 & 1 & -3 \\ -1 & 2 & 0 & 1 \\ -3 & 0 & 1 & 0 \\ 1 & 1 & 0 & 0 \end{vmatrix} \qquad \begin{vmatrix} + & - & + & - \\ - & + & - & + \\ + & - & + & - \\ - & + & - & + \end{vmatrix}$$

4×4 determinant 4×4 sign array

43. Use expansion by minors to evaluate the preceding 4×4 determinant by expanding it across row 1.

44. Evaluate the preceding determinant by expanding it down column 4.

45. Use expansion by minors down column 3 to evaluate the preceding determinant.

46. Evaluate the preceding determinant by expanding it across row 4.

We begin this section with a look at how determinants can be used to solve a system of linear equations in two variables. The method we use is called *Cramer's rule.* We state it here as a theorem without proof.

THEOREM (CRAMER'S RULE)

The solution to the system

$$a_1x + b_1y = c_1$$
$$a_2x + b_2y = c_2$$

is given by

$$x = \frac{D_x}{D}, \qquad y = \frac{D_y}{D}$$

where

$$D = \begin{vmatrix} a_1 & b_1 \\ a_2 & b_2 \end{vmatrix} \qquad D_x = \begin{vmatrix} c_1 & b_1 \\ c_2 & b_2 \end{vmatrix} \qquad D_y = \begin{vmatrix} a_1 & c_1 \\ a_2 & c_2 \end{vmatrix} \qquad (D \neq 0)$$

The determinant D is made up of the coefficients of x and y in the original system. The determinants D_x and D_y are found by replacing the coefficients of x or y by the constant terms in the original system. Notice also that Cramer's rule does not apply if $D = 0$. In this case the equations are either inconsistent or dependent.

EXAMPLE 1   Use Cramer's rule to solve

$$2x - 3y = 4$$

$$4x + 5y = 3$$

Solution   We begin by calculating the determinants D, D_x, and D_y.

$$D = \begin{vmatrix} 2 & -3 \\ 4 & 5 \end{vmatrix} = 2(5) - 4(-3) = 22$$

$$D_x = \begin{vmatrix} 4 & -3 \\ 3 & 5 \end{vmatrix} = 4(5) - 3(-3) = 29$$

$$D_y = \begin{vmatrix} 2 & 4 \\ 4 & 3 \end{vmatrix} = 2(3) - 4(4) = -10$$

$$x = \frac{D_x}{D} = \frac{29}{22} \qquad \text{and} \qquad y = \frac{D_y}{D} = \frac{-10}{22} = -\frac{5}{11}$$

The solution set for the system is $\{(\frac{29}{22}, -\frac{5}{11})\}$.

Cramer's rule can also be applied to systems of linear equations in three variables.

THEOREM (ALSO CRAMER'S RULE)

The solution set to the system

$$a_1 x + b_1 y + c_1 z = d_1$$
$$a_2 x + b_2 y + c_2 z = d_2$$
$$a_3 x + b_3 y + c_3 z = d_3$$

is given by

$$x = \frac{D_x}{D}, \qquad y = \frac{D_y}{D}, \qquad \text{and} \qquad z = \frac{D_z}{D}$$

where

$$D = \begin{vmatrix} a_1 & b_1 & c_1 \\ a_2 & b_2 & c_2 \\ a_3 & b_3 & c_3 \end{vmatrix} \qquad D_x = \begin{vmatrix} d_1 & b_1 & c_1 \\ d_2 & b_2 & c_2 \\ d_3 & b_3 & c_3 \end{vmatrix} \qquad (D \neq 0)$$

$$D_y = \begin{vmatrix} a_1 & d_1 & c_1 \\ a_2 & d_2 & c_2 \\ a_3 & d_3 & c_3 \end{vmatrix} \qquad D_z = \begin{vmatrix} a_1 & b_1 & d_1 \\ a_2 & b_2 & d_2 \\ a_3 & b_3 & d_3 \end{vmatrix}$$

Again, the determinant D consists of the coefficients of x, y, and z in the original system. The determinants D_x, D_y, and D_z are found by replacing the coefficients of x, y, and z, respectively, with the constant terms from the original system. If $D = 0$, there is no unique solution to the system.

EXAMPLE 2   Use Cramer's rule to solve

$$x + y + z = 6$$
$$2x - y + z = 3$$
$$x + 2y - 3z = -4$$

Solution   This is the same system used in Example 1 in Section 4.2, so we can compare Cramer's rule with our previous methods of solving a system in three variables. We begin by setting up and evaluating D, D_x, D_y, and D_z. (Recall that there are a number of ways to evaluate a 3×3 determinant. Since we have four of these determinants, we can use both Methods 1 and 2 from the previous section.) We evaluate D using Method 1 from Section 4.3.

$$D = \begin{vmatrix} 1 & 1 & 1 \\ 2 & -1 & 1 \\ 1 & 2 & -3 \end{vmatrix} \begin{matrix} 1 & 1 \\ 2 & -1 \\ 1 & 2 \end{matrix}$$

$$= 3 + 1 + 4 - (-1) - (2) - (-6) = 13$$

We evaluate D_x using Method 2 from Section 4.3 and expanding across row 1:

$$D_x = \begin{vmatrix} 6 & 1 & 1 \\ 3 & -1 & 1 \\ -4 & 2 & -3 \end{vmatrix} = 6 \begin{vmatrix} -1 & 1 \\ 2 & -3 \end{vmatrix} - 1 \begin{vmatrix} 3 & 1 \\ -4 & -3 \end{vmatrix} + 1 \begin{vmatrix} 3 & -1 \\ -4 & 2 \end{vmatrix}$$

$$= 6(1) - 1(-5) + 1(2)$$

$$= 13$$

Find D_y by expanding across row 2:

$$D_y = \begin{vmatrix} 1 & 6 & 1 \\ 2 & 3 & 1 \\ 1 & -4 & -3 \end{vmatrix} = -2 \begin{vmatrix} 6 & 1 \\ -4 & -3 \end{vmatrix} + 3 \begin{vmatrix} 1 & 1 \\ 1 & -3 \end{vmatrix} - 1 \begin{vmatrix} 1 & 6 \\ 1 & -4 \end{vmatrix}$$

$$= -2(-14) + 3(-4) - 1(-10)$$

$$= 26$$

Find D_z by expanding down column 1:

$$D_z = \begin{vmatrix} 1 & 1 & 6 \\ 2 & -1 & 3 \\ 1 & 2 & -4 \end{vmatrix} = 1 \begin{vmatrix} -1 & 3 \\ 2 & -4 \end{vmatrix} - 2 \begin{vmatrix} 1 & 6 \\ 2 & -4 \end{vmatrix} + 1 \begin{vmatrix} 1 & 6 \\ -1 & 3 \end{vmatrix}$$

$$= 1(-2) - 2(-16) + 1(9)$$

$$= 39$$

$$x = \frac{D_x}{D} = \frac{13}{13} = 1 \qquad y = \frac{D_y}{D} = \frac{26}{13} = 2 \qquad z = \frac{D_z}{D} = \frac{39}{13} = 3$$

The solution set is $\{(1, 2, 3)\}$.

Note: We are solving each of these determinants by expanding about different rows or columns just to show the different ways these determinants can be evaluated.

EXAMPLE 3   Use Cramer's rule to solve

$$x + y = -1$$
$$2x - z = 3$$
$$y + 2z = -1$$

Solution   It is helpful to rewrite the system using zeros for the coefficients of those variables not shown:

$$x + y + 0z = -1$$
$$2x + 0y - z = 3$$
$$0x + y + 2z = -1$$

The four determinants used in Cramer's rule are

$$D = \begin{vmatrix} 1 & 1 & 0 \\ 2 & 0 & -1 \\ 0 & 1 & 2 \end{vmatrix} = -3$$

$$D_x = \begin{vmatrix} -1 & 1 & 0 \\ 3 & 0 & -1 \\ -1 & 1 & 2 \end{vmatrix} = -6$$

$$D_y = \begin{vmatrix} 1 & -1 & 0 \\ 2 & 3 & -1 \\ 0 & -1 & 2 \end{vmatrix} = 9$$

$$D_z = \begin{vmatrix} 1 & 1 & -1 \\ 2 & 0 & 3 \\ 0 & 1 & -1 \end{vmatrix} = -3$$

$$x = \frac{D_x}{D} = \frac{-6}{-3} = 2 \qquad y = \frac{D_y}{D} = \frac{9}{-3} = -3 \qquad z = \frac{D_z}{D} = \frac{-3}{-3} = 1$$

The solution set is $\{(2, -3, 1)\}$.

Finally, we should mention the possible situations that can occur when the determinant D is 0, when we are using Cramer's rule.

If $D = 0$ and at least one of the other determinants, D_x or D_y (or D_z), is not 0, then the system is inconsistent. In this case there is no solution to the system.

On the other hand, if $D = 0$ and both D_x and D_y (and D_z in a system of three equations in three variables) are 0, then the system is a dependent one.

A Note on the History of Cramer's Rule   Cramer's rule is named after the Swiss mathematician *Gabriel Cramer* (1704–1752). Cramer's rule appeared in the appendix of an algebraic work of his classifying curves, but the basic idea behind his now-famous rule was formulated earlier by Leibniz and Chinese mathematicians. It was actually Cramer's superior notation that helped to popularize the technique.

Cramer has a respectable reputation as a mathematician, but he does not rank with the great mathematicians of his time, although through his extensive travels he met many of them, such as the Bernoullis, Euler, and D'Alembert.

Cramer had very broad interests. He wrote on philosophy, law, and government, as well as mathematics; served in public office; and was an expert on cathedrals, often instructing workers about their repair and coordinating excavations to recover cathedral archives. Cramer never married, and a fall from a carriage eventually led to his death.

Getting Ready for Class

After reading through the preceding section, respond in your own words and in complete sentences.

A. When applying Cramer's rule, when will you use 2 × 2 determinants?

B. Why would it be impossible to use Cramer's rule if the determinant $D = 0$?

C. When applying Cramer's rule to solve a system of two linear equations in two variables, how many numbers should you obtain? How do these numbers relate to the system?

D. What will happen when you apply Cramer's rule to a system of equations made up of two parallel lines?

PROBLEM SET 4.4

Solve each of the following systems using Cramer's rule.

1. $2x - 3y = 3$
$4x - 2y = 10$

2. $3x + y = -2$
$-3x + 2y = -4$

3. $5x - 2y = 4$
$-10x + 4y = 1$

4. $-4x + 3y = -11$
$5x + 4y = 6$

5. $4x - 7y = 3$
$5x + 2y = -3$

6. $3x - 4y = 7$
$6x - 2y = 5$

7. $9x - 8y = 4$
$2x + 3y = 6$

8. $4x - 7y = 10$
$-3x + 2y = -9$

9. $x + y + z = 4$
$x - y - z = 2$
$2x + 2y - z = 2$

10. $-x + y + 3z = 6$
$x + y + 2z = 7$
$2x + 3y + z = 4$

11. $x + y - z = 2$
$-x + y + z = 3$
$x + y + z = 4$

12. $-x - y + z = 1$
$x - y + z = 3$
$x + y - z = 4$

13. $3x - y + 2z = 4$
$6x - 2y + 4z = 8$
$x - 5y + 2z = 1$

14. $2x - 3y + z = 1$
$3x - y - z = 4$
$4x - 6y + 2z = 3$

15. $2x - y + 3z = 4$
$x - 5y - 2z = 1$
$-4x - 2y + z = 3$

16. $4x - y + 5z = 1$
$2x + 3y + 4z = 5$
$x + y + 3z = 2$

17. $-x - 7y = 1$
$x + 3z = 11$
$2y + z = 0$

18. $x + y = 2$
$-x + 3z = 0$
$2y + z = 3$

19. $x - y = 2$
$3x + z = 11$
$y - 2z = -3$

20. $4x + 5y = -1$
$2y + 3z = -5$
$x + 2z = -1$

Applying the Concepts

21. Break-Even Point If a company has fixed costs of $100 per week and each item it produces costs $10 to manufacture, then the total cost y per week to produce x items is

$$y = 10x + 100$$

If the company sells each item it manufactures for $12, then the total amount of money y the company brings in for selling x items is

$$y = 12x$$

Use Cramer's rule to solve the system

$$y = 10x + 100$$

$$y = 12x$$

for x to find the number of items the company must sell per week in order to break even.

22. Break-Even Point Suppose a company has fixed costs of $200 per week and each item it produces costs $20 to manufacture.

(a) Write an equation that gives the total cost per week y to manufacture x items.

(b) If each item sells for $25, write an equation that gives the total amount of money y the company brings in for selling x items.

(c) Use Cramer's rule to find the number of items the company must sell each week to break even.

23. Health Insurance For years between 1980 and 1991, the number (in millions) of U.S. residents without health insurance, y, may be approximated by the equation

$$y = 0.98x - 1,915.8$$

where x represents the year, and $1980 \leq x \leq 1991$. To determine the year in which 30 million U.S. residents were without health insurance, we solve the system of equations made up of the equation above and the equation $y = 30$. Solve this system using Cramer's rule. (When you obtain an answer, you will need to round it to the nearest year.)

24. Price Index From 1970 to 1990, the price index of dental care, d, may be closely approximated by the equation

$$d = 6x - 11,780$$

where x is the year, and $1970 \leq x \leq 1990$. Determine when the price index for dental care reached 120 by forming a system of equations using the equation above along with the equation $d = 120$. Solve this system using Cramer's rule.

Review Problems

The following problems review material we covered in Section 3.6.

Let $f(x) = \frac{1}{2}x + 3$ and $g(x) = x^2 - 4$, and find

25. $f(0)$ **26.** $g(0)$

27. $g(2)$ **28.** $f(2)$

29. $f(-4)$ **30.** $g(-6)$

31. $f[g(2)]$ **32.** $g[f(2)]$

Extending the Concepts

Solve for x and y using Cramer's rule. Your answers will contain the constants a and b.

33. $ax + by = -1$
$\quad\ bx + ay = 1$

34. $ax + y = b$
$\quad\ bx + y = a$

35. $a^2x + by = 1$
$\quad\ b^2x + ay = 1$

36. $ax + by = a$
$\quad\ bx + ay = a$

37. Name the system of equations for which Cramer's rule yields the following determinants.

$$D = \begin{vmatrix} 1 & 2 \\ 3 & 4 \end{vmatrix} \qquad D_x = \begin{vmatrix} 1 & 2 \\ 0 & 4 \end{vmatrix}$$

38. Name the system of equations for which Cramer's rule yields the following determinants.

$$D = \begin{vmatrix} 1 & 3 & 2 \\ -1 & 0 & 4 \\ 2 & 5 & -1 \end{vmatrix} \qquad D_y = \begin{vmatrix} 1 & 1 & 2 \\ -1 & 3 & 4 \\ 2 & 5 & -1 \end{vmatrix}$$

4.5 Applications

Many times word problems involve more than one unknown quantity. If a problem is stated in terms of two unknowns and we represent each unknown quantity with a different variable, then we must write the relationships between the variables with two equations. The two equations written in terms of the two variables form a system of linear equations that we solve using the methods developed in this chapter. If we find a problem that relates three unknown quantities, then we need three equations in order to form a linear system we can solve.

Here is our Blueprint for Problem Solving, modified to fit the application problems that you will find in this section.

Blueprint for Problem Solving Using a System of Equations

Step 1: ***Read*** the problem, and then mentally ***list*** the items that are known and the items that are unknown.

Step 2: ***Assign variables*** to each of the unknown items. That is, let $x =$ one of the unknown items and $y =$ the other unknown item (and $z =$ the third unknown item, if there is a third one). Then ***translate*** the other ***information*** in the problem to expressions involving the two (or three) variables.

Step 3: ***Reread*** the problem, and then ***write a system of equations,*** using the items and variables listed in steps 1 and 2, that describes the situation.

Step 4: ***Solve the system*** found in step 3.

Step 5: ***Write your answers*** using complete sentences.

Step 6: ***Reread*** the problem, and ***check*** your solution with the original words in the problem.

EXAMPLE 1   One number is 2 more than 3 times another. Their sum is 26. Find the two numbers.

Solution   Applying the steps from our Blueprint, we have:

Step 1: ***Read and list.***
We know that we have two numbers, whose sum is 26. One of them is 2 more than 3 times the other. The unknown quantities are the two numbers.

Step 2: ***Assign variables and translate information.***
Let $x =$ one of the numbers. Then $y =$ the other number.

Step 3: ***Write a system of equations.***
The first sentence in the problem translates into $y = 3x + 2$. The second sentence gives us a second equation: $x + y = 26$. Together, these two equations give us the following system of equations:

$$x + y = 26$$

$$y = 3x + 2$$

Step 4: ***Solve the system.***
Substituting the expression for y from the second equation into the first and solving for x yields

$$x + (3x + 2) = 26$$

$$4x + 2 = 26$$

$$4x = 24$$

$$x = 6$$

Using $x = 6$ in $y = 3x + 2$ gives the second number:

$$y = 3(6) + 2$$
$$y = 20$$

Step 5: Write answers.
The two numbers are 6 and 20.

Step 6: Reread and check.
The sum of 6 and 20 is 26, and 20 is 2 more than 3 times 6.

EXAMPLE 2 Suppose 850 tickets were sold for a game for a total of $1,100. If adult tickets cost $1.50 and children's tickets cost $1.00, how many of each kind of ticket were sold?

Solution

Step 1: Read and list.
The total number of tickets sold is 850. The total income from tickets is $1,100. Adult tickets are $1.50 each. Children's tickets are $1.00 each. We don't know how many of each type of ticket have been sold.

Step 2: Assign variables and translate information.
We let $x =$ the number of adult tickets and $y =$ the number of children's tickets.

Step 3: Write a system of equations.
The total number of tickets sold is 850, giving us our first equation.

$$x + y = 850$$

Since each adult ticket costs $1.50 and each children's ticket costs $1.00 and the total amount of money paid for tickets was $1,100, a second equation is

$$1.50x + 1.00y = 1,100$$

The same information can also be obtained by summarizing the problem with a table. One such table follows. Notice that the two equations we obtained previously are given by the two rows of the table.

	Adult Tickets	Children's Tickets	Total
Number	x	y	850
Value	$1.50x$	$1.00y$	1,100

Whether we use a table to summarize the information in the problem, or just talk our way through the problem, the system of equations that describes the situation is

$$x + y = 850$$
$$1.50x + 1.00y = 1,100$$

Step 4: Solve the system.

If we multiply the second equation by 10 to clear it of decimals, we have the system

$$x + y = 850$$

$$15x + 10y = 11{,}000$$

Multiplying the first equation by -10 and adding the result to the second equation eliminates the variable y from the system:

$$\begin{array}{r} -10x - 10y = -8{,}500 \\ 15x + 10y = 11{,}000 \\ \hline 5x = 2{,}500 \\ x = 500 \end{array}$$

The number of adult tickets sold was 500. To find the number of children's tickets, we substitute $x = 500$ into $x + y = 850$ to get

$$500 + y = 850$$

$$y = 350$$

Step 5: Write answers.

The number of children's tickets is 350, and the number of adult tickets is 500.

Step 6: Reread and check.

The total number of tickets is $350 + 500 = 850$. The amount of money from selling the two types of tickets is

$$\begin{array}{l} 350 \text{ children's tickets at } \$1.00 \text{ each is } 350(1.00) = \$350 \\ 500 \text{ adult tickets at } \$1.50 \text{ each is } 500(1.50) = \$750 \\ \hline \end{array}$$

The total income from ticket sales is \$1,100

EXAMPLE 3 Suppose a person invests a total of \$10,000 in two accounts. One account earns 8% annually, and the other earns 9% annually. If the total interest earned from both accounts in a year is \$860, how much was invested in each account?

Solution

Step 1: Read and list.

The total investment is \$10,000 split between two accounts. One account earns 8% annually, and the other earns 9% annually. The interest from both accounts is \$860 in 1 year. We don't know how much is in each account.

Step 2: Assign variables and translate information.

We let x equal the amount invested at 9% and y be the amount invested at 8%.

Step 3: Write a system of equations.

Since the total investment is \$10,000, one relationship between x and y can be written as

$$x + y = 10,000$$

The total interest earned from both accounts is $860. The amount of interest earned on x dollars at 9% is 0.09x, while the amount of interest earned on y dollars at 8% is 0.08y. This relationship is represented by the equation

$$0.09x + 0.08y = 860$$

The two equations we have just written can also be found by first summarizing the information from the problem in a table. Again, the two rows of the table yield the two equations just written. Here is the table.

	Dollars at 9%	**Dollars at 8%**	**Total**
Number	x	y	10,000
Interest	0.09x	0.08y	860

The system of equations that describes this situation is given by

$$x + \quad y = 10,000$$

$$0.09x + 0.08y = \quad 860$$

Step 4: Solve the system.

Multiplying the second equation by 100 will clear it of decimals. The system that results after doing so is

$$x + \ y = 10,000$$

$$9x + 8y = 86,000$$

We can eliminate y from this system by multiplying the first equation by -8 and adding the result to the second equation.

$$-8x - 8y = -80,000$$
$$\underline{9x + 8y = \quad 86,000}$$
$$x \qquad = \quad 6,000$$

The amount of money invested at 9% is $6,000. Since the total investment was $10,000, the amount invested at 8% must be $4,000.

Step 5: Write answers.

The amount invested at 8% is $4,000, and the amount invested at 9% is $6,000.

Step 6: Reread and check.

The total investment is $4,000 + $6,000 = $10,000. The amount of interest earned from the two accounts is

In 1 year, $4,000 invested at 8% earns 0.08(4,000) = $320
In 1 year, $6,000 invested at 9% earns 0.09(6,000) = $540

The total interest from the two accounts is $860

EXAMPLE 4   How much 20% alcohol solution and 50% alcohol solution must be mixed to get 12 gallons of 30% alcohol solution?

Solution   To solve this problem, we must first understand that a 20% alcohol solution is 20% alcohol and 80% water.

Step 1: *Read and list.*

We will mix two solutions to obtain 12 gallons of solution that is 30% alcohol. One of the solutions is 20% alcohol and the other 50% alcohol. We don't know how much of each solution we need.

Step 2: *Assign variables and translate information.*

Let x = the number of gallons of 20% alcohol solution needed, and y = the number of gallons of 50% alcohol solution needed.

Step 3: *Write a system of equations.*

Since we must end up with a total of 12 gallons of solution, one equation for the system is

$$x + y = 12$$

The amount of alcohol in the x gallons of 20% solution is $0.20x$, while the amount of alcohol in the y gallons of 50% solution is $0.50y$. Since the total amount of alcohol in the 20% and 50% solutions must add up to the amount of alcohol in the 12 gallons of 30% solution, the second equation in our system can be written as

$$0.20x + 0.50y = 0.30(12)$$

Again, let's make a table that summarizes the information we have to this point in the problem.

	20% Solution	50% Solution	Final Solution
Total number of gallons	x	y	12
Gallons of alcohol	$0.20x$	$0.50y$	$0.30(12)$

20% alcohol   50% alcohol

12 gal
30% alcohol

Our system of equations is

$$x + \quad y = 12$$

$$0.20x + 0.50y = 0.30(12) = 3.6$$

Step 4: Solve the system.

Multiplying the second equation by 10 gives us an equivalent system:

$$x + y = 12$$

$$2x + 5y = 36$$

Multiplying the top equation by -2 to eliminate the x-variable, we have

$$
\begin{array}{rcr}
-2x - 2y &=& -24 \\
2x + 5y &=& 36 \\
\hline
3y &=& 12 \\
y &=& 4
\end{array}
$$

Substituting $y = 4$ into $x + y = 12$, we solve for x:

$$x + 4 = 12$$

$$x = 8$$

Step 5: Write answers.

It takes 8 gallons of 20% alcohol solution and 4 gallons of 50% alcohol solution to produce 12 gallons of 30% alcohol solution.

Step 6: Reread and check.

If we mix 8 gallons of 20% solution and 4 gallons of 50% solution, we end up with a total of 12 gallons of solution. To check the percentages we look for the total amount of alcohol in the two initial solutions and in the final solution.

In the initial solutions

The amount of alcohol in 8 gallons of 20% solution is $0.20(8) = 1.6$ gallons
The amount of alcohol in 4 gallons of 50% solution is $0.50(4) = 2.0$ gallons

The total amount of alcohol in the initial solutions is 3.6 gallons

In the final solution

The amount of alcohol in 12 gallons of 30% solution is $0.30(12) = 3.6$ gallons

EXAMPLE 5

It takes 2 hours for a boat to travel 28 miles downstream (with the current). The same boat can travel 18 miles upstream (against the current) in 3 hours. What is the speed of the boat in still water, and what is the speed of the current of the river?

Solution

Step 1: *Read and list.*

A boat travels 18 miles upstream and 28 miles downstream. The trip upstream takes 3 hours. The trip downstream takes 2 hours. We don't know the speed of the boat or the speed of the current.

Step 2: *Assign variables and translate information.*

Let x = the speed of the boat in still water and let y = the speed of the current. The average speed (rate) of the boat upstream is $x - y$, since it is traveling against the current. The rate of the boat downstream is $x + y$, since the boat is traveling with the current.

Step 3: *Write a system of equations.*

Putting the information into a table, we have

	d (distance, miles)	r (rate, mph)	t (time, h)
Upstream	18	$x - y$	3
Downstream	28	$x + y$	2

The formula for the relationship between distance d, rate r, and time t is $d = rt$ (the rate equation). Since $d = r \cdot t$, the system we need to solve the problem is

$$18 = (x - y) \cdot 3$$
$$28 = (x + y) \cdot 2$$

which is equivalent to

$$6 = x - y$$
$$14 = x + y$$

Step 4: *Solve the system.*

Adding the two equations, we have

$$20 = 2x$$
$$x = 10$$

Substituting $x = 10$ into $14 = x + y$, we see that

$$y = 4$$

Step 5: *Write answers.*

The speed of the boat in still water is 10 miles per hour; the speed of the current is 4 miles per hour.

Step 6: *Reread and check.*

The boat travels at $10 + 4 = 14$ miles per hour downstream, so in 2 hours it will travel $14 \cdot 2 = 28$ miles. The boat travels at $10 - 4 = 6$ miles per hour upstream, so in 3 hours it will travel $6 \cdot 3 = 18$ miles.

EXAMPLE 6   A coin collection consists of 14 coins with a total value of \$1.35. If the coins are nickels, dimes, and quarters, and the number of nickels is 3 less than twice the number of dimes, how many of each coin is there in the collection?

Solution   This problem will require three variables and three equations.

Step 1: *Read and list.*

We have 14 coins with a total value of \$1.35. The coins are nickels, dimes, and quarters. The number of nickels is 3 less than twice the number of dimes. We do not know how many of each coin we have.

Step 2: *Assign variables and translate information.*

Since we have three types of coins, we will have to use three variables. Let's let $x =$ the number of nickels, $y =$ the number of dimes, and $z =$ the number of quarters.

Step 3: *Write a system of equations.*

Since the total number of coins is 14, our first equation is

$$x + y + z = 14$$

Since the number of nickels is 3 less than twice the number of dimes, a second equation is

$$x = 2y - 3 \qquad \text{which is equivalent to} \qquad x - 2y = -3$$

Our last equation is obtained by considering the value of each coin and the total value of the collection. Let's write the equation in terms of cents, so we won't have to clear it of decimals later.

$$5x + 10y + 25z = 135$$

Here is our system, with the equations numbered for reference:

$$
\begin{aligned}
x + \ y + \ z &= \ 14 \qquad &(1)\\
x - \ 2y \quad\ &= -3 \qquad &(2)\\
5x + 10y + 25z &= 135 \qquad &(3)
\end{aligned}
$$

Step 4: *Solve the system.*

Let's begin by eliminating x from the first and second equations, and the first and third equations. Adding -1 times the second equation to the first equation gives us an equation in only y and z. We call this equation (4).

$$3y + z = 17 \qquad (4)$$

Adding -5 times equation (1) to equation (3) gives us

$$5y + 20z = 65 \qquad (5)$$

We can eliminate z from equations (4) and (5) by adding -20 times (4) to (5). Here is the result:

$$-55y = -275$$

$$y = 5$$

Substituting $y = 5$ into equation (4) gives us $z = 2$. Substituting $y = 5$ and $z = 2$ into equation (1) gives us $x = 7$.

Step 5: Write answers.
The collection consists of 7 nickels, 5 dimes, and 2 quarters.

Step 6: Reread and check.
The total number of coins is $7 + 5 + 2 = 14$. The number of nickels, 7, is 3 less than twice the number of dimes, 5. To find the total value of the collection, we have

The value of the 7 nickels is $7(0.05) = \$0.35$
The value of the 5 dimes is $5(0.10) = \$0.50$
The value of the 2 quarters is $2(0.25) = \$0.50$

The total value of the collection is $\$1.35$

If you go on to take a chemistry class, you may see the next example (or one much like it).

EXAMPLE 7   In a chemistry lab, students record the temperature of water at room temperature and find that it is 77° on the Fahrenheit temperature scale and 25° on the Celsius temperature scale. The water is then heated until it boils. The temperature of the boiling water is 212°F and 100°C. Assume that the relationship

77°F

between the two temperature scales is a linear one, then use the preceding data to find the formula that gives the Celsius temperature C in terms of the Fahrenheit temperature F.

Solution   The data is summarized in Table 1.

TABLE I Corresponding Temperatures	
In Degrees Fahrenheit	**In Degrees Celsius**
77	25
212	100

If we assume the relationship is linear, then the formula that relates the two temperature scales can be written in slope-intercept form as

$$C = mF + b$$

Substituting $C = 25$ and $F = 77$ into this formula gives us

$$25 = 77m + b$$

Substituting $C = 100$ and $F = 212$ into the formula yields

$$100 = 212m + b$$

Together, the two equations form a system of equations, which we can solve using the addition method.

$$
\begin{array}{ll}
25 = 77m + b \xrightarrow{\text{Times } -1} & -25 = -77m - b \\
100 = 212m + b \xrightarrow{\text{No change}} & 100 = 212m + b \\
\hline
& 75 = 135m
\end{array}
$$

$$m = \frac{75}{135} = \frac{5}{9}$$

To find the value of b, we substitute $m = \frac{5}{9}$ into $25 = 77m + b$ and solve for b.

$$25 = 77\left(\frac{5}{9}\right) + b$$

$$25 = \frac{385}{9} + b$$

$$b = 25 - \frac{385}{9} = \frac{225}{9} - \frac{385}{9} = -\frac{160}{9}$$

The equation that gives C in terms of F is

$$C = \frac{5}{9}F - \frac{160}{9}$$

Note: Although we have solved each system of equations in this section by the addition method or the substitution method, we could have used Cramer's rule also. Cramer's rule applies to any system of linear equations with a unique solution.

 Getting Ready for Class

After reading through the preceding section, respond in your own words and in complete sentences.

A. If you were to apply the Blueprint for Problem Solving from Section 2.3 to the examples in this section, what would be the first step?

B. If you were to apply the Blueprint for Problem Solving from Section 2.3 to the examples in this section, what would be the last step?

C. When working application problems involving boats moving in rivers, how does the current of the river affect the speed of the boat?

D. Write an application problem for which the solution depends on solving the system of equations:

$$x + \quad y = 1,000$$
$$0.05x + 0.06y = \quad 55$$

PROBLEM SET 4.5

Number Problems

1. One number is 3 more than twice another. The sum of the numbers is 18. Find the two numbers.

2. The sum of two numbers is 32. One of the numbers is 4 less than 5 times the other. Find the two numbers.

3. The difference of two numbers is 6. Twice the smaller is 4 more than the larger. Find the two numbers.

4. The larger of two numbers is 5 more than twice the smaller. If the smaller is subtracted from the larger, the result is 12. Find the two numbers.

5. The sum of three numbers is 8. Twice the smallest is 2 less than the largest, while the sum of the largest and smallest is 5. Use a linear system in three variables to find the three numbers.

6. The sum of three numbers is 14. The largest is 4 times the smallest, while the sum of the smallest and twice the largest is 18. Use a linear system in three variables to find the three numbers.

Ticket and Interest Problems

7. A total of 925 tickets were sold for a game for a total of $1,150. If adult tickets sold for $2.00 and children's tickets sold for $1.00, how many of each kind of ticket were sold?

8. If tickets for a show cost $2.00 for adults and $1.50 for children, how many of each kind of ticket were sold if a total of 300 tickets were sold for $525?

9. Mr. Jones has $20,000 to invest. He invests part at 6% and the rest at 7%. If he earns $1,280 in interest after 1 year, how much did he invest at each rate?

10. A man invests $17,000 in two accounts. One account earns 5% interest per year and the other 6.5%. If his total interest after 1 year is $970, how much did he invest at each rate?

11. Susan invests twice as much money at 7.5% as she does at 6%. If her total interest after 1 year is $840, how much does she have invested at each rate?

12. A woman earns $1,350 in interest from two accounts in 1 year. If she has three times as much invested at 7% as she does at 6%, how much does she have in each account?

13. A man invests $2,200 in three accounts that pay 6%, 8%, and 9% in annual interest, respectively. He has three times as much invested at 9% as he does at 6%. If his total interest for the year is $178, how much is invested at each rate?

14. A student has money in three accounts that pay 5%, 7%, and 8% in annual interest. She has three times as much invested at 8% as she does at 5%. If the total amount she has invested is $1,600 and her interest for the year comes to $115, how much money does she have in each account?

Mixture Problems

15. How many gallons of 20% alcohol solution and 50% alcohol solution must be mixed to get 9 gallons of 30% alcohol solution?

16. How many ounces of 30% hydrochloric acid solution and 80% hydrochloric acid solution must be mixed to get 10 ounces of 50% hydrochloric acid solution?

17. A mixture of 16% disinfectant solution is to be made from 20% and 14% disinfectant solutions. How much of each solution should be used if 15 gallons of the 16% solution are needed?

18. How much 25% antifreeze and 50% antifreeze should be combined to give 40 gallons of 30% antifreeze?

Rate Problems

19. It takes a boat 2 hours to travel 24 miles downstream and 3 hours to travel 18 miles upstream. What is the speed of the boat in still water? What is the speed of the current of the river?

20. A boat on a river travels 20 miles downstream in only 2 hours. It takes the same boat 6 hours to travel 12 miles upstream. What are the speed of the boat and the speed of the current?

21. An airplane flying with the wind can cover a certain distance in 2 hours. The return trip against the wind takes $2\frac{1}{2}$ hours. How fast is the plane and what is the speed of the air, if the distance is 600 miles?

22. An airplane covers a distance of 1,500 miles in 3 hours when it flies with the wind and $3\frac{1}{3}$ hours when it flies against the wind. What is the speed of the plane in still air?

Coin Problems

23. Bob has 20 coins totaling $1.40. If he has only dimes and nickels, how many of each coin does he have?

24. If Amy has 15 coins totaling $2.70, and the coins are quarters and dimes, how many of each coin does she have?

25. A collection of nickels, dimes, and quarters consists of 9 coins with a total value of $1.20. If the number of dimes is equal to the number of nickels, find the number of each type of coin.

26. A coin collection consists of 12 coins with a total value of $1.20. If the collection consists only of nickels, dimes, and quarters, and the number of dimes is two more than twice the number of nickels, how many of each type of coin are in the collection?

Additional Problems

27. Price and Demand   A manufacturing company finds that they can sell 300 items if the price per item is $2.00, and 400 items if the price is $1.50 per item. If the relationship between the number of items sold x and the price per item p is a linear one, find a formula that gives x in terms of p. Then use the formula to find the number of items they will sell if the price per item is $3.00.

28. Price and Demand   A company manufactures and sells bracelets. They have found from past experience that they can sell 300 bracelets each week if the price per bracelet is $2.00, but only 150 bracelets are sold if the price is $2.50 per bracelet. If the relationship between the number of bracelets sold x and the price per bracelet p is a linear one, find a formula that gives x in terms of p. Then use the formula to find the number of bracelets they will sell at $3.00 each.

29. Garbage Rates   Five Cities Garbage charges a flat monthly fee for their services plus a certain amount for each bag of trash they pick up. A customer notices that the January bill for picking up 5 bags of trash was $25.60, while the February bill for 7 bags of trash was $27.10. Assume the relationship between the total monthly charges C and the number of bags of trash picked up x is a linear relationship. Use the data to find the formula that gives C in terms of x. Then use the formula to predict the cost for picking up 12 bags of trash in one month.

30. **Bottled Water Rates** A bottled water company charges a flat fee each month for the use of their water dispenser plus a certain amount for each gallon of water delivered. Suppose that the company delivers 10 gallons of water in March and the March bill is $18. Then, in April, 15 gallons of water are delivered for a total charge of $23.50. Assume the relationship between the total monthly charges C and the number of gallons of water delivered x is a linear relationship. Use the data to find the formula that gives C in terms of x. Then use the formula to predict the cost if 20 gallons of water are delivered in 1 month.

31. **Height of a Ball** A ball is tossed into the air so that the height after 1, 3, and 5 seconds is as given in the following table. If the relationship between the height of the ball h and the time t is quadratic, then the relationship can be written as

$$h = at^2 + bt + c$$

Use the information in the table to write a system of three equations in three variables a, b, and c. Solve the system to find the exact relationship between h and t. Then find the maximum height attained by the ball.

t (sec)	h (ft)
1	128
3	128
5	0

32. **Height of a Ball** A ball is tossed into the air and its height above the ground after 1, 3, and 4 seconds is recorded as shown in the following table. The relationship between the height of the ball h and the time t is quadratic and can be written as

$$h = at^2 + bt + c$$

Use the information in the table to write a system of three equations in three variables a, b, and c.

Solve the system to find the exact relationship between the variables h and t. Then find the maximum height attained by the ball.

t (sec)	h (ft)
1	96
3	64
4	0

33. **Metal Alloys** Metal workers solve systems of equations when forming metal alloys. If a certain metal alloy is 40% copper and another alloy is 60% copper, then a system of equations may be written to determine the amount of each alloy necessary to make 50 pounds of a metal alloy that is 55% copper. Write the system and determine this amount.

34. **Cell Phones** In 1990 5.8 million people bought cellular telephones, and in 1993 13.8 million people bought them. The relationship between the years 1990 through 1993 and the sales during these years was a linear one.
    (a) Find an equation relating sales, S, to the year, x, for this time period.
    (b) Use the equation from part (a) to determine the number of people who bought cellular telephones in 1992.

Review Problems

The following problems review material we covered in Section 3.7.

35. If y varies directly with the square of x, and y is 75 when x is 5, find y when x is 7.

36. Suppose y varies directly with the cube of x. If y is 16 when x is 2, find y when x is 3.

37. Suppose y varies inversely with x. If y is 10 when x is 25, find x when y is 5.

38. If y varies inversely with the cube of x, and y is 2 when x is 2, find y when x is 4.

39. Suppose z varies jointly with x and the square of y. If z is 40 when x is 5 and y is 2, find z when x is 2 and y is 5.

40. Suppose z varies jointly with x and the cube of y. If z is 48 when x is 3 and y is 2, find z when x is 4 and y is $\frac{1}{2}$.

Extending the Concepts

41. High School Dropout Rate   The high school dropout rates for males and females over the years 1967 to 1991 are shown in the following table.

Year	Female Dropout Rate (%)	Male Dropout Rate (%)
1967	18	16
1970	16	14
1975	15	13
1980	13	16
1985	12	15
1991	13	14

Plotting the years along the horizontal axis and the dropout rates along the vertical axis, draw a line graph for these data. Draw the female line dashed and the male line solid for easier reading and comparison.

42. High School Dropout Rate   Refer to the data in the preceding exercise to answer the following questions.

(a) Using the slope, determine the time interval when the decline in the female dropout rate was the steepest.

(b) Using the slope, determine the time interval when the increase in the male dropout rate was the steepest.

(c) What appears unusual about the time period from 1975 to 1980?

(d) Are the dropout rates for males and females generally increasing or generally decreasing?

(e) Do the dropout rates for males or females show the most consistent pattern?

43. High School Dropout Rate   Refer to the data about high school dropout rates for males and females.

(a) Write a linear equation for the dropout rate of males, M, based on the year, x, for the years 1975 and 1980.

(b) Write a linear equation for the dropout rate of females, F, based on the year, x, for the years 1975 and 1980.

(c) Using your results from parts (a) and (b), determine when the dropout rates for males and females were the same.

CHAPTER 4   SUMMARY

Examples

1. The solution to the system

$$x + 2y = 4$$
$$x - y = 1$$

is the ordered pair (2, 1). It is the only ordered pair that satisfies both equations.

Systems of Linear Equations [4.1, 4.2]

A system of linear equations consists of two or more linear equations considered simultaneously. The solution set to a linear system in two variables is the set of ordered pairs that satisfies both equations. The solution set to a linear system in three variables consists of all the ordered triples that satisfy each equation in the system.

2. We can eliminate the y-variable from the system in Example 1 by multiplying both sides of the second equation by 2 and adding the result to the first equation:

$$x + 2y = 4 \xrightarrow{\text{No change}} x + 2y = 4$$
$$x - y = 1 \xrightarrow{\text{Times 2}} \underline{2x - 2y = 2}$$
$$3x \qquad = 6$$
$$x = 2$$

Substituting $x = 2$ into either of the original two equations gives $y = 1$. The solution is (2, 1).

To Solve a System by the Addition Method [4.1]

Step 1: Look the system over to decide which variable will be easier to eliminate.

Step 2: Use the multiplication property of equality on each equation separately if necessary to ensure that the coefficients of the variable to be eliminated are opposites.

Step 3: Add the left and right sides of the system produced in step 2, and solve the resulting equation.

Step 4: Substitute the solution from step 3 back into any equation with both x- and y-variables, and solve.

Step 5: Check your solution in both equations if necessary.

3. We can apply the substitution method to the system in Example 1 by first solving the second equation for x to get

$$x = y + 1$$

Substituting this expression for x into the first equation we have

$$y + 1 + 2y = 4$$
$$3y + 1 = 4$$
$$3y = 3$$
$$y = 1$$

Using $y = 1$ in either of the original equations gives $x = 2$.

To Solve a System by the Substitution Method [4.1]

Step 1: Solve either of the equations for one of the variables (this step is not necessary if one of the equations has the correct form already).

Step 2: Substitute the results of step 1 into the other equation, and solve.

Step 3: Substitute the results of step 2 into an equation with both x- and y-variables, and solve. (The equation produced in step 1 is usually a good one to use.)

Step 4: Check your solution if necessary.

4. If the two lines are parallel, then the system will be inconsistent, and the solution is $\varnothing$. If the two lines coincide, then the system is dependent.

Inconsistent and Dependent Equations [4.1, 4.2]

Two linear equations that have no solutions in common are said to be *inconsistent*, while two linear equations that have all their solutions in common are said to be *dependent*.

5. $\begin{vmatrix} 3 & 4 \\ -2 & 5 \end{vmatrix} = 15 - (-8) = 23$

2 × 2 Determinants [4.3]
The value of a 2 × 2 determinant is as follows:

$$\begin{vmatrix} a & c \\ b & d \end{vmatrix} = ad - bc$$

6. Expanding $\begin{vmatrix} 1 & 3 & -2 \\ 2 & 0 & 1 \\ 4 & -1 & 1 \end{vmatrix}$

across the first row gives us

$1\begin{vmatrix} 0 & 1 \\ -1 & 1 \end{vmatrix} - 3\begin{vmatrix} 2 & 1 \\ 4 & 1 \end{vmatrix} - 2\begin{vmatrix} 2 & 0 \\ 4 & -1 \end{vmatrix}$

$= 1(1) - 3(-2) - 2(-2)$

$= 11$

3 × 3 Determinants [4.3]
The value of a 3 × 3 determinant is given by

$$\begin{vmatrix} a_1 & b_1 & c_1 \\ a_2 & b_2 & c_2 \\ a_3 & b_3 & c_3 \end{vmatrix} = a_1b_2c_3 + a_3b_1c_2 + a_2b_3c_1 - a_3b_2c_1 - a_1b_3c_2 - a_2b_1c_3$$

There are two methods of finding the six products in the expansion of a 3 × 3 determinant. One method involves a cross-multiplication scheme. The other method involves expanding the determinant by minors.

7. For the system $\begin{matrix} x + y = 6 \\ 3x - 2y = -2 \end{matrix}$

we have

$D = \begin{vmatrix} 1 & 1 \\ 3 & -2 \end{vmatrix} = -5$

$D_x = \begin{vmatrix} 6 & 1 \\ -2 & -2 \end{vmatrix} = -10$

$x = \dfrac{-10}{-5} = 2$

$D_y = \begin{vmatrix} 1 & 6 \\ 3 & -2 \end{vmatrix} = -20$

$y = \dfrac{-20}{-5} = 4$

Cramer's Rule for a Linear System in Two Variables [4.4]
The solution to the system

$$a_1x + b_1y = c_1$$
$$a_2x + b_2y = c_2$$

is given by

$$x = \frac{D_x}{D} \quad \text{and} \quad y = \frac{D_y}{D} \quad (D \neq 0)$$

where

$$D = \begin{vmatrix} a_1 & b_1 \\ a_2 & b_2 \end{vmatrix}, \quad D_x = \begin{vmatrix} c_1 & b_1 \\ c_2 & b_2 \end{vmatrix}, \quad D_y = \begin{vmatrix} a_1 & c_1 \\ a_2 & c_2 \end{vmatrix}$$

8. For the system

$\begin{matrix} x + y = -1 \\ 2x - z = 3 \\ y + 2z = -1 \end{matrix}$

$D = \begin{vmatrix} 1 & 1 & 0 \\ 2 & 0 & -1 \\ 0 & 1 & 2 \end{vmatrix} = -3$

$D_x = \begin{vmatrix} -1 & 1 & 0 \\ 3 & 0 & -1 \\ -1 & 1 & 2 \end{vmatrix} = -6$

$x = \dfrac{-6}{-3} = 2$

Cramer's Rule for a Linear System in Three Variables [4.4]
The solution to the system

$$a_1x + b_1y + c_1z = d_1$$
$$a_2x + b_2y + c_2z = d_2$$
$$a_3x + b_3y + c_3z = d_3$$

is given by

$$x = \frac{D_x}{D}, \quad y = \frac{D_y}{D}, \quad \text{and} \quad z = \frac{D_z}{D} \quad (D \neq 0)$$

where

$$D = \begin{vmatrix} a_1 & b_1 & c_1 \\ a_2 & b_2 & c_2 \\ a_3 & b_3 & c_3 \end{vmatrix} \quad D_x = \begin{vmatrix} d_1 & b_1 & c_1 \\ d_2 & b_2 & c_2 \\ d_3 & b_3 & c_3 \end{vmatrix}$$

$$D_y = \begin{vmatrix} 1 & -1 & 0 \\ 2 & 3 & -1 \\ 0 & -1 & 2 \end{vmatrix} = 9 \qquad D_y = \begin{vmatrix} a_1 & d_1 & c_1 \\ a_2 & d_2 & c_2 \\ a_3 & d_3 & c_3 \end{vmatrix} \qquad D_z = \begin{vmatrix} a_1 & b_1 & d_1 \\ a_2 & b_2 & d_2 \\ a_3 & b_3 & d_3 \end{vmatrix}$$

$$y = \frac{9}{-3} = -3$$

$$D_z = \begin{vmatrix} 1 & 1 & -1 \\ 2 & 0 & 3 \\ 0 & 1 & -1 \end{vmatrix} = -3$$

$$z = \frac{-3}{-3} = 1$$

CHAPTER 4 REVIEW

Solve each system using the addition method. [4.1]

1. $x + y = 4$
$2x - y = 14$

2. $3x + y = 2$
$2x + y = 0$

3. $2x - 4y = 5$
$-x + 2y = 3$

4. $5x - 2y = 7$
$3x + y = 2$

5. $6x - 5y = -5$
$3x + y = 1$

6. $6x + 4y = 8$
$9x + 6y = 12$

7. $3x - 7y = 2$
$-4x + 6y = -6$

8. $6x + 5y = 9$
$4x + 3y = 6$

9. $-7x + 4y = -1$
$5x - 3y = 0$

10. $\frac{1}{2}x - \frac{3}{4}y = -4$
$\frac{1}{4}x + \frac{3}{2}y = 13$

11. $\frac{2}{3}x - \frac{1}{6}y = 0$
$\frac{4}{3}x + \frac{5}{6}y = 14$

12. $-\frac{1}{2}x + \frac{1}{3}y = -\frac{13}{6}$
$\frac{4}{5}x + \frac{3}{4}y = \frac{9}{10}$

Solve each system by the substitution method. [4.1]

13. $x + y = 2$
$y = x - 1$

14. $2x - 3y = 5$
$y = 2x - 7$

15. $x + y = 4$
$2x + 5y = 2$

16. $x + y = 3$
$2x + 5y = -6$

17. $3x + 7y = 6$
$x = -3y + 4$

18. $5x - y = 4$
$y = 5x - 3$

Solve each system. [4.2]

19. $x + y + z = 6$
$x - y - 3z = -8$
$x + y - 2z = -6$

20. $3x + 2y + z = 4$
$2x - 4y + z = -1$
$x + 6y + 3z = -4$

21. $5x + 8y - 4z = -7$
$7x + 4y + 2z = -2$
$3x - 2y + 8z = 8$

22. $5x - 3y - 6z = 5$
$4x - 6y - 3z = 4$
$-x + 9y + 9z = 7$

23. $5x - 2y + z = 6$
$-3x + 4y - z = 2$
$6x - 8y + 2z = -4$

24. $4x - 6y + 8z = 4$
$5x + y - 2z = 4$
$6x - 9y + 12z = 6$

25. $2x - y = 5$
$3x - 2z = -2$
$5y + z = -1$

26. $x - y = 2$
$y - z = -3$
$x - z = -1$

Evaluate each determinant. [4.3]

27. $\begin{vmatrix} 2 & 3 \\ -5 & 4 \end{vmatrix}$

28. $\begin{vmatrix} 3 & 0 \\ 5 & -1 \end{vmatrix}$

29. $\begin{vmatrix} 1 & 0 \\ -7 & -3 \end{vmatrix}$

30. $\begin{vmatrix} 3 & 0 & 2 \\ -1 & 4 & 0 \\ 2 & 0 & 0 \end{vmatrix}$

31. $\begin{vmatrix} 3 & -1 & 0 \\ 0 & 2 & -4 \\ 6 & 0 & 2 \end{vmatrix}$

32. $\begin{vmatrix} -3 & -2 & 0 \\ 0 & -4 & 2 \\ 5 & 1 & 1 \end{vmatrix}$

Solve for x. [4.3]

33. $\begin{vmatrix} 2 & 3x \\ -1 & 2x \end{vmatrix} = 4$ 　　　　**34.** $\begin{vmatrix} 4x & 1 \\ 3 & x \end{vmatrix} = -4x$

Use Cramer's rule to solve each system. [4.4]

35. $3x - 5y = 4$
$7x - 2y = 3$

36. $7x - 5y = 8$
$4x + 3y = 2$

37. $3x - 6y = 9$
$2x - 4y = 6$

38. $6x - 9y = 5$
$7x + 3y = 4$

39. $-6x + 3y = 7$
$5x - 8y = -2$

40. $2x - y + 3z = 4$
$5x + 2y - z = 3$
$-x - 3y + 2z = 1$

41. $4x - 5y = -3$
$2x + 3z = 4$
$3y - z = 8$

42. $2x - 4y = 2$
$4x - 2z = 3$
$4y - z = 2$

Use a system of equations to solve each application problem. In each case be sure to show the system used. [4.5]

43. Ticket Prices Tickets for the show cost $2.00 for adults and $1.50 for children. How many adult tickets and how many children's tickets were sold if a total of 127 tickets were sold for $214?

44. Coin Collection John has 20 coins totaling $3.20. If he has only dimes and quarters, how many of each coin does he have?

45. Investments Ms. Jones invests money in two accounts, one of which pays 12% per year, while the other pays 15% per year. If her total investment is $12,000 and the interest after 1 year is $1,650, how much is invested in each account?

46. Speed It takes a boat on a river 2 hours to travel 28 miles downstream and 3 hours to travel 30 miles upstream. What is the speed of the boat and the current of the river?

CHAPTER 4 **PROJECTS**

SYSTEMS OF EQUATIONS

GROUP PROJECT

COMPARING PHONE BILLS

Number of People: 2–3

Time Needed: 20–30 minutes

Equipment: Paper, pencil, and graphing calculator

Background: In the group project in Chapter 3, we introduced step functions, or greatest integer functions. We saw that on a graphing calculator, the notation int(X) gives the greatest integer less than or equal to x. In other words, it "rounds down." For the calculator to "round up," we use the notation $-int(-X)$. Figures 1 and 2 show the graphs of $y = int(x)$ and $y = -int(-x)$.

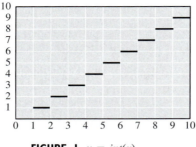

FIGURE 1 $y = int(x)$

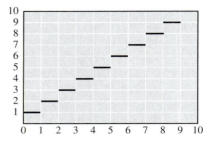

FIGURE 2 $y = -int(-x)$

Procedure: We are going to compare two phone rates. Plan A is from the company we have already used. They charge $.50 per phone call plus $.35 per minute. Fractions of a minute are rounded up to the next full minute. Plan B is from a company we are thinking of changing to. Plan B charges $.50 per call, plus $.40 cents per minute. They calculate time on 6-second intervals, rounding up to the next tenth of a minute.

1. Find the step functions that give the cost of a call as a function of time for each plan.
2. Sketch the graph of each function on the given grid, and then fill in the following table, using the table feature of your graphing calculator.

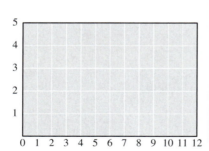

FIGURE 3

Time (min)	Plan A Cost ($)	Plan B Cost ($)
0.1		
0.8		
2.3		
2.5		
4.8		

3. In general for a call lasting 5 minutes or less, when will Plan A cost less and when will Plan B cost less?
4. What is the longest call you can make and still have equal cost for the two plans?
5. What is the longest call you can make on Plan B that is still less expensive than Plan A?

RESEARCH PROJECT

ZENO'S PARADOXES

Zeno of Elea was born at about the same time that Pythagoras died. He is responsible for three paradoxes that have come to be known as Zeno's paradoxes. One of the three has to do with a race between Achilles and a tortoise. Achilles is much faster than the tortoise, but the tortoise has a head start. According to Zeno's method of reasoning, Achilles can never pass the tortoise because each time he reaches the place where the tortoise was, the tortoise is gone. Research Zeno's paradox concerning Achilles and the tortoise. Put your findings into essay form that begins with a definition for the word "paradox." Then use Zeno's method of reasoning to describe a race between Achilles and the tortoise, if Achilles runs at 10 miles per hour, the tortoise at 1 mile per hour, and the tortoise has a 1-mile head start. Next, use the methods shown in this chapter to find the distance at which Achilles reaches the tortoise and the time at which Achilles reaches the tortoise. Conclude your essay by summarizing what you have done and showing how the two results you have obtained form a paradox.

CHAPTER 4   TEST

Solve the following systems by the addition method. [4.1]

1. $2x - 5y = -8$
$3x + y = 5$

2. $4x - 7y = -2$
$-5x + 6y = -3$

3. $\dfrac{1}{3}x - \dfrac{1}{6}y = 3$
$-\dfrac{1}{5}x + \dfrac{1}{4}y = 0$

Solve the following systems by the substitution method. [4.1]

4. $2x - 5y = 14$
$ y = 3x + 8$

5. $6x - 3y = 0$
$ x + 2y = 5$

6. Solve the system. [4.2]

$2x - y + z = 9$
$x + y - 3z = -2$
$3x + y - z = 6$

Evaluate each determinant. [4.3]

7. $\begin{vmatrix} 3 & -5 \\ -4 & 2 \end{vmatrix}$

8. $\begin{vmatrix} 1 & 0 & -3 \\ 2 & 1 & 0 \\ 0 & 5 & 4 \end{vmatrix}$

Use Cramer's rule to solve. [4.4]

9. $5x - 4y = 2$
$-2x + y = 3$

10. $2x + 4y = 3$
$-4x - 8y = -6$

11. $2x - y + 3z = 2$
$x - 4y - z = 6$
$3x - 2y + z = 4$

Solve each word problem. [4.5]

12. Number Problem   A number is 1 less than twice another. Their sum is 14. Find the two numbers.

13. Investing   John invests twice as much money at 6% as he does at 5%. If his investments earn a total of $680 in 1 year, how much does he have invested at each rate?

14. Ticket Cost   There were 750 tickets sold for a basketball game for a total of $1,090. If adult tickets cost $2.00 and children's tickets cost $1.00, how many of each kind were sold?

15. Mixture Problem   How much 30% alcohol solution and 70% alcohol solution must be mixed to get 16 gallons of 60% solution?

16. Speed of a Boat   A boat can travel 20 miles downstream in 2 hours. The same boat can travel 18 miles upstream in 3 hours. What is the speed of the boat in still water, and what is the speed of the current?

17. Coin Problem   A collection of nickels, dimes, and quarters consists of 15 coins with a total value of $1.10. If the number of nickels is 1 less than 4 times the number of dimes, how many of each coin are contained in the collection?

18. Nutrition   For a woman of average height and weight between the ages of 19 and 22, the Food and Nutrition Board of the National Academy of Sciences has determined the Recommended Daily Allowance (RDA) of ascorbic acid to be 45 mg (milligrams). They also determined the RDA for niacin to be 14 mg for the same woman.

Each ounce of cereal I contains 10 mg of ascorbic acid and 4 mg of niacin, while each ounce of cereal II contains 15 mg of ascorbic acid and 2 mg of niacin. If cereals are combined, how many ounces of each cereal must the average woman between the ages of 19 and 22 consume in order to have the RDAs for both ascorbic acid and niacin?

	Cereal I	Cereal II	Recommended Daily Allowance (RDA)
Ascorbic acid	10 mg	15 mg	45 mg
Niacin	4 mg	2 mg	14 mg

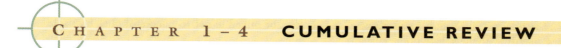

Simplify.

1. $2^4 - 4(5 - 12 + 4)$

2. $2(7x - 4) + 7(2x - 3)$

3. $18 - 7 - 5 - (-3)$ **4.** $-6(2x + 3) + 15x$

5. $\frac{3}{5} \cdot \frac{8}{9} \cdot 10$ **6.** $\frac{2}{3} + \frac{5}{9}$

Solve.

7. $-\frac{4}{7}a - 3 = 9$

8. $-5 - 2(4x - 3) = -5$

9. $\frac{3}{5}(2x - 3) + \frac{4}{5} = 5$ **10.** $|a| - 3 = 7$

11. $|2y + 3| = |2y - 7|$

Solve each inequality and graph the solution.

12. $4y - 1 \geq 11$ or $4y - 1 \leq -11$

13. $-2(4x - 4) \geq -4(3x + 1)$

14. $|6x + 18| - 7 > 5$

Graph on a rectangular coordinate system.

15. $-4x + 3y = 12$ **16.** $y = -3$

17. Find the slope of the line you graphed in Problem 16.

Factor completely.

18. 594

Solve each system.

19. $3x + y = 2$
$-6x + 2y = -4$

20. $3x - y = 2$
$-6x + 2y = 4$

21. $4x - 8y = 6$
$6x - 12y = 6$

22. $3x - 2y = 5$
$y = 3x - 7$

23. $-\frac{1}{5}x + \frac{4}{3}y = \frac{14}{15}$
$\frac{1}{3}x - \frac{1}{4}y = \frac{5}{12}$

24. $2x + 3y - 8z = 2$
$3x - y + 2z = 10$
$4x + y + 8z = 16$

Evaluate each determinant.

25. $\begin{vmatrix} -5 & -2 \\ 0 & -6 \end{vmatrix}$ **26.** $\begin{vmatrix} 1 & 2 \\ 2 & 1 \end{vmatrix}$

27. $\begin{vmatrix} 3 & 5 & -1 \\ 0 & 4 & 2 \\ -6 & 0 & 7 \end{vmatrix}$

28. Solve for x: $\begin{vmatrix} 6 & 2x \\ -2 & 3x \end{vmatrix} = 5$

Solve using Cramer's rule.

29. $3x - 5y = 1$
$2x + 7y = -1$

30. $5x - 2y = 3$
$-4x + 3z = -2$
$3y - z = 6$

31. Solve $3 - 2t < 7$. Write your answer with interval notation.

32. Solve for W if $P = 2L + 2W$ and $P = 20$ and $L = 7$.

33. Give the next number of the sequence and state whether it is arithmetic or geometric: -5, -1, 3, . . .

34. Translate into an inequality: x is between -7 and 7 inclusive.

35. For the set $\{-2, -\sqrt{2}, -1, 0, \sqrt{3}, 2.4\}$ list all elements belonging to the set of irrational numbers.

36. State the properties that justify: $x + (4 + y) = (x + y) + 4$

37. Give the opposite and reciprocal of $\frac{7}{3}$.

38. 32 is 5% of what number?

39. Indicate whether the graph you drew in Problem 15 is the graph of a function.

40. Let $f(x) = x^2 - 2x$ and $g(x) = x + 5$. Find: $f(-1) + g(4)$.

41. Let $f(x) = 2x^2 - x + 3$ and $g(x) = x + 5$. Find: $f[g(-4)]$.

42. Find the slope of the line through $(-3, -2)$ and $(-5, 3)$.

43. Find y if the line through $(-2, -8)$ and $(1, y)$ has a slope of 3.

44. Find the slope and y-intercept of $2x - 3y = 12$.

45. Give the equation of a line with slope $-\frac{3}{4}$ and y-intercept $= 2$.

46. Find the equation of the line through $(-8, -2)$ and $(1, -3)$.

47. **Variation** Suppose z varies jointly with x and the cube of y. If z is -108 when x is 4 and y is 3, find x when z is 8 and y is -2.

48. **Variation** The power P (in watts) in an electric circuit varies directly with the square of the current I (in amperes). If P is 30 when I is 2, find P when I is 6.

49. **Mixture** How many ounces of 30% HCl (hydrochloric acid) solution and 70% HCl solution must be mixed to get 15 ounces of 50% HCl solution?

50. **Geometry** Find all three angles in a triangle if the smallest angle is one-fifth the largest angle and the remaining angle is 12 degrees more than the smallest angle.

Exponents and Polynomials

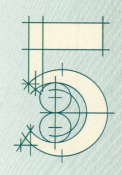

(Mitch Kezar/Tony Stone Images)

INTRODUCTION

If you go on to take a business course or an economics course, you will find your-self spending lots of time with the three expressions that form the mathematical foundation of business: profit, revenue, and cost. Many times these expressions are given as polynomials, the topic of this chapter. The relationship between the three equations is known as the profit equation:

$$\text{Profit} = \text{Revenue} - \text{Cost}$$

$$P(x) = R(x) - C(x)$$

The table and graphs shown below were produced on a graphing calculator. They give numerical and graphical descriptions of revenue, profit, and cost for a company that manufactures and sells prerecorded videotapes according to the equations

$$R(x) = 11.5x - 0.05x^2 \qquad \text{and} \qquad C(x) = 200 + 2x$$

TABLE I	Revenue, Cost, and Profit on a Graphing Calculator		
Number of Videotapes X	Revenue Y_1	Cost Y_2	Profit Y_3
0	0	200	-200
50	450	300	150
100	650	400	250
150	600	500	100
200	300	600	-300

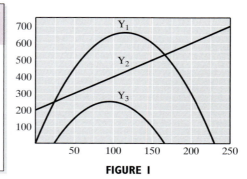

FIGURE I

By studying the material in this chapter, you will get a head start on learning the equations and relationships that are emphasized in business and economics.

STUDY SKILLS

The study skills for this chapter cover the way you approach new situations in mathematics. The first study skill is a point of view you hold about your natural instincts for what does and doesn't work in mathematics. The second study skill gives you a way of testing your instincts.

1. **Don't Let Your Intuition Fool You** As you become more experienced and more successful in mathematics you will be able to trust your mathematical intuition. For now, though, it can get in the way of your success. For example, if you ask some students to "subtract 3 from -5" they will answer -2 or 2. Both answers are incorrect, even though they may seem intuitively true. Likewise, some students will expand $(a + b)^2$ and arrive at $a^2 + b^2$, which is incorrect. In both cases, intuition leads directly to the wrong answer.

2. **Test Properties of Which You Are Unsure** From time to time you will be in a situation where you would like to apply a property or rule, but you are not sure it is true. You can always test a property or statement by substituting numbers for variables. For instance, I always have students that rewrite $(x + 3)^2$ as $x^2 + 9$, thinking that the two expressions are equivalent. The fact that the two expressions are not equivalent becomes obvious when we substitute 10 for x in each one.

$$\text{When } x = 10, \text{ the expression } (x + 3)^2 \text{ is } (10 + 3)^2 = 13^2 = 169$$

$$\text{When } x = 10, \text{ the expression } x^2 + 9 = 10^2 + 9 = 100 + 9 = 109$$

When you test the equivalence of expressions by substituting numbers for the variable, make it easy on yourself by choosing numbers that are easy to work with, such as 10. Don't try to verify the equivalence of expressions by substituting 0, 1, or 2 for the variable, as using these numbers will occasionally give you false results.

It is not good practice to trust your intuition or instincts in every new situation in algebra. If you have any doubt about the generalizations you are making, test them by replacing variables with numbers and simplifying.

Properties of Exponents

The figure shows a square and a cube, each with a side of length 1.5 centimeters. To find the area of the square, we raise 1.5 to the second power: 1.5^2. To find the volume of the cube, we raise 1.5 to the third power: 1.5^3.

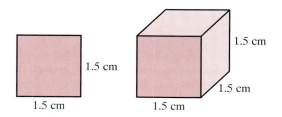

1.5 cm
1.5 cm
1.5 cm
1.5 cm
1.5 cm

Because the area of the square is 1.5^2, we say second powers are *squares*; that is, x^2 is read "*x* squared." Likewise, since the volume of the cube is 1.5^3, we say third powers are *cubes*; that is, x^3 is read "*x* cubed." Exponents and the vocabulary associated with them are topics we will study in this section.

PROPERTIES OF EXPONENTS

In this section, we will be concerned with the simplification of expressions that involve exponents. We begin by making some generalizations about exponents, based on specific examples.

EXAMPLE 1 Write the product $x^3 \cdot x^4$ with a single exponent.

Solution
$$x^3 \cdot x^4 = (x \cdot x \cdot x)(x \cdot x \cdot x \cdot x)$$
$$= (x \cdot x \cdot x \cdot x \cdot x \cdot x \cdot x)$$
$$= x^7 \qquad \textit{Notice: } 3 + 4 = 7$$

We can generalize this result into the first property of exponents.

PROPERTY 1 FOR EXPONENTS

If a is a real number and r and s are integers, then

$$a^r \cdot a^s = a^{r+s}$$

EXAMPLE 2 Write $(5^3)^2$ with a single exponent.

Solution
$$(5^3)^2 = 5^3 \cdot 5^3$$
$$= 5^6 \qquad \textit{Notice: } 3 \cdot 2 = 6$$

Generalizing this result, we have a second property of exponents.

PROPERTY 2 FOR EXPONENTS

If a is a real number and r and s are integers, then

$$(a^r)^s = a^{r \cdot s}$$

A third property of exponents arises when we have the product of two or more numbers raised to an integer power.

EXAMPLE 3 Expand $(3x)^4$ and then multiply.

Solution

$$
\begin{aligned}
(3x)^4 &= (3x)(3x)(3x)(3x) \\
&= (3 \cdot 3 \cdot 3 \cdot 3)(x \cdot x \cdot x \cdot x) \\
&= 3^4 \cdot x^4 \qquad \textit{Notice: } \text{The exponent 4 distributes over the} \\
&\qquad\qquad\qquad\qquad \text{product } 3x. \\
&= 81x^4
\end{aligned}
$$

Generalizing Example 3 we have Property 3 for exponents.

PROPERTY 3 FOR EXPONENTS

If a and b are any two real numbers and r is an integer, then

$$(ab)^r = a^r \cdot b^r$$

Here are some examples that use combinations of the first three properties of exponents to simplify expressions involving exponents.

EXAMPLES Simplify each expression using the properties of exponents.

4. $\begin{aligned}[t](-3x^2)(5x^4) &= -3(5)(x^2 \cdot x^4) \qquad &\text{Commutative and associative} \\ &= -15x^6 &\text{Property 1 for exponents}\end{aligned}$

5. $\begin{aligned}[t](-2x^2)^3(4x^5) &= (-2)^3(x^2)^3(4x^5) \qquad &\text{Property 3} \\ &= -8x^6 \cdot (4x^5) &\text{Property 2} \\ &= (-8 \cdot 4)(x^6 \cdot x^5) &\text{Commutative and associative} \\ &= -32x^{11} &\text{Property 1}\end{aligned}$

6. $\begin{aligned}[t](x^2)^4(x^2y^3)^2(y^4)^3 &= x^8 \cdot x^4 \cdot y^6 \cdot y^{12} \qquad &\text{Properties 2 and 3} \\ &= x^{12}y^{18} &\text{Property 1}\end{aligned}$

The next property of exponents deals with negative integer exponents.

PROPERTY 4 FOR EXPONENTS

If a is any nonzero real number and r is a positive integer, then

$$a^{-r} = \frac{1}{a^r}$$

Note: This property is actually a definition. That is, we are defining negative-integer exponents as indicating reciprocals. Doing so gives us a way to write an expression with a negative exponent as an equivalent expression with a positive exponent.

EXAMPLES   Write with positive exponents, and then simplify.

7. $5^{-2} = \dfrac{1}{5^2} = \dfrac{1}{25}$

8. $(-2)^{-3} = \dfrac{1}{(-2)^3} = \dfrac{1}{-8} = -\dfrac{1}{8}$

9. $\left(\dfrac{3}{4}\right)^{-2} = \dfrac{1}{\left(\frac{3}{4}\right)^2} = \dfrac{1}{\frac{9}{16}} = \dfrac{16}{9}$

If we generalize the result in Example 9, we have the following extension of Property 4,

$$\left(\frac{a}{b}\right)^{-r} = \left(\frac{b}{a}\right)^{r}$$

which indicates that raising a fraction to a negative power is equivalent to raising the reciprocal of the fraction to the positive power.

Property 3 indicated that exponents distribute over products. Since division is defined in terms of multiplication, we can expect that exponents will distribute over quotients as well. Property 5 is the formal statement of this fact.

PROPERTY 5 FOR EXPONENTS

If a and b are any two real numbers with $b \neq 0$, and r is an integer, then

$$\left(\frac{a}{b}\right)^{r} = \frac{a^r}{b^r}$$

Proof of Property 5

$$\left(\frac{a}{b}\right)^r = \underbrace{\left(\frac{a}{b}\right)\left(\frac{a}{b}\right)\left(\frac{a}{b}\right)\cdots\left(\frac{a}{b}\right)}_{r \text{ factors}}$$

$$= \frac{a\cdot a\cdot a\cdots a \; \leftarrow r \text{ factors}}{b\cdot b\cdot b\cdots b \; \leftarrow r \text{ factors}}$$

$$= \frac{a^r}{b^r}$$

Since multiplication with the same base resulted in addition of exponents, it seems reasonable to expect division with the same base to result in subtraction of exponents.

PROPERTY 6 FOR EXPONENTS

If a is any nonzero real number, and r and s are any two integers, then

$$\frac{a^r}{a^s} = a^{r-s}$$

Notice again that we have specified r and s to be any integers. Our definition of negative exponents is such that the properties of exponents hold for all integer exponents, whether positive or negative integers. Here is a proof of Property 6.

Proof of Property 6

Our proof is centered on the fact that division by a number is equivalent to multiplication by the reciprocal of the number.

$$\frac{a^r}{a^s} = a^r \cdot \frac{1}{a^s} \qquad \text{Dividing by } a^s \text{ is equivalent to multiplying by } \frac{1}{a^s}.$$

$$= a^r a^{-s} \qquad \text{Property 4}$$

$$= a^{r+(-s)} \qquad \text{Property 1}$$

$$= a^{r-s} \qquad \text{Definition of subtraction}$$

EXAMPLES   Apply Property 6 to each expression, and then simplify the result. All answers that contain exponents should contain positive exponents only.

10. $\dfrac{2^8}{2^3} = 2^{8-3} = 2^5 = 32$

11. $\dfrac{x^2}{x^{18}} = x^{2-18} = x^{-16} = \dfrac{1}{x^{16}}$

12. $\dfrac{a^6}{a^{-8}} = a^{6-(-8)} = a^{14}$

13. $\dfrac{m^{-5}}{m^{-7}} = m^{-5-(-7)} = m^2$

Let's complete our list of properties by looking at how the numbers 0 and 1 behave when used as exponents.

We can use the original definition for exponents when the number 1 is used as an exponent.

$$a^1 = \underbrace{a}_{\text{1 factor}}$$

For 0 as an exponent, consider the expression $\dfrac{3^4}{3^4}.$ Since $3^4 = 81,$ we have

$$\frac{3^4}{3^4} = \frac{81}{81} = 1$$

On the other hand, since we have the quotient of two expressions with the same base, we can subtract exponents.

$$\frac{3^4}{3^4} = 3^{4-4} = 3^0$$

Hence, 3^0 must be the same as 1.

Summarizing these results, we have our last property for exponents.

PROPERTY 7 FOR EXPONENTS

If a is any real number, then

$$a^1 = a$$

and

$$a^0 = 1 \qquad (\text{as long as } a \neq 0)$$

EXAMPLES   Simplify.

14. $(2x^2y^4)^0 = 1$

15. $(2x^2y^4)^1 = 2x^2y^4$

Here are some examples that use many of the properties of exponents. There are a number of ways to proceed on problems like these. You should use the method that works best for you.

E X A M P L E S   Simplify.

16. $\dfrac{(x^3)^{-2}(x^4)^5}{(x^{-2})^7} = \dfrac{x^{-6}x^{20}}{x^{-14}}$ Property 2

$\qquad\qquad\qquad = \dfrac{x^{14}}{x^{-14}}$ Property 1

$\qquad\qquad\qquad = x^{28}$ Property 6: $x^{14-(-14)} = x^{28}$

17. $\dfrac{6a^5b^{-6}}{12a^3b^{-9}} = \dfrac{6}{12}\cdot\dfrac{a^5}{a^3}\cdot\dfrac{b^{-6}}{b^{-9}}$ Write as separate fractions.

$\qquad\qquad\qquad = \dfrac{1}{2}\,a^2b^3$ Property 6

Note: This last answer can also be written as $\frac{a^2b^3}{2}$. Either answer is correct.

18. $\dfrac{(4x^{-5}y^3)^2}{(x^4y^{-6})^{-3}} = \dfrac{16x^{-10}y^6}{x^{-12}y^{18}}$ Properties 2 and 3

$\qquad\qquad\qquad = 16x^2y^{-12}$ Property 6

$\qquad\qquad\qquad = 16x^2\cdot\dfrac{1}{y^{12}}$ Property 4

$\qquad\qquad\qquad = \dfrac{16x^2}{y^{12}}$ Multiplication

SCIENTIFIC NOTATION

Scientific notation is a way in which to write very large or very small numbers in a more manageable form. Here is the definition.

> **DEFINITION**   A number is written in **scientific notation** if it is written as the product of a number between 1 and 10 and an integer power of 10. A number written in scientific notation has the form
>
> $$n \times 10^r$$
>
> where $1 \le n < 10$ and $r =$ an integer.

E X A M P L E 19   Write 376,000 in scientific notation.

Solution   We must rewrite 376,000 as the product of a number between 1 and 10 and a power of 10. To do so, we move the decimal point five places to the left so that it appears between the 3 and the 7. Then we multiply this number by 10^5. The number that results has the same value as our original number and is written in scientific notation.

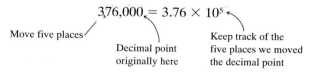

$$376{,}000 = 3.76 \times 10^5$$

Move five places

Decimal point
originally here

Keep track of the
five places we moved
the decimal point

If a number written in expanded form is greater than or equal to 10, then when the number is written in scientific notation the exponent on 10 will be positive. A number that is less than one and greater than zero will have a negative exponent when written in scientific notation.

EXAMPLE 20   Write 4.52×10^3 in expanded form.

Solution   Since 10^3 is 1,000, we can think of this as simply a multiplication problem. That is,

$$4.52 \times 10^3 = 4.52 \times 1{,}000 = 4{,}520$$

On the other hand, we can think of the exponent 3 as indicating the number of places we need to move the decimal point in order to write our number in expanded form. Since our exponent is positive 3, we move the decimal point three places to the right.

$$4.52 \times 10^3 = 4{,}520$$

The following table lists some additional examples of numbers written in expanded form and in scientific notation. In each case, note the relationship between the number of places the decimal point is moved and the exponent on 10.

Number Written in Expanded Form		Number Written in Scientific Notation
376,000	=	3.76×10^5
49,500	=	4.95×10^4
3,200	=	3.2×10^3
591	=	5.91×10^2
46	=	4.6×10^1
8	=	8×10^0
0.47	=	4.7×10^{-1}
0.093	=	9.3×10^{-2}
0.00688	=	6.88×10^{-3}
0.0002	=	2×10^{-4}
0.000098	=	9.8×10^{-5}

Calculator Note: Some scientific calculators have a key that allows you to enter numbers in scientific notation. The key is labeled

$$\boxed{\text{EXP}} \quad \text{or} \quad \boxed{\text{EE}} \quad \text{or} \quad \boxed{\text{SCI}}$$

To enter the number 3.45×10^6 you would first enter the decimal number, then press the scientific notation key, and finally, enter the exponent:

$$3.45 \ \boxed{\text{EXP}} \ 6$$

To enter 6.2×10^{-27} you would use the following sequence:

$$6.2 \ \boxed{\text{EXP}} \ 27 \ \boxed{+/-}$$

We can use the properties of exponents to do arithmetic with numbers written in scientific notation. Here are some examples.

EXAMPLES   Simplify each expression and write all answers in scientific notation.

21. $(2 \times 10^8)(3 \times 10^{-3}) = (2)(3) \times (10^8)(10^{-3})$
$$= 6 \times 10^5$$

22. $\dfrac{4.8 \times 10^9}{2.4 \times 10^{-3}} = \dfrac{4.8}{2.4} \times \dfrac{10^9}{10^{-3}}$
$$= 2 \times 10^{9-(-3)}$$
$$= 2 \times 10^{12}$$

23. $\dfrac{(6.8 \times 10^5)(3.9 \times 10^{-7})}{7.8 \times 10^{-4}} = \dfrac{(6.8)(3.9)}{7.8} \times \dfrac{(10^5)(10^{-7})}{10^{-4}}$
$$= 3.4 \times 10^2$$

Calculator Note: On a scientific calculator with a scientific notation key, you would use the following sequence of keys to do Example 22:

$$4.8 \ \boxed{\text{EXP}} \ 9 \ \boxed{\div} \ 2.4 \ \boxed{\text{EXP}} \ 3 \ \boxed{+/-} \ \boxed{=}$$

 Getting Ready for Class

After reading through the preceding section, respond in your own words and in complete sentences.

A. Explain the difference between -2^4 and $(-2)^4$.

B. Explain the difference between 2^5 and 2^{-5}.

C. If a positive base is raised to a negative exponent, can the result be a negative number?

D. State Property 1 for exponents in your own words.

PROBLEM SET 5.1

Evaluate each of the following.

1. 4^2 **2.** $(-4)^2$ **3.** -4^2

4. $-(-4)^2$ **5.** -0.3^3 **6.** $(-0.3)^3$

7. 2^5 **8.** 2^4 **9.** $(\frac{1}{2})^3$

10. $(\frac{3}{4})^2$ **11.** $(-\frac{5}{6})^2$ **12.** $(-\frac{7}{8})^2$

Use the properties of exponents to simplify each of the following as much as possible.

13. $x^5 \cdot x^4$ **14.** $x^6 \cdot x^3$ **15.** $(2^3)^2$

16. $(3^2)^2$ **17.** $(-\frac{2}{3}x^2)^3$ **18.** $(-\frac{3}{5}x^4)^3$

19. $-3a^2(2a^4)$ **20.** $5a^7(-4a^6)$

Write each of the following with positive exponents. Then simplify as much as possible.

21. 3^{-2} **22.** $(-5)^{-2}$ **23.** $(-2)^{-5}$

24. 2^{-5} **25.** $(\frac{3}{4})^{-2}$ **26.** $(\frac{3}{5})^{-2}$

27. $(\frac{1}{3})^{-2} + (\frac{1}{2})^{-3}$ **28.** $(\frac{1}{2})^{-2} + (\frac{1}{3})^{-3}$ ← invert Fraction

Simplify each expression. Write all answers with positive exponents only. (Assume all variables are nonzero.)

29. $x^{-4}x^7$ **30.** $x^{-3}x^8$

31. $(a^2b^{-5})^3$ **32.** $(a^4b^{-3})^3$

33. $(5y^4)^{-3}(2y^{-2})^3$ **34.** $(3y^5)^{-2}(2y^{-4})^3$

35. $(\frac{1}{2}x^3)(\frac{2}{3}x^4)(\frac{3}{5}x^{-7})$ **36.** $(\frac{1}{7}x^{-3})(\frac{7}{8}x^{-5})(\frac{8}{9}x^8)$

37. $(4a^5b^2)(2b^{-5}c^2)(3a^7c^4)$

38. $(3a^{-2}c^3)(5b^{-6}c^5)(4a^6b^{-2})$

39. $(2x^2y^{-5})^3(3x^{-4}y^2)^{-4}$ **40.** $(4x^{-4}y^9)^{-2}(5x^4y^{-3})^2$

Use the properties of exponents to simplify each expression. Write all answers with positive exponents only. (Assume all variables are nonzero.)

41. $\dfrac{x^{-1}}{x^9}$ **42.** $\dfrac{x^{-3}}{x^5}$ **43.** $\dfrac{a^4}{a^{-6}}$

44. $\dfrac{a^5}{a^{-2}}$ **45.** $\dfrac{t^{-10}}{t^{-4}}$ **46.** $\dfrac{t^{-8}}{t^{-5}}$

47. $\left(\dfrac{x^5}{x^3}\right)^6$ **48.** $\left(\dfrac{x^7}{x^4}\right)^5$ **49.** $\dfrac{(x^5)^6}{(x^3)^4}$

50. $\dfrac{(x^7)^3}{(x^4)^5}$ **51.** $\dfrac{(x^{-2})^3(x^3)^{-2}}{x^{10}}$

52. $\dfrac{(x^{-4})^3(x^3)^{-4}}{x^{10}}$ **53.** $\dfrac{5a^8b^3}{20a^5b^{-4}}$ **54.** $\dfrac{7a^6b^{-2}}{21a^2b^{-5}}$

55. $\dfrac{(3x^{-2}y^8)^4}{(9x^4y^{-3})^2}$ **56.** $\dfrac{(6x^{-3}y^{-5})^2}{(3x^{-4}y^{-3})^4}$

57. $\left(\dfrac{8x^2y}{4x^4y^{-3}}\right)^4$ **58.** $\left(\dfrac{5x^4y^5}{10xy^{-2}}\right)^3$

59. $\left(\dfrac{x^{-5}y^2}{x^{-3}y^5}\right)^{-2}$ **60.** $\left(\dfrac{x^{-8}y^{-3}}{x^{-5}y^6}\right)^{-1}$

61. $\left(\dfrac{ab^{-3}c^{-2}}{a^{-3}b^0c^{-5}}\right)^{-1}$ **62.** $\left(\dfrac{a^3b^2c^1}{a^{-1}b^{-2}c^{-3}}\right)^{-2}$

Write each number in scientific notation.

63. 378,000   **64.** 3,780,000

65. 4,900   **66.** 490

67. 0.00037   **68.** 0.000037

69. 0.00495   **70.** 0.0495

Write each number in expanded form.

71. 5.34×10^3 **72.** 5.34×10^2

73. 7.8×10^6 **74.** 7.8×10^4

75. 3.44×10^{-3} **76.** 3.44×10^{-5}

77. 4.9×10^{-1} **78.** 4.9×10^{-2}

Use the properties of exponents to simplify each of the following expressions. Write all answers in scientific notation.

79. $(4 \times 10^{10})(2 \times 10^{-6})$

80. $(3 \times 10^{-12})(3 \times 10^4)$

81. $\dfrac{8 \times 10^{14}}{4 \times 10^5}$ **82.** $\dfrac{6 \times 10^8}{2 \times 10^3}$

83. $\dfrac{(5 \times 10^6)(4 \times 10^{-8})}{8 \times 10^4}$

84. $\dfrac{(6 \times 10^{-7})(3 \times 10^9)}{5 \times 10^6}$

85. $\dfrac{(2.4 \times 10^{-3})(3.6 \times 10^{-7})}{(4.8 \times 10^6)(1 \times 10^{-9})}$

86. $\dfrac{(7.5 \times 10^{-6})(1.5 \times 10^9)}{(1.8 \times 10^4)(2.5 \times 10^{-2})}$

87. The number 237×10^4 is not written in scientific notation because 237 is larger than 10. Write 237×10^4 in scientific notation.

88. Write 46.2×10^{-3} in scientific notation.

Applying the Concepts

89. Large Numbers If you are 20 years old, you have been alive for more than 630,000,000 seconds. Write this last number in scientific notation.

90. Large Numbers Use the information from Problem 89 to give the approximate number of seconds you have lived if you are 40 years old. Write your answer in scientific notation.

91. Very Large Numbers The mass of the Earth is approximately 5.98×10^{24} kilograms. If this number were written in expanded form, how many zeros would it contain?

92. Very Small Numbers The mass of a single hydrogen atom is approximately 1.67×10^{-27} kilogram. If this number were written in expanded form, how many digits would there be to the right of the decimal point?

93. Very Large Numbers A light-year, the distance light travels in 1 year, is approximately 5.9×10^{12} miles. The Andromeda galaxy is approximately 1.7×10^{6} light-years from our galaxy. Find the distance in miles between our galaxy and the Andromeda galaxy.

94. Very Large Numbers The distance from the Earth to the Sun is approximately 9.3×10^{7} miles.

If light travels 1.2×10^{7} miles in 1 minute, how many minutes does it take the light from the sun to reach the Earth?

95. National Debt The U.S. national debt increased in June 1998 by $34,312,982,073. Round this number to the nearest hundred million, and write it in scientific notation.

96. National Debt The new total of U.S. national debt in June 1998 was $5,530,020,864,838. Round this number to the nearest hundred billion, and write it in scientific notation.

97. Astronomy The following table shows the minimum miles each planet is from Earth, and distances are shown in millions of miles. (Minimum miles are specified because planet orbits are not perfectly circular — these distances are the closest each planet comes to the Earth.)

Planet	Minimum Distance From Earth (in millions of miles)
Mercury	48
Venus	24
Mars	34
Jupiter	366
Saturn	743
Uranus	1,604
Neptune	2,676
Pluto	2,668

Express each of these distances in miles using scientific notation.

98. Mass of the Moon As stated in an earlier exercise, the mass of the Earth is 5.98×10^{24} kilograms. If the mass of the Moon is 1.23% of the mass of the Earth, what is the mass of the Moon? Write your answer in scientific notation.

99. Volume of the Planets Recall that the volume of a sphere, V, is found by the formula

$$V = \frac{4}{3} \pi r^3$$

where r is the radius of the sphere. The following table shows the mean radius of each of the planets.
(a) Use the information in this table to find the approximate volume, in cubic miles, of each planet. Express your answers in scientific notation, rounded to three places to the right of the decimal point, and use $\pi = 3.14$.

Planet	Mean Radius (in miles)
Mercury	1,516
Venus	3,761
Earth	3,959
Mars	2,106
Jupiter	43,441
Saturn	36,184
Uranus	15,759
Neptune	15,301
Pluto	707

(b) Why are the answers in part (a) only approximate and not exact?

100. Volume of the Sun The approximate volume of the Sun is 3.3876×10^{17} cubic miles. Using the fact from the previous exercise that the mean radius of the Earth is 3,959 miles, how many times larger is the volume of the Sun than the Earth? Express your answer in scientific notation.

Review Problems

The problems that follow review material we covered in Section 4.1.

Solve each system by the addition method.

101. $4x + 3y = 10$
    $2x + y = 4$

102. $3x - 5y = -2$
    $2x - 3y = 1$

103. $4x + 5y = 5$
    $\frac{6}{5}x + y = 2$

104. $4x + 2y = -2$
    $\frac{1}{2}x + y = 0$

Solve each system by the substitution method.

105. $x + y = 3$
    $y = x + 3$

106. $x + y = 6$
    $y = x - 4$

107. $2x - 3y = -6$
    $y = 3x - 5$

108. $7x - y = 24$
    $x = 2y + 9$

Extending the Concepts

Assume all variable exponents represent positive integers, and simplify each expression.

109. $x^{m+2} \cdot x^{-2m} \cdot x^{m-5}$

110. $x^{m-4}x^{m+9}x^{-2m}$

111. $(y^m)^2(y^{-3})^m(y^{m+3})$

112. $(y^m)^{-4}(y^3)^m(y^{m+6})$

113. $\dfrac{x^{n+2}}{x^{n-3}}$

114. $\dfrac{x^{n-3}}{x^{n-7}}$

5.2   ## Polynomials, Sums, and Differences

The following chart is from a company that duplicates videotapes. It shows the revenue and cost to duplicate a 30-minute video. From the chart you can see that 300 copies will bring in $900 in revenue, with a cost of $600. The profit is the difference between revenue and cost, or $300.

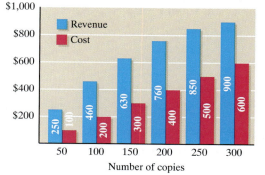

Revenue and Cost to Duplicate a 30-Minute Video

■ Revenue
■ Cost

Number of copies

The relationship between profit, revenue, and cost is one application of the polynomials we will study in this section. Let's begin with a definition that we will use to build polynomials.

POLYNOMIALS IN GENERAL

> **DEFINITION** A **term**, or **monomial**, is a constant or the product of a constant and one or more variables raised to whole-number exponents.

The following are monomials, or terms:

$$-16 \qquad 3x^2y \qquad -\frac{2}{5}a^3b^2c \qquad xy^2z$$

The numerical part of each monomial is called the *numerical coefficient,* or just *coefficient* for short. For the preceding terms, the coefficients are -16, 3, $-\frac{2}{5}$, and 1. Notice that the coefficient for xy^2z is understood to be 1.

> **DEFINITION** A **polynomial** is any finite sum of terms. Since subtraction can be written in terms of addition, finite differences are also included in this definition.

The following are polynomials:

$$2x^2 - 6x + 3 \qquad -5x^2y + 2xy^2 \qquad 4a - 5b + 6c + 7d$$

Polynomials can be classified further according to the number of terms present. If a polynomial consists of two terms, it is said to be a *binomial.* If it has three terms, it is called a *trinomial.* And, as stated, a polynomial with only one term is said to be a *monomial.*

> **DEFINITION** The **degree** of a polynomial with one variable is the highest power to which the variable is raised in any one term.

EXAMPLES

1. $6x^2 + 2x - 1$ A trinomial of degree 2
2. $5x - 3$ A binomial of degree 1
3. $7x^6 - 5x^3 + 2x - 4$ A polynomial of degree 6
4. $-7x^4$ A monomial of degree 4
5. 15 A monomial of degree 0

Polynomials in one variable are usually written in decreasing powers of the variable. When this is the case, the coefficient of the first term is called the *leading coefficient.* In Example 1, the leading coefficient is 6. In Example 2, it is 5. The leading coefficient in Example 3 is 7.

> **DEFINITION**   Two or more terms that differ only in their numerical coefficients are called **similar,** or **like,** terms. Since similar terms differ only in their coefficients, they have identical variable parts.

ADDITION AND SUBTRACTION OF POLYNOMIALS

To add two polynomials, we simply apply the commutative and associative properties to group similar terms together and then use the distributive property as we have in the following example.

EXAMPLE 6   Add $5x^2 - 4x + 2$ and $3x^2 + 9x - 6$.

Solution

$$(5x^2 - 4x + 2) + (3x^2 + 9x - 6)$$

$$= (5x^2 + 3x^2) + (-4x + 9x) + (2 - 6) \qquad \text{Commutative and associative properties}$$

$$= (5 + 3)x^2 + (-4 + 9)x + (2 - 6) \qquad \text{Distributive property}$$

$$= 8x^2 + 5x + (-4)$$

$$= 8x^2 + 5x - 4$$

EXAMPLE 7   Find the sum of $-8x^3 + 7x^2 - 6x + 5$ and $10x^3 + 3x^2 - 2x - 6$.

Solution   We can add the two polynomials using the method of Example 6, or we can arrange similar terms in columns and add vertically. Using the column method, we have

$$\begin{array}{r} -8x^3 + 7x^2 - 6x + 5 \\ \underline{10x^3 + 3x^2 - 2x - 6} \\ 2x^3 + 10x^2 - 8x - 1 \end{array}$$

To find the difference of two polynomials, we need to use the fact that the opposite of a sum is the sum of the opposites. That is,

$$-(a + b) = -a + (-b)$$

One way to remember this is to observe that $-(a + b)$ is equivalent to $-1(a + b) = (-1)a + (-1)b = -a + (-b)$.

If a negative sign directly precedes the parentheses surrounding a polynomial, we may remove the parentheses and the preceding negative sign by changing the sign of each term within the parentheses. For example:

$$-(3x + 4) = -3x + (-4) = -3x - 4$$
$$-(5x^2 - 6x + 9) = -5x^2 + 6x - 9$$
$$-(-x^2 + 7x - 3) = x^2 - 7x + 3$$

E X A M P L E 8   Subtract $(9x^2 - 3x + 5) - (4x^2 + 2x - 3)$.

Solution   We subtract by adding the opposite of each term in the polynomial that follows the subtraction sign:

$(9x^2 - 3x + 5) - (4x^2 + 2x - 3)$

$= 9x^2 - 3x + 5 + (-4x^2) + (-2x) + 3$ The opposite of a sum is the sum of the opposites.

$= (9x^2 - 4x^2) + (-3x - 2x) + (5 + 3)$ Commutative and associative properties

$= 5x^2 - 5x + 8$ Combine similar terms.

E X A M P L E 9   Subtract $4x^2 - 9x + 1$ from $-3x^2 + 5x - 2$.

Solution   Again, to subtract, we add the opposite:

$$(-3x^2 + 5x - 2) - (4x^2 - 9x + 1) = -3x^2 + 5x - 2 - 4x^2 + 9x - 1$$
$$= (-3x^2 - 4x^2) + (5x + 9x) + (-2 - 1)$$
$$= -7x^2 + 14x - 3$$

E X A M P L E 10   Simplify $4x - 3[2 - (3x + 4)]$.

Solution   Removing the innermost parentheses first, we have

$$4x - 3[2 - (3x + 4)] = 4x - 3(2 - 3x - 4)$$
$$= 4x - 3(-3x - 2)$$
$$= 4x + 9x + 6$$
$$= 13x + 6$$

E X A M P L E 11   Simplify $(2x + 3) - [(3x + 1) - (x - 7)]$.

Solution

$$(2x + 3) - [(3x + 1) - (x - 7)] = (2x + 3) - (3x + 1 - x + 7)$$
$$= (2x + 3) - (2x + 8)$$
$$= 2x + 3 - 2x - 8$$
$$= -5$$

In the example that follows we will find the value of a polynomial for a given value of the variable.

EXAMPLE 12   Find the value of $5x^3 - 3x^2 + 4x - 5$ when x is 2.

Solution   We begin by substituting 2 for x in the original polynomial:

When $x = 2$

the polynomial $5x^3 - 3x^2 + 4x - 5$

becomes $5 \cdot 2^3 - 3 \cdot 2^2 + 4 \cdot 2 - 5 = 5 \cdot 8 - 3 \cdot 4 + 4 \cdot 2 - 5$

$$= 40 - 12 + 8 - 5$$

$$= 31$$

POLYNOMIALS AND FUNCTION NOTATION

Example 12 can be restated using function notation by calling the polynomial $P(x)$ and asking for $P(2)$. The solution would look like this:

If $P(x) = 5x^3 - 3x^2 + 4x - 5$

then $P(2) = 5 \cdot 2^3 - 3 \cdot 2^2 + 4 \cdot 2 - 5$

$$= 31$$

Our next example is stated in terms of function notation.

As we mentioned in the introduction to this chapter, three functions that occur very frequently in business and economics classes are profit, revenue, and cost functions. If a company manufactures and sells x items, then the revenue $R(x)$ is the total amount of money obtained by selling all x items. The cost $C(x)$ is the total amount of money it costs the company to manufacture the x items. The profit $P(x)$ obtained by selling all x items is the difference between the revenue and the cost and is given by the equation

$$P(x) = R(x) - C(x)$$

EXAMPLE 13   A company produces and sells copies of an accounting program for home computers. The total weekly cost (in dollars) to produce x copies of the program is $C(x) = 8x + 500$. Find its weekly profit if the total revenue obtained from selling all x programs is $R(x) = 35x - 0.1x^2$. How much profit will the company make if it produces and sells 100 programs a week? That is, find $P(100)$.

Solution   Using the equation $P(x) = R(x) - C(x)$ and the information given in the problem, we have

$$P(x) = R(x) - C(x)$$

$$= 35x - 0.1x^2 - (8x + 500)$$

$$= 35x - 0.1x^2 - 8x - 500$$

$$= -500 + 27x - 0.1x^2$$

If the company produces and sells 100 copies of the program, its weekly profit is

$$P(100) = -500 + 27(100) - 0.1(100)^2$$
$$= -500 + 27(100) - 0.1(10,000)$$
$$= -500 + 2,700 - 1,000$$
$$= 1,200$$

The weekly profit is $1,200.

Using TECHNOLOGY

GRAPHING CALCULATORS

Your graphing calculator allows you access to a function list in which you can input formulas for a number of functions. There are many ways to use the function list. For the problems in this section, we can use the function list to evaluate a polynomial. Let's use the profit function from Example 13 to illustrate. To input $P(x) = -500 + 27x - 0.1x^2$, we assign $P(x)$ to the first function in our function list, which is Y_1. Call up the function list for your calculator, and enter the profit function this way:

$$Y_1 = -500 + 27X - 0.1X^2$$

Now that you have an expression for Y_1, you can evaluate it by substituting a value for X. To do so, quit the function list and store the number 100 into the variable X:

100 → X ENTER   or   100 STO X ENTER

The display shows 100. Next, display the variable Y_1 (on many calculators, individual variables are on the Y-VAR menu), then press ENTER:

Y_1 ENTER

The display will show 1,200.

Now that you have the profit function stored in function Y_1, you can evaluate it for any value of X. Find the weekly profit if 50, 100, 150, 200, 250, and 300 programs are sold.

CLASSIFYING POLYNOMIAL FUNCTIONS

Any equation in two variables, such as $y = x^2 - 2x + 3$, in which the right side is a polynomial in one variable, is a function. (Such an equation is a function because it is impossible to obtain two different values of y for the same value of x.) Polynomial functions have the form

$$f(x) = \text{a polynomial in the variable } x$$

Some of the equations we have worked with in the past are polynomial functions. For example, all of the linear equations in two variables, except those with vertical lines for graphs, are polynomial functions. Each of the functions in the following groups is a polynomial function.

CONSTANT FUNCTIONS

Any function that can be written in the form

$$f(x) = c$$

where c is a real number, is called a **constant function.** The graph of every constant function is a horizontal line.

EXAMPLE 14   The function $f(x) = 3$ is an example of a constant function. Since all ordered pairs belonging to f have a y-coordinate of 3, the graph is the horizontal line given by $y = 3$. Remember, y and $f(x)$ are equivalent — that is, $y = f(x)$.

LINEAR FUNCTIONS

Any function that can be written in the form

$$f(x) = ax + b$$

where a and b are real numbers, $a \neq 0$, is called a **linear function.** The graph of every linear function is a straight line. In the past we have written linear functions in the form

$$y = mx + b$$

EXAMPLE 15   The function $f(x) = 2x - 3$ is an example of a linear function. The graph of this function is a straight line with slope 2 and y-intercept -3.

QUADRATIC FUNCTIONS

A **quadratic function** is any function that can be written in the form

$$f(x) = ax^2 + bx + c$$

where a, b, and c are real numbers and $a \neq 0$. We graphed the quadratic function $y = x^2$ in Section 3.5.

EXAMPLE 16   The function $f(x) = 2x^2 - 4x - 6$ is an example of a quadratic function.

> ## CUBIC FUNCTIONS
>
> A **cubic function** is any function that can be written in the form
>
> $$f(x) = ax^3 + bx^2 + cx + d$$
>
> where a, b, c, and d are real numbers and $a \neq 0$.

EXAMPLE 17 Graph the cubic function $y = \frac{1}{2}x^3$.

Solution The graph is shown in Figure 1. The table next to the graph shows some ordered pairs that satisfy the equation.

x	y
-3	$-\frac{27}{2}$
-2	-4
-1	$-\frac{1}{2}$
0	0
1	$\frac{1}{2}$
2	4
3	$\frac{27}{2}$

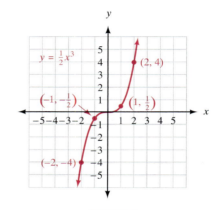

FIGURE 1

Getting Ready for Class

After reading through the preceding section, respond in your own words and in complete sentences.

A. Is $3x^2 + 2x - \frac{1}{x}$ a polynomial? Explain.

B. What are similar terms?

C. Explain in words how you subtract one polynomial from another.

D. What is revenue?

PROBLEM SET 5.2

Identify those of the following that are monomials, binomials, or trinomials. Give the degree of each, and name the leading coefficient.

1. $5x^2 - 3x + 2$

2. $2x^2 + 4x - 1$

3. $3x - 5$

4. $5y + 3$

5. $8a^2 + 3a - 5$

6. $9a^2 - 8a - 4$

7. $4x^3 - 6x^2 + 5x - 3$

8. $9x^4 + 4x^3 - 2x^2 + x$

9. $-\frac{3}{4}$

10. -16

11. $4x - 5 + 6x^3$

12. $9x + 2 + 3x^3$

Simplify each of the following by combining similar terms.

13. $(4x + 2) + (3x - 1)$

14. $(8x - 5) + (-5x + 4)$

15. $2x^2 - 3x + 10x - 15$

16. $6x^2 - 4x - 15x + 10$

17. $12a^2 + 8ab - 15ab - 10b^2$

18. $28a^2 - 8ab + 7ab - 2b^2$

19. $(5x^2 - 6x + 1) - (4x^2 + 7x - 2)$

20. $(11x^2 - 8x) - (4x^2 - 2x - 7)$

21. $(\frac{1}{2}x^2 - \frac{1}{3}x - \frac{1}{6}) - (\frac{1}{4}x^2 + \frac{7}{12}x) + (\frac{1}{3}x - \frac{1}{12})$

22. $(\frac{2}{3}x^2 - \frac{1}{2}x) - (\frac{1}{4}x^2 + \frac{1}{6}x + \frac{1}{12}) - (\frac{1}{2}x^2 + \frac{1}{4})$

23. $(y^3 - 2y^2 - 3y + 4) - (2y^3 - y^2 + y - 3)$

24. $(8y^3 - 3y^2 + 7y + 2) -$
$\qquad (-4y^3 + 6y^2 - 5y - 8)$

25. $(5x^3 - 4x^2) - (3x + 4) +$
$\qquad (5x^2 - 7) - (3x^3 + 6)$

26. $(x^3 - x) - (x^2 + x) + (x^3 - 1) - (-3x + 2)$

27. $(\frac{4}{7}x^2 - \frac{1}{7}xy + \frac{1}{14}y^2) - (\frac{1}{2}x^2 - \frac{2}{7}xy - \frac{9}{14}y^2)$

28. $(\frac{1}{5}x^2 - \frac{1}{2}xy + \frac{1}{10}y^2) - (-\frac{3}{10}x^2 + \frac{2}{5}xy - \frac{1}{2}y^2)$

29. $(3a^3 + 2a^2b + ab^2 - b^3) -$
$\qquad (6a^3 - 4a^2b + 6ab^2 - b^3)$

30. $(a^3 - 3a^2b + 3ab^2 - b^3) -$
$\qquad (a^3 + 3a^2b + 3ab^2 + b^3)$

31. Subtract $2x^2 - 4x$ from $2x^2 - 7x$.

32. Subtract $-3x + 6$ from $-3x + 9$.

33. Find the sum of $x^2 - 6xy + y^2$ and $2x^2 - 6xy - y^2$.

34. Find the sum of $9x^3 - 6x^2 + 2$ and $3x^2 - 5x + 4$.

35. Subtract $-8x^5 - 4x^3 + 6$ from $9x^5 - 4x^3 - 6$.

36. Subtract $4x^4 - 3x^3 - 2x^2$ from $2x^4 + 3x^3 + 4x^2$.

37. Find the sum of $11a^2 + 3ab + 2b^2$, $9a^2 - 2ab + b^2$, and $-6a^2 - 3ab + 5b^2$.

38. Find the sum of $a^2 - ab - b^2$, $a^2 + ab - b^2$, and $a^2 + 2ab + b^2$.

Simplify each of the following. Begin by working on the innermost parentheses.

39. $-[2 - (4 - x)]$

40. $-[-3 - (x - 6)]$

41. $-5[-(x - 3) - (x + 2)]$

42. $-6[(2x - 5) - 3(8x - 2)]$

43. $4x - 5[3 - (x - 4)]$

44. $x - 7[3x - (2 - x)]$

45. $-(3x - 4y) - [(4x + 2y) - (3x + 7y)]$

46. $(8x - y) - [-(2x + y) - (-3x - 6y)]$

47. $4a - \{3a + 2[a - 5(a + 1) + 4]\}$

48. $6a - \{-2a - 6[2a + 3(a - 1) - 6]\}$

49. Find the value of $2x^2 - 3x - 4$ when x is 2.

50. Find the value of $4x^2 + 3x - 2$ when x is -1.

51. If $P(x) = \frac{3}{2}x^2 - \frac{3}{4}x + 1$, find $P(12)$.

52. If $P(x) = \frac{2}{5}x^2 - \frac{1}{10}x + 2$, find $P(10)$.

53. If $Q(x) = x^3 - x^2 + x - 1$, find $Q(-2)$.

54. If $Q(x) = x^3 + x^2 + x + 1$, find $Q(-2)$.

55. Let $a = 3$ in each of the following expressions, and then simplify each one.

$\qquad (a + 4)^2 \qquad a^2 + 16 \qquad a^2 + 8a + 16$

56. Let $a = 2$ in each of the following expressions, and then simplify each one.

$\qquad (2a - 3)^2 \qquad 4a^2 - 9 \qquad 4a^2 - 12a + 9$

Sketch the graph of each of the following functions. Identify each as a constant function, linear function, quadratic function, or cubic function.

57. $f(x) = x^2 - 3$

58. $g(x) = 2x^2$

59. $g(x) = 4x - 1$

60. $f(x) = 3x + 2$

61. $f(x) = 5$

62. $f(x) = -3$

63. $y = x^3$

64. $y = 2x^3$

65. $y = x^3 - 2$

66. $y = x^3 + 3$

Applying the Concepts

Problems 67–72 may be solved using a graphing calculator.

67. Height of an Object If an object is propelled straight up into the air with a velocity of 128 feet/second, then its height $h(t)$ above the ground t seconds later is given by the formula

$$h(t) = -16t^2 + 128t$$

Find the height after 3 seconds and after 5 seconds. [Find $h(3)$ and $h(5)$.]

68. **Height of an Object** The formula for the height of an object that has been thrown straight up with a velocity of 64 feet/second is

$$h(t) = -16t^2 + 64t$$

Find the height after 1 second and after 3 seconds. [Find $h(1)$ and $h(3)$.]

69. **Profits** The total cost (in dollars) for a company to manufacture and sell x items per week is $C(x) = 60x + 300$. If the revenue brought in by selling all x items is $R(x) = 100x - 0.5x^2$, find the weekly profit. How much profit will be made by producing and selling 60 items each week?

70. **Profits** The total cost (in dollars) for a company to produce and sell x items per week is $C(x) = 200x + 1,600$. If the revenue brought in by selling all x items is $R(x) = 300x - 0.6x^2$, find the weekly profit. How much profit will be made by producing and selling 50 items each week?

71. **Profits** Suppose it costs a company selling patterns $C(x) = 800 + 6.5x$ dollars to produce and sell x patterns a month. If the revenue obtained by selling x patterns is $R(x) = 10x - 0.002x^2$, what is the profit equation? How much profit will be made if 1,000 patterns are produced and sold in May?

72. **Profits** Suppose a company manufactures and sells x picture frames each month with a total cost of $C(x) = 1,200 + 3.5x$ dollars. If the revenue obtained by selling x frames is $R(x) = 9x - 0.003x^2$, find the profit equation. How much profit will be made if 1,000 frames are manufactured and sold in June?

Review Problems

The problems that follow review material we covered in Section 5.1. Reviewing these problems will help you with the next section.

Simplify each expression.

73. $4x^3(5x^2)$ 74. $4x^3(-3x)$
75. $2a^2b(ab^2)$ 76. $2a^2b(-6a^2b)$

Extending the Concepts

77. **Reading Graphs** The graph of two polynomial functions are given in Figures 2 and 3, below. Use the graphs to solve a–h, on the next page.

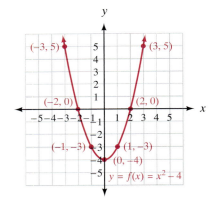

FIGURE 2

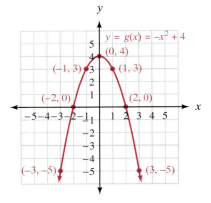

FIGURE 3

(a) $f(-3)$ (b) $f(0)$ (c) $f(1)$

(d) $g(-1)$ (e) $g(0)$ (f) $g(2)$

(g) $f[g(2)]$ (h) $g[f(2)]$

78. Profit, Revenue, and Cost The following line graphs show the monthly revenue and monthly cost, $R(x)$ and $C(x)$, for a local business.

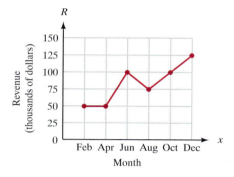

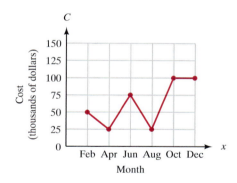

Use the preceding line graphs to evaluate each of the following functions. Remember, $P(x) = R(x) - C(x)$.

(a) R(February) C(February) P(February)

(b) R(June) C(June) P(June)

(c) R(October) C(October) P(October)

(d) Construct a line graph of $P(x)$ using the following template.

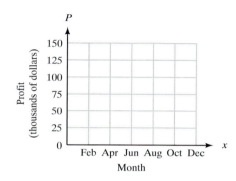

5.3   Multiplication of Polynomials

In the previous section we found the relationship between profit, revenue, and cost to be

$$P(x) = R(x) - C(x)$$

Revenue itself can be broken down further by another formula common in the business world. The revenue obtained from selling all x items is the product of the number of items sold and the price per item. That is,

$$\text{Revenue} = (\text{number of items sold})(\text{price of each item})$$

For example, if 100 items are sold for \$9 each, the revenue is $100(9) = \$900$. Likewise, if 500 items are sold for \$11 each, then the revenue is $500(11) = \$5,500$. In general, if x is the number of items sold and p is the selling price of each item, then we can write

$$R = xp$$

Many times, x and p are polynomials, which means that the expression xp is the product of two polynomials. In this section we learn how to multiply polynomials, and in so doing, increase our understanding of the equations and formulas that describe business applications.

EXAMPLE 1   Find the product of $4x^3$ and $5x^2 - 3x + 1$.

Solution   To multiply, we apply the distributive property:

$$4x^3(5x^2 - 3x + 1) = 4x^3(5x^2) + 4x^3(-3x) + 4x^3(1) \qquad \text{Distributive property}$$
$$= 20x^5 - 12x^4 + 4x^3$$

Notice that we multiply coefficients and add exponents.

EXAMPLE 2   Multiply $2x - 3$ and $x + 5$.

Solution   Distributing the $2x - 3$ across the sum $x + 5$ gives us

$$(2x - 3)(x + 5) = (2x - 3)x + (2x - 3)5 \qquad \text{Distributive property}$$
$$= 2x(x) + (-3)x + 2x(5) + (-3)5 \qquad \text{Distributive property}$$
$$= 2x^2 - 3x + 10x - 15$$
$$= 2x^2 + 7x - 15 \qquad \text{Combine like terms.}$$

Notice the third line in this example. It consists of all possible products of terms in the first binomial and those of the second binomial. We can generalize this into a rule for multiplying two polynomials.

RULE

To multiply two polynomials, multiply each term in the first polynomial by each term in the second polynomial.

Multiplying polynomials can be accomplished by a method that looks very similar to long multiplication with whole numbers.

E X A M P L E 3   Multiply $(2x - 3y)$ and $(3x^2 - xy + 4y^2)$ vertically.

Solution

$$
\begin{array}{r}
3x^2 - \quad xy + \ 4y^2 \\
2x - \ 3y \\
\hline
6x^3 - \quad 2x^2y + \ 8xy^2 \\
- \ 9x^2y + \quad 3xy^2 - 12y^3 \\
\hline
6x^3 - 11x^2y + 11xy^2 - 12y^3
\end{array}
$$

Multiply $(3x^2 - xy + 4y^2)$ by $2x$.
Multiply $(3x^2 - xy + 4y^2)$ by $-3y$.
Add similar terms.

MULTIPLYING BINOMIALS — THE FOIL METHOD

Consider the product of $(2x - 5)$ and $(3x - 2)$. Distributing $(3x - 2)$ over $2x$ and -5, we have

$$
\begin{aligned}
(2x - 5)(3x - 2) &= (2x)(3x - 2) + (-5)(3x - 2) \\
&= (2x)(3x) + (2x)(-2) + (-5)(3x) + (-5)(-2) \\
&= 6x^2 - 4x - 15x + 10 \\
&= 6x^2 - 19x + 10
\end{aligned}
$$

Looking closely at the second and third lines, we notice the following:

1. $6x^2$ comes from multiplying the *first* terms in each binomial:

$$(2x - 5)(3x - 2) \qquad 2x(3x) = 6x^2 \qquad \textit{First } \text{terms}$$

2. $-4x$ comes from multiplying the *outside* terms in the product:

$$(2x - 5)(3x - 2) \qquad 2x(-2) = -4x \qquad \textit{Outside } \text{terms}$$

3. $-15x$ comes from multiplying the *inside* terms in the product:

$$(2x - 5)(3x - 2) \qquad -5(3x) = -15x \qquad \textit{Inside } \text{terms}$$

4. 10 comes from multiplying the *last* two terms in the product:

$$(2x - 5)(3x - 2) \qquad -5(-2) = 10 \qquad \textit{Last } \text{terms}$$

Once we know where the terms in the answer come from, we can reduce the number of steps used in finding the product:

$$(2x - 5)(3x - 2) = 6x^2 - \quad 4x \quad - \quad 15x + 10 = 6x^2 - 19x + 10$$
$$\qquad\qquad\qquad\quad \text{First} \quad \text{Outside} \quad \text{Inside} \quad \text{Last}$$

EXAMPLES   Multiply using the FOIL method.

4. $(4a - 5b)(3a + 2b) = \underset{F}{12a^2} + \underset{O}{8ab} - \underset{I}{15ab} - \underset{L}{10b^2}$

$ = 12a^2 - 7ab - 10b^2$

5. $(3 - 2t)(4 + 7t) = \underset{F}{12} + \underset{O}{21t} - \underset{I}{8t} - \underset{L}{14t^2}$

$ = 12 + 13t - 14t^2$

6. $(2x + \tfrac{1}{2})(4x - \tfrac{1}{2}) = \underset{F}{8x^2} - \underset{O}{x} + \underset{I}{2x} - \underset{L}{\tfrac{1}{4}}$

$\phantom{(2x + \tfrac{1}{2})(4x - \tfrac{1}{2})} = 8x^2 + x - \tfrac{1}{4}$

7. $(a^5 + 3)(a^5 - 7) = \underset{F}{a^{10}} - \underset{O}{7a^5} + \underset{I}{3a^5} - \underset{L}{21}$

$ = a^{10} - 4a^5 - 21$

8. $(2x + 3)(5y - 4) = \underset{F}{10xy} - \underset{O}{8x} + \underset{I}{15y} - \underset{L}{12}$

THE SQUARE OF A BINOMIAL

EXAMPLE 9   Find $(4x - 6)^2$.

Solution   Applying the definition of exponents and then the FOIL method, we have

$$(4x - 6)^2 = (4x - 6)(4x - 6)$$

$$= \underset{F}{16x^2} - \underset{O}{24x} - \underset{I}{24x} + \underset{L}{36}$$

$$= 16x^2 - 48x + 36$$

This example is the square of a binomial. This type of product occurs frequently enough in algebra that we have special formulas for it. Here are the formulas for binomial squares:

$$(a + b)^2 = (a + b)(a + b) = a^2 + ab + ab + b^2 = a^2 + 2ab + b^2$$

$$(a - b)^2 = (a - b)(a - b) = a^2 - ab - ab + b^2 = a^2 - 2ab + b^2$$

Observing the results in both cases, we have the following rule.

> **RULE**
>
> The square of a binomial is the sum of the square of the first term, twice the product of the two terms, and the square of the last term. Or:
>
> $$(a + b)^2 = \underset{\substack{\text{Square} \\ \text{of} \\ \text{first} \\ \text{term}}}{a^2} + \underset{\substack{\text{Twice the} \\ \text{product} \\ \text{of the} \\ \text{two terms}}}{2ab} + \underset{\substack{\text{Square} \\ \text{of} \\ \text{last} \\ \text{term}}}{b^2}$$
>
> $$(a - b)^2 = a^2 - 2ab + b^2$$

EXAMPLES Use the preceding formulas to expand each binomial square.

10. $(x + 7)^2 = x^2 + 2(x)(7) + 7^2 = x^2 + 14x + 49$

11. $(3t - 5)^2 = (3t)^2 - 2(3t)(5) + 5^2 = 9t^2 - 30t + 25$

12. $(4x + 2y)^2 = (4x)^2 + 2(4x)(2y) + (2y)^2 = 16x^2 + 16xy + 4y^2$

13. $(5 - a^3)^2 = 5^2 - 2(5)(a^3) + (a^3)^2 = 25 - 10a^3 + a^6$

PRODUCTS RESULTING IN THE DIFFERENCE OF TWO SQUARES

Another frequently occurring kind of product is found when multiplying two binomials that differ only in the sign between their terms.

EXAMPLE 14 Multiply $(3x - 5)$ and $(3x + 5)$.

Solution Applying the FOIL method, we have

$$(3x - 5)(3x + 5) = 9x^2 + 15x - 15x - 25 \qquad \text{Two middle terms add to 0.}$$
$$\ \ \, \text{F}\quad\ \text{O}\quad\ \ \text{I}\quad\ \text{L}$$
$$= 9x^2 - 25$$

The outside and inside products in Example 14 are opposites and therefore add to 0. Here it is in general:

$$(a - b)(a + b) = a^2 + ab - ab - b^2 \qquad \text{Two middle terms add to 0.}$$
$$= a^2 - b^2$$

> **RULE**
>
> To multiply two binomials that differ only in the sign between their two terms, simply subtract the square of the second term from the square of the first term:
>
> $$(a + b)(a - b) = a^2 - b^2$$

The expression $a^2 - b^2$ is called the *difference of two squares*.

EXAMPLES   Find the following products.

15. $(x - 5)(x + 5) = x^2 - 25$

16. $(2a - 3)(2a + 3) = 4a^2 - 9$

17. $(x^2 + 4)(x^2 - 4) = x^4 - 16$

18. $(x^3 - 2a)(x^3 + 2a) = x^6 - 4a^2$

MORE ABOUT FUNCTION NOTATION

From the introduction to this chapter, we know that the revenue obtained from selling x items at p dollars per item is

$$R = \text{Revenue} = xp \qquad \text{(The number of items} \times \text{price per item)}$$

For example, if a store sells 100 items at \$4.50 per item, the revenue is $100(4.50) = \$450$. If we have an equation that gives the relationship between x and p, then we can write the revenue in terms of x or in terms of p. With function notation, we would write the revenue as either $R(x)$ or $R(p)$, where

$R(x)$ is the revenue function that gives the revenue R in terms of the number of items x.

$R(p)$ is the revenue function that gives the revenue R in terms of the price per item p.

With function notation we can see exactly which variables we want our formulas written in terms of.

In the next two examples, we will use function notation to combine a number of problems we have worked previously.

EXAMPLE 19   A company manufactures and sells prerecorded videotapes. They find that they can sell x videotapes each day at p dollars per tape, according to the equation $x = 230 - 20p$. Find $R(x)$ and $R(p)$.

Solution   The notation $R(p)$ tells us we are to write the revenue equation in terms of the variable p. To do so, we use the formula $R(p) = xp$ and substitute $230 - 20p$ for x to obtain

$$R(p) = xp = (230 - 20p)p = 230p - 20p^2$$

The notation $R(x)$ indicates that we are to write the revenue equation in terms of the variable x. We need to solve the equation $x = 230 - 20p$ for p. Let's begin by interchanging the two sides of the equation:

$$230 - 20p = x$$

$$-20p = -230 + x \qquad \text{Add } -230 \text{ to each side.}$$

$$p = \frac{-230 + x}{-20} \qquad \text{Divide each side by } -20.$$

$$p = 11.5 - 0.05x \qquad \frac{230}{20} = 11.5 \text{ and } \frac{1}{20} = 0.05$$

Now we can find $R(x)$ by substituting $11.5 - 0.05x$ for p in the formula $R(x) = xp$:

$$R(x) = xp = x(11.5 - 0.05x) = 11.5x - 0.05x^2$$

Our two revenue functions are actually equivalent. To offer some justification for this, suppose that the company decides to sell each tape for $5. The equation $x = 230 - 20p$ indicates that, at $5 per tape, they will sell $x = 230 - 20(5) = 230 - 100 = 130$ tapes per day. To find the revenue from selling the tapes for $5 each, we use $R(p)$ with $p = 5$:

$$\text{If} \qquad p = 5$$

$$\text{then} \qquad R(p) = R(5)$$
$$= 230(5) - 20(5)^2$$
$$= 1,150 - 500$$
$$= \$650$$

On the other hand, to find the revenue from selling 130 tapes, we use $R(x)$ with $x = 130$:

$$\text{If} \qquad x = 130$$

$$\text{then} \qquad R(x) = R(130)$$
$$= 11.5(130) - 0.05(130)^2$$
$$= 1,495 - 845$$
$$= \$650$$

EXAMPLE 20   Suppose the daily cost function for the videotapes in Example 19 is $C(x) = 200 + 2x$. Find the profit function $P(x)$ and then find $P(130)$.

Solution   Since profit is equal to the difference of the revenue and the cost, we have

$$P(x) = R(x) - C(x)$$
$$= 11.5x - 0.05x^2 - (200 + 2x)$$
$$= -0.05x^2 + 9.5x - 200$$

Notice that we used the formula for $R(x)$ from Example 19 instead of the formula for $R(p)$. We did so because we were asked to find $P(x)$, meaning we want the profit P only in terms of the variable x.

Next, we use the formula we just obtained to find $P(130)$:

$$P(130) = -0.05(130)^2 + 9.5(130) - 200$$
$$= -0.05(16,900) + 9.5(130) - 200$$
$$= -845 + 1,235 - 200$$
$$= \$190$$

Since $P(130) = \$190$, the company will make a profit of $190 per day by selling 130 tapes per day.

Using TECHNOLOGY

GRAPHING CALCULATORS

More About Example 20

We can visualize the three functions given in Example 20 if we set up the functions list and graphing window on our calculator this way:

$$Y_1 = 11.5X - 0.05X^2 \qquad \text{This gives the graph of } R(x).$$

$$Y_2 = 200 + 2X \qquad \text{This gives the graph of } C(x).$$

$$Y_3 = Y_1 - Y_2 \qquad \text{This gives the graph of } P(x).$$

Window: X from 0 to 250, Y from 0 to 750

The graphs in Figure 1 are similar to what you will obtain using the functions list and window shown here. Next, find the value of $P(x)$ when $R(x)$ and $C(x)$ intersect.

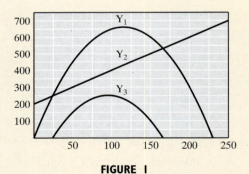

FIGURE 1

Getting Ready for Class

After reading through the preceding section, respond in your own words and in complete sentences.

A. Describe how the distributive property is used to multiply a monomial and a polynomial.

B. Describe how you would use the FOIL method to multiply two binomials.

C. Explain why $(x + 3)^2 \neq x^2 + 9$.

D. When will the product of two binomials result in a binomial?

PROBLEM SET 5.3

Multiply the following by applying the distributive property.

1. $2x(6x^2 - 5x + 4)$ **2.** $-3x(5x^2 - 6x - 4)$

3. $-3a^2(a^3 - 6a^2 + 7)$ **4.** $4a^3(3a^2 - a + 1)$

5. $2a^2b(a^3 - ab + b^3)$

6. $5a^2b^2(8a^2 - 2ab + b^2)$

Multiply the following vertically.

7. $(x - 5)(x + 3)$ **8.** $(x + 4)(x + 6)$

9. $(2x^2 - 3)(3x^2 - 5)$ **10.** $(3x^2 + 4)(2x^2 - 5)$

11. $(x + 3)(x^2 + 6x + 5)$

12. $(x - 2)(x^2 - 5x + 7)$

13. $(a - b)(a^2 + ab + b^2)$

14. $(a + b)(a^2 - ab + b^2)$

15. $(2x + y)(4x^2 - 2xy + y^2)$

16. $(x - 3y)(x^2 + 3xy + 9y^2)$

17. $(2a - 3b)(a^2 + ab + b^2)$

18. $(5a - 2b)(a^2 - ab - b^2)$

Multiply the following using the FOIL method.

19. $(x - 2)(x + 3)$ **20.** $(x + 2)(x - 3)$

21. $(2a + 3)(3a + 2)$ **22.** $(5a - 4)(2a + 1)$

23. $(5 - 3t)(4 + 2t)$ **24.** $(7 - t)(6 - 3t)$

25. $(x^3 + 3)(x^3 - 5)$ **26.** $(x^3 + 4)(x^3 - 7)$

27. $(5x - 6y)(4x + 3y)$ **28.** $(6x - 5y)(2x - 3y)$

29. $(3t + \frac{1}{3})(6t - \frac{2}{3})$ **30.** $(5t - \frac{1}{5})(10t + \frac{3}{5})$

Find the following special products.

31. $(5x + 2y)^2$ **32.** $(3x - 4y)^2$

33. $(5 - 3t^3)^2$ **34.** $(7 - 2t^4)^2$

35. $(2a + 3b)(2a - 3b)$ **36.** $(6a - 1)(6a + 1)$

37. $(3r^2 + 7s)(3r^2 - 7s)$

38. $(5r^2 - 2s)(5r^2 + 2s)$

39. $(\frac{1}{3}x - \frac{2}{5})(\frac{1}{3}x + \frac{2}{5})$

40. $(\frac{3}{4}x - \frac{1}{7})(\frac{3}{4}x + \frac{1}{7})$

Find the following products.

41. $(x - 2)^3$ **42.** $(4x + 1)^3$

43. $(x - \frac{1}{2})^3$ **44.** $(x + \frac{1}{4})^3$

45. $3(x - 1)(x - 2)(x - 3)$

46. $2(x + 1)(x + 2)(x + 3)$

47. $(b^2 + 8)(a^2 + 1)$ **48.** $(b^2 + 1)(a^4 - 5)$

49. $(x + 1)^2 + (x + 2)^2 + (x + 3)^2$

50. $(x - 1)^2 + (x - 2)^2 + (x - 3)^2$

51. $(2x + 3)^2 - (2x - 3)^2$

52. $(x - 3)^3 - (x + 3)^3$

53. Multiply $(x + y - 4)(x + y + 5)$ by first writing it like this:

$$[(x + y) - 4][(x + y) + 5]$$

and then applying the FOIL method.

54. Multiply $(x - 5 - y)(x - 5 + y)$ by first writing it like this:

$$[(x - 5) - y][(x - 5) + y]$$

and then applying the FOIL method.

55. Let $a = 2$ and $b = 3$, and evaluate each of the following expressions.

$$a^4 - b^4 \qquad (a - b)^4 \qquad (a^2 + b^2)(a + b)(a - b)$$

56. Let $a = 2$ and $b = 3$, and evaluate each of the following expressions.

$$a^3 + b^3 \qquad (a + b)^3 \qquad a^3 + 3a^2b + 3ab^2 + b^3$$

Applying the Concepts

57. Revenue   A store selling art supplies finds that it can sell x sketch pads per week at p dollars each, according to the formula $x = 900 - 300p$. Write formulas for $R(p)$ and $R(x)$. Then find the revenue obtained by selling the pads for $1.60 each.

58. Revenue   A company selling diskettes for home computers finds that it can sell x diskettes per day at p dollars per diskette, according to the formula $x = 800 - 100p$. Write formulas for $R(p)$ and $R(x)$. Then find the revenue obtained by selling the diskettes for $3.80 each.

59. Revenue A company sells an inexpensive accounting program for home computers. If it can sell x programs per week at p dollars per program, according to the formula $x = 350 - 10p$, find formulas for $R(p)$ and $R(x)$. How much will the weekly revenue be if it sells 65 programs?

60. Revenue A company sells boxes of greeting cards through the mail. It finds that it can sell x boxes of cards each week at p dollars per box, according to the formula $x = 1,475 - 250p$. Write formulas for $R(p)$ and $R(x)$. What revenue will it bring in each week if it sells 200 boxes of cards?

61. Profit If the cost to produce the x programs in Problem 59 is $C(x) = 5x + 500$, find $P(x)$ and $P(60)$.

62. Profit If the cost to produce the x diskettes in Problem 58 is $C(x) = 2x + 200$, find $P(x)$ and $P(40)$.

63. Interest If you deposit $100 in an account with an interest rate r that is compounded annually, then the amount of money in that account at the end of 4 years is given by the formula $A = 100(1 + r)^4$. Expand the right side of this formula.

64. Interest If you deposit P dollars in an account with an annual interest rate r that is compounded twice a year, then at the end of a year the amount of money in that account is given by the formula

$$A = P\left(1 + \frac{r}{2}\right)^2$$

Expand the right side of this formula.

Review Problems

The following problems review material from Section 4.2.

Solve each system.

65. $\begin{aligned} x + y + z &= 6 \\ 2x - y + z &= 3 \\ x + 2y - 3z &= -4 \end{aligned}$ **66.** $\begin{aligned} x + y + z &= 6 \\ x - y + 2z &= 7 \\ 2x - y - z &= 0 \end{aligned}$

67. $\begin{aligned} 3x + 4y &= 15 \\ 2x - 5z &= -3 \\ 4y - 3z &= 9 \end{aligned}$ **68.** $\begin{aligned} x + 3y &= 5 \\ 6y + z &= 12 \\ x - 2z &= -10 \end{aligned}$

Extending the Concepts

69. Multiplication and Area The following diagram shows a square that is divided into four parts: two smaller squares and two smaller rectangles. The diagram can be used to justify the formula $(a + b)^2 = a^2 + 2ab + b^2$. The area of the complete figure is $(a + b)^2$. Copy the diagram and then find the area of each of the four smaller parts, and show how this justifies our formula.

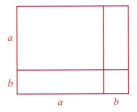

70. Multiplication and Volume The following diagram shows a cube that is divided into eight smaller parts: two smaller cubes and six smaller rectangular solids. The diagram can be used to justify the formula

$$(a + b)^3 = a^3 + 3a^2b + 3ab^2 + b^3$$

The volume of the complete figure is $(a + b)^3$.
(a) What is the volume of the cube in the upper left part of the diagram?
(b) Find the volume of the smaller cube in the back on lower right side.
(c) What is the volume of the green rectangular solid? How many of these solids are there?
(d) What is the volume of the red rectangular solid? How many of these solids are there?
(e) Add the volumes of all eight smaller parts to justify the preceding formula.

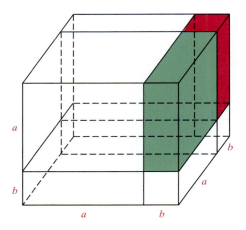

71. Multiplication and Area An area model may be used to multiply polynomials. The areas of the following rectangles can be combined to form the product $(2x + 3)(x + 2)$. Fill in all the missing areas, and then combine them to form the product $(2x + 3)(x + 2)$.

72. Multiplication and Area The following diagram contains nine areas, each one of which represents a term in the expansion of $(x + y + z)^2$. Fill in all the missing areas, and then combine the results to arrive at the expanded expression for $(x + y + z)^2$.

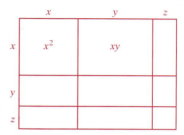

5.4   The Greatest Common Factor and Factoring by Grouping

In general, factoring is the reverse of multiplication. The following diagram illustrates the relationship between factoring and multiplication:

Multiplication

Factors $\longrightarrow 3 \cdot 7 = 21 \longleftarrow$ Product

Factoring

Reading from left to right, we say the product of 3 and 7 is 21. Reading in the other direction, from right to left, we say 21 factors into 3 times 7. Or, 3 and 7 are factors of 21.

> **DEFINITION**   The **greatest common factor** for a polynomial is the largest monomial that divides (is a factor of) each term of the polynomial.

Note: The term *largest monomial,* as used here, refers to the monomial with the largest integer exponents whose coefficient has the greatest absolute value.

The greatest common factor for the polynomial $25x^5 + 20x^4 - 30x^3$ is $5x^3$ since it is the largest monomial that is a factor of each term. We can apply the distributive property and write

$$25x^5 + 20x^4 - 30x^3 = 5x^3(5x^2) + 5x^3(4x) - 5x^3(6)$$
$$= 5x^3(5x^2 + 4x - 6)$$

The last line is written in factored form.

EXAMPLE 1   Factor the greatest common factor from

$$8a^3 - 8a^2 - 48a$$

Solution   The greatest common factor is $8a$. It is the largest monomial that divides each term of our polynomial. We can write each term in our polynomial as the product of $8a$ and another monomial. Then, we apply the distributive property to factor $8a$ from each term:

$$8a^3 - 8a^2 - 48a = 8a(a^2) - 8a(a) - 8a(6)$$
$$= 8a(a^2 - a - 6)$$

EXAMPLE 2   Factor the greatest common factor from

$$16a^5b^4 - 24a^2b^5 - 8a^3b^3$$

Solution   The largest monomial that divides each term is $8a^2b^3$. We write each term of the original polynomial in terms of $8a^2b^3$ and apply the distributive property to write the polynomial in factored form:

$$16a^5b^4 - 24a^2b^5 - 8a^3b^3 = 8a^2b^3(2a^3b) - 8a^2b^3(3b^2) - 8a^2b^3(a)$$
$$= 8a^2b^3(2a^3b - 3b^2 - a)$$

EXAMPLE 3   Factor the greatest common factor from

$$5x^2(a + b) - 6x(a + b) - 7(a + b)$$

Solution   The greatest common factor is $a + b$. Factoring this from each term, we have

$$5x^2(a + b) - 6x(a + b) - 7(a + b) = (a + b)(5x^2 - 6x - 7)$$

EXAMPLE 4   A company manufacturing prerecorded videotapes finds that the total daily revenue for selling x tapes is given by

$$R(x) = 11.5x - 0.05x^2$$

Factor x from each term on the right side of the equation to find the formula that gives the price p in terms of x.

Solution   We begin by factoring x from the right side of the equation:

$$\text{If} \qquad R(x) = 11.5x - 0.05x^2$$
$$\text{then} \qquad R(x) = x(11.5 - 0.05x)$$

Because R is always xp, the quantity in parentheses must be p. The price it should charge if it wants to sell x items per day is therefore

$$p = 11.5 - 0.05x$$

FACTORING BY GROUPING

Many polynomials have no greatest common factor other than the number 1. Some of these can be factored using the distributive property if those terms with a common factor are grouped together.

For example, the polynomial $5x + 5y + x^2 + xy$ can be factored by noticing that the first two terms have a 5 in common, whereas the last two have an x in common.

Applying the distributive property, we have

$$5x + 5y + x^2 + xy = 5(x + y) + x(x + y)$$

This last expression can be thought of as having two terms, $5(x + y)$ and $x(x + y)$, each of which has a common factor $(x + y)$. We apply the distributive property again to factor $(x + y)$ from each term:

$$5(x + y) + x(x + y)$$
$$= (x + y)(5 + x)$$

EXAMPLE 5   Factor $a^2b^2 + b^2 + 8a^2 + 8$.

Solution   The first two terms have b^2 in common; the last two have 8 in common:

$$a^2b^2 + b^2 + 8a^2 + 8 = b^2(a^2 + 1) + 8(a^2 + 1)$$
$$= (a^2 + 1)(b^2 + 8)$$

EXAMPLE 6   Factor $15 - 5y^4 - 3x^3 + x^3y^4$.

Solution   Let's try factoring a 5 from the first two terms and an $-x^3$ from the last two terms:

$$15 - 5y^4 - 3x^3 + x^3y^4 = 5(3 - y^4) - x^3(3 - y^4)$$
$$= (3 - y^4)(5 - x^3)$$

EXAMPLE 7   Factor by grouping $x^3 + 2x^2 + 9x + 18$.

Solution   We begin by factoring x^2 from the first two terms and 9 from the second two terms:

$$x^3 + 2x^2 + 9x + 18 = x^2(x + 2) + 9(x + 2)$$
$$= (x + 2)(x^2 + 9)$$

Getting Ready for Class

After reading through the preceding section, respond in your own words and in complete sentences.

A. What is the greatest common factor for a polynomial?

B. After factoring a polynomial, how can you check your result?

C. When would you try to factor by grouping?

D. What is the relationship between multiplication and factoring?

PROBLEM SET 5.4

Factor the greatest common factor from each of the following. (The answers in the back of the book all show greatest common factors whose coefficients are positive.)

1. $10x^3 - 15x^2$

2. $12x^5 + 18x^7$

3. $9y^6 + 18y^3$

4. $24y^4 - 8y^2$

5. $9a^2b - 6ab^2$

6. $30a^3b^4 + 20a^4b^3$

7. $21xy^4 + 7x^2y^2$

8. $14x^6y^3 - 6x^2y^4$

9. $3a^2 - 21a + 30$

10. $3a^2 - 3a - 6$

11. $4x^3 - 16x^2 - 20x$

12. $2x^3 - 14x^2 + 20x$

13. $10x^4y^2 + 20x^3y^3 - 30x^2y^4$

14. $6x^4y^2 + 18x^3y^3 - 24x^2y^4$

15. $-x^2y + xy^2 - x^2y^2$

16. $-x^3y^2 - x^2y^3 - x^2y^2$

17. $4x^3y^2z - 8x^2y^2z^2 + 6xy^2z^3$

18. $7x^4y^3z^2 - 21x^2y^2z^2 - 14x^2y^3z^4$

19. $20a^2b^2c^2 - 30ab^2c + 25a^2bc^2$

20. $8a^3bc^5 - 48a^2b^4c + 16ab^3c^5$

21. $5x(a - 2b) - 3y(a - 2b)$

22. $3a(x - y) - 7b(x - y)$

23. $3x^2(x + y)^2 - 6y^2(x + y)^2$

24. $10x^3(2x - 3y) - 15x^2(2x - 3y)$

25. $2x^2(x + 5) + 7x(x + 5) + 6(x + 5)$

26. $2x^2(x + 2) + 13x(x + 2) + 15(x + 2)$

Factor each of the following by grouping.

27. $3xy + 3y + 2ax + 2a$

28. $5xy^2 + 5y^2 + 3ax + 3a$

29. $x^2y + x + 3xy + 3$

30. $x^3y^3 + 2x^3 + 5x^2y^3 + 10x^2$

31. $3xy^2 - 6y^2 + 4x - 8$

32. $8x^2y - 4x^2 + 6y - 3$

33. $x^2 - ax - bx + ab$

34. $ax - x^2 - bx + ab$

35. $ab + 5a - b - 5$

36. $x^2 - xy - ax + ay$

37. $a^4b^2 + a^4 - 5b^2 - 5$

38. $2a^2 - a^2b - bc^2 + 2c^2$

39. $x^3 + 3x^2 - 4x - 12$

40. $x^3 + 5x^2 - 4x - 20$

41. $x^3 + 2x^2 - 25x - 50$

42. $x^3 + 4x^2 - 9x - 36$

43. $2x^3 + 3x^2 - 8x - 12$

44. $3x^3 + 2x^2 - 27x - 18$

45. $4x^3 + 12x^2 - 9x - 27$

46. $9x^3 + 18x^2 - 4x - 8$

47. The greatest common factor of the binomial $3x - 9$ is 3. The greatest common factor of the binomial $6x - 2$ is 2. What is the greatest common factor of their product, $(3x - 9)(6x - 2)$, when it has been multiplied out?

48. The greatest common factors of the binomials $5x - 10$ and $2x + 4$ are 5 and 2, respectively. What is the greatest common factor of their product, $(5x - 10)(2x + 4)$, when it has been multiplied out?

Applying the Concepts

49. Investing   If P dollars are placed in a savings account in which the rate of interest r is compounded yearly, then at the end of 1 year the amount of money in the account can be written as $P + Pr$. At the end of 2 years the amount of money in the account is

$$P + Pr + (P + Pr)r$$

Use factoring by grouping to show that this last expression can be written as $P(1 + r)^2$.

50. Investing   At the end of 3 years, the amount of money in the savings account in Problem 49 will be

$$P(1 + r)^2 + P(1 + r)^2 r$$

Use factoring to show that this last expression can be written as $P(1 + r)^3$.

Use Example 4 as a guide in solving the next four problems.

51. Price   A company manufacturing prerecorded videotapes finds that the total daily revenue R for selling x tapes at p dollars per tape is given by

$$R(x) = 11.5x - 0.05x^2$$

Factor x from each term on the right side of the equation to find the formula that gives the price p in terms of x. Then, use it to find the price they should charge if they want to sell 125 videotapes per day.

52. Price   A company producing diskettes for home computers finds that the total daily revenue for selling x diskettes at p dollars per diskette is given by

$$R(x) = 8x - 0.01x^2$$

Use the fact that $R = xp$ and your knowledge of factoring to find a formula that gives the price p in terms of x. Then, use it to find the price they should charge if they want to sell 420 diskettes per day.

53. Price   The weekly revenue equation for a company selling an inexpensive accounting program for home computers is given by the equation

$$R(x) = 35x - 0.1x^2$$

where x is the number of programs they sell per week. What price p should they charge if they want to sell 65 programs per week?

54. Price   The weekly revenue equation for a small mail-order company selling boxes of greeting cards is

$$R(x) = 5.9x - 0.004x^2$$

where x is the number of boxes they sell per week. What price p should they charge if they want to sell 200 boxes each week?

Review Problems

The problems that follow review material we covered in Sections 4.3 and 5.3. Reviewing the problems from Section 5.3 will help you understand the next section.

Multiply using the FOIL method. [5.3]

55. $(x + 2)(x + 3)$ **56.** $(x - 2)(x - 3)$
57. $(2y + 5)(3y - 7)$ **58.** $(2y - 5)(3y + 7)$
59. $(4 - 3a)(5 - a)$ **60.** $(4 - 3a)(5 + a)$

Evaluate each determinant. [4.3]

61. $\begin{vmatrix} 3 & 5 \\ -6 & 2 \end{vmatrix}$ **62.** $\begin{vmatrix} -2 & 0 \\ 0 & -1 \end{vmatrix}$

63. $\begin{vmatrix} 1 & -2 & 3 \\ 0 & 4 & -1 \\ 2 & -4 & 6 \end{vmatrix}$ **64.** $\begin{vmatrix} 2 & 0 & 0 \\ 0 & -3 & 0 \\ 0 & 0 & 4 \end{vmatrix}$

Extending the Concepts

Assume n is a positive integer and factor each expression by factoring out the greatest common factor.

65. $x^{n+2} + x^{n+1} + x^n$ **66.** $x^{n+4} + x^{n+2} + x^n$
67. $x^{n+5} + x^{n+4} + x^{n+3}$ **68.** $x^{n+3} + x^{n+2} + x^{n+1}$

FACTORING TRINOMIALS WITH A LEADING COEFFICIENT OF 1

In Section 5.3 we multiplied binomials:

$$(x - 2)(x + 3) = x^2 + x - 6$$

$$(x + 5)(x + 2) = x^2 + 7x + 10$$

In each case the product of two binomials is a trinomial. The first term in the resulting trinomial is obtained by multiplying the first term in each binomial. The middle term comes from adding the product of the two inside terms with the product of the two outside terms. The last term is the product of the last term in each binomial.

In general,

$$(x + a)(x + b) = x^2 + ax + bx + ab$$
$$= x^2 + (a + b)x + ab$$

Writing this as a factoring problem, we have

$$x^2 + (a + b)x + ab = (x + a)(x + b)$$

To factor a trinomial with a leading coefficient of 1, we simply find the two numbers a and b whose sum is the coefficient of the middle term and whose product is the constant term.

E X A M P L E 1   Factor $x^2 + 2x - 15$.

Solution   Again the leading coefficient is 1. We need two integers whose product is -15 and whose sum is $+2$. The integers are $+5$ and -3.

$$x^2 + 2x - 15 = (x + 5)(x - 3)$$

If a trinomial is factorable, then its factors are unique. For instance, in the preceding example we found factors of $x + 5$ and $x - 3$. These are the only two factors for $x^2 + 2x - 15$. There is no other pair of binomials whose product is $x^2 + 2x - 15$.

E X A M P L E 2   Factor $x^2 - xy - 12y^2$.

Solution   We need two numbers whose product is $-12y^2$ and whose sum is $-y$. The numbers are $-4y$ and $3y$:

$$x^2 - xy - 12y^2 = (x - 4y)(x + 3y)$$

Checking this result gives

$$(x - 4y)(x + 3y) = x^2 + 3xy - 4xy - 12y^2$$
$$= x^2 - xy - 12y^2$$

EXAMPLE 3   Factor $x^2 - 8x + 6$.

Solution   Since there is no pair of integers whose product is 6 and whose sum is -8, the trinomial $x^2 - 8x + 6$ is not factorable. We say it is a *prime polynomial*.

EXAMPLE 4   Factor $3x^4 - 15x^3y - 18x^2y^2$.

Solution   The leading coefficient is not 1. Each term is divisible by $3x^2$, however. Factoring this out to begin with, we have

$$3x^4 - 15x^3y - 18x^2y^2 = 3x^2(x^2 - 5xy - 6y^2)$$

Factoring the resulting trinomial as in the previous examples gives

$$3x^2(x^2 - 5xy - 6y^2) = 3x^2(x - 6y)(x + y)$$

FACTORING OTHER TRINOMIALS BY TRIAL AND ERROR

We want to turn our attention now to trinomials with leading coefficients other than 1 and with no greatest common factor other than 1.

Suppose we want to factor $3x^2 - x - 2$. The factors will be a pair of binomials. The product of the first terms will be $3x^2$, and the product of the last terms will be -2. We can list all the possible factors along with their products as follows.

Possible Factors	First Term	Middle Term	Last Term
$(x + 2)(3x - 1)$	$3x^2$	$+5x$	-2
$(x - 2)(3x + 1)$	$3x^2$	$-5x$	-2
$(x + 1)(3x - 2)$	$3x^2$	$+x$	-2
$(x - 1)(3x + 2)$	$3x^2$	$-x$	-2

From the last line we see that the factors of $3x^2 - x - 2$ are $(x - 1)(3x + 2)$. That is,

$$3x^2 - x - 2 = (x - 1)(3x + 2)$$

To factor trinomials with leading coefficients other than 1, when the greatest common factor is 1, we must use trial and error or list all the possible factors. In either case the idea is this: Look only at pairs of binomials whose products give the correct first and last terms, then look for the combination that will give the correct middle term.

EXAMPLE 5   Factor $2x^2 + 13xy + 15y^2$.

Solution   Listing all possible factors the product of whose first terms is $2x^2$ and the product of whose last terms is $+15y^2$ yields

Possible Factors	Middle Term of Product
$(2x - 5y)(x - 3y)$	$-11xy$
$(2x - 3y)(x - 5y)$	$-13xy$
$(2x + 5y)(x + 3y)$	$+11xy$
$(2x + 3y)(x + 5y)$	$+13xy$
$(2x + 15y)(x + y)$	$+17xy$
$(2x - 15y)(x - y)$	$-17xy$

The fourth line has the correct middle term:

$$2x^2 + 13xy + 15y^2 = (2x + 3y)(x + 5y)$$

Actually, we did not need to check the first two pairs of possible factors in the preceding list. All the signs in the trinomial $2x^2 + 13xy + 15y^2$ are positive. The binomial factors must then be of the form $(ax + b)(cx + d)$, where $a, b, c,$ and d are all positive.

There are other ways to reduce the number of possible factors to consider. For example, if we were to factor the trinomial $2x^2 - 11x + 12$, we would not have to consider the pair of possible factors $(2x - 4)(x - 3)$. If the original trinomial has no greatest common factor other than 1, then neither of its binomial factors will either. The trinomial $2x^2 - 11x + 12$ has a greatest common factor of 1, but the possible factor $2x - 4$ has a greatest common factor of 2: $2x - 4 = 2(x - 2)$. Therefore, we do not need to consider $2x - 4$ as a possible factor.

EXAMPLE 6   Factor $12x^4 + 17x^2 + 6$.

Solution   This is a trinomial in x^2:

$$12x^4 + 17x^2 + 6 = (4x^2 + 3)(3x^2 + 2)$$

EXAMPLE 7   Factor $2x^2(x - 3) - 5x(x - 3) - 3(x - 3)$.

Solution   We begin by factoring out the greatest common factor $(x - 3)$. Then we factor the trinomial that remains.

$$2x^2(x - 3) - 5x(x - 3) - 3(x - 3) = (x - 3)(2x^2 - 5x - 3)$$
$$= (x - 3)(2x + 1)(x - 3)$$
$$= (x - 3)^2(2x + 1)$$

ANOTHER METHOD OF FACTORING TRINOMIALS

As an alternative to the trial-and-error method of factoring trinomials, we present the following method. The new method does not require as much trial and error. To use this new method, we must rewrite our original trinomial in such a way that the factoring by grouping method can be applied.

Here are the steps we use to factor $ax^2 + bx + c$.

Step 1: Form the product ac.

Step 2: Find a pair of numbers whose product is ac and whose sum is b.

Step 3: Rewrite the polynomial to be factored so that the middle term bx is written as the sum of two terms whose coefficients are the two numbers found in step 2.

Step 4: Factor by grouping.

EXAMPLE 8   Factor $3x^2 - 10x - 8$ using these steps.

Solution   The trinomial $3x^2 - 10x - 8$ has the form $ax^2 + bx + c$, where $a = 3, b = -10$, and $c = -8$.

Step 1: The product ac is $3(-8) = -24$.

Step 2: We need to find two numbers whose product is -24 and whose sum is -10. Let's list all the pairs of numbers whose product is -24 to find the pair whose sum is -10.

Product	Sum
$1(-24) = -24$	$1 + (-24) = -23$
$-1(24) = -24$	$-1 + 24\quad = 23$
$2(-12) = -24$	$2 + (-12) = -10$
$-2(12) = -24$	$-2 + 12\quad = 10$
$3(-8) = -24$	$3 + (-8)\ = -5$
$-3(8) = -24$	$-3 + 8\quad = 5$
$4(-6) = -24$	$4 + (-6)\ = -2$
$-4(6) = -24$	$-4 + 6\quad = 2$

As you can see, of all the pairs of numbers whose product is -24, only 2 and -12 have a sum of -10.

Step 3: We now rewrite our original trinomial so the middle term $-10x$ is written as the sum of $-12x$ and $2x$:

$$3x^2 - 10x - 8 = 3x^2 - 12x + 2x - 8$$

Step 4: Factoring by grouping, we have

$$3x^2 - 12x + 2x - 8 = 3x(x - 4) + 2(x - 4)$$
$$= (x - 4)(3x + 2)$$

You can see that this method works by multiplying $x - 4$ and $3x + 2$ to get

$$3x^2 - 10x - 8$$

EXAMPLE 9   Factor $9x^2 + 15x + 4$.

Solution   In this case $a = 9$, $b = 15$, and $c = 4$. The product ac is $9 \cdot 4 = 36$. Listing all the pairs of numbers whose product is 36 with their corresponding sums, we have

Product	Sum
$1(36) = 36$	$1 + 36 = 37$
$2(18) = 36$	$2 + 18 = 20$
$3(12) = 36$	$3 + 12 = 15$
$4(9) = 36$	$4 + 9 = 13$
$6(6) = 36$	$6 + 6 = 12$

Notice we list only positive numbers since both the product and sum we are looking for are positive. The numbers 3 and 12 are the numbers we are looking for. Their product is 36, and their sum is 15. We now rewrite the original polynomial $9x^2 + 15x + 4$ with the middle term written as $3x + 12x$. We then factor by grouping:

$$9x^2 + 15x + 4 = 9x^2 + 3x + 12x + 4$$
$$= 3x(3x + 1) + 4(3x + 1)$$
$$= (3x + 1)(3x + 4)$$

The polynomial $9x^2 + 15x + 4$ factors into the product

$$(3x + 1)(3x + 4)$$

EXAMPLE 10   Factor $8x^2 - 2x - 15$.

Solution   The product ac is $8(-15) = -120$. There are many pairs of numbers whose product is -120. We are looking for the pair whose sum is also -2. The numbers are -12 and 10. Writing $-2x$ as $-12x + 10x$ and then factoring by grouping, we have

$$8x^2 - 2x - 15 = 8x^2 - 12x + 10x - 15$$
$$= 4x(2x - 3) + 5(2x - 3)$$
$$= (2x - 3)(4x + 5)$$

Getting Ready for Class

After reading through the preceding section, respond in your own words and in complete sentences.

A. What is a prime polynomial?

B. When factoring polynomials, what should you look for first?

C. How can you check to see that you have factored a trinomial correctly?

D. Describe how to determine the binomial factors of $6x^2 + 5x - 25$.

PROBLEM SET 5.5

Factor each of the following trinomials.

1. $x^2 + 7x + 12$ **2.** $x^2 - 7x + 12$

3. $x^2 - x - 12$ **4.** $x^2 + x - 12$

5. $y^2 + y - 6$ **6.** $y^2 - y - 6$

7. $16 - 6x - x^2$ **8.** $3 + 2x - x^2$

9. $12 + 8x + x^2$ **10.** $15 - 2x - x^2$

Factor completely by first factoring out the greatest common factor and then factoring the trinomial that remains.

11. $3a^2 - 21a + 30$ **12.** $3a^2 - 3a - 6$

13. $4x^3 - 16x^2 - 20x$ **14.** $2x^3 - 14x^2 + 20x$

Factor.

15. $x^2 + 3xy + 2y^2$ **16.** $x^2 - 5xy - 24y^2$

17. $a^2 + 3ab - 18b^2$ **18.** $a^2 - 8ab - 9b^2$

19. $x^2 - 2xa - 48a^2$ **20.** $x^2 + 14xa + 48a^2$

21. $x^2 - 12xb + 36b^2$ **22.** $x^2 + 10xb + 25b^2$

Factor completely. Be sure to factor out the greatest common factor first if it is other than 1.

23. $3x^2 - 6xy - 9y^2$ **24.** $5x^2 + 25xy + 20y^2$

25. $2a^5 + 4a^4b + 4a^3b^2$

26. $3a^4 - 18a^3b + 27a^2b^2$

27. $10x^4y^2 + 20x^3y^3 - 30x^2y^4$

28. $6x^4y^2 + 18x^3y^3 - 24x^2y^4$

29. $2x^2 + 7x - 15$ **30.** $2x^2 - 7x - 15$

31. $2x^2 + x - 15$ **32.** $2x^2 - x - 15$

33. $2x^2 - 13x + 15$ **34.** $2x^2 + 13x + 15$

35. $2x^2 - 11x + 15$ **36.** $2x^2 + 11x + 15$

37. $2x^2 + 7x + 15$ **38.** $2x^2 + x + 15$

39. $2 + 7a + 6a^2$ **40.** $2 - 7a + 6a^2$

41. $60y^2 - 15y - 45$ **42.** $72y^2 + 60y - 72$

43. $6x^4 - x^3 - 2x^2$ **44.** $3x^4 + 2x^3 - 5x^2$

45. $40r^3 - 120r^2 + 90r$

46. $40r^3 + 200r^2 + 250r$

47. $4x^2 - 11xy - 3y^2$

48. $3x^2 + 19xy - 14y^2$

49. $10x^2 - 3xa - 18a^2$

50. $9x^2 + 9xa - 10a^2$

51. $18a^2 + 3ab - 28b^2$

52. $6a^2 - 7ab - 5b^2$

53. $600 + 800t - 800t^2$

54. $200 - 600t - 350t^2$

55. $9y^4 + 9y^3 - 10y^2$ **56.** $4y^5 + 7y^4 - 2y^3$

57. $24a^2 - 2a^3 - 12a^4$ **58.** $60a^2 + 65a^3 - 20a^4$

59. $8x^4y^2 - 2x^3y^3 - 6x^2y^4$

60. $8x^4y^2 - 47x^3y^3 - 6x^2y^4$

61. $300x^4 + 1{,}000x^2 + 300$

62. $600x^4 - 100x^2 - 200$

63. $20a^4 + 37a^2 + 15$ **64.** $20a^4 + 13a^2 - 15$

65. $9 + 3r^2 - 12r^4$ **66.** $2 - 4r^2 - 30r^4$

Factor each of the following by first factoring out the greatest common factor, and then factoring the trinomial that remains.

67. $2x^2(x + 5) + 7x(x + 5) + 6(x + 5)$

68. $2x^2(x + 2) + 13x(x + 2) + 15(x + 2)$

69. $x^2(2x + 3) + 7x(2x + 3) + 10(2x + 3)$

70. $2x^2(x + 1) + 7x(x + 1) + 6(x + 1)$

71. What polynomial, when factored, gives $(3x + 5y)(3x - 5y)$?

72. What polynomial, when factored, gives $(7x + 2y)(7x - 2y)$?

73. One factor of the trinomial $a^2 + 260a + 2{,}500$ is $a + 10$. What is the other factor?

74. One factor of the trinomial $a^2 - 75a - 2{,}500$ is $a + 25$. What is the other factor?

75. Factor the right side of the equation $y = 4x^2 + 18x - 10$, and then use the result to find y when x is $\frac{1}{2}$, when x is -5, and when x is 2.

76. Factor the right side of the equation $y = 9x^2 + 33x - 12$, and use the result to find y when x is $\frac{1}{3}$, when x is -4, and when x is 3.

Applying the Concepts

77. Height of a Bullet A bullet is fired into the air with an initial upward velocity of 80 feet per second from the top of a building 96 feet high. The equation that gives the height of the bullet at any time t is

$$h(t) = 96 + 80t - 16t^2$$

Factor the right side of this equation, and then find h when t is 6 seconds and when t is 3 seconds. [Find $h(6)$ and $h(3)$.]

78. Height of an Arrow   An arrow is shot into the air with an upward velocity of 16 feet per second from a hill 32 feet high. The equation that gives the height of the arrow at any time t is

$$h(t) = 32 + 16t - 16t^2$$

Factor the right side of this equation, and then find h when t is 2 seconds and when t is 1 second. [Find $h(2)$ and $h(1)$.]

Review Problems

The following problems review material we covered in Section 5.3. Reviewing these problems will help you with the next section.

Multiply.

79. $(2x - 3)(2x + 3)$ **80.** $(4 - 5x)(4 + 5x)$

81. $(2x - 3)^2$ **82.** $(4 - 5x)^2$

83. $(2x - 3)(4x^2 + 6x + 9)$

84. $(2x + 3)(4x^2 - 6x + 9)$

Extending the Concepts

Factor completely.

85. $8x^6 + 26x^3y^2 + 15y^4$ **86.** $24x^4 + 6x^2y^3 - 45y^6$

87. $3x^2 + 295x - 500$ **88.** $3x^2 + 594x - 1,200$

89. $\frac{1}{8}x^2 + x + 2$ **90.** $\frac{1}{9}x^2 + x + 2$

91. $2x^2 + 1.5x + 0.25$ **92.** $6x^2 + 2x + 0.16$

93. Factoring and Area   The following area model gives us a way to visualize the factorization of the trinomial $x^2 + 5x + 6$. Construct a similar diagram that will allow you to visualize the factorization of $x^2 + 5x + 4$.

94. Factoring and Area   Refer to the area model for factoring trinomials mentioned in the preceding exercise. Write the factoring problem represented by each of the following diagrams.

(a)

(b)

To find the area of the large square in the margin, we can square the length of its side, giving us $(a + b)^2$. On the other hand, we can add the areas of the four smaller figures to arrive at the same result.

Since the area of the large square is the same whether we find it by squaring a side or by adding the four smaller areas, we can write the following relationship:

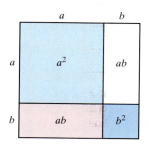

$$(a + b)^2 = a^2 + 2ab + b^2$$

This is the formula for the square of a binomial that we found in Section 5.3. The figure gives us a geometric interpretation for one of the special multiplication formulas. We begin this section by looking at the special multiplication formulas from a factoring perspective.

PERFECT SQUARE TRINOMIALS

We previously listed some special products found in multiplying polynomials. Two of the formulas looked like this:

$$(a + b)^2 = a^2 + 2ab + b^2$$
$$(a - b)^2 = a^2 - 2ab + b^2$$

If we exchange the left and right sides of each formula, we have two special formulas for factoring:

$$a^2 + 2ab + b^2 = (a + b)^2$$
$$a^2 - 2ab + b^2 = (a - b)^2$$

The left side of each formula is called a **perfect square trinomial.** The right sides are binomial squares. Perfect square trinomials can always be factored using the usual methods for factoring trinomials. However, if we notice that the first and last terms of a trinomial are perfect squares, it is wise to see whether the trinomial factors as a binomial square before attempting to factor by the usual method.

EXAMPLE 1   Factor: $x^2 - 6x + 9$

Solution   Since the first and last terms are perfect squares, we attempt to factor according to the preceding formulas:

$$x^2 - 6x + 9 = (x - 3)^2$$

If we expand $(x - 3)^2$, we have $x^2 - 6x + 9$, indicating that we have factored correctly.

EXAMPLES   Factor each of the following perfect square trinomials.

2. $16a^2 + 40ab + 25b^2 = (4a + 5b)^2$

3. $49 - 14t + t^2 = (7 - t)^2$

321

4. $9x^4 - 12x^2 + 4 = (3x^2 - 2)^2$

5. $(y + 3)^2 + 10(y + 3) + 25 = [(y + 3) + 5]^2 = (y + 8)^2$

EXAMPLE 6 Factor: $8x^2 - 24xy + 18y^2$

Solution We begin by factoring the greatest common factor 2 from each term:

$$8x^2 - 24xy + 18y^2 = 2(4x^2 - 12xy + 9y^2)$$
$$= 2(2x - 3y)^2$$

THE DIFFERENCE OF TWO SQUARES

Recall the formula that results in the difference of two squares:

$$(a - b)(a + b) = a^2 - b^2$$

Writing this as a factoring formula, we have

$$a^2 - b^2 = (a - b)(a + b)$$

EXAMPLES Each of the following is the difference of two squares. Use the formula $a^2 - b^2 = (a - b)(a + b)$ to factor each one.

7. $x^2 - 25 = x^2 - 5^2 = (x - 5)(x + 5)$

8. $49 - t^2 = 7^2 - t^2 = (7 - t)(7 + t)$

9. $81a^2 - 25b^2 = (9a)^2 - (5b)^2 = (9a - 5b)(9a + 5b)$

10. $4x^6 - 1 = (2x^3)^2 - 1^2 = (2x^3 - 1)(2x^3 + 1)$

11. $x^2 - \frac{4}{9} = x^2 - (\frac{2}{3})^2 = (x - \frac{2}{3})(x + \frac{2}{3})$

As our next example shows, the difference of two fourth powers can be factored as the difference of two squares.

EXAMPLE 12 Factor: $16x^4 - 81y^4$

Solution The first and last terms are perfect squares. We factor according to the preceding formula:

$$16x^4 - 81y^4 = (4x^2)^2 - (9y^2)^2$$
$$= (4x^2 - 9y^2)(4x^2 + 9y^2)$$

Notice that the first factor is also the difference of two squares. Factoring completely, we have

$$16x^4 - 81y^4 = (2x - 3y)(2x + 3y)(4x^2 + 9y^2)$$

Here is another example of the difference of two squares.

EXAMPLE 13   Factor: $(x - 3)^2 - 25$

Solution   This example has the form $a^2 - b^2$, where a is $x - 3$ and b is 5. We factor it according to the formula for the difference of two squares:

$$
\begin{aligned}
(x - 3)^2 - 25 &= (x - 3)^2 - 5^2 &&\text{Write 25 as } 5^2. \\
&= [(x - 3) - 5][(x - 3) + 5] &&\text{Factor.} \\
&= (x - 8)(x + 2) &&\text{Simplify.}
\end{aligned}
$$

Notice in this example that we could have expanded $(x - 3)^2$, subtracted 25, and then factored to obtain the same result:

$$
\begin{aligned}
(x - 3)^2 - 25 &= x^2 - 6x + 9 - 25 &&\text{Expand } (x - 3)^2. \\
&= x^2 - 6x - 16 &&\text{Simplify.} \\
&= (x - 8)(x + 2) &&\text{Factor.}
\end{aligned}
$$

EXAMPLE 14   Factor: $x^2 - 10x + 25 - y^2$

Solution   Notice that the first three terms form a perfect square trinomial. That is, $x^2 - 10x + 25 = (x - 5)^2$. If we replace the first three terms with $(x - 5)^2$, the expression that results has the form $a^2 - b^2$. We can then factor as we did in Example 13:

$$
\begin{aligned}
&x^2 - 10x + 25 - y^2 \\
&= (x^2 - 10x + 25) - y^2 &&\text{Group first three terms together.} \\
&= (x - 5)^2 - y^2 &&\text{This has the form } a^2 - b^2. \\
&= [(x - 5) - y][(x - 5) + y] &&\text{Factoring according to the formula} \\
&&&\quad a^2 - b^2 = (a - b)(a + b) \\
&= (x - 5 - y)(x - 5 + y) &&\text{Simplify.}
\end{aligned}
$$

We could check this result by multiplying the two factors together. (You may want to do that to convince yourself that we have the correct result.)

EXAMPLE 15   Factor completely: $x^3 + 2x^2 - 9x - 18$

Solution   We use factoring by grouping to begin and then factor the difference of two squares:

$$
\begin{aligned}
x^3 + 2x^2 - 9x - 18 &= x^2(x + 2) - 9(x + 2) \\
&= (x + 2)(x^2 - 9) \\
&= (x + 2)(x - 3)(x + 3)
\end{aligned}
$$

THE SUM AND DIFFERENCE OF TWO CUBES

Here are the formulas for factoring the sum and difference of two cubes:

$$a^3 + b^3 = (a + b)(a^2 - ab + b^2)$$
$$a^3 - b^3 = (a - b)(a^2 + ab + b^2)$$

EXAMPLE 16 Verify the two formulas.

Solution We verify the formulas by multiplying the right sides and comparing the results with the left sides:

$$
\begin{array}{r}
a^2 - ab + b^2 \\
a + b \\
\hline
a^3 - a^2b + ab^2 \\
a^2b - ab^2 + b^3 \\
\hline
a^3 \qquad\qquad + b^3
\end{array}
\qquad
\begin{array}{r}
a^2 + ab + b^2 \\
a - b \\
\hline
a^3 + a^2b + ab^2 \\
- a^2b - ab^2 - b^3 \\
\hline
a^3 \qquad\qquad - b^3
\end{array}
$$

Here are some examples using the formulas for factoring the sum and difference of two cubes.

EXAMPLE 17 Factor: $64 + t^3$

Solution The first term is the cube of 4 and the second term is the cube of t. Therefore,

$$64 + t^3 = 4^3 + t^3$$
$$= (4 + t)(16 - 4t + t^2)$$

EXAMPLE 18 Factor: $27x^3 + 125y^3$

Solution Writing both terms as perfect cubes, we have

$$27x^3 + 125y^3 = (3x)^3 + (5y)^3$$
$$= (3x + 5y)(9x^2 - 15xy + 25y^2)$$

EXAMPLE 19 Factor: $a^3 - \frac{1}{8}$

Solution The first term is the cube of a, while the second term is the cube of $\frac{1}{2}$:

$$a^3 - \frac{1}{8} = a^3 - \left(\frac{1}{2}\right)^3$$
$$= (a - \frac{1}{2})(a^2 + \frac{1}{2}a + \frac{1}{4})$$

EXAMPLE 20   Factor: $x^6 - y^6$

Solution   We have a choice of how we initially want to write the two terms. We can write the expression as the difference of two squares, $(x^3)^2 - (y^3)^2$, or as the difference of two cubes, $(x^2)^3 - (y^2)^3$. It is better to start with the difference of two squares if we have a choice:

$$x^6 - y^6 = (x^3)^2 - (y^3)^2$$
$$= (x^3 - y^3)(x^3 + y^3)$$
$$= (x - y)(x^2 + xy + y^2)(x + y)(x^2 - xy + y^2)$$

Factor completely, using the sum and difference of two cubes.

Try this example again, writing the first line as the difference of two cubes instead of the difference of two squares. It will become apparent why it is better to use the difference of two squares.

FACTORING: A GENERAL REVIEW

We end this section by reviewing all the different methods of factoring we have covered. To begin, here is a list of the steps that can be used to factor polynomials of any type.

TO FACTOR A POLYNOMIAL

Step 1: If the polynomial has a greatest common factor other than 1, then factor out the greatest common factor.

Step 2: If the polynomial has two terms (a binomial), then see if it is the difference of two squares, or the sum or difference of two cubes, and then factor accordingly. (*Note:* If it is the *sum* of two squares, it will not factor.)

Step 3: If the polynomial has three terms (a trinomial), then either it is a perfect square trinomial, which will factor into the square of a binomial, or it is not a perfect square trinomial, in which case you must use the trial-and-error method developed in Section 5.5.

Step 4: If the polynomial has more than three terms, try to factor it by grouping.

Step 5: As a final check, see if any of the factors you have written can be factored further. If you have overlooked a common factor, you can catch it here.

Here are some examples illustrating how we use these five steps. There are no new factoring problems here. The problems are all similar to the problems you have seen before, but they are not grouped according to type.

EXAMPLE 21   Factor: $2x^5 - 8x^3$

Solution   First we check to see if the greatest common factor is other than 1. Since the greatest common factor is $2x^3$, we begin by factoring it out. Once we have done this, we notice that the binomial that remains is the difference of two squares, which we factor according to the formula $a^2 - b^2 = (a + b)(a - b)$.

$$2x^5 - 8x^3 = 2x^3(x^2 - 4) \qquad \text{Factor out the greatest common}$$
$$\text{factor, } 2x^3.$$
$$= 2x^3(x + 2)(x - 2) \qquad \text{Factor the difference of two squares.}$$

EXAMPLE 22   Factor: $3x^4 - 18x^3 + 27x^2$

Solution   Step 1 is to factor out the greatest common factor, $3x^2$. After we have done this, we notice that the trinomial that remains is a perfect square trinomial, which will factor as the square of a binomial:

$$3x^4 - 18x^3 + 27x^2 = 3x^2(x^2 - 6x + 9) \qquad \text{Factor out } 3x^2.$$
$$= 3x^2(x - 3)^2 \qquad x^2 - 6x + 9 \text{ is the}$$
$$\text{square of } x - 3.$$

EXAMPLE 23   Factor: $y^3 + 25y$

Solution   We begin by factoring out the y that is common to both terms. The binomial that remains after we have done this is the sum of two squares, which does not factor. So, after the first step, we are finished:

$$y^3 + 25y = y(y^2 + 25)$$

EXAMPLE 24   Factor: $6a^2 - 11a + 4$

Solution   Here we have a trinomial that does not have a greatest common factor other than 1. Since it is not a perfect square trinomial, we factor it by trial and error. Without showing all the different possibilities, here is the answer:

$$6a^2 - 11a + 4 = (3a - 4)(2a - 1)$$

EXAMPLE 25   Factor: $2x^4 + 16x$

Solution   This binomial has a greatest common factor of $2x$. The binomial that remains after the $2x$ has been factored from each term is the sum of two cubes, which we factor according to the formula $a^3 + b^3 = (a + b)(a^2 - ab + b^2)$.

$$2x^4 + 16x = 2x(x^3 + 8) \qquad \text{Factor } 2x \text{ from each term.}$$
$$= 2x(x + 2)(x^2 - 2x + 4) \qquad \text{The sum of two cubes}$$

EXAMPLE 26   Factor: $2ab^5 + 8ab^4 + 2ab^3$

Solution   The greatest common factor is $2ab^3$. We begin by factoring it from each term. After that, we find that the trinomial that remains cannot be factored further:

$$2ab^5 + 8ab^4 + 2ab^3 = 2ab^3(b^2 + 4b + 1)$$

EXAMPLE 27   Factor: $4x^2 - 6x + 2ax - 3a$

Solution   This polynomial has four terms, so we factor by grouping:

$$4x^2 - 6x + 2ax - 3a = 2x(2x - 3) + a(2x - 3)$$
$$= (2x - 3)(2x + a)$$

 # Getting Ready for Class

After reading through the preceding section, respond in your own words and in complete sentences.

A. In what cases can you factor a binomial?

B. What is a perfect square trinomial?

C. How do you know when you've factored completely?

D. If a polynomial has four terms, what method of factoring should you try?

PROBLEM SET 5.6

Factor each perfect square trinomial.

1. $x^2 - 6x + 9$

2. $x^2 + 10x + 25$

3. $a^2 - 12a + 36$

4. $36 - 12a + a^2$

5. $25 - 10t + t^2$

6. $64 + 16t + t^2$

7. $4y^4 - 12y^2 + 9$

8. $9y^4 + 12y^2 + 4$

9. $16a^2 + 40ab + 25b^2$

10. $25a^2 - 40ab + 16b^2$

11. $\frac{1}{25} + \frac{1}{10}t^2 + \frac{1}{16}t^4$

12. $\frac{1}{9} - \frac{1}{3}t^3 + \frac{1}{4}t^6$

13. $(x + 2)^2 + 6(x + 2) + 9$

14. $(x + 5)^2 + 4(x + 5) + 4$

Factor completely.

15. $49x^2 - 64y^2$

16. $81x^2 - 49y^2$

17. $4a^2 - \frac{1}{4}$

18. $25a^2 - \frac{1}{25}$

19. $x^2 - \frac{9}{25}$

20. $x^2 - \frac{25}{36}$

21. $25 - t^2$

22. $64 - t^2$

23. $16a^4 - 81$

24. $81a^4 - 16b^4$

25. $x^2 - 10x + 25 - y^2$

26. $x^2 - 6x + 9 - y^2$

27. $a^2 + 8a + 16 - b^2$

28. $a^2 + 12a + 36 - b^2$

29. $x^3 + 2x^2 - 25x - 50$

30. $x^3 + 4x^2 - 9x - 36$

31. $2x^3 + 3x^2 - 8x - 12$

32. $3x^3 + 2x^2 - 27x - 18$

33. $4x^3 + 12x^2 - 9x - 27$

34. $9x^3 + 18x^2 - 4x - 8$

Factor each of the following as the sum or difference of two cubes.

35. $x^3 - y^3$ **36.** $x^3 + y^3$

37. $a^3 + 8$ **38.** $a^3 - 8$

39. $y^3 - 1$ **40.** $y^3 + 1$

41. $10r^3 - 1{,}250$ **42.** $10r^3 + 1{,}250$

43. $64 + 27a^3$ **44.** $27 - 64a^3$

45. $t^3 + \frac{1}{27}$ **46.** $t^3 - \frac{1}{27}$

Factor each of the following polynomials completely, if possible. That is, once you are finished factoring, none of the factors you obtain should be factorable.

47. $x^2 - 81$ **48.** $x^2 - 18x + 81$

49. $x^2 + 2x - 15$ **50.** $15x^2 + 13x - 6$

51. $x^2y^2 + 2y^2 + x^2 + 2$ **52.** $21y^2 - 25y - 4$

53. $2a^3b + 6a^2b + 2ab$ **54.** $6a^2 - ab - 15b^2$

55. $x^2 + x + 1$

56. $x^2y + 3y + 2x^2 + 6$

57. $12a^2 - 75$ **58.** $18a^2 - 50$

59. $25 - 10t + t^2$ **60.** $t^2 + 4t + 4 - y^2$

61. $4x^3 + 16xy^2$ **62.** $16x^2 + 49y^2$

63. $x^3 + 5x^2 - 9x - 45$

64. $x^3 + 5x^2 - 16x - 80$

65. $x^2 + 49$ **66.** $16 - x^4$

67. $x^2(x - 3) - 14x(x - 3) + 49(x - 3)$

68. $x^2 + 3ax - 2bx - 6ab$

69. $8 - 14x - 15x^2$ **70.** $5x^4 + 14x^2 - 3$

71. $r^2 - \frac{1}{25}$ **72.** $27 - r^3$

73. $49x^2 + 9y^2$

74. $12x^4 - 62x^3 + 70x^2$

75. $100x^2 - 100x - 600$

76. $100x^2 - 100x - 1{,}200$

77. $3x^4 - 14x^2 - 5$ **78.** $8 - 2x - 15x^2$

79. $24a^5b - 3a^2b$

80. $18a^4b^2 - 24a^3b^3 + 8a^2b^4$

81. $64 - r^3$ **82.** $r^2 - \frac{1}{9}$

83. $20x^4 - 45x^2$ **84.** $16x^3 + 16x^2 + 3x$

85. $16x^5 - 44x^4 + 30x^3$ **86.** $16x^2 + 16x - 1$

87. $y^6 - 1$ **88.** $25y^7 - 16y^5$

89. $50 - 2a^2$ **90.** $4a^2 + 2a + \frac{1}{4}$

91. $x^2 - 4x + 4 - y^2$

92. $x^2 - 12x + 36 - b^2$

93. Find two values of b that will make $9x^2 + bx + 25$ a perfect square trinomial.

94. Find a value of c that will make $49x^2 - 42x + c$ a perfect square trinomial.

Applying the Concepts

Compound Interest If $100 is invested in an account with an annual interest rate of r compounded twice a year, then the amount of money in the account at the end of the year is given by the formula

$$A = 100\left(1 + r + \frac{r^2}{4}\right)$$

If the annual interest rate is 12%, find the amount of money in the account at the end of 1 year

95. . . . without factoring the right side of the formula.

96. . . . by first factoring the right side of the formula completely.

97. Volume Between Two Cubes Recall that the volume of a cube is $V = (\text{side})^3$. A cube with side of length r inches is placed inside a cube of side p inches. Write an expression, in factored form, for the volume of the space between the two cubes.

98. Binomial Squares From Area The area model at the beginning of this section gives a visual interpretation of the factoring problem $a^2 + 2ab + b^2$. Using that diagram as a reference, draw a similar diagram for the factorization of $4x^2 + 4xy + y^2$.

99. Binomial Squares and Area Refer to the preceding exercise. The following diagram is the start of an area model for $x^2 - 2xy + y^2 = (x - y)^2$. Finish labeling this area model.

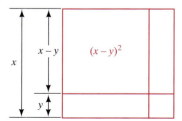

100. Genetics The Austrian monk *Gregor Mendel* (1822–1884) showed that genes from both the mother and father determine what traits are inherited by a new generation. The Punnett square

shown here is used by biologists to study the probabilities associated with inherited traits. Explain how a Punnett square is modeled on the square of a binomial.

Male genes

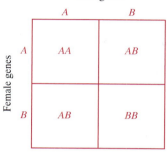

Female genes

Review Problems

The following problems review material we covered in Section 4.4.

Solve each system by using Cramer's rule.

101. $4x - 7y = 3$
$5x + 2y = -3$

102. $9x - 8y = 4$
$2x + 3y = 6$

103. $3x + 4y = 15$
$2x - 5z = -3$
$4y - 3z = 9$

104. $x + 3y = 5$
$6y + z = 12$
$x - 2z = -10$

5.7   Solving Equations by Factoring

In this section we will use our knowledge of factoring to solve equations. Most of the equations we will solve in this section are *quadratic equations*. Here is the definition of a quadratic equation.

> **DEFINITION**   Any equation that can be written in the form
> $$ax^2 + bx + c = 0$$
> where a, b, and c are constants and a is not 0 ($a \neq 0$), is called a *quadratic equation*. The form $ax^2 + bx + c = 0$ is called *standard form* for quadratic equations.

Each of the following is a quadratic equation:
$$2x^2 = 5x + 3 \qquad 5x^2 = 75 \qquad 4x^2 - 3x + 2 = 0$$

Note: The third equation is clearly a quadratic equation since it is in standard form. (Notice that a is 4, b is -3, and c is 2.) The first two equations are also quadratic because they could be put in the form $ax^2 + bx + c = 0$ by using the addition property of equality.

Notation   For a quadratic equation written in standard form, the first term ax^2 is called the *quadratic term;* the second term bx is the *linear term;* and the last term c is called the *constant term.*

In the past we have noticed that the number 0 is a special number. There is another property of 0 that is the key to solving quadratic equations. It is called the *zero-factor property.*

ZERO-FACTOR PROPERTY

For all real numbers r and s,

$$r \cdot s = 0 \quad \text{if and only if} \quad r = 0 \quad \text{or} \quad s = 0 \quad \text{(or both)}$$

EXAMPLE 1 Solve $x^2 - 2x - 24 = 0$.

Solution We begin by factoring the left side as $(x - 6)(x + 4)$ and get

$$(x - 6)(x + 4) = 0$$

Now both $(x - 6)$ and $(x + 4)$ represent real numbers. We notice that their product is 0. By the zero-factor property, one or both of them must be 0:

$$x - 6 = 0 \quad \text{or} \quad x + 4 = 0$$

We have used factoring and the zero-factor property to rewrite our original second-degree equation as two first-degree equations connected by the word *or*. Completing the solution, we solve the two first-degree equations:

$$x - 6 = 0 \quad \text{or} \quad x + 4 = 0$$
$$x = 6 \quad \text{or} \quad x = -4$$

We check our solutions in the original equation as follows:

Check $x = 6$	Check $x = -4$
$6^2 - 2(6) - 24 \stackrel{?}{=} 0$	$(-4)^2 - 2(-4) - 24 \stackrel{?}{=} 0$
$36 - 12 - 24 \stackrel{?}{=} 0$	$16 + 8 - 24 \stackrel{?}{=} 0$
$0 = 0$	$0 = 0$

In both cases the result is a true statement, which means that both 6 and -4 are solutions to the original equation.

Although the next equation is not quadratic, the method we use is similar.

EXAMPLE 2 Solve $\frac{1}{3}x^3 = \frac{5}{6}x^2 + \frac{1}{2}x$.

Solution We can simplify our work if we clear the equation of fractions. Multiplying both sides by the LCD, 6, we have

$$\mathbf{6} \cdot \frac{1}{3}x^3 = \mathbf{6} \cdot \frac{5}{6}x^2 + \mathbf{6} \cdot \frac{1}{2}x$$
$$2x^3 = 5x^2 + 3x$$

Next we add $-5x^2$ and $-3x$ to each side so that the right side will become 0.

$$2x^3 - 5x^2 - 3x = 0 \qquad \text{Standard form}$$

We factor the left side and then use the zero-factor property to set each factor to 0.

$$x(2x^2 - 5x - 3) = 0 \qquad \text{Factor out the greatest common factor.}$$

$$x(2x + 1)(x - 3) = 0 \qquad \text{Continue factoring.}$$

$$x = 0 \quad \text{or} \quad 2x + 1 = 0 \quad \text{or} \quad x - 3 = 0 \qquad \text{Zero-factor property}$$

Solving each of the resulting equations, we have

$$x = 0 \quad \text{or} \quad x = -\frac{1}{2} \quad \text{or} \quad x = 3$$

To generalize the preceding example, here are the steps used in solving a quadratic equation by factoring.

TO SOLVE AN EQUATION BY FACTORING

Step 1: Write the equation in standard form.
Step 2: Factor the left side.
Step 3: Use the zero-factor property to set each factor equal to 0.
Step 4: Solve the resulting linear equations.

EXAMPLE 3   Solve $100x^2 = 300x$.

Solution   We begin by writing the equation in standard form and factoring:

$$100x^2 = 300x$$

$$100x^2 - 300x = 0 \qquad \text{Standard form}$$

$$100x(x - 3) = 0 \qquad \text{Factor.}$$

Using the zero-factor property to set each factor to 0, we have

$$100x = 0 \quad \text{or} \quad x - 3 = 0$$

$$x = 0 \quad \text{or} \quad x = 3$$

The two solutions are 0 and 3.

EXAMPLE 4   Solve $(x - 2)(x + 1) = 4$.

Solution   We begin by multiplying the two factors on the left side. (Notice that it would be incorrect to set each of the factors on the left side equal to 4. The fact that the product is 4 does not imply that either of the factors must be 4.)

$$(x - 2)(x + 1) = 4$$

$$x^2 - x - 2 = 4 \qquad \text{Multiply the left side.}$$

$$x^2 - x - 6 = 0 \qquad \text{Standard form}$$

$$(x - 3)(x + 2) = 0 \qquad \text{Factor.}$$

$$x - 3 = 0 \quad \text{or} \quad x + 2 = 0 \qquad \text{Zero-factor property}$$

$$x = 3 \quad \text{or} \qquad x = -2$$

E X A M P L E   5   Solve for x: $x^3 + 2x^2 - 9x - 18 = 0$

Solution   We factored the left side of this equation in Section 5.6. We start with factoring by grouping.

$$x^3 + 2x^2 - 9x - 18 = 0$$

$$x^2(x + 2) - 9(x + 2) = 0$$

$$(x + 2)(x^2 - 9) = 0$$

$$(x + 2)(x - 3)(x + 3) = 0 \qquad \text{The difference of two squares}$$

$$x + 2 = 0 \quad \text{or} \quad x - 3 = 0 \quad \text{or} \quad x + 3 = 0 \qquad \text{Set factors to 0.}$$

$$x = -2 \quad \text{or} \quad x = 3 \quad \text{or} \quad x = -3$$

We have three solutions: -2, 3, and -3.

E X A M P L E   6   The sum of the squares of two consecutive integers is 25. Find the two integers.

Solution   We apply the Blueprint for Problem Solving to solve this application problem. Remember, step 1 in the blueprint is done mentally.

Step 1: Read and list.

Known items: Two consecutive integers. If we add their squares, the result is 25.

Unknown items: The two integers

Step 2: Assign a variable and translate information.

Let $x =$ the first integer; then $x + 1 =$ the next consecutive integer.

Step 3: Reread and write an equation.

Since the sum of the squares of the two integers is 25, the equation that describes the situation is

$$x^2 + (x + 1)^2 = 25$$

Step 4: Solve the equation.

$$x^2 + (x + 1)^2 = 25$$

$$x^2 + (x^2 + 2x + 1) = 25$$

$$2x^2 + 2x - 24 = 0$$

$$x^2 + x - 12 = 0$$

$$(x + 4)(x - 3) = 0$$

$$x = -4 \quad \text{or} \quad x = 3$$

Step 5: *Write the answer.*

If $x = -4$, then $x + 1 = -3$. If $x = 3$, then $x + 1 = 4$. The two integers are -4 and -3, or the two integers are 3 and 4.

Step 6: *Reread and check.*

The two integers in each pair are consecutive integers, and the sum of the squares of either pair is 25.

Another application of quadratic equations involves the Pythagorean theorem, an important theorem from geometry. The theorem gives the relationship between the sides of any right triangle (a triangle with a 90-degree angle). We state it here without proof.

PYTHAGOREAN THEOREM

In any right triangle, the square of the longest side (hypotenuse) is equal to the sum of the squares of the other two sides (legs).

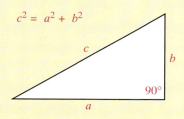

$$c^2 = a^2 + b^2$$

EXAMPLE 7   The lengths of the three sides of a right triangle are given by three consecutive integers. Find the lengths of the three sides.

Solution

Step 1: *Read and list.*

Known items: A right triangle. The three sides are three consecutive integers.

Unknown items: The three sides

Step 2: *Assign a variable and translate information.*

Let $x =$ first integer (shortest side)

Then $x + 1 =$ next consecutive integer

$x + 2 =$ last consecutive integer (longest side)

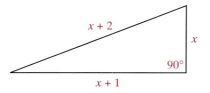

Step 3: *Reread and write an equation.*
By the Pythagorean theorem, we have
$$(x + 2)^2 = (x + 1)^2 + x^2$$

Step 4: *Solve the equation.*
$$x^2 + 4x + 4 = x^2 + 2x + 1 + x^2$$
$$x^2 - 2x - 3 = 0$$
$$(x - 3)(x + 1) = 0$$
$$x = 3 \quad \text{or} \quad x = -1$$

Step 5: *Write the answer.*
Since x is the length of a side in a triangle, it must be a positive number. Therefore, $x = -1$ cannot be used.
The shortest side is 3. The other two sides are 4 and 5.

Step 6: *Reread and check.*
The three sides are given by consecutive integers. The square of the longest side is equal to the sum of the squares of the two shorter sides.

Our next two examples involve formulas that are quadratic.

EXAMPLE 8   If an object is projected into the air with an initial vertical velocity of v feet/second, its height h, in feet, above the ground after t seconds will be given by
$$h = vt - 16t^2$$

Find t if $v = 64$ feet/second and $h = 48$ feet.

Solution   Substituting $v = 64$ and $h = 48$ into the preceding formula, we have
$$48 = 64t - 16t^2$$

which is a quadratic equation. We write it in standard form and solve by factoring:
$$16t^2 - 64t + 48 = 0$$
$$t^2 - 4t + 3 = 0 \qquad \text{Divide each side by 16.}$$
$$(t - 1)(t - 3) = 0$$
$$t - 1 = 0 \quad \text{or} \quad t - 3 = 0$$
$$t = 1 \quad \text{or} \quad t = 3$$

Here is how we interpret our results: If an object is projected upward with an initial vertical velocity of 64 feet/second, it will be 48 feet above the ground after 1 second and after 3 seconds. That is, it passes 48 feet going up and also coming down.

EXAMPLE 9   A manufacturer of small portable radios knows that the number of radios she can sell each week is related to the price of the radios by the equation $x = 1,300 - 100p$, where x is the number of radios and p is the price per radio. What price should she charge for each radio if she wants the weekly revenue to be $4,000?

Solution   The formula for total revenue is $R = xp$. Since we want R in terms of p, we substitute $1,300 - 100p$ for x in the equation $R = xp$:

$$\text{If} \qquad R = xp$$
$$\text{and} \qquad x = 1,300 - 100p$$
$$\text{then} \qquad R = (1,300 - 100p)p$$

We want to find p when R is 4,000. Substituting 4,000 for R in the formula gives us

$$4,000 = (1,300 - 100p)p$$
$$4,000 = 1,300p - 100p^2$$

which is a quadratic equation. To write it in standard form, we add $100p^2$ and $-1,300p$ to each side, giving us

$$100p^2 - 1,300p + 4,000 = 0$$
$$p^2 - 13p + 40 = 0 \qquad \text{Divide each side by 100.}$$
$$(p - 5)(p - 8) = 0$$
$$p - 5 = 0 \qquad \text{or} \qquad p - 8 = 0$$
$$p = 5 \qquad \text{or} \qquad p = 8$$

If she sells the radios for $5 each or for $8 each she will have a weekly revenue of $4,000.

Getting Ready for Class

After reading through the preceding section, respond in your own words and in complete sentences.

A. Explain the Pythagorean theorem in words.

B. What is the first step in solving an equation by factoring?

C. Describe the zero-factor property in your own words.

D. Write an application problem for which the solution depends on solving the equation $x^2 + (x + 1)^2 = 313$.

PROBLEM SET 5.7

Solve each equation.

1. $x^2 - 5x - 6 = 0$ **2.** $x^2 + 5x - 6 = 0$

3. $x^3 - 5x^2 + 6x = 0$ **4.** $x^3 + 5x^2 + 6x = 0$

5. $3y^2 + 11y - 4 = 0$ **6.** $3y^2 - y - 4 = 0$

7. $60x^2 - 130x + 60 = 0$

8. $90x^2 + 60x - 80 = 0$

9. $\frac{1}{10}t^2 - \frac{5}{2} = 0$ **10.** $\frac{2}{7}t^2 - \frac{7}{2} = 0$

11. $100x^4 = 400x^3 + 2{,}100x^2$

12. $100x^4 = -400x^3 + 2{,}100x^2$

13. $\frac{1}{5}y^2 - 2 = -\frac{3}{10}y$ **14.** $\frac{1}{2}y^2 + \frac{5}{3} = \frac{17}{6}y$

15. $9x^2 - 12x = 0$ **16.** $4x^2 + 4x = 0$

17. $0.02r + 0.01 = 0.15r^2$

18. $0.02r - 0.01 = -0.08r^2$

19. $9a^3 = 16a$ **20.** $16a^3 = 25a$

21. $-100x = 10x^2$ **22.** $800x = 100x^2$

23. $(x + 6)(x - 2) = -7$

24. $(x - 7)(x + 5) = -20$

25. $(y - 4)(y + 1) = -6$

26. $(y - 6)(y + 1) = -12$

27. $(x + 1)^2 = 3x + 7$ **28.** $(x + 2)^2 = 9x$

29. $(2r + 3)(2r - 1) = -(3r + 1)$

30. $(3r + 2)(r - 1) = -(7r - 7)$

31. $x^3 + 3x^2 - 4x - 12 = 0$

32. $x^3 + 5x^2 - 4x - 20 = 0$

33. $x^3 + 2x^2 - 25x - 50 = 0$

34. $x^3 + 4x^2 - 9x - 36 = 0$

35. $2x^3 + 3x^2 - 8x - 12 = 0$

36. $3x^3 + 2x^2 - 27x - 18 = 0$

37. $4x^3 + 12x^2 - 9x - 27 = 0$

38. $9x^3 + 18x^2 - 4x - 8 = 0$

Applying the Concepts

39. Consecutive Integers   The sum of the squares of two consecutive odd integers is 34. Find the two integers.

40. Consecutive Integers   The sum of the squares of two consecutive even integers is 100. Find the two integers.

41. Consecutive Integers   The square of the sum of two consecutive integers is 81. Find the two integers.

42. Consecutive Integers   Find two consecutive even integers whose sum squared is 100.

43. A 24-foot ladder is leaning against a building. The base of the ladder is 7 feet from the side of the building. How high does the ladder reach along the side of the building?

7 ft

44. Noreen wants to place her 13-foot ramp against the side of her house so that the top of the ramp rests on a ledge that is 5 feet above the ground. How far will the base of the ramp be from the house?

5 ft

45. Geometry   The lengths of the three sides of a right triangle are given by three consecutive even integers. Find the lengths of the three sides.

46. Geometry The longest side of a right triangle is 3 less than twice the shortest side. The third side measures 12 inches. Find the length of the shortest side.

47. Geometry The length of a rectangle is 2 feet more than 3 times the width. If the area is 16 square feet, find the width and the length.

48. Geometry The length of a rectangle is 4 yards more than twice the width. If the area is 70 square yards, find the width and the length.

49. Geometry The base of a triangle is 2 inches more than 4 times the height. If the area is 36 square inches, find the base and the height.

50. Geometry The height of a triangle is 4 feet less than twice the base. If the area is 48 square feet, find the base and the height.

51. Projectile Motion If an object is thrown straight up into the air with an initial velocity of 32 feet per second, then its height above the ground at any time t is given by the formula $h = 32t - 16t^2$. Find the times at which the object is on the ground by letting $h = 0$ in the equation and solving for t.

52. Projectile Motion An object is projected into the air with an initial velocity of 64 feet per second. Its height at any time t is given by the formula $h = 64t - 16t^2$. Find the times at which the object is on the ground.

The formula $h = vt - 16t^2$ gives the height h, in feet, of an object projected into the air with an initial vertical velocity v, in feet per second, after t seconds.

53. Projectile Motion If an object is propelled upward with an initial velocity of 48 feet per second, at what times will it reach a height of 32 feet above the ground?

54. Projectile Motion If an object is propelled upward into the air with an initial velocity of 80 feet per second, at what times will it reach a height of 64 feet above the ground?

55. Projectile Motion An object is projected into the air with a vertical velocity of 24 feet per second. At what times will the object be on the ground? (It is on the ground when h is 0.)

56. Projectile Motion An object is projected into the air with a vertical velocity of 20 feet per second. At what times will the object be on the ground?

57. Height of a Bullet A bullet is fired into the air with an initial upward velocity of 80 feet per second from the top of a building 96 feet high. The equation that gives the height of the bullet at any time t is $h = 96 + 80t - 16t^2$. At what times will the bullet be 192 feet in the air?

58. Height of an Arrow An arrow is shot into the air with an upward velocity of 48 feet per second from a hill 32 feet high. The equation that gives the height of the arrow at any time t is $h = 32 + 48t - 16t^2$. Find the times at which the arrow will be 64 feet above the ground.

59. Price and Revenue A company that manufactures typewriter ribbons knows that the number of ribbons x it can sell each week is related to the price per ribbon p by the equation $x = 1{,}200 - 100p$. At what price should it sell the ribbons if it wants the weekly revenue to be $3,200? (*Remember:* The equation for revenue is $R = xp$.)

60. Price and Revenue A company manufactures diskettes for home computers. It knows from past experience that the number of diskettes x it can sell each day is related to the price per diskette p by the equation $x = 800 - 100p$. At what price should it sell its diskettes if it wants the daily revenue to be $1,200?

61. Price and Revenue The relationship between the number of calculators x a company sells per day and the price of each calculator p is given by the equation $x = 1{,}700 - 100p$. At what price should the calculators be sold if the daily revenue is to be $7,000?

62. Price and Revenue The relationship between the number of pencil sharpeners x a company can sell each week and the price of each sharpener p is given by the equation $x = 1{,}800 - 100p$. At what price should the sharpeners be sold if the weekly revenue is to be $7,200?

63. Area of a Square The area of a square is quadrupled by increasing its side 5 feet. What was the length of the original side?

64. Skydiving The distance an object falls, s, is related to the time it falls by the formula $s = 16t^2$, where s is in feet and t is in seconds. Skydivers typically jump from about 10,000 feet and fall half that distance before opening their chute. How long (in seconds) does the skydiver fall before opening his chute?

Review Problems

The following problems review material from Section 4.5. They are taken from the book *Algebra for the Practical Man,* written by J. E. Thompson and published by D. Van Nostrand Company in 1931.

65. A man spent $112.80 for 108 geese and ducks, each goose costing 14 dimes and each duck 6 dimes. How many of each did he buy?

66. If 15 pounds of tea and 10 pounds of coffee together cost $15.50, while 25 pounds of tea and 13 pounds of coffee at the same prices cost $24.55, find the price per pound of each.

67. A number of oranges at the rate of 3 for $0.10 and apples at $0.15 a dozen cost, together, $6.80. Five times as many oranges and $\frac{1}{4}$ as many apples at the same rates would have cost $25.45. How many of each were bought?

68. An estate is divided among three persons, A, B, and C. A's share is 3 times that of B, and B's share is twice that of C. If A receives $9,000 more than C, how much does each receive?

CHAPTER 5   SUMMARY

Examples

1. These expressions illustrate the properties of exponents.

(a) $x^2 \cdot x^3 = x^{2+3} = x^5$

(b) $(x^2)^3 = x^{2 \cdot 3} = x^6$

(c) $(3x)^2 = 3^2 \cdot x^2 = 9x^2$

(d) $2^{-3} = \dfrac{1}{2^3} = \dfrac{1}{8}$

(e) $(\tfrac{x}{5})^2 = \dfrac{x^2}{5^2} = \dfrac{x^2}{25}$

(f) $\dfrac{x^7}{x^5} = x^{7-5} = x^2$

(g) $3^1 = 3$
$3^0 = 1$

Properties of Exponents [5.1]

If a and b represent real numbers and r and s represent integers, then

1. $a^r \cdot a^s = a^{r+s}$

2. $(a^r)^s = a^{r \cdot s}$

3. $(ab)^r = a^r \cdot b^r$

4. $a^{-r} = \dfrac{1}{a^r} \qquad (a \neq 0)$

5. $\left(\dfrac{a}{b}\right)^r = \dfrac{a^r}{b^r} \qquad (b \neq 0)$

6. $\dfrac{a^r}{a^s} = a^{r-s} \qquad (a \neq 0)$

7. $a^1 = a$
$a^0 = 1 \qquad (a \neq 0)$

2. $49{,}800{,}000 = 4.98 \times 10^7$
$0.00462 = 4.62 \times 10^{-3}$

Scientific Notation [5.1]

A number is written in scientific notation when it is written as the product of a number between 1 and 10 and an integer power of 10. That is, when it has the form

$$n \times 10^r$$

where $1 \leq n < 10$ and $r =$ an integer.

3. $(3x^2 + 2x - 5)$
$\qquad + (4x^2 - 7x + 2)$
$= 7x^2 - 5x - 3$

Addition of Polynomials [5.2]

To add two polynomials, simply combine the coefficients of similar terms.

4. $-(2x^2 - 8x - 9)$
$= -2x^2 + 8x + 9$

Negative Sign Preceding Parentheses [5.2]

If there is a negative sign directly preceding the parentheses surrounding a polynomial, we may remove the parentheses and preceding negative sign by changing the sign of each term within the parentheses. (This procedure is actually just another application of the distributive property.)

5. $(3x - 5)(x + 2)$
$= 3x^2 + 6x - 5x - 10$
$= 3x^2 + x - 10$

Multiplication of Polynomials [5.3]

To multiply two polynomials, multiply each term in the first by each term in the second.

6. The following are examples of the three special products:

$(x + 3)^2 = x^2 + 6x + 9$

$(5 - x)^2 = 25 - 10x + x^2$

$(x + 7)(x - 7) = x^2 - 49$

Special Products [5.3]

$$(a + b)^2 = a^2 + 2ab + b^2$$

$$(a - b)^2 = a^2 - 2ab + b^2$$

$$(a + b)(a - b) = a^2 - b^2$$

7. A company makes x items each week and sells them for p dollars each, according to the equation $p = 35 - 0.1x$. Then, the revenue is

$$R = x(35 - 0.1x) = 35x - 0.1x^2$$

If the total cost to make all x items is $C = 8x + 500$, then the profit gained by selling the x items is

$$P = 35x - 0.1x^2 - (8x + 500)$$
$$= -500 + 27x - 0.1x^2$$

Business Applications [5.2, 5.3, 5.4]

If a company manufactures and sells x items at p dollars per item, then the revenue R is given by the formula

$$R = xp$$

If the total cost to manufacture all x items is C, then the profit obtained from selling all x items is

$$P = R - C$$

8. The greatest common factor of $10x^5 - 15x^4 + 30x^3$ is $5x^3$. Factoring it out of each term, we have

$$5x^3(2x^2 - 3x + 6)$$

Greatest Common Factor [5.4]

The greatest common factor of a polynomial is the largest monomial (the monomial with the largest coefficient and highest exponent) that divides each term of the polynomial. The first step in factoring a polynomial is to factor the greatest common factor (if it is other than 1) out of each term.

9. $x^2 + 5x + 6 = (x + 2)(x + 3)$

$x^2 - 5x + 6 = (x - 2)(x - 3)$

$x^2 + x - 6 = (x - 2)(x + 3)$

$x^2 - x - 6 = (x + 2)(x - 3)$

Factoring Trinomials [5.5]

We factor a trinomial by writing it as the product of two binomials. (This refers to trinomials whose greatest common factor is 1.) Each factorable trinomial has a unique set of factors. Finding the factors is sometimes a matter of trial and error.

10. Here are some polynomials that have been factored this way.

$x^2 + 6x + 9 = (x + 3)^2$

$x^2 - 6x + 9 = (x - 3)^2$

$x^2 - 9 = (x + 3)(x - 3)$

$x^3 - 27 = (x - 3)$
$\qquad (x^2 + 3x + 9)$

$x^3 + 27 = (x + 3)$
$\qquad (x^2 - 3x + 9)$

Special Factoring [5.6]

$a^2 + 2ab + b^2 = (a + b)^2$ Perfect square trinomials

$a^2 - 2ab + b^2 = (a - b)^2$

$a^2 - b^2 = (a - b)(a + b)$ Difference of two squares

$a^3 - b^3 = (a - b)(a^2 + ab + b^2)$ Difference of two cubes

$a^3 + b^3 = (a + b)(a^2 - ab + b^2)$ Sum of two cubes

11. Factor completely.

(a) $3x^3 - 6x^2 = 3x^2(x - 2)$

(b) $x^2 - 9 = (x + 3)(x - 3)$
$x^3 - 8 = (x - 2)$
$\qquad (x^2 + 2x + 4)$
$x^3 + 27 = (x + 3)$
$\qquad (x^2 - 3x + 9)$

To Factor Polynomials in General [5.6]

Step 1: If the polynomial has a greatest common factor other than 1, then factor out the greatest common factor.

Step 2: If the polynomial has two terms (it is a binomial), then see if it is the difference of two squares, or the sum or difference of two cubes, and then factor accordingly. Remember, if it is the sum of two squares it will not factor.

(c) $x^2 - 6x + 9 = (x - 3)^2$

$6x^2 - 7x - 5$
$$= (2x + 1)(3x - 5)$$

(d) $x^2 + ax + bx + ab$
$$= x(x + a) + b(x + a)$$
$$= (x + a)(x + b)$$

Step 3: If the polynomial has three terms (a trinomial), then it is either a perfect square trinomial, which will factor into the square of a binomial, or it is not a perfect square trinomial, in which case you use one of the methods developed in Section 5.5.

Step 4: If the polynomial has more than three terms, then try to factor it by grouping.

Step 5: As a final check, see if any of the factors you have written can be factored further. If you have overlooked a common factor, you can catch it here.

12. Solve $x^2 - 5x = -6$.

$$x^2 - 5x + 6 = 0$$
$$(x - 3)(x - 2) = 0$$
$x - 3 = 0$ or $x - 2 = 0$
$x = 3$ or $x = 2$

To Solve an Equation by Factoring [5.7]

Step 1: Write the equation in standard form.
Step 2: Factor the left side.
Step 3: Use the zero-factor property to set each factor equal to zero.
Step 4: Solve the resulting linear equations.

COMMON MISTAKES

When we subtract one polynomial from another, it is common to forget to add the opposite of each term in the second polynomial. For example,

$$(6x - 5) - (3x + 4) = 6x - 5 - 3x + 4 \qquad \text{Mistake}$$

This mistake occurs if the negative sign outside the second set of parentheses is not distributed over all terms inside the parentheses. To avoid this mistake, remember: The opposite of a sum is the sum of the opposites, or,

$$-(3x + 4) = -3x + (-4)$$

CHAPTER 5   REVIEW

Simplify each of the following. [5.1]

1. $x^3 \cdot x^7$ **2.** $(5x^3)^2$

3. $(2x^3y)^2(-2x^4y^2)^3$

Write with positive exponents, and then simplify. [5.1]

4. 2^{-3} **5.** $\left(\frac{2}{3}\right)^{-2}$

6. $2^{-2} + 4^{-1}$

Write in scientific notation. [5.1]

7. 34,500,000   **8.** 0.00357

Write in expanded form. [5.1]

9. 4.45×10^4 **10.** 4.45×10^{-4}

Simplify each expression. All answers should contain positive exponents only. (Assume all variables are nonnegative.) [5.1]

11. $\dfrac{a^{-4}}{a^5}$ **12.** $\dfrac{(4x^2)(-3x^3)^2}{(12x^{-2})^2}$

13. $\dfrac{x^n x^{3n}}{x^{4n-2}}$

Simplify each expression as much as possible. Write all answers in scientific notation. [5.1]

14. $(2 \times 10^3)(4 \times 10^{-5})$ 15. $\dfrac{(600{,}000)(0.000008)}{(4{,}000)(3{,}000{,}000)}$

[handwritten: 4×10^{-10}]

Simplify by combining similar terms. [5.2]

16. $(6x^2 - 3x + 2) - (4x^2 + 2x - 5)$
17. $(x^3 - x) - (x^2 + x) + (x^3 - 3) - (x^2 + 1)$
18. Subtract $2x^2 - 3x + 1$ from $3x^2 - 5x - 2$.
19. Simplify $-3[2x - 4(3x + 1)]$. [5.2]
20. Find the value of $2x^2 - 3x + 1$ when x is -2. [5.2]

Multiply. [5.3]

21. $3x(4x^2 - 2x + 1)$
22. $2a^2b^3(a^2 + 2ab + b^2)$
23. $(6 - y)(3 - y)$ 24. $(2x^2 - 1)(3x^2 + 4)$
25. $2t(t + 1)(t - 3)$
26. $(x + 3)(x^2 - 3x + 9)$
27. $(2x - 3)(4x^2 + 6x + 9)$
28. $(a^2 - 2)^2$ 29. $(3x + 5)^2$
30. $(4x - 3y)^2$ 31. $\left(x - \dfrac{1}{3}\right)\left(x + \dfrac{1}{3}\right)$
32. $(2a + b)(2a - b)$ 33. $(x - 1)^3$
34. $(x^m + 2)(x^m - 2)$

Factor out the greatest common factor. [5.4]

35. $6x^4y - 9xy^4 + 18x^3y^3$
36. $4x^2(x + y)^2 - 8y^2(x + y)^2$

Factor by grouping. [5.4, 5.6]

37. $8x^2 + 10 - 4x^2y - 5y$

38. $x^3 + 8b^2 - x^3y^2 - 8y^2b^2$

Factor completely. [5.4, 5.5]

39. $x^2 - 5x + 6$ 40. $2x^3 + 4x^2 - 30x$
41. $20a^2 - 41ab + 20b^2$ 42. $6x^4 - 11x^3 - 10x^2$
43. $24x^2y - 6xy - 45y$

Factor completely. [5.6]

44. $x^4 - 16$ 45. $3a^4 + 18a^2 + 27$
46. $a^3 - 8$
47. $5x^3 + 30x^2y + 45xy^2$ 48. $3a^3b - 27ab^3$
49. $x^2 - 10x + 25 - y^2$ 50. $36 - 25a^2$
51. $x^3 + 4x^2 - 9x - 36$

Solve each equation. [5.7]

52. $x^2 + 5x + 6 = 0$ 53. $\dfrac{5}{6}y^2 = \dfrac{1}{4}y + \dfrac{1}{3}$
54. $9x^2 - 25 = 0$
55. $5x^2 = -10x$ 56. $(x + 2)(x - 5) = 8$
57. $x^3 + 4x^2 - 9x - 36 = 0$

Solve each application. In each case be sure to show the equation used. [5.7]

58. **Consecutive Numbers** The product of two consecutive even integers is 80. Find the two integers.
59. **Consecutive Numbers** The sum of the squares of two consecutive integers is 41. Find the two integers.
60. **Geometry** The lengths of the three sides of a right triangle are given by three consecutive integers. Find the three sides.
61. **Geometry** The lengths of the three sides of a right triangle are given by three consecutive even integers. Find the three sides.

CHAPTER 5   PROJECTS

EXPONENTS AND POLYNOMIALS

GROUP PROJECT

DISCOVERING PASCAL'S TRIANGLE

Number of People: 3

Time Needed: 20 minutes

Equipment: Paper and pencils

Background: The triangular array of numbers shown here is known as Pascal's triangle, after the French philosopher Blaise Pascal (1623–1662).

$$
\begin{array}{ccccccc}
 & & & 1 & & & \\
 & & 1 & & 1 & & \\
 & 1 & & 2 & & 1 & \\
 1 & & 3 & & 3 & & 1 \\
1 & & 4 & & 6 & & 4 & & 1 \\
1 & & 5 & & 10 & & 10 & & 5 & & 1
\end{array}
$$

Procedure: Look at Pascal's triangle and discover how the numbers in each row of the triangle are obtained from the numbers in the row above it.

1. Once you have discovered how to extend the triangle, write the next two rows.

2. Pascal's triangle can be linked to the Fibonacci sequence by rewriting Pascal's triangle so that the 1's on the left side of the triangle line up under one another, and the other columns are equally spaced to the right of the first column. Rewrite Pascal's triangle as indicated and then look along the diagonals of the new array until you discover how the Fibonacci sequence can be obtained from it.

3. The diagram at the left shows Pascal's triangle as written in Japan in 1781. Use your knowledge of Pascal's triangle to translate the numbers written in Japanese into our number system. Then write down the Japanese numbers from 1 to 20.

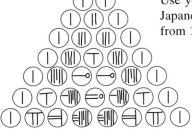

Pascal's triangle in Japan

RESEARCH PROJECT

BINOMIAL EXPANSIONS

The title on the following diagram is *Binomial Expansions* because each line gives the expansion of the binomial $x + y$ raised to a whole-number power.

Binomial Expansions

$$(x + y)^0 = \quad 1$$
$$(x + y)^1 = \quad x + y$$
$$(x + y)^2 = \quad x^2 + 2xy + y^2$$
$$(x + y)^3 = x^3 + 3x^2y + 3xy^2 + y^3$$
$$(x + y)^4 =$$
$$(x + y)^5 =$$

The fourth row in the diagram was completed by expanding $(x + y)^3$ using the methods developed in this chapter. Next, complete the diagram by expanding the binomials $(x + y)^4$ and $(x + y)^5$ using the multiplication procedures you have learned in this chapter. Finally, study the completed diagram until you see patterns that will allow you to continue the diagram one more row without using multiplication. (One pattern that you will see is Pascal's triangle, which we mentioned in the preceding group project.) When you are finished, write an essay in which you describe what you have done and the results you have obtained.

CHAPTER 5   TEST

Simplify. All answers should contain positive exponents only. (Assume all variables are nonnegative.) [5.1]

1. $x^4 \cdot x^7 \cdot x^{-3}$

2. 2^{-5}

3. $\left(\frac{3}{4}\right)^{-2}$

4. $(2x^2y)^3(2x^3y^4)^2$

5. $\dfrac{a^{-5}}{a^{-7}}$

6. $\dfrac{x^{n+1}}{x^{n-5}}$

7. $\dfrac{(2ab^3)^{-2}(a^4b^{-3})}{(a^{-4}b^3)^4(2a^{-2}b^2)^{-3}}$

Write each number in scientific notation. [5.1]

8. 6,530,000

9. 0.00087

Perform the indicated operations, and write your answers in scientific notation. [5.1]

10. $(2.9 \times 10^{12})(3 \times 10^{-5})$

11. $\dfrac{(6 \times 10^{-4})(4 \times 10^9)}{8 \times 10^{-3}}$

Simplify. [5.2]

12. $\left(\frac{3}{4}x^3 - x^2 - \frac{3}{2}\right) - \left(\frac{1}{4}x^2 + 2x - \frac{1}{2}\right)$

13. $3 - 4[2x - 3(x + 6)]$

Multiply. [5.3]

14. $(3y - 7)(2y + 5)$

15. $(2x - 5)(x^2 + 4x - 3)$

16. $(8 - 3t^3)^2$

17. $(1 - 6y)(1 + 6y)$

18. $2x(x - 3)(2x + 5)$

19. $\left(5t^2 - \frac{1}{2}\right)\left(2t^2 + \frac{1}{5}\right)$

Factor completely. [5.4, 5.5, 5.6]

20. $x^2 + x - 12$

21. $12x^4 + 26x^2 - 10$

Factor completely.

22. $16a^4 - 81y^4$

23. $7ax^2 - 14ay - b^2x^2 + 2b^2y$

24. $t^3 + \frac{1}{8}$

25. $4a^5b - 24a^4b^2 - 64a^3b^3$

26. $x^2 - 10x + 25 - b^2$

27. $81 - x^4$

Solve each equation. [5.7]

28. $\frac{1}{5}x^2 = \frac{1}{3}x + \frac{2}{15}$ **29.** $100x^3 = 500x^2$

30. $(x + 1)(x + 2) = 12$

31. $x^3 + 2x^2 - 16x - 32 = 0$

32. Number Problem   One integer is 3 more than another. The sum of their squares is 29. Find the two integers. [5.7]

33. Geometry   The longest side of a right triangle is 4 inches more than the shortest side. The third side is 2 inches more than the shortest side. Find the length of each side. [5.7]

Profit, Revenue, and Cost   A company making ceramic coffee cups finds that it can sell x cups per week at p dollars each, according to the formula $p = 25 - 0.2x$. If the total cost to produce and sell x coffee cups is $C = 2x + 100$, find [5.2, 5.3]

34. . . . an equation for the revenue that gives the revenue in terms of x.

35. . . . the profit equation.

36. . . . the revenue brought in by selling 100 coffee cups.

37. . . . the cost of producing 100 coffee cups.

38. . . . the profit obtained by making and selling 100 coffee cups.

Simplify.

1. $15 - 12 \div 4 - 3 \cdot 2$

2. $6(11 - 13)^3 - 5(8 - 11)^2$

3. $\left(\dfrac{2}{5}\right)^{-2}$

4. $4(3x - 2) + 3(2x + 5)$

5. $5 - 3[2x - 4(x - 2)]$

6. $(3y + 2)^2 - (3y - 2)^2$

7. $(2x + 3)(x^2 - 4x + 2)$

Solve.

8. $-4y - 2 = 6y + 8$

9. $-6 + 2(2x + 3) = 0$

10. $3x^2 = 17x - 10$ **11.** $|2x - 3| + 7 = 1$

Solve each system.

12. $\begin{aligned} 2x - 5y &= -7 \\ -3x + 4y &= 0 \end{aligned}$ **13.** $\begin{aligned} 8x + 6y &= 4 \\ 12x + 9y &= 8 \end{aligned}$

14. $\begin{aligned} 2x + y &= 3 \\ y &= -2x + 3 \end{aligned}$ **15.** $\begin{aligned} 2x + y &= 8 \\ 4y - z &= -9 \\ 3x - 2z &= -6 \end{aligned}$

Solve each inequality, and graph the solution.

16. $3 < \dfrac{1}{4}x + 4 < 5$ **17.** $|2x - 7| \le 3$

18. $|2x - 7| - 5 \ge 6$

19. Solve $-3t \ge 12$. Write your answer with interval notation.

20. Solve for x: $ax - 4 = bx + 9$.

21. Write the first five terms of the sequence:

$$a_n = \frac{2n}{n + 3}.$$

22. Give the opposite and reciprocal of -7.

23. Add $-\dfrac{2}{9}$ to the product of -6 and $\dfrac{7}{54}$.

24. Translate into an inequality: x is between -5 and 5.

25. For the set $\{-1, 0, \sqrt{2}, 2.35, \sqrt{3}, 4\}$ list all elements belonging to the set of *rational numbers*.

26. If $a = \{0, 2, 3, 5, 9\}$ and $B = \{1, 2, 3, 6\}$, find $A \cup B$.

27. Identify the hypothesis and conclusion: If x is a multiple of 12, then x is divisible by 3.

28. Write as positive exponents and simplify:

$$\left(\tfrac{2}{13}\right)^{-2} - \left(\tfrac{2}{5}\right)^{-2}$$

29. Simplify and write answer with positive exponents:

$$(5x^{-3}y^2z^{-2})(6x^{-5}y^4z^{-3})$$

30. Write 0.000469 in scientific notation.

31. Simplify and write in scientific notation:

$$\frac{(7 \times 10^{-5})(21 \times 10^{-6})}{3 \times 10^{-12}}$$

32. Specify the domain and range for the relation $\{(-1, 3), (2, -10), (3, 3)\}$. Is the relation also a function?

33. State the domain: $y = -\sqrt{5 - x}$.

Graph on a rectangular coordinate system.

34. $0.03x + 0.04y = 0.04$ **35.** $3x - y < -2$

36. Find the slope of the line through $\left(\tfrac{2}{3}, -\tfrac{1}{2}\right)$ and $\left(\tfrac{1}{2}, -\tfrac{5}{6}\right)$.

37. Find the slope and y-intercept of $3x - 5y = 15$.

38. Give the equation of a line with slope $-\tfrac{2}{3}$ and y-intercept $= -3$.

39. Write the equation of the line in slope-intercept form if the slope is $\tfrac{7}{3}$ and $(6, 8)$ is a point on the line.

40. Find the equation of the line through $(1, 4)$ and $(-1, -2)$.

Factor completely.

41. $16y^2 + 2y + \tfrac{1}{16}$

42. $6a^2x + 2x + 3a^2y^2 + y^2$

43. $x^3 - 8$ **44.** $16a^4 - 81b^4$

Use Cramer's rule to solve each system.

45. $\begin{aligned} 7x + 9y &= 2 \\ -5x + 3y &= 1 \end{aligned}$ **46.** $\begin{aligned} 5x - 3y + z &= 2 \\ 3x - y + 4z &= 3 \\ 2x + 3y - z &= -1 \end{aligned}$

Evaluate each determinant.

47. $\begin{vmatrix} 1 & 8 \\ -2 & 7 \end{vmatrix}$

48. $\begin{vmatrix} 1 & 3 & 5 \\ -3 & -1 & -1 \\ 0 & 1 & 0 \end{vmatrix}$

49. Speed   A boat travels 36 miles down a river in 3 hours. If it takes the boat 9 hours to travel the same distance going up the river, what is the speed of the boat? What is the speed of the current of the river?

50. Concentric Circles   Recall that the area of a circle is $A = \pi r^2$. Two circles are concentric if they have the same center. Find the area between two concentric circles in which the larger circle has radius a and the smaller circle has radius b. Express your answer in a form in which π is a factor.

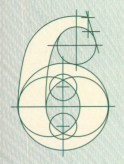

Rational Expressions

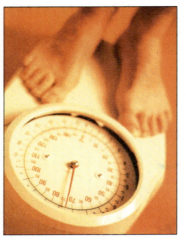

(Jon Bradley/Tony Stone Images)

INTRODUCTION

If you have ever put yourself on a weight-loss diet, you know that you lose more weight at the beginning of the diet than you do later on. If we let $W(x)$ represent a person's weight after x weeks on the diet, then the rational function

$$W(x) = \frac{80(2x + 15)}{x + 6}$$

is a mathematical model of the person's weekly progress on a diet intended to take them from 200 pounds to about 160 pounds. Rational functions are good models for quantities that fall off rapidly to begin with, and then level off over time. Table 1 is a table of values for this function, while Figure 1 shows the graph of this function.

TABLE I	Weekly Weight Loss
Weeks Since Starting Diet	**Weight (nearest pound)**
0	200
4	184
8	177
12	173
16	171
20	169
24	168

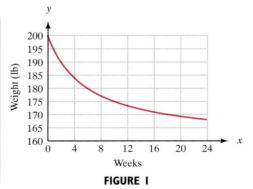

FIGURE I

As you progress through this chapter you will acquire an intuitive feel for these types of functions, and as a result, you will see why they are good models for situations such as dieting.

STUDY SKILLS

This is the last chapter in which we will mention study skills. You know by now what works best for you and what you have to do to achieve your goals for this course. From now on it is simply a matter of sticking with the things that work for you and avoiding the things that do not. It seems simple, but as with anything that takes effort, it is up to you to see that you maintain the skills that get you where you want to be in the course.

If you intend to take more classes in mathematics, and you want to ensure your success in those classes, then you can work toward this goal: *Become the type of student who can learn mathematics on his or her own.* Most people who have degrees in mathematics were students who could learn mathematics on their own. This doesn't mean that you have to learn it all on your own; it simply means that if you have to, you can learn it on your own. Attaining this goal gives you independence and puts you in control of your success in any math class you take.

We will begin this section with the definition of a rational expression. We will then state the two basic properties associated with rational expressions and go on to apply one of the properties to reduce rational expressions to lowest terms.

Recall from Chapter 1 that a *rational number* is any number that can be expressed as the ratio of two integers:

$$\text{Rational numbers} = \left\{ \frac{a}{b} \middle| a \text{ and } b \text{ are integers, } b \neq 0 \right\}$$

A rational expression is defined similarly as any expression that can be written as the ratio of two polynomials:

$$\text{Rational expressions} = \left\{ \frac{P}{Q} \middle| P \text{ and } Q \text{ are polynomials, } Q \neq 0 \right\}$$

Some examples of rational expressions are

$$\frac{2x - 3}{x + 5} \qquad \frac{x^2 - 5x - 6}{x^2 - 1} \qquad \frac{a - b}{b - a}$$

BASIC PROPERTIES

For rational expressions, multiplying the numerator and denominator by the same nonzero expression may change the form of the rational expression, but it will always produce an expression equivalent to the original one. The same is true when dividing the numerator and denominator by the same nonzero quantity.

PROPERTIES OF RATIONAL EXPRESSIONS

If P, Q, and K are polynomials with $Q \neq 0$ and $K \neq 0$, then

$$\frac{P}{Q} = \frac{PK}{QK} \qquad \text{and} \qquad \frac{P}{Q} = \frac{P/K}{Q/K}$$

REDUCING TO LOWEST TERMS

The fraction $\frac{6}{8}$ can be written in lowest terms as $\frac{3}{4}$. The process is shown here:

$$\frac{6}{8} = \frac{3 \cdot \overset{1}{\cancel{2}}}{4 \cdot \underset{1}{\cancel{2}}} = \frac{3}{4}$$

Reducing $\frac{6}{8}$ to $\frac{3}{4}$ involves dividing the numerator and denominator by 2, the factor they have in common. Before dividing out the common factor 2, we must notice that the common factor *is* 2! (This may not be obvious since we are very familiar with the numbers 6 and 8 and therefore do not have to put much thought into finding what number divides both of them.)

We reduce rational expressions to lowest terms by first factoring the numerator and denominator and then dividing both numerator and denominator by any factors they have in common.

EXAMPLE 1 Reduce $\dfrac{x^2 - 9}{x - 3}$ to lowest terms.

Solution Factoring, we have

$$\frac{x^2 - 9}{x - 3} = \frac{(x + 3)(x - 3)}{x - 3}$$

The numerator and denominator have the factor $x - 3$ in common. Dividing the numerator and denominator by $x - 3$, we have

$$\frac{(x + 3)\cancel{(x - 3)}}{\cancel{x - 3}} = \frac{x + 3}{1} = x + 3$$

Note: The lines drawn through the $(x - 3)$ in the numerator and denominator indicate that we have divided through by $(x - 3)$. As the problems become more involved, these lines will help keep track of which factors have been divided out and which have not.

For the problem in Example 1, there is an implied restriction on the variable x: It cannot be 3. If x were 3, the expression $(x^2 - 9)/(x - 3)$ would become $0/0$, an expression that we cannot associate with a real number. For all problems involving rational expressions, we restrict the variable to only those values that result in a nonzero denominator. When we state the relationship

$$\frac{x^2 - 9}{x - 3} = x + 3$$

we are assuming that it is true for all values of x except $x = 3$.

Here are some other examples of reducing rational expressions to lowest terms.

EXAMPLES Reduce to lowest terms.

2. $\dfrac{y^2 - 5y - 6}{y^2 - 1} = \dfrac{(y - 6)\cancel{(y + 1)}}{(y - 1)\cancel{(y + 1)}}$ Factor numerator and denominator.

$\qquad = \dfrac{y - 6}{y - 1}$ Divide out common factor $y + 1$.

3. $\dfrac{2a^3 - 16}{4a^2 - 12a + 8} = \dfrac{2(a^3 - 8)}{4(a^2 - 3a + 2)}$ Factor numerator and denominator.

$\qquad = \dfrac{2\cancel{(a - 2)}(a^2 + 2a + 4)}{4\cancel{(a - 2)}(a - 1)}$

$\qquad = \dfrac{a^2 + 2a + 4}{2(a - 1)}$ Divide out common factor $2(a - 2)$.

4. $\dfrac{x^2 - 3x + ax - 3a}{x^2 - ax - 3x + 3a} = \dfrac{x(x-3) + a(x-3)}{x(x-a) - 3(x-a)}$ Factor numerator

$\qquad\qquad\qquad\qquad = \dfrac{\cancel{(x-3)}(x+a)}{(x-a)\cancel{(x-3)}}$ and denominator.

$\qquad\qquad\qquad\qquad = \dfrac{x+a}{x-a}$ Divide out common factor $x-3$.

The answer to Example 4 is $(x+a)/(x-a)$. The problem cannot be reduced further. It is a fairly common mistake to attempt to divide out an x or an a in this last expression. Remember, we can divide out only the factors common to the numerator and denominator of a rational expression. For the last expression in Example 4, neither the numerator nor the denominator can be factored further; x is not a factor of the numerator or the denominator, and neither is a. The expression is in lowest terms.

The next example involves what we call a trick. The trick is to reverse the order of the terms in a difference by factoring -1 from each term in either the numerator or the denominator. The next examples illustrate how this is done.

EXAMPLE 5   Reduce to lowest terms: $\dfrac{a-b}{b-a}$

Solution   The relationship between $a - b$ and $b - a$ is that they are opposites. We can show this fact by factoring -1 from each term in the numerator:

$\qquad \dfrac{a-b}{b-a} = \dfrac{-1(-a+b)}{b-a}$ Factor -1 from each term in the numerator.

$\qquad\qquad = \dfrac{-1(b-a)}{b-a}$ Reverse the order of the terms in the numerator.

$\qquad\qquad = -1$ Divide out common factor $b-a$.

EXAMPLE 6   Reduce to lowest terms: $\dfrac{x^2 - 25}{5 - x}$

Solution   We begin by factoring the numerator:

$$\dfrac{x^2 - 25}{5 - x} = \dfrac{(x-5)(x+5)}{5-x}$$

The factors $x - 5$ and $5 - x$ are similar but are not exactly the same. We can reverse the order of either by factoring -1 from it. That is: $5 - x = -1(-5 + x) = -1(x - 5)$.

$$\dfrac{(x-5)(x+5)}{5-x} = \dfrac{\cancel{(x-5)}(x+5)}{-1\cancel{(x-5)}}$$

$$= \dfrac{x+5}{-1}$$

$$= -(x+5)$$

APPLICATIONS

Ratios   You may recall from previous math classes that the ratio of a to b is the same as the fraction $\frac{a}{b}$. To illustrate, if the ratio of men to women in a math class is 3 to 2, then $\frac{3}{5}$ of the class are men, and $\frac{2}{5}$ of the class are women. Here are two ratios that are used frequently in mathematics:

1. The number π is defined as the ratio of the circumference of a circle to the diameter of a circle. That is,

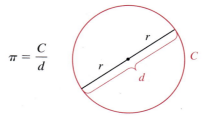

$$\pi = \frac{C}{d}$$

Multiplying both sides of this formula by d, we have the more common form $C = \pi d$.

Note: Since the diameter of a circle is twice the radius, the formula for circumference is sometimes written $C = 2\pi r$. Neither formula should be confused with the formula for the area of a circle, which is $A = \pi r^2$.

2. The *average speed* of a moving object is defined to be the ratio of distance to time. If you drive your car for 5 hours and travel a distance of 200 miles, then your average rate of speed is

$$\text{Average speed} = \frac{200 \text{ miles}}{5 \text{ hours}} = 40 \text{ miles per hour}$$

The formula we use for the relationship between average speed r, distance d, and time t is

$$r = \frac{d}{t}$$

This formula is sometimes called the *rate equation*. Multiplying both sides by t, we have the equivalent form of the rate equation, $d = rt$.

Our next example involves both the formula for the circumference of a circle and the rate equation.

EXAMPLE 7   The first Ferris wheel was designed and built by George Ferris in 1893. The diameter of the wheel was 250 feet. It had 36 carriages, equally spaced around the wheel, each of which held a maximum of 40 people. One trip around the wheel took 20 minutes. Find the average speed of a rider on the first Ferris wheel. (Use 3.14 as an approximation for π.)

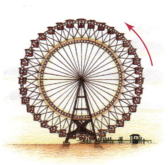

Solution The distance traveled is the circumference of the wheel, which is

$$C = 250\pi = 250(3.14) = 785 \text{ feet}$$

To find the average speed, we divide the distance traveled by the amount of time it took to go once around the wheel.

$$r = \frac{d}{t} = \frac{785 \text{ feet}}{20 \text{ minutes}} = 39.3 \text{ feet per minute (to the nearest tenth)}$$

Later in this chapter we will convert this ratio into an equivalent ratio that gives the speed of the rider in miles per hour.

   The Ferris wheel in Example 7 has a circumference of 785 feet. In the next example we graph the relationship between the average speed of a person riding this wheel and the amount of time it takes the wheel to complete one revolution.

EXAMPLE 8   A Ferris wheel has a circumference of 785 feet. If one complete revolution of the wheel takes from 10 to 30 minutes, then the relationship between the average speed of a rider on the wheel and the amount of time it takes the wheel to complete one revolution is given by the function

$$r(t) = \frac{785}{t} \qquad 10 \le t \le 30$$

where $r(t)$ is the average speed (in feet per minute) and t is the amount of time (in minutes) it takes the wheel to complete one revolution. Graph the function.

Solution Since the variables r and t represent speed and time, both must be positive quantities. The graph of this function will therefore lie in the first quadrant only. The following table displays the values of t and $r(t)$ found from the function, along with the graph of the function (Figure 1). (Some of the numbers in the table have been rounded to the nearest tenth.)

Time to Complete 1 Revolution t	Speed (ft/min) $r(t)$
10	78.5
15	52.3
20	39.3
25	31.4
30	26.2

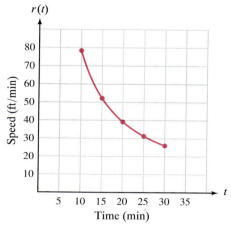

FIGURE 1

Using TECHNOLOGY

MORE ABOUT EXAMPLE 8

If we use a graphing calculator to graph the equation in Example 8, it is not necessary to construct the table first. In fact, if we graph

$$Y_1 = 785/X \qquad \text{Window:} \quad \text{X from 0 to 40, Y from 0 to 90}$$

we can use the Trace and Zoom features together to produce the numbers in the table next to Figure 1. Graph the preceding equation, and zoom in on the point with x-coordinate 20 until you are convinced that the table values for x and y are correct.

RATIONAL FUNCTIONS

The function shown in Example 8 is called a *rational function* because the right side, $785/t$, is a rational expression (the numerator, 785, is a polynomial of degree 0). We can extend our knowledge of rational expressions to functions with the following definition:

> **DEFINITION**   A **rational function** is any function that can be written in the form
>
> $$f(x) = \frac{P(x)}{Q(x)}$$
>
> where $P(x)$ and $Q(x)$ are polynomials and $Q(x) \neq 0$.

EXAMPLE 9 For the rational function $f(x) = \dfrac{x-4}{x-2}$, find $f(0), f(-4), f(4),$ $f(-2)$, and $f(2)$.

Solution To find these function values, we substitute the given value of x into the rational expression, and then simplify if possible.

$$f(0) = \frac{0-4}{0-2} = \frac{-4}{-2} = 2 \qquad f(-2) = \frac{-2-4}{-2-2} = \frac{-6}{-4} = \frac{3}{2}$$

$$f(-4) = \frac{-4-4}{-4-2} = \frac{-8}{-6} = \frac{4}{3} \qquad f(2) = \frac{2-4}{2-2} = \frac{-2}{0} \quad \text{Undefined}$$

$$f(4) = \frac{4-4}{4-2} = \frac{0}{2} = 0$$

Because the rational function in Example 9 is not defined when x is 2, the domain of that function does not include 2. We have more to say about the domain of a rational function next.

THE DOMAIN OF A RATIONAL FUNCTION

In Example 8 the domain of the rational function is specified as $10 \le t \le 30$, and the function is defined for all values of t in that domain. If the domain of a rational function is not specified, it is assumed to be all real numbers for which the function is defined. That is, the domain of the rational function

$$f(x) = \frac{P(x)}{Q(x)}$$

is all x for which $Q(x)$ is nonzero. For example:

The domain for $r(t) = \dfrac{785}{t}$, $10 \le t \le 30$, is $\{t \mid 10 \le t \le 30\}$.

The domain for $f(x) = \dfrac{x-4}{x-2}$ is $\{x \mid x \ne 2\}$.

The domain for $g(x) = \dfrac{x^2+5}{x+1}$ is $\{x \mid x \ne -1\}$.

The domain for $h(x) = \dfrac{x}{x^2-9}$ is $\{x \mid x \ne -3, x \ne 3\}$.

Notice that, for these functions, $f(2), g(-1), h(-3),$ and $h(3)$ are all undefined, and that is why the domains are written as shown.

EXAMPLE 10 Graph the equation $y = \dfrac{x^2-9}{x-3}$. How is this graph different from the graph of $y = x + 3$?

Solution   We know from the discussion in Example 1 that

$$y = \frac{x^2 - 9}{x - 3} = \frac{(x + 3)(x - 3)}{x - 3} = x + 3$$

This relationship is true for all x except $x = 3$, because the rational expressions with $x - 3$ in the denominator are undefined when x is 3. However, for all other values of x, the expressions

$$\frac{x^2 - 9}{x - 3} \qquad \text{and} \qquad x + 3$$

are equal. Therefore, the graphs of

$$y = \frac{x^2 - 9}{x - 3} \qquad \text{and} \qquad y = x + 3$$

will be the same except when x is 3. In the first equation, there is no value of y to correspond to $x = 3$. In the second equation, $y = x + 3$, so y is 6 when x is 3.

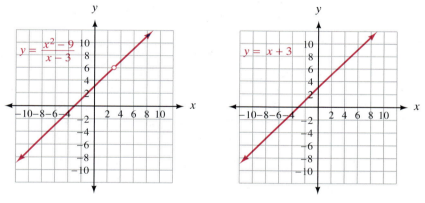

FIGURE 2 Graphs of $y = \dfrac{x^2 - 9}{x - 3}$ (left) and $y = x + 3$ (right)

Now you can see the difference in the graphs of the two equations. To show that there is no y value for $x = 3$ in the graph on the left in Figure 2, we draw an open circle at that point on the line.

Notice that the two graphs shown in Figure 2 are both graphs of functions. Suppose we use function notation to designate them as follows:

$$f(x) = \frac{x^2 - 9}{x - 3} \qquad \text{and} \qquad g(x) = x + 3$$

The two functions, f and g, are equivalent except when $x = 3$, because $f(3)$ is undefined, while $g(3) = 6$. The domain of the function f is all real numbers except $x = 3$, while the domain for g is all real numbers, with no restrictions.

Getting Ready for Class

After reading through the preceding section, respond in your own words and in complete sentences.

A. What is a rational expression?

B. Explain how to determine if a rational expression is in "lowest terms."

C. When is a rational expression undefined?

D. Explain the process we use to reduce a rational expression or a fraction to lowest terms.

PROBLEM SET 6.1

Reduce each fraction to lowest terms.

1. $-\dfrac{12}{36}$

2. $-\dfrac{45}{60}$

3. $\dfrac{2a^2b^3}{4a^2}$

4. $\dfrac{3a^3b^2}{6b^2}$

5. $-\dfrac{24x^3y^5}{16x^4y^2}$

6. $-\dfrac{36x^6y^8}{24x^3y^9}$

7. $\dfrac{144a^2b^3c^4}{56a^4b^3c^2}$

8. $\dfrac{108a^5b^2c^5}{27a^2b^5c^2}$

Reduce each rational expression to lowest terms.

9. $\dfrac{x^2 - 16}{6x + 24}$

10. $\dfrac{5x + 25}{x^2 - 25}$

11. $\dfrac{12x - 9y}{3x^2 + 3xy}$

12. $\dfrac{x^3 - xy^2}{4x + 4y}$

13. $\dfrac{a^4 - 81}{a - 3}$

14. $\dfrac{a + 4}{a^2 - 16}$

15. $\dfrac{a^2 - 4a - 12}{a^2 + 8a + 12}$

16. $\dfrac{a^2 - 7a + 12}{a^2 - 9a + 20}$

17. $\dfrac{20y^2 - 45}{10y^2 - 5y - 15}$

18. $\dfrac{54y^2 - 6}{18y^2 - 60y + 18}$

19. $\dfrac{20x^2 - 93x + 34}{4x^2 - 9x - 34}$

20. $\dfrac{15x^2 - 59x + 52}{5x^2 - 33x + 52}$

21. $\dfrac{12y - 2xy - 2x^2y}{6y - 4xy - 2x^2y}$

22. $\dfrac{250a + 100ax + 10ax^2}{50a - 2ax^2}$

23. $\dfrac{(x - 3)^2(x + 2)}{(x + 2)^2(x - 3)}$

24. $\dfrac{(x - 4)^3(x + 3)}{(x + 3)^2(x - 4)}$

25. $\dfrac{a^3 + b^3}{a^2 - b^2}$

26. $\dfrac{a^2 - b^2}{a^3 - b^3}$

27. $\dfrac{8x^4 - 8x}{4x^4 + 4x^3 + 4x^2}$

28. $\dfrac{6x^5 - 48x^2}{12x^3 + 24x^2 + 48x}$

29. $\dfrac{6x^2 + 7xy - 3y^2}{6x^2 + xy - y^2}$

30. $\dfrac{4x^2 - y^2}{4x^2 - 8xy - 5y^2}$

31. $\dfrac{ax + 2x + 3a + 6}{ay + 2y - 4a - 8}$

32. $\dfrac{ax - x - 5a + 5}{ax + x - 5a - 5}$

33. $\dfrac{x^2 + bx - 3x - 3b}{x^2 - 2bx - 3x + 6b}$

34. $\dfrac{x^2 - 3ax - 2x + 6a}{x^2 - 3ax + 2x - 6a}$

35. $\dfrac{x^3 + 3x^2 - 4x - 12}{x^2 + x - 6}$

36. $\dfrac{x^3 + 5x^2 - 4x - 20}{x^2 + 7x + 10}$

37. $\dfrac{3x^3 + 21x^2 + 36x}{3x^4 + 12x^3 - 27x^2 - 108x}$

38. $\dfrac{2x^4 + 14x^3 + 20x^2}{2x^5 + 4x^4 - 50x^3 - 100x^2}$

39. $\dfrac{4x^4 - 25}{6x^3 - 4x^2 + 15x - 10}$

40. $\dfrac{16x^4 - 49}{8x^3 - 12x^2 + 14x - 21}$

Refer to Examples 5 and 6 in this section, and reduce the following to lowest terms.

41. $\dfrac{x - 4}{4 - x}$ **42.** $\dfrac{6 - x}{x - 6}$

43. $\dfrac{y^2 - 36}{6 - y}$ **44.** $\dfrac{1 - y}{y^2 - 1}$

45. $\dfrac{1 - 9a^2}{9a^2 - 6a + 1}$ **46.** $\dfrac{1 - a^2}{a^2 - 2a + 1}$

Reduce each rational expression to lowest terms; then subtract.

47. $\dfrac{28x^2 - 41x + 15}{7x - 5} - \dfrac{12x^2 - 41x + 24}{3x - 8}$

48. $\dfrac{42x^2 + 47x - 55}{7x - 5} - \dfrac{18x^2 - 15x - 88}{3x - 8}$

49. $\dfrac{x^3 - 8}{x - 2} - \dfrac{x^3 + 8}{x + 2}$ **50.** $\dfrac{x^4 - 16}{x + 2} - \dfrac{x^4 - 16}{x - 2}$

51. If $g(x) = \dfrac{x + 3}{x - 1}$, find $g(0)$, $g(-3)$, $g(3)$, $g(-1)$, and $g(1)$, if possible.

52. If $g(x) = \dfrac{x - 2}{x - 1}$, find $g(0)$, $g(-2)$, $g(2)$, $g(-1)$, and $g(1)$, if possible.

53. If $h(t) = \dfrac{t - 3}{t + 1}$, find $h(0)$, $h(-3)$, $h(3)$, $h(-1)$, and $h(1)$, if possible.

54. If $h(t) = \dfrac{t - 2}{t + 1}$, find $h(0)$, $h(-2)$, $h(2)$, $h(-1)$, and $h(1)$, if possible.

State the domain for each rational function.

55. $f(x) = \dfrac{x - 3}{x - 1}$ **56.** $f(x) = \dfrac{x + 4}{x - 2}$

57. $g(x) = \dfrac{x^2 - 4}{x - 2}$ **58.** $g(x) = \dfrac{x^2 - 9}{x - 3}$

59. $h(t) = \dfrac{t - 4}{t^2 - 16}$ **60.** $h(t) = \dfrac{t - 5}{t^2 - 25}$

Let $f(x) = \dfrac{x^2 - 4}{x - 2}$ and $g(x) = x + 2$, and evaluate the following expressions, if possible.

61. $f(0)$ and $g(0)$ **62.** $f(1)$ and $g(1)$

63. $f(2)$ and $g(2)$ **64.** $f(3)$ and $g(3)$

Let $f(x) = \dfrac{x^2 - 1}{x - 1}$ and $g(x) = x + 1$, and evaluate the following expressions, if possible.

65. $f(0)$ and $g(0)$ **66.** $f(1)$ and $g(1)$

67. $f(2)$ and $g(2)$ **68.** $f(3)$ and $g(3)$

69. Graph the equation $y = \dfrac{x^2 - 4}{x - 2}$. Then explain how this graph is different from the graph of $y = x + 2$.

70. Graph the equation $y = \dfrac{x^2 - 1}{x - 1}$. Then explain how this graph is different from the graph of $y = x + 1$.

Applying the Concepts

71. Diet The following rational function is the one we mentioned in the introduction to this chapter. The quantity $W(x)$ is the weight (in pounds) of the person after x weeks of dieting. Use the function to fill in the table. Then compare your results with the graph in the chapter introduction.

$$W(x) = \dfrac{80(2x + 15)}{x + 6}$$

Weeks x	Weight (lb) $W(x)$
0	
1	
4	
12	
24	

72. Drag Racing The following rational function gives the speed $V(x)$, in miles per hour, of a dragster at each second x during a quarter-mile race.

Use the function to fill in the table.

$$V(x) = \frac{340x}{x + 3}$$

Time (sec) x	Speed (mi/hr) V(x)
0	
1	
2	
3	
4	
5	
6	

For Problems 73–76, round all answers to the nearest tenth.

73. Average Speed A jogger covers 3.5 miles in 29.75 minutes. Find the average speed of the jogger in miles per minute.

74. Average Speed A bullet fired from a gun travels a distance of 4,750 feet in 3.2 seconds. Find the average speed of the bullet in feet per second.

75. Fuel Consumption A pickup truck travels 175.8 miles on 16.3 gallons of gas. Give the average rate of fuel consumption of the truck in miles per gallon.

76. Fuel Consumption A luxury car travels 200 miles on 16.5 gallons of gas. Give the average fuel consumption of the car in miles per gallon.

For Problems 77 and 78, use 3.14 as an approximation for π. Round answers to the nearest tenth.

77. Average Speed A person riding a Ferris wheel with a diameter of 65 feet travels once around the wheel in 30 seconds. What is the average speed of the rider in feet per second?

78. Average Speed A person riding a Ferris wheel with a diameter of 102 feet travels once around the wheel in 3.5 minutes. What is the average speed of the rider in feet per minute?

The abbreviation "rpm" stands for revolutions per minute. If a point on a circle rotates at 300 rpm, then it rotates through one complete revolution 300 times every minute. The length of time it takes to rotate once around the circle is $\frac{1}{300}$ minute. Use 3.14 as an approximation for π.

79. Average Speed A $3\frac{1}{2}$-inch diskette, when placed in the disk drive of a computer, rotates at 300 rpm (1 revolution takes $\frac{1}{300}$ minute). Find the average speed of a point 2 inches from the center of the diskette. Then find the average speed of a point 1.5 inches from the center of the diskette.

80. Average Speed A 5-inch fixed disk in a computer rotates at 3,600 rpm. Find the average speed of a point 2 inches from the center of the disk. Then find the average speed of a point 1.5 inches from the center.

81. Average Speed The Ferris wheel in Problem 77 has a circumference of 204 feet (to the nearest foot). If a ride on the wheel takes from 20 to 50 seconds, then the relationship between the average speed of a rider and the amount of time it takes to complete one revolution is given by the function

$$r(t) = \frac{204}{t} \qquad 20 \le t \le 50$$

where $r(t)$ is in feet per second and t is in seconds.
(a) State the domain for this function.
(b) Graph the function.

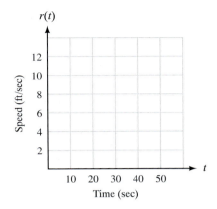

82. Average Speed The Ferris wheel in Problem 78 has a circumference of 320 feet (to the nearest foot). If a ride on the wheel takes from 3 to 5 minutes, then the relationship between the average speed of a rider and the amount of time it takes to complete one revolution is given by the function

$$r(t) = \frac{320}{t} \qquad 3 \le t \le 5$$

where $r(t)$ is in feet per minute and t is in minutes.
(a) State the domain for this function.
(b) Graph the function.

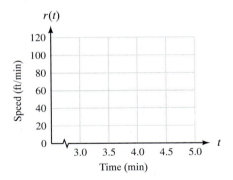

83. Intensity of Light The relationship between the intensity of light that falls on a surface from a 100-watt light bulb and the distance from that surface is given by the rational function

$$I(d) = \frac{120}{d^2} \qquad \text{for } 1 \le d \le 6$$

where $I(d)$ is the intensity of light (in lumens per square foot) and d is the distance (in feet) from the light bulb to the surface.
(a) State the domain for this function.

(b) Use the template in Figure 3 to graph this function.

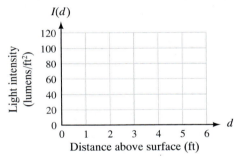

FIGURE 3 Template for graphing a rational function

84. Average Speed If it takes Maria t minutes to run a mile, then her average speed $s(t)$ is given by the rational function

$$s(t) = \frac{60}{t} \qquad \text{for } 6 \le t \le 12$$

where $s(t)$ is in miles per hour and t is in minutes.

(a) State the domain for this function.

(b) Use the template in Figure 4 to graph this function.

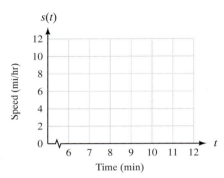

FIGURE 4 Template for graphing a rational function

85. The following is a graph of a person's weight loss (in pounds) after x weeks of dieting:

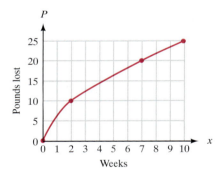

(a) Find the ratio of weight lost after 2 weeks to weight lost after 10 weeks.

(b) Find the ratio of weight lost after 7 weeks to weight lost after 10 weeks.

86. **Traveling Time**   Suppose you travel 220 miles on the freeway to get to a relative's house. If the speed limit has been changed from 55 miles per hour to 70 miles per hour, how much time will you save with the new speed limit?

The traffic capacity of a road can be approximated with the formula

$$N = \frac{1,760V}{I}$$

where N is the traffic capacity in vehicles per hour, V is the speed of the vehicles in miles per hour, and I is the distance, in yards, from the front of one vehicle to the front of the next vehicle for vehicles in the same lane. Use this formula to solve the following problems.

87. **Traffic Capacity**   Suppose $V = 70$ miles per hour and $I = 40$ yards. Under these conditions, what is the traffic capacity of a four-lane highway?

88. **Traffic Capacity**   Suppose $V = 75$ miles per hour and $I = 42.5$ yards. What is the traffic capacity of a four-lane highway under these conditions?

Review Problems

The following problems review material we covered in Section 5.2. Reviewing these problems will help you with the next section.

Subtract as indicated.

89. Subtract $x^2 + 2x + 1$ from $4x^2 - 5x + 5$.

90. Subtract $3x^2 - 5x + 2$ from $7x^2 + 6x + 4$.

91. Subtract $10x - 20$ from $10x - 11$.

92. Subtract $-6x - 18$ from $-6x + 5$.

93. Subtract $4x^3 - 8x^2$ from $4x^3$.

94. Subtract $2x^2 + 6x$ from $2x^2$.

Extending the Concepts

95. The graphs of two rational functions are given in Figures 5 and 6. Use the graphs to find the following.

(a) $f(2)$

(b) $f(-1)$

(c) $f(0)$

(d) $g(3)$

(e) $g(6)$

(f) $g(-1)$

(g) $f(g(6))$

(h) $g(f(-2))$

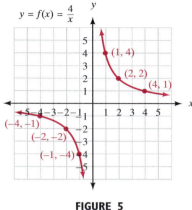

$y = f(x) = \dfrac{4}{x}$

(1, 4)
(2, 2)
(4, 1)
(−4, −1)
(−2, −2)
(−1, −4)

FIGURE 5

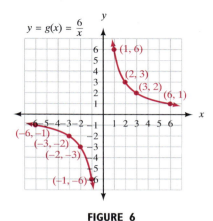

$y = g(x) = \dfrac{6}{x}$

(1, 6)
(2, 3)
(3, 2)
(6, 1)
(−6, −1)
(−3, −2)
(−2, −3)
(−1, −6)

FIGURE 6

6.2 **Division of Polynomials**

First Bank of San Luis Obispo charges $2.00 per month and $0.15 per check for a regular checking account. So, if you write x checks in one month, the total monthly cost of the checking account will be $C(x) = 2.00 + 0.15x$. From this formula we see that the more checks we write in a month, the more we pay for the account. But it is also true that the more checks we write in a month, the lower the cost per check. To find the cost per check, we use the average cost function. To find the average cost function, we divide the total cost by the number of checks written.

$$\text{Average Cost} = \overline{C}(x) = \frac{C(x)}{x} = \frac{2.00 + 0.15x}{x}$$

This last expression gives us the average cost per check for each of the x checks written. To work with this last expression, we need to know something about division with polynomials, and that is what we will cover in this section.

We begin this section by considering division of a polynomial by a monomial. This is the simplest kind of polynomial division. The rest of the section is devoted to division of a polynomial by a polynomial. This kind of division is similar to long division with whole numbers.

DIVIDING A POLYNOMIAL BY A MONOMIAL

To divide a polynomial by a monomial, we use the definition of division and apply the distributive property. The following example illustrates the procedure.

EXAMPLE 1   Divide $\dfrac{10x^5 - 15x^4 + 20x^3}{5x^2}$.

Solution

$$= (10x^5 - 15x^4 + 20x^3) \cdot \dfrac{1}{5x^2}$$

Dividing by $5x^2$ is the same as multiplying by $\dfrac{1}{5x^2}$.

$$= 10x^5 \cdot \dfrac{1}{5x^2} - 15x^4 \cdot \dfrac{1}{5x^2} + 20x^3 \cdot \dfrac{1}{5x^2}$$

Distributive property

$$= \dfrac{10x^5}{5x^2} - \dfrac{15x^4}{5x^2} + \dfrac{20x^3}{5x^2}$$

Multiplying by $\dfrac{1}{5x^2}$ is the same as dividing by $5x^2$.

$$= 2x^3 - 3x^2 + 4x$$

Divide coefficients and subtract exponents.

Notice that division of a polynomial by a monomial is accomplished by dividing each term of the polynomial by the monomial. The first two steps are usually not shown in a problem like this. They are part of Example 1 to justify distributing $5x^2$ under all three terms of the polynomial $10x^5 - 15x^4 + 20x^3$.

Here are some more examples of this kind of division.

EXAMPLES   Divide. Write all results with positive exponents.

2. $\dfrac{8x^3y^5 - 16x^2y^2 + 4x^4y^3}{-2x^2y} = \dfrac{8x^3y^5}{-2x^2y} + \dfrac{-16x^2y^2}{-2x^2y} + \dfrac{4x^4y^3}{-2x^2y}$

$$= -4xy^4 + 8y - 2x^2y^2$$

3. $\dfrac{10a^4b^2 + 8ab^3 - 12a^3b + 6ab}{4a^2b^2} = \dfrac{10a^4b^2}{4a^2b^2} + \dfrac{8ab^3}{4a^2b^2} - \dfrac{12a^3b}{4a^2b^2} + \dfrac{6ab}{4a^2b^2}$

$$= \dfrac{5a^2}{2} + \dfrac{2b}{a} - \dfrac{3a}{b} + \dfrac{3}{2ab}$$

Notice in Example 3 that the result is not a polynomial because of the last three terms. If we were to write each as a product, some of the variables would have negative exponents. For example, the second term would be

$$\dfrac{2b}{a} = 2a^{-1}b$$

The divisor in each of the preceding examples was a monomial. We now want to turn our attention to division of polynomials in which the divisor has two or more terms.

DIVIDING A POLYNOMIAL BY A POLYNOMIAL

EXAMPLE 4   Divide: $\dfrac{x^2 - 6xy - 7y^2}{x + y}$

Solution   In this case, we can factor the numerator and perform division by simply dividing out common factors, just like we did in the previous section:

$$\frac{x^2 - 6xy - 7y^2}{x + y} = \frac{\cancel{(x + y)}(x - 7y)}{\cancel{x + y}}$$

$$= x - 7y$$

The diagram in Figure 1 is an important diagram from calculus. Although it may look complicated, the point of it is simple: The slope of the line passing through the points P and Q is given by the formula

$$\text{Slope of line through } PQ = m = \frac{f(x) - f(a)}{x - a}$$

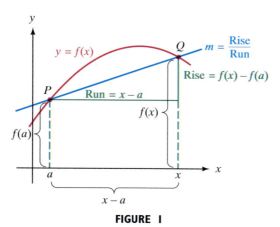

FIGURE I

If $f(x)$ is a polynomial, then the expression $\dfrac{f(x) - f(a)}{x - a}$ will be a rational expression. In our next example, we explore this situation.

EXAMPLE 5   If $f(x) = x^2 - 4$, find $\dfrac{f(x) - f(a)}{x - a}$ and simplify.

Solution   Since $f(x) = x^2 - 4$ and $f(a) = a^2 - 4$, we have

$$\frac{f(x) - f(a)}{x - a} = \frac{(x^2 - 4) - (a^2 - 4)}{x - a}$$

$$= \frac{x^2 - 4 - a^2 + 4}{x - a}$$

$$= \frac{x^2 - a^2}{x - a}$$

$$= \frac{(x + a)\cancel{(x - a)}}{\cancel{x - a}} \qquad \text{Factor and divide out common factor.}$$

$$= x + a$$

For the type of division shown in Examples 4 and 5, the denominator must be a factor of the numerator. When the denominator is not a factor of the numerator, or in the case where we can't factor the numerator, the method used in Examples 4 and 5 won't work. We need to develop a new method for these cases. Since this new method is very similar to long division with whole numbers, we will review the method of long division here.

EXAMPLE 6   Divide $25\overline{)4{,}628}$.

Solution

$$\begin{array}{r} 1 \quad \leftarrow \text{Estimate 25 into 46.} \\ 25\overline{)4{,}628} \\ 2\,5 \quad \leftarrow \text{Multiply } 1 \times 25 = 25. \\ \overline{2\,1} \quad \leftarrow \text{Subtract } 46 - 25 = 21. \end{array}$$

$$\begin{array}{r} 1 \quad\quad \\ 25\overline{)4{,}628} \\ 2\,5\downarrow \quad \\ \overline{2\,12} \quad \leftarrow \text{Bring down the 2.} \end{array}$$

These are the four basic steps in long division: estimate, multiply, subtract, and bring down the next term. To complete the problem, we simply perform the same four steps:

$$\begin{array}{r} 18 \quad \leftarrow \text{8 is the estimate.} \\ 25\overline{)4{,}628} \\ 2\,5 \\ \overline{2\,12} \\ 2\,00\downarrow \leftarrow \text{Multiply to get 200.} \\ \overline{128} \leftarrow \text{Subtract to get 12, then bring down the 8.} \end{array}$$

One more time:

$$
\begin{array}{r}
185 \leftarrow 5 \text{ is the estimate.} \\
25\overline{)4{,}628} \\
\underline{25} \\
2\,12 \\
\underline{2\,00\downarrow} \\
128 \\
\underline{125} \leftarrow \text{Multiply to get 125.} \\
3 \leftarrow \text{Subtract to get 3.}
\end{array}
$$

Since 3 is less than 25 and we have no more terms to bring down, we have our answer:

$$\frac{4{,}628}{25} = 185 + \frac{3}{25}$$

To check our answer, we multiply 185 by 25 and then add 3 to the result:

$$25(185) + 3 = 4{,}625 + 3 = 4{,}628$$

Long division with polynomials is very similar to long division with whole numbers. Both use the same four basic steps: estimate, multiply, subtract, and bring down the next term. We use long division with polynomials when the denominator has two or more terms and is not a factor of the numerator. Here is an example.

EXAMPLE 7 Divide $\dfrac{2x^2 - 7x + 9}{x - 2}$.

Solution

$$
\begin{array}{r}
2x \\
x - 2\overline{)2x^2 - 7x + 9} \leftarrow \text{Estimate } 2x^2 \div x = 2x. \\
\underline{-+} \\
\cancel{2x^2} \cancel{-} 4x \leftarrow \text{Multiply } 2x(x - 2) = 2x^2 - 4x. \\
-3x \leftarrow \text{Subtract } (2x^2 - 7x) - (2x^2 - 4x) = -3x.
\end{array}
$$

$$
\begin{array}{r}
2x \\
x - 2\overline{)2x^2 - 7x + 9} \\
\underline{-+} \\
\cancel{2x^2} \cancel{-} 4x \downarrow \\
-3x + 9 \leftarrow \text{Bring down the 9.}
\end{array}
$$

Notice we change the signs on $2x^2 - 4x$ and add in the subtraction step. Subtracting a polynomial is equivalent to adding its opposite.

We repeat the four steps.

$$
\require{enclose}
\begin{array}{r}
2x - 3 \\
x - 2 \enclose{longdiv}{2x^2 - 7x + 9}
\end{array}
\qquad \leftarrow -3 \text{ is the estimate: } -3x \div x = -3.
$$

$$
\begin{array}{r}
2x - 3 \\
\cancel{-}\; \cancel{+} \\
\cancel{2x^2}\; \cancel{4x} \\
\hline
-3x + 9 \\
+ \quad - \\
\cancel{3x}\; \cancel{6} \\
\hline
3
\end{array}
$$

$\leftarrow$ Multiply $-3(x - 2) = -3x + 6$.

$\leftarrow$ Subtract $(-3x + 9) - (-3x + 6) = 3$.

Since we have no other term to bring down, we have our answer:

$$\frac{2x^2 - 7x + 9}{x - 2} = 2x - 3 + \frac{3}{x - 2}$$

To check, we multiply $(2x - 3)(x - 2)$ to get $2x^2 - 7x + 6$; then, adding the remainder 3 to this result, we have $2x^2 - 7x + 9$.

In setting up a division problem involving two polynomials, we must remember two things: (1) Both polynomials should be in decreasing powers of the variable, and (2) neither should skip any powers from the highest power down to the constant term. If there are any missing terms, they can be filled in using a coefficient of 0.

EXAMPLE 8　Divide $2x - 4)\overline{4x^3 - 6x - 11}$.

Solution　Since the trinomial is missing a term in x^2, we can fill it in with $0x^2$:

$$4x^3 - 6x - 11 = 4x^3 + 0x^2 - 6x - 11$$

Adding $0x^2$ does not change our original problem.

$$
\begin{array}{r}
2x^2 + 4x + 5 \\
2x - 4 \enclose{longdiv}{4x^3 + 0x^2 - 6x - 11}
\end{array}
$$

Notice: Adding the $0x^2$ term gives us a column in which to write $-8x^2$.

$$
\begin{array}{r}
\cancel{4x^3}\; \cancel{8x^2} \\
\hline
8x^2 - 6x \\
8x^2\; 16x \\
\hline
10x - 11 \\
10x\; 20 \\
\hline
9
\end{array}
$$

$$\frac{4x^3 - 6x - 11}{2x - 4} = 2x^2 + 4x + 5 + \frac{9}{2x - 4}$$

To check this result, we multiply $2x - 4$ and $2x^2 + 4x + 5$:

$$
\begin{array}{r}
2x^2 + 4x + 5 \\
2x - 4 \\
\hline
4x^3 + 8x^2 + 10x \\
- 8x^2 - 16x - 20 \\
\hline
4x^3 \qquad - 6x - 20
\end{array}
$$

Adding 9 (the remainder) to this result gives us the polynomial $4x^3 - 6x - 11$. Our answer checks.

For our next example, let's do Example 4 again, but this time use long division.

EXAMPLE 9   Divide $\dfrac{x^2 - 6xy - 7y^2}{x + y}$.

Solution

$$
\begin{array}{r}
x - 7y \\
x + y \overline{)\, x^2 - 6xy - 7y^2} \\
\end{array}
$$

In this case, the remainder is 0, and we have

$$\frac{x^2 - 6xy - 7y^2}{x + y} = x - 7y$$

which is easy to check since

$$(x + y)(x - 7y) = x^2 - 6xy - 7y^2$$

EXAMPLE 10   Factor $x^3 + 9x^2 + 26x + 24$ completely if $x + 2$ is one of its factors.

Solution   Since $x + 2$ is one of the factors of the polynomial we are trying to factor, it must divide that polynomial evenly — that is, without a remainder. Therefore, we begin by dividing the polynomial by $x + 2$:

$$\begin{array}{r}
x^2 + 7x\ + 12 \\
x + 2{\overline{)}\ \ x^3 + 9x^2 + 26x + 24}
\end{array}$$

$$\begin{array}{r}
\ \ \ {\not-}\ x^3\ {\not-}\ 2x^2 \downarrow \\
+\ 7x^2 + 26x \\
{\not-}\ 7x^2\ {\not-}\ 14x \downarrow \\
+\ 12x + 24 \\
{\not-}\ 12x\ {\not-}\ 24 \\
0
\end{array}$$

Now we know that the polynomial we are trying to factor is equal to the product of $x + 2$ and $x^2 + 7x + 12$. To factor completely, we simply factor $x^2 + 7x + 12$:

$$x^3 + 9x^2 + 26x + 24 = (x + 2)(x^2 + 7x + 12)$$
$$= (x + 2)(x + 3)(x + 4)$$

Getting Ready for Class

After reading through the preceding section, respond in your own words and in complete sentences.

A. What are the four steps used in long division with polynomials?

B. What does it mean to have a remainder of 0?

C. When must long division be performed, and when can factoring be used to divide polynomials?

D. What property of real numbers is the key to dividing a polynomial by a monomial?

PROBLEM SET 6.2

Find the following quotients.

1. $\dfrac{4x^3 - 8x^2 + 6x}{2x}$

2. $\dfrac{6x^3 + 12x^2 - 9x}{3x}$

8. $\dfrac{-9x^5 + 10x^3 - 12x}{-6x^4}$

3. $\dfrac{10x^4 + 15x^3 - 20x^2}{-5x^2}$

4. $\dfrac{12x^5 - 18x^4 - 6x^3}{6x^3}$

9. $\dfrac{28a^3b^5 + 42a^4b^3}{7a^2b^2}$

10. $\dfrac{a^2b + ab^2}{ab}$

5. $\dfrac{8y^5 + 10y^3 - 6y}{4y^3}$

6. $\dfrac{6y^4 - 3y^3 + 18y^2}{9y^2}$

11. $\dfrac{10x^3y^2 - 20x^2y^3 - 30x^3y^3}{-10x^2y}$

7. $\dfrac{5x^3 - 8x^2 - 6x}{-2x^2}$

12. $\dfrac{9x^4y^4 + 18x^3y^4 - 27x^2y^4}{-9xy^3}$

Divide by factoring numerators and then dividing out common factors.

13. $\dfrac{x^2 - x - 6}{x - 3}$

14. $\dfrac{x^2 - x - 6}{x + 2}$

15. $\dfrac{2a^2 - 3a - 9}{2a + 3}$

16. $\dfrac{2a^2 + 3a - 9}{2a - 3}$

17. $\dfrac{5x^2 - 14xy - 24y^2}{x - 4y}$

18. $\dfrac{5x^2 - 26xy - 24y^2}{5x + 4y}$

19. $\dfrac{x^3 - y^3}{x - y}$

20. $\dfrac{x^3 + 8}{x + 2}$

21. $\dfrac{y^4 - 16}{y - 2}$

22. $\dfrac{y^4 - 81}{y - 3}$

23. $\dfrac{x^3 + 2x^2 - 25x - 50}{x - 5}$

24. $\dfrac{x^3 + 2x^2 - 25x - 50}{x + 5}$

25. $\dfrac{4x^3 + 12x^2 - 9x - 27}{x + 3}$

26. $\dfrac{9x^3 + 18x^2 - 4x - 8}{x + 2}$

Divide using the long division method.

27. $\dfrac{x^2 - 5x - 7}{x + 2}$

28. $\dfrac{x^2 + 4x - 8}{x - 3}$

29. $\dfrac{6x^2 + 7x - 18}{3x - 4}$

30. $\dfrac{8x^2 - 26x - 9}{2x - 7}$

31. $\dfrac{2x^3 - 3x^2 - 4x + 5}{x + 1}$

32. $\dfrac{3x^3 - 5x^2 + 2x - 1}{x - 2}$

33. $\dfrac{2y^3 - 9y^2 - 17y + 39}{2y - 3}$

34. $\dfrac{3y^3 - 19y^2 + 17y + 4}{3y - 4}$

35. $\dfrac{2x^3 - 9x^2 + 11x - 6}{2x^2 - 3x + 2}$

36. $\dfrac{6x^3 + 7x^2 - x + 3}{3x^2 - x + 1}$

37. $\dfrac{6y^3 - 8y + 5}{2y - 4}$

38. $\dfrac{9y^3 - 6y^2 + 8}{3y - 3}$

39. $\dfrac{a^4 - 2a + 5}{a - 2}$

40. $\dfrac{a^4 + a^3 - 1}{a + 2}$

41. $\dfrac{y^4 - 16}{y - 2}$

42. $\dfrac{y^4 - 81}{y - 3}$

43. $\dfrac{x^4 + x^3 - 3x^2 - x + 2}{x^2 + 3x + 2}$

44. $\dfrac{2x^4 + x^3 + 4x - 3}{2x^2 - x + 3}$

45. Factor $x^3 + 6x^2 + 11x + 6$ completely if one of its factors is $x + 3$.

46. Factor $x^3 + 10x^2 + 29x + 20$ completely if one of its factors is $x + 4$.

47. Factor $x^3 + 5x^2 - 2x - 24$ completely if one of its factors is $x + 3$.

48. Factor $x^3 + 3x^2 - 10x - 24$ completely if one of its factors is $x + 2$.

49. Problems 21 and 41 are the same problem. Are the two answers you obtained equivalent?

50. Problems 22 and 42 are the same problem. Are the two answers you obtained equivalent?

51. Find $P(-2)$ if $P(x) = x^2 - 5x - 7$. Compare it with the remainder in Problem 27.

52. Find $P(3)$ if $P(x) = x^2 + 4x - 8$. Compare it with the remainder in Problem 28.

Applying the Concepts

53. The Factor Theorem The factor theorem of algebra states that if $x - a$ is a factor of a polynomial, $P(x)$, then $P(a) = 0$. Verify the following.
(a) That $x - 2$ is a factor of $P(x) = x^3 - 3x^2 + 5x - 6$, and that $P(2) = 0$
(b) That $x - 5$ is a factor of $P(x) = x^4 - 5x^3 - x^2 + 6x - 5$, and that $P(5) = 0$

54. The Remainder Theorem The remainder theorem of algebra states that if a polynomial, $P(x)$, is divided by $x - a$, then the remainder is $P(a)$. Verify the remainder theorem by showing that when $P(x) = x^2 - x + 3$ is divided by $x - 2$ the remainder is 5, and that $P(2) = 5$.

55. Checking Account First Bank of San Luis Obispo charges $2.00 per month and $0.15 per check for a regular checking account. As we mentioned in the introduction to this section, the total monthly cost of this account is

$C(x) = 2.00 + 0.15x$. To find the average cost of each of the x checks, we divide the total cost by the number of checks written. That is,

$$\overline{C}(x) = \frac{C(x)}{x}$$

(a) Use the total cost function to fill in the following table.

x	1	5	10	15	20
$C(x)$					

(b) Find the formula for the average cost function, $\overline{C}(x)$.
(c) Use the average cost function to fill in the following table.

x	1	5	10	15	20
$\overline{C}(x)$					

(d) What happens to the average cost as more items are produced?
(e) Assume that you write at least 1 check a month, but never more than 20 checks per month, and graph both $y = C(x)$ and $y = \overline{C}(x)$ on the same set of axes.
(f) Give the domain and range of each of the functions you graphed in part (e).

56. Average Cost A company that manufactures computer diskettes uses the function $C(x) = 200 + 2x$ to represent the daily cost of producing x diskettes.
(a) Find the average cost function, $\overline{C}(x)$.
(b) Use the average cost function to fill in the following table:

x	1	5	10	20	50
$\overline{C}(x)$					

(c) What happens to the average cost as more items are produced?
(d) Graph the function $y = \overline{C}(x)$ for $x > 0$.

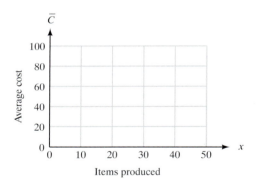

(e) What is the domain of this function?
(f) What is the range of this function?

Review Problems

The problems that follow review material we covered in Sections 1.4 and 5.1. Reviewing the problems from Section 1.4 will help you with the next section.

Divide. [1.4]

57. $\dfrac{3}{5} \div \dfrac{2}{7}$ **58.** $\dfrac{2}{7} \div \dfrac{3}{5}$

59. $\dfrac{3}{4} \div \dfrac{6}{11}$ **60.** $\dfrac{6}{8} \div \dfrac{3}{5}$

61. $\dfrac{4}{9} \div 8$ **62.** $\dfrac{3}{7} \div 6$

63. $8 \div \dfrac{1}{4}$ **64.** $12 \div \dfrac{2}{3}$

Write each expression with positive exponents, and simplify as much as possible. [5.1]

65. $\left(\dfrac{1}{3}\right)^{-2} + \left(\dfrac{1}{2}\right)^{-3}$ **66.** $\left(\dfrac{1}{2}\right)^{-3} - \left(\dfrac{1}{3}\right)^{-3}$

Simplify, and write your answers with positive exponents only. [5.1]

67. $(9x^{-4}y^9)^{-2}(3x^2y^{-1})^4$ **68.** $(4x^4y^{-3})^2(2x^{-6}y^4)^{-3}$

Extending the Concepts

Divide.

69. $\dfrac{4x^5 - x^4 - 20x^3 + 8x^2 - 15}{x^2 - 5}$

70. $\dfrac{4x^5 + 2x^4 + x^3 - 20x^2 - 10x - 5}{x^3 - 5}$

71. $\dfrac{0.5x^3 - 0.3x^2 + 0.22x + 0.06}{x + 0.2}$

72. $\dfrac{0.6x^3 - 1.1x^2 - 0.1x + 0.6}{0.3x + 0.2}$

73. $\dfrac{3x^2 + x - 10}{2x + 4}$

75. $\dfrac{2x^2 + \frac{1}{3}x + \frac{5}{3}}{3x - 1}$

74. $\dfrac{2x^2 - x - 10}{3x + 6}$

76. $\dfrac{x^2 + \frac{3}{5}x + \frac{8}{5}}{5x - 2}$

6.3 Multiplication and Division of Rational Expressions

If you have ever taken a home videotape to be duplicated, you know the amount you pay for the duplication service depends on the number of copies you have made: The more copies you have made, the lower the charge per copy. The following demand function gives the price (in dollars) per tape $p(x)$ a company charges for making x copies of a 30-minute videotape. As you can see, it is a rational function.

$$p(x) = \frac{2(x + 60)}{x + 5}$$

The graph in Figure 1 shows this function from $x = 0$ to $x = 100$. As you can see, the more copies that are made, the lower the price per copy.

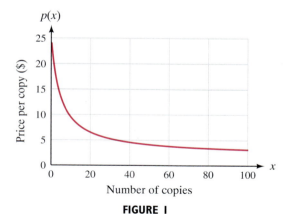

FIGURE 1

If we were interested in finding the revenue function for this situation, we would multiply the number of copies made x by the price per copy $p(x)$. This involves multiplication with a rational expression, which is one of the topics we cover in this section.

In Section 6.1 we found the process of reducing rational expressions to lowest terms to be the same process used in reducing fractions to lowest terms. The similarity also holds for the process of multiplication or division of rational expressions.

Multiplication with fractions is the simplest of the four basic operations. To multiply two fractions we simply multiply numerators and multiply denominators. That is, if a, b, c, and d are real numbers, with $b \neq 0$ and $d \neq 0$, then

$$\frac{a}{b} \cdot \frac{c}{d} = \frac{ac}{bd}$$

EXAMPLE 1 Multiply $\frac{6}{7} \cdot \frac{14}{18}$.

Solution

$$\frac{6}{7} \cdot \frac{14}{18} = \frac{6(14)}{7(18)} \qquad \text{Multiply numerators and denominators.}$$

$$= \frac{2 \cdot 3(2 \cdot 7)}{7(2 \cdot 3 \cdot 3)} \qquad \text{Factor.}$$

$$= \frac{2}{3} \qquad \text{Divide out common factors.}$$

Our next example is similar to some of the problems we worked in Chapter 5. We multiply fractions whose numerators and denominators are monomials by multiplying numerators and multiplying denominators and then reducing to lowest terms. Here is how it looks.

EXAMPLE 2 Multiply $\frac{8x^3}{27y^8} \cdot \frac{9y^3}{12x^2}$.

Solution We multiply numerators and denominators without actually carrying out the multiplication:

$$\frac{8x^3}{27y^8} \cdot \frac{9y^3}{12x^2} = \frac{8 \cdot 9x^3y^3}{27 \cdot 12x^2y^8} \qquad \begin{array}{l} \text{Multiply numerators.} \\ \text{Multiply denominators.} \end{array}$$

$$= \frac{4 \cdot 2 \cdot 9x^3y^3}{9 \cdot 3 \cdot 4 \cdot 3x^2y^8} \qquad \text{Factor coefficients.}$$

$$= \frac{2x}{9y^5} \qquad \text{Divide out common factors.}$$

The product of two rational expressions is the product of their numerators over the product of their denominators.

Once again, we should mention that the little slashes we have drawn through the factors are simply used to denote the factors we have divided out of the numerator and denominator.

EXAMPLE 3 Multiply $\dfrac{x-3}{x^2-4}\cdot\dfrac{x+2}{x^2-6x+9}$.

Solution We begin by multiplying numerators and denominators. We then factor all polynomials and divide out factors common to the numerator and denominator:

$$\frac{x-3}{x^2-4}\cdot\frac{x+2}{x^2-6x+9}=\frac{(x-3)(x+2)}{(x^2-4)(x^2-6x+9)} \qquad \text{Multiply.}$$

$$=\frac{\cancel{(x-3)}\cancel{(x+2)}}{\cancel{(x+2)}(x-2)\cancel{(x-3)}(x-3)} \qquad \text{Factor.}$$

$$=\frac{1}{(x-2)(x-3)} \qquad \begin{array}{l}\text{Divide out}\\ \text{common factors.}\end{array}$$

The first two steps can be combined to save time. We can perform the multiplication and factoring steps together.

EXAMPLE 4 Multiply $\dfrac{2y^2-4y}{2y^2-2}\cdot\dfrac{y^2-2y-3}{y^2-5y+6}$.

Solution

$$\frac{2y^2-4y}{2y^2-2}\cdot\frac{y^2-2y-3}{y^2-5y+6}=\frac{2y\cancel{(y-2)}\cancel{(y-3)}\cancel{(y+1)}}{2\cancel{(y+1)}(y-1)\cancel{(y-3)}\cancel{(y-2)}}$$

$$=\frac{y}{y-1}$$

Notice in both of the preceding examples that we did not actually multiply the polynomials as we did in Chapter 5. It would be senseless to do that since we would then have to factor each of the resulting products to reduce them to lowest terms.

The quotient of two rational expressions is the product of the first and the reciprocal of the second. That is, we find the quotient of two rational expressions the same way we find the quotient of two fractions. Here is an example that reviews division with fractions.

EXAMPLE 5 Divide $\dfrac{6}{8}\div\dfrac{3}{5}$.

Solution

$$\frac{6}{8}\div\frac{3}{5}=\frac{6}{8}\cdot\frac{5}{3} \qquad \text{Write division in terms of multiplication.}$$

$$=\frac{6(5)}{8(3)} \qquad \text{Multiply numerators and denominators.}$$

$$= \frac{2 \cdot \cancel{3}(5)}{2 \cdot 2 \cdot 2\cancel{(3)}} \qquad \text{Factor.}$$

$$= \frac{5}{4} \qquad \text{Divide out common factors.}$$

To divide one rational expression by another, we use the definition of division to multiply by the reciprocal of the expression that follows the division symbol.

EXAMPLE 6   Divide $\dfrac{8x^3}{5y^2} \div \dfrac{4x^2}{10y^6}$.

Solution   First we rewrite the problem in terms of multiplication. Then we multiply.

$$\frac{8x^3}{5y^2} \div \frac{4x^2}{10y^6} = \frac{8x^3}{5y^2} \cdot \frac{10y^6}{4x^2}$$

$$= \frac{\overset{2}{\cancel{8}} \cdot \overset{2}{\cancel{10}} x^3 y^6}{\cancel{4} \cdot \cancel{5} x^2 y^2}$$

$$= 4xy^4$$

EXAMPLE 7   Divide $\dfrac{x^2 - y^2}{x^2 - 2xy + y^2} \div \dfrac{x^3 + y^3}{x^3 - x^2y}$.

Solution   We begin by writing the problem as the product of the first and the reciprocal of the second and then proceed as in the previous two examples:

$$\frac{x^2 - y^2}{x^2 - 2xy + y^2} \div \frac{x^3 + y^3}{x^3 - x^2y} \qquad \text{Multiply by the reciprocal of the divisor.}$$

$$= \frac{x^2 - y^2}{x^2 - 2xy + y^2} \cdot \frac{x^3 - x^2y}{x^3 + y^3}$$

$$= \frac{\cancel{(x-y)}\cancel{(x+y)}(x^2)\cancel{(x-y)}}{\cancel{(x-y)}\cancel{(x-y)}\cancel{(x+y)}(x^2 - xy + y^2)} \qquad \text{Factor and multiply.}$$

$$= \frac{x^2}{x^2 - xy + y^2} \qquad \text{Divide out common factors.}$$

Here are some more examples of multiplication and division with rational expressions.

EXAMPLE 8   Perform the indicated operations.

$$\frac{a^2 - 8a + 15}{a + 4} \cdot \frac{a + 2}{a^2 - 5a + 6} \div \frac{a^2 - 3a - 10}{a^2 + 2a - 8}$$

Solution   First we rewrite the division as multiplication by the reciprocal. Then we proceed as usual.

$$\frac{a^2 - 8a + 15}{a + 4} \cdot \frac{a + 2}{a^2 - 5a + 6} \div \frac{a^2 - 3a - 10}{a^2 + 2a - 8}$$

Change division to multiplication by the reciprocal.

$$= \frac{(a^2 - 8a + 15)(a + 2)(a^2 + 2a - 8)}{(a + 4)(a^2 - 5a + 6)(a^2 - 3a - 10)}$$

$$= \frac{(a - 5)(a - 3)(a + 2)(a + 4)(a - 2)}{(a + 4)(a - 3)(a - 2)(a - 5)(a + 2)}$$ Factor.

$$= 1$$ Divide out common factors.

Our next example involves factoring by grouping. As you may have noticed, working the problems in this chapter gives you a very detailed review of factoring.

EXAMPLE 9   Multiply $\dfrac{xa + xb + ya + yb}{xa - xb - ya + yb} \cdot \dfrac{xa + xb - ya - yb}{xa - xb + ya - yb}$.

Solution   We will factor each polynomial by grouping, which takes two steps.

$$\frac{xa + xb + ya + yb}{xa - xb - ya + yb} \cdot \frac{xa + xb - ya - yb}{xa - xb + ya - yb}$$

$$= \frac{x(a + b) + y(a + b)}{x(a - b) - y(a - b)} \cdot \frac{x(a + b) - y(a + b)}{x(a - b) + y(a - b)}$$

$$= \frac{(a + b)(x + y)(a + b)(x - y)}{(a - b)(x - y)(a - b)(x + y)}$$

$$= \frac{(a + b)^2}{(a - b)^2}$$

Factor by grouping.

EXAMPLE 10   Multiply $(4x^2 - 36) \cdot \dfrac{12}{4x + 12}$.

Solution   We can think of $4x^2 - 36$ as having a denominator of 1. Thinking of it in this way allows us to proceed as we did in the previous examples.

$$(4x^2 - 36) \cdot \frac{12}{4x + 12}$$

$$= \frac{4x^2 - 36}{1} \cdot \frac{12}{4x + 12}$$ Write $4x^2 - 36$ with denominator 1.

$$= \frac{4(x - 3)(x + 3)12}{4(x + 3)}$$ Factor.

$$= 12(x - 3)$$ Divide out common factors.

EXAMPLE 11 Multiply $3(x-2)(x-1) \cdot \dfrac{5}{x^2 - 3x + 2}$.

Solution This problem is very similar to the problem in Example 10. Writing the first rational expression with a denominator of 1, we have

$$\frac{3(x-2)(x-1)}{1} \cdot \frac{5}{x^2 - 3x + 2} = \frac{3\cancel{(x-2)}\cancel{(x-1)}5}{\cancel{(x-2)}\cancel{(x-1)}}$$

$$= 3 \cdot 5$$

$$= 15$$

 Getting Ready for Class

After reading through the preceding section, respond in your own words and in complete sentences.

A. Summarize the steps used to multiply fractions.

B. What is the first step in multiplying two rational expressions?

C. Why is factoring important when multiplying and dividing rational expressions?

D. How is division with rational expressions different than multiplication of rational expressions?

PROBLEM SET 6.3

Perform the indicated operations involving fractions.

1. $\dfrac{2}{9} \cdot \dfrac{3}{4}$

2. $\dfrac{5}{6} \cdot \dfrac{7}{8}$

3. $\dfrac{3}{4} \div \dfrac{1}{3}$

4. $\dfrac{3}{8} \div \dfrac{5}{4}$

5. $\dfrac{3}{7} \cdot \dfrac{14}{24} \div \dfrac{1}{2}$

6. $\dfrac{6}{5} \cdot \dfrac{10}{36} \div \dfrac{3}{4}$

7. $\dfrac{10x^2}{5y^2} \cdot \dfrac{15y^3}{2x^4}$

8. $\dfrac{8x^3}{7y^4} \cdot \dfrac{14y^6}{16x^2}$

9. $\dfrac{11a^2b}{5ab^2} \div \dfrac{22a^3b^2}{10ab^4}$

10. $\dfrac{8ab^3}{9a^2b} \div \dfrac{16a^2b^2}{18ab^3}$

11. $\dfrac{6x^2}{5y^3} \cdot \dfrac{11z^2}{2x^2} \div \dfrac{33z^5}{10y^8}$

12. $\dfrac{4x^3}{7y^2} \cdot \dfrac{6z^5}{5x^6} \div \dfrac{24z^2}{35x^6}$

Perform the indicated operations. Be sure to write all answers in lowest terms.

13. $\dfrac{x^2 - 9}{x^2 - 4} \cdot \dfrac{x - 2}{x - 3}$

14. $\dfrac{x^2 - 16}{x^2 - 25} \cdot \dfrac{x - 5}{x - 4}$

15. $\dfrac{y^2 - 1}{y + 2} \cdot \dfrac{y^2 + 5y + 6}{y^2 + 2y - 3}$

16. $\dfrac{y - 1}{y^2 - y - 6} \cdot \dfrac{y^2 + 5y + 6}{y^2 - 1}$

17. $\dfrac{3x - 12}{x^2 - 4} \cdot \dfrac{x^2 + 6x + 8}{x - 4}$

18. $\dfrac{x^2 + 5x + 1}{4x - 4} \cdot \dfrac{x - 1}{x^2 + 5x + 1}$

Perform the indicated operations. Be sure to write all answers in lowest terms.

19. $\dfrac{5x + 2y}{25x^2 - 5xy - 6y^2} \cdot \dfrac{20x^2 - 7xy - 3y^2}{4x + y}$

20. $\dfrac{7x + 3y}{42x^2 - 17xy - 15y^2} \cdot \dfrac{12x^2 - 4xy - 5y^2}{2x + y}$

21. $\dfrac{a^2 - 5a + 6}{a^2 - 2a - 3} \div \dfrac{a - 5}{a^2 + 3a + 2}$

22. $\dfrac{a^2 + 7a + 12}{a - 5} \div \dfrac{a^2 + 9a + 18}{a^2 - 7a + 10}$

23. $\dfrac{4t^2 - 1}{6t^2 + t - 2} \div \dfrac{8t^3 + 1}{27t^3 + 8}$

24. $\dfrac{9t^2 - 1}{6t^2 + 7t - 3} \div \dfrac{27t^3 + 1}{8t^3 + 27}$

25. $\dfrac{2x^2 - 5x - 12}{4x^2 + 8x + 3} \div \dfrac{x^2 - 16}{2x^2 + 7x + 3}$

26. $\dfrac{x^2 - 2x + 1}{3x^2 + 7x - 20} \div \dfrac{x^2 + 3x - 4}{3x^2 - 2x - 5}$

27. $\dfrac{6a^2b + 2ab^2 - 20b^3}{4a^2b - 16b^3} \cdot \dfrac{10a^2 - 22ab + 4b^2}{27a^3 - 125b^3}$

28. $\dfrac{12a^2b - 3ab^2 - 42b^3}{9a^2 - 36b^2} \cdot \dfrac{6a^2 - 15ab + 6b^2}{8a^3b - b^4}$

29. $\dfrac{360x^3 - 490x}{36x^2 + 84x + 49} \cdot \dfrac{30x^2 + 83x + 56}{150x^3 + 65x^2 - 280x}$

30. $\dfrac{490x^2 - 640}{49x^2 - 112x + 64} \cdot \dfrac{28x^2 - 95x + 72}{56x^3 - 62x^2 - 144x}$

31. $\dfrac{x^5 - x^2}{5x^5 - 5x} \cdot \dfrac{10x^4 - 10x^2}{2x^4 + 2x^3 + 2x^2}$

32. $\dfrac{2x^4 - 16x}{3x^6 - 48x^2} \cdot \dfrac{6x^5 + 24x^3}{4x^4 + 8x^3 + 16x^2}$

33. $\dfrac{a^2 - 16b^2}{a^2 - 8ab + 16b^2} \cdot \dfrac{a^2 - 9ab + 20b^2}{a^2 - 7ab + 12b^2}$

$\div \dfrac{a^2 - 25b^2}{a^2 - 6ab + 9b^2}$

34. $\dfrac{a^2 - 6ab + 9b^2}{a^2 - 4b^2} \cdot \dfrac{a^2 - 5ab + 6b^2}{(a - 3b)^2}$

$\div \dfrac{a^2 - 9b^2}{a^2 - ab - 6b^2}$

35. $\dfrac{2y^2 - 7y - 15}{42y^2 - 29y - 5} \cdot \dfrac{12y^2 - 16y + 5}{7y^2 - 36y + 5}$

$\div \dfrac{4y^2 - 9}{49y^2 - 1}$

36. $\dfrac{8y^2 + 18y - 5}{21y^2 - 16y + 3} \cdot \dfrac{35y^2 - 22y + 3}{6y^2 + 17y + 5}$

$\div \dfrac{16y^2 - 1}{9y^2 - 1}$

37. $\dfrac{xy - 2x + 3y - 6}{xy + 2x - 4y - 8} \cdot \dfrac{xy + x - 4y - 4}{xy - x + 3y - 3}$

38. $\dfrac{ax + bx + 2a + 2b}{ax - 3a + bx - 3b} \cdot \dfrac{ax - bx - 3a + 3b}{ax - bx - 2a + 2b}$

39. $\dfrac{xy^2 - y^2 + 4xy - 4y}{xy - 3y + 4x - 12}$

$\div \dfrac{xy^3 + 2xy^2 + y^3 + 2y^2}{xy^2 - 3y^2 + 2xy - 6y}$

40. $\dfrac{4xb - 8b + 12x - 24}{xb^2 + 3b^2 + 3xb + 9b}$

$\div \dfrac{4xb - 8b - 8x + 16}{xb^2 + 3b^2 - 2xb - 6b}$

41. $\dfrac{2x^3 + 10x^2 - 8x - 40}{x^3 + 4x^2 - 9x - 36} \cdot \dfrac{x^2 + x - 12}{2x^2 + 14x + 20}$

42. $\dfrac{x^3 + 2x^2 - 9x - 18}{x^4 + 3x^3 - 4x^2 - 12x} \cdot \dfrac{x^3 + 5x^2 + 6x}{x^2 - x - 6}$

Use the method shown in Example 10 to find the following products.

43. $(3x - 6) \cdot \dfrac{x}{x - 2}$

44. $(4x + 8) \cdot \dfrac{x}{x + 2}$

45. $(x^2 - 25) \cdot \dfrac{2}{x - 5}$

46. $(x^2 - 49) \cdot \dfrac{5}{x + 7}$

47. $(x^2 - 3x + 2) \cdot \dfrac{3}{3x - 3}$

48. $(x^2 - 3x + 2) \cdot \dfrac{-1}{x - 2}$

49. $(y - 3)(y - 4)(y + 3) \cdot \dfrac{-1}{y^2 - 9}$

50. $(y + 1)(y + 4)(y - 1) \cdot \dfrac{3}{y^2 - 1}$

51. $a(a + 5)(a - 5) \cdot \dfrac{a + 1}{a^2 + 5a}$

52. $a(a + 3)(a - 3) \cdot \dfrac{a - 1}{a^2 - 3a}$

Applying the Concepts

At the beginning of this section we introduced the demand equation shown here. Use it to work Problems 53–56.

$$p(x) = \dfrac{2(x + 60)}{x + 5}$$

53. Demand Equation Use the demand equation to fill in the table. Then compare your results with the graph shown in Figure 1 of this section.

Number of Copies x	Price per Copy ($) $p(x)$
1	
10	
20	
50	
100	

54. Demand Equation To find the revenue for selling 50 copies of a tape, we multiply the price per tape by 50. Find the revenue for selling 50 tapes.

55. Revenue Find the revenue for selling 100 tapes.

56. Revenue Find the revenue equation $R(x)$.

57. Thinking Critically Replace each question mark with either multiplication or division to make a true statement.

(a) $\dfrac{a^2 b}{a^3} ? \dfrac{ab^2}{b^3} ? \dfrac{a^4}{b} = \dfrac{b}{a^4}$

(b) $\dfrac{a^2 b}{a^3} ? \dfrac{ab^2}{b^3} ? \dfrac{a^4}{b} = \dfrac{a^4}{b}$

(c) $\dfrac{a^2 b}{a^3} ? \dfrac{ab^2}{b^3} ? \dfrac{a^4}{b} = a^2 b$

58. Area The following box has a square top. The front face of the box has an area of $A = x^3 - 2x^2 - 2x - 3$. The height of the box is $h = x^2 + x + 1$. Find a formula for the area of the top square in terms of x.

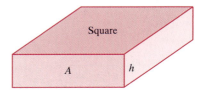

Square

A h

Surface Area of a Cylinder The surface area of the cylinder in the figure is defined as the area of its two circular bases and the lateral, or side, area. The surface area may be found by the formula

$$A = 2\pi r^2 + 2\pi rh$$

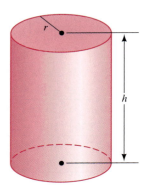

59. If the surface area is 6π, find an expression for $(r^2 + rh)$.

60. If the surface area is 6π, and $h = 2$, find r.

Review Problems

The following problems review material we covered in Sections 5.3 and 1.3.

Multiply. [5.3]

61. $2x^2(5x^3 + 4x - 3)$ **62.** $3x^3(7x^2 - 4x - 8)$

63. $(3a - 1)(4a + 5)$ **64.** $(6a - 3)(2a + 1)$

65. $(3x + 7)(4y - 2)$ **66.** $(x + 2a)(2 - 3b)$

67. $(3 - t^2)^2$ **68.** $(2 - t^3)^2$

69. $3(x + 1)(x + 2)(x + 3)$

70. $4(x - 1)(x - 2)(x - 3)$

Combine. [1.3]

71. $\dfrac{3}{14} + \dfrac{7}{30}$ **72.** $\dfrac{5}{12} + \dfrac{7}{18}$

73. $\dfrac{4}{15} - \dfrac{5}{21}$ **74.** $\dfrac{3}{14} - \dfrac{5}{22}$

This section is concerned with addition and subtraction of rational expressions. In the first part of this section we will look at addition of expressions that have the same denominator. In the second part of this section we will look at addition of expressions that have different denominators.

ADDITION AND SUBTRACTION WITH THE SAME DENOMINATOR

To add two expressions that have the same denominator, we simply add numerators and put the sum over the common denominator. Since the process we use to add and subtract rational expressions is the same process used to add and subtract fractions, we will begin with an example involving fractions.

EXAMPLE 1 Add $\frac{4}{9} + \frac{2}{9}$.

Solution We add fractions with the same denominator by using the distributive property. Here is a detailed look at the steps involved.

$$\frac{4}{9} + \frac{2}{9} = 4\left(\frac{1}{9}\right) + 2\left(\frac{1}{9}\right)$$
$$= (4 + 2)\left(\frac{1}{9}\right) \qquad \text{Distributive property}$$
$$= 6\left(\frac{1}{9}\right)$$
$$= \frac{6}{9}$$
$$= \frac{2}{3} \qquad \text{Divide numerator and denominator}$$
$$\text{by common factor 3.}$$

Note that the important thing about the fractions in this example is that they each have a denominator of 9. If they did not have the same denominator, we could not have written them as two terms with a factor of $\frac{1}{9}$ in common. Without the $\frac{1}{9}$ common to each term, we couldn't apply the distributive property. And without the distributive property, we would not have been able to add the two fractions.

In the examples that follow, we will not show all the steps we showed in Example 1. The steps are shown in Example 1 so that you will see why both fractions must have the same denominator before we can add them. In practice we simply add numerators and place the result over the common denominator.

We add and subtract rational expressions with the same denominator by combining numerators and writing the result over the common denominator. Then we reduce the result to lowest terms, if possible. Example 2 shows this process in detail. If you see the similarities between operations on rational numbers and operations on rational expressions, this chapter will look like an extension of rational numbers rather than a completely new set of topics.

EXAMPLE 2   Add $\dfrac{x}{x^2 - 1} + \dfrac{1}{x^2 - 1}$.

Solution   Since the denominators are the same, we simply add numerators:

$$\frac{x}{x^2 - 1} + \frac{1}{x^2 - 1} = \frac{x + 1}{x^2 - 1} \qquad \text{Add numerators.}$$

$$= \frac{x + 1}{(x - 1)(x + 1)} \qquad \text{Factor denominator.}$$

$$= \frac{1}{x - 1} \qquad \text{Divide out common factor } x + 1.$$

Our next example involves subtraction of rational expressions. Pay careful attention to what happens to the signs of the terms in the numerator of the second expression when we subtract it from the first expression.

EXAMPLE 3   Subtract $\dfrac{2x - 5}{x - 2} - \dfrac{x - 3}{x - 2}$.

Solution   Since each expression has the same denominator, we simply subtract the numerator in the second expression from the numerator in the first expression and write the difference over the common denominator $x - 2$. We must be careful, however, that we subtract both terms in the second numerator. To ensure that we do, we will enclose that numerator in parentheses.

$$\frac{2x - 5}{x - 2} - \frac{x - 3}{x - 2} = \frac{2x - 5 - (x - 3)}{x - 2} \qquad \text{Subtract numerators.}$$

$$= \frac{2x - 5 - x + 3}{x - 2} \qquad \text{Remove parentheses.}$$

$$= \frac{x - 2}{x - 2} \qquad \begin{array}{l}\text{Combine similar terms} \\ \text{in the numerator.}\end{array}$$

$$= 1 \qquad \text{Reduce (or divide).}$$

Note the $+3$ in the numerator of the second step. It is a very common mistake to write this as -3, by forgetting to subtract both terms in the numerator of the second expression. Whenever the expression we are subtracting has two or more terms in its numerator, we have to watch for this mistake.

Next we consider addition and subtraction of fractions and rational expressions that have different denominators.

ADDITION AND SUBTRACTION WITH DIFFERENT DENOMINATORS

Before we look at an example of addition of fractions with different denominators, we need to review the definition for the least common denominator.

> **DEFINITION**   The **least common denominator**, abbreviated LCD, for a set of denominators is the smallest expression that is divisible by each of the denominators.

The first step in combining two fractions is to find the LCD. Once we have the common denominator, we rewrite each fraction as an equivalent fraction with the common denominator. After that, we simply add or subtract as we did in our first three examples.

Example 4 is a review of the step-by-step procedure used to add two fractions with different denominators.

EXAMPLE   4   Add $\frac{3}{14} + \frac{7}{30}$.

Solution

   Step 1:   Find the LCD.

To do this, we first factor both denominators into prime factors.

$$\text{Factor 14:} \qquad 14 = 2 \cdot 7$$

$$\text{Factor 30:} \qquad 30 = 2 \cdot 3 \cdot 5$$

Since the LCD must be divisible by 14, it must have factors of $2 \cdot 7$. It must also be divisible by 30 and, therefore, have factors of $2 \cdot 3 \cdot 5$. We do not need to repeat the 2 that appears in both the factors of 14 and those of 30. Therefore,

$$\text{LCD} = 2 \cdot 3 \cdot 5 \cdot 7 = 210$$

   Step 2:   Change to equivalent fractions.

Since we want each fraction to have a denominator of 210 and at the same time keep its original value, we multiply each by 1 in the appropriate form.
   Change $\frac{3}{14}$ to a fraction with denominator 210:

$$\frac{3}{14} \cdot \frac{\mathbf{15}}{\mathbf{15}} = \frac{45}{210}$$

Change $\frac{7}{30}$ to a fraction with denominator 210:

$$\frac{7}{30} \cdot \frac{\mathbf{7}}{\mathbf{7}} = \frac{49}{210}$$

   Step 3:   Add numerators of equivalent fractions found in step 2:

$$\frac{45}{210} + \frac{49}{210} = \frac{94}{210}$$

Step 4:   Reduce to lowest terms if necessary:

$$\frac{94}{210} = \frac{47}{105}$$

The main idea in adding fractions is to write each fraction again with the LCD for a denominator. In doing so, we must be sure not to change the value of either of the original fractions.

EXAMPLE 5   Add $\dfrac{-2}{x^2 - 2x - 3} + \dfrac{3}{x^2 - 9}$.

Solution

Step 1:   Factor each denominator and build the LCD from the factors:

$$\left.\begin{array}{l} x^2 - 2x - 3 = (x - 3)(x + 1) \\ x^2 - 9 = (x - 3)(x + 3) \end{array}\right\} \text{LCD} = (x - 3)(x + 3)(x + 1)$$

Step 2:   Change each rational expression to an equivalent expression that has the LCD for a denominator:

$$\frac{-2}{x^2 - 2x - 3} = \frac{-2}{(x - 3)(x + 1)} \cdot \frac{(x + 3)}{(x + 3)} = \frac{-2x - 6}{(x - 3)(x + 3)(x + 1)}$$

$$\frac{3}{x^2 - 9} = \frac{3}{(x - 3)(x + 3)} \cdot \frac{(x + 1)}{(x + 1)} = \frac{3x + 3}{(x - 3)(x + 3)(x + 1)}$$

Step 3:   Add numerators of the rational expressions found in step 2:

$$\frac{-2x - 6}{(x - 3)(x + 3)(x + 1)} + \frac{3x + 3}{(x - 3)(x + 3)(x + 1)} = \frac{x - 3}{(x - 3)(x + 3)(x + 1)}$$

Step 4:   Reduce to lowest terms by dividing out the common factor $x - 3$:

$$= \frac{1}{(x + 3)(x + 1)}$$

EXAMPLE 6   Subtract $\dfrac{x + 4}{2x + 10} - \dfrac{5}{x^2 - 25}$.

Solution   We begin by factoring each denominator:

$$\frac{x + 4}{2x + 10} - \frac{5}{x^2 - 25} = \frac{x + 4}{2(x + 5)} - \frac{5}{(x + 5)(x - 5)}$$

The LCD is $2(x + 5)(x - 5)$. Completing the problem we have

$$= \frac{x + 4}{2(x + 5)} \cdot \frac{(x - 5)}{(x - 5)} - \frac{5}{(x + 5)(x - 5)} \cdot \frac{2}{2}$$

$$= \frac{x^2 - x - 20}{2(x + 5)(x - 5)} - \frac{10}{2(x + 5)(x - 5)}$$

$$= \frac{x^2 - x - 30}{2(x + 5)(x - 5)}$$

To see if this expression will reduce, we factor the numerator into $(x - 6)(x + 5)$.

$$= \frac{(x - 6)\cancel{(x + 5)}}{2\cancel{(x + 5)}(x - 5)}$$

$$= \frac{x - 6}{2(x - 5)}$$

EXAMPLE 7   Subtract $\dfrac{2x - 2}{x^2 + 4x + 3} - \dfrac{x - 1}{x^2 + 5x + 6}$.

Solution   We factor each denominator and build the LCD from those factors:

$$\frac{2x - 2}{x^2 + 4x + 3} - \frac{x - 1}{x^2 + 5x + 6}$$

$$= \frac{2x - 2}{(x + 3)(x + 1)} - \frac{x - 1}{(x + 3)(x + 2)}$$

$$= \frac{2x - 2}{(x + 3)(x + 1)} \cdot \frac{(x + 2)}{(x + 2)} - \frac{x - 1}{(x + 3)(x + 2)} \cdot \frac{(x + 1)}{(x + 1)} \qquad \begin{array}{l} \text{The LCD is} \\ (x + 1)(x + 2) \\ (x + 3). \end{array}$$

$$= \frac{2x^2 + 2x - 4}{(x + 1)(x + 2)(x + 3)} - \frac{x^2 - 1}{(x + 1)(x + 2)(x + 3)} \qquad \begin{array}{l} \text{Multiply out} \\ \text{each numerator.} \end{array}$$

$$= \frac{(2x^2 + 2x - 4) - (x^2 - 1)}{(x + 1)(x + 2)(x + 3)} \left.\begin{array}{l} \\ \\ \\ \\ \\ \\ \end{array}\right\} \quad \begin{array}{l} \text{Subtract} \\ \text{numerators.} \end{array}$$

$$= \frac{x^2 + 2x - 3}{(x + 1)(x + 2)(x + 3)}$$

$$= \frac{\cancel{(x + 3)}(x - 1)}{(x + 1)(x + 2)\cancel{(x + 3)}} \qquad \begin{array}{l} \text{Factor numer-} \\ \text{ator to see if} \\ \text{we can reduce.} \end{array}$$

$$= \frac{x - 1}{(x + 1)(x + 2)} \qquad \text{Reduce.}$$

EXAMPLE 8   Add $\dfrac{x^2}{x - 7} + \dfrac{6x + 7}{7 - x}$.

Solution   In Section 6.1 we were able to reverse the terms in a factor such as $7 - x$ by factoring -1 from each term. In a problem like this, the same result can be obtained by multiplying the numerator and denominator by -1:

$$\frac{x^2}{x - 7} + \frac{6x + 7}{7 - x} \cdot \frac{-1}{-1} = \frac{x^2}{x - 7} + \frac{-6x - 7}{x - 7}$$

$$= \frac{x^2 - 6x - 7}{x - 7} \qquad \text{Add numerators.}$$

$$= \frac{\cancel{(x - 7)}(x + 1)}{\cancel{(x - 7)}} \qquad \text{Factor numerator.}$$

$$= x + 1 \qquad \text{Divide out } x - 7.$$

For our next example we will look at a problem in which we combine a whole number and a rational expression.

EXAMPLE 9 Subtract $2 - \dfrac{9}{3x + 1}$.

Solution To subtract these two expressions, we think of 2 as a rational expression with a denominator of 1.

$$2 - \frac{9}{3x + 1} = \frac{2}{1} - \frac{9}{3x + 1}$$

The LCD is $3x + 1$. Multiplying the numerator and denominator of the first expression by $3x + 1$ gives us a rational expression equivalent to 2, but with a denominator of $3x + 1$.

$$\frac{2}{1} \cdot \frac{(3x + 1)}{(3x + 1)} - \frac{9}{3x + 1} = \frac{6x + 2 - 9}{3x + 1}$$

$$= \frac{6x - 7}{3x + 1}$$

The numerator and denominator of this last expression do not have any factors in common other than 1, so the expression is in lowest terms.

EXAMPLE 10 Write an expression for the sum of a number and twice its reciprocal. Then, simplify that expression.

Solution If x is the number, then its reciprocal is $\dfrac{1}{x}$. Twice its reciprocal is $\dfrac{2}{x}$. The sum of the number and twice its reciprocal is

$$x + \frac{2}{x}$$

To combine these two expressions, we think of the first term x as a rational expression with a denominator of 1. The least common denominator is x:

$$x + \frac{2}{x} = \frac{x}{1} + \frac{2}{x}$$

$$= \frac{x}{1} \cdot \frac{x}{x} + \frac{2}{x}$$

$$= \frac{x^2 + 2}{x}$$

Getting Ready for Class

After reading through the preceding section, respond in your own words and in complete sentences.

A. Briefly describe how you would add two rational expressions that have the same denominator.

B. Why is factoring important in finding a least common denominator?

C. What is the last step in adding or subtracting two rational expressions?

D. Explain how you would change the fraction $\dfrac{5}{x-3}$ to an equivalent fraction with denominator $x^2 - 9$.

PROBLEM SET 6.4

Combine the following fractions.

1. $\dfrac{3}{4} + \dfrac{1}{2}$ **2.** $\dfrac{5}{6} + \dfrac{1}{3}$ **3.** $\dfrac{2}{5} - \dfrac{1}{15}$

4. $\dfrac{5}{8} - \dfrac{1}{4}$ **5.** $\dfrac{5}{6} + \dfrac{7}{8}$ **6.** $\dfrac{3}{4} + \dfrac{2}{3}$

7. $\dfrac{9}{48} - \dfrac{3}{54}$ **8.** $\dfrac{6}{28} - \dfrac{5}{42}$

9. $\dfrac{3}{4} - \dfrac{1}{8} + \dfrac{2}{3}$ **10.** $\dfrac{1}{3} - \dfrac{5}{6} + \dfrac{5}{12}$

Combine the following rational expressions. Reduce all answers to lowest terms.

11. $\dfrac{x}{x+3} + \dfrac{3}{x+3}$ **12.** $\dfrac{5x}{5x+2} + \dfrac{2}{5x+2}$

13. $\dfrac{4}{y-4} - \dfrac{y}{y-4}$ **14.** $\dfrac{8}{y+8} + \dfrac{y}{y+8}$

15. $\dfrac{x}{x^2-y^2} - \dfrac{y}{x^2-y^2}$ **16.** $\dfrac{x}{x^2-y^2} + \dfrac{y}{x^2-y^2}$

17. $\dfrac{2x-3}{x-2} - \dfrac{x-1}{x-2}$ **18.** $\dfrac{2x-4}{x+2} - \dfrac{x-6}{x+2}$

19. $\dfrac{1}{a} + \dfrac{2}{a^2} - \dfrac{3}{a^3}$ **20.** $\dfrac{3}{a} + \dfrac{2}{a^2} - \dfrac{1}{a^3}$

21. $\dfrac{7x-2}{2x+1} - \dfrac{5x-3}{2x+1}$ **22.** $\dfrac{7x-1}{3x+2} - \dfrac{4x-3}{3x+2}$

Combine the following rational expressions. Reduce all answers to lowest terms.

23. $\dfrac{2}{t^2} - \dfrac{3}{2t}$ **24.** $\dfrac{5}{3t} - \dfrac{4}{t^2}$

25. $\dfrac{3x+1}{2x-6} - \dfrac{x+2}{x-3}$ **26.** $\dfrac{x+1}{x-2} - \dfrac{4x+7}{5x-10}$

27. $\dfrac{6x+5}{5x-25} - \dfrac{x+2}{x-5}$ **28.** $\dfrac{4x+2}{3x+12} - \dfrac{x-2}{x+4}$

29. $\dfrac{x+1}{2x-2} - \dfrac{2}{x^2-1}$ **30.** $\dfrac{x+7}{2x+12} + \dfrac{6}{x^2-36}$

31. $\dfrac{1}{a-b} - \dfrac{3ab}{a^3-b^3}$ **32.** $\dfrac{1}{a+b} + \dfrac{3ab}{a^3+b^3}$

33. $\dfrac{1}{2y-3} - \dfrac{18y}{8y^3-27}$ **34.** $\dfrac{1}{3y-2} - \dfrac{18y}{27y^3-8}$

35. $\dfrac{x}{x^2-5x+6} - \dfrac{3}{3-x}$

36. $\dfrac{x}{x^2+4x+4} - \dfrac{2}{2+x}$

37. $\dfrac{2}{4t-5} + \dfrac{9}{8t^2-38t+35}$

38. $\dfrac{3}{2t-5} + \dfrac{21}{8t^2-14t-15}$

39. $\dfrac{1}{a^2-5a+6} + \dfrac{3}{a^2-a-2}$

40. $\dfrac{-3}{a^2+a-2} + \dfrac{5}{a^2-a-6}$

41. $\dfrac{1}{8x^3-1} - \dfrac{1}{4x^2-1}$

42. $\dfrac{1}{27x^3 - 1} - \dfrac{1}{9x^2 - 1}$

43. $\dfrac{4}{4x^2 - 9} - \dfrac{6}{8x^2 - 6x - 9}$

44. $\dfrac{9}{9x^2 + 6x - 8} - \dfrac{6}{9x^2 - 4}$

45. $\dfrac{4a}{a^2 + 6a + 5} - \dfrac{3a}{a^2 + 5a + 4}$

46. $\dfrac{3a}{a^2 + 7a + 10} - \dfrac{2a}{a^2 + 6a + 8}$

47. $\dfrac{2x - 1}{x^2 + x - 6} - \dfrac{x + 2}{x^2 + 5x + 6}$

48. $\dfrac{4x + 1}{x^2 + 5x + 4} - \dfrac{x + 3}{x^2 + 4x + 3}$

49. $\dfrac{2x - 8}{3x^2 + 8x + 4} + \dfrac{x + 3}{3x^2 + 5x + 2}$

50. $\dfrac{5x + 3}{2x^2 + 5x + 3} - \dfrac{3x + 9}{2x^2 + 7x + 6}$

51. $\dfrac{2}{x^2 + 5x + 6} - \dfrac{4}{x^2 + 4x + 3} + \dfrac{3}{x^2 + 3x + 2}$

52. $\dfrac{-5}{x^2 + 3x - 4} + \dfrac{5}{x^2 + 2x - 3} + \dfrac{1}{x^2 + 7x + 12}$

53. $\dfrac{2x + 8}{x^2 + 5x + 6} - \dfrac{x + 5}{x^2 + 4x + 3} - \dfrac{x - 1}{x^2 + 3x + 2}$

54. $\dfrac{2x + 11}{x^2 + 9x + 20} - \dfrac{x + 1}{x^2 + 7x + 12} - \dfrac{x + 6}{x^2 + 8x + 15}$

55. $2 + \dfrac{3}{2x + 1}$ **56.** $3 - \dfrac{2}{2x + 3}$

57. $5 + \dfrac{2}{4 - t}$ **58.** $7 + \dfrac{3}{5 - t}$

59. $x - \dfrac{4}{2x + 3}$ **60.** $x - \dfrac{5}{3x + 4} + 1$

61. $\dfrac{x}{x + 2} + \dfrac{1}{2x + 4} - \dfrac{3}{x^2 + 2x}$

62. $\dfrac{x}{x + 3} + \dfrac{7}{3x + 9} - \dfrac{2}{x^2 + 3x}$

63. $\dfrac{1}{x} + \dfrac{x}{2x + 4} - \dfrac{2}{x^2 + 2x}$

64. $\dfrac{1}{x} + \dfrac{x}{3x + 9} - \dfrac{3}{x^2 + 3x}$

Applying the Concepts

65. Optometry The formula

$$P = \dfrac{1}{a} + \dfrac{1}{b}$$

is used by optometrists to help determine how strong to make the lenses for a pair of eyeglasses. If a is 10 and b is 0.2, find the corresponding value of P.

66. Optometry Show that the formula in Problem 65 can be written

$$P = \dfrac{a + b}{ab}$$

Then let $a = 10$ and $b = 0.2$ in this new form of the formula to find P.

67. Comparing Expressions Show that the expressions $(x + y)^{-1}$ and $x^{-1} + y^{-1}$ are not equal when $x = 3$ and $y = 4$.

68. Comparing Expressions Show that the expressions $(x + y)^{-1}$ and $x^{-1} + y^{-1}$ are not equal. (Begin by writing each with positive exponents only.)

69. Elliptical Orbits Consider two objects, A and B, that move in the same direction along an elliptical path at constant but different velocities.

It can be shown that the time, T, it takes for the two objects to meet can be found from the formula

$$\dfrac{1}{T} = \dfrac{1}{t_A} - \dfrac{1}{t_B}$$

where t_A = time required for object A to orbit, and t_B = time required for object B to orbit.
(a) If $t_A = 24$ months and $t_B = 30$ months, when will these two objects meet?
(b) If $t_A = t_B$ what can one conclude?

70. Average Velocity If a car travels at a constant velocity, v_1, for 10 miles and then at a constant, but different velocity, v_2, for the next 10 miles, it can be shown that the car's average velocity, v_{avg}, over these 20 miles satisfies the equation

$$\frac{2}{v_{avg}} = \frac{1}{v_1} + \frac{1}{v_2}$$

Find the average velocity of a car that travels a constant 45 miles per hour for 10 miles and then increases to a constant 60 miles per hour for the next 10 miles.

71. Number Problem   Write an expression for the sum of a number and 4 times its reciprocal. Then, simplify that expression.

72. Number Problem   Write an expression for the sum of a number and 3 times its reciprocal. Then, simplify that expression.

73. Number Problem   Write an expression for the sum of the reciprocals of two consecutive integers. Then, simplify that expression.

74. Number Problem   Write an expression for the sum of the reciprocals of two consecutive even integers. Then, simplify that expression.

Review Problems

The problems that follow review material we covered in Section 5.1 on scientific notation.

Write each number in scientific notation.

75. 54,000   **76.** 768,000

77. 0.00034   **78.** 0.0359

Write each number in expanded form.

79. 6.44×10^3

80. 2.5×10^2

81. 6.44×10^{-3}

82. 2.5×10^{-2}

Simplify each expression as much as possible. Write all answers in scientific notation.

83. $(3 \times 10^8)(4 \times 10^{-5})$

84. $\dfrac{8 \times 10^{-3}}{4 \times 10^{-6}}$

Extending the Concepts

Simplify.

85. $\left(1 - \dfrac{1}{x}\right)\left(1 - \dfrac{1}{x+1}\right)\left(1 - \dfrac{1}{x+2}\right)\left(1 - \dfrac{1}{x+3}\right)$

86. $\left(1 + \dfrac{1}{x}\right)\left(1 + \dfrac{1}{x+1}\right)\left(1 + \dfrac{1}{x+2}\right)\left(1 + \dfrac{1}{x+3}\right)$

6.5 Complex Fractions

The quotient of two fractions or two rational expressions is called a *complex fraction*. This section is concerned with the simplification of complex fractions.

EXAMPLE 1   Simplify $\dfrac{\frac{3}{4}}{\frac{5}{8}}$.

Solution   There are generally two methods that can be used to simplify complex fractions.

Method 1   We can multiply the numerator and denominator of the complex fraction by the LCD for both of the fractions, which in this case is 8.

$$\frac{\frac{3}{4}}{\frac{5}{8}} = \frac{\frac{3}{4} \cdot \mathbf{8}}{\frac{5}{8} \cdot \mathbf{8}} = \frac{6}{5}$$

Method 2 Instead of dividing by $\frac{5}{8}$ we can multiply by $\frac{8}{5}$.

$$\frac{\dfrac{3}{4}}{\dfrac{5}{8}} = \frac{3}{4} \cdot \frac{8}{5} = \frac{24}{20} = \frac{6}{5}$$

Here are some examples of complex fractions involving rational expressions. Most can be solved using either of the two methods shown in Example 1.

EXAMPLE 2 Simplify $\dfrac{\dfrac{1}{x} + \dfrac{1}{y}}{\dfrac{1}{x} - \dfrac{1}{y}}$.

Solution This problem is most easily solved using Method 1. We begin by multiplying both the numerator and denominator by the quantity xy, which is the LCD for all the fractions:

$$\frac{\dfrac{1}{x} + \dfrac{1}{y}}{\dfrac{1}{x} - \dfrac{1}{y}} = \frac{\left(\dfrac{1}{x} + \dfrac{1}{y}\right) \cdot xy}{\left(\dfrac{1}{x} - \dfrac{1}{y}\right) \cdot xy}$$

$$= \frac{\dfrac{1}{x}(xy) + \dfrac{1}{y}(xy)}{\dfrac{1}{x}(xy) - \dfrac{1}{y}(xy)} \qquad \text{Apply the distributive property to distribute } xy \text{ over both terms in the numerator and denominator.}$$

$$= \frac{y + x}{y - x}$$

EXAMPLE 3 Simplify $\dfrac{\dfrac{x - 2}{x^2 - 9}}{\dfrac{x^2 - 4}{x + 3}}$.

Solution Applying Method 2, we have

$$\frac{\dfrac{x - 2}{x^2 - 9}}{\dfrac{x^2 - 4}{x + 3}} = \frac{x - 2}{x^2 - 9} \cdot \frac{x + 3}{x^2 - 4}$$

$$= \frac{\cancel{(x-2)}(x+3)}{\cancel{(x+3)}(x-3)(x+2)\cancel{(x-2)}}$$

$$= \frac{1}{(x-3)(x+2)}$$

EXAMPLE 4 Simplify $\dfrac{1 - \dfrac{4}{x^2}}{1 - \dfrac{1}{x} - \dfrac{6}{x^2}}$.

Solution   The simplest way to simplify this complex fraction is to multiply the numerator and denominator by the LCD, x^2:

$$\frac{1 - \dfrac{4}{x^2}}{1 - \dfrac{1}{x} - \dfrac{6}{x^2}} = \frac{x^2\left(1 - \dfrac{4}{x^2}\right)}{x^2\left(1 - \dfrac{1}{x} - \dfrac{6}{x^2}\right)} \qquad \text{Multiply numerator and denominator by } x^2.$$

$$= \frac{x^2 \cdot 1 - x^2 \cdot \dfrac{4}{x^2}}{x^2 \cdot 1 - x^2 \cdot \dfrac{1}{x} - x^2 \cdot \dfrac{6}{x^2}} \qquad \text{Distributive property}$$

$$= \frac{x^2 - 4}{x^2 - x - 6} \qquad \text{Simplify.}$$

$$= \frac{(x - 2)\cancel{(x+2)}}{(x - 3)\cancel{(x+2)}} \qquad \text{Factor.}$$

$$= \frac{x - 2}{x - 3} \qquad \text{Reduce.}$$

EXAMPLE 5 Simplify $2 - \dfrac{3}{x + \frac{1}{3}}$.

Solution   First we simplify the expression that follows the subtraction sign.

$$2 - \frac{3}{x + \dfrac{1}{3}} = 2 - \frac{3 \cdot 3}{3\left(x + \dfrac{1}{3}\right)} = 2 - \frac{9}{3x + 1}$$

Now we subtract by rewriting the first term, 2, with the LCD, $3x + 1$.

$$2 - \frac{9}{3x + 1} = \frac{2}{1} \cdot \frac{3x + 1}{3x + 1} - \frac{9}{3x + 1}$$

$$= \frac{6x + 2 - 9}{3x + 1} = \frac{6x - 7}{3x + 1}$$

Getting Ready for Class

After reading through the preceding section, respond in your own words and in complete sentences.

A. What is a complex fraction?

B. Explain how a least common denominator can be used to simplify a complex fraction.

C. Explain how some complex fractions can be converted to division problems. When is it more efficient to convert a complex fraction to a division problem of rational expressions?

D. Which method of simplifying complex fractions do you prefer? Why?

PROBLEM SET 6.5

Simplify each of the following as much as possible.

1. $\dfrac{\dfrac{3}{4}}{\dfrac{2}{3}}$

2. $\dfrac{\dfrac{5}{9}}{\dfrac{7}{12}}$

3. $\dfrac{\dfrac{1}{3} - \dfrac{1}{4}}{\dfrac{1}{2} + \dfrac{1}{8}}$

4. $\dfrac{\dfrac{1}{6} - \dfrac{1}{3}}{\dfrac{1}{4} - \dfrac{1}{8}}$

5. $\dfrac{3 + \dfrac{2}{5}}{1 - \dfrac{3}{7}}$

6. $\dfrac{2 + \dfrac{5}{6}}{1 - \dfrac{7}{8}}$

7. $\dfrac{\dfrac{1}{x}}{1 + \dfrac{1}{x}}$

8. $\dfrac{1 - \dfrac{1}{x}}{\dfrac{1}{x}}$

9. $\dfrac{1 + \dfrac{1}{a}}{1 - \dfrac{1}{a}}$

10. $\dfrac{1 - \dfrac{2}{a}}{1 - \dfrac{3}{a}}$

11. $\dfrac{\dfrac{1}{x} - \dfrac{1}{y}}{\dfrac{1}{x} + \dfrac{1}{y}}$

12. $\dfrac{\dfrac{1}{x} + \dfrac{2}{y}}{\dfrac{2}{x} + \dfrac{1}{y}}$

13. $\dfrac{\dfrac{x - 5}{x^2 - 4}}{\dfrac{x^2 - 25}{x + 2}}$

14. $\dfrac{\dfrac{3x + 1}{x^2 - 49}}{\dfrac{9x^2 - 1}{x - 7}}$

15. $\dfrac{\dfrac{4a}{2a^3 + 2}}{\dfrac{8a}{4a + 4}}$

16. $\dfrac{\dfrac{2a}{3a^3 - 3}}{\dfrac{4a}{6a - 6}}$

17. $\dfrac{1 - \dfrac{9}{x^2}}{1 - \dfrac{1}{x} - \dfrac{6}{x^2}}$

18. $\dfrac{4 - \dfrac{1}{x^2}}{4 + \dfrac{4}{x} + \dfrac{1}{x^2}}$

19. $\dfrac{2 + \dfrac{5}{a} - \dfrac{3}{a^2}}{2 - \dfrac{5}{a} + \dfrac{2}{a^2}}$

20. $\dfrac{3 + \dfrac{5}{a} - \dfrac{2}{a^2}}{3 - \dfrac{10}{a} + \dfrac{3}{a^2}}$

21. $\dfrac{2 + \dfrac{3}{x} - \dfrac{18}{x^2} - \dfrac{27}{x^3}}{2 + \dfrac{9}{x} + \dfrac{9}{x^2}}$

22. $\dfrac{3 + \dfrac{5}{x} - \dfrac{12}{x^2} - \dfrac{20}{x^3}}{3 + \dfrac{11}{x} + \dfrac{10}{x^2}}$

23. $\dfrac{1 + \dfrac{1}{x + 3}}{1 - \dfrac{1}{x + 3}}$

24. $\dfrac{1 + \dfrac{1}{x - 2}}{1 - \dfrac{1}{x - 2}}$

25. $\dfrac{1 + \dfrac{1}{x + 3}}{1 + \dfrac{7}{x - 3}}$

26. $\dfrac{1 + \dfrac{1}{x - 2}}{1 - \dfrac{3}{x + 2}}$

27. $\dfrac{1 - \dfrac{1}{a + 1}}{1 + \dfrac{1}{a - 1}}$

28. $\dfrac{\dfrac{1}{a - 1} + 1}{\dfrac{1}{a + 1} - 1}$

29. $\dfrac{\dfrac{1}{x + 3} + \dfrac{1}{x - 3}}{\dfrac{1}{x + 3} - \dfrac{1}{x - 3}}$

30. $\dfrac{\dfrac{1}{x + a} + \dfrac{1}{x - a}}{\dfrac{1}{x + a} - \dfrac{1}{x - a}}$

31. $\dfrac{\dfrac{y + 1}{y - 1} + \dfrac{y - 1}{y + 1}}{\dfrac{y + 1}{y - 1} - \dfrac{y - 1}{y + 1}}$

32. $\dfrac{\dfrac{y - 1}{y + 1} - \dfrac{y + 1}{y - 1}}{\dfrac{y - 1}{y + 1} + \dfrac{y + 1}{y - 1}}$

33. $1 - \dfrac{x}{1 - \dfrac{1}{x}}$

34. $x - \dfrac{1}{x - \dfrac{1}{2}}$

35. $1 + \dfrac{1}{1 + \dfrac{1}{1 + 1}}$

36. $1 - \dfrac{1}{1 - \dfrac{1}{1 - \frac{1}{2}}}$

37. $\dfrac{1 - \dfrac{1}{x + \frac{1}{2}}}{1 + \dfrac{1}{x + \frac{1}{2}}}$

38. $\dfrac{2 + \dfrac{1}{x - \frac{1}{3}}}{2 - \dfrac{1}{x - \frac{1}{3}}}$

Applying the Concepts

39. Optics   The formula $f = \dfrac{ab}{a + b}$ is used in optics to find the focal length of a lens. Show that the formula $f = (a^{-1} + b^{-1})^{-1}$ is equivalent to the preceding formula by rewriting it without the negative exponents and then simplifying the results.

40. Optics   Show that the expression $(a^{-1} - b^{-1})^{-1}$ can be simplified to $\dfrac{ab}{b - a}$ by first writing it without the negative exponents and then simplifying the result.

41. Doppler Effect   The change in the pitch of a sound (such as a train whistle) as an object passes is called the Doppler effect, named after C. J. Doppler (1803–1853). A person will *hear* a sound with a frequency, h, according to the formula

$$h = \dfrac{f}{1 + \dfrac{v}{s}}$$

where f is the actual frequency of the sound being produced, s is the speed of sound (about 740 miles per hour), and v is the velocity of the moving object.

(a) Examine this fraction, and then explain why h and f approach the same value as v becomes smaller and smaller.

(b) Solve this formula for v.

42. Mean of Two Numbers   The mean of two numbers is a point on the number line midway between them.

(a) Given two numbers, a and b, determine a formula for the mean of a and b.

(b) Given two numbers, a and b, determine a formula for the mean of their reciprocals.

(c) If $a = 4$ and $b = 6$, find the mean of their reciprocals.

43. Work Problem   A water storage tank has two drains. It can be shown that the time it takes to empty the tank if both drains are open is given by the formula

$$\dfrac{1}{\dfrac{1}{a} + \dfrac{1}{b}}$$

where $a = $ time it takes for the first drain to empty the tank, and $b = $ time for the second drain to empty the tank.

(a) Simplify this complex fraction.

(b) Find the amount of time needed to empty the tank using both drains if, used alone, the first drain empties the tank in 4 hours and the second drain can empty the tank in 3 hours.

44. Investing   If an amount of money, P, is invested for 1 year at 6% interest, compounded n times per year, then the amount of money in the account at the end of 1 year may be found from the formula

$$A = P\left(1 + \dfrac{0.06}{n}\right)^n$$

(a) Determine A if $n = 10$ and if $n = 11$; if $n = 50$ and if $n = 51$; if $n = 100$ and if

$n = 101$. (Round your answers to six decimal places to the right of the decimal point.)

(b) Use your results from part (a) to determine a limiting value of the ratio A_n/A_{n+1}.

Review Problems

The following problems review material we covered in Sections 2.1 and 5.7. Reviewing these problems will help you with the next section.

Solve each equation.

45. $10 - 2(x + 3) = x + 1$

46. $15 - 3(x - 1) = x - 2$

47. $x^2 - x - 12 = 0$

48. $3x^2 + x - 10 = 0$

49. $(x + 1)(x - 6) = -12$

50. $(x + 1)(x - 4) = -6$

6.6   Equations Involving Rational Expressions

The first step in solving an equation that contains one or more rational expressions is to find the LCD for all denominators in the equation. We then multiply both sides of the equation by the LCD to clear the equation of all fractions. That is, after we have multiplied through by the LCD, each term in the resulting equation will have a denominator of 1.

EXAMPLE 1   Solve $\frac{x}{2} - 3 = \frac{2}{3}$.

Solution   The LCD for 2 and 3 is 6. Multiplying both sides by 6, we have

$$6\left(\frac{x}{2} - 3\right) = 6\left(\frac{2}{3}\right)$$

$$6\left(\frac{x}{2}\right) - 6(3) = 6\left(\frac{2}{3}\right)$$

$$3x - 18 = 4$$

$$3x = 22$$

$$x = \frac{22}{3}$$

Multiplying both sides of an equation by the LCD clears the equation of fractions because the LCD has the property that all the denominators divide it evenly.

EXAMPLE 2   Solve $\frac{6}{a - 4} = \frac{3}{8}$.

Solution   The LCD for $a - 4$ and 8 is $8(a - 4)$. Multiplying both sides by this quantity yields

$$8(a - 4) \cdot \frac{6}{a - 4} = 8(a - 4) \cdot \frac{3}{8}$$

$$48 = (a - 4) \cdot 3$$

$$48 = 3a - 12$$

$$60 = 3a$$

$$20 = a$$

The solution set is $\{20\}$, which checks in the original equation.

When we multiply both sides of an equation by an expression containing the variable, we must be sure to check our solutions. The multiplication property of equality does not allow multiplication by 0. If the expression we multiply by contains the variable, then it has the possibility of being 0. In the last example we multiplied both sides by $8(a - 4)$. This gives a restriction $a \neq 4$ for any solution we come up with.

EXAMPLE 3   Solve $\dfrac{x}{x - 2} + \dfrac{2}{3} = \dfrac{2}{x - 2}$.

Solution   The LCD is $3(x - 2)$. We are assuming $x \neq 2$ when we multiply both sides of the equation by $3(x - 2)$:

$$3(x - 2) \cdot \left(\frac{x}{x - 2} + \frac{2}{3} \right) = 3(x - 2) \cdot \frac{2}{x - 2}$$

$$3x + (x - 2) \cdot 2 = 3 \cdot 2$$

$$3x + 2x - 4 = 6$$

$$5x - 4 = 6$$

$$5x = 10$$

$$x = 2$$

The only possible solution is $x = 2$. Checking this value back in the original equation gives

$$\frac{2}{2 - 2} + \frac{2}{3} \overset{?}{=} \frac{2}{2 - 2}$$

$$\frac{2}{0} + \frac{2}{3} \overset{?}{=} \frac{2}{0}$$

The first and last terms are undefined. The proposed solution, $x = 2$, does not check in the original equation. The solution set is the empty set. There is no solution to the original equation.

Note: In the process of solving the equation, we multiplied both sides by $3(x - 2)$, solved for x, and got $x = 2$ for our solution. But when x is 2, the quantity $3(x - 2) = 3(2 - 2) = 3(0) = 0$, which means we multiplied both sides of our equation by 0, which is not allowed under the multiplication property of equality.

When the proposed solution to an equation is not actually a solution, it is called an *extraneous* solution. In the last example, $x = 2$ is an extraneous solution.

EXAMPLE 4 Solve $\dfrac{5}{x^2 - 3x + 2} - \dfrac{1}{x - 2} = \dfrac{1}{3x - 3}$.

Solution Writing the equation again with the denominators in factored form, we have

$$\frac{5}{(x - 2)(x - 1)} - \frac{1}{x - 2} = \frac{1}{3(x - 1)}$$

The LCD is $3(x - 2)(x - 1)$. Multiplying through by the LCD, we have

$$3(x - 2)(x - 1)\frac{5}{(x - 2)(x - 1)} - 3(x - 2)(x - 1) \cdot \frac{1}{(x - 2)}$$

$$= 3(x - 2)(x - 1) \cdot \frac{1}{3(x - 1)}$$

$$3 \cdot 5 - 3(x - 1) \cdot 1 = (x - 2) \cdot 1$$

$$15 - 3x + 3 = x - 2$$

$$-3x + 18 = x - 2$$

$$-4x + 18 = -2$$

$$-4x = -20$$

$$x = 5$$

Note: We can check the proposed solution in any of the equations obtained before multiplying through by the LCD. We cannot check the proposed solution in an equation obtained *after* multiplying through by the LCD since, if we have multiplied by 0, the resulting equations will not be equivalent to the original one.

Checking the proposed solution $x = 5$ in the original equation yields a true statement. Try it and see.

EXAMPLE 5 Solve $3 + \dfrac{1}{x} = \dfrac{10}{x^2}$.

Solution   To clear the equation of denominators, we multiply both sides by x^2:

$$x^2\left(3 + \frac{1}{x}\right) = x^2\left(\frac{10}{x^2}\right)$$

$$3(x^2) + \left(\frac{1}{x}\right)(x^2) = \left(\frac{10}{x^2}\right)(x^2)$$

$$3x^2 + x = 10$$

Rewrite in standard form, and solve:

$$3x^2 + x - 10 = 0$$

$$(3x - 5)(x + 2) = 0$$

$$3x - 5 = 0 \quad \text{or} \quad x + 2 = 0$$

$$x = \frac{5}{3} \quad \text{or} \quad x = -2$$

The solution set is $\{-2, \frac{5}{3}\}$. Both solutions check in the original equation. Remember: We have to check *all solutions* any time we multiply both sides of the equation by an expression that contains the variable, just to be sure we haven't multiplied by 0.

EXAMPLE 6   Solve $\dfrac{y - 4}{y^2 - 5y} = \dfrac{2}{y^2 - 25}$.

Solution   Factoring each denominator, we find the LCD is $y(y - 5)(y + 5)$. Multiplying each side of the equation by the LCD clears the equation of denominators and leads us to our possible solutions:

$$y(y - 5)(y + 5) \cdot \frac{y - 4}{y(y - 5)} = \frac{2}{(y - 5)(y + 5)} \cdot y(y - 5)(y + 5)$$

$$(y + 5)(y - 4) = 2y$$

$$y^2 + y - 20 = 2y \qquad \text{Multiply out the left side.}$$

$$y^2 - y - 20 = 0 \qquad \text{Add } -2y \text{ to each side.}$$

$$(y - 5)(y + 4) = 0$$

$$y - 5 = 0 \quad \text{or} \quad y + 4 = 0$$

$$y = 5 \quad \text{or} \quad y = -4$$

The two possible solutions are 5 and -4. If we substitute -4 for y in the original equation, we find that it leads to a true statement. It is therefore a solution. On the other hand, if we substitute 5 for y in the original equation, we find that both sides of the equation are undefined. The only solution to our original equation is $y = -4$. The other possible solution $y = 5$ is extraneous.

EXAMPLE 7 Solve for y: $x = \dfrac{y - 4}{y - 2}$

Solution To solve for y, we first multiply each side by $y - 2$ to obtain

$$x(y - 2) = y - 4$$

$$xy - 2x = y - 4 \qquad \text{Distributive property}$$

$$xy - y = 2x - 4 \qquad \text{Collect all terms containing } y \text{ on the left side.}$$

$$y(x - 1) = 2x - 4 \qquad \text{Factor } y \text{ from each term on the left side.}$$

$$y = \dfrac{2x - 4}{x - 1} \qquad \text{Divide each side by } x - 1.$$

EXAMPLE 8 Solve the formula $\dfrac{1}{x} = \dfrac{1}{b} + \dfrac{1}{a}$ for x.

Solution We begin by multiplying both sides by the least common denominator xab. As you can see from our previous examples, multiplying both sides of an equation by the LCD is equivalent to multiplying each term of both sides by the LCD:

$$xab \cdot \dfrac{1}{x} = \dfrac{1}{b} \cdot xab + \dfrac{1}{a} \cdot xab$$

$$ab = xa + xb$$

$$ab = (a + b)x \qquad \text{Factor } x \text{ from the right side.}$$

$$\dfrac{ab}{a + b} = x$$

We know we are finished because the variable we were solving for is alone on one side of the equation and does not appear on the other side.

GRAPHING RATIONAL FUNCTIONS

We graphed simple rational functions in Section 3.7 when we graphed the inverse variation statement $y = \dfrac{1}{x}$. In the first section of this chapter we looked at a graph that was a little more complicated when we graphed $r(t) = \dfrac{785}{t}$. Our next example continues our investigation of the graphs of rational functions.

EXAMPLE 9 Graph the rational function $f(x) = \dfrac{6}{x - 2}$.

Solution Unlike the graphs in Section 3.7, this graph will cross the y-axis. To find the y-intercept, we let x equal 0.

$$\text{When } x = 0: \qquad y = \dfrac{6}{0 - 2} = \dfrac{6}{-2} = -3 \qquad y\text{-intercept}$$

The graph will not cross the x-axis. If it did, we would have a solution to the equation

$$0 = \frac{6}{x - 2}$$

which has no solution because there is no number to divide 6 by to obtain 0.

The graph of our equation is shown in Figure 1 along with a table giving values of x and y that satisfy the equation. Notice that y is undefined when x is 2. This means that the graph will not cross the vertical line $x = 2$. (If it did, there would be a value of y for $x = 2$.) The line $x = 2$ is called a *vertical asymptote* of the graph. The graph will get very close to the vertical asymptote, but will never touch or cross it.

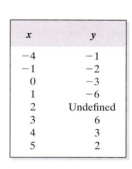

x	y
-4	-1
-1	-2
0	-3
1	-6
2	Undefined
3	6
4	3
5	2

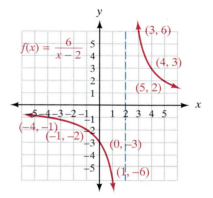

FIGURE 1 The graph of $f(x) = \dfrac{6}{x - 2}$

If you were to graph $y = \dfrac{6}{x}$ on the coordinate system in Figure 1, you would see that the graph of $y = \dfrac{6}{x - 2}$ is the graph of $y = \dfrac{6}{x}$ with all points shifted 2 units to the right.

Using TECHNOLOGY

MORE ABOUT EXAMPLE 9

We know the graph of $f(x) = \dfrac{6}{x - 2}$ will not cross the vertical asymptote $x = 2$ because replacing x with 2 in the equation gives us an undefined expression, meaning there is no value of y to associate with $x = 2$. We can use a graphing calculator to explore the behavior of this function when x gets closer and closer to 2 by using the table function on the calculator. We want to put our own val-

(continued)

(Using Technology, continued)

ues for X into the table, so we set the independent variable to Ask. (On a TI-82/83, use the TBLSET key to set up the table.) To see how the function behaves as x gets close to 2, we let X take on values of 1.9, 1.99, and 1.999. Then we move to the other side of 2 and let X become 2.1, 2.01, and 2.001.

Table Setup	*Y Variables*
Table minimum = 0	$Y_1 = 6/(x - 2)$
Table increment = 1	
Independent variable: Ask	
Dependent variable: Auto	

The table will look like this:

X	Y_1
1.9	-60
1.99	-600
1.999	-6000
2.1	60
2.01	600
2.001	6000

Again, the calculator asks us for a table increment. Because we are inputting the *x*-values ourselves, the increment value does not matter.

As you can see, the values in the table support the shape of the curve in Figure 1 around the vertical asymptote $x = 2$.

EXAMPLE 10   Graph: $g(x) = \dfrac{6}{x + 2}$

Solution   The only difference between this equation and the equation in Example 9 is in the denominator. This graph will have the same shape as the graph in Example 9, but the vertical asymptote will be $x = -2$ instead of $x = 2$. Figure 2 shows the graph.

Notice that the graphs shown in Figures 1 and 2 are both graphs of functions because no vertical line will cross either graph in more than one place. Notice the similarities and differences in our two functions,

$$f(x) = \frac{6}{x - 2} \qquad \text{and} \qquad g(x) = \frac{6}{x + 2}$$

and their graphs. The vertical asymptotes shown in Figures 1 and 2 correspond to the fact that both $f(2)$ and $g(-2)$ are undefined. The domain for the function f is all real numbers except $x = 2$, while the domain for g is all real numbers except $x = -2$.

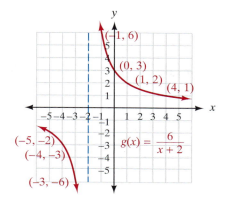

FIGURE 2 The graph of $g(x) = \dfrac{6}{x + 2}$

 ## Getting Ready for Class

After reading through the preceding section, respond in your own words and in complete sentences.

A. Explain how a least common denominator can be used to simplify an equation.

B. What is an extraneous solution?

C. How does the location of the vertical asymptote in the graph of a rational function relate to the equation of the function?

D. What is the last step in solving an equation that contains rational expressions?

PROBLEM SET 6.6

Solve each of the following equations.

1. $\dfrac{x}{5} + 4 = \dfrac{5}{3}$

2. $\dfrac{x}{5} = \dfrac{x}{2} - 9$

3. $\dfrac{a}{3} + 2 = \dfrac{4}{5}$

4. $\dfrac{a}{4} + \dfrac{1}{2} = \dfrac{2}{3}$

5. $\dfrac{y}{2} + \dfrac{y}{4} + \dfrac{y}{6} = 3$

6. $\dfrac{y}{3} - \dfrac{y}{6} + \dfrac{y}{2} = 1$

7. $\dfrac{5}{2x} = \dfrac{1}{x} + \dfrac{3}{4}$

8. $\dfrac{1}{2a} = \dfrac{2}{a} - \dfrac{3}{8}$

9. $\dfrac{1}{x} = \dfrac{1}{3} - \dfrac{2}{3x}$

10. $\dfrac{5}{2x} = \dfrac{2}{x} - \dfrac{1}{12}$

11. $\dfrac{2x}{x - 3} + 2 = \dfrac{2}{x - 3}$

12. $\dfrac{2}{x + 5} = \dfrac{2}{5} - \dfrac{x}{x + 5}$

13. $1 - \dfrac{1}{x} = \dfrac{12}{x^2}$

14. $2 + \dfrac{5}{x} = \dfrac{3}{x^2}$

15. $y - \dfrac{4}{3y} = -\dfrac{1}{3}$

16. $\dfrac{y}{2} - \dfrac{4}{y} = -\dfrac{7}{2}$

17. $\dfrac{x + 2}{x + 1} = \dfrac{1}{x + 1} + 2$

18. $\dfrac{x + 6}{x + 3} = \dfrac{3}{x + 3} + 2$

19. $\dfrac{3}{a - 2} = \dfrac{2}{a - 3}$

20. $\dfrac{5}{a + 1} = \dfrac{4}{a + 2}$

21. $6 - \dfrac{5}{x^2} = \dfrac{7}{x}$

22. $10 - \dfrac{3}{x^2} = -\dfrac{1}{x}$

23. $\dfrac{1}{x - 1} - \dfrac{1}{x + 1} = \dfrac{3x}{x^2 - 1}$

24. $\dfrac{5}{x-1} + \dfrac{2}{x-1} = \dfrac{4}{x+1}$

25. $\dfrac{2}{x-3} + \dfrac{x}{x^2-9} = \dfrac{4}{x+3}$

26. $\dfrac{2}{x+5} + \dfrac{3}{x+4} = \dfrac{2x}{x^2+9x+20}$

27. $\dfrac{3}{2} - \dfrac{1}{x-4} = \dfrac{-2}{2x-8}$

28. $\dfrac{2}{x} - \dfrac{1}{x+1} = \dfrac{-2}{5x+5}$

29. $\dfrac{t-4}{t^2-3t} = \dfrac{-2}{t^2-9}$

30. $\dfrac{t+3}{t^2-2t} = \dfrac{10}{t^2-4}$

31. $\dfrac{3}{y-4} - \dfrac{2}{y+1} = \dfrac{5}{y^2-3y-4}$

32. $\dfrac{1}{y+2} - \dfrac{2}{y-3} = \dfrac{-2y}{y^2-y-6}$

33. $\dfrac{2}{1+a} = \dfrac{3}{1-a} + \dfrac{5}{a}$

34. $\dfrac{1}{a+3} - \dfrac{a}{a^2-9} = \dfrac{2}{3-a}$

35. $\dfrac{3}{2x-6} - \dfrac{x+1}{4x-12} = 4$

36. $\dfrac{2x-3}{5x+10} + \dfrac{3x-2}{4x+8} = 1$

37. $\dfrac{y+2}{y^2-y} - \dfrac{6}{y^2-1} = 0$

38. $\dfrac{y+3}{y^2-y} - \dfrac{8}{y^2-1} = 0$

39. $\dfrac{4}{2x-6} - \dfrac{12}{4x+12} = \dfrac{12}{x^2-9}$

40. $\dfrac{1}{x+2} + \dfrac{1}{x-2} = \dfrac{4}{x^2-4}$

41. $\dfrac{2}{y^2-7y+12} - \dfrac{1}{y^2-9} = \dfrac{4}{y^2-y-12}$

42. $\dfrac{1}{y^2+5y+4} + \dfrac{3}{y^2-1} = \dfrac{-1}{y^2+3y-4}$

43. Solve the equation $6x^{-1} + 4 = 7$ by multiplying both sides by x. (*Remember:* $x^{-1} \cdot x = x^{-1} \cdot x^1 = x^0 = 1$.)

44. Solve the equation $3x^{-1} - 5 = 2x^{-1} - 3$ by multiplying both sides by x.

45. Solve the equation $1 + 5x^{-2} = 6x^{-1}$ by multiplying both sides by x^2.

46. Solve the equation $1 + 3x^{-2} = 4x^{-1}$ by multiplying both sides by x^2.

47. Solve the formula $\dfrac{1}{x} = \dfrac{1}{b} - \dfrac{1}{a}$ for x.

48. Solve $\dfrac{1}{x} = \dfrac{1}{a} - \dfrac{1}{b}$ for x.

49. Solve for R in the formula $\dfrac{1}{R} = \dfrac{1}{R_1} + \dfrac{1}{R_2}$.

50. Solve for R in the formula

$$\dfrac{1}{R} = \dfrac{1}{R_1} + \dfrac{1}{R_2} + \dfrac{1}{R_3}$$

Solve for y.

51. $x = \dfrac{y-3}{y-1}$

52. $x = \dfrac{y-2}{y-3}$

53. $x = \dfrac{2y+1}{3y+1}$

54. $x = \dfrac{3y+2}{5y+1}$

Graph each function. Show the vertical asymptote.

55. $f(x) = \dfrac{1}{x-3}$

56. $f(x) = \dfrac{1}{x+3}$

57. $f(x) = \dfrac{4}{x+2}$

58. $f(x) = \dfrac{4}{x-2}$

59. $g(x) = \dfrac{2}{x-4}$

60. $g(x) = \dfrac{2}{x+4}$

61. $g(x) = \dfrac{6}{x+1}$

62. $g(x) = \dfrac{6}{x-1}$

Let $f(x) = \dfrac{1}{x-3}$ and $g(x) = \dfrac{1}{x+3}$, and evaluate the following.

63. $f(0)$ and $f(6)$

64. $g(0)$ and $g(-6)$

65. $f(1)$ and $f(5)$

66. $g(1)$ and $g(-7)$

67. Give the domain for the function f as defined above.

68. Give the domain for the function g as defined above.

Let $f(x) = \dfrac{4}{x+2}$ and $g(x) = \dfrac{4}{x-2}$, and evaluate the following.

69. $f(0)$ and $f(-4)$ **70.** $g(0)$ and $g(4)$

71. $f(2)$ and $f(-6)$ **72.** $g(1)$ and $g(-5)$

73. Give the domain for the function f as defined above.

74. Give the domain for the function g as defined above.

Review Problems

The following problems review material we covered in Sections 2.3 and 5.7. Reviewing these problems will get you ready for the next section. In each case, be sure to show the equation used.

75. Number Problem   Twice the sum of a number and 3 is 16. Find the number.

76. Number Problem   The sum of two consecutive odd integers is 48. Find the two integers.

77. Geometry   The length of a rectangle is 3 less than twice the width. The perimeter is 42 meters. Find the length and width.

78. Geometry   The smallest angle in a triangle is one fourth as large as the largest angle. The third angle is 9 degrees more than the smallest angle. Find the measure of all three angles.

79. Consecutive Integers   The sum of the squares of two consecutive integers is 61. Find the integers.

80. Consecutive Integers   The square of the sum of two consecutive integers is 121. Find the two integers.

81. Geometry   The lengths of the sides of a right triangle are given by three consecutive integers. Find the lengths of the three sides.

82. Geometry   The longest side of a right triangle is 8 inches more than the shortest side. The other side is 7 inches more than the shortest side. Find the lengths of the three sides.

83. Geometry   From plane geometry and the principle of similar triangles, the relationship between y_1, y_2, and h shown in Figure 3 can be expressed as

$$\frac{1}{h} = \frac{1}{y_1} + \frac{1}{y_2}$$

Two poles are 12 feet high and 8 feet high. If a cable is attached to the top of each one and stretched to the bottom of the other, what is the height above the ground at which the two wires will meet?

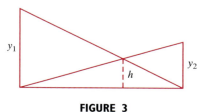

FIGURE 3

84. Geometry   We can use the diagram in Figure 4 to develop the formula

$$\frac{1}{h} = \frac{1}{y_1} + \frac{1}{y_2}$$

by taking the following steps:

Step 1   By similar triangles, $\dfrac{y}{h} = \dfrac{x+y}{?}$.

Step 2   Solve the proportion from step 1 for h.

Step 3   Solve the proportion from step 2 for $1/h$.

Step 4   Write the right side of the proportion from step 3 as two fractions.

Step 5   Use the fact that $y \cdot y_1 = x \cdot y_2$, and substitute in the proportion from step 4 to obtain the desired result.

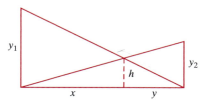

FIGURE 4

85. An Identity   An identity is an equation that is true for any value of the variable for which the expression is defined. Verify that the following expression is an identity by simplifying the left side of the expression.

$$\frac{2}{x-y} - \frac{1}{y-x} = \frac{3}{x-y}$$

86. Batting Average In baseball, a player's batting average is the ratio of the number of his hits to the number of his official at-bats. There is a dispute about who had the highest batting average in 1910. Although many sources list Ty Cobb of the Detroit Tigers as having the highest batting average of .385, some baseball researchers claim that his average was actually .382, and that he was second that year to Napoleon Lajoie of the Cleveland Indians, who had an average of .383. Assuming that both Cobb and Lajoie each had about 550 official at-bats in 1910, and that the researchers are correct and Cobb hit "only" .382 that year, how many more hits than Cobb did Lajoie have?

87. Kayak Race In a kayak race, the participants must paddle a kayak 450 meters down a river and then return 450 meters up the river to the starting point (Figure 5). Susan has correctly deduced that the total time t (in seconds) depends on the speed c (in meters per second) of the water according to the following expression:

$$t = \frac{450}{v + c} + \frac{450}{v - c}$$

where v is the speed of the kayak relative to the water (the speed of the kayak in still water).
(a) Fill in the following table.

Time t (sec)	Speed of Kayak Relative to the Water v (m/sec)	Current of the River c (m/sec)
240		1
300		2
	4	3
	3	1
540	3	
	3	3

(b) If the kayak race were conducted in the still waters of a lake, do you think that the total time of a given participant would be greater

than, equal to, or smaller than the time in the river? Justify your answer.
(c) Suppose Peter can drive his kayak at 4.1 meters per second and that the speed of the current is 4.1 meters per second. What will happen when Peter makes the turn and tries to come back up the river? How does this situation show up in the equation for total time?

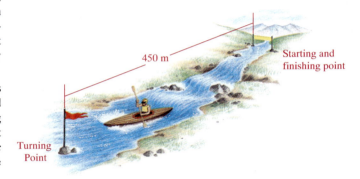

450 m

Starting and finishing point

Turning Point

FIGURE 5

88. Harmonic Mean A number, h, is the harmonic mean of two numbers, n_1 and n_2, if $1/h$ is the mean (average) of $1/n_1$ and $1/n_2$.
(a) Write an equation relating the harmonic mean, h, to two numbers, n_1 and n_2.
(b) Find the harmonic mean of 3 and 5.

Extending the Concepts

Solve each equation.

89. $\dfrac{12}{x} + \dfrac{8}{x^2} - \dfrac{75}{x^3} - \dfrac{50}{x^4} = 0$

90. $\dfrac{45}{x} + \dfrac{18}{x^2} - \dfrac{80}{x^3} - \dfrac{32}{x^4} = 0$

91. $\dfrac{1}{x^3} - \dfrac{1}{3x^2} - \dfrac{1}{4x} + \dfrac{1}{12} = 0$

92. $\dfrac{1}{x^3} - \dfrac{1}{2x^2} - \dfrac{1}{9x} + \dfrac{1}{18} = 0$

We begin this section with some application problems, the solutions to which involve equations that contain rational expressions. As you will see, the solutions to the examples show only the essential steps from our Blueprint for Problem Solving. Recall that step 1 was done mentally; we read the problem and mentally list the items that are known and the items that are unknown. This is an essential part of problem solving. Now that you have had experience with application problems, however, you are doing step 1 automatically.

Also in this section we will look at a method of solving conversion problems that is called *unit analysis*. With unit analysis, we can convert expressions with units of feet per minute to equivalent expressions in miles per hour. This method of converting between different units of measure is used often in chemistry, physics, and engineering classes.

EXAMPLE 1   One number is twice another. The sum of their reciprocals is 2. Find the numbers.

Solution   Let $x =$ the smaller number. The larger number is $2x$. Their reciprocals are $\frac{1}{x}$ and $\frac{1}{2x}$. The equation that describes the situation is

$$\frac{1}{x} + \frac{1}{2x} = 2$$

Multiplying both sides by the LCD $2x$, we have

$$2x \cdot \frac{1}{x} + 2x \cdot \frac{1}{2x} = 2x(2)$$

$$2 + 1 = 4x$$

$$3 = 4x$$

$$x = \frac{3}{4}$$

The smaller number is $\frac{3}{4}$. The larger is $2\left(\frac{3}{4}\right) = \frac{6}{4} = \frac{3}{2}$. Adding their reciprocals, we have

$$\frac{4}{3} + \frac{2}{3} = \frac{6}{3} = 2$$

The sum of the reciprocals of $\frac{3}{4}$ and $\frac{3}{2}$ is 2.

EXAMPLE 2   The speed of a boat in still water is 20 miles per hour. It takes the same amount of time for the boat to travel 3 miles downstream (with the current) as it does to travel 2 miles upstream (against the current). Find the speed of the current.

405

Solution   The following table will be helpful in finding the equation necessary to solve this problem.

	d (distance)	*r* (rate)	*t* (time)
Upstream			
Downstream			

If we let $x =$ the speed of the current, the speed (rate) of the boat upstream is $20 - x$ since it is traveling against the current. The rate downstream is $20 + x$ since the boat is then traveling with the current. The distance traveled upstream is 2 miles, and the distance traveled downstream is 3 miles. Putting the information given here into the table, we have

	d	*r*	*t*
Upstream	2	$20 - x$	
Downstream	3	$20 + x$	

To fill in the last two spaces in the table we must use the relationship $d = r \cdot t$. Since we know the spaces to be filled in are in the time column, we solve the equation $d = r \cdot t$ for t and get

$$t = \frac{d}{r}$$

$$\frac{40 - 2x}{20 \, / \, x}$$

The completed table then is

	d	*r*	*t*
Upstream	2	$20 - x$	$\dfrac{2}{20 - x}$
Downstream	3	$20 + x$	$\dfrac{3}{20 + x}$

Reading the problem again, we find that the time moving upstream is equal to the time moving downstream, or

$$\frac{2}{20 - x} = \frac{3}{20 + x}$$

Multiplying both sides by the LCD $(20 - x)(20 + x)$ gives

$$(20 + x) \cdot 2 = 3(20 - x)$$
$$40 + 2x = 60 - 3x$$
$$5x = 20$$
$$x = 4$$

The speed of the current is 4 miles per hour.

E X A M P L E   3   The current of a river is 3 miles per hour. It takes a motorboat a total of 3 hours to travel 12 miles upstream and return 12 miles downstream. What is the speed of the boat in still water?

Solution   This time we let $x =$ the speed of the boat in still water. Then, we fill in as much of the table as possible using the information given in the problem. For instance, since we let $x =$ the speed of the boat in still water, the rate upstream (against the current) must be $x - 3$. The rate downstream (with the current) is $x + 3$.

	d	*r*	*t*
Upstream	12	$x - 3$	
Downstream	12	$x + 3$	

The last two boxes can be filled in using the relationship

$$t = \frac{d}{r}$$

	d	*r*	*t*
Upstream	12	$x - 3$	$\dfrac{12}{x - 3}$
Downstream	12	$x + 3$	$\dfrac{12}{x + 3}$

The total time for the trip up and back is 3 hours:

$$\text{Time upstream} + \text{Time downstream} = \text{Total time}$$

$$\frac{12}{x-3} + \frac{12}{x+3} = 3$$

Multiplying both sides by $(x-3)(x+3)$, we have

$$12(x+3) + 12(x-3) = 3(x^2-9)$$
$$12x + 36 + 12x - 36 = 3x^2 - 27$$
$$3x^2 - 24x - 27 = 0$$
$$x^2 - 8x - 9 = 0 \qquad\qquad \text{Divide both sides by 3.}$$
$$(x-9)(x+1) = 0$$
$$x = 9 \quad\text{or}\quad x = -1$$

The speed of the motorboat in still water is 9 miles per hour. (We don't use $x = -1$ because the speed of the motorboat cannot be a negative number.)

EXAMPLE 4 An inlet pipe can fill a pool in 10 hours, while the drain can empty it in 12 hours. If the pool is empty and both the inlet pipe and drain are open, how long will it take to fill the pool?

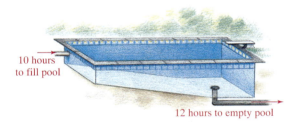

10 hours to fill pool

12 hours to empty pool

Solution It is helpful to think in terms of how much work is done by each pipe in 1 hour.

Let $x =$ the time it takes to fill the pool with both pipes open.

If the inlet pipe can fill the pool in 10 hours, then in 1 hour it is $\frac{1}{10}$ full. If the outlet pipe empties the pool in 12 hours, then in 1 hour it is $\frac{1}{12}$ empty. If the pool can be filled in x hours with both the inlet pipe and the drain open, then in 1 hour it is $\frac{1}{x}$ full when both pipes are open.

Here is the equation:

In 1 hour

$$\left[\begin{array}{c}\text{Amount filled by}\\\text{inlet pipe}\end{array}\right] - \left[\begin{array}{c}\text{Amount emptied by}\\\text{the drain}\end{array}\right] = \left[\begin{array}{c}\text{Fraction of pool filled}\\\text{with both pipes open}\end{array}\right]$$

$$\frac{1}{10} \qquad - \qquad \frac{1}{12} \qquad = \qquad \frac{1}{x}$$

Multiplying through by $60x$, we have

$$60x \cdot \frac{1}{10} - 60x \cdot \frac{1}{12} = 60x \cdot \frac{1}{x}$$

$$6x - 5x = 60$$

$$x = 60$$

It takes 60 hours to fill the pool if both the inlet pipe and the drain are open.

UNIT ANALYSIS

In the 1950s the United States had a spy plane, the U-2, that could fly at an altitude of 65,000 feet. Do you know how many miles are in 65,000 feet?

65,000 ft

We can solve problems like this by using a method called *unit analysis*. With unit analysis, we analyze the units we are given and the units for which we are asked, and then multiply by the appropriate *conversion factor*. Since 1 mile is 5,280 feet, the conversion factor we use is

$$\frac{1 \text{ mile}}{5,280 \text{ feet}}$$

which is the number 1. Multiplying 65,000 feet by this conversion factor we have the following:

$$65,000 \text{ feet} = \frac{65,000 \text{ feet}}{1} \cdot \frac{1 \text{ mile}}{5,280 \text{ feet}}$$

We treat the units common to the numerator and denominator in the same way we treat factors common to the numerator and denominator: We divide out common units, just as we divide out common factors. In the preceding expression we have feet common to the numerator and denominator. Dividing them out leaves us with miles only. Here is the complete problem.

$$65{,}000 \text{ feet} = \frac{65{,}000 \text{ feet}}{1} \cdot \frac{1 \text{ mile}}{5{,}280 \text{ feet}}$$

$$= \frac{65{,}000}{5{,}280} \text{ mile}$$

$$= 12.3 \text{ miles, to the nearest tenth of a mile}$$

The key to solving a problem like this one lies in choosing the appropriate conversion factor. The fact that 1 mile = 5,280 feet yields two conversion factors, each of which is equal to the number 1. They are

$$\frac{1 \text{ mile}}{5{,}280 \text{ feet}} \quad \text{and} \quad \frac{5{,}280 \text{ feet}}{1 \text{ mile}}$$

The conversion factor we choose depends on the units we are given and the units with which we want to end up. Multiplying any expression by either of the two conversion factors leaves the value of the original expression unchanged because each of the conversion factors is simply the number 1.

EXAMPLE 5 In Section 6.1 we found a rider on the first Ferris wheel was traveling at approximately 39.3 feet per minute. Convert 39.3 feet per minute to miles per hour.

Solution   We know that 5,280 feet = 1 mile and 60 minutes = 1 hour. Therefore, we have the following conversion factors, each of which is equal to 1.

$$\frac{5{,}280 \text{ feet}}{1 \text{ mile}} \qquad \frac{1 \text{ mile}}{5{,}280 \text{ feet}} \qquad \frac{60 \text{ minutes}}{1 \text{ hour}} \qquad \frac{1 \text{ hour}}{60 \text{ minutes}}$$

The conversion factors we choose to multiply by are the ones that will allow us to divide out the units we are converting from and leave us with the units we are converting to. Specifically, we want to get rid of feet and be left with miles. Likewise, we want to get rid of minutes and be left with hours. Here is the conversion process that will accomplish these goals:

$$39.3 \text{ feet per minute} = \frac{39.3 \text{ feet}}{1 \text{ minute}} \cdot \frac{1 \text{ mile}}{5,280 \text{ feet}} \cdot \frac{60 \text{ minutes}}{1 \text{ hour}}$$

$$= \frac{39.3 \cdot 60 \text{ miles}}{5,280 \text{ hours}}$$

$$= 0.45 \text{ mile per hour, to the nearest hundredth}$$

EXAMPLE 6   In 1993, a ski resort in Vermont advertised their new high-speed chair lift as "the world's fastest chair lift, with a speed of 1,100 feet per second." Show why the speed cannot be correct.

Solution   To solve this problem, we can convert feet per second into miles per hour, a unit of measure we are more familiar with on an intuitive level.

$$1,100 \text{ feet per second} = \frac{1,100 \text{ feet}}{1 \text{ second}} \cdot \frac{1 \text{ mile}}{5,280 \text{ feet}} \cdot \frac{60 \text{ seconds}}{1 \text{ minute}} \cdot \frac{60 \text{ minutes}}{1 \text{ hour}}$$

$$= \frac{1,100 \cdot 60 \cdot 60 \text{ miles}}{5,280 \text{ hours}}$$

$$= 750 \text{ miles per hour}$$

Obviously, there is a mistake in the advertisement.

More About Graphing Rational Functions

We continue our investigation of the graphs of rational functions by considering the graph of a rational function with binomials in the numerator and denominator.

EXAMPLE 7   Graph the rational function $y = \dfrac{x - 4}{x - 2}$.

Solution   In addition to making a table to find some points on the graph, we can analyze the graph as follows:

1. The graph will have a *y*-intercept of 2, because when $x = 0$, $y = \frac{-4}{-2} = 2$.

2. To find the *x*-intercept, we let $y = 0$ to get

$$0 = \frac{x - 4}{x - 2}$$

The only way this expression can be 0 is if the numerator is 0, which happens when $x = 4$. (If you want to solve this equation, multiply both sides by $x - 2$. You will get the same solution, $x - 4$.)

3. The graph will have a vertical asymptote at $x = 2$, since $x = 2$ will make the denominator of the function 0, meaning *y* is undefined when *x* is 2.

4. The graph will have a *horizontal asymptote* at $y = 1$ because for very large values of *x*, $\frac{x - 4}{x - 2}$ is very close to 1. The larger *x* is, the closer $\frac{x - 4}{x - 2}$ is to 1. The same is true for very small values of *x*, such as $-1{,}000$ and $-10{,}000$.

Putting this information together with the ordered pairs in the table next to the figure, we have the graph shown in Figure 1.

x	y
-1	$\frac{5}{3}$
0	2
1	3
2	Undefined
3	-1
4	0
5	$\frac{1}{3}$

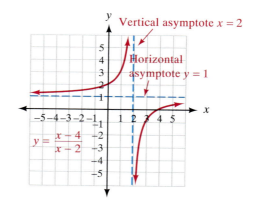

FIGURE 1

 Using TECHNOLOGY

MORE ABOUT EXAMPLE 7

In the previous section we used technology to explore the graph of a rational function around a vertical asymptote. This time, we are going to explore the graph near the horizontal asymptote. In Figure 1, the horizontal asymptote is at $y = 1$. To show that the graph approaches this line as *x* becomes very large, we use the table function on our graphing calculator, with X taking values of 100,

1,000, and 10,000. To show that the graph approaches the line $y = 1$ on the left side of the coordinate system, we let X become -100, $-1,000$, and $-10,000$.

Table Setup   *Y Variables*

Table minimum = 0 $Y_1 = (x - 4)/(x - 2)$
Table increment = 1
Independent variable: Ask
Dependent variable: Auto

The table will look like this:

X	Y_1	
100	.97959	
1000	.998	
10000	.9998	
-100	1.0196	
-1000	1.002	
-10000	1.0002	

As you can see, as x becomes very large in the positive direction, the graph approaches the line $y = 1$ from below. As x becomes very small in the negative direction, the graph approaches the line $y = 1$ from above.

Getting Ready for Class

After reading through the preceding section, respond in your own words and in complete sentences.

A. Briefly list the steps in the Blueprint for Problem Solving that you have used previously to solve application problems.

B. Write an application problem for which the solution depends on solving the equation $\frac{1}{2} + \frac{1}{3} = \frac{1}{x}$.

C. What is a conversion factor in unit analysis?

D. What conversion factors would you use to convert feet per second to miles per hour?

PROBLEM SET 6.7

Solve each of the following word problems. Be sure to show the equation in each case.

Number Problems

1. One number is 3 times another. The sum of their reciprocals is $\frac{20}{3}$. Find the numbers.

2. One number is 3 times another. The sum of their reciprocals is $\frac{4}{9}$. Find the numbers.

3. The sum of a number and its reciprocal is $\frac{10}{3}$. Find the number.

4. The sum of a number and twice its reciprocal is $\frac{27}{5}$. Find the number.

5. The sum of the reciprocals of two consecutive integers is $\frac{7}{12}$. Find the two integers.

6. Find two consecutive even integers, the sum of whose reciprocals is $\frac{3}{4}$.

7. If a certain number is added to the numerator and denominator of $\frac{7}{9}$, the result is $\frac{5}{6}$. Find the number.

8. Find the number you would add to both the numerator and denominator of $\frac{8}{11}$ so that the result would be $\frac{6}{7}$.

Rate Problems

9. The speed of a boat in still water is 5 miles per hour. If the boat travels 3 miles downstream in the same amount of time it takes to travel 1.5 miles upstream, what is the speed of the current?

10. A boat, which moves at 18 miles per hour in still water, travels 14 miles downstream in the same amount of time it takes to travel 10 miles upstream. Find the speed of the current.

11. The current of a river is 2 miles per hour. A boat travels to a point 8 miles upstream and back again in 3 hours. What is the speed of the boat in still water?

12. A motorboat travels at 4 miles per hour in still water. It goes 12 miles upstream and 12 miles back again in a total of 8 hours. Find the speed of the current of the river.

13. Train A has a speed 15 miles per hour greater than that of train B. If train A travels 150 miles in the same time train B travels 120 miles, what are the speeds of the two trains?

14. A train travels 30 miles per hour faster than a car. If the train covers 120 miles in the same time the car covers 80 miles, what is the speed of each of them?

15. A small airplane flies 810 miles from Los Angeles to Portland, OR, with an average speed of 270 miles per hour. An hour and a half after the plane leaves, a Boeing 747 leaves Los Angeles for Portland. Both planes arrive in Portland at the same time. What was the average speed of the 747?

16. Lou leaves for a cross-country excursion on a bicycle traveling at 20 miles per hour. His friends are driving the trip and will meet him at several rest stops along the way. The first stop is scheduled 30 miles from the original starting point. If the people driving leave 15 minutes after Lou from the same place, how fast will they have to drive to reach the first rest stop at the same time as Lou?

17. A tour bus leaves Sacramento every Friday evening at 5:00 P.M. for a 270-mile trip to Las Vegas. This week, however, the bus leaves at 5:30 P.M. In order to arrive in Las Vegas on time, the driver drives 6 miles per hour faster than usual. What is the bus's usual speed?

18. A bakery delivery truck leaves the bakery at 5:00 A.M. each morning on its 140-mile route. One day the driver gets a late start and does not leave the bakery until 5:30 A.M. In order to finish her route on time the driver drives 5 miles per hour faster than usual. At what speed does she usually drive?

Work Problems

19. A water tank can be filled by an inlet pipe in 8 hours. It takes twice that long for the outlet pipe to empty the tank. How long will it take to fill the tank if both pipes are open?

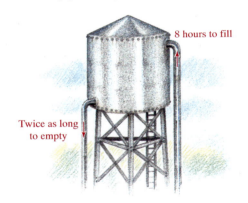

8 hours to fill

Twice as long to empty

20. A sink can be filled from the faucet in 5 minutes. It takes only 3 minutes to empty the sink when the drain is open. If the sink is full and both the faucet and the drain are open, how long will it take to empty the sink?

21. It takes 10 hours to fill a pool with the inlet pipe. It can be emptied in 15 hours with the outlet pipe. If the pool is half full to begin with, how long will it take to fill it from there if both pipes are open?

10 hours to fill pool

15 hours to empty pool

22. A sink is one-quarter full when both the faucet and the drain are opened. The faucet alone can fill the sink in 6 minutes, while it takes 8 minutes to empty it with the drain. How long will it take to fill the remaining three quarters of the sink?

23. A sink has two faucets: one for hot water and one for cold water. The sink can be filled by a cold-water faucet in 3.5 minutes. If both faucets are open, the sink is filled in 2.1 minutes. How long does it take to fill the sink with just the hot-water faucet open?

24. A water tank is being filled by two inlet pipes. Pipe A can fill the tank in $4\frac{1}{2}$ hours, but both pipes together can fill the tank in 2 hours. How long does it take to fill the tank using only pipe B?

Unit Analysis Problems

Give your answers to the following problems to the nearest tenth.

25. The South Coast Shopping Mall in Costa Mesa, California, covers an area of 2,224,750 square feet. If 1 acre = 43,560 square feet, how many acres does the South Coast Shopping Mall cover?

26. The relationship between liters and cubic inches, both of which are measures of volume, is 0.0164 liters = 1 cubic inch. If a Ford Mustang has a motor with a displacement of 4.9 liters, what is the displacement in cubic inches?

27. The Forest chair lift at the Northstar ski resort in Lake Tahoe is 5,750 feet long. If a ride on this chair lift takes 11 minutes, what is the average speed of the lift in miles per hour?

28. The Bear Paw chair lift at the Northstar ski resort in Lake Tahoe is 790 feet long. If a ride on this chair lift takes 2.2 minutes, what is the average speed of the lift in miles per hour?

29. A sprinter runs 100 meters in 10.8 seconds. What is the sprinter's average speed in miles per hour? (1 meter = 3.28 feet)

30. A runner covers 400 meters in 49.8 seconds. What is the average speed of the runner in miles per hour?

31. A person riding a Ferris wheel with a diameter of 65 feet travels once around the wheel in 30 seconds. What is the average speed of the rider in miles per hour?

32. A person riding a Ferris wheel with a diameter of 102 feet travels once around the wheel in 3.5 minutes. What is the average speed of the rider in miles per hour?

33. A $3\frac{1}{2}$-inch diskette, when placed in the disk drive of a computer, rotates at 300 rpm (meaning one revolution takes 1/300 minute). Find the average speed of a point 1.5 inches from the center of the diskette in miles per hour.

34. A 5-inch fixed disk in a computer rotates at 3,600 rpm. Find the average speed of a point 2 inches from the center of the disk in miles per hour.

35. **Unit Analysis** An ad similar to the one shown in Figure 1 appeared recently in a national magazine. The ad indicates that 3,241,440 minutes is 6 years. Is this statement correct?

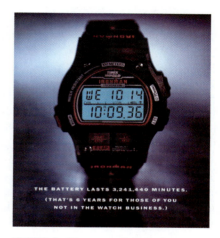

FIGURE 1 *(Courtesy of Timex Corp)*

36. **Unit Analysis and Thomas Jefferson** In addition to being our third president, Thomas Jefferson was also this nation's first Secretary of State. One of his official duties in 1791 was to summarize the units presently in use in the new United States. The units of capacity that he listed in his official report were as follows:

Measures of Capacity

Let the bushel be divided into 10 pottles;
Each pottle into 10 demi-pints;
Each demi-pint into 10 metres, which will be of a cubic inch each.

(a) Change 8 bushels into demi-pints.
(b) Change 76 metres into pottles.
(c) What should the number of bushels be multiplied by in order to convert bushels into metres?

Miscellaneous Problems

37. Rhind Papyrus Nearly 4,000 years ago, Egyptians worked mathematical exercises involving reciprocals. The *Rhind Papyrus* contains a wealth of such problems, and one of them is as follows:

A quantity and its two thirds are added together, one third of this is added, then one third of the sum is taken, and the result is 10.

Write an equation and solve this exercise.

38. Photography For clear photographs, a camera must be properly focused. Professional photographers use a mathematical relationship relating the distance from the camera lens to the object being photographed, *a*; the distance from the lens to the film, *b*; and the focal length of the lens, *f*. These quantities, *a*, *b*, and *f*, are related by the equation

$$\frac{1}{a} + \frac{1}{b} = \frac{1}{f}$$

A camera has a focal length of 3 inches. If the lens is 5 inches from the film, how far should the lens be placed from the object being photographed for the camera to be perfectly focused?

The Periodic Table If you take a chemistry class, you will work with the Periodic Table of Elements. Figure 2 shows three of the elements listed in the periodic table. As you can see, the bottom number in each figure is the molecular weight of the element. In chemistry, a *mole* is the amount of a substance that will give the weight in grams equal to the molecular weight. For example, 1 mole of lead is 207.2 grams.

Name	Lead	Carbon	Sulfur
Atomic number	82	6	16
Symbol	Pb	C	S
Atomic weight	207.2	12.01	32.07

FIGURE 2

39. Chemistry For the element carbon, 1 mole = 12.01 grams.
(a) To the nearest gram, how many grams of carbon are in 2.5 moles of carbon?
(b) How many moles of carbon are in 39 grams of carbon? Round to the nearest hundredth.

40. Chemistry For the element sulfur, 1 mole = 32.07 grams.
(a) How many grams of sulfur are in 3 moles of sulfur?
(b) How many moles of sulfur are found in 80.2 grams of sulfur?

Graph each rational function. In each case, show the vertical asymptote, the horizontal asymptote, and any intercepts that exist.

41. $f(x) = \dfrac{x - 3}{x - 1}$ **42.** $f(x) = \dfrac{x + 4}{x - 2}$

43. $f(x) = \dfrac{x + 3}{x - 1}$ **44.** $f(x) = \dfrac{x - 2}{x - 1}$

45. $g(x) = \dfrac{x - 3}{x + 1}$ **46.** $g(x) = \dfrac{x - 2}{x + 1}$

Review Problems

The problems that follow review material we covered in Sections 6.3, 6.4, 6.5, and 6.6. Reviewing these problems will help clarify the different methods we have used in this chapter.

Perform the indicated operations. [6.3, 6.4]

47. $\dfrac{2a + 10}{a^3} \cdot \dfrac{a^2}{3a + 15}$

48. $\dfrac{4a + 8}{a^2 - a - 6} \div \dfrac{a^2 + 7a + 12}{a^2 - 9}$

49. $(x^2 - 9)\left(\dfrac{x + 2}{x + 3}\right)$ **50.** $\dfrac{1}{x + 4} + \dfrac{8}{x^2 - 16}$

51. $\dfrac{2x - 7}{x - 2} - \dfrac{x - 5}{x - 2}$ **52.** $2 + \dfrac{25}{5x - 1}$

Simplify each expression. [6.5]

53. $\dfrac{\dfrac{1}{x} - \dfrac{1}{3}}{\dfrac{1}{x} + \dfrac{1}{3}}$ **54.** $\dfrac{1 - \dfrac{9}{x^2}}{1 - \dfrac{1}{x} - \dfrac{6}{x^2}}$

Solve each equation. [6.6]

55. $\dfrac{x}{x - 3} + \dfrac{3}{2} = \dfrac{3}{x - 3}$ **56.** $1 - \dfrac{3}{x} = \dfrac{-2}{x^2}$

Examples

1. $\dfrac{3}{4}$ is a rational number. $\dfrac{x-3}{x^2-9}$ is a rational expression.

Rational Numbers and Expressions [6.1]

A *rational number* is any number that can be expressed as the ratio of two integers:

$$\text{Rational numbers} = \left\{\dfrac{a}{b}\,\middle|\, a \text{ and } b \text{ are integers}, b \neq 0\right\}$$

A *rational expression* is any quantity that can be expressed as the ratio of two polynomials:

$$\text{Rational expressions} = \left\{\dfrac{P}{Q}\,\middle|\, P \text{ and } Q \text{ are polynomials}, Q \neq 0\right\}$$

Properties of Rational Expressions [6.1]

If P, Q, and K are polynomials with $Q \neq 0$ and $K \neq 0$, then

$$\dfrac{P}{Q} = \dfrac{PK}{QK} \quad \text{and} \quad \dfrac{P}{Q} = \dfrac{P/K}{Q/K}$$

which is to say that multiplying or dividing the numerator and denominator of a rational expression by the same nonzero quantity always produces an equivalent rational expression.

2. $\dfrac{x-3}{x^2-9} = \dfrac{\cancel{x-3}}{\cancel{(x-3)}(x+3)}$

$\qquad = \dfrac{1}{x+3}$

Reducing to Lowest Terms [6.1]

To reduce a rational expression to lowest terms, we first factor the numerator and denominator and then divide the numerator and denominator by any factors they have in common.

3. $\dfrac{15x^3 - 20x^2 + 10x}{5x}$

$\quad = 3x^2 - 4x + 2$

Dividing a Polynomial by a Monomial [6.2]

To divide a polynomial by a monomial, divide each term of the polynomial by the monomial.

4.
$$
\begin{array}{r}
x - 2 \\
x-3\overline{)\,x^2 - 5x + 8}\\
\underline{\mp x^2 \mp 3x}\\
-2x + 8\\
\underline{\mp 2x \mp 6}\\
2
\end{array}
$$

Long Division With Polynomials [6.2]

If division with polynomials cannot be accomplished by dividing out factors common to the numerator and denominator, then we use a process similar to long division with whole numbers. The steps in the process are estimate, multiply, subtract, and bring down the next term.

5. $\dfrac{x+1}{x^2-4} \cdot \dfrac{x+2}{3x+3}$

$= \dfrac{\cancel{(x+1)}\cancel{(x+2)}}{(x-2)\cancel{(x+2)}(3)\cancel{(x+1)}}$

$= \dfrac{1}{3(x-2)}$

Multiplication [6.3]

To multiply two rational numbers or rational expressions, multiply numerators and multiply denominators. In symbols,

$$\frac{P}{Q} \cdot \frac{R}{S} = \frac{PR}{QS} \qquad (Q \neq 0 \text{ and } S \neq 0)$$

In practice, we don't really multiply, but rather, we factor and then divide out common factors.

6. $\dfrac{x^2-y^2}{x^3+y^3} \div \dfrac{x-y}{x^2-xy+y^2}$

$= \dfrac{x^2-y^2}{x^3+y^3} \cdot \dfrac{x^2-xy+y^2}{x-y}$

$= \dfrac{\cancel{(x+y)}\cancel{(x-y)}\cancel{(x^2-xy+y^2)}}{\cancel{(x+y)}\cancel{(x^2-xy+y^2)}\cancel{(x-y)}}$

$= 1$

Division [6.3]

To divide one rational expression by another, we use the definition of division to rewrite our division problem as an equivalent multiplication problem. Instead of dividing by a rational expression, we multiply by its reciprocal. In symbols,

$$\frac{P}{Q} \div \frac{R}{S} = \frac{P}{Q} \cdot \frac{S}{R} = \frac{PS}{QR} \qquad (Q \neq 0, S \neq 0, R \neq 0)$$

7. The LCD for $\dfrac{2}{x-3}$ and $\dfrac{3}{5}$ is $5(x-3)$.

Least Common Denominator [6.4]

The *least common denominator,* LCD, for a set of denominators is the smallest quantity divisible by each of the denominators.

8. $\dfrac{2}{x-3} + \dfrac{3}{5}$

$= \dfrac{2}{x-3} \cdot \dfrac{5}{5} + \dfrac{3}{5} \cdot \dfrac{x-3}{x-3}$

$= \dfrac{3x+1}{5(x-3)}$

Addition and Subtraction [6.4]

If P, Q, and R represent polynomials, $R \neq 0$, then

$$\frac{P}{R} + \frac{Q}{R} = \frac{P+Q}{R} \qquad \text{and} \qquad \frac{P}{R} - \frac{Q}{R} = \frac{P-Q}{R}$$

When adding or subtracting rational expressions with different denominators, we must find the LCD for all denominators and change each rational expression to an equivalent expression that has the LCD.

9. $\dfrac{\dfrac{1}{x} + \dfrac{1}{y}}{\dfrac{1}{x} - \dfrac{1}{y}} = \dfrac{xy\left(\dfrac{1}{x} + \dfrac{1}{y}\right)}{xy\left(\dfrac{1}{x} - \dfrac{1}{y}\right)}$

$= \dfrac{y+x}{y-x}$

Complex Fractions [6.5]

A rational expression that contains, in its numerator or denominator, other rational expressions is called a *complex fraction.* One method of simplifying a complex fraction is to multiply the numerator and denominator by the LCD for all denominators.

10. Solve $\dfrac{x}{2} + 3 = \dfrac{1}{3}$.

$6\left(\dfrac{x}{2}\right) + 6 \cdot 3 = 6 \cdot \dfrac{1}{3}$

$3x + 18 = 2$

$x = -\dfrac{16}{3}$

Equations Involving Rational Expressions [6.6]

To solve an equation involving rational expressions, we first find the LCD for all denominators appearing on either side of the equation. We then multiply both sides by the LCD to clear the equation of all fractions and solve as usual.

COMMON MISTAKES

1. Attempting to divide the numerator and denominator of a rational expression by a quantity that is not a factor of both. Like this:

$$\frac{x^2 - \overset{3}{\cancel{9}}x + \overset{2}{\cancel{20}}}{x^2 - \cancel{3}x - \cancel{10}} \quad \text{Mistake}$$
$$\underset{1 \quad\quad 1}{}$$

This makes no sense at all. The numerator and denominator must be factored completely before any factors they have in common can be recognized:

$$\frac{x^2 - 9x + 20}{x^2 - 3x - 10} = \frac{(\cancel{x - 5})(x - 4)}{(\cancel{x - 5})(x + 2)}$$
$$= \frac{x - 4}{x + 2}$$

2. Forgetting to check solutions to equations involving rational expressions. When we multiply both sides of an equation by a quantity containing the variable, we must be sure to check for extraneous solutions (see Section 6.6).

CHAPTER 6   REVIEW

Reduce to lowest terms. [6.1]

1. $\dfrac{125x^4yz^3}{35x^2y^4z^3}$

2. $\dfrac{a^3 - ab^2}{4a + 4b}$

3. $\dfrac{x^2 - 25}{x^2 + 10x + 25}$

4. $\dfrac{ax + x - 5a - 5}{ax - x - 5a + 5}$

Divide. If the denominator is a factor of the numerator, you may want to factor the numerator and divide out the common factor. [6.2]

5. $\dfrac{12x^3 + 8x^2 + 16x}{4x^2}$

6. $\dfrac{27a^2b^3 - 15a^3b^2 + 21a^4b^4}{-3a^2b^2}$

7. $\dfrac{x^{6n} - x^{5n}}{x^{3n}}$

8. $\dfrac{x^2 - x - 6}{x - 3}$

9. $\dfrac{5x^2 - 14xy - 24y^2}{x - 4y}$

10. $\dfrac{y^4 - 16}{y - 2}$

11. $\dfrac{8x^2 - 26x - 9}{2x - 7}$

12. $\dfrac{2y^3 - 9y^2 - 17y + 39}{2y - 3}$

Multiply and divide as indicated. [6.3]

13. $\dfrac{3}{4} \cdot \dfrac{12}{15} \div \dfrac{1}{3}$

14. $\dfrac{15x^2y}{8xy^2} \div \dfrac{10xy}{4x}$

15. $\dfrac{x^3 - 1}{x^4 - 1} \cdot \dfrac{x^2 - 1}{x^2 + x + 1}$

16. $\dfrac{a^2 + 5a + 6}{a + 1} \cdot \dfrac{a + 5}{a^2 + 2a - 3} \div \dfrac{a^2 + 7a + 10}{a^2 - 1}$

17. $\dfrac{ax + bx + 2a + 2b}{ax - 3a + bx - 3b} \div \dfrac{ax - bx - 2a + 2b}{ax - bx - 3a + 3b}$

18. $(4x^2 - 9) \cdot \dfrac{x + 3}{2x + 3}$

Add and subtract as indicated. [6.4]

19. $\dfrac{3}{5} - \dfrac{1}{10} + \dfrac{8}{15}$

20. $\dfrac{5}{x - 5} - \dfrac{x}{x - 5}$

21. $\dfrac{1}{x} + \dfrac{1}{x^2} + \dfrac{1}{x^3}$

22. $\dfrac{8}{y^2 - 16} - \dfrac{7}{y^2 - y - 12}$

23. $\dfrac{x - 2}{x^2 + 5x + 4} - \dfrac{x - 4}{2x^2 + 12x + 16}$

24. $3 + \dfrac{4}{5x - 2}$

Simplify each complex fraction. [6.5]

25. $\dfrac{1 + \dfrac{2}{3}}{1 - \dfrac{2}{3}}$

26. $\dfrac{\dfrac{4a}{2a^3 + 2}}{\dfrac{8a}{4a + 4}}$

27. $1 + \dfrac{1}{x + \dfrac{1}{x}}$

28. $\dfrac{1 - \dfrac{9}{x^2}}{1 - \dfrac{1}{x} - \dfrac{6}{x^2}}$

Solve each equation. [6.6]

29. $\dfrac{3}{x - 1} = \dfrac{3}{5}$

30. $\dfrac{x + 1}{3} + \dfrac{x - 3}{4} = \dfrac{1}{6}$

31. $\dfrac{5}{y + 1} = \dfrac{4}{y + 2}$

32. $\dfrac{x + 6}{x + 3} - 2 = \dfrac{3}{x + 3}$

33. $\dfrac{4}{x^2 - x - 12} + \dfrac{1}{x^2 - 9} = \dfrac{2}{x^2 - 7x + 12}$

34. $\dfrac{a + 4}{a^2 + 5a} = \dfrac{-2}{a^2 - 25}$

Graph each rational function. [6.6]

35. $f(x) = \dfrac{6}{x + 2}$

36. $g(x) = \dfrac{x - 3}{x + 2}$

37. Distance, Rate, and Time A car makes a 120-mile trip 10 miles per hour faster than a truck. The truck takes 2 hours longer to make the trip. What are the speeds of the car and the truck? [6.7]

38. Average Speed A jogger covers 3.5 miles in 28 minutes. Find the average speed of the jogger in miles per hour. [6.7]

39. Unit Analysis The speed of sound is 1,088 feet per second. Convert the speed of sound to miles per hour. Round your answer to the nearest whole number. [6.7]

CHAPTER 6 PROJECTS

RATIONAL EXPRESSIONS

GROUP PROJECT

Number of People: 3

Time Needed: 10–15 minutes

Equipment: Pencil and paper

Procedure: The four problems shown below all involve the same rational expressions. Many times, students who have worked problems successfully on their homework have trouble when they take a test on rational expressions because the problems are mixed up and do not have similar instructions. Noticing similarities and differences between the type of problems involving rational expressions can help with this situation.

Problem 1: Add: $\dfrac{-2}{x^2 - 2x - 3} + \dfrac{3}{x^2 - 9}$

Problem 2: Divide: $\dfrac{-2}{x^2 - 2x - 3} \div \dfrac{3}{x^2 - 9}$

Problem 3: Solve: $\dfrac{-2}{x^2 - 2x - 3} + \dfrac{3}{x^2 - 9} = -1$

Problem 4: Simplify: $\dfrac{\dfrac{-2}{x^2 - 2x - 3}}{\dfrac{3}{x^2 - 9}}$

1. Which of the problems above does not require the use of a least common denominator?
2. Which two problems involve multiplying by the least common denominator?
3. Which of the problems will have an answer that is one or two numbers, but no variables?
4. Work each of the four problems.

RESEARCH PROJECT

FERRIS WHEELS AND *THE THIRD MAN*

Among the large Ferris wheels built around the turn of the century was one built in Vienna in 1897. It is the only one of those large wheels that is still in operation today. Known as the Riesenrad, it has a diameter of 197 feet and can carry a total of 800 people. A brochure that gives some statistics associated with the Riesenrad indicates that passengers riding it travel at 2 feet 6 inches per second. You can check the accuracy of this number by watching the movie *The Third*

(Hulton Getty/Tony Stone Images)   *(Selznick Films/The Kobal Collection)*

Man. In the movie, Orson Welles rides the Riesenrad through one complete revolution. Play *The Third Man* on a VCR, so you can view the Riesenrad in operation. Use the pause button and the timer on the VCR to time how long it takes Orson Welles to ride once around the wheel. Then calculate his average speed during the ride. Use your results to either prove or disprove the claim that passengers travel at 2 feet 6 inches per second on the Riesenrad. When you have finished, write your procedures and results in essay form.

CHAPTER 6 TEST

Reduce to lowest terms. [6.1]

1. $\dfrac{x^2 - y^2}{x - y}$

2. $\dfrac{2x^2 - 5x + 3}{2x^2 - x - 3}$

Divide. [6.2]

3. $\dfrac{24x^3y + 12x^2y^2 - 16xy^3}{4xy}$

4. $\dfrac{2x^3 - 9x^2 + 10}{2x - 1}$

Multiply and divide as indicated. [6.3]

5. $\dfrac{a^2 - 16}{5a - 15} \cdot \dfrac{10(a - 3)^2}{a^2 - 7a + 12}$

6. $\dfrac{a^4 - 81}{a^2 + 9} \div \dfrac{a^2 - 8a + 15}{4a - 20}$

7. $\dfrac{x^3 - 8}{2x^2 - 9x + 10} \div \dfrac{x^2 + 2x + 4}{2x^2 + x - 15}$

Add and subtract as indicated. [6.4]

8. $\dfrac{4}{21} + \dfrac{6}{35}$

9. $\dfrac{3}{4} - \dfrac{1}{2} + \dfrac{5}{8}$

10. $\dfrac{a}{a^2 - 9} + \dfrac{3}{a^2 - 9}$

11. $\dfrac{1}{x} + \dfrac{2}{x - 3}$

12. $\dfrac{4x}{x^2 + 6x + 5} - \dfrac{3x}{x^2 + 5x + 4}$

13. $\dfrac{2x + 8}{x^2 + 4x + 3} - \dfrac{x + 4}{x^2 + 5x + 6}$

Simplify each complex fraction. [6.5]

14. $\dfrac{3 - \dfrac{1}{a + 3}}{3 + \dfrac{1}{a + 3}}$

15. $\dfrac{1 - \dfrac{9}{x^2}}{1 + \dfrac{1}{x} - \dfrac{6}{x^2}}$

Solve each of the following equations. [6.6]

16. $\dfrac{1}{x} + 3 = \dfrac{4}{3}$

17. $\dfrac{x}{x - 3} + 3 = \dfrac{3}{x - 3}$

18. $\dfrac{y + 3}{2y} + \dfrac{5}{y - 1} = \dfrac{1}{2}$

19. $1 - \dfrac{1}{x} = \dfrac{6}{x^2}$

20. Graph $f(x) = \dfrac{x + 4}{x - 1}$.

Solve the following applications. Be sure to show the equation in each case. [6.7]

21. **Number Problem** What number must be subtracted from the denominator of $\frac{10}{23}$ to make the result $\frac{1}{3}$?

22. **Speed of a Boat** The current of a river is 2 miles per hour. It takes a motorboat a total of 3 hours to travel 8 miles upstream and return 8 miles downstream. What is the speed of the boat in still water?

23. **Filling a Pool** An inlet pipe can fill a pool in 10 hours, and the drain can empty it in 15 hours. If the pool is half full and both the inlet pipe and the drain are left open, how long will it take to fill the pool the rest of the way?

24. Unit Analysis The top of Mount Whitney, the highest point in California, is 14,494 feet above sea level. Give this height in miles to the nearest tenth of a mile.

25. Unit Analysis A bullet fired from a gun travels a distance of 4,750 feet in 3.2 seconds. Find the average speed of the bullet in miles per hour. Round to the nearest whole number.

Simplify.

1. $11 - (-9) - 7 - (-5)$

2. $\left(\dfrac{5}{6}\right)^{-2}$

3. $\dfrac{x^{-5}}{x^{-8}}$

4. $\left(\dfrac{x^{-6}y^3}{x^{-3}y^{-4}}\right)^{-1}$

5. $-3(5x + 4) + 12x$

6. $\dfrac{\dfrac{2a}{3a^3 - 3}}{\dfrac{4a}{6a - 6}}$

7. $(x + 3)^2 - (x - 3)^2$

8. $\begin{vmatrix} 3 & 4 \\ 1 & 2 \end{vmatrix}$

9. Subtract $\frac{3}{4}$ from the product of -3 and $\frac{5}{12}$.

10. Write in symbols: The difference of $5a$ and $7b$ is greater than their sum.

11. If $A = \{0, 2, 3, 5, 9\}$ and $B = \{0, 1, 3, 5, 7\}$, find $\{x \mid x \notin A$ and $x \in B\}$

12. Give the next number of the sequence, and state whether arithmetic or geometric: $-3, 2, 7, \ldots$

13. State the properties that justify: $(3x - 4) + 2y = (3x + 2y) - 4$

Subtract.

14. $\dfrac{6}{y^2 - 9} - \dfrac{5}{y^2 - y - 6}$

15. $\dfrac{y}{x^2 - y^2} - \dfrac{x}{x^2 - y^2}$

Multiply.

16. $\left(4t^2 + \dfrac{1}{3}\right)\left(3t^2 - \dfrac{1}{4}\right)$

17. $\dfrac{x^4 - 16}{x^3 - 8} \cdot \dfrac{x^2 + 2x + 4}{x^2 + 4}$

Divide.

18. $\dfrac{10x^{3n} - 15x^{4n}}{5x^n}$

19. $\dfrac{a^4 + a^3 - 1}{a + 2}$

Solve.

20. $-\dfrac{3}{5}a + 3 = 15$

21. $7y - 6 = 2y + 9$

22. $\dfrac{2}{5}(15x - 2) - \dfrac{1}{5} = 5$

23. $|a| - 5 = 7$

24. $x^3 - 3x^2 - 25x + 75 = 0$

25. $\dfrac{3}{y - 2} = \dfrac{2}{y - 3}$

26. $2 - \dfrac{11}{x} = -\dfrac{12}{x^2}$

Solve each system.

27. $5x - 2y = -1$
     $y = 3x + 2$

28. $5x - 8y = 4$
     $3x + 2y = -1$

29. $-5x + 3y = 1$
     $\frac{5}{3}x - y = 2$

30. $x - y + z = -4$
     $-4x - 3y - 2z = 2$
     $-5x + 4y + z = 2$

Use Cramer's rule to solve.

31. $7x - 9y = 2$
     $-3x + 11y = 1$

32. $x - 2y = 1$
     $3x + 5z = 8$
     $4y - 7z = -3$

Solve each inequality, and graph the solution.

33. $-3(3x - 1) \le -2(3x - 3)$

34. $|2x + 3| - 4 < 1$

35. Find the slope of the line $x + y = -3$.

Graph on a rectangular coordinate system.

36. $6x - 5y = 30$

37. $f(x) = \dfrac{2}{x - 1}$

38. $3x - y < 4$

Factor completely.

39. 168

40. $x^2 - 3x - 70$

41. $x^2 + 10x + 25 - y^2$

42. Reduce to lowest terms: $\dfrac{x^3 + 2x^2 - 9x - 18}{x^2 - x - 6}$

43. Write $9,270,000.00$ in scientific notation.

44. 18 is 8% of what number?

45. Let $f(x) = x^2 - 2x$ and $g(x) = x + 5$. Find: $f(-2) - g(3)$.

46. Find the slope of any perpendicular to the line through $(-6, -1)$ and $(-3, -5)$.

47. Find the equation of the line through $(-2, 6)$ and $(-2, 3)$.

48. Joint Variation Suppose z varies jointly with x and the cube of y. If z is -48 when x is 3 and y is 2, find z when x is 2 and y is 3.

49. Geometry The height of a triangle is 5 feet less than 2 times the base. If the area is 75 square feet, find the base and height.

50. Geometry Find all three angles in a triangle if the smallest angle is one-sixth the largest angle and the remaining angle is 20 degrees more than the smallest angle.

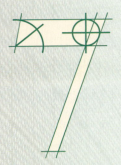

Rational Exponents and Roots

(Chip Porter / Tony Stone Images)

INTRODUCTION

Ecology and conservation are topics that interest most college students. If our rivers and oceans are to be preserved for future generations, we need to work to eliminate pollution from our waters. If a river is flowing at 1 meter per second and a pollutant is entering the river at a constant rate, the shape of the pollution plume can often be modeled by the simple equation

$$y = \sqrt{x}$$

The following table and graph were produced from the equation.

TABLE I Width of a Pollutant Plume	
Distance From Source (meters) x	Width of Plume (meters) y
0	0
1	1
4	2
9	3
16	4

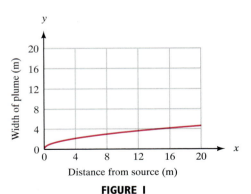

FIGURE I

To visualize how Figure 1 models the pollutant plume, imagine that the river is flowing from left to right, parallel to the x-axis, with the x-axis as one of its banks. The pollutant is entering the river from the bank at (0, 0).

By modeling pollution with mathematics, we can use our knowledge of mathematics to help control and eliminate pollution.

Figure 1 shows a square in which each of the four sides is 1 inch long. To find the square of the length of the diagonal c, we apply the Pythagorean theorem:

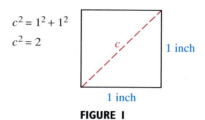

$$c^2 = 1^2 + 1^2$$
$$c^2 = 2$$

1 inch

1 inch

FIGURE 1

Because we know that c is positive and that its square is 2, we call c the *positive square root* of 2, and we write $c = \sqrt{2}$. Associating numbers, such as $\sqrt{2}$, with the diagonal of a square or rectangle allows us to analyze some interesting items from geometry. One particularly interesting geometric object that we will study in this section is shown in Figure 2. It is constructed from a right triangle, and the length of the diagonal is found from the Pythagorean theorem. We will come back to this figure at the end of this section.

The Golden Rectangle

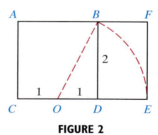

FIGURE 2

In Chapter 5 we developed notation (exponents) to give us the square, cube, or any other power of a number. For instance, if we wanted the square of 3, we wrote $3^2 = 9$. If we wanted the cube of 3, we wrote $3^3 = 27$. In this section we will develop notation that will take us in the reverse direction, that is, from the square of a number, say 25, back to the original number, 5.

DEFINITION   If x is a nonnegative real number, then the expression $\sqrt{x}$ is called the **positive square root** of x and is such that

$$(\sqrt{x})^2 = x$$

In words: $\sqrt{x}$ is the positive number we square to get x.

The negative square root of x, $-\sqrt{x}$, is defined in a similar manner.

EXAMPLE 1   The positive square root of 64 is 8 because 8 is the positive number with the property $8^2 = 64$. The negative square root of 64 is -8 since -8 is the negative number whose square is 64. We can summarize both of these facts by saying

$$\sqrt{64} = 8 \qquad \text{and} \qquad -\sqrt{64} = -8$$

The higher roots, cube roots, fourth roots, and so on, are defined by definitions similar to that of square roots.

Note: It is a common mistake to assume that an expression like $\sqrt{25}$ indicates both square roots, 5 and -5. The expression $\sqrt{25}$ indicates only the positive square root of 25, which is 5. If we want the negative square root, we must use a negative sign: $-\sqrt{25} = -5$.

DEFINITION   If x is a real number and n is a positive integer, then

Positive square root of x, $\sqrt{x}$, is such that $(\sqrt{x})^2 = x$ $\qquad x \geq 0$

Cube root of x, $(\sqrt[3]{x})^3 = x$

Positive fourth root of x, $\sqrt[4]{x}$, is such that $(\sqrt[4]{x})^4 = x$ $\qquad x \geq 0$

Fifth root of x, $\sqrt[5]{x}$, is such that $(\sqrt[5]{x})^5 = x$

$$\vdots \qquad\qquad \vdots$$

The **nth root of x**, $\sqrt[n]{x}$, is such that $(\sqrt[n]{x})^n = x$ $\qquad x \geq 0$ if n is even

---

Note: We have restricted the even roots in this definition to nonnegative numbers. Even roots of negative numbers exist, but are not represented by real numbers. That is, $\sqrt{-4}$ is not a real number since there is no real number whose square is -4.

The following is a table of the most common roots used in this book. Any of the roots that are unfamiliar should be memorized.

Square Roots		Cube Roots	Fourth Roots
$\sqrt{0} = 0$	$\sqrt{49} = 7$	$\sqrt[3]{0} = 0$	$\sqrt[4]{0} = 0$
$\sqrt{1} = 1$	$\sqrt{64} = 8$	$\sqrt[3]{1} = 1$	$\sqrt[4]{1} = 1$
$\sqrt{4} = 2$	$\sqrt{81} = 9$	$\sqrt[3]{8} = 2$	$\sqrt[4]{16} = 2$
$\sqrt{9} = 3$	$\sqrt{100} = 10$	$\sqrt[3]{27} = 3$	$\sqrt[4]{81} = 3$
$\sqrt{16} = 4$	$\sqrt{121} = 11$	$\sqrt[3]{64} = 4$	
$\sqrt{25} = 5$	$\sqrt{144} = 12$	$\sqrt[3]{125} = 5$	
$\sqrt{36} = 6$	$\sqrt{169} = 13$		

Notation   An expression like $\sqrt[3]{8}$ that involves a root is called a *radical expression*. In the expression $\sqrt[3]{8}$, the 3 is called the *index*, the $\sqrt{}$ is the *radical sign*, and 8 is called the *radicand.* The index of a radical must be a positive integer greater than 1. If no index is written, it is assumed to be 2.

ROOTS AND NEGATIVE NUMBERS

When dealing with negative numbers and radicals, the only restriction concerns negative numbers under even roots. We can have negative signs in front of radicals and negative numbers under odd roots and still obtain real numbers. Here are some examples to help clarify this. In the last section of this chapter we will see how to deal with even roots of negative numbers.

EXAMPLES   Simplify each expression, if possible.

2. $\sqrt[3]{-8} = -2$ because $(-2)^3 = -8$.

3. $\sqrt{-4}$ is not a real number since there is no real number whose square is -4.

4. $-\sqrt{25} = -5$ is the negative square root of 25.

5. $\sqrt[5]{-32} = -2$ because $(-2)^5 = -32$.

6. $\sqrt[4]{-81}$ is not a real number since there is no real number we can raise to the fourth power and obtain -81.

VARIABLES UNDER A RADICAL

From the preceding examples it is clear that we must be careful that we do not try to take an even root of a negative number. For this reason, we will assume that all variables appearing under a radical sign represent nonnegative numbers.

EXAMPLES   Assume all variables represent nonnegative numbers, and simplify each expression as much as possible.

7. $\sqrt{25a^4b^6} = 5a^2b^3$ because $(5a^2b^3)^2 = 25a^4b^6$.

8. $\sqrt[3]{x^6y^{12}} = x^2y^4$ because $(x^2y^4)^3 = x^6y^{12}$.

9. $\sqrt[4]{81r^8s^{20}} = 3r^2s^5$ because $(3r^2s^5)^4 = 81r^8s^{20}$.

RATIONAL NUMBERS AS EXPONENTS

We will now develop a second kind of notation involving exponents that will allow us to designate square roots, cube roots, and so on in another way.

Consider the equation $x = 8^{1/3}$. Although we have not encountered fractional exponents before, let's assume that all the properties of exponents hold in this case. Cubing both sides of the equation, we have

$$x^3 = (8^{1/3})^3$$

$$x^3 = 8^{(1/3)(3)}$$

$$x^3 = 8^1$$

$$x^3 = 8$$

The last line tells us that x is the number whose cube is 8. It must be true, then, that x is the cube root of 8, $x = \sqrt[3]{8}$. Since we started with $x = 8^{1/3}$, it follows that

$$8^{1/3} = \sqrt[3]{8}$$

It seems reasonable, then, to define fractional exponents as indicating roots. Here is the formal definition.

DEFINITION   If x is a real number and n is a positive integer greater than 1, then

$$x^{1/n} = \sqrt[n]{x} \qquad (x \geq 0 \text{ when } n \text{ is even})$$

In words: The quantity $x^{1/n}$ is the nth root of x.

With this definition we have a way of representing roots with exponents. Here are some examples.

EXAMPLES   Write each expression as a root and then simplify, if possible.

10. $8^{1/3} = \sqrt[3]{8} = 2$

11. $36^{1/2} = \sqrt{36} = 6$

12. $-25^{1/2} = -\sqrt{25} = -5$

13. $(-25)^{1/2} = \sqrt{-25}$, which is not a real number

14. $\left(\dfrac{4}{9}\right)^{1/2} = \sqrt{\dfrac{4}{9}} = \dfrac{2}{3}$

The properties of exponents developed in Chapter 5 were applied to integer exponents only. We will now extend these properties to include rational exponents also. We do so without proof.

PROPERTIES OF EXPONENTS

If a and b are real numbers and r and s are rational numbers, and a and b are nonnegative whenever r and s indicate even roots, then

1. $a^r \cdot a^s = a^{r+s}$ **4.** $a^{-r} = \dfrac{1}{a^r}$ $(a \neq 0)$

2. $(a^r)^s = a^{rs}$ **5.** $\left(\dfrac{a}{b}\right)^r = \dfrac{a^r}{b^r}$ $(b \neq 0)$

3. $(ab)^r = a^r b^r$ **6.** $\dfrac{a^r}{a^s} = a^{r-s}$ $(a \neq 0)$

Sometimes rational exponents can simplify our work with radicals. Here are Examples 8 and 9 again, but this time we will work them using rational exponents.

EXAMPLES   Write each radical with a rational exponent and then simplify.

15. $\sqrt[3]{x^6 y^{12}} = (x^6 y^{12})^{1/3}$

$\qquad = (x^6)^{1/3}(y^{12})^{1/3}$

$\qquad = x^2 y^4$

16. $\sqrt[4]{81 r^8 s^{20}} = (81 r^8 s^{20})^{1/4}$

$\qquad = 81^{1/4}(r^8)^{1/4}(s^{20})^{1/4}$

$\qquad = 3 r^2 s^5$

So far, the numerators of all the rational exponents we have encountered have been 1. The next theorem extends the work we can do with rational exponents to rational exponents with numerators other than 1.

We can extend our properties of exponents with the following theorem.

THEOREM 7.1

If a is a nonnegative real number, m is an integer, and n is a positive integer, then

$$a^{m/n} = (a^{1/n})^m = (a^m)^{1/n}$$

Proof   We can prove Theorem 7.1 using the properties of exponents. Since $\frac{m}{n} = m\left(\frac{1}{n}\right)$ we have

$$a^{m/n} = a^{m(1/n)} \qquad \qquad a^{m/n} = a^{(1/n)(m)}$$

$$\qquad = (a^m)^{1/n} \qquad \qquad \quad = (a^{1/n})^m$$

Here are some examples that illustrate how we use this theorem.

EXAMPLES   Simplify as much as possible.

17. $8^{2/3} = (8^{1/3})^2$ Theorem 7.1

$\qquad = 2^2$ Definition of fractional exponents

$\qquad = 4$ The square of 2 is 4.

Note: On a scientific calculator, Example 17 would look like this:

$$8 \;\boxed{y^x}\; \boxed{(}\; 2 \;\boxed{\div}\; 3 \;\boxed{)}\; \boxed{=}$$

18. $25^{3/2} = (25^{1/2})^3$ Theorem 7.1

        $= 5^3$ Definition of fractional exponents

        $= 125$ The cube of 5 is 125.

19. $9^{-3/2} = (9^{1/2})^{-3}$ Theorem 7.1

        $= 3^{-3}$ Definition of fractional exponents

        $= \dfrac{1}{3^3}$ Property 4 for exponents

        $= \dfrac{1}{27}$ The cube of 3 is 27.

20. $\left(\dfrac{27}{8}\right)^{-4/3} = \left[\left(\dfrac{27}{8}\right)^{1/3}\right]^{-4}$ Theorem 7.1

        $= \left(\dfrac{3}{2}\right)^{-4}$ Definition of fractional exponents

        $= \left(\dfrac{2}{3}\right)^{4}$ Property 4 for exponents

        $= \dfrac{16}{81}$ The fourth power of $\frac{2}{3}$ is $\frac{16}{81}$.

The following examples show the application of the properties of exponents to rational exponents.

EXAMPLES   Assume all variables represent positive quantities, and simplify as much as possible.

21. $x^{1/3} \cdot x^{5/6} = x^{1/3 + 5/6}$ Property 1

        $= x^{2/6 + 5/6}$ LCD is 6.

        $= x^{7/6}$ Add fractions.

22. $(y^{2/3})^{3/4} = y^{(2/3)(3/4)}$ Property 2

        $= y^{1/2}$ Multiply fractions: $\frac{2}{3} \cdot \frac{3}{4} = \frac{6}{12} = \frac{1}{2}$

23. $\dfrac{z^{1/3}}{z^{1/4}} = z^{1/3 - 1/4}$ Property 6

        $= z^{4/12 - 3/12}$ LCD is 12.

        $= z^{1/12}$ Subtract fractions.

24. $\left(\dfrac{a^{-1/3}}{b^{1/2}}\right)^6 = \dfrac{(a^{-1/3})^6}{(b^{1/2})^6}$ Property 5

        $= \dfrac{a^{-2}}{b^3}$ Property 2

        $= \dfrac{1}{a^2 b^3}$ Property 4

25. $\dfrac{(x^{-3}y^{1/2})^4}{x^{10}y^{3/2}} = \dfrac{(x^{-3})^4(y^{1/2})^4}{x^{10}y^{3/2}}$ Property 3

$\phantom{\dfrac{(x^{-3}y^{1/2})^4}{x^{10}y^{3/2}}} = \dfrac{x^{-12}y^2}{x^{10}y^{3/2}}$ Property 2

$\phantom{\dfrac{(x^{-3}y^{1/2})^4}{x^{10}y^{3/2}}} = x^{-22}y^{1/2}$ Property 6

$\phantom{\dfrac{(x^{-3}y^{1/2})^4}{x^{10}y^{3/2}}} = \dfrac{y^{1/2}}{x^{22}}$ Property 4

FACTS FROM
Geometry

The Pythagorean Theorem (Again) and the Golden Rectangle

Now that we have had some experience working with square roots, we can rewrite the Pythagorean theorem using a square root. If triangle *ABC* is a right triangle with $C = 90°$, then the length of the longest side is the *positive square root* of the sum of the squares of the other two sides (see Figure 3).

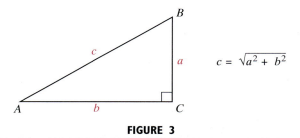

$$c = \sqrt{a^2 + b^2}$$

FIGURE 3

In the introduction to this chapter we mentioned the golden rectangle. Its origins can be traced back over 2,000 years to the Greek civilization that produced Pythagoras, Socrates, Plato, Aristotle, and Euclid. The most important mathematical work to come from that Greek civilization was Euclid's *Elements,* an elegantly written summary of all that was known about geometry at that time in history. Euclid's *Elements,* according to Howard Eves, an authority on the history of mathematics, exercised a greater influence on scientific thinking than any other work. Here is how we construct a golden rectangle from a square of side 2, using the same method that Euclid used in his *Elements.*

CONSTRUCTING A GOLDEN RECTANGLE
FROM A SQUARE OF SIDE 2

Step 1:   Draw a square with a side of length 2. Connect the midpoint of side *CD* to corner *B*. (Note that we have labeled the midpoint of segment *CD* with the letter *O*.)

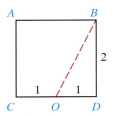

Step 2:   Drop the diagonal from step 1 down so it aligns with side *CD*.

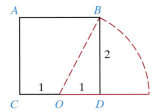

Step 3:   Form rectangle *ACEF.* This is a golden rectangle.

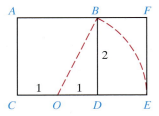

All golden rectangles are constructed from squares. Every golden rectangle, no matter how large or small it is, will have the same shape. To associate a number with the shape of the golden rectangle, we use the ratio of its length to its width. This ratio is called the *golden ratio*. To calculate the golden ratio, we must first find the length of the diagonal we used to construct the golden rectangle. Figure 4 shows the golden rectangle that we constructed from a square of side 2. The length of the diagonal *OB* is found by applying the Pythagorean theorem to triangle *OBD*.

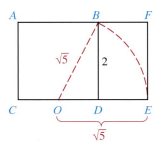

FIGURE 4

The length of segment OE is equal to the length of diagonal OB; both are $\sqrt{5}$. Since the distance from C to O is 1, the length CE of the golden rectangle is $1 + \sqrt{5}$. Now we can find the golden ratio:

$$\text{Golden ratio} = \frac{\text{length}}{\text{width}} = \frac{CE}{EF} = \frac{1 + \sqrt{5}}{2}$$

Using TECHNOLOGY

GRAPHING CALCULATORS—A WORD OF CAUTION

Some graphing calculators give surprising results when evaluating expressions such as $(-8)^{2/3}$. As you know from reading this section, the expression $(-8)^{2/3}$ simplifies to 4, either by taking the cube root first and then squaring the result, or by squaring the base first and then taking the cube root of the result. Here are three different ways to evaluate this expression on your calculator:

1. $(-8)\char`^(2/3)$ To evaluate $(-8)^{2/3}$
2. $((-8)\char`^2)\char`^(1/3)$ To evaluate $((-8)^2)^{1/3}$
3. $((-8)\char`^(1/3))\char`^2$ To evaluate $((-8)^{1/3})^2$

Note any differences in the results.
    Next, graph each of the following functions, one at a time.

1. $Y_1 = X^{2/3}$ **2.** $Y_2 = (X^2)^{1/3}$ **3.** $Y_3 = (X^{1/3})^2$

The correct graph is shown in Figure 5. Note which of your graphs match the correct graph.

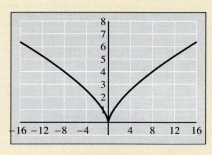

FIGURE 5

Different calculators evaluate exponential expressions in different ways. You should use the method (or methods) that gave you the correct graph.

Getting Ready for Class

After reading through the preceding section, respond in your own words and in complete sentences.

A. Every real number has two square roots. Explain the notation we use to tell them apart. Use the square roots of 3 for examples.

B. Explain why a square root of -4 is not a real number.

C. We use the notation $\sqrt{2}$ to represent the positive square root of 2. Explain why there isn't a simpler way to express the positive square root of 2.

D. For the expression $a^{m/n}$, explain the significance of the numerator m and the significance of the denominator n in the exponent.

PROBLEM SET 7.1

Find each of the following roots, if possible.

1. $\sqrt{144}$

2. $-\sqrt{144}$

3. $\sqrt{-144}$

4. $\sqrt{-49}$

5. $-\sqrt{49}$

6. $\sqrt{49}$

7. $\sqrt[3]{-27}$

8. $-\sqrt[3]{27}$

9. $\sqrt[4]{16}$

10. $-\sqrt[4]{16}$

11. $\sqrt[4]{-16}$

12. $-\sqrt[4]{-16}$

13. $\sqrt{0.04}$

14. $\sqrt{0.81}$

15. $\sqrt[3]{0.008}$

16. $\sqrt[3]{0.125}$

Simplify each expression. Assume all variables represent nonnegative numbers.

17. $\sqrt{36a^8}$

18. $\sqrt{49a^{10}}$

19. $\sqrt[3]{27a^{12}}$

20. $\sqrt[3]{8a^{15}}$

21. $\sqrt[3]{x^3y^6}$

22. $\sqrt[3]{x^6y^3}$

23. $\sqrt[5]{32x^{10}y^5}$

24. $\sqrt[5]{32x^5y^{10}}$

25. $\sqrt[4]{16a^{12}b^{20}}$

26. $\sqrt[4]{81a^{24}b^8}$

Use the definition of rational exponents to write each of the following with the appropriate root. Then simplify.

27. $36^{1/2}$

28. $49^{1/2}$

29. $-9^{1/2}$

30. $-16^{1/2}$

31. $8^{1/3}$

32. $-8^{1/3}$

33. $(-8)^{1/3}$

34. $-27^{1/3}$

35. $32^{1/5}$

36. $81^{1/4}$

37. $\left(\dfrac{81}{25}\right)^{1/2}$

38. $\left(\dfrac{9}{16}\right)^{1/2}$

39. $\left(\dfrac{64}{125}\right)^{1/3}$

40. $\left(\dfrac{8}{27}\right)^{1/3}$

Use Theorem 7.1 to simplify each of the following as much as possible.

41. $27^{2/3}$

42. $8^{4/3}$

43. $25^{3/2}$

44. $9^{3/2}$

45. $16^{3/4}$

46. $81^{3/4}$

Simplify each expression. Remember, negative exponents give reciprocals.

47. $27^{-1/3}$

48. $9^{-1/2}$

49. $81^{-3/4}$

50. $4^{-3/2}$

51. $\left(\dfrac{25}{36}\right)^{-1/2}$

52. $\left(\dfrac{16}{49}\right)^{-1/2}$

53. $\left(\dfrac{81}{16}\right)^{-3/4}$

54. $\left(\dfrac{27}{8}\right)^{-2/3}$

55. $16^{1/2} + 27^{1/3}$

56. $25^{1/2} + 100^{1/2}$

57. $8^{-2/3} + 4^{-1/2}$

58. $49^{-1/2} + 25^{-1/2}$

Use the properties of exponents to simplify each of the following as much as possible. Assume all bases are positive.

59. $x^{3/5} \cdot x^{1/5}$

60. $x^{3/4} \cdot x^{5/4}$

61. $(a^{3/4})^{4/3}$

62. $(a^{2/3})^{3/4}$

63. $\dfrac{x^{1/5}}{x^{3/5}}$

64. $\dfrac{x^{2/7}}{x^{5/7}}$

65. $\dfrac{x^{5/6}}{x^{2/3}}$

66. $\dfrac{x^{7/8}}{x^{8/7}}$

67. $(x^{3/5}y^{5/6}z^{1/3})^{3/5}$

68. $(x^{3/4}y^{1/8}z^{5/6})^{4/5}$

69. $\dfrac{a^{3/4}b^2}{a^{7/8}b^{1/4}}$

70. $\dfrac{a^{1/3}b^4}{a^{3/5}b^{1/3}}$

71. $\dfrac{(y^{2/3})^{3/4}}{(y^{1/3})^{3/5}}$

72. $\dfrac{(y^{5/4})^{2/5}}{(y^{1/4})^{4/3}}$

73. $\left(\dfrac{a^{-1/4}}{b^{1/2}}\right)^8$

74. $\left(\dfrac{a^{-1/5}}{b^{1/3}}\right)^{15}$

75. $\dfrac{(r^{-2}s^{1/3})^6}{r^8 s^{3/2}}$

76. $\dfrac{(r^{-5}s^{1/2})^4}{r^{12}s^{5/2}}$

77. $\dfrac{(25a^6b^4)^{1/2}}{(8a^{-9}b^3)^{-1/3}}$

78. $\dfrac{(27a^3b^6)^{1/3}}{(81a^8b^{-4})^{1/4}}$

79. Show that the expression $(a^{1/2} + b^{1/2})^2$ is not equal to $a + b$ by replacing a with 9 and b with 4 in both expressions and then simplifying each.

80. Show that the statement $(a^2 + b^2)^{1/2} = a + b$ is not, in general, true by replacing a with 3 and b with 4 and then simplifying both sides.

81. You may have noticed, if you have been using a calculator to find roots, that you can find the fourth root of a number by pressing the square root button twice. Written in symbols, this fact looks like this:

$$\sqrt{\sqrt{a}} = \sqrt[4]{a} \qquad (a \geq 0)$$

Show that this statement is true by rewriting each side with exponents instead of radical notation and then simplifying the left side.

82. Show that the following statement is true by rewriting each side with exponents instead of radical notation and then simplifying the left side.

$$\sqrt[3]{\sqrt{a}} = \sqrt[6]{a} \qquad (a \geq 0)$$

Applying the Concepts

83. Maximum Speed The maximum speed (v) that an automobile can travel around a curve of radius r without skidding is given by the equation

$$v = \left(\dfrac{5r}{2}\right)^{1/2}$$

where v is in miles per hour and r is measured in feet. What is the maximum speed a car can travel around a curve with a radius of 250 feet without skidding?

84. Relativity The equation

$$L = \left(1 - \dfrac{v^2}{c^2}\right)^{1/2}$$

gives the relativistic length of a 1-foot ruler traveling with velocity v. Find L if

$$\dfrac{v}{c} = \dfrac{3}{5}$$

85. Golden Ratio The golden ratio is the ratio of the length to the width in any golden rectangle. The exact value of this number is $\dfrac{1 + \sqrt{5}}{2}$. Use a calculator to find a decimal approximation to this number and round it to the nearest thousandth.

86. Golden Ratio The reciprocal of the golden ratio is $\dfrac{2}{1 + \sqrt{5}}$. Find a decimal approximation to this number that is accurate to the nearest thousandth.

87. Sequences Find the next term in the following sequence. Then explain how this sequence is related to the Fibonacci sequence.

$$\dfrac{3}{2}, \dfrac{5}{3}, \dfrac{8}{5}, \ldots$$

88. Sequences Write the first ten terms in the sequence shown in Problem 87. Then find a decimal approximation to each of the ten terms, rounding each to the nearest thousandth.

89. Chemistry Figure 6 shows part of a model of a magnesium oxide (MgO) crystal. Each corner of the square is at the center of one oxygen ion (O^{2-}), and the center of the middle ion is at the center of the square. The radius for each oxygen ion is 60 picometers (pm), and the radius for each magnesium ion (Mg^{2+}) is 150 picometers.

(a) Find the length of the side of the square. Write your answer in picometers.

(b) Find the length of the diagonal of the square. Write your answer in picometers.

(c) If 1 meter is 10^{12} picometers, give the length of the diagonal of the square in meters.

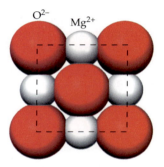

FIGURE 6 *(Susan M. Young)*

90. Chemistry Figure 7 shows part of a model of crystallized aluminum. Each corner of the square is at the center of one aluminum atom, and the center of the middle atom is at the center of the square. The radius for each aluminum atom is 143 picometers (pm).
  (a) Find the length of the side of the square. Write your answer in picometers.
  (b) If 1 meter is 10^{12} picometers, give the length of the side of the square in meters.

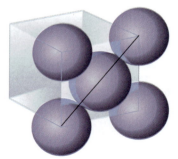

FIGURE 7 *(Susan M. Young)*

91. Geometry The length of each side of the cube shown in Figure 8 is 1 inch.
  (a) Find the length of the diagonal CH.
  (b) Find the length of the diagonal CF.

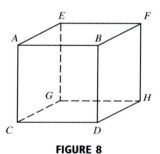

FIGURE 8

92. Chemistry Figure 9 shows part of a model of a crystal containing three atoms. The endpoints of the diagonal of the cube are at the centers of the two outside atoms, and the center of the cube is at the center of the middle atom. The radius of each atom is 100 picometers.
  (a) Find the length of the diagonal of the cube.
  (b) Find the length of the side of the cube.

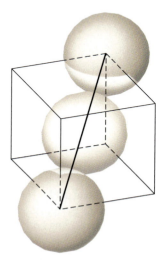

FIGURE 9 *(Susan M. Young)*

93. Comparing Graphs Identify the graph with the correct equation.
  (a) $y = x$
  (b) $y = x^2$
  (c) $y = x^{2/3}$
  (d) What are the two points of intersection of all three graphs?

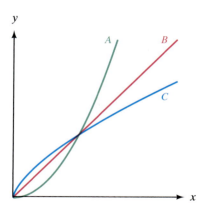

94. Falling Objects The time t it takes an object to fall d feet is given by the equation

$$t = \frac{1}{4}\sqrt{d}$$

(a) The Sears Tower in Chicago is 1,450 feet tall. How long would it take a penny to fall to the ground from the top of the Sears Tower?

(b) An object took 30 seconds to fall to the ground. From what distance must it have been dropped?

Review Problems

The problems that follow review material we covered in Section 5.3. Reviewing these problems will help you understand the next section.

Multiply.

95. $x^2(x^4 - x)$ **96.** $5x^2(2x^3 - x)$

97. $(x - 3)(x + 5)$ **98.** $(x - 2)(x + 2)$

99. $(x^2 - 5)^2$ **100.** $(x^2 + 5)^2$

101. $(x - 3)(x^2 + 3x + 9)$

102. $(x + 3)(x^2 - 3x + 9)$

Extending the Concepts

Graph the equation $y = x^{3/4}$, and then use the Trace and Zoom features to approximate each of the following to the nearest tenth.

103. $2^{3/4}$ **104.** $3^{3/4}$

105. $10^{3/4}$ **106.** $16^{3/4}$

Graph $y = x^{3/4}$ and $y = x^{4/3}$ on the same coordinate system, using a window with X from -4 to 4 and Y from -1 to 4. Use the graphs to answer the following questions.

107. Where do the two graphs intersect?

108. For what values of x is $x^{3/4} \geq x^{4/3}$?

Carbon-14 dating is used extensively in science to find the age of fossils. If at one time a fossilized substance contains an amount A of carbon-14, then t years later the amount of carbon-14 it contains is given by the formula

$$A \cdot 2^{-t/5,600}$$

Use this formula and a calculator to solve the following problems.

109. A fossilized substance contains 3 micrograms of carbon-14. How much carbon-14 will be left at the end of

(a) 5,000 years? (b) 10,000 years?

(c) 56,000 years? (d) 112,000 years?

110. A fossilized substance contains 5 micrograms of carbon-14. How much carbon-14 will be left at the end of

(a) 500 years? (b) 5,000 years?

(c) 56,000 years? (d) 112,000 years?

Suppose you purchased ten silver proof coin sets in 1997 for $21 each, for a total investment of $210. Three years later, in 2000, you find that each set is worth $30, which means that your ten sets have a total value of $300.

United States Mint Proof Set

You can calculate the annual rate of return on this investment using a formula that involves rational exponents. The annual rate of return will tell you at what interest rate you would have to invest your original $210 in order for it to be worth $300, 3 years later. As you will see at the end of this section, the annual rate of return on this investment is 12.6%, which is a good return on your money.

In this section we will look at multiplication, division, factoring, and simplification of some expressions that resemble polynomials but contain rational exponents. The problems in this section will be of particular interest to you if you are planning to take either an engineering calculus class or a business calculus class. As was the case in the previous section, we will assume all variables represent nonnegative real numbers. That way, we will not have to worry about the possibility of introducing undefined terms — even roots of negative numbers — into any of our examples. Let's begin this section with a look at multiplication of expressions containing rational exponents.

EXAMPLE   1   Multiply $x^{2/3}(x^{4/3} - x^{1/3})$.

Solution   Applying the distributive property and then simplifying the resulting terms, we have:

$$
\begin{aligned}
x^{2/3}(x^{4/3} - x^{1/3}) &= x^{2/3}x^{4/3} - x^{2/3}x^{1/3} &&\text{Distributive property}\\
&= x^{6/3} - x^{3/3} &&\text{Add exponents.}\\
&= x^2 - x &&\text{Simplify.}
\end{aligned}
$$

EXAMPLE   2   Multiply $(x^{2/3} - 3)(x^{2/3} + 5)$.

Solution   Applying the FOIL method, we multiply as if we were multiplying two binomials:

$$(x^{2/3} - 3)(x^{2/3} + 5) = x^{2/3}x^{2/3} + 5x^{2/3} - 3x^{2/3} - 15$$
$$= x^{4/3} + 2x^{2/3} - 15$$

EXAMPLE 3   Multiply $(3a^{1/3} - 2b^{1/3})(4a^{1/3} - b^{1/3})$.

Solution   Again, we use the FOIL method to multiply:

$$(3a^{1/3} - 2b^{1/3})(4a^{1/3} - b^{1/3}) = 3a^{1/3}4a^{1/3} - 3a^{1/3}b^{1/3} - 2b^{1/3}4a^{1/3} + 2b^{1/3}b^{1/3}$$
$$= 12a^{2/3} - 11a^{1/3}b^{1/3} + 2b^{2/3}$$

EXAMPLE 4   Expand $(t^{1/2} - 5)^2$.

Solution   We can use the definition of exponents and the FOIL method:

$$(t^{1/2} - 5)^2 = (t^{1/2} - 5)(t^{1/2} - 5)$$
$$= t^{1/2}t^{1/2} - 5t^{1/2} - 5t^{1/2} + 25$$
$$= t - 10t^{1/2} + 25$$

We can obtain the same result by using the formula for the square of a binomial, $(a - b)^2 = a^2 - 2ab + b^2$.

$$(t^{1/2} - 5)^2 = (t^{1/2})^2 - 2t^{1/2} \cdot 5 + 5^2$$
$$= t - 10t^{1/2} + 25$$

EXAMPLE 5   Multiply $(x^{3/2} - 2^{3/2})(x^{3/2} + 2^{3/2})$.

Solution   This product has the form $(a - b)(a + b)$, which will result in the difference of two squares, $a^2 - b^2$:

$$(x^{3/2} - 2^{3/2})(x^{3/2} + 2^{3/2}) = (x^{3/2})^2 - (2^{3/2})^2$$
$$= x^3 - 2^3$$
$$= x^3 - 8$$

EXAMPLE 6   Multiply $(a^{1/3} - b^{1/3})(a^{2/3} + a^{1/3}b^{1/3} + b^{2/3})$.

Solution   We can find this product by multiplying in columns:

$$
\begin{array}{l}
a^{2/3} + a^{1/3}b^{1/3} + b^{2/3} \\
\underline{ a^{1/3} - b^{1/3}} \\
a + a^{2/3}b^{1/3} + a^{1/3}b^{2/3} \\
\underline{ - a^{2/3}b^{1/3} - a^{1/3}b^{2/3} - b} \\
a - b
\end{array}
$$

The product is $a - b$.

Our next example involves division with expressions that contain rational exponents. As you will see, this kind of division is very similar to division of a polynomial by a monomial (as shown previously in Section 6.2).

EXAMPLE 7 Divide $\dfrac{15x^{2/3}y^{1/3} - 20x^{4/3}y^{2/3}}{5x^{1/3}y^{1/3}}$.

Solution We can approach this problem in the same way we approached division by a monomial. We simply divide each term in the numerator by the term in the denominator:

$$\frac{15x^{2/3}y^{1/3} - 20x^{4/3}y^{2/3}}{5x^{1/3}y^{1/3}} = \frac{15x^{2/3}y^{1/3}}{5x^{1/3}y^{1/3}} - \frac{20x^{4/3}y^{2/3}}{5x^{1/3}y^{1/3}}$$
$$= 3x^{1/3} - 4xy^{1/3}$$

The next three examples involve factoring. In the first example, we are told what to factor from each term of an expression.

EXAMPLE 8 Factor $3(x - 2)^{1/3}$ from $12(x - 2)^{4/3} - 9(x - 2)^{1/3}$, and then simplify, if possible.

Solution This solution is similar to factoring out the greatest common factor:

$$12(x - 2)^{4/3} - 9(x - 2)^{1/3} = 3(x - 2)^{1/3}[4(x - 2) - 3]$$
$$= 3(x - 2)^{1/3}(4x - 11)$$

Although an expression containing rational exponents is not a polynomial — remember, a polynomial must have exponents that are whole numbers — we are going to treat the expressions that follow as if they were polynomials.

EXAMPLE 9 Factor $x^{2/3} - 3x^{1/3} - 10$ as if it were a trinomial.

Solution We can think of $x^{2/3} - 3x^{1/3} - 10$ as if it is a trinomial in which the variable is $x^{1/3}$. To see this, replace $x^{1/3}$ with y to get

$$y^2 - 3y - 10$$

Since this trinomial in y factors as $(y - 5)(y + 2)$, we can factor our original expression similarly:

$$x^{2/3} - 3x^{1/3} - 10 = (x^{1/3} - 5)(x^{1/3} + 2)$$

Remember, with factoring, we can always multiply our factors to check that we have factored correctly.

EXAMPLE 10   Factor $6x^{2/5} + 11x^{1/5} - 10$ as if it were a trinomial.

Solution   We can think of the expression in question as a trinomial in $x^{1/5}$.

$$6x^{2/5} + 11x^{1/5} - 10 = (3x^{1/5} - 2)(2x^{1/5} + 5)$$

In our next example, we combine two expressions by applying the methods we used to add and subtract fractions or rational expressions in Chapter 6.

EXAMPLE 11   Subtract $(x^2 + 4)^{1/2} - \dfrac{x^2}{(x^2 + 4)^{1/2}}$.

Solution   To combine these two expressions, we need to find a least common denominator, change to equivalent fractions, and subtract numerators. The least common denominator is $(x^2 + 4)^{1/2}$.

$$(x^2 + 4)^{1/2} - \frac{x^2}{(x^2 + 4)^{1/2}} = \frac{(x^2 + 4)^{1/2}}{1} \cdot \frac{(x^2 + 4)^{1/2}}{(x^2 + 4)^{1/2}} - \frac{x^2}{(x^2 + 4)^{1/2}}$$

$$= \frac{x^2 + 4 - x^2}{(x^2 + 4)^{1/2}}$$

$$= \frac{4}{(x^2 + 4)^{1/2}}$$

EXAMPLE 12   If you purchase an investment for P dollars and t years later it is worth A dollars, then the annual rate of return r on that investment is given by the formula

$$r = \left(\frac{A}{P}\right)^{1/t} - 1$$

United States Mint Proof Set

Find the annual rate of return on a coin collection that was purchased for $210 and sold 3 years later for $300.

Solution   Using $A = 300$, $P = 210$, and $t = 3$ in the formula, we have

$$r = \left(\frac{300}{210}\right)^{1/3} - 1$$

The easiest way to simplify this expression is with a calculator.

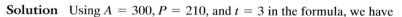

Allowing three decimal places, the result is 0.126. The annual return on the coin collection is approximately 12.6%. To do as well with a savings account, we would have to invest the original $210 in an account that paid 12.6%, compounded annually.

Getting Ready for Class

After reading through the preceding section, respond in your own words and in complete sentences.

A. To multiply two expressions with fractional exponents, should the bases be the same, or should the exponents be the same?

B. Is it possible to multiply two expressions with fractional exponents and end up with an expression containing only integer components? Support your answer with examples.

C. Write an application modeled by the equation $r = (\frac{1,000}{600})^{1/8} - 1$.

D. When can you use the FOIL method with expressions that contain rational exponents?

PROBLEM SET 7.2

Multiply. (Assume all variables in this problem set represent nonnegative real numbers.)

1. $x^{2/3}(x^{1/3} + x^{4/3})$ **2.** $x^{2/5}(x^{3/5} - x^{8/5})$

3. $a^{1/2}(a^{3/2} - a^{1/2})$ **4.** $a^{1/4}(a^{3/4} + a^{7/4})$

5. $2x^{1/3}(3x^{8/3} - 4x^{5/3} + 5x^{2/3})$

6. $5x^{1/2}(4x^{5/2} + 3x^{3/2} + 2x^{1/2})$

7. $4x^{1/2}y^{3/5}(3x^{3/2}y^{-3/5} - 9x^{-1/2}y^{7/5})$

8. $3x^{4/5}y^{1/3}(4x^{6/5}y^{-1/3} - 12x^{-4/5}y^{5/3})$

9. $(x^{2/3} - 4)(x^{2/3} + 2)$ **10.** $(x^{2/3} - 5)(x^{2/3} + 2)$

11. $(a^{1/2} - 3)(a^{1/2} - 7)$ **12.** $(a^{1/2} - 6)(a^{1/2} - 2)$

13. $(4y^{1/3} - 3)(5y^{1/3} + 2)$

14. $(5y^{1/3} - 2)(4y^{1/3} + 3)$

15. $(5x^{2/3} + 3y^{1/2})(2x^{2/3} + 3y^{1/2})$

16. $(4x^{2/3} - 2y^{1/2})(5x^{2/3} - 3y^{1/2})$

17. $(t^{1/2} + 5)^2$ **18.** $(t^{1/2} - 3)^2$

19. $(x^{3/2} + 4)^2$ **20.** $(x^{3/2} - 6)^2$

21. $(a^{1/2} - b^{1/2})^2$ **22.** $(a^{1/2} + b^{1/2})^2$

23. $(2x^{1/2} - 3y^{1/2})^2$ **24.** $(5x^{1/2} + 4y^{1/2})^2$

25. $(a^{1/2} - 3^{1/2})(a^{1/2} + 3^{1/2})$

26. $(a^{1/2} - 5^{1/2})(a^{1/2} + 5^{1/2})$

27. $(x^{3/2} + y^{3/2})(x^{3/2} - y^{3/2})$

28. $(x^{5/2} + y^{5/2})(x^{5/2} - y^{5/2})$

29. $(t^{1/2} - 2^{3/2})(t^{1/2} + 2^{3/2})$

30. $(t^{1/2} - 5^{3/2})(t^{1/2} + 5^{3/2})$

31. $(2x^{3/2} + 3^{1/2})(2x^{3/2} - 3^{1/2})$

32. $(3x^{1/2} + 2^{3/2})(3x^{1/2} - 2^{3/2})$

33. $(x^{1/3} + y^{1/3})(x^{2/3} - x^{1/3}y^{1/3} + y^{2/3})$

34. $(x^{1/3} - y^{1/3})(x^{2/3} + x^{1/3}y^{1/3} + y^{2/3})$

35. $(a^{1/3} - 2)(a^{2/3} + 2a^{1/3} + 4)$

36. $(a^{1/3} + 3)(a^{2/3} - 3a^{1/3} + 9)$

37. $(2x^{1/3} + 1)(4x^{2/3} - 2x^{1/3} + 1)$

38. $(3x^{1/3} - 1)(9x^{2/3} + 3x^{1/3} + 1)$

39. $(t^{1/4} - 1)(t^{1/4} + 1)(t^{1/2} + 1)$

40. $(t^{1/4} - 2)(t^{1/4} + 2)(t^{1/2} + 4)$

Divide.

41. $\dfrac{18x^{3/4} + 27x^{1/4}}{9x^{1/4}}$ **42.** $\dfrac{25x^{1/4} + 30x^{3/4}}{5x^{1/4}}$

43. $\dfrac{12x^{2/3}y^{1/3} - 16x^{1/3}y^{2/3}}{4x^{1/3}y^{1/3}}$

44. $\dfrac{12x^{4/3}y^{1/3} - 18x^{1/3}y^{4/3}}{6x^{1/3}y^{1/3}}$

45. $\dfrac{21a^{7/5}b^{3/5} - 14a^{2/5}b^{8/5}}{7a^{2/5}b^{3/5}}$

46. $\dfrac{24a^{9/5}b^{3/5} - 16a^{4/5}b^{8/5}}{8a^{4/5}b^{3/5}}$

47. Factor $3(x - 2)^{1/2}$ from
$12(x - 2)^{3/2} - 9(x - 2)^{1/2}$.

48. Factor $4(x + 1)^{1/3}$ from
$4(x + 1)^{4/3} + 8(x + 1)^{1/3}$.

49. Factor $5(x - 3)^{7/5}$ from
$5(x - 3)^{12/5} - 15(x - 3)^{7/5}$.

50. Factor $6(x + 3)^{8/7}$ from
$6(x + 3)^{15/7} - 12(x + 3)^{8/7}$.

51. Factor $3(x + 1)^{1/2}$ from
$9x(x + 1)^{3/2} + 6(x + 1)^{1/2}$.

52. Factor $4x(x + 1)^{1/2}$ from
$4x^2(x + 1)^{1/2} + 8x(x + 1)^{3/2}$.

Factor each of the following as if it were a trinomial.

53. $x^{2/3} - 5x^{1/3} + 6$
54. $x^{2/3} - x^{1/3} - 6$
55. $a^{2/5} - 2a^{1/5} - 8$
56. $a^{2/5} + 2a^{1/5} - 8$
57. $2y^{2/3} - 5y^{1/3} - 3$
58. $3y^{2/3} + 5y^{1/3} - 2$
59. $9t^{2/5} - 25$
60. $16t^{2/5} - 49$
61. $4x^{2/7} + 20x^{1/7} + 25$
62. $25x^{2/7} - 20x^{1/7} + 4$

Simplify each of the following to a single fraction.

63. $\dfrac{3}{x^{1/2}} + x^{1/2}$
64. $\dfrac{2}{x^{1/2}} - x^{1/2}$

65. $x^{2/3} + \dfrac{5}{x^{1/3}}$
66. $x^{3/4} - \dfrac{7}{x^{1/4}}$

67. $\dfrac{3x^2}{(x^3 + 1)^{1/2}} + (x^3 + 1)^{1/2}$

68. $\dfrac{x^3}{(x^2 - 1)^{1/2}} + 2x(x^2 - 1)^{1/2}$

69. $\dfrac{x^2}{(x^2 + 4)^{1/2}} - (x^2 + 4)^{1/2}$

70. $\dfrac{x^5}{(x^2 - 2)^{1/2}} + 4x^3(x^2 - 2)^{1/2}$

Use a calculator to find approximations to each of the following. Round your answers for Problems 75 and 76 to three places past the decimal point.

71. $16^{0.25}$
72. $81^{0.25}$

73. $9^{1.5}$
74. $32^{0.4}$

75. $\left(\dfrac{1}{2}\right)^{1/5}$
76. $\left(\dfrac{1}{2}\right)^{1/10}$

Applying the Concepts

77. Investing   A coin collection is purchased as an investment for $500 and sold 4 years later for $900. Find the annual rate of return on the investment.

78. Investing   An investor buys stock in a company for $800. Five years later, the same stock is worth $1,600. Find the annual rate of return on the stocks.

79. Investing   Find the annual rate of return on a home that is purchased for $60,000 and is sold 5 years later for $80,000.

80. Investing   Find the annual rate of return on a home that is purchased for $75,000 and is sold 10 years later for $150,000.

Review Problems

The problems that follow review material we covered in Sections 6.1 and 6.2.

Reduce to lowest terms. [6.1]

81. $\dfrac{x^2 - 9}{x^4 - 81}$
82. $\dfrac{6 - a - a^2}{3 - 2a - a^2}$

Divide. [6.2]

83. $\dfrac{15x^2y - 20x^4y^2}{5xy}$
84. $\dfrac{12x^3y^2 - 24x^2y^3}{6xy}$

Divide using long division. [6.2]

85. $\dfrac{10x^2 + 7x - 12}{2x + 3}$
86. $\dfrac{6x^2 - x - 35}{2x - 5}$

87. $\dfrac{x^3 - 125}{x - 5}$
88. $\dfrac{x^3 + 64}{x + 4}$

7.3 Simplified Form for Radicals

Earlier in this chapter we showed how the Pythagorean theorem can be used to construct a golden rectangle. In a similar manner, the Pythagorean theorem can be used to construct the attractive spiral shown here.

The Spiral of Roots

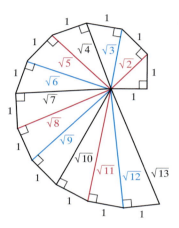

This spiral is called the Spiral of Roots because each of the diagonals is the positive square root of one of the positive integers. At the end of this section, we will use the Pythagorean theorem and some of the material in this section to construct this spiral.

In this section we will use radical notation instead of rational exponents. We will begin by stating two properties of radicals. Following this, we will give a definition for simplified form for radical expressions. The examples in this section show how we use the properties of radicals to write radical expressions in simplified form.

There are two properties of radicals. For these two properties, we will assume a and b are nonnegative real numbers whenever n is an even number.

PROPERTY 1 FOR RADICALS

$$\sqrt[n]{ab} = \sqrt[n]{a}\,\sqrt[n]{b}$$

In words: The nth root of a product is the product of the nth roots.

Proof of Property 1

$\sqrt[n]{ab} = (ab)^{1/n}$	Definition of fractional exponents
$= a^{1/n}b^{1/n}$	Exponents distribute over products.
$= \sqrt[n]{a}\,\sqrt[n]{b}$	Definition of fractional exponents

Note: There is no property for radicals that says the *n*th root of a sum is the sum of the *n*th roots. That is,

$$\sqrt[n]{a + b} \neq \sqrt[n]{a} + \sqrt[n]{b}$$

PROPERTY 2 FOR RADICALS

$$\sqrt[n]{\frac{a}{b}} = \frac{\sqrt[n]{a}}{\sqrt[n]{b}} \qquad (b \neq 0)$$

In words: The *n*th root of a quotient is the quotient of the *n*th roots.

The proof of Property 2 is similar to the proof of Property 1.

These two properties of radicals allow us to change the form of and simplify radical expressions without changing their value.

SIMPLIFIED FORM FOR RADICAL EXPRESSIONS

A radical expression is in *simplified form* if

1. None of the factors of the radicand (the quantity under the radical sign) can be written as powers greater than or equal to the index — that is, no perfect squares can be factors of the quantity under a square root sign, no perfect cubes can be factors of what is under a cube root sign, and so forth;

2. There are no fractions under the radical sign; and

3. There are no radicals in the denominator.

Satisfying the first condition for simplified form actually amounts to taking as much out from under the radical sign as possible. The following examples illustrate the first condition for simplified form.

EXAMPLE 1   Write $\sqrt{50}$ in simplified form.

Solution   The largest perfect square that divides 50 is 25. We write 50 as $25 \cdot 2$ and apply Property 1 for radicals:

$$\sqrt{50} = \sqrt{25 \cdot 2} \qquad 50 = 25 \cdot 2$$
$$= \sqrt{25}\,\sqrt{2} \qquad \text{Property 1}$$
$$= 5\sqrt{2} \qquad \sqrt{25} = 5$$

We have taken as much as possible out from under the radical sign — in this case, factoring 25 from 50 and then writing $\sqrt{25}$ as 5.

EXAMPLE 2 Write in simplified form: $\sqrt{48x^4y^3}$, where $x, y \geq 0$

Solution The largest perfect square that is a factor of the radicand is $16x^4y^2$. Applying Property 1 again, we have

$$\sqrt{48x^4y^3} = \sqrt{16x^4y^2 \cdot 3y}$$
$$= \sqrt{16x^4y^2}\,\sqrt{3y}$$
$$= 4x^2y\sqrt{3y}$$

EXAMPLE 3 Write $\sqrt[3]{40a^5b^4}$ in simplified form.

Solution We now want to factor the largest perfect cube from the radicand. We write $40a^5b^4$ as $8a^3b^3 \cdot 5a^2b$ and proceed as we did in Examples 1 and 2.

$$\sqrt[3]{40a^5b^4} = \sqrt[3]{8a^3b^3 \cdot 5a^2b}$$
$$= \sqrt[3]{8a^3b^3}\,\sqrt[3]{5a^2b}$$
$$= 2ab\sqrt[3]{5a^2b}$$

Here are some further examples concerning the first condition for simplified form.

EXAMPLES Write each expression in simplified form.

4. $\sqrt{12x^7y^6} = \sqrt{4x^6y^6 \cdot 3x}$
$\quad\quad\quad\quad\; = \sqrt{4x^6y^6}\,\sqrt{3x}$
$\quad\quad\quad\quad\; = 2x^3y^3\sqrt{3x}$

5. $\sqrt[3]{54a^6b^2c^4} = \sqrt[3]{27a^6c^3 \cdot 2b^2c}$
$\quad\quad\quad\quad\quad\; = \sqrt[3]{27a^6c^3}\,\sqrt[3]{2b^2c}$
$\quad\quad\quad\quad\quad\; = 3a^2c\sqrt[3]{2b^2c}$

The second property of radicals is used to simplify a radical that contains a fraction.

EXAMPLE 6 Simplify $\sqrt{\frac{3}{4}}$.

Solution Applying Property 2 for radicals, we have

$$\sqrt{\frac{3}{4}} = \frac{\sqrt{3}}{\sqrt{4}} \quad\quad \text{Property 2}$$

$$= \frac{\sqrt{3}}{2} \quad\quad \sqrt{4} = 2$$

The last expression is in simplified form because it satisfies all three conditions for simplified form.

EXAMPLE 7 Write $\sqrt{\frac{5}{6}}$ in simplified form.

Solution Proceeding as in Example 6, we have

$$\sqrt{\frac{5}{6}} = \frac{\sqrt{5}}{\sqrt{6}}$$

The resulting expression satisfies the second condition for simplified form since neither radical contains a fraction. It does, however, violate Condition 3 since it has a radical in the denominator. Getting rid of the radical in the denominator is called *rationalizing the denominator* and is accomplished, in this case, by multiplying the numerator and denominator by $\sqrt{6}$:

$$\frac{\sqrt{5}}{\sqrt{6}} = \frac{\sqrt{5}}{\sqrt{6}} \cdot \frac{\sqrt{6}}{\sqrt{6}}$$

$$= \frac{\sqrt{30}}{\sqrt{6^2}}$$

$$= \frac{\sqrt{30}}{6}$$

EXAMPLES Rationalize the denominator.

8. $\dfrac{4}{\sqrt{3}} = \dfrac{4}{\sqrt{3}} \cdot \dfrac{\sqrt{3}}{\sqrt{3}}$

$\qquad = \dfrac{4\sqrt{3}}{\sqrt{3^2}}$

$\qquad = \dfrac{4\sqrt{3}}{3}$

9. $\dfrac{2\sqrt{3x}}{\sqrt{5y}} = \dfrac{2\sqrt{3x}}{\sqrt{5y}} \cdot \dfrac{\sqrt{5y}}{\sqrt{5y}}$

$\qquad = \dfrac{2\sqrt{15xy}}{\sqrt{(5y)^2}}$

$\qquad = \dfrac{2\sqrt{15xy}}{5y}$

When the denominator involves a cube root, we must multiply by a radical that will produce a perfect cube under the cube root sign in the denominator, as our next example illustrates.

EXAMPLE 10 Rationalize the denominator in $\dfrac{7}{\sqrt[3]{4}}$.

$2\sqrt[3]{2}$

Solution Since $4 = 2^2$, we can multiply both numerator and denominator by $\sqrt[3]{2}$ and obtain $\sqrt[3]{2^3}$ in the denominator.

$$\frac{7}{\sqrt[3]{4}} = \frac{7}{\sqrt[3]{2^2}}$$

$$= \frac{7}{\sqrt[3]{2^2}} \cdot \frac{\sqrt[3]{2}}{\sqrt[3]{2}}$$

$$= \frac{7\sqrt[3]{2}}{\sqrt[3]{2^3}}$$

$$= \frac{7\sqrt[3]{2}}{2}$$

EXAMPLE 11 Simplify $\sqrt{\dfrac{12x^5y^3}{5z}}$. $\dfrac{2x^2y\sqrt{3xy}}{\sqrt{5z}} = \dfrac{2x^2y\sqrt{15xyz}}{5z}$

Solution We use Property 2 to write the numerator and denominator as two separate radicals:

$$\sqrt{\frac{12x^5y^3}{5z}} = \frac{\sqrt{12x^5y^3}}{\sqrt{5z}}$$

Simplifying the numerator, we have

$$\frac{\sqrt{12x^5y^3}}{\sqrt{5z}} = \frac{\sqrt{4x^4y^2}\,\sqrt{3xy}}{\sqrt{5z}}$$

$$= \frac{2x^2y\sqrt{3xy}}{\sqrt{5z}}$$

To rationalize the denominator, we multiply the numerator and denominator by $\sqrt{5z}$:

$$\frac{2x^2y\sqrt{3xy}}{\sqrt{5z}} \cdot \frac{\sqrt{5z}}{\sqrt{5z}} = \frac{2x^2y\sqrt{15xyz}}{\sqrt{(5z)^2}}$$

$$= \frac{2x^2y\sqrt{15xyz}}{5z}$$

THE SQUARE ROOT OF A PERFECT SQUARE

So far in this chapter, we have assumed that all our variables are nonnegative when they appear under a square root symbol. There are times, however, when this is not the case.

Consider the following two statements:

$$\sqrt{3^2} = \sqrt{9} = 3 \qquad \text{and} \qquad \sqrt{(-3)^2} = \sqrt{9} = 3$$

Whether we operate on 3 or -3, the result is the same: Both expressions simplify to 3. The other operation we have worked with in the past that produces the same result is absolute value. That is,

$$|3| = 3 \quad \text{and} \quad |-3| = 3$$

This leads us to the next property of radicals.

PROPERTY 3 FOR RADICALS

If a is a real number, then $\sqrt{a^2} = |a|$.

The result of this discussion and Property 3 is simply this:

If we know a is positive, then $\sqrt{a^2} = a$.
If we know a is negative, then $\sqrt{a^2} = |a|$.
If we don't know if a is positive or negative, then $\sqrt{a^2} = |a|$.

EXAMPLES   Simplify each expression. Do *not* assume the variables represent positive numbers.

12. $\sqrt{9x^2} = 3|x|$
13. $\sqrt{x^3} = |x|\sqrt{x}$
14. $\sqrt{x^2 - 6x + 9} = \sqrt{(x-3)^2} = |x-3|$
15. $\sqrt{x^3 - 5x^2} = \sqrt{x^2(x-5)} = |x|\sqrt{x-5}$

As you can see, we must use absolute value symbols when we take a square root of a perfect square, unless we know the base of the perfect square is a positive number. The same idea holds for higher even roots, but not for odd roots. With odd roots, no absolute value symbols are necessary.

EXAMPLES   Simplify each expression.

16. $\sqrt[3]{(-2)^3} = \sqrt[3]{-8} = -2$
17. $\sqrt[3]{(-5)^3} = \sqrt[3]{-125} = -5$

We can extend this discussion to all roots as follows:

EXTENDING PROPERTY 3 FOR RADICALS

If a is a real number, then

$$\sqrt[n]{a^n} = |a| \qquad \text{if} \qquad n \text{ is even}$$

$$\sqrt[n]{a^n} = a \qquad \text{if} \qquad n \text{ is odd}$$

THE SPIRAL OF ROOTS

In order to visualize the square roots of the positive integers, we can construct the spiral of roots that we mentioned in the introduction to this chapter. To begin, we draw two line segments, each of length 1, at right angles to each other. Then we use the Pythagorean theorem to find the length of the diagonal. Figure 1 illustrates this procedure.

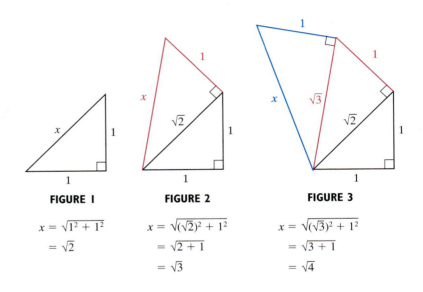

FIGURE 1	FIGURE 2	FIGURE 3
$x = \sqrt{1^2 + 1^2}$	$x = \sqrt{(\sqrt{2})^2 + 1^2}$	$x = \sqrt{(\sqrt{3})^2 + 1^2}$
$\quad = \sqrt{2}$	$\quad = \sqrt{2 + 1}$	$\quad = \sqrt{3 + 1}$
	$\quad = \sqrt{3}$	$\quad = \sqrt{4}$

Next, we construct a second triangle by connecting a line segment of length 1 to the end of the first diagonal so that the angle formed is a right angle. We find the length of the second diagonal using the Pythagorean theorem. Figure 2 illustrates this procedure. Continuing to draw new triangles by connecting line segments of length 1 to the end of each new diagonal, so that the angle formed is a right angle, the spiral of roots begins to appear (Figure 3).

THE SPIRAL OF ROOTS AND FUNCTION NOTATION

Looking over the diagrams and calculations in the preceding discussion, we see that each diagonal in the spiral of roots is found by using the length of the previous diagonal.

First diagonal: $\sqrt{1^2 + 1^2} = \sqrt{2}$

Second diagonal: $\sqrt{(\sqrt{2})^2 + 1^2} = \sqrt{3}$

Third diagonal: $\sqrt{(\sqrt{3})^2 + 1^2} = \sqrt{4}$

Fourth diagonal: $\sqrt{(\sqrt{4})^2 + 1^2} = \sqrt{5}$

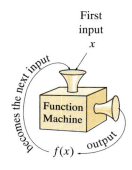

First input

x

becomes the next input

Function Machine

$f(x)$ — output

A process like this one, in which the answer to one calculation is used to find the answer to the next calculation, is called a *recursive* process. In this particular case, we can use function notation to model the process. If we let x represent the length of any diagonal, then the length of the next diagonal is given by

$$f(x) = \sqrt{x^2 + 1}$$

To begin the process of finding the diagonals, we let $x = 1$:

$$f(1) = \sqrt{1^2 + 1} = \sqrt{2}$$

To find the next diagonal, we substitute $\sqrt{2}$ for x to obtain

$$f[f(1)] = f(\sqrt{2}) = \sqrt{(\sqrt{2})^2 + 1} = \sqrt{3}$$
$$f(f[f(1)]) = f(\sqrt{3}) = \sqrt{(\sqrt{3})^2 + 1} = \sqrt{4}$$

We can describe this process of finding the diagonals of the spiral of roots very concisely this way:

$$f(1), f[f(1)], f(f[f(1)]), \ \ldots \qquad \text{where } f(x) = \sqrt{x^2 + 1}$$

This sequence of function values is a special case of a general category of similar sequences that are closely connected to *fractals* and *chaos,* two topics in mathematics that are currently receiving a good deal of attention.

Using TECHNOLOGY

As our preceding discussion indicates, the length of each diagonal in the spiral of roots is used to calculate the length of the next diagonal. The `ANS` key on a graphing calculator can be used very effectively in a situation like this. To begin, we store the number 1 in the variable ANS. Next, we key in the formula used to produce each diagonal using ANS for the variable. After that, it is simply a matter of pressing `ENTER`, as many times as we like, to produce the lengths of as many diagonals as we like. Here is a summary of what we do:

Enter This	Display Shows
1 `ENTER`	1.000
$\sqrt{\ }$ (ANS^2 + 1) `ENTER`	1.414
`ENTER`	1.732
`ENTER`	2.000
`ENTER`	2.236

If you continue to press the `ENTER` key, you will produce decimal approximations for as many of the diagonals in the spiral of roots as you like.

Getting Ready for Class

After reading through the preceding section, respond in your own words and in complete sentences.

A. Explain why this statement is false: "The square root of a sum is the sum of the square roots."

B. What is simplified form for an expression that contains a square root?

C. Why is it not necessarily true that $\sqrt{a^2} = a$?

D. What does it mean to rationalize the denominator in an expression?

PROBLEM SET 7.3

Use Property 1 for radicals to write each of the following expressions in simplified form. (Assume all variables are nonnegative through Problem 70.)

1. $\sqrt{8}$ **2.** $\sqrt{32}$ **3.** $\sqrt{98}$

4. $\sqrt{75}$ **5.** $\sqrt{288}$ **6.** $\sqrt{128}$

7. $\sqrt{80}$ **8.** $\sqrt{200}$ **9.** $\sqrt{48}$

10. $\sqrt{27}$ **11.** $\sqrt{675}$ **12.** $\sqrt{972}$

13. $\sqrt[3]{54}$ **14.** $\sqrt[3]{24}$ **15.** $\sqrt[3]{128}$

16. $\sqrt[3]{162}$ **17.** $\sqrt[3]{432}$ **18.** $\sqrt[3]{1,536}$

19. $\sqrt[5]{64}$ **20.** $\sqrt[4]{48}$ **21.** $\sqrt{18x^3}$

22. $\sqrt{27x^5}$ **23.** $\sqrt[4]{32y^7}$ **24.** $\sqrt[5]{32y^7}$

25. $\sqrt[3]{40x^4y^7}$ **26.** $\sqrt[3]{128x^6y^2}$

27. $\sqrt{48a^2b^3c^4}$ **28.** $\sqrt{72a^4b^3c^2}$

29. $\sqrt[3]{48a^2b^3c^4}$ **30.** $\sqrt[3]{72a^4b^3c^2}$

31. $\sqrt[5]{64x^8y^{12}}$ **32.** $\sqrt[4]{32x^9y^{10}}$

33. $\sqrt[5]{243x^7y^{10}z^5}$ **34.** $\sqrt[5]{64x^8y^4z^{11}}$

Substitute the given numbers into the expression $\sqrt{b^2 - 4ac}$, and then simplify.

35. $a = 2, b = -6, c = 3$

36. $a = 6, b = 7, c = -5$

37. $a = 1, b = 2, c = 6$

38. $a = 2, b = 5, c = 3$

39. $a = \dfrac{1}{2}, b = -\dfrac{1}{2}, c = -\dfrac{5}{4}$

40. $a = \dfrac{7}{4}, b = -\dfrac{3}{4}, c = -2$

Rationalize the denominator in each of the following expressions.

41. $\dfrac{2}{\sqrt{3}}$ **42.** $\dfrac{3}{\sqrt{2}}$ **43.** $\dfrac{5}{\sqrt{6}}$

44. $\dfrac{7}{\sqrt{5}}$ **45.** $\sqrt{\dfrac{1}{2}}$ **46.** $\sqrt{\dfrac{1}{3}}$

47. $\sqrt{\dfrac{1}{5}}$ **48.** $\sqrt{\dfrac{1}{6}}$ **49.** $\dfrac{4}{\sqrt[3]{2}}$

50. $\dfrac{5}{\sqrt[3]{3}}$ **51.** $\dfrac{2}{\sqrt[3]{9}}$ **52.** $\dfrac{3}{\sqrt[3]{4}}$

53. $\sqrt[4]{\dfrac{3}{2x^2}}$ **54.** $\sqrt[4]{\dfrac{5}{3x^2}}$ **55.** $\sqrt[4]{\dfrac{8}{y}}$

56. $\sqrt[4]{\dfrac{27}{y}}$ **57.** $\sqrt[3]{\dfrac{4x}{3y}}$ **58.** $\sqrt[3]{\dfrac{7x}{6y}}$

59. $\sqrt[3]{\dfrac{2x}{9y}}$ **60.** $\sqrt[3]{\dfrac{5x}{4y}}$ **61.** $\sqrt[4]{\dfrac{1}{8x^3}}$

62. $\sqrt[4]{\dfrac{8}{9x^3}}$

Write each of the following in simplified form.

63. $\sqrt{\dfrac{27x^3}{5y}}$ **64.** $\sqrt{\dfrac{12x^5}{7y}}$ **65.** $\sqrt{\dfrac{75x^3y^2}{2z}}$

66. $\sqrt{\dfrac{50x^2y^3}{3z}}$ **67.** $\sqrt[3]{\dfrac{16a^4b^3}{9c}}$ **68.** $\sqrt[3]{\dfrac{54a^5b^4}{25c^2}}$

69. $\sqrt[3]{\dfrac{8x^3y^6}{9z}}$ **70.** $\sqrt[3]{\dfrac{27x^6y^3}{2z^2}}$

Simplify each expression. Do *not* assume the variables represent positive numbers.

$4a^2(a^2 + 4a + 4)$
$4a^2(a+2)^2$

71. $\sqrt{25x^2}$

72. $\sqrt{49x^2}$

73. $\sqrt{27x^3y^2}$

74. $\sqrt{40x^3y^2}$

75. $\sqrt{x^2 - 10x + 25}$

76. $\sqrt{x^2 - 16x + 64}$

77. $\sqrt{4x^2 + 12x + 9}$

78. $\sqrt{16x^2 + 40x + 25}$

79. $\sqrt{4a^4 + 16a^3 + 16a^2}$

80. $\sqrt{9a^4 + 18a^3 + 9a^2}$

81. $\sqrt{4x^3 - 8x^2}$

82. $\sqrt{18x^3 - 9x^2}$ $2x\sqrt{x-2}$

83. Show that the statement $\sqrt{a + b} = \sqrt{a} + \sqrt{b}$ is not true by replacing a with 9 and b with 16 and simplifying both sides.

84. Find a pair of values for a and b that will make the statement $\sqrt{a + b} = \sqrt{a} + \sqrt{b}$ true.

Applying the Concepts

85. Diagonal Distance The distance d between opposite corners of a rectangular room with length l and width w is given by

$$d = \sqrt{l^2 + w^2}$$

How far is it between opposite corners of a living room that measures 10 by 15 feet?

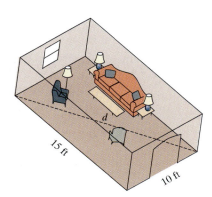

86. Radius of a Sphere The radius r of a sphere with volume V can be found by using the formula

$$r = \sqrt[3]{\dfrac{3V}{4\pi}}$$

Find the radius of a sphere with volume 9 cubic feet. Write your answer in simplified form. (Use $\frac{22}{7}$ for π.)

Volume

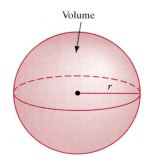

87. Spiral of Roots Construct your own spiral of roots by using a ruler. Draw the first triangle by using two 1-inch lines. The first diagonal will have a length of $\sqrt{2}$ inches. Each new triangle will be formed by drawing a 1-inch line segment at the end of the previous diagonal so that the angle formed is 90°.

88. Spiral of Roots Construct a spiral of roots by using line segments of length 2 inches. The length of the first diagonal will be $2\sqrt{2}$ inches. The length of the second diagonal will be $2\sqrt{3}$ inches.

89. Spiral of Roots If $f(x) = \sqrt{x^2 + 1}$, find the first six terms in the following sequence. Use your results to predict the value of the 10th term and the 100th term.

$$f(1), f[f(1)], f(f[f(1)]), \ \cdots$$

90. Spiral of Roots If $f(x) = \sqrt{x^2 + 4}$, find the first six terms in the following sequence. Use your results to predict the value of the 10th term and the 100th term. (The numbers in this sequence are the lengths of the diagonals of the spiral you drew in Problem 88.)

$$f(2), f[f(2)], f(f[f(2)]), \ \cdots$$

Review Problems

The following problems review material we covered in Section 6.3.

Perform the indicated operations.

91. $\dfrac{8xy^3}{9x^2y} \div \dfrac{16x^2y^2}{18xy^3}$

92. $\dfrac{25x^2}{5y^4} \cdot \dfrac{30y^3}{2x^5}$

93. $\dfrac{12a^2 - 4a - 5}{2a + 1} \cdot \dfrac{7a + 3}{42a^2 - 17a - 15}$

94. $\dfrac{20a^2 - 7a - 3}{4a + 1} \cdot \dfrac{25a^2 - 5a - 6}{5a + 2}$

95. $\dfrac{8x^3 + 27}{27x^3 + 1} \div \dfrac{6x^2 + 7x - 3}{9x^2 - 1}$

96. $\dfrac{27x^3 + 8}{8x^3 + 1} \div \dfrac{6x^2 + x - 2}{4x^2 - 1}$

Extending the Concepts

Factor each radicand into the product of prime factors. Then simplify each radical.

97. $\sqrt[3]{8{,}640}$ **98.** $\sqrt{8{,}640}$

99. $\sqrt[3]{10{,}584}$ **100.** $\sqrt{10{,}584}$

Assume a is a positive number, and rationalize each denominator.

101. $\dfrac{1}{\sqrt[10]{a^3}}$ **102.** $\dfrac{1}{\sqrt[12]{a^7}}$

103. $\dfrac{1}{\sqrt[20]{a^{11}}}$ **104.** $\dfrac{1}{\sqrt[15]{a^{13}}}$

105. Show that the two expressions $\sqrt{x^2 + 1}$ and $x + 1$ are not, in general, equal to each other by graphing $y = \sqrt{x^2 + 1}$ and $y = x + 1$ in the same viewing window.

106. Show that the two expressions $\sqrt{x^2 + 9}$ and $x + 3$ are not, in general, equal to each other by graphing $y = \sqrt{x^2 + 9}$ and $y = x + 3$ in the same viewing window.

107. Approximately how far apart are the graphs in Problem 105 when $x = 2$?

108. Approximately how far apart are the graphs in Problem 106 when $x = 2$?

109. For what value of x are the expressions $\sqrt{x^2 + 1}$ and $x + 1$ equal?

110. For what value of x are the expressions $\sqrt{x^2 + 9}$ and $x + 3$ equal?

111. Hero's Formula   Hero's formula for the area of a triangle is

$$A = \sqrt{s(s - a)(s - b)(s - c)}$$

in which a, b, and c are the lengths of the sides of the triangle and s is one-half the perimeter of the triangle.

(a) Write a formula to find s in terms of a, b, and c.

(b) Use the result from part (a) to find the area of a triangle that has sides of 5, 6, and 7.

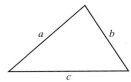

112. Brahmagupta's Formula   A cyclic quadrilateral is a quadrilateral that has each of its vertices on a circle. The area of a cyclic quadrilateral can be found with Brahmagupta's formula

$$A = \sqrt{(s - a)(s - b)(s - c)(s - d)}$$

where a, b, c, and d are the lengths of the sides of the cyclic quadrilateral and s is one-half the perimeter of the quadrilateral.

(a) Write a formula for s in terms of a, b, c, and d.

(b) Find the area of a cyclic quadrilateral with sides of 4, 6, 9, and 3.

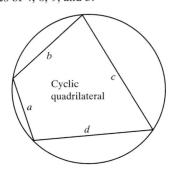

In Chapter 5 we found that we could add similar terms when combining polynomials. The same idea applies to addition and subtraction of radical expressions.

> **DEFINITION** Two radicals are said to be **similar radicals** if they have the same index and the same radicand.

The expressions $5\sqrt[3]{7}$ and $-8\sqrt[3]{7}$ are similar since the index is 3 in both cases and the radicands are 7. The expressions $3\sqrt[4]{5}$ and $7\sqrt[3]{5}$ are not similar because they have different indices, and the expressions $2\sqrt[5]{8}$ and $3\sqrt[5]{9}$ are not similar because the radicands are not the same.

We add and subtract radical expressions in the same way we add and subtract polynomials — by combining similar terms under the distributive property.

EXAMPLE 1 Combine $5\sqrt{3} - 4\sqrt{3} + 6\sqrt{3}$.

Solution All three radicals are similar. We apply the distributive property to get

$$5\sqrt{3} - 4\sqrt{3} + 6\sqrt{3} = (5 - 4 + 6)\sqrt{3}$$
$$= 7\sqrt{3}$$

EXAMPLE 2 Combine $3\sqrt{8} + 5\sqrt{18}$.

Solution The two radicals do not seem to be similar. We must write each in simplified form before applying the distributive property.

$$3\sqrt{8} + 5\sqrt{18} = 3\sqrt{4 \cdot 2} + 5\sqrt{9 \cdot 2}$$
$$= 3\sqrt{4}\,\sqrt{2} + 5\sqrt{9}\,\sqrt{2}$$
$$= 3 \cdot 2\sqrt{2} + 5 \cdot 3\sqrt{2}$$
$$= 6\sqrt{2} + 15\sqrt{2}$$
$$= (6 + 15)\sqrt{2}$$
$$= 21\sqrt{2}$$

The result of Example 2 can be generalized to the following rule for sums and differences of radical expressions.

> **RULE**
>
> To add or subtract radical expressions, put each in simplified form, and apply the distributive property if possible. We can add only similar radicals. We must write each expression in simplified form for the radicals before we can tell if the radicals are similar.

EXAMPLE 3   Combine $7\sqrt{75xy^3} - 4y\sqrt{12xy}$, where $x, y \geq 0$.

Solution   We write each expression in simplified form and combine similar radicals:

$$
\begin{aligned}
7\sqrt{75xy^3} - 4y\sqrt{12xy} &= 7\sqrt{25y^2}\,\sqrt{3xy} - 4y\sqrt{4}\,\sqrt{3xy} \\
&= 35y\sqrt{3xy} - 8y\sqrt{3xy} \\
&= (35y - 8y)\sqrt{3xy} \\
&= 27y\sqrt{3xy}
\end{aligned}
$$

EXAMPLE 4   Combine $10\sqrt[3]{8a^4b^2} + 11a\sqrt[3]{27ab^2}$.

Solution   Writing each radical in simplified form and combining similar terms, we have

$$
\begin{aligned}
10\sqrt[3]{8a^4b^2} + 11a\sqrt[3]{27ab^2} &= 10\sqrt[3]{8a^3}\,\sqrt[3]{ab^2} + 11a\sqrt[3]{27}\,\sqrt[3]{ab^2} \\
&= 20a\sqrt[3]{ab^2} + 33a\sqrt[3]{ab^2} \\
&= 53a\sqrt[3]{ab^2}
\end{aligned}
$$

EXAMPLE 5   Combine $\dfrac{\sqrt{3}}{2} + \dfrac{1}{\sqrt{3}}$.

Solution   We begin by writing the second term in simplified form.

$$
\begin{aligned}
\frac{\sqrt{3}}{2} + \frac{1}{\sqrt{3}} &= \frac{\sqrt{3}}{2} + \frac{1}{\sqrt{3}} \cdot \frac{\sqrt{3}}{\sqrt{3}} \\
&= \frac{\sqrt{3}}{2} + \frac{\sqrt{3}}{3} \\
&= \frac{1}{2}\sqrt{3} + \frac{1}{3}\sqrt{3} \\
&= \left(\frac{1}{2} + \frac{1}{3}\right)\sqrt{3} \\
&= \frac{5}{6}\sqrt{3} = \frac{5\sqrt{3}}{6}
\end{aligned}
$$

EXAMPLE 6 Construct a golden rectangle from a square of side 4. Then show that the ratio of the length to the width is the golden ratio $\dfrac{1 + \sqrt{5}}{2}$.

Solution   Figure 1 shows the golden rectangle constructed from a square of side 4.

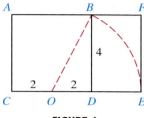

FIGURE I

The length of the diagonal OB is found from the Pythagorean theorem.

$$OB = \sqrt{2^2 + 4^2} = \sqrt{4 + 16} = \sqrt{20} = 2\sqrt{5}$$

The ratio of the length to the width for the rectangle is the golden ratio.

$$\text{Golden ratio} = \frac{CE}{EF} = \frac{2 + 2\sqrt{5}}{4} = \frac{2(1 + \sqrt{5})}{2 \cdot 2} = \frac{1 + \sqrt{5}}{2}$$

As you can see, showing that the ratio of length to width in this rectangle is the golden ratio depends on our ability to write $\sqrt{20}$ as $2\sqrt{5}$ and our ability to reduce to lowest terms by factoring and then dividing out the common factor 2 from the numerator and denominator.

 # Getting Ready for Class

After reading through the preceding section, respond in your own words and in complete sentences.

A. What are similar radicals?

B. When can we add two radical expressions?

C. What is the first step when adding or subtracting expressions containing radicals?

D. What is the golden ratio, and where does it come from?

PROBLEM SET 7.4

Combine the following expressions. (Assume any variables under an even root are nonnegative.)

1. $3\sqrt{5} + 4\sqrt{5}$

2. $6\sqrt{3} - 5\sqrt{3}$

3. $3x\sqrt{7} - 4x\sqrt{7}$

4. $6y\sqrt{a} + 7y\sqrt{a}$

5. $5\sqrt[3]{10} - 4\sqrt[3]{10}$

6. $6\sqrt[4]{2} + 9\sqrt[4]{2}$

7. $8\sqrt[5]{6} - 2\sqrt[5]{6} + 3\sqrt[5]{6}$

8. $7\sqrt[6]{7} - \sqrt[6]{7} + 4\sqrt[6]{7}$

9. $3x\sqrt{2} - 4x\sqrt{2} + x\sqrt{2}$

10. $5x\sqrt{6} - 3x\sqrt{6} - 2x\sqrt{6}$

11. $\sqrt[3]{20} - \sqrt[4]{80} + \sqrt[3]{45}$

12. $\sqrt[2]{8} - \sqrt[4]{32} - \sqrt{18}$

13. $4\sqrt{8} - 2\sqrt{50} - 5\sqrt{72}$

14. $\sqrt{48} - 3\sqrt{27} + 2\sqrt{75}$

15. $5x\sqrt{8} + 3\sqrt{32x^2} - 5\sqrt{50x^2}$

16. $2\sqrt{50x^2} - 8x\sqrt{18} - 3\sqrt{72x^2}$

17. $5\sqrt[3]{16} - 4\sqrt[3]{54}$

18. $\sqrt[3]{81} + 3\sqrt[3]{24}$

19. $\sqrt[3]{x^4y^2} + 7x\sqrt[3]{xy^2}$

20. $2\sqrt[3]{x^8y^6} - 3y^2\sqrt[3]{8x^8}$

21. $5a^2\sqrt{27ab^3} - 6b\sqrt{12a^5b}$

22. $9a\sqrt{20a^3b^2} + 7b\sqrt{45a^5}$

23. $b\sqrt[3]{24a^5b} + 3a\sqrt[3]{81a^2b^4}$

24. $7\sqrt[3]{a^4b^3c^2} - 6ab\sqrt[3]{ac^2}$

25. $5x\sqrt[4]{3y^5} + y\sqrt[4]{243x^4y} + \sqrt[4]{48x^4y^5}$

26. $x\sqrt[4]{5xy^8} + y\sqrt[4]{405x^5y^4} + y^2\sqrt[4]{80x^5}$

27. $\dfrac{\sqrt{2}}{2} + \dfrac{1}{\sqrt{2}}$

28. $\dfrac{\sqrt{3}}{3} + \dfrac{1}{\sqrt{3}}$

29. $\dfrac{\sqrt{5}}{3} + \dfrac{1}{\sqrt{5}}$

30. $\dfrac{\sqrt{6}}{2} + \dfrac{1}{\sqrt{6}}$

31. $\sqrt{x} - \dfrac{1}{\sqrt{x}}$

32. $\sqrt{x} + \dfrac{1}{\sqrt{x}}$

33. $\dfrac{\sqrt{18}}{6} + \sqrt{\dfrac{1}{2}} + \dfrac{\sqrt{2}}{2}$

34. $\dfrac{\sqrt{12}}{6} + \sqrt{\dfrac{1}{3}} + \dfrac{\sqrt{3}}{3}$

35. $\sqrt{6} - \sqrt{\dfrac{2}{3}} + \sqrt{\dfrac{1}{6}}$

36. $\sqrt{15} - \sqrt{\dfrac{3}{5}} + \sqrt{\dfrac{5}{3}}$

37. $\sqrt[3]{25} + \dfrac{3}{\sqrt[3]{5}}$

38. $\sqrt[4]{8} + \dfrac{1}{\sqrt[4]{2}}$

39. Use a calculator to find a decimal approximation for $\sqrt{12}$ and for $2\sqrt{3}$.

40. Use a calculator to find decimal approximations for $\sqrt{50}$ and $5\sqrt{2}$.

41. Use a calculator to find a decimal approximation for $\sqrt{8} + \sqrt{18}$. Is it equal to the decimal approximation for $\sqrt{26}$ or $\sqrt{50}$?

42. Use a calculator to find a decimal approximation for $\sqrt{3} + \sqrt{12}$. Is it equal to the decimal approximation for $\sqrt{15}$ or $\sqrt{27}$?

Each of the following statements is false. Correct the right side of each one to make the statement true.

43. $3\sqrt{2x} + 5\sqrt{2x} = 8\sqrt{4x}$

44. $5\sqrt{3} - 7\sqrt{3} = -2\sqrt{9}$

45. $\sqrt{9 + 16} = 3 + 4$

46. $\sqrt{36 + 64} = 6 + 8$

Applying the Concepts

47. Golden Rectangle Construct a golden rectangle from a square of side 8. Then show that the ratio of the length to the width is the golden ratio $\dfrac{1 + \sqrt{5}}{2}$.

48. Golden Rectangle Construct a golden rectangle from a square of side 10. Then show that the ratio of the length to the width is the golden ratio $\dfrac{1 + \sqrt{5}}{2}$.

49. Golden Rectangle Use a ruler to construct a golden rectangle from a square of side 1 inch. Then show that the ratio of the length to the width is the golden ratio.

50. Golden Rectangle Use a ruler to construct a golden rectangle from a square of side $\frac{2}{3}$ inch. Then show that the ratio of the length to the width is the golden ratio.

51. Golden Rectangle To show that all golden rectangles have the same ratio of length to width, construct a golden rectangle from a square of side $2x$. Then show that the ratio of the length to the width is the golden ratio.

52. Golden Rectangle To show that all golden rectangles have the same ratio of length to width, construct a golden rectangle from a square of side x. Then show that the ratio of the length to the width is the golden ratio.

53. Isosceles Right Triangles A triangle is isosceles if it has two equal sides, and a triangle is a right triangle if it has a right angle in it. Sketch an isosceles right triangle, and find the ratio of the hypotenuse to a leg.

54. Equilateral Triangles A triangle is equilateral if it has three equal sides. The triangle in the figure is equilateral with each side of length $2x$. Find the ratio of the height to a side.

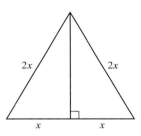

55. Pyramids The following solid is called a regular square pyramid because its base is a square and all eight edges are the same length, 5. It is also true that the vertex, V, is directly above the center of the base.

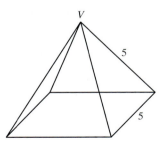

(a) Find the ratio of a diagonal of the base to the length of a side.

(b) Find the ratio of the area of the base to the diagonal of the base.

(c) Find the ratio of the area of the base to the perimeter of the base.

56. Pyramids Refer to the diagram of a square pyramid in Problem 55. Find the ratio of the height h of the pyramid to the altitude a.

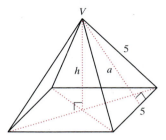

Review Problems

The following problems review material we covered in Section 6.4.

Add and subtract as indicated.

57. $\dfrac{2a - 4}{a + 2} - \dfrac{a - 6}{a + 2}$ **58.** $\dfrac{2a - 3}{a - 2} - \dfrac{a - 1}{a - 2}$

59. $3 + \dfrac{4}{3 - t}$ **60.** $6 + \dfrac{2}{5 - t}$

61. $\dfrac{3}{2x - 5} - \dfrac{39}{8x^2 - 14x - 15}$

62. $\dfrac{2}{4x - 5} + \dfrac{9}{8x^2 - 38x + 35}$

63. $\dfrac{1}{x - y} - \dfrac{3xy}{x^3 - y^3}$

64. $\dfrac{1}{x + y} + \dfrac{3xy}{x^3 + y^3}$

We have worked with the golden rectangle more than once in this chapter. The following is one such golden rectangle.

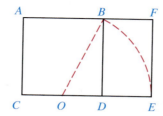

By now you know that in any golden rectangle constructed from a square (of any size), the ratio of the length to the width will be

$$\frac{1 + \sqrt{5}}{2}$$

which we call the golden ratio. What is interesting is that the smaller rectangle on the right, *BFED*, is also a golden rectangle. We will use the mathematics developed in this section to confirm this fact.

In this section we will look at multiplication and division of expressions that contain radicals. As you will see, multiplication of expressions that contain radicals is very similar to multiplication of polynomials. The division problems in this section are just an extension of the work we did previously when we rationalized denominators.

EXAMPLE 1 Multiply $(3\sqrt{5})(2\sqrt{7})$.

Solution We can rearrange the order and grouping of the numbers in this product by applying the commutative and associative properties. Following this, we apply Property 1 for radicals and multiply:

$$(3\sqrt{5})(2\sqrt{7}) = (3 \cdot 2)(\sqrt{5}\,\sqrt{7}) \quad \text{Commutative and associative properties}$$
$$= (3 \cdot 2)(\sqrt{5 \cdot 7}) \quad \text{Property 1 for radicals}$$
$$= 6\sqrt{35} \quad \text{Multiplication}$$

In practice, it is not necessary to show the first two steps.

EXAMPLE 2 Multiply $\sqrt{3}(2\sqrt{6} - 5\sqrt{12})$.

Solution Applying the distributive property, we have

$$\sqrt{3}(2\sqrt{6} - 5\sqrt{12}) = \sqrt{3} \cdot 2\sqrt{6} - \sqrt{3} \cdot 5\sqrt{12}$$
$$= 2\sqrt{18} - 5\sqrt{36}$$

Writing each radical in simplified form gives

$$2\sqrt{18} - 5\sqrt{36} = 2\sqrt{9}\sqrt{2} - 5\sqrt{36}$$
$$= 6\sqrt{2} - 30$$

EXAMPLE 3   Multiply $(\sqrt{3} + \sqrt{5})(4\sqrt{3} - \sqrt{5})$.

Solution   The same principle that applies when multiplying two binomials applies to this product. We must multiply each term in the first expression by each term in the second one. Any convenient method can be used. Let's use the FOIL method.

$$
\overset{\text{F}\qquad\quad\text{O}\qquad\quad\text{I}\qquad\quad\text{L}}{(\sqrt{3} + \sqrt{5})(4\sqrt{3} - \sqrt{5}) = \sqrt{3}\cdot 4\sqrt{3} - \sqrt{3}\sqrt{5} + \sqrt{5}\cdot 4\sqrt{3} - \sqrt{5}\sqrt{5}}
$$
$$= 4\cdot 3 - \sqrt{15} + 4\sqrt{15} - 5$$
$$= 12 + 3\sqrt{15} - 5$$
$$= 7 + 3\sqrt{15}$$

EXAMPLE 4   Expand and simplify $(\sqrt{x} + 3)^2$.

Solution 1   We can write this problem as a multiplication problem and proceed as we did in Example 3:

$$(\sqrt{x} + 3)^2 = (\sqrt{x} + 3)(\sqrt{x} + 3)$$
$$
\overset{\text{F}\qquad\ \ \text{O}\qquad\ \ \text{I}\qquad\ \ \text{L}}{= \sqrt{x}\cdot\sqrt{x} + 3\sqrt{x} + 3\sqrt{x} + 3\cdot 3}
$$
$$= x + 3\sqrt{x} + 3\sqrt{x} + 9$$
$$= x + 6\sqrt{x} + 9$$

Solution 2   We can obtain the same result by applying the formula for the square of a sum: $(a + b)^2 = a^2 + 2ab + b^2$.

$$(\sqrt{x} + 3)^2 = (\sqrt{x})^2 + 2(\sqrt{x})(3) + 3^2$$
$$= x + 6\sqrt{x} + 9$$

EXAMPLE 5   Expand $(3\sqrt{x} - 2\sqrt{y})^2$ and simplify the result.

Solution   Let's apply the formula for the square of a difference, $(a - b)^2 = a^2 - 2ab + b^2$.

$$(3\sqrt{x} - 2\sqrt{y})^2 = (3\sqrt{x})^2 - 2(3\sqrt{x})(2\sqrt{y}) + (2\sqrt{y})^2$$
$$= 9x - 12\sqrt{xy} + 4y$$

EXAMPLE 6   Expand and simplify $(\sqrt{x+2} - 1)^2$.

Solution   Applying the formula $(a - b)^2 = a^2 - 2ab + b^2$, we have

$$(\sqrt{x+2} - 1)^2 = (\sqrt{x+2})^2 - 2\sqrt{x+2}(1) + 1^2$$
$$= x + 2 - 2\sqrt{x+2} + 1$$
$$= x + 3 - 2\sqrt{x+2}$$

EXAMPLE 7   Multiply $(\sqrt{6} + \sqrt{2})(\sqrt{6} - \sqrt{2})$.

Solution   We notice the product is of the form $(a + b)(a - b)$, which always gives the difference of two squares, $a^2 - b^2$:

$$(\sqrt{6} + \sqrt{2})(\sqrt{6} - \sqrt{2}) = (\sqrt{6})^2 - (\sqrt{2})^2$$
$$= 6 - 2$$
$$= 4$$

In Example 7 the two expressions $(\sqrt{6} + \sqrt{2})$ and $(\sqrt{6} - \sqrt{2})$ are called *conjugates*. In general, the conjugate of $\sqrt{a} + \sqrt{b}$ is $\sqrt{a} - \sqrt{b}$. If a and b are integers, multiplying conjugates of this form always produces a rational number. That is, if a and b are positive integers, then

$$(\sqrt{a} + \sqrt{b})(\sqrt{a} - \sqrt{b}) = \sqrt{a}\sqrt{a} - \sqrt{a}\sqrt{b} + \sqrt{a}\sqrt{b} - \sqrt{b}\sqrt{b}$$
$$= a - \sqrt{ab} + \sqrt{ab} - b$$
$$= a - b$$

which is rational if a and b are rational.

Division with radical expressions is the same as rationalizing the denominator. In Section 7.3 we were able to divide $\sqrt{3}$ by $\sqrt{2}$ by rationalizing the denominator:

$$\frac{\sqrt{3}}{\sqrt{2}} = \frac{\sqrt{3}}{\sqrt{2}} \cdot \frac{\sqrt{2}}{\sqrt{2}} = \frac{\sqrt{6}}{2}$$

We can accomplish the same result with expressions such as

$$\frac{6}{\sqrt{5} - \sqrt{3}}$$

by multiplying the numerator and denominator by the conjugate of the denominator.

EXAMPLE 8   Divide $\dfrac{6}{\sqrt{5} - \sqrt{3}}$. (Rationalize the denominator.)

Solution   Since the product of two conjugates is a rational number, we multiply the numerator and denominator by the conjugate of the denominator.

$$\frac{6}{\sqrt{5} - \sqrt{3}} = \frac{6}{\sqrt{5} - \sqrt{3}} \cdot \frac{(\sqrt{5} + \sqrt{3})}{(\sqrt{5} + \sqrt{3})}$$

$$= \frac{6\sqrt{5} + 6\sqrt{3}}{(\sqrt{5})^2 - (\sqrt{3})^2}$$

$$= \frac{6\sqrt{5} + 6\sqrt{3}}{5 - 3}$$

$$= \frac{6\sqrt{5} + 6\sqrt{3}}{2}$$

The numerator and denominator of this last expression have a factor of 2 in common. We can reduce the lowest terms by factoring 2 from the numerator and then dividing both the numerator and denominator by 2:

$$= \frac{\cancel{2}(3\sqrt{5} + 3\sqrt{3})}{\cancel{2}}$$

$$= 3\sqrt{5} + 3\sqrt{3}$$

EXAMPLE   9   Rationalize the denominator $\dfrac{\sqrt{5} - 2}{\sqrt{5} + 2}$.

Solution   To rationalize the denominator, we multiply the numerator and denominator by the conjugate of the denominator:

$$\frac{\sqrt{5} - 2}{\sqrt{5} + 2} = \frac{\sqrt{5} - 2}{\sqrt{5} + 2} \cdot \frac{(\sqrt{5} - 2)}{(\sqrt{5} - 2)}$$

$$= \frac{5 - 2\sqrt{5} - 2\sqrt{5} + 4}{(\sqrt{5})^2 - 2^2}$$

$$= \frac{9 - 4\sqrt{5}}{5 - 4}$$

$$= \frac{9 - 4\sqrt{5}}{1}$$

$$= 9 - 4\sqrt{5}$$

EXAMPLE   10   A golden rectangle constructed from a square of side 2 is shown in Figure 1. Show that the smaller rectangle *BDEF* is also a golden rectangle by finding the ratio of its length to its width.

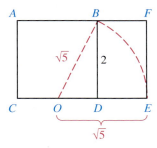

FIGURE 1

Solution   First, find expressions for the length and width of the smaller rectangle.

$$\text{Length} = EF = 2$$

$$\text{Width} = DE = \sqrt{5} - 1$$

Next, we find the ratio of length to width.

$$\text{Ratio of length to width} = \frac{EF}{DE} = \frac{2}{\sqrt{5} - 1}$$

To show that the small rectangle is a golden rectangle, we must show that the ratio of length to width is the golden ratio. We do so by rationalizing the denominator.

$$\frac{2}{\sqrt{5} - 1} = \frac{2}{\sqrt{5} - 1} \cdot \frac{\sqrt{5} + 1}{\sqrt{5} + 1}$$

$$= \frac{2(\sqrt{5} + 1)}{5 - 1}$$

$$= \frac{2(\sqrt{5} + 1)}{4}$$

$$= \frac{\sqrt{5} + 1}{2} \qquad \text{Divide out common factor 2.}$$

Since addition is commutative, this last expression is the golden ratio. Therefore, the small rectangle in Figure 1 is a golden rectangle.

Getting Ready for Class

After reading through the preceding section, respond in your own words and in complete sentences.

A. Explain why $(\sqrt{5} + \sqrt{2})^2 \neq 5 + 2$.

B. Explain in words how you would rationalize the denominator in the expression $\dfrac{\sqrt{3}}{\sqrt{5} - \sqrt{2}}$.

C. What are conjugates?

D. What result is guaranteed when multiplying radical expressions that are conjugates?

PROBLEM SET 7.5

Multiply. (Assume all expressions appearing under a square root symbol represent nonnegative numbers throughout this problem set.)

1. $\sqrt{6}\,\sqrt{3}$

2. $\sqrt{6}\,\sqrt{2}$

3. $(2\sqrt{3})(5\sqrt{7})$

4. $(3\sqrt{5})(2\sqrt{7})$

5. $(4\sqrt{6})(2\sqrt{15})(3\sqrt{10})$

6. $(4\sqrt{35})(2\sqrt{21})(5\sqrt{15})$

7. $(3\sqrt[3]{3})(6\sqrt[3]{9})$

8. $(2\sqrt[3]{2})(6\sqrt[3]{4})$

9. $\sqrt{3}(\sqrt{2} - 3\sqrt{3})$

10. $\sqrt{2}(5\sqrt{3} + 4\sqrt{2})$

11. $6\sqrt[3]{4}(2\sqrt[3]{2} + 1)$

12. $7\sqrt[3]{5}(3\sqrt[3]{25} - 2)$

13. $(\sqrt{3} + \sqrt{2})(3\sqrt{3} - \sqrt{2})$

14. $(\sqrt{5} - \sqrt{2})(3\sqrt{5} + 2\sqrt{2})$

15. $(\sqrt{x} + 5)(\sqrt{x} - 3)$

16. $(\sqrt{x} + 4)(\sqrt{x} + 2)$

17. $(3\sqrt{6} + 4\sqrt{2})(\sqrt{6} + 2\sqrt{2})$

18. $(\sqrt{7} - 3\sqrt{3})(2\sqrt{7} - 4\sqrt{3})$

19. $(\sqrt{3} + 4)^2$

20. $(\sqrt{5} - 2)^2$

21. $(\sqrt{x} - 3)^2$

22. $(\sqrt{x} + 4)^2$

23. $(2\sqrt{a} - 3\sqrt{b})^2$

24. $(5\sqrt{a} - 2\sqrt{b})^2$

25. $(\sqrt{x - 4} + 2)^2$

26. $(\sqrt{x - 3} + 2)^2$

27. $(\sqrt{x - 5} - 3)^2$

28. $(\sqrt{x - 3} - 4)^2$

29. $(\sqrt{3} - \sqrt{2})(\sqrt{3} + \sqrt{2})$

30. $(\sqrt{5} - \sqrt{2})(\sqrt{5} + \sqrt{2})$

31. $(\sqrt{a} + 7)(\sqrt{a} - 7)$

32. $(\sqrt{a} + 5)(\sqrt{a} - 5)$

33. $(5 - \sqrt{x})(5 + \sqrt{x})$

34. $(3 - \sqrt{x})(3 + \sqrt{x})$

35. $(\sqrt{x - 4} + 2)(\sqrt{x - 4} - 2)$

36. $(\sqrt{x + 3} + 5)(\sqrt{x + 3} - 5)$

37. $(\sqrt{3} + 1)^3$

38. $(\sqrt{5} - 2)^3$

Rationalize the denominator in each of the following.

39. $\dfrac{\sqrt{2}}{\sqrt{6} - \sqrt{2}}$

40. $\dfrac{\sqrt{5}}{\sqrt{5} + \sqrt{3}}$

41. $\dfrac{\sqrt{5}}{\sqrt{5} + 1}$

42. $\dfrac{\sqrt{7}}{\sqrt{7} - 1}$

43. $\dfrac{\sqrt{x}}{\sqrt{x} - 3}$

44. $\dfrac{\sqrt{x}}{\sqrt{x} + 2}$

45. $\dfrac{\sqrt{5}}{2\sqrt{5} - 3}$

46. $\dfrac{\sqrt{7}}{3\sqrt{7} - 2}$

47. $\dfrac{3}{\sqrt{x} - \sqrt{y}}$

48. $\dfrac{2}{\sqrt{x} + \sqrt{y}}$

49. $\dfrac{\sqrt{6} + \sqrt{2}}{\sqrt{6} - \sqrt{2}}$

50. $\dfrac{\sqrt{5} - \sqrt{3}}{\sqrt{5} + \sqrt{3}}$

51. $\dfrac{\sqrt{7} - 2}{\sqrt{7} + 2}$

52. $\dfrac{\sqrt{11} + 3}{\sqrt{11} - 3}$

53. $\dfrac{\sqrt{a} + \sqrt{b}}{\sqrt{a} - \sqrt{b}}$

54. $\dfrac{\sqrt{a} - \sqrt{b}}{\sqrt{a} + \sqrt{b}}$

55. $\dfrac{\sqrt{x} + 2}{\sqrt{x} - 2}$

56. $\dfrac{\sqrt{x} - 3}{\sqrt{x} + 3}$

57. $\dfrac{2\sqrt{3} - \sqrt{7}}{3\sqrt{3} + \sqrt{7}}$

58. $\dfrac{5\sqrt{6} + 2\sqrt{2}}{\sqrt{6} - \sqrt{2}}$

59. $\dfrac{3\sqrt{x} + 2}{1 + \sqrt{x}}$

60. $\dfrac{5\sqrt{x} - 1}{2 + \sqrt{x}}$

61. Show that the product $(\sqrt[3]{2} + \sqrt[3]{3})(\sqrt[3]{4} - \sqrt[3]{6} + \sqrt[3]{9})$ is 5.

62. Show that the product $(\sqrt[3]{x} + 2)(\sqrt[3]{x^2} - 2\sqrt[3]{x} + 4)$ is $x + 8$.

Each of the following statements below is false. Correct the right side of each one to make it true.

63. $5(2\sqrt{3}) = 10\sqrt{15}$

64. $3(2\sqrt{x}) = 6\sqrt{3x}$

65. $(\sqrt{x} + 3)^2 = x + 9$

66. $(\sqrt{x} - 7)^2 = x - 49$

67. $(5\sqrt{3})^2 = 15$

68. $(3\sqrt{5})^2 = 15$

Applying the Concepts

69. Gravity If an object is dropped from the top of a 100-foot building, the amount of time t (in seconds) that it takes for the object to be h feet from the ground is given by the formula

$$t = \frac{\sqrt{100 - h}}{4}$$

How long does it take before the object is 50 feet from the ground? How long does it take to reach the ground? (When it is on the ground, h is 0.)

70. Gravity Use the formula given in Problem 69 to determine h if t is 1.25 seconds.

71. Golden Rectangle Rectangle *ACEF* in Figure 2 is a golden rectangle. If side *AC* is 6 inches, show that the smaller rectangle *BDEF* is also a golden rectangle.

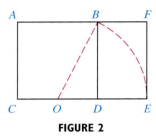

FIGURE 2

72. Golden Rectangle Rectangle *ACEF* in Figure 2 is a golden rectangle. If side *AC* is 1 inch, show that the smaller rectangle *BDEF* is also a golden rectangle.

73. Golden Rectangle If side *AC* in Figure 2 is 2*x*, show that rectangle *BDEF* is a golden rectangle.

74. Golden Rectangle If side *AC* in Figure 2 is *x*, show that rectangle *BDEF* is a golden rectangle.

Review Problems

The problems that follow review material we covered in Section 6.5.

Simplify each complex fraction.

75. $\dfrac{\dfrac{1}{4} - \dfrac{1}{3}}{\dfrac{1}{2} + \dfrac{1}{6}}$

76. $\dfrac{\dfrac{1}{8} - \dfrac{1}{3}}{\dfrac{1}{4} - \dfrac{1}{3}}$

77. $\dfrac{1 - \dfrac{2}{y}}{1 + \dfrac{2}{y}}$

78. $\dfrac{1 + \dfrac{3}{y}}{1 - \dfrac{3}{y}}$

79. $\dfrac{4 + \dfrac{4}{x} + \dfrac{1}{x^2}}{4 - \dfrac{1}{x^2}}$

80. $\dfrac{1 - \dfrac{1}{x} - \dfrac{6}{x^2}}{1 - \dfrac{9}{x^2}}$

7.6 Equations With Radicals

This section is concerned with solving equations that involve one or more radicals. The first step in solving an equation that contains a radical is to eliminate the radical from the equation. To do so, we need an additional property.

SQUARING PROPERTY OF EQUALITY

If both sides of an equation are squared, the solutions to the original equation are solutions to the resulting equation.

We will never lose solutions to our equations by squaring both sides. We may, however, introduce *extraneous solutions*. Extraneous solutions satisfy the equation obtained by squaring both sides of the original equation, but do not satisfy the original equation.

We know that if two real numbers *a* and *b* are equal, then so are their squares:

$$\text{If} \qquad a = b$$
$$\text{then} \qquad a^2 = b^2$$

On the other hand, extraneous solutions are introduced when we square opposites. That is, even though opposites are not equal, their squares are. For example,

$$5 = -5 \qquad \text{A false statement}$$

$$(5)^2 = (-5)^2 \qquad \text{Square both sides.}$$

$$25 = 25 \qquad \text{A true statement}$$

We are free to square both sides of an equation any time it is convenient. We must be aware, however, that doing so may introduce extraneous solutions. We must, therefore, check all our solutions in the original equation if at any time we square both sides of the original equation.

EXAMPLE 1   Solve for x: $\sqrt{3x + 4} = 5$

Solution   We square both sides and proceed as usual:

$$\sqrt{3x + 4} = 5$$

$$(\sqrt{3x + 4})^2 = 5^2$$

$$3x + 4 = 25$$

$$3x = 21$$

$$x = 7$$

Checking $x = 7$ in the original equation, we have

$$\sqrt{3(7) + 4} \stackrel{?}{=} 5$$

$$\sqrt{21 + 4} = 5$$

$$\sqrt{25} = 5$$

$$5 = 5$$

The solution $x = 7$ satisfies the original equation.

EXAMPLE 2   Solve $\sqrt{4x - 7} = -3$.

Solution   Squaring both sides, we have

$$\sqrt{4x - 7} = -3$$

$$(\sqrt{4x - 7})^2 = (-3)^2$$

$$4x - 7 = 9$$

$$4x = 16$$

$$x = 4$$

Checking $x = 4$ in the original equation gives

$$\sqrt{4(4) - 7} = -3$$
$$\sqrt{16 - 7} = -3$$
$$\sqrt{9} = -3$$
$$3 = -3$$

The solution $x = 4$ produces a false statement when checked in the original equation. Since $x = 4$ was the only possible solution, there is no solution to the original equation. The possible solution $x = 4$ is an extraneous solution. It satisfies the equation obtained by squaring both sides of the original equation, but does not satisfy the original equation.

Note: The fact that there is no solution to the equation in Example 2 was obvious to begin with. Notice that the left side of the equation is the *positive* square root of $4x - 7$, which must be a positive number or 0. The right side of the equation is -3. Since we cannot have a number that is either positive or zero equal to a negative number, there is no solution to the equation.

EXAMPLE 3   Solve $\sqrt{5x - 1} + 3 = 7$.

Solution   We must isolate the radical on the left side of the equation. If we attempt to square both sides without doing so, the resulting equation will also contain a radical. Adding -3 to both sides, we have

$$\sqrt{5x - 1} + 3 = 7$$
$$\sqrt{5x - 1} = 4$$

We can now square both sides and proceed as usual:

$$(\sqrt{5x - 1})^2 = 4^2$$
$$5x - 1 = 16$$
$$5x = 17$$
$$x = \frac{17}{5}$$

Checking $x = \frac{17}{5}$, we have

$$\sqrt{5\left(\frac{17}{5}\right) - 1} + 3 = 7$$
$$\sqrt{17 - 1} + 3 = 7$$
$$\sqrt{16} + 3 = 7$$
$$4 + 3 = 7$$
$$7 = 7$$

E X A M P L E   4   Solve $t + 5 = \sqrt{t + 7}$.

Solution   This time, squaring both sides of the equation results in a quadratic equation:

$$(t + 5)^2 = (\sqrt{t + 7})^2 \qquad \text{Square both sides.}$$

$$t^2 + 10t + 25 = t + 7$$

$$t^2 + 9t + 18 = 0 \qquad \text{Standard form}$$

$$(t + 3)(t + 6) = 0 \qquad \text{Factor the left side.}$$

$$t + 3 = 0 \quad \text{or} \quad t + 6 = 0 \qquad \text{Set factors equal to 0.}$$

$$t = -3 \quad \text{or} \quad t = -6$$

We must check each solution in the original equation:

Check $t = -3$	Check $t = -6$
$-3 + 5 \overset{?}{=} \sqrt{-3 + 7}$	$-6 + 5 \overset{?}{=} \sqrt{-6 + 7}$
$2 = \sqrt{4}$	$-1 = \sqrt{1}$
$2 = 2$ A true statement	$-1 = 1$ A false statement

Since $t = -6$ does not check, your only solution is $t = -3$.

E X A M P L E   5   Solve $\sqrt{x - 3} = \sqrt{x} - 3$.

Solution   We begin by squaring both sides. Note carefully what happens when we square the right side of the equation, and compare the square of the right side with the square of the left side. You must convince yourself that these results are correct. (The note on the next page will help if you are having trouble convincing yourself that what is written below is true.)

$$(\sqrt{x - 3})^2 = (\sqrt{x} - 3)^2$$

$$x - 3 = x - 6\sqrt{x} + 9$$

Now we still have a radical in our equation, so we will have to square both sides again. Before we do, though, let's isolate the remaining radical.

$$x - 3 = x - 6\sqrt{x} + 9$$

$$-3 = -6\sqrt{x} + 9 \qquad \text{Add } -x \text{ to each side.}$$

$$-12 = -6\sqrt{x} \qquad \text{Add } -9 \text{ to each side.}$$

$$2 = \sqrt{x} \qquad \text{Divide each side by } -6.$$

$$4 = x \qquad \text{Square each side.}$$

Our only possible solution is $x = 4$, which we check in our original equation as follows:

$$\sqrt{4 - 3} = \sqrt{4} - 3$$

$$\sqrt{1} = 2 - 3$$

$$1 = -1 \qquad \text{A false statement}$$

Substituting 4 for x in the original equation yields a false statement. Since 4 was our only possible solution, there is no solution to our equation.

Note: It is very important that you realize that the square of $(\sqrt{x} - 3)$ is not $x + 9$. Remember, when we square a difference with two terms, we use the formula

$$(a - b)^2 = a^2 - 2ab + b^2$$

Applying this formula to $(\sqrt{x} - 3)^2$, we have

$$(\sqrt{x} - 3)^2 = (\sqrt{x})^2 - 2(\sqrt{x})(3) + 3^2$$
$$= x - 6\sqrt{x} + 9$$

Here is another example of an equation for which we must apply our squaring property twice before all radicals are eliminated.

EXAMPLE 6   Solve $\sqrt{x + 1} = 1 - \sqrt{2x}$.

Solution   This equation has two separate terms involving radical signs. Squaring both sides gives

$$
\begin{array}{ll}
x + 1 = 1 - 2\sqrt{2x} + 2x & \\
-x = -2\sqrt{2x} & \text{Add } -2x \text{ and } -1 \text{ to both sides.} \\
x^2 = 4(2x) & \text{Square both sides.} \\
x^2 - 8x = 0 & \text{Standard form}
\end{array}
$$

Our equation is a quadratic equation in standard form. To solve for x, we factor the left side and set each factor equal to 0:

$$
\begin{array}{lll}
x(x - 8) = 0 & & \text{Factor left side.} \\
x = 0 \quad \text{or} \quad x - 8 = 0 & & \text{Set factors equal to 0.} \\
\phantom{x = 0 \quad \text{or} \quad} x = 8 & &
\end{array}
$$

Since we squared both sides of our equation, we have the possibility that one or both of the solutions are extraneous. We must check each one in the original equation:

Check $x = 8$	Check $x = 0$
$\sqrt{8 + 1} \stackrel{?}{=} 1 - \sqrt{2 \cdot 8}$	$\sqrt{0 + 1} \stackrel{?}{=} 1 - \sqrt{2 \cdot 0}$
$\sqrt{9} = 1 - \sqrt{16}$	$\sqrt{1} = 1 - \sqrt{0}$
$3 = 1 - 4$	$1 = 1 - 0$
$3 = -3$ A false statement	$1 = 1$ A true statement

Since $x = 8$ does not check, it is an extraneous solution. Our only solution is $x = 0$.

EXAMPLE 7   Solve $\sqrt{x + 1} = \sqrt{x + 2} - 1$.

Solution   Squaring both sides we have

$$(\sqrt{x + 1})^2 = (\sqrt{x + 2} - 1)^2$$
$$x + 1 = x + 2 - 2\sqrt{x + 2} + 1$$

Once again we are left with a radical in our equation. Before we square each side again, we must isolate the radical on the right side of the equation.

$x + 1 = x + 3 - 2\sqrt{x + 2}$	Simplify the right side.
$1 = 3 - 2\sqrt{x + 2}$	Add $-x$ to each side.
$-2 = -2\sqrt{x + 2}$	Add -3 to each side.
$1 = \sqrt{x + 2}$	Divide each side by -2.
$1 = x + 2$	Square both sides.
$-1 = x$	Add -2 to each side.

Checking our only possible solution, $x = -1$, in our original equation, we have

$$\sqrt{-1 + 1} \stackrel{?}{=} \sqrt{-1 + 2} - 1$$
$$\sqrt{0} = \sqrt{1} - 1$$
$$0 = 1 - 1$$
$$0 = 0 \qquad \text{A true statement}$$

Our solution checks.

It is also possible to raise both sides of an equation to powers greater than 2. We only need to check for extraneous solutions when we raise both sides of an equation to an even power. Raising both sides of an equation to an odd power will not produce extraneous solutions.

EXAMPLE 8   Solve $\sqrt[3]{4x + 5} = 3$.

Solution   Cubing both sides we have

$$(\sqrt[3]{4x + 5})^3 = 3^3$$
$$4x + 5 = 27$$
$$4x = 22$$
$$x = \frac{22}{4}$$
$$x = \frac{11}{2}$$

We do not need to check $x = \frac{11}{2}$ since we raised both sides to an odd power.

We end this section by looking at graphs of some equations that contain radicals.

EXAMPLE 9 Graph $y = \sqrt{x}$ and $y = \sqrt[3]{x}$.

Solution The graphs are shown in Figures 1 and 2. Notice that the graph of $y = \sqrt{x}$ appears in the first quadrant only, because in the equation $y = \sqrt{x}$, x and y cannot be negative.

The graph of $y = \sqrt[3]{x}$ appears in Quadrants 1 and 3 since the cube root of a positive number is also a positive number, and the cube root of a negative number is a negative number. That is, when x is positive, y will be positive, and when x is negative, y will be negative.

The graphs of both equations will contain the origin, since $y = 0$ when $x = 0$ in both equations.

x	y
-4	Undefined
-1	Undefined
0	0
1	1
4	2
9	3
16	4

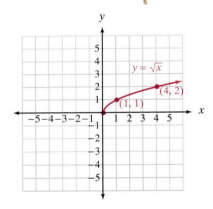

FIGURE 1

x	y
-27	-3
-8	-2
-1	-1
0	0
1	1
8	2
27	3

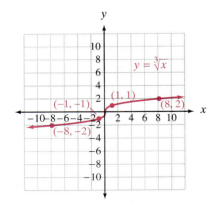

FIGURE 2

Getting Ready for Class

After reading through the preceding section, respond in your own words and in complete sentences.

A. What is the squaring property of equality?

B. Under what conditions do we obtain extraneous solutions to equations that contain radical expressions?

C. If we have raised both sides of an equation to a power, when is it not necessary to check for extraneous solutions?

D. When will you need to apply the squaring property of equality twice in the process of solving an equation containing radicals?

PROBLEM·SET 7.6

Solve each of the following equations.

1. $\sqrt{2x + 1} = 3$

2. $\sqrt{3x + 1} = 4$

3. $\sqrt{4x + 1} = -5$

4. $\sqrt{6x + 1} = -5$

5. $\sqrt{2y - 1} = 3$

6. $\sqrt{3y - 1} = 2$

7. $\sqrt{5x - 7} = -1$

8. $\sqrt{8x + 3} = -6$

9. $\sqrt{2x - 3} - 2 = 4$

10. $\sqrt{3x + 1} - 4 = 1$

11. $\sqrt{4a + 1} + 3 = 2$

12. $\sqrt{5a - 3} + 6 = 2$

13. $\sqrt[4]{3x + 1} = 2$

14. $\sqrt[4]{4x + 1} = 3$

15. $\sqrt[3]{2x - 5} = 1$

16. $\sqrt[3]{5x + 7} = 2$

17. $\sqrt[3]{3a + 5} = -3$

18. $\sqrt[3]{2a + 7} = -2$

19. $\sqrt{y - 3} = y - 3$

20. $\sqrt{y + 3} = y - 3$

21. $\sqrt{a + 2} = a + 2$

22. $\sqrt{a + 10} = a - 2$

23. $\sqrt{2x + 4} = \sqrt{1 - x}$

24. $\sqrt{3x + 4} = -\sqrt{2x + 3}$

25. $\sqrt{4a + 7} = -\sqrt{a + 2}$

26. $\sqrt{7a - 1} = \sqrt{2a + 4}$

27. $\sqrt[4]{5x - 8} = \sqrt[4]{4x - 1}$

28. $\sqrt[4]{6x + 7} = \sqrt[4]{x + 2}$

29. $x + 1 = \sqrt{5x + 1}$

30. $x - 1 = \sqrt{6x + 1}$

31. $t + 5 = \sqrt{2t + 9}$

32. $t + 7 = \sqrt{2t + 13}$

33. $\sqrt{y - 8} = \sqrt{8 - y}$

34. $\sqrt{2y + 5} = \sqrt{5y + 2}$

35. $\sqrt[3]{3x + 5} = \sqrt[3]{5 - 2x}$

36. $\sqrt[3]{4x + 9} = \sqrt[3]{3 - 2x}$

The following equations will require that you square both sides twice before all the radicals are eliminated. Solve each equation using the methods shown in Examples 5, 6, and 7.

37. $\sqrt{x - 8} = \sqrt{x} - 2$

38. $\sqrt{x + 3} = \sqrt{x} - 3$

39. $\sqrt{x + 1} = \sqrt{x} + 1$

40. $\sqrt{x - 1} = \sqrt{x} - 1$

41. $\sqrt{x + 8} = \sqrt{x - 4} + 2$

42. $\sqrt{x + 5} = \sqrt{x - 3} + 2$

43. $\sqrt{x - 5} - 3 = \sqrt{x - 8}$

44. $\sqrt{x - 3} - 4 = \sqrt{x - 3}$

45. $\sqrt{x + 4} = 2 - \sqrt{2x}$ $x+4 = 4 - 2\sqrt{2x} + v$

46. $\sqrt{5x + 1} = 1 + \sqrt{5x}$

47. $\sqrt{2x + 4} = \sqrt{x + 3} + 1$

48. $\sqrt{2x - 1} = \sqrt{x - 4} + 2$

Applying the Concepts

49. Solving a Formula Solve the following formula for h:

$$t = \frac{\sqrt{100 - h}}{4}$$

50. Solving a Formula Solve the following formula for h:

$$t = \sqrt{\frac{2h - 40t}{g}}$$

51. Pendulum Clock The length of time (T) in seconds it takes the pendulum of a clock to swing through one complete cycle is given by the formula

$$T = 2\pi\sqrt{\frac{L}{32}}$$

where L is the length, in feet, of the pendulum, and π is approximately $\frac{22}{7}$. How long must the pendulum be if one complete cycle takes 2 seconds?

1 sec

52. Pendulum Clock Solve the formula in Problem 51 for L.

53. Similar Rectangles Two rectangles are similar if their vertices lie along the same diagonal, as shown in the following diagram.

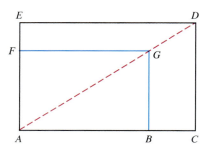

Rectangle ABGF is similar to rectangle ACDE.

If two rectangles are similar, then corresponding sides are in proportion, which means that $\frac{ED}{DC} = \frac{FG}{GB}$. Use these facts in the following diagram to express the length of the larger rectangle w in terms of x.

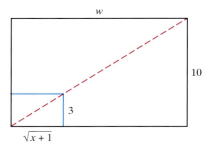

54. Volume Recall that the volume of a box V can be found from the formula $V = $ (length)(width)(height). Find the volume of the following box in terms of x.

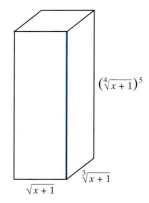

Pollution A long straight river, 100 meters wide, is flowing at 1 meter per second. A pollutant is entering the river at a constant rate, from one of its banks. As the pollutant disperses in the water, it forms a plume that is modeled by the equation $y = \sqrt{x}$. Use this information to answer the following questions.

Pollution source

55. How wide is the plume 25 meters down river from the source of the pollution?

56. How wide is the plume 100 meters down river from the source of the pollution?

57. How far down river from the source of the pollution does the plume reach halfway across the river?

58. How far down the river from the source of the pollution does the plume reach the other side of the river?

59. For the situation described in the instructions and modeled by the equation $y = \sqrt{x}$, what is the range of values that y can assume?

60. If the river was moving at 2 meters per second, would the plume be larger or smaller 100 meters downstream from the source?

Graph each equation.

61. $y = 2\sqrt{x}$ 　　　　 **62.** $y = -2\sqrt{x}$

63. $y = \sqrt{x} - 2$ 　　　 **64.** $y = \sqrt{x} + 2$

65. $y = \sqrt{x-2}$ 　　　 **66.** $y = \sqrt{x+2}$

67. $y = 3\sqrt[3]{x}$ 　　　　 **68.** $y = -3\sqrt[3]{x}$

69. $y = \sqrt[3]{x} + 3$ 　　　 **70.** $y = \sqrt[3]{x} - 3$

71. $y = \sqrt[3]{x+3}$ 　　　 **72.** $y = \sqrt[3]{x-3}$

Review Problems

The problems that follow review material we covered in Section 7.5. Reviewing these problems will help you understand the next section.

Multiply.

73. $\sqrt{2}(\sqrt{3} - \sqrt{2})$ 　　　 **74.** $(\sqrt{x} - 4)(\sqrt{x} + 5)$

75. $(\sqrt{x} + 5)^2$ 　　　　 **76.** $(\sqrt{5} + \sqrt{3})(\sqrt{5} - \sqrt{3})$

Rationalize the denominator.

77. $\dfrac{\sqrt{x}}{\sqrt{x} + 3}$ 　　　 **78.** $\dfrac{\sqrt{5} - \sqrt{3}}{\sqrt{5} + \sqrt{3}}$

Extending the Concepts

Solve each equation.

79. $\dfrac{x}{3\sqrt{2x-3}} - \dfrac{1}{\sqrt{2x-3}} = \dfrac{1}{3}$

80. $\dfrac{x}{5\sqrt{2x+10}} + \dfrac{1}{\sqrt{2x+10}} = \dfrac{1}{5}$

81. $x + 1 = \sqrt[3]{4x + 4}$

82. $x - 1 = \sqrt[3]{4x - 4}$

Solve for y in terms of x.

83. $y + 2 = \sqrt{x^2 + (y-2)^2}$

84. $y + \dfrac{1}{2} = \sqrt{x^2 + \left(y - \dfrac{1}{2}\right)^2}$

85. Use your Y variables list, or write a program, to graph the family of curves $Y = \sqrt{X} + B$ for $B = -3, -2, -1, 0, 1, 2,$ and 3.

86. Use your Y variables list, or write a program, to graph the family of curves $Y = \sqrt{X + B}$ for $B = -3, -2, -1, 0, 1, 2,$ and 3.

87. Summarize the results of Problem 85 by giving a written description of the effect of b on the graph of $y = \sqrt{x} + b$.

88. Summarize the results of Problem 86 by giving a written description of the effect of b on the graph of $y = \sqrt{x + b}$.

89. Use your Y variables list, or write a program, to graph the family of curves $Y = \sqrt[3]{X} + B$ for $B = -3, -2, -1, 0, 1, 2,$ and 3.

90. Use your Y variables list, or write a program, to graph the family of curves $Y = \sqrt[3]{X + B}$ for $B = -3, -2, -1, 0, 1, 2,$ and 3.

91. Summarize the results of Problem 89 by giving a written description of the effect of b on the graph of $y = \sqrt[3]{x} + b$.

92. Summarize the results of Problem 90 by giving a written description of the effect of b on the graph of $y = \sqrt[3]{x + b}$.

93. Use your Y variables list, or write a program, to graph the family of curves $Y = A\sqrt{X}$ for $A = -3, -2, -1, 0, 1, 2,$ and 3.

94. Use your Y variables list, or write a program, to graph the family of curves $Y = A\sqrt{X}$ for $A = \frac{1}{4}, \frac{1}{3}, \frac{1}{2}, 1, 2,$ and 3.

95. Summarize the results of Problems 93 and 94 by giving a written description of the effect of a on the graph of $y = a\sqrt{x}$.

The equation $x^2 = -9$ has no real number solutions since the square of a real number is always positive. We have been unable to work with square roots of negative numbers like $\sqrt{-25}$ and $\sqrt{-16}$ for the same reason. Complex numbers allow us to expand our work with radicals to include square roots of negative numbers and to solve equations like $x^2 = -9$ and $x^2 = -64$. Our work with complex numbers is based on the following definition.

DEFINITION The **number** i is such that $i = \sqrt{-1}$ (which is the same as saying $i^2 = -1$).

The number i, as we have defined it here, is not a real number. Because of the way we have defined i, we can use it to simplify square roots of negative numbers.

SQUARE ROOTS OF NEGATIVE NUMBERS

If a is a positive number, then $\sqrt{-a}$ can always be written as $i\sqrt{a}$. That is,

$$\sqrt{-a} = i\sqrt{a} \qquad \text{if } a \text{ is a positive number}$$

To justify our rule, we simply square the quantity $i\sqrt{a}$ to obtain $-a$. Here is what it looks like when we do so:

$$(i\sqrt{a})^2 = i^2 \cdot (\sqrt{a})^2$$
$$= -1 \cdot a$$
$$= -a$$

Here are some examples that illustrate the use of our new rule.

EXAMPLES Write each square root in terms of the number i.

1. $\sqrt{-25} = i\sqrt{25} = i \cdot 5 = 5i$
2. $\sqrt{-49} = i\sqrt{49} = i \cdot 7 = 7i$
3. $\sqrt{-12} = i\sqrt{12} = i \cdot 2\sqrt{3} = 2i\sqrt{3}$
4. $\sqrt{-17} = i\sqrt{17}$

Note: In Examples 3 and 4 we wrote i before the radical simply to avoid confusion. If we were to write the answer to 3 as $2\sqrt{3}i$, some people would think the i was under the radical sign, and it is not.

If we assume all the properties of exponents hold when the base is i, we can write any power of i as i, -1, $-i$, or 1. Using the fact that $i^2 = -1$, we have

$$i^1 = i$$

$$i^2 = -1$$

$$i^3 = i^2 \cdot i = -1(i) = -i$$

$$i^4 = i^2 \cdot i^2 = -1(-1) = 1$$

Since $i^4 = 1$, i^5 will simplify to i, and we will begin repeating the sequence i, -1, $-i$, 1 as we simplify higher powers of i: Any power of i simplifies to i, -1, $-i$, or 1. The easiest way to simplify higher powers of i is to write them in terms of i^2. For instance, to simplify i^{21}, we would write it as

$$(i^2)^{10} \cdot i \qquad \text{because } 2 \cdot 10 + 1 = 21$$

Then, since $i^2 = -1$, we have

$$(-1)^{10} \cdot i = 1 \cdot i = i$$

EXAMPLES   Simplify as much as possible.

5. $i^{30} = (i^2)^{15} = (-1)^{15} = -1$

6. $i^{11} = (i^2)^5 \cdot i = (-1)^5 \cdot i = (-1)i = -i$

7. $i^{40} = (i^2)^{20} = (-1)^{20} = 1$

> **DEFINITION**   A **complex number** is any number that can be put in the form
>
> $$a + bi$$
>
> where a and b are real numbers and $i = \sqrt{-1}$. The form $a + bi$ is called *standard form* for complex numbers. The number a is called the *real part* of the complex number. The number b is called the *imaginary part* of the complex number.
>
> Every real number is also a complex number. The real number 8, for example, can be written as $8 + 0i$; therefore, 8 is also considered a complex number.

EQUALITY FOR COMPLEX NUMBERS

Two complex numbers are equal if and only if their real parts are equal and their imaginary parts are equal. That is, for real numbers a, b, c, and d,

$$a + bi = c + di \qquad \text{if and only if} \qquad a = c \quad \text{and} \quad b = d$$

EXAMPLE 8 Find x and y if $3x + 4i = 12 - 8yi$.

Solution Since the two complex numbers are equal, their real parts are equal, and their imaginary parts are equal:

$$3x = 12 \qquad \text{and} \qquad 4 = -8y$$
$$x = 4 \qquad\qquad y = -\frac{1}{2}$$

EXAMPLE 9 Find x and y if $(4x - 3) + 7i = 5 + (2y - 1)i$.

Solution The real parts are $4x - 3$ and 5. The imaginary parts are 7 and $2y - 1$:

$$4x - 3 = 5 \qquad \text{and} \qquad 7 = 2y - 1$$
$$4x = 8 \qquad\qquad 8 = 2y$$
$$x = 2 \qquad\qquad y = 4$$

ADDITION AND SUBTRACTION OF COMPLEX NUMBERS

To add two complex numbers, add their real parts and add their imaginary parts. That is, if a, b, c, and d are real numbers, then

$$(a + bi) + (c + di) = (a + c) + (b + d)i$$

If we assume that the commutative, associative, and distributive properties hold for the number i, then the definition of addition is simply an extension of these properties.

We define subtraction in a similar manner. If a, b, c, and d are real numbers, then

$$(a + bi) - (c + di) = (a - c) + (b - d)i$$

EXAMPLES Add or subtract as indicated.

10. $(3 + 4i) + (7 - 6i) = (3 + 7) + (4 - 6)i = 10 - 2i$
11. $(7 + 3i) - (5 + 6i) = (7 - 5) + (3 - 6)i = 2 - 3i$
12. $(5 - 2i) - (9 - 4i) = (5 - 9) + (-2 + 4)i = -4 + 2i$

MULTIPLICATION OF COMPLEX NUMBERS

Since complex numbers have the same form as binomials, we find the product of two complex numbers the same way we find the product of two binomials.

EXAMPLE 13 Multiply $(3 - 4i)(2 + 5i)$.

Solution Multiplying each term in the second complex number by each term in the first, we have

$$\overset{\text{F}\qquad\text{O}\qquad\text{I}\qquad\text{L}}{(3 - 4i)(2 + 5i) = 3 \cdot 2 + 3 \cdot 5i - 2 \cdot 4i - 5i(4i)}$$

$$= 6 + 15i - 8i - 20i^2$$

Combining similar terms and using the fact that $i^2 = -1$, we can simplify as follows:

$$6 + 15i - 8i - 20i^2 = 6 + 7i - 20(-1)$$

$$= 6 + 7i + 20$$

$$= 26 + 7i$$

The product of the complex numbers $3 - 4i$ and $2 + 5i$ is the complex number $26 + 7i$.

EXAMPLE 14 Multiply $2i(4 - 6i)$.

Solution Applying the distributive property gives us

$$2i(4 - 6i) = 2i \cdot 4 - 2i \cdot 6i$$

$$= 8i - 12i^2$$

$$= 12 + 8i$$

EXAMPLE 15 Expand $(3 + 5i)^2$.

Solution We treat this like the square of a binomial. Remember, $(a + b)^2 = a^2 + 2ab + b^2$:

$$(3 + 5i)^2 = 3^2 + 2(3)(5i) + (5i)^2$$

$$= 9 + 30i + 25i^2$$

$$= 9 + 30i - 25$$

$$= -16 + 30i$$

EXAMPLE 16 Multiply $(2 - 3i)(2 + 3i)$.

Solution This product has the form $(a - b)(a + b)$, which we know results in the difference of two squares, $a^2 - b^2$:

$$(2 - 3i)(2 + 3i) = 2^2 - (3i)^2$$
$$= 4 - 9i^2$$
$$= 4 + 9$$
$$= 13$$

The product of the two complex numbers $2 - 3i$ and $2 + 3i$ is the real number 13. The two complex numbers $2 - 3i$ and $2 + 3i$ are called complex conjugates. The fact that their product is a real number is very useful.

> **DEFINITION** The complex numbers $a + bi$ and $a - bi$ are called **complex conjugates.** One important property they have is that their product is the real number $a^2 + b^2$. Here's why:
>
> $$(a + bi)(a - bi) = a^2 - (bi)^2$$
> $$= a^2 - b^2i^2$$
> $$= a^2 - b^2(-1)$$
> $$= a^2 + b^2$$

DIVISION WITH COMPLEX NUMBERS

The fact that the product of two complex conjugates is a real number is the key to division with complex numbers.

EXAMPLE 17 Divide $\dfrac{2 + i}{3 - 2i}$.

Solution We want a complex number in standard form that is equivalent to the quotient $\dfrac{2 + i}{3 - 2i}$. We need to eliminate i from the denominator. Multiplying the numerator and denominator by $3 + 2i$ will give us what we want:

$$\frac{2 + i}{3 - 2i} = \frac{2 + i}{3 - 2i} \cdot \frac{(3 + 2i)}{(3 + 2i)}$$

$$= \frac{6 + 4i + 3i + 2i^2}{9 - 4i^2}$$

$$= \frac{6 + 7i - 2}{9 + 4}$$

$$= \frac{4 + 7i}{13}$$

$$= \frac{4}{13} + \frac{7}{13}i$$

Dividing the complex number $2 + i$ by $3 - 2i$ gives the complex number $\frac{4}{13} + \frac{7}{13}i$.

EXAMPLE 18   Divide $\dfrac{7 - 4i}{i}$.

Solution   The conjugate of the denominator is $-i$. Multiplying numerator and denominator by this number, we have

$$\frac{7 - 4i}{i} = \frac{7 - 4i}{i} \cdot \frac{-i}{-i}$$

$$= \frac{-7i + 4i^2}{-i^2}$$

$$= \frac{-7i + 4(-1)}{-(-1)}$$

$$= -4 - 7i$$

Getting Ready for Class

After reading through the preceding section, respond in your own words and in complete sentences.

A. What is the number i?

B. What is a complex number?

C. What kind of number will always result when we multiply complex conjugates?

D. Explain how to divide complex numbers.

PROBLEM SET 7.7

Write the following in terms of i, and simplify as much as possible.

1. $\sqrt{-36}$ **2.** $\sqrt{-49}$ **3.** $-\sqrt{-25}$

4. $-\sqrt{-81}$ **5.** $\sqrt{-72}$ **6.** $\sqrt{-48}$

7. $-\sqrt{-12}$ **8.** $-\sqrt{-75}$

Write each of the following as i, -1, $-i$, or 1.

9. i^{28} **10.** i^{31} **11.** i^{26}

12. i^{37} **13.** i^{75} **14.** i^{42}

Find x and y so that each of the following equations is true.

15. $2x + 3yi = 6 - 3i$ **16.** $4x - 2yi = 4 + 8i$

17. $2 - 5i = -x + 10yi$ **18.** $4 + 7i = 6x - 14yi$

19. $2x + 10i = -16 - 2yi$

20. $4x - 5i = -2 + 3yi$

21. $(2x - 4) - 3i = 10 - 6yi$

22. $(4x - 3) - 2i = 8 + yi$

23. $(7x - 1) + 4i = 2 + (5y + 2)i$

24. $(5x + 2) - 7i = 4 + (2y + 1)i$

Combine the following complex numbers.

25. $(2 + 3i) + (3 + 6i)$ **26.** $(4 + i) + (3 + 2i)$

27. $(3 - 5i) + (2 + 4i)$ **28.** $(7 + 2i) + (3 - 4i)$

29. $(5 + 2i) - (3 + 6i)$ **30.** $(6 + 7i) - (4 + i)$

31. $(3 - 5i) - (2 + i)$

32. $(7 - 3i) - (4 + 10i)$

33. $[(3 + 2i) - (6 + i)] + (5 + i)$

34. $[(4 - 5i) - (2 + i)] + (2 + 5i)$

35. $[(7 - i) - (2 + 4i)] - (6 + 2i)$

36. $[(3 - i) - (4 + 7i)] - (3 - 4i)$

37. $(3 + 2i) - [(3 - 4i) - (6 + 2i)]$

38. $(7 - 4i) - [(-2 + i) - (3 + 7i)]$

39. $(4 - 9i) + [(2 - 7i) - (4 + 8i)]$

40. $(10 - 2i) - [(2 + i) - (3 - i)]$

Find the following products.

41. $3i(4 + 5i)$ **42.** $2i(3 + 4i)$

43. $6i(4 - 3i)$ **44.** $11i(2 - i)$

45. $(3 + 2i)(4 + i)$ **46.** $(2 - 4i)(3 + i)$

47. $(4 + 9i)(3 - i)$ **48.** $(5 - 2i)(1 + i)$

49. $(1 + i)^3$ **50.** $(1 - i)^3$

51. $(2 - i)^3$ **52.** $(2 + i)^3$

53. $(2 + 5i)^2$ **54.** $(3 + 2i)^2$

55. $(1 - i)^2$ **56.** $(1 + i)^2$

57. $(3 - 4i)^2$ **58.** $(6 - 5i)^2$

59. $(2 + i)(2 - i)$ **60.** $(3 + i)(3 - i)$

61. $(6 - 2i)(6 + 2i)$ **62.** $(5 + 4i)(5 - 4i)$

63. $(2 + 3i)(2 - 3i)$ **64.** $(2 - 7i)(2 + 7i)$

65. $(10 + 8i)(10 - 8i)$ **66.** $(11 - 7i)(11 + 7i)$

Find the following quotients. Write all answers in standard form for complex numbers.

67. $\dfrac{2 - 3i}{i}$ **68.** $\dfrac{3 + 4i}{i}$

69. $\dfrac{5 + 2i}{-i}$ **70.** $\dfrac{4 - 3i}{-i}$ **71.** $\dfrac{4}{2 - 3i}$

72. $\dfrac{3}{4 - 5i}$ **73.** $\dfrac{6}{-3 + 2i}$ **74.** $\dfrac{-1}{-2 - 5i}$

75. $\dfrac{2 + 3i}{2 - 3i}$ **76.** $\dfrac{4 - 7i}{4 + 7i}$ **77.** $\dfrac{5 + 4i}{3 + 6i}$

78. $\dfrac{2 + i}{5 - 6i}$

Applying the Concepts

79. Electric Circuits   Complex numbers may be applied to electrical circuits. Electrical engineers use the fact that resistance R to electrical flow of the electrical current I and the voltage V are related by the formula $V = RI$. (Voltage is measured in volts, resistance in ohms, and current in amperes.) Find the resistance to electrical flow in a circuit that has a voltage $V = (80 + 20i)$ volts and current $I = (-6 + 2i)$ amps.

80. Electric Circuits   Refer to the information about electrical circuits in Problem 79, and find the current in a circuit that has a resistance of $(4 + 10i)$ ohms and a voltage of $(5 - 7i)$ volts.

Review Problems

The following problems review material we covered in Sections 6.6 and 6.7.

Solve each equation. [6.6]

81. $\dfrac{t}{3} - \dfrac{1}{2} = -1$

82. $\dfrac{x}{x - 2} + \dfrac{2}{3} = \dfrac{2}{x - 2}$

83. $2 + \dfrac{5}{y} = \dfrac{3}{y^2}$ **84.** $1 - \dfrac{1}{y} = \dfrac{12}{y^2}$

Solve each application problem. [6.7]

85. The sum of a number and its reciprocal is $\frac{41}{20}$. Find the number.

86. It takes an inlet pipe 8 hours to fill a tank. The drain can empty the tank in 6 hours. If the tank is full and both the inlet pipe and drain are open, how long will it take to drain the tank?

Extending the Concepts

87. Show that $-i$ and $\dfrac{1}{i}$ (the opposite and the reciprocal of i) are the same number.

88. Show that i^{2n+1} is the same as i for all positive even integers n.

89. Show that $x = 1 + i$ is a solution to the equation $x^2 - 2x + 2 = 0$.

90. Show that $x = 1 - i$ is a solution to the equation $x^2 - 2x + 2 = 0$.

91. Show that $x = 2 + i$ is a solution to the equation $x^3 - 11x + 20 = 0$.

92. Show that $x = 2 - i$ is a solution to the equation $x^3 - 11x + 20 = 0$.

Examples

1. The number 49 has two square roots, 7 and -7. They are written like this:

$$\sqrt{49} = 7 \qquad -\sqrt{49} = -7$$

Square Roots [7.1]

Every positive real number x has two square roots. The *positive square root* of x is written $\sqrt{x}$, and the *negative square root* of x is written $-\sqrt{x}$. Both the positive and the negative square roots of x are numbers we square to get x. That is,

$$\text{and} \quad \left.\begin{array}{l} (\sqrt{x})^2 = x \\ (-\sqrt{x})^2 = x \end{array}\right\} \quad \text{for } x \geq 0$$

2. $\sqrt[3]{8} = 2$
$\sqrt[3]{-27} = -3$

Higher Roots [7.1]

Consider the expression $\sqrt[n]{a}$: n is the *index*, a is the *radicand*, and $\sqrt{}$ is the *radical sign*. The expression $\sqrt[n]{a}$ is such that

$$(\sqrt[n]{a})^n = a \qquad a \geq 0 \text{ when } n \text{ is even}$$

3. $25^{1/2} = \sqrt{25} = 5$
$8^{2/3} = (\sqrt[3]{8})^2 = 2^2 = 4$
$9^{3/2} = (\sqrt{9})^3 = 3^3 = 27$

Rational Exponents [7.1, 7.2]

Rational exponents are used to indicate roots. The relationship between rational exponents and roots is as follows:

$$a^{1/n} = \sqrt[n]{a} \qquad \text{and} \qquad a^{m/n} = (a^{1/n})^m = (a^m)^{1/n}$$

$$a \geq 0 \text{ when } n \text{ is even}$$

4. $\sqrt{4 \cdot 5} = \sqrt{4}\,\sqrt{5} = 2\sqrt{5}$
$\sqrt{\dfrac{7}{9}} = \dfrac{\sqrt{7}}{\sqrt{9}} = \dfrac{\sqrt{7}}{3}$

Properties of Radicals [7.3]

If a and b are nonnegative real numbers whenever n is even, then

1. $\sqrt[n]{ab} = \sqrt[n]{a}\,\sqrt[n]{b}$

2. $\sqrt[n]{\dfrac{a}{b}} = \dfrac{\sqrt[n]{a}}{\sqrt[n]{b}} \qquad (b \neq 0)$

5. $\sqrt{\dfrac{4}{5}} = \dfrac{\sqrt{4}}{\sqrt{5}}$

$= \dfrac{2}{\sqrt{5}} \cdot \dfrac{\sqrt{5}}{\sqrt{5}}$

$= \dfrac{2\sqrt{5}}{5}$

Simplified Form for Radicals [7.3]

A radical expression is said to be in *simplified form*

1. If there is no factor of the radicand that can be written as a power greater than or equal to the index;

2. If there are no fractions under the radical sign; and

3. If there are no radicals in the denominator.

6. $5\sqrt{3} - 7\sqrt{3} = (5 - 7)\sqrt{3}$
$\qquad = -2\sqrt{3}$
$\sqrt{20} + \sqrt{45} = 2\sqrt{5} + 3\sqrt{5}$
$\qquad = (2 + 3)\sqrt{5}$
$\qquad = 5\sqrt{5}$

Addition and Subtraction of Radical Expressions [7.4]

We add and subtract radical expressions by using the distributive property to combine similar radicals. Similar radicals are radicals with the same index and the same radicand.

7. $(\sqrt{x} + 2)(\sqrt{x} + 3)$
$= \sqrt{x}\,\sqrt{x} + 3\sqrt{x} + 2\sqrt{x} + 2 \cdot 3$
$= x + 5\sqrt{x} + 6$

Multiplication of Radical Expressions [7.5]
We multiply radical expressions in the same way that we multiply polynomials. We can use the distributive property and the FOIL method.

8. $\dfrac{3}{\sqrt{2}} = \dfrac{3}{\sqrt{2}} \cdot \dfrac{\sqrt{2}}{\sqrt{2}} = \dfrac{3\sqrt{2}}{2}$

$\dfrac{3}{\sqrt{5} - \sqrt{3}} = \dfrac{3}{\sqrt{5} - \sqrt{3}} \cdot \dfrac{\sqrt{5} + \sqrt{3}}{\sqrt{5} + \sqrt{3}}$

$= \dfrac{3\sqrt{5} + 3\sqrt{3}}{5 - 3}$

$= \dfrac{3\sqrt{5} + 3\sqrt{3}}{2}$

Rationalizing the Denominator [7.3, 7.5]
When a fraction contains a square root in the denominator, we rationalize the denominator by multiplying numerator and denominator by

1. The square root itself if there is only one term in the denominator, or
2. The conjugate of the denominator if there are two terms in the denominator.

Rationalizing the denominator can also be called division of radical expressions.

9. $\sqrt{2x + 1} = 3$
$(\sqrt{2x + 1})^2 = 3^2$
$2x + 1 = 9$
$x = 4$

Squaring Property of Equality [7.6]
We may square both sides of an equation any time it is convenient to do so, as long as we check all resulting solutions in the original equation.

10. $3 + 4i$ is a complex number.

Addition

$(3 + 4i) + (2 - 5i) = 5 - i$

Multiplication

$(3 + 4i)(2 - 5i)$
$= 6 - 15i + 8i - 20i^2$
$= 6 - 7i + 20$
$= 26 - 7i$

Division

$\dfrac{2}{3 + 4i} = \dfrac{2}{3 + 4i} \cdot \dfrac{3 - 4i}{3 - 4i}$

$= \dfrac{6 - 8i}{9 + 16}$

$= \dfrac{6}{25} - \dfrac{8}{25}i$

Complex Numbers [7.7]
A *complex number* is any number that can be put in the form

$$a + bi$$

where a and b are real numbers and $i = \sqrt{-1}$. The *real part* of the complex number is a, and b is the *imaginary part*.

If a, b, c, and d are real numbers, then we have the following definitions associated with complex numbers:

1. Equality

$a + bi = c + di$ if and only if $a = c$ and $b = d$

2. Addition and subtraction

$(a + bi) + (c + di) = (a + c) + (b + d)i$
$(a + bi) - (c + di) = (a - c) + (b - d)i$

3. Multiplication

$(a + bi)(c + di) = (ac - bd) + (ad + bc)i$

4. Division is similar to rationalizing the denominator.

Simplify each expression as much as possible. [7.1]

1. $\sqrt{49}$ **2.** $(-27)^{1/3}$ **3.** $16^{1/4}$

4. $9^{3/2}$ **5.** $\sqrt[5]{32x^{15}y^{10}}$ **6.** $8^{-4/3}$

Use the properties of exponents to simplify each expression. Assume all bases represent positive numbers. [7.1]

7. $x^{2/3} \cdot x^{4/3}$ **8.** $(a^{2/3}b^{4/3})^3$ **9.** $\dfrac{a^{3/5}}{a^{1/4}}$

10. $\dfrac{a^{2/3}b^3}{a^{1/4}b^{1/3}}$

Multiply. [7.2]

11. $(3x^{1/2} + 5y^{1/2})(4x^{1/2} - 3y^{1/2})$

12. $(a^{1/3} - 5)^2$

13. Divide: $\dfrac{28x^{5/6} + 14x^{7/6}}{7x^{1/3}}$ (Assume $x > 0$.) [7.2]

14. Factor $2(x - 3)^{1/4}$ from $8(x - 3)^{5/4} - 2(x - 3)^{1/4}$. [7.2]

15. Simplify $x^{3/4} + \dfrac{5}{x^{1/4}}$ into a single fraction. (Assume $x > 0$.) [7.2]

Write each expression in simplified form for radicals. (Assume all variables represent nonnegative numbers.) [7.3]

16. $\sqrt{12}$ **17.** $\sqrt{50}$ **18.** $\sqrt[3]{16}$

19. $\sqrt{18x^2}$ **20.** $\sqrt{80a^3b^4c^2}$ **21.** $\sqrt[4]{32a^4b^5c^6}$

Rationalize the denominator in each expression. [7.3]

22. $\dfrac{3}{\sqrt{2}}$ **23.** $\dfrac{6}{\sqrt[3]{2}}$

Write each expression in simplified form. (Assume all variables represent positive numbers.) [7.3]

24. $\sqrt{\dfrac{48x^3}{7y}}$ **25.** $\sqrt[3]{\dfrac{40x^2y^3}{3z}}$

Combine the following expressions. (Assume all variables represent positive numbers.) [7.4]

26. $5x\sqrt{6} + 2x\sqrt{6} - 9x\sqrt{6}$

27. $\sqrt{12} + \sqrt{3}$ **28.** $\dfrac{3}{\sqrt{5}} + \sqrt{5}$

29. $3\sqrt{8} - 4\sqrt{72} + 5\sqrt{50}$

30. $3b\sqrt{27a^5b} + 2a\sqrt{3a^3b^3}$

31. $2x\sqrt[3]{xy^3z^2} - 6y\sqrt[3]{x^4z^2}$

Multiply. [7.5]

32. $\sqrt{2}(\sqrt{3} - 2\sqrt{2})$ **33.** $(\sqrt{x} - 2)(\sqrt{x} - 3)$

Rationalize the denominator. (Assume $x, y > 0$.) [7.5]

34. $\dfrac{3}{\sqrt{5} - 2}$ **35.** $\dfrac{\sqrt{7} + \sqrt{5}}{\sqrt{7} - \sqrt{5}}$ **36.** $\dfrac{3\sqrt{7}}{3\sqrt{7} - 4}$

Solve each equation. [7.6]

37. $\sqrt{4a + 1} = 1$ **38.** $\sqrt[3]{3x - 8} = 1$

39. $\sqrt{3x + 1} - 3 = 1$ **40.** $\sqrt{x + 4} = \sqrt{x} - 2$

Graph each equation. [7.6]

41. $y = 3\sqrt{x}$ **42.** $y = \sqrt[3]{x} + 2$

Write each of the following as i, -1, $-i$, or 1. [7.7]

43. i^{24} **44.** i^{27}

Find x and y so that each of the following equations is true. [7.7]

45. $3 - 4i = -2x + 8yi$

46. $(3x + 2) - 8i = -4 + 2yi$

Combine the following complex numbers. [7.7]

47. $(3 + 5i) + (6 - 2i)$

48. $(2 + 5i) - [(3 + 2i) + (6 - i)]$

Multiply. [7.7]

49. $3i(4 + 2i)$ **50.** $(2 + 3i)(4 + i)$

51. $(4 + 2i)^2$ **52.** $(4 + 3i)(4 - 3i)$

Divide. Write all answers in standard form for complex numbers. [7.7]

53. $\dfrac{3 + i}{i}$ **54.** $\dfrac{-3}{2 + i}$

55. Construction The roof of a house shown in Figure 1 is to extend up 13.5 feet above the ceiling, which is 36 feet across. Find the length of one side of the roof.

FIGURE 1

56. Surveying A surveyor is attempting to find the distance across a pond. From a point on one side of the pond he walks 25 yards to the end of the pond and then makes a 90-degree turn and walks another 60 yards before coming to a point directly across the pond from the point at which he started. What is the distance across the pond? (See Figure 2.)

FIGURE 2

C H A P T E R 7 **PROJECTS**

RATIONAL EXPONENTS AND ROOTS

G R O U P P R O J E C T

CONSTRUCTING THE SPIRAL OF ROOTS

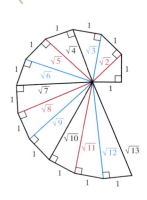

Number of People: 3

Time Needed: 20 minutes

Equipment: Two sheets of graph paper (4 or 5 squares per inch) and pencils

Background: The spiral of roots gives us a way to visualize the positive square roots of the counting numbers, and in so doing, we see many line segments whose lengths are irrational numbers.

Procedure: You are to construct a spiral of roots from a line segment 1 inch long. The graph paper you have contains either 4 or 5 squares per inch, allowing you to accurately draw 1-inch line segments. Since the lines on the graph paper are perpendicular to one another, if you are careful, you can also use the graph paper to connect one line segment to another so that they form a right angle.

1. Fold one of the pieces of graph paper so it can be used as a ruler.

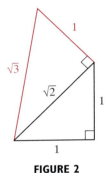

FIGURE 1

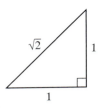

FIGURE 2

2. Use the folded paper to draw a line segment 1 inch long, just to the right of the middle of the unfolded paper. On the end of this segment, attach another segment of 1-inch length at a right angle to the first one. Connect the endpoints of the segments to form a right triangle. Label each of the sides of this triangle. When you are finished, your work should resemble Figure 1.

3. On the end of the hypotenuse of the triangle, attach a 1-inch line segment so that the two segments form a right angle. (Use the folded paper to do this.) Draw the hypotenuse of this triangle. Label all the sides of this second triangle. Your work should resemble Figure 2.

4. Continue to draw a new right triangle by attaching 1-inch line segments at right angles to the previous 1-inch line segment. Label all the sides of each triangle.

5. Stop when you have drawn a hypotenuse $\sqrt{8}$ inches long.

R E S E A R C H   P R O J E C T

CONNECTIONS

Although it may not look like it, the three items shown here are related very closely to one another. Your job is to find the connection.

A Continued Fraction *The Fibonacci Sequence* *The Golden Rectangle*

$$1 + \cfrac{1}{1 + \cfrac{1}{1 + \cfrac{1}{1 + \cdots}}}$$

1, 1, 2, 3, 5, . . .

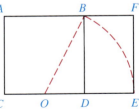

Step 1: The dots in the continued fraction indicate that the pattern shown continues indefinitely. This means that there is no way for us to simplify this expression, as we have simplified the expressions in this chapter. However, we can begin to understand the continued fraction, and what it simplifies to, by working with the following sequence of expressions. Simplify each expression. Write each answer as a fraction, in lowest terms.

$$1 + \cfrac{1}{1+1} \qquad 1 + \cfrac{1}{1+\cfrac{1}{1+1}} \qquad 1 + \cfrac{1}{1+\cfrac{1}{1+\cfrac{1}{1+1}}} \qquad 1 + \cfrac{1}{1+\cfrac{1}{1+\cfrac{1}{1+\cfrac{1}{1+1}}}}$$

Step 2: Compare the fractional answers to step 1 with the numbers in the Fibonacci sequence. Based on your observation, give the answer to the following problem, without actually doing any arithmetic.

$$1 + \cfrac{1}{1+\cfrac{1}{1+\cfrac{1}{1+\cfrac{1}{1+\cfrac{1}{1+1}}}}}$$

Step 3: Continue the sequence of simplified fractions you have written in steps 1 and 2, until you have nine numbers in the sequence. Convert each of these numbers to a decimal, accurate to four places past the decimal point.

Step 4: Find a decimal approximation to the golden ratio $\dfrac{1+\sqrt5}{2}$, accurate to four places past the decimal point.

Step 5: Compare the results in steps 3 and 4, and then make a conjecture about what number the continued fraction would simplify to, if it was actually possible to simplify it.

C H A P T E R 7   **TEST**

Simplify each of the following. (Assume all variable bases are positive integers and all variable exponents are positive real numbers throughout this test.) [7.1]

1. $27^{-2/3}$

2. $\left(\dfrac{25}{49}\right)^{-1/2}$

3. $a^{3/4} \cdot a^{-1/3}$

4. $\dfrac{(x^{2/3}y^{-3})^{1/2}}{(x^{3/4}y^{1/2})^{-1}}$

5. $\sqrt{49x^8y^{10}}$

6. $\sqrt[5]{32x^{10}y^{20}}$

7. $\dfrac{(36a^8b^4)^{1/2}}{(27a^9b^6)^{1/3}}$

8. $\dfrac{(x^ny^{1/n})^n}{(x^{1/n}y^n)^{n^2}}$

Multiply. [7.2]

9. $2a^{1/2}(3a^{3/2} - 5a^{1/2})$ **10.** $(4a^{3/2} - 5)^2$

Factor. [7.2]

11. $3x^{2/3} + 5x^{1/3} - 2$ **12.** $9x^{2/3} - 49$

Combine. [7.4]

13. $\dfrac{4}{x^{1/2}} + x^{1/2}$

14. $\dfrac{x^2}{(x^2 - 3)^{1/2}} - (x^2 - 3)^{1/2}$

Write in simplified form. [7.3]

15. $\sqrt{125x^3y^5}$

16. $\sqrt[3]{40x^7y^8}$

17. $\sqrt{\dfrac{2}{3}}$

18. $\sqrt{\dfrac{12a^4b^3}{5c}}$

Combine. [7.4]

19. $3\sqrt{12} - 4\sqrt{27}$

20. $\sqrt[3]{24a^3b^3} - 5a\sqrt[3]{3b^3}$

Multiply. [7.5]

21. $(\sqrt{x} + 7)(\sqrt{x} - 4)$ **22.** $(3\sqrt{2} - \sqrt{3})^2$

Rationalize the denominator. [7.5]

23. $\dfrac{5}{\sqrt{3} - 1}$ **24.** $\dfrac{\sqrt{x} - \sqrt{2}}{\sqrt{x} + \sqrt{2}}$

Solve for x. [7.6]

25. $\sqrt{3x + 1} = x - 3$ **26.** $\sqrt[3]{2x + 7} = -1$

27. $\sqrt{x + 3} = \sqrt{x + 4} - 1$

Graph. [7.6]

28. $y = \sqrt{x} - 2$ **29.** $y = \sqrt[3]{x} + 3$

30. Solve for x and y so that the following equation is true [7.7]:

$$(2x + 5) - 4i = 6 - (y - 3)i$$

Perform the indicated operations. [7.7]

31. $(3 + 2i) - [(7 - i) - (4 + 3i)]$

32. $(2 - 3i)(4 + 3i)$

33. $(5 - 4i)^2$ **34.** $\dfrac{2 - 3i}{2 + 3i}$

35. Show that i^{38} can be written as -1. [7.7]

Simplify.

1. $33 - 22 - (-11) + 1$
2. $12 - 20 \div 4 - 3 \cdot 2$
3. $-6 + 5[3 - 2(-4 - 1)]$
4. $3(2x + 5) + 4(4x - 1)$
5. $(2y^{-3})^{-1}(4y^{-3})^2$
6. $8^{2/3}$
7. $\sqrt{72y^5}$
8. $\sqrt{15} - \sqrt{\dfrac{3}{5}}$
9. Give the opposite and reciprocal of 12.
10. Specify the domain and range for the relation $\{(-1, 2), (3, -1), (-1, 0)\}$. Is this relation also a function?
11. Write 41,500.00 in scientific notation.
12. Let $f(x) = x^2 - 3x$ and $g(x) = x - 1$. Find: $f(3) - g(0)$

Factor completely.

13. $625a^4 - 16b^4$
14. $24a^4 + 10a^2 - 6$

Reduce to lowest terms.

15. $\dfrac{246}{861}$
16. $\dfrac{28xy^3z^2}{14x^2y^3z}$
17. $\dfrac{x^2 - 9x + 20}{x^2 - 7x + 12}$

Multiply.

18. $\dfrac{6}{7} \cdot \dfrac{21}{35} \cdot 5$
19. $(2x - 5y)(3x - 2y)$
20. $(9x^2 - 25) \cdot \dfrac{x + 5}{3x - 5}$
21. $(\sqrt{x} - 2)^2$

Divide.

22. $\dfrac{18a^4b^2 - 9a^2b^2 + 27a^2b^4}{-9a^2b^2}$
23. $\dfrac{27x^3y^2}{13x^2y^4} \div \dfrac{9xy}{26y}$
24. $\dfrac{\sqrt{x} + \sqrt{y}}{\sqrt{x} - \sqrt{y}}$

Subtract.

25. $\left(\dfrac{2}{3}x^3 + \dfrac{1}{6}x^2 + \dfrac{1}{2}\right) - \left(\dfrac{1}{4}x^2 - \dfrac{1}{3}x + \dfrac{1}{12}\right)$
26. $\dfrac{3}{x^2 + 8x + 15} - \dfrac{1}{x^2 + 7x + 12} - \dfrac{1}{x^2 + 9x + 20}$

Solve.

27. $3y - 8 = -4y + 6$
28. $\dfrac{2}{3}(9x - 2) + \dfrac{1}{3} = 4$
29. $|3x - 1| - 2 = 6$
30. $x^3 + 2x^2 - 16x - 32 = 0$
31. $\dfrac{x + 2}{x + 1} - 2 = \dfrac{1}{x + 1}$
32. $\sqrt[3]{8 - 3x} = -1$
33. $(x + 3)^2 - 3(x + 3) - 70 = 0$
34. $\sqrt{y + 7} - \sqrt{y + 2} = 1$
35. Solve for L if $P = 2L + 2W$, $P = 18$ and $W = 3$.

Solve each inequality, and graph the solution.

36. $4y - 2 \geq 10$ or $4y - 2 \leq -10$
37. $|5x - 4| \geq 6$

Solve each system.

38. $\begin{aligned} -7x + 8y &= 7 \\ 6x - 5y &= -19 \end{aligned}$
39. $\begin{aligned} 4x + 9y &= 2 \\ y &= 2x - 12 \end{aligned}$
40. $\begin{aligned} x + y + z &= -2 \\ 2x + y - 3z &= 1 \\ -2x - 3y + 4z &= 9 \end{aligned}$
41. $\begin{aligned} x + y &= -2 \\ y + 10z &= -1 \\ 2x - 13z &= 5 \end{aligned}$
42. Evaluate $\begin{vmatrix} 1 & 0 & 2 \\ 0 & 5 & -1 \\ 4 & 3 & 0 \end{vmatrix}$.

Graph on a rectangular coordinate system.

43. $y = \dfrac{1}{3}x - 3$
44. $y = \sqrt{x} + 3$
45. $y = \dfrac{x - 2}{x + 1}$
46. Find y if the line through $(-1, 2)$ and $(-3, y)$ has a slope of -2.
47. Find the slope and y-intercept of $4x - 5y = 15$.
48. Give the equation of a line with slope $-\dfrac{2}{5}$ and y-intercept $= 5$.
49. Write the equation of the line in slope-intercept form if the slope is $\dfrac{2}{3}$ and $(9, 3)$ is a point on the line.
50. **Geometry** The length of a rectangle is 4 feet more than twice the width. The perimeter is 68 feet. Find the dimensions.

Quadratic Functions

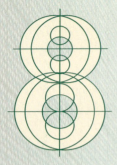

(Jerry Yulsman/The Image Bank)

INTRODUCTION

If you have been to the circus or the county fair recently, you may have witnessed one of the more spectacular acts, the human cannonball. The human cannonball shown in the photograph will reach a height of 70 feet, and travel a distance of 160 feet, before landing in a safety net. In this chapter we use this information to derive the equation

$$y = -\frac{7}{640}(x - 80)^2 + 70 \quad \text{for } 0 \le x \le 160$$

which describes the path flown by this particular cannonball. The table and graph below were constructed from this equation.

TABLE I	Path of a Human Cannonball
x (feet)	y (nearest foot)
0	0
40	53
80	70
120	53
160	0

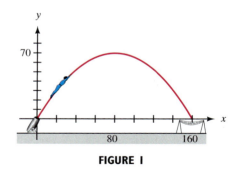

FIGURE I

All objects that are projected into the air, whether they are basketballs, bullets, arrows, or coins, follow parabolic paths like the one shown in Figure 1. Studying the material in this chapter will give you a more mathematical hold on the world around you.

Table 2 is taken from the trail map given to skiers at the Northstar at Tahoe Ski Resort in Lake Tahoe, California. The table gives the length of each chair lift at Northstar, along with the change in elevation from the beginning of the lift to the end of the lift.

Right triangles are good mathematical models for chair lifts. In this chapter we will use our knowledge of right triangles, along with the new material developed in the chapter, to solve problems involving chair lifts and a variety of other examples.

TABLE 2   From the Trail Map for the Northstar at Tahoe Ski Resort

Lift Information

Lift	Vertical Rise (feet)	Length (feet)
Big Springs Gondola	480	4,100
Bear Paw Double	120	790
Echo Triple	710	4,890
Aspen Express Quad	900	5,100
Forest Double	1,170	5,750
Lookout Double	960	4,330
Comstock Express Quad	1,250	5,900
Rendezvous Triple	650	2,900
Schaffer Camp Triple	1,860	6,150
Chipmunk Tow Lift	28	280
Bear Cub Tow Lift	120	750

8.1   Completing the Square

In this section we will develop the first of our new methods of solving quadratic equations. The new method is called *completing the square*. Completing the square on a quadratic equation allows us to obtain solutions, regardless of whether the equation can be factored. Before we solve equations by completing the square, we need to learn how to solve equations by taking square roots of both sides.

Consider the equation

$$x^2 = 16$$

We could solve it by writing it in standard form, factoring the left side, and proceeding as we did in Chapter 5. We can shorten our work considerably, however, if we simply notice that x must be either the positive square root of 16 or the negative square root of 16. That is,

$$\text{If} \quad x^2 = 16$$

$$\text{then} \quad x = \sqrt{16} \quad \text{or} \quad x = -\sqrt{16}$$

$$x = 4 \quad \text{or} \quad x = -4$$

We can generalize this result into a theorem as follows.

THEOREM 8.1

If $a^2 = b$, where b is a real number, then $a = \sqrt{b}$ or $a = -\sqrt{b}$.

Notation   The expression $a = \sqrt{b}$ or $a = -\sqrt{b}$ can be written in shorthand form as $a = \pm \sqrt{b}$. The symbol $\pm$ is read "plus or minus."

We can apply Theorem 8.1 to some fairly complicated quadratic equations.

EXAMPLE   I   Solve $(2x - 3)^2 = 25$.

Solution

$$(2x - 3)^2 = 25$$

$$2x - 3 = \pm \sqrt{25} \qquad \text{Theorem 8.1}$$

$$2x - 3 = \pm 5 \qquad \sqrt{25} = 5$$

$$2x = 3 \pm 5 \qquad \text{Add 3 to both sides.}$$

$$x = \frac{3 \pm 5}{2} \qquad \text{Divide both sides by 2.}$$

The last equation can be written as two separate statements:

$$x = \frac{3 + 5}{2} \quad \text{or} \quad x = \frac{3 - 5}{2}$$

$$= \frac{8}{2} \qquad\qquad\quad = \frac{-2}{2}$$

$$= 4 \quad \text{or} \quad = -1$$

The solution set is $\{4, -1\}$.

Notice that we could have solved the equation in Example 1 by expanding the left side, writing the resulting equation in standard form, and then factoring. The problem would look like this:

$$(2x - 3)^2 = 25 \qquad \text{Original equation}$$
$$4x^2 - 12x + 9 = 25 \qquad \text{Expand the left side.}$$
$$4x^2 - 12x - 16 = 0 \qquad \text{Add } -25 \text{ to each side.}$$
$$4(x^2 - 3x - 4) = 0 \qquad \text{Begin factoring.}$$
$$4(x - 4)(x + 1) = 0 \qquad \text{Factor completely.}$$
$$x - 4 = 0 \quad \text{or} \quad x + 1 = 0 \qquad \text{Set variable factors equal to 0.}$$
$$x = 4 \quad \text{or} \quad x = -1$$

As you can see, solving the equation by factoring leads to the same two solutions.

EXAMPLE 2   Solve for x: $(3x - 1)^2 = -12$

Solution

$$(3x - 1)^2 = -12$$
$$3x - 1 = \pm\sqrt{-12} \qquad \text{Theorem 8.1}$$
$$3x - 1 = \pm 2i\sqrt{3} \qquad \sqrt{-12} = 2i\sqrt{3}$$
$$3x = 1 \pm 2i\sqrt{3} \qquad \text{Add 1 to both sides.}$$
$$x = \frac{1 \pm 2i\sqrt{3}}{3} \qquad \text{Divide both sides by 3.}$$

The solution set is $\left\{ \dfrac{1 + 2i\sqrt{3}}{3}, \dfrac{1 - 2i\sqrt{3}}{3} \right\}$.

Both solutions are complex. Here is a check of the first solution:

When
$$x = \frac{1 + 2i\sqrt{3}}{3}$$

the equation
$$(3x - 1)^2 = -12$$

becomes
$$\left(3 \cdot \frac{1 + 2i\sqrt{3}}{3} - 1 \right)^2 \overset{?}{=} -12$$

or
$$(1 + 2i\sqrt{3} - 1)^2 \overset{?}{=} -12$$
$$(2i\sqrt{3})^2 \overset{?}{=} -12$$
$$4 \cdot i^2 \cdot 3 \overset{?}{=} -12$$
$$12(-1) \overset{?}{=} -12$$
$$-12 = -12$$

Note: We cannot solve the equation in Example 2 by factoring. If we expand the left side and write the resulting equation in standard form, we are left with a quadratic equation that does not factor:

$$(3x - 1)^2 = -12 \qquad \text{Equation from Example 2}$$

$$9x^2 - 6x + 1 = -12 \qquad \text{Expand the left side.}$$

$$9x^2 - 6x + 13 = 0 \qquad \text{Standard form, but not factorable}$$

EXAMPLE 3   Solve $x^2 + 6x + 9 = 12$.

Solution   We can solve this equation as we have the equations in Examples 1 and 2 if we first write the left side as $(x + 3)^2$.

$$x^2 + 6x + 9 = 12 \qquad \text{Original equation}$$

$$(x + 3)^2 = 12 \qquad \text{Write } x^2 + 6x + 9 \text{ as } (x + 3)^2.$$

$$x + 3 = \pm 2\sqrt{3} \qquad \text{Theorem 8.1}$$

$$x = -3 \pm 2\sqrt{3} \qquad \text{Add } -3 \text{ to each side.}$$

We have two irrational solutions: $-3 + 2\sqrt{3}$ and $-3 - 2\sqrt{3}$. What is important about this problem, however, is the fact that the equation was easy to solve because the left side was a perfect square trinomial.

METHOD OF COMPLETING THE SQUARE

The method of completing the square is simply a way of transforming any quadratic equation into an equation of the form found in the preceding three examples.

The key to understanding the method of completing the square lies in recognizing the relationship between the last two terms of any perfect square trinomial whose leading coefficient is 1.

Consider the following list of perfect square trinomials and their corresponding binomial squares:

$$x^2 - 6x + 9 = (x - 3)^2$$

$$x^2 + 8x + 16 = (x + 4)^2$$

$$x^2 - 10x + 25 = (x - 5)^2$$

$$x^2 + 12x + 36 = (x + 6)^2$$

In each case the leading coefficient is 1. A more important observation comes from noticing the relationship between the linear and constant terms (middle and last terms) in each trinomial. Observe that the constant term in each case is the square of half the coefficient of x in the middle term. For example, in the last expression, the constant term 36 is the square of half of 12, where 12 is the coefficient of x in the middle term. (Notice also that the second terms in all the binomials on the right side are half the coefficients of the middle terms of the trinomials on the left side.) We can use these observations to build our own perfect square trinomials and, in doing so, solve some quadratic equations.

Consider the following equation:

$$x^2 + 6x = 3$$

We can think of the left side as having the first two terms of a perfect square trinomial. We need only add the correct constant term. If we take half the coefficient of x, we get 3. If we then square this quantity, we have 9. Adding the 9 to both sides, the equation becomes

$$x^2 + 6x + \mathbf{9} = 3 + \mathbf{9}$$

The left side is the perfect square $(x + 3)^2$; the right side is 12:

$$(x + 3)^2 = 12$$

The equation is now in the correct form. We can apply Theorem 8.1 and finish the solution:

$$(x + 3)^2 = 12$$
$$x + 3 = \pm\sqrt{12} \qquad \text{Theorem 8.1}$$
$$x + 3 = \pm 2\sqrt{3}$$
$$x = -3 \pm 2\sqrt{3}$$

The solution set is $\{-3 + 2\sqrt{3}, -3 - 2\sqrt{3}\}$. The method just used is called *completing the square,* since we complete the square on the left side of the original equation by adding the appropriate constant term.

E X A M P L E   4   Solve by completing the square: $x^2 + 5x - 2 = 0$

Solution   We must begin by adding 2 to both sides. (The left side of the equation, as it is, is not a perfect square, because it does not have the correct constant term. We will simply "move" that term to the other side and use our own constant term.)

$$x^2 + 5x = 2 \qquad \text{Add 2 to each side.}$$

We complete the square by adding the square of half the coefficient of the linear term to both sides:

$$x^2 + 5x + \frac{25}{4} = 2 + \frac{25}{4} \qquad \text{Half of 5 is } \tfrac{5}{2}, \text{ the square of which is } \tfrac{25}{4}.$$

$$\left(x + \frac{5}{2}\right)^2 = \frac{33}{4} \qquad\qquad 2 + \frac{25}{4} = \frac{8}{4} + \frac{25}{4} = \frac{33}{4}$$

$$x + \frac{5}{2} = \pm\sqrt{\frac{33}{4}} \qquad\qquad \text{Theorem 8.1}$$

$$x + \frac{5}{2} = \pm\frac{\sqrt{33}}{2} \qquad\qquad \text{Simplify the radical.}$$

$$x = -\frac{5}{2} \pm \frac{\sqrt{33}}{2} \qquad\qquad \text{Add } -\frac{5}{2} \text{ to both sides.}$$

$$x = \frac{-5 \pm \sqrt{33}}{2}$$

The solution set is $\left\{\dfrac{-5 + \sqrt{33}}{2}, \dfrac{-5 - \sqrt{33}}{2}\right\}$.

We can use a calculator to get decimal approximations to these solutions. If $\sqrt{33} \approx 5.74$, then

$$\frac{-5 + 5.74}{2} = 0.37$$

$$\frac{-5 - 5.74}{2} = -5.37$$

EXAMPLE 5   Solve for x: $3x^2 - 8x + 7 = 0$

Solution

$$3x^2 - 8x + 7 = 0$$

$$3x^2 - 8x = -7 \qquad \text{Add } -7 \text{ to both sides.}$$

We cannot complete the square on the left side because the leading coefficient is not 1. We take an extra step and divide both sides by 3:

$$\frac{3x^2}{3} - \frac{8x}{3} = -\frac{7}{3}$$

$$x^2 - \frac{8}{3}x = -\frac{7}{3}$$

Half of $\frac{8}{3}$ is $\frac{4}{3}$, the square of which is $\frac{16}{9}$:

$$x^2 - \frac{8}{3}x + \mathbf{\frac{16}{9}} = -\frac{7}{3} + \mathbf{\frac{16}{9}} \qquad \text{Add } \mathbf{\frac{16}{9}} \text{ to both sides.}$$

$$\left(x - \frac{4}{3}\right)^2 = -\frac{5}{9} \qquad \text{Simplify right side.}$$

$$x - \frac{4}{3} = \pm\sqrt{-\frac{5}{9}} \qquad \text{Theorem 8.1}$$

$$x - \frac{4}{3} = \pm\frac{i\sqrt{5}}{3} \qquad\quad \sqrt{-\frac{5}{9}} = \frac{\sqrt{-5}}{3} = \frac{i\sqrt{5}}{3}$$

$$x = \frac{4}{3} \pm \frac{i\sqrt{5}}{3} \qquad \text{Add } \frac{4}{3} \text{ to both sides.}$$

$$x = \frac{4 \pm i\sqrt{5}}{3}$$

The solution set is $\left\{ \dfrac{4 + i\sqrt{5}}{3}, \dfrac{4 - i\sqrt{5}}{3} \right\}$.

TO SOLVE A QUADRATIC EQUATION BY COMPLETING THE SQUARE

To summarize the method used in the preceding two examples, we list the following steps:

Step 1: Write the equation in the form $ax^2 + bx = c$.

Step 2: If the leading coefficient is not 1, divide both sides by the coefficient so that the resulting equation has a leading coefficient of 1. That is, if $a \neq 1$, then divide both sides by a.

Step 3: Add the square of half the coefficient of the linear term to both sides of the equation.

Step 4: Write the left side of the equation as the square of a binomial, and simplify the right side if possible.

Step 5: Apply Theorem 8.1, and solve as usual.

FACTS FROM
Geometry

More Special Triangles

The triangles shown in Figures 1 and 2 occur frequently in mathematics.

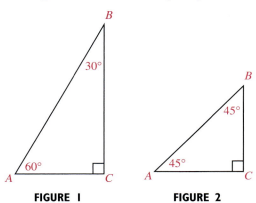

FIGURE 1   **FIGURE 2**

Note that both of the triangles are right triangles. We refer to the triangle in Figure 1 as a $30°{-}60°{-}90°$ triangle, and the triangle in Figure 2 as a $45°{-}45°{-}90°$ triangle.

E X A M P L E   6   If the shortest side in a 30°–60°–90° triangle is 1 inch, find the lengths of the other two sides.

Solution   In Figure 3 triangle ABC is a 30°–60°–90° triangle in which the shortest side AC is 1 inch long. Triangle DBC is also a 30°–60°–90° triangle in which the shortest side DC is 1 inch long.

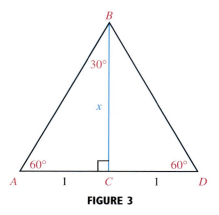

FIGURE 3

Notice that the large triangle ABD is an equilateral triangle because each of its interior angles is 60°. Each side of triangle ABD is 2 inches long. Side AB in triangle ABC is therefore 2 inches. To find the length of side BC, we use the Pythagorean theorem.

$$BC^2 + AC^2 = AB^2$$
$$x^2 + 1^2 = 2^2$$
$$x^2 + 1 = 4$$
$$x^2 = 3$$
$$x = \sqrt{3} \text{ inches}$$

Note that we write only the positive square root because x is the length of a side in a triangle and is therefore a positive number.

E X A M P L E   7   Table 2 in the introduction to this section gives the vertical rise of the Forest Double chair lift as 1,170 feet and the length of the chair lift as 5,750 feet. To the nearest foot, find the horizontal distance covered by a person riding this lift.

Solution   Figure 4 is a model of the Forest Double chair lift. A rider gets on the lift at point A and exits at point B. The length of the lift is AB.

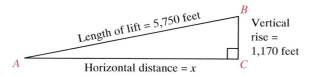

FIGURE 4

To find the horizontal distance covered by a person riding the chair lift, we use the Pythagorean theorem.

$$5{,}750^2 = x^2 + 1{,}170^2$$ Pythagorean theorem

$$33{,}062{,}500 = x^2 + 1{,}368{,}900$$ Simplify squares.

$$x^2 = 33{,}062{,}500 - 1{,}368{,}900$$ Solve for x^2.

$$x^2 = 31{,}693{,}600$$ Simplify the right side.

$$x = \sqrt{31{,}693{,}600}$$ Theorem 8.1

$$= 5{,}630 \text{ feet} \qquad \text{to the nearest foot}$$

A rider getting on the lift at point A and riding to point B will cover a horizontal distance of approximately 5,630 feet.

Getting Ready for Class

After reading through the preceding section, respond in your own words and in complete sentences.

A. What kind of equation do we solve using the method of completing the square?

B. Explain in words how you would complete the square on $x^2 - 16x = 4$.

C. What is the relationship between the shortest side and the longest side in a $30°-60°-90°$ triangle?

D. What two expressions together are equivalent to $x = \pm 4$?

PROBLEM SET 8.1

Solve the following equations.

1. $x^2 = 25$ **2.** $x^2 = 16$

3. $a^2 = -9$ **4.** $a^2 = -49$

5. $y^2 = \dfrac{3}{4}$ **6.** $y^2 = \dfrac{5}{9}$

7. $x^2 + 12 = 0$ **8.** $x^2 + 8 = 0$

9. $4a^2 - 45 = 0$ **10.** $9a^2 - 20 = 0$

11. $(2y - 1)^2 = 25$ **12.** $(3y + 7)^2 = 1$

13. $(2a + 3)^2 = -9$ **14.** $(3a - 5)^2 = -49$

15. $(5x + 2)^2 = -8$ **16.** $(6x - 7)^2 = -75$

17. $x^2 + 8x + 16 = -27$

18. $x^2 - 12x + 36 = -8$

19. $4a^2 - 12a + 9 = -4$

20. $9a^2 - 12a + 4 = -9$

Simplify the left side of each equation, and then solve for x.

21. $(x + 5)^2 + (x - 5)^2 = 52$

22. $(2x + 1)^2 + (2x - 1)^2 = 10$

23. $(2x + 3)^2 + (2x - 3)^2 = 26$

24. $(3x + 2)^2 + (3x - 2)^2 = 26$

25. $(3x + 4)(3x - 4) - (x + 2)(x - 2) = -4$

26. $(5x + 2)(5x - 2) - (x + 3)(x - 3) = 29$

Copy each of the following, and fill in the blanks so that the left side of each is a perfect square trinomial. That is, complete the square.

27. $x^2 + 12x + \underline{\quad} = (x + \underline{\quad})^2$

28. $x^2 + 6x + \underline{\quad} = (x + \underline{\quad})^2$

29. $x^2 - 4x + \underline{} = (x - \underline{})^2$

30. $x^2 - 2x + \underline{} = (x - \underline{})^2$

31. $a^2 - 10a + \underline{} = (a - \underline{})^2$

32. $a^2 - 8a + \underline{} = (a - \underline{})^2$

33. $x^2 + 5x + \underline{} = (x + \underline{})^2$

34. $x^2 + 3x + \underline{} = (x + \underline{})^2$

35. $y^2 - 7y + \underline{} = (y - \underline{})^2$

36. $y^2 - y + \underline{} = (y - \underline{})^2$

Solve each of the following quadratic equations by completing the square.

37. $x^2 + 4x = 12$

38. $x^2 - 2x = 8$

39. $x^2 + 12x = -27$

40. $x^2 - 6x = 16$

41. $a^2 - 2a + 5 = 0$

42. $a^2 + 10a + 22 = 0$

43. $y^2 - 8y + 1 = 0$

44. $y^2 + 6y - 1 = 0$

45. $x^2 - 5x - 3 = 0$

46. $x^2 - 5x - 2 = 0$

47. $2x^2 - 4x - 8 = 0$

48. $3x^2 - 9x - 12 = 0$

49. $3t^2 - 8t + 1 = 0$

50. $5t^2 + 12t - 1 = 0$

51. $4x^2 - 3x + 5 = 0$

52. $7x^2 - 5x + 2 = 0$

Applying the Concepts

53. Geometry If the shortest side in a $30°-60°-90°$ triangle is $\frac{1}{2}$ inch long, find the lengths of the other two sides.

54. Geometry If the shortest side in a $30°-60°-90°$ triangle is 3 feet long, find the lengths of the other two sides.

55. Geometry If the length of the shortest side of a $30°-60°-90°$ triangle is x, find the lengths of the other two sides in terms of x.

56. Geometry If the length of the longest side of a $30°-60°-90°$ triangle is x, find the lengths of the other two sides in terms of x.

57. Geometry If the length of the shorter sides of a $45°-45°-90°$ triangle is 1 inch, find the length of the hypotenuse.

58. Geometry If the length of the shorter sides of a $45°-45°-90°$ triangle is 3 feet, find the length of the hypotenuse.

59. Geometry If the length of the hypotenuse of a $45°-45°-90°$ triangle is 1 inch, find the length of the shorter sides.

60. Geometry If the length of the hypotenuse of a $45°-45°-90°$ triangle is 2 feet, find the length of the shorter sides.

61. Geometry If the length of the shorter sides of a $45°-45°-90°$ triangle is x, find the length of the hypotenuse, in terms of x.

62. Geometry If the length of the hypotenuse of a $45°-45°-90°$ triangle is x, find the length of the shorter sides, in terms of x.

63. Chair Lift Use Table 2 from the introduction to this section to find the horizontal distance covered by a person riding the Bear Paw Double chair lift. Round your answer to the nearest foot.

64. Chair Lift Use Table 2 from the introduction to this section to find the horizontal distance covered by a person riding the Big Springs Gondola lift. Round your answer to the nearest foot.

65. Chair Lift Using a right triangle to model the Forest Double chair lift, find the slope of the lift to the nearest hundredth.

66. Chair Lift Using a right triangle to model the Echo Triple chair lift, find the slope of the lift to the nearest hundredth.

67. Length of an Escalator An escalator in a department store is to carry people a vertical distance of 20 feet between floors. How long is the escalator if it makes an angle of 45° with the ground? (See Figure 5.)

FIGURE 5

68. Dimensions of a Tent A two-person tent is to be made so that the height at the center is 4 feet. If the sides of the tent are to meet the ground at an angle of 60°, and the tent is to be 6 feet in length, how many square feet of material will be needed to make the tent? (Figure 6; assume that the tent has a floor and is closed at both ends.) Give your answer to the nearest tenth of a square foot.

FIGURE 6

69. Interest Rate   Suppose a deposit of $3,000 in a savings account that paid an annual interest rate r (compounded yearly) is worth $3,456 after 2 years. Using the formula $A = P(1 + r)^t$, we have

$$3{,}456 = 3{,}000(1 + r)^2$$

Solve for r to find the annual interest rate.

70. Special Triangles   In Figure 7, triangle ABC has angles 45° and 30°, and height x. Find the lengths of sides AB, BC, and AC, in terms of x.

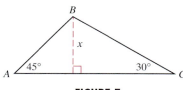

FIGURE 7

71. Special Triangles   In Figure 8, AB has a length of 12 inches. Find the length of AD.

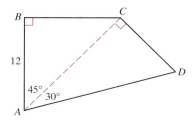

FIGURE 8

72. Wheelchair Access   Most local laws state that ramps for wheelchairs may not have more than a 1/10 slope; that is, for every foot in the vertical direction the ramp must travel at least 10 feet in the horizontal

direction. A wheelchair ramp will take the place of four steps that are 8 inches deep and 8 inches high (Figure 9). What is the minimum length of such a ramp?

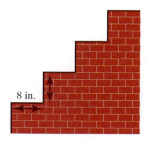

FIGURE 9

Review Problems

The problems that follow review material we covered in Section 7.3. Reviewing these problems will help you with the next section.

Write each of the following in simplified form for radicals.

73. $\sqrt{45}$ **74.** $\sqrt{24}$ **75.** $\sqrt{27y^5}$

76. $\sqrt{8y^3}$ **77.** $\sqrt[3]{54x^6y^5}$ **78.** $\sqrt[3]{16x^9y^7}$

79. Simplify $\sqrt{b^2 - 4ac}$ when $a = 6$, $b = 7$, and $c = -5$.

80. Simplify $\sqrt{b^2 - 4ac}$ when $a = 2$, $b = -6$, and $c = 3$.

Rationalize the denominator.

81. $\dfrac{3}{\sqrt{2}}$ **82.** $\dfrac{5}{\sqrt{3}}$ **83.** $\dfrac{2}{\sqrt[3]{4}}$ **84.** $\dfrac{3}{\sqrt[3]{2}}$

Extending the Concepts

Solve for x.

85. $(x + a)^2 + (x - a)^2 = 10a^2$

86. $(ax + 1)^2 + (ax - 1)^2 = 10$

Assume a is a positive number and solve for x by completing the square on x.

87. $x^2 + 2ax = -a^2$ **88.** $x^2 + 2ax = -4a^2$

89. $x^2 + 2ax = 0$ **90.** $x^2 + ax = 0$

Assume p and q are positive numbers and solve for x by completing the square on x.

91. $x^2 + px + q = 0$ **92.** $x^2 - px + q = 0$

The Quadratic Formula

In this section we will use the method of completing the square from the preceding section to derive the quadratic formula. The *quadratic formula* is a very useful tool in mathematics. It allows us to solve all types of quadratic equations.

THE QUADRATIC THEOREM

For any quadratic equation in the form $ax^2 + bx + c = 0$, where $a \neq 0$, the two solutions are

$$x = \frac{-b + \sqrt{b^2 - 4ac}}{2a} \qquad \text{and} \qquad x = \frac{-b - \sqrt{b^2 - 4ac}}{2a}$$

Proof We will prove the quadratic theorem by completing the square on $ax^2 + bx + c = 0$:

$$ax^2 + bx + c = 0$$

$$ax^2 + bx = -c \qquad \text{Add } -c \text{ to both sides.}$$

$$x^2 + \frac{b}{a}x = -\frac{c}{a} \qquad \text{Divide both sides by } a.$$

To complete the square on the left side, we add the square of $\frac{1}{2}$ of $\frac{b}{a}$ to both sides. $\left(\frac{1}{2} \text{ of } \frac{b}{a} \text{ is } \frac{b}{2a}.\right)$

$$x^2 + \frac{b}{a}x + \left(\frac{b}{2a}\right)^2 = -\frac{c}{a} + \left(\frac{b}{2a}\right)^2$$

We now simplify the right side as a separate step. We square the second term and combine the two terms by writing each with the least common denominator $4a^2$:

$$-\frac{c}{a} + \left(\frac{b}{2a}\right)^2 = -\frac{c}{a} + \frac{b^2}{4a^2} = \frac{4a}{4a}\left(\frac{-c}{a}\right) + \frac{b^2}{4a^2} = \frac{-4ac + b^2}{4a^2}$$

It is convenient to write this last expression as

$$\frac{b^2 - 4ac}{4a^2}$$

Continuing with the proof, we have

$$x^2 + \frac{b}{a}x + \left(\frac{b}{2a}\right)^2 = \frac{b^2 - 4ac}{4a^2}$$

$$\left(x + \frac{b}{2a}\right)^2 = \frac{b^2 - 4ac}{4a^2} \qquad \begin{array}{l}\text{Write left side as a} \\ \text{binomial square.}\end{array}$$

$$x + \frac{b}{2a} = \pm \frac{\sqrt{b^2 - 4ac}}{2a} \qquad \text{Theorem 8.1}$$

$$x = -\frac{b}{2a} \pm \frac{\sqrt{b^2 - 4ac}}{2a} \qquad \text{Add } -\frac{b}{2a} \text{ to both sides.}$$

$$= \frac{-b \pm \sqrt{b^2 - 4ac}}{2a}$$

Our proof is now complete. What we have is this: If our equation is in the form $ax^2 + bx + c = 0$ (standard form), where $a \neq 0$, the two solutions are always given by the formula

$$x = \frac{-b \pm \sqrt{b^2 - 4ac}}{2a}$$

This formula is known as the *quadratic formula*. If we substitute the coefficients a, b, and c of any quadratic equation in standard form into the formula, we need only perform some basic arithmetic to arrive at the solution set.

E X A M P L E   I   Use the quadratic formula to solve $6x^2 + 7x - 5 = 0$.

Solution   Using $a = 6$, $b = 7$, and $c = -5$ in the formula

$$x = \frac{-b \pm \sqrt{b^2 - 4ac}}{2a}$$

we have

$$x = \frac{-7 \pm \sqrt{49 - 4(6)(-5)}}{2(6)}$$

or

$$x = \frac{-7 \pm \sqrt{49 + 120}}{12}$$

$$= \frac{-7 \pm \sqrt{169}}{12}$$

$$= \frac{-7 \pm 13}{12}$$

We separate the last equation into the two statements

$$x = \frac{-7 + 13}{12} \qquad \text{or} \qquad x = \frac{-7 - 13}{12}$$

$$x = \frac{1}{2} \qquad \text{or} \qquad x = -\frac{5}{3}$$

The solution set is $\{\frac{1}{2}, -\frac{5}{3}\}$.

Whenever the solutions to a quadratic equation are rational numbers, as they are in Example 1, it means that the original equation was solvable by factoring. To illustrate, let's solve the equation from Example 1 again, but this time by factoring:

$$6x^2 + 7x - 5 = 0 \qquad \text{Equation in standard form}$$

$$(3x + 5)(2x - 1) = 0 \qquad \text{Factor the left side.}$$

$$3x + 5 = 0 \qquad \text{or} \qquad 2x - 1 = 0 \qquad \text{Set factors equal to 0.}$$

$$x = -\frac{5}{3} \qquad \text{or} \qquad x = \frac{1}{2}$$

When an equation can be solved by factoring, then factoring is usually the faster method of solution. It is best to try to factor first, and then if you have trouble factoring, go to the quadratic formula. It always works.

EXAMPLE 2   Solve $\dfrac{x^2}{3} - x = -\dfrac{1}{2}$.

Solution   Multiplying through by 6 and writing the result in standard form, we have

$$2x^2 - 6x + 3 = 0$$

the left side of which is not factorable. Therefore, we use the quadratic formula with $a = 2$, $b = -6$, and $c = 3$. The two solutions are given by

$$x = \frac{-(-6) \pm \sqrt{36 - 4(2)(3)}}{2(2)}$$

$$= \frac{6 \pm \sqrt{12}}{4}$$

$$= \frac{6 \pm 2\sqrt{3}}{4} \qquad \sqrt{12} = \sqrt{4 \cdot 3} = \sqrt{4}\,\sqrt{3} = 2\sqrt{3}$$

We can reduce this last expression to lowest terms by factoring 2 from the numerator and denominator and then dividing the numerator and denominator by 2:

$$x = \frac{2(3 \pm \sqrt{3})}{2 \cdot 2} = \frac{3 \pm \sqrt{3}}{2}$$

EXAMPLE 3   Solve $\dfrac{1}{x + 2} - \dfrac{1}{x} = \dfrac{1}{3}$.

Solution   To solve this equation, we must first put it in standard form. To do so, we must clear the equation of fractions by multiplying each side by the LCD for all the denominators, which is $3x(x + 2)$. Multiplying both sides by the LCD, we have

$$3x(x + 2)\left(\frac{1}{x + 2} - \frac{1}{x}\right) = \frac{1}{3} \cdot 3x(x + 2) \qquad \text{Multiply each by the LCD.}$$

$$3x(\cancel{x + 2}) \cdot \frac{1}{\cancel{x + 2}} - 3x(x + 2) \cdot \frac{1}{\cancel{x}} = \frac{1}{\cancel{3}} \cdot \cancel{3}x(x + 2)$$

$$3x - 3(x + 2) = x(x + 2)$$

$$3x - 3x - 6 = x^2 + 2x \qquad \text{Multiplication}$$

$$-6 = x^2 + 2x \qquad \text{Simplify left side.}$$

$$0 = x^2 + 2x + 6 \qquad \text{Add 6 to each side.}$$

Since the right side of our last equation is not factorable, we use the quadratic formula. From our last equation, we have $a = 1$, $b = 2$, and $c = 6$. Using these numbers for a, b, and c in the quadratic formula gives us

$$x = \frac{-2 \pm \sqrt{4 - 4(1)(6)}}{2(1)}$$

$$= \frac{-2 \pm \sqrt{4 - 24}}{2} \qquad \text{Simplify inside the radical.}$$

$$= \frac{-2 \pm \sqrt{-20}}{2} \qquad 4 - 24 = -20$$

$$= \frac{-2 \pm 2i\sqrt{5}}{2} \qquad \sqrt{-20} = i\sqrt{20} = i\sqrt{4}\,\sqrt{5} = 2i\sqrt{5}$$

$$= \frac{2(-1 \pm i\sqrt{5})}{2} \qquad \text{Factor 2 from the numerator.}$$

$$= -1 \pm i\sqrt{5} \qquad \text{Divide numerator and denominator by 2.}$$

Since neither of the two solutions, $-1 + i\sqrt{5}$ nor $-1 - i\sqrt{5}$, will make any of the denominators in our original equation 0, they are both solutions.

Although the equation in our next example is not a quadratic equation, we solve it by using both factoring and the quadratic formula.

EXAMPLE 4   Solve $27t^3 - 8 = 0$.

Solution   It would be a mistake to add 8 to each side of this equation and then take the cube root of each side because we would lose two of our solutions. Instead, we factor the left side, and then set the factors equal to 0:

$$27t^3 - 8 = 0 \qquad \text{Equation in standard form}$$

$$(3t - 2)(9t^2 + 6t + 4) = 0 \qquad \text{Factor as the difference of two cubes.}$$

$$3t - 2 = 0 \quad \text{or} \quad 9t^2 + 6t + 4 = 0 \qquad \text{Set each factor equal to 0.}$$

The first equation leads to a solution of $t = \frac{2}{3}$. The second equation does not factor, so we use the quadratic formula with $a = 9$, $b = 6$, and $c = 4$:

$$t = \frac{-6 \pm \sqrt{36 - 4(9)(4)}}{2(9)}$$

$$= \frac{-6 \pm \sqrt{36 - 144}}{18}$$

$$= \frac{-6 \pm \sqrt{-108}}{18}$$

$$= \frac{-6 \pm 6i\sqrt{3}}{18} \qquad\qquad \sqrt{-108} = i\sqrt{36 \cdot 3} = 6i\sqrt{3}$$

$$= \frac{6(-1 \pm i\sqrt{3})}{6 \cdot 3} \qquad\qquad \text{Factor 6 from the numerator and denominator.}$$

$$= \frac{-1 \pm i\sqrt{3}}{3} \qquad\qquad \text{Divide out common factor 6.}$$

The three solutions to our original equation are

$$\frac{2}{3}, \qquad \frac{-1 + i\sqrt{3}}{3}, \qquad \text{and} \qquad \frac{-1 - i\sqrt{3}}{3}$$

EXAMPLE 5   If an object is thrown downward with an initial velocity of 20 feet per second, the distance $s(t)$, in feet, it travels in t seconds is given by the function $s(t) = 20t + 16t^2$. How long does it take the object to fall 40 feet?

Solution   We let $s(t) = 40$, and solve for t:

When $s(t) = 40$

the function $s(t) = 20t + 16t^2$

becomes $40 = 20t + 16t^2$

or $16t^2 + 20t - 40 = 0$

$4t^2 + 5t - 10 = 0$ Divide by 4.

Using the quadratic formula, we have

$$t = \frac{-5 \pm \sqrt{25 - 4(4)(-10)}}{2(4)}$$

$$= \frac{-5 \pm \sqrt{185}}{8}$$

$$= \frac{-5 + \sqrt{185}}{8} \qquad \text{or} \qquad t = \frac{-5 - \sqrt{185}}{8}$$

The second solution is impossible since it is a negative number and time t must be positive. It takes

$$t = \frac{-5 + \sqrt{185}}{8} \qquad \text{or approximately} \qquad \frac{-5 + 13.60}{8} \approx 1.08 \text{ seconds}$$

for the object to fall 40 feet.

Recall from Chapter 5 that the relationship between profit, revenue, and cost is given by the formula

$$P(x) = R(x) - C(x)$$

where $P(x)$ is the profit, $R(x)$ is the total revenue, and $C(x)$ is the total cost of producing and selling x items.

EXAMPLE 6 A company produces and sells copies of an accounting program for home computers. The total weekly cost (in dollars) to produce x copies of the program is $C(x) = 8x + 500$, and the weekly revenue for selling all x copies of the program is $R(x) = 35x - 0.1x^2$. How many programs must be sold each week for the weekly profit to be $1,200?

Solution   Substituting the given expressions for $R(x)$ and $C(x)$ in the equation $P(x) = R(x) - C(x)$, we have a polynomial in x that represents the weekly profit $P(x)$:

$$\begin{aligned}
P(x) &= R(x) - C(x) \\
&= 35x - 0.1x^2 - (8x + 500) \\
&= 35x - 0.1x^2 - 8x - 500 \\
&= -500 + 27x - 0.1x^2
\end{aligned}$$

Setting this expression equal to 1,200, we have a quadratic equation to solve that gives us the number of programs x that need to be sold each week to bring in a profit of $1,200:

$$1,200 = -500 + 27x - 0.1x^2$$

We can write this equation in standard form by adding the opposite of each term on the right side of the equation to both sides of the equation. Doing so produces the following equation:

$$0.1x^2 - 27x + 1,700 = 0$$

Applying the quadratic formula to this equation with $a = 0.1$, $b = -27$, and $c = 1,700$, we have

$$\begin{aligned}
x &= \frac{27 \pm \sqrt{(-27)^2 - 4(0.1)(1,700)}}{2(0.1)} \\
&= \frac{27 \pm \sqrt{729 - 680}}{0.2} \\
&= \frac{27 \pm \sqrt{49}}{0.2} \\
&= \frac{27 \pm 7}{0.2}
\end{aligned}$$

Writing this last expression as two separate expressions, we have our two solutions:

$$x = \frac{27 + 7}{0.2} \quad \text{or} \quad x = \frac{27 - 7}{0.2}$$

$$= \frac{34}{0.2} \qquad\qquad = \frac{20}{0.2}$$

$$= 170 \qquad\qquad\quad = 100$$

The weekly profit will be \$1,200 if the company produces and sells 100 programs or 170 programs.

What is interesting about the equation we solved in Example 6 is that it has rational solutions, meaning it could have been solved by factoring. But looking back at the equation, factoring does not seem like a reasonable method of solution because the coefficients are either very large or very small. So, there are times when using the quadratic formula is a faster method of solution, even though the equation you are solving is factorable.

Using TECHNOLOGY

GRAPHING CALCULATORS

More About Example 5

We can solve the problem discussed in Example 5 by graphing the function $Y_1 = 20X + 16X^2$ in a window with X from 0 to 2 (because X is taking the place of t and we know t is a positive quantity) and Y from 0 to 50 (because we are looking for X when Y_1 is 40). Graphing Y_1 gives a graph similar to the graph in Figure 1. Using the Zoom and Trace features at $Y_1 = 40$ gives us X = 1.08 to the nearest hundredth, matching the results we obtained by solving the original equation algebraically.

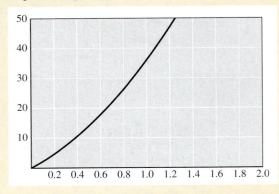

FIGURE 1

(continued)

More About Example 6

To visualize the functions in Example 6, we set up our calculator this way:

$$Y_1 = 35X - .1X^2 \qquad \text{Revenue function}$$
$$Y_2 = 8X + 500 \qquad \text{Cost function}$$
$$Y_3 = Y_1 - Y_2 \qquad \text{Profit function}$$

Window: X from 0 to 350, Y from 0 to 3,500

Graphing these functions produces graphs similar to the ones shown in Figure 2. The lowest graph is the graph of the profit function. Using the Zoom and Trace features on the lowest graph at $Y_3 = 1{,}200$ produces two corresponding values of X, 170 and 100, which match the results in Example 6.

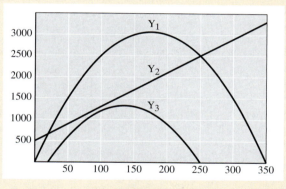

FIGURE 2

We will continue this discussion of the relationship between graphs of functions and solutions to equations in the Using Technology material in the next section.

Getting Ready for Class

After reading through the preceding section, respond in your own words and in complete sentences.

A. What is the quadratic formula?

B. Under what circumstances should the quadratic formula be applied?

C. When would the quadratic formula result in complex solutions?

D. When will the quadratic formula result in only one solution?

PROBLEM SET 8.2

Solve each equation. Use factoring or the quadratic formula, whichever is appropriate. (Try factoring first. If you have any difficulty factoring, then go right to the quadratic formula.)

1. $x^2 + 5x + 6 = 0$

2. $x^2 + 5x - 6 = 0$

3. $a^2 - 4a + 1 = 0$

4. $a^2 + 4a + 1 = 0$

5. $\frac{1}{6}x^2 - \frac{1}{2}x + \frac{1}{3} = 0$

6. $\frac{1}{4}x^2 + \frac{1}{4}x - \frac{1}{2} = 0$

7. $\frac{x^2}{2} + 1 = \frac{2x}{3}$

8. $\frac{x^2}{2} + \frac{2}{3} = -\frac{2x}{3}$

9. $y^2 - 5y = 0$

10. $2y^2 + 10y = 0$

11. $30x^2 + 40x = 0$

12. $50x^2 - 20x = 0$

13. $\frac{2t^2}{3} - t = -\frac{1}{6}$

14. $\frac{t^2}{3} - \frac{t}{2} = -\frac{3}{2}$

15. $0.01x^2 + 0.06x - 0.08 = 0$

16. $0.02x^2 - 0.03x + 0.05 = 0$

17. $2x + 3 = -2x^2$

18. $2x - 3 = 3x^2$

19. $100x^2 - 200x + 100 = 0$

20. $100x^2 - 600x + 900 = 0$

21. $\frac{1}{2}r^2 = \frac{1}{6}r - \frac{2}{3}$

22. $\frac{1}{4}r^2 = \frac{2}{5}r + \frac{1}{10}$

23. $(x - 3)(x - 5) = 1$

24. $(x - 3)(x + 1) = -6$

25. $(x + 3)^2 + (x - 8)(x - 1) = 16$

26. $(x - 4)^2 + (x + 2)(x + 1) = 9$

27. $\frac{x^2}{3} - \frac{5x}{6} = \frac{1}{2}$

28. $\frac{x^2}{6} + \frac{5}{6} = -\frac{x}{3}$

Multiply both sides of each equation by its LCD. Then solve the resulting equation.

29. $\frac{1}{x + 1} - \frac{1}{x} = \frac{1}{2}$

30. $\frac{1}{x + 1} + \frac{1}{x} = \frac{1}{3}$

31. $\frac{1}{y - 1} + \frac{1}{y + 1} = 1$

32. $\frac{2}{y + 2} + \frac{3}{y - 2} = 1$

33. $\frac{1}{x + 2} + \frac{1}{x + 3} = 1$

34. $\frac{1}{x + 3} + \frac{1}{x + 4} = 1$

35. $\frac{6}{r^2 - 1} - \frac{1}{2} = \frac{1}{r + 1}$

36. $2 + \frac{5}{r - 1} = \frac{12}{(r - 1)^2}$

Solve each equation. In each case you will have three solutions.

37. $x^3 - 8 = 0$

38. $x^3 - 27 = 0$

39. $8a^3 + 27 = 0$

40. $27a^3 + 8 = 0$

41. $125t^3 - 1 = 0$

42. $64t^3 + 1 = 0$

Each of the following equations has three solutions. Look for the greatest common factor; then use the quadratic formula to find all solutions.

43. $2x^3 + 2x^2 + 3x = 0$

44. $6x^3 - 4x^2 + 6x = 0$

45. $3y^4 = 6y^3 - 6y^2$

46. $4y^4 = 16y^3 - 20y^2$

47. $6t^5 + 4t^4 = -2t^3$

48. $8t^5 + 2t^4 = -10t^3$

49. One solution to a quadratic equation is $\frac{-3 + 2i}{5}$. What is the other solution?

50. One solution to a quadratic equation is $\frac{-2 + 3i\sqrt{2}}{5}$. What is the other solution?

Applying the Concepts

51. Falling Object   An object is thrown downward with an initial velocity of 5 feet per second. The relationship between the distance s it travels and time t is given by $s = 5t + 16t^2$. How long does it take the object to fall 74 feet?

52. Falling Object   The distance an object falls from rest is given by the equation $s = 16t^2$, where $s =$

distance and t = time. How long does it take an object dropped from a 100-foot cliff to hit the ground?

53. Ball Toss   A ball is thrown upward with an initial velocity of 20 feet per second. The equation that gives the height h of the object at any time t is $h = 20t - 16t^2$. At what times will the object be 4 feet off the ground?

54. Coin Toss   A coin is tossed upward with an initial velocity of 32 feet per second from a height of 16 feet above the ground. The equation giving the object's height h at any time t is $h = 16 + 32t - 16t^2$. Does the object ever reach a height of 32 feet?

55. Profit   The total cost (in dollars) for a company to manufacture and sell x items per week is $C = 60x + 300$, whereas the revenue brought in by selling all x items is $R = 100x - 0.5x^2$. How many items must be sold to obtain a weekly profit of $300?

56. Profit   The total cost (in dollars) for a company to produce and sell x items per week is $C = 200x + 1,600$, whereas the revenue brought in by selling all x items is $R = 300x - 0.5x^2$. How many items must be sold in order for the weekly profit to be $2,150?

57. Profit   Suppose it costs a company selling patterns $C = 800 + 6.5x$ dollars to produce and sell x patterns a month. If the revenue obtained by selling x patterns is $R = 10x - 0.002x^2$, how many patterns must it sell each month if it wants a monthly profit of $700?

58. Profit   Suppose a company manufactures and sells x picture frames each month with a total cost of $C = 1,200 + 3.5x$ dollars. If the revenue obtained by selling x frames is $R = 9x - 0.002x^2$, find the number of frames it must sell each month if its monthly profit is to be $2,300.

59. Photograph Cropping   The following figure shows a photographic image on a 10.5-centimeter by 8.2-centimeter background. The overall area of the background is to be reduced to 80% of its original area by cutting off (cropping) equal strips on all four sides. What is the width of the strip that is cut from each side?

(American Museum of Natural History)

60. Area of a Garden   A garden measures 20.3 meters by 16.4 meters. In order to double the area of the garden, strips of equal width are added to all four sides.
(a) Draw a diagram that illustrates these conditions.
(b) What are the new overall dimensions of the garden?

61. Area and Perimeter   A rectangle has a perimeter of 20 yards and an area of 15 square yards.
(a) Write two equations that state these facts in terms of the rectangle's length, l, and its width, w.
(b) Solve the two equations from part (a) to determine the actual length and width of the rectangle.
(c) Explain why two answers are possible to part (b).

62. Population Size   Writing in 1829, former President James Madison made some predictions about the growth of the population of the United States. The populations he predicted fit the equation

$$y = 0.029x^2 - 1.39x + 42$$

where y is the population in millions of people x years from 1829.
(a) Use the equation to determine the approximate year President Madison would have predicted that the U.S. population would reach 100,000,000.
(b) If the U.S. population in 1990 was approximately 200,000,000, were President Madison's predictions accurate in the long term? Explain why or why not.

Review Problems

The problems that follow review material we covered in Sections 6.2 and 7.1. Reviewing these problems will help you with the next section.

Divide, using long division. [6.2]

63. $\dfrac{8y^2 - 26y - 9}{2y - 7}$

64. $\dfrac{6y^2 + 7y - 18}{3y - 4}$

65. $\dfrac{x^3 + 9x^2 + 26x + 24}{x + 2}$

66. $\dfrac{x^3 + 6x^2 + 11x + 6}{x + 3}$

Simplify each expression. (Assume $x, y > 0$.) [7.1]

67. $25^{1/2}$

68. $8^{1/3}$

69. $\left(\dfrac{9}{25}\right)^{3/2}$

70. $\left(\dfrac{16}{81}\right)^{3/4}$

71. $8^{-2/3}$

72. $4^{-3/2}$

73. $\dfrac{(49x^8y^{-4})^{1/2}}{(27x^{-3}y^9)^{-1/3}}$

74. $\dfrac{(x^{-2}y^{1/3})^6}{x^{-10}y^{3/2}}$

Extending the Concepts

So far, all the equations we have solved have had coefficients that were rational numbers. Here are some equations that have irrational coefficients and some that have complex coefficients. Solve each equation. (Remember, $i^2 = -1$.)

75. $x^2 + \sqrt{3}x - 6 = 0$

76. $x^2 - \sqrt{5}x - 5 = 0$

77. $\sqrt{2}x^2 + 2x - \sqrt{2} = 0$

78. $\sqrt{7}x^2 + 2\sqrt{2}x - \sqrt{7} = 0$

79. $x^2 + ix + 2 = 0$

80. $x^2 + 3ix - 2 = 0$

81. $ix^2 + 3x + 4i = 0$

82. $4ix^2 + 5x + 9i = 0$

8.3 Additional Items Involving Solutions to Equations

In this section we will do two things. First, we will define the discriminant and use it to find the kind of solutions a quadratic equation has without solving the equation. Second, we will use the zero-factor property to build equations from their solutions.

THE DISCRIMINANT

The quadratic formula

$$x = \frac{-b \pm \sqrt{b^2 - 4ac}}{2a}$$

gives the solutions to any quadratic equation in standard form. There are times, when working with quadratic equations, when it is important only to know what kind of solutions the equation has.

> **DEFINITION** The expression under the radical in the quadratic formula is called the **discriminant**:
>
> $$\text{Discriminant} = D = b^2 - 4ac$$

The discriminant indicates the number and type of solutions to a quadratic equation, when the original equation has integer coefficients. For example, if we were to use the quadratic formula to solve the equation $2x^2 + 2x + 3 = 0$, we would find the discriminant to be

$$b^2 - 4ac = 2^2 - 4(2)(3) = -20$$

Since the discriminant appears under a square root symbol, we have the square root of a negative number in the quadratic formula. Our solutions would therefore be complex numbers. Similarly, if the discriminant were 0, the quadratic formula would yield

$$x = \frac{-b \pm \sqrt{0}}{2a} = \frac{-b \pm 0}{2a} = \frac{-b}{2a}$$

and the equation would have one rational solution, the number $\dfrac{-b}{2a}$.

The following table gives the relationship between the discriminant and the type of solutions to the equation.

For the equation $ax^2 + bx + c = 0$ where a, b, and c are integers and $a \neq 0$:

If the Discriminant $b^2 - 4ac$ Is	Then the Equation Will Have
Negative	Two complex solutions containing i
Zero	One rational solution
A positive number that is also a perfect square	Two rational solutions
A positive number that is not a perfect square	Two irrational solutions

In the second and third cases, when the discriminant is 0 or a positive perfect square, the solutions are rational numbers. The quadratic equations in these two cases are the ones that can be factored.

EXAMPLES   For each equation, give the number and kind of solutions.

1. $x^2 - 3x - 40 = 0$

Solution   Using $a = 1$, $b = -3$, and $c = -40$ in $b^2 - 4ac$, we have $(-3)^2 - 4(1)(-40) = 9 + 160 = 169$.

The discriminant is a perfect square. The equation therefore has two rational solutions.

2. $2x^2 - 3x + 4 = 0$

Solution   Using $a = 2$, $b = -3$, and $c = 4$, we have

$$b^2 - 4ac = (-3)^2 - 4(2)(4) = 9 - 32 = -23$$

The discriminant is negative, implying the equation has two complex solutions that contain i.

3. $4x^2 - 12x + 9 = 0$

Solution   Using $a = 4$, $b = -12$, and $c = 9$, the discriminant is

$$b^2 - 4ac = (-12)^2 - 4(4)(9) = 144 - 144 = 0$$

Since the discriminant is 0, the equation will have one rational solution.

4. $x^2 + 6x = 8$

Solution   We must first put the equation in standard form by adding -8 to each side. If we do so, the resulting equation is

$$x^2 + 6x - 8 = 0$$

Now we identify a, b, and c as 1, 6, and -8, respectively:

$$b^2 - 4ac = 6^2 - 4(1)(-8) = 36 + 32 = 68$$

The discriminant is a positive number, but not a perfect square. The equation will therefore have two irrational solutions.

EXAMPLE 5   Find an appropriate k so that the equation $4x^2 - kx = -9$ has exactly one rational solution.

Solution   We begin by writing the equation in standard form:

$$4x^2 - kx + 9 = 0$$

Using $a = 4$, $b = -k$, and $c = 9$, we have

$$b^2 - 4ac = (-k)^2 - 4(4)(9)$$
$$= k^2 - 144$$

An equation has exactly one rational solution when the discriminant is 0. We set the discriminant equal to 0 and solve:

$$k^2 - 144 = 0$$
$$k^2 = 144$$
$$k = \pm 12$$

Choosing k to be 12 or -12 will result in an equation with one rational solution.

BUILDING EQUATIONS FROM THEIR SOLUTIONS

Suppose we know that the solutions to an equation are $x = 3$ and $x = -2$. We can find equations with these solutions by using the zero-factor property. First, let's write our solutions as equations with 0 on the right side:

If	$x = 3$	First solution
then	$x - 3 = 0$	Add -3 to each side.

$$\text{and if} \qquad x = -2 \qquad \text{Second solution}$$

$$\text{then} \qquad x + 2 = 0 \qquad \text{Add 2 to each side.}$$

Now, since both $x - 3$ and $x + 2$ are 0, their product must be 0 also. We can therefore write

$$(x - 3)(x + 2) = 0 \qquad \text{Zero-factor property}$$

$$x^2 - x - 6 = 0 \qquad \text{Multiply out the left side.}$$

Many other equations have 3 and -2 as solutions. For example, any constant multiple of $x^2 - x - 6 = 0$, such as $5x^2 - 5x - 30 = 0$, also has 3 and -2 as solutions. Similarly, any equation built from positive integer powers of the factors $x - 3$ and $x + 2$ will also have 3 and -2 as solutions. One such equation is

$$(x - 3)^2(x + 2) = 0$$

$$(x^2 - 6x + 9)(x + 2) = 0$$

$$x^3 - 4x^2 - 3x + 18 = 0$$

In mathematics we distinguish between the solutions to this last equation and those to the equation $x^2 - x - 6 = 0$ by saying $x = 3$ is a solution of *multiplicity 2* in the equation $x^3 - 4x^2 - 3x + 18 = 0$, and a solution of *multiplicity 1* in the equation $x^2 - x - 6 = 0$.

EXAMPLE   6   Find an equation that has solutions $t = 5$, $t = -5$, and $t = 3$.

Solution   First, we use the given solutions to write equations that have 0 on their right sides:

$$\text{If} \qquad t = 5 \qquad t = -5 \qquad t = 3$$

$$\text{then} \qquad t - 5 = 0 \qquad t + 5 = 0 \qquad t - 3 = 0$$

Since $t - 5$, $t + 5$, and $t - 3$ are all 0, their product is also 0 by the zero-factor property. An equation with solutions of 5, -5, and 3 is

$$(t - 5)(t + 5)(t - 3) = 0 \qquad \text{Zero-factor property}$$

$$(t^2 - 25)(t - 3) = 0 \qquad \text{Multiply first two binomials.}$$

$$t^3 - 3t^2 - 25t + 75 = 0 \qquad \text{Complete the multiplication.}$$

The last line gives us an equation with solutions of 5, -5, and 3. Remember, many other equations have these same solutions.

EXAMPLE   7   Find an equation with solutions $x = -\frac{2}{3}$ and $x = \frac{4}{5}$.

Solution   The solution $x = -\frac{2}{3}$ can be rewritten as $3x + 2 = 0$ as follows:

$$x = -\frac{2}{3} \qquad \text{The first solution}$$

$$3x = -2 \qquad \text{Multiply each side by 3.}$$

$$3x + 2 = 0 \qquad \text{Add 2 to each side.}$$

Similarly, the solution $x = \frac{4}{5}$ can be rewritten as $5x - 4 = 0$:

$$x = \frac{4}{5} \qquad \text{The second solution}$$

$$5x = 4 \qquad \text{Multiply each side by 5.}$$

$$5x - 4 = 0 \qquad \text{Add } -4 \text{ to each side.}$$

Since both $3x + 2$ and $5x - 4$ are 0, their product is 0 also, giving us the equation we are looking for:

$$(3x + 2)(5x - 4) = 0 \qquad \text{Zero-factor property}$$

$$15x^2 - 2x - 8 = 0 \qquad \text{Multiplication}$$

 Using **TECHNOLOGY**

GRAPHING CALCULATORS

Solving Equations

Now that we have explored the relationship between equations and their solutions, we can look at how a graphing calculator can be used in the solution process. To begin, let's solve the equation $x^2 = x + 2$ using techniques from algebra: writing it in standard form, factoring, and then setting each factor equal to 0.

$$x^2 - x - 2 = 0 \qquad \text{Standard form}$$

$$(x - 2)(x + 1) = 0 \qquad \text{Factor.}$$

$$x - 2 = 0 \quad \text{or} \quad x + 1 = 0 \qquad \text{Set each factor equal to 0.}$$

$$x = 2 \quad \text{or} \quad x = -1 \qquad \text{Solve.}$$

Our original equation, $x^2 = x + 2$, has two solutions: $x = 2$ and $x = -1$. To solve the equation using a graphing calculator, we need to associate it with an equation (or equations) in two variables. One way to do this is to associate the left side with the equation $y = x^2$ and the right side of the equation with $y = x + 2$. To do so, we set up the functions list in our calculator this way:

$$Y_1 = X^2$$

$$Y_2 = X + 2$$

Window: X from -5 to 5, Y from -5 to 5

Graphing these functions in this window will produce a graph similar to the one shown in Figure 1.

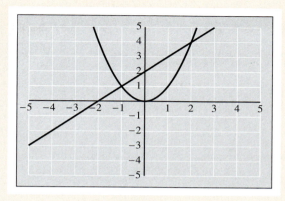

FIGURE 1

If we use the Trace feature to find the coordinates of the points of intersection, we find that the two curves intersect at $(-1, 1)$ and $(2, 4)$. We note that the x-coordinates of these two points match the solutions to the equation $x^2 = x + 2$, which we found using algebraic techniques. This makes sense because if two graphs intersect at a point (x, y), then the coordinates of that point satisfy both equations. If a point (x, y) satisfies both $y = x^2$ and $y = x + 2$, then, for that particular point, $x^2 = x + 2$. From this we conclude that the x-coordinates of the points of intersection are solutions to our original equation. Here is a summary of what we have discovered:

Conclusion 1   If the graphs of two functions $y = f(x)$ and $y = g(x)$ intersect in the coordinate plane, then the x-coordinates of the points of intersection are solutions to the equation $f(x) = g(x)$.

A second method of solving our original equation $x^2 = x + 2$ graphically requires the use of one function instead of two. To begin, we write the equation in standard form as $x^2 - x - 2 = 0$. Next, we graph the function $y = x^2 - x - 2$. The x-intercepts of the graph are the points with y-coordinates of 0. They therefore satisfy the equation $0 = x^2 - x - 2$, which is equivalent to our original equation. The graph in Figure 2 shows $Y_1 = X^2 - X - 2$ in a window with X from -5 to 5 and Y from -5 to 5.

Using the Trace feature, we find that the x-intercepts of the graph are $x = -1$ and $x = 2$, which match the solutions to our original equation $x^2 = x + 2$. We can summarize the relationship between solutions to an equation and the intercepts of its associated graph this way:

Conclusion 2   If $y = f(x)$ is a function, then any x-intercept on the graph of $y = f(x)$ is a solution to the equation $f(x) = 0$.

(continued)

(continued)

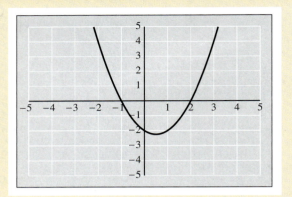

FIGURE 2

Solving Equations With Complex Solutions

There are limitations to using a graph to solve an equation. To illustrate, suppose we solve the equation $4x^3 - 6x^2 + 2x - 3 = 0$ by factoring:

$$4x^3 - 6x^2 + 2x - 3 = 0 \qquad \text{Original equation}$$

$$2x^2(2x - 3) + (2x - 3) = 0$$

$$(2x^2 + 1)(2x - 3) = 0 \qquad \text{Factor by grouping.}$$

$$2x^2 + 1 = 0 \qquad \text{or} \qquad 2x - 3 = 0 \qquad \text{Set factors equal to 0.}$$

$$2x^2 = -1 \qquad \text{or} \qquad 2x = 3$$

$$x^2 = -\frac{1}{2} \qquad \text{or} \qquad x = \frac{3}{2} \qquad \text{Solve the resulting equations.}$$

$$x = \pm\frac{1}{\sqrt{2}}\,i$$

$$= \pm\frac{\sqrt{2}}{2}\,i$$

We have three solutions:

$$-\frac{\sqrt{2}}{2}\,i, \qquad \frac{\sqrt{2}}{2}\,i, \qquad \frac{3}{2}$$

Figure 3 shows the graph of $y = 4x^3 - 6x^2 + 2x - 3$. As you can see, the graph crosses the x-axis exactly once at $x = \frac{3}{2}$, which we expect. The rectangular coordinate system consists of ordered pairs of *real numbers,* so our complex solutions cannot appear on the graph.

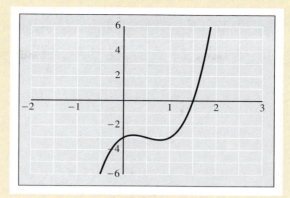

FIGURE 3

Every cubic equation will have at least one real number solution. If we are interested in exact values for all solutions to our equation, including complex solutions as well as irrational solutions, we can take our one real solution and, with the aid of long division and factoring or the quadratic formula, find the other solutions. But how do we find the one real solution in the first place? We use a graphing calculator.

Suppose we want to solve the equation $x^3 = 15x + 4$ completely. We can use a graphing calculator to graph the function $y = x^3 - 15x - 4$ and note that it crosses the x-axis at $x = 4$. The graph is shown in Figure 4.

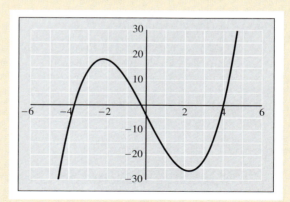

FIGURE 4

If we zoom and trace on the other two solutions, we find they are approximated by -3.732 and -0.268 to the nearest thousandth. To find exact values for these last two solutions, we reason that if $x = 4$ is a solution to our equation, then $x - 4$ must divide $x^3 - 15x - 4$ without a remainder. Here is the

(continued)

(continued)

division problem:

$$x - 4 \overline{\smash{)}\begin{aligned} x^2 + 4x + 1 \\[-2pt] x^3 + 0x^2 - 15x - 4 \end{aligned}}$$

$$\begin{aligned}
\underline{x^3 - 4x^2} \phantom{{}- 15x - 4} \\
4x^2 - 15x \phantom{{}- 4} \\
\underline{4x^2 - 16x} \phantom{{}- 4} \\
x - 4 \\
\underline{x - 4} \\
0
\end{aligned}$$

We find that $x^3 - 15x - 4$ factors into $(x - 4)(x^2 + 4x + 1)$. This allows us to find all solutions to our original equation:

$$x^3 - 15x - 4 = 0$$

$$(x - 4)(x^2 + 4x + 1) = 0$$

$$x - 4 = 0 \qquad \text{or} \qquad x^2 + 4x + 1 = 0$$

$$x = 4 \qquad \text{or} \qquad x = \frac{-4 \pm \sqrt{16 - 4(1)(1)}}{2(1)}$$

$$= \frac{-4 \pm \sqrt{12}}{2}$$

$$= \frac{-4 \pm 2\sqrt{3}}{2}$$

$$= -2 \pm \sqrt{3}$$

The three exact solutions to our equation are 4, $-2 + \sqrt{3}$, and $-2 - \sqrt{3}$.

Getting Ready for Class

After reading through the preceding section, respond in your own words and in complete sentences.

A. What is the discriminant?

B. What kind of solutions do we get to a quadratic equation when the discriminant is negative?

C. What does it mean for a solution to have multiplicity 3?

D. When will a quadratic equation have two rational solutions?

PROBLEM SET 8.3

Use the discriminant to find the number and kind of solutions for each of the following equations.

1. $x^2 - 6x + 5 = 0$

2. $x^2 - x - 12 = 0$

3. $4x^2 - 4x = -1$

4. $9x^2 + 12x = -4$

5. $x^2 + x - 1 = 0$

6. $x^2 - 2x + 3 = 0$

7. $2y^2 = 3y + 1$

8. $3y^2 = 4y - 2$

9. $x^2 - 9 = 0$

10. $4x^2 - 81 = 0$

11. $5a^2 - 4a = 5$

12. $3a = 4a^2 - 5$

Determine k so that each of the following has exactly one real solution.

13. $x^2 - kx + 25 = 0$

14. $x^2 + kx + 25 = 0$

15. $x^2 = kx - 36$

16. $x^2 = kx - 49$

17. $4x^2 - 12x + k = 0$

18. $9x^2 + 30x + k = 0$

19. $kx^2 - 40x = 25$

20. $kx^2 - 2x = -1$

21. $3x^2 - kx + 2 = 0$

22. $5x^2 + kx + 1 = 0$

For each of the following problems, find an equation that has the given solutions.

23. $x = 5, x = 2$

24. $x = -5, x = -2$

25. $t = -3, t = 6$

26. $t = -4, t = 2$

27. $y = 2, y = -2, y = 4$

28. $y = 1, y = -1, y = 3$

29. $x = \frac{1}{2}, x = 3$

30. $x = \frac{1}{3}, x = 5$

31. $t = -\frac{3}{4}, t = 3$

32. $t = -\frac{4}{5}, t = 2$

33. $x = 3, x = -3, x = \frac{5}{6}$

34. $x = 5, x = -5, x = \frac{2}{3}$

35. $a = -\frac{1}{2}, a = \frac{3}{5}$

36. $a = -\frac{1}{3}, a = \frac{4}{7}$

37. $x = -\frac{2}{3}, x = \frac{2}{3}, x = 1$

38. $x = -\frac{4}{5}, x = \frac{4}{5}, x = -1$

39. $x = 2, x = -2, x = 3, x = -3$

40. $x = 1, x = -1, x = 5, x = -5$

41. Find an equation that has a solution of $x = 3$ of multiplicity 1 and a solution $x = -5$ of multiplicity 2.

42. Find an equation that has a solution of $x = 5$ of multiplicity 1 and a solution $x = -3$ of multiplicity 2.

43. Find an equation that has solutions $x = 3$ and $x = -3$, both of multiplicity 2.

44. Find an equation that has solutions $x = 4$ and $x = -4$, both of multiplicity 2.

45. Find all solutions to $x^3 + 6x^2 + 11x + 6 = 0$ if $x = -3$ is one of its solutions.

46. Find all solutions to $x^3 + 10x^2 + 29x + 20 = 0$ if $x = -4$ is one of its solutions.

47. One solution to $y^3 + 5y^2 - 2y - 24 = 0$ is $y = -3$. Find all solutions.

48. One solution to $y^3 + 3y^2 - 10y - 24 = 0$ is $y = -2$. Find all solutions.

49. If $x = 3$ is one solution to $x^3 - 5x^2 + 8x = 6$, find the other solutions.

50. If $x = 2$ is one solution to $x^3 - 6x^2 + 13x = 10$, find the other solutions.

51. Find all solutions to $t^3 = 13t^2 - 65t + 125$ if $t = 5$ is one of the solutions.

52. Find all solutions to $t^3 = 8t^2 - 25t + 26$ if $t = 2$ is one of the solutions.

Review Problems

Problems 53–62 review material we covered in Section 7.2. Reviewing these problems will help you with the next section.

Multiply.

53. $a^4(a^{3/2} - a^{1/2})$

54. $(a^{1/2} - 5)(a^{1/2} + 3)$

55. $(x^{3/2} - 3)^2$

56. $(x^{1/2} - 8)(x^{1/2} + 8)$

Divide.

57. $\dfrac{30x^{3/4} - 25x^{5/4}}{5x^{1/4}}$

58. $\dfrac{45x^{5/3}y^{7/3} - 36x^{8/3}y^{4/3}}{9x^{2/3}y^{1/3}}$

59. Factor $5(x - 3)^{1/2}$ from

$$10(x - 3)^{3/2} - 15(x - 3)^{1/2}$$

60. Factor $2(x + 1)^{1/3}$ from

$$8(x + 1)^{4/3} - 2(x + 1)^{1/3}$$

Factor each of the following as if they were trinomials.

61. $2x^{2/3} - 11x^{1/3} + 12$

62. $9x^{2/3} + 12x^{1/3} + 4$

Extending the Concepts

63. Find all solutions to $x^4 + x^3 - x^2 + x - 2 = 0$ if $x = -2$ is one solution.

64. Find all solutions to $x^4 - x^3 + 2x^2 - 4x - 8 = 0$ if $x = 2$ is one solution.

65. Find all solutions to $x^3 + 3ax^2 + 3a^2x + a^3 = 0$ if $x = -a$ is one solution.

66. If $x = -2a$ is one solution to $x^3 + 6ax^2 + 12a^2x + 8a^3 = 0$, find the other solutions.

Find all solutions to the following equations. Solve using algebra and by graphing. If rounding is necessary, round to the nearest hundredth.

67. $x^2 = 4x + 5$

68. $4x^2 = 8x + 5$

69. $x^2 - 1 = 2x$

70. $4x^2 - 1 = 4x$

Find all solutions to each equation. If rounding is necessary, round to the nearest hundredth.

71. $2x^3 - x^2 - 2x + 1 = 0$

72. $3x^3 - 2x^2 - 3x + 2 = 0$

73. $2x^3 + 2 = x^2 + 4x$

74. $3x^3 - 9x = 2x^2 - 6$

Each of the following equations has only one real solution. Find it by graphing. Then use long division or the quadratic formula to find the remaining solutions.

75. $3x^3 - 8x^2 + 10x - 4 = 0$

76. $10x^3 + 6x^2 + x - 2 = 0$

8.4 Equations Quadratic in Form

We are now in a position to put our knowledge of quadratic equations to work to solve a variety of equations.

EXAMPLE 1 Solve $(x + 3)^2 - 2(x + 3) - 8 = 0$.

Solution We can see that this equation is quadratic in form by replacing $x + 3$ with another variable, say, y. Replacing $x + 3$ with y we have

$$y^2 - 2y - 8 = 0$$

We can solve this equation by factoring the left side and then setting each factor equal to 0.

$$y^2 - 2y - 8 = 0$$

$$(y - 4)(y + 2) = 0 \qquad \text{Factor.}$$

$$y - 4 = 0 \quad \text{or} \quad y + 2 = 0 \qquad \text{Set factors to 0.}$$

$$y = 4 \quad \text{or} \quad y = -2$$

Since our original equation was written in terms of the variable x, we would like our solutions in terms of x also. Replacing y with $x + 3$ and then solving for x we have

$$x + 3 = 4 \quad \text{or} \quad x + 3 = -2$$

$$x = 1 \quad \text{or} \quad x = -5$$

The solutions to our original equation are 1 and -5.

The method we have just shown lends itself well to other types of equations that are quadratic in form, as we will see. In this example, however, there is another method that works just as well. Let's solve our original equation again, but this time, let's begin by expanding $(x + 3)^2$ and $2(x + 3)$.

$$(x + 3)^2 - 2(x + 3) - 8 = 0$$

$$x^2 + 6x + 9 - 2x - 6 - 8 = 0 \qquad \text{Multiply.}$$

$$x^2 + 4x - 5 = 0 \qquad \text{Combine similar terms.}$$

$$(x - 1)(x + 5) = 0 \qquad \text{Factor.}$$

$$x - 1 = 0 \quad \text{or} \quad x + 5 = 0 \qquad \text{Set factors to 0.}$$

$$x = 1 \quad \text{or} \quad x = -5$$

As you can see, either method produces the same result.

EXAMPLE 2   Solve $4x^4 + 7x^2 = 2$.

Solution   This equation is quadratic in x^2. We can make it easier to look at by using the substitution $y = x^2$. (The choice of the letter y is arbitrary. We could just as easily use the substitution $m = x^2$.) Making the substitution $y = x^2$ and then solving the resulting equation we have

$$4y^2 + 7y = 2$$

$$4y^2 + 7y - 2 = 0 \qquad \text{Standard form}$$

$$(4y - 1)(y + 2) = 0 \qquad \text{Factor.}$$

$$4y - 1 = 0 \quad \text{or} \quad y + 2 = 0 \qquad \text{Set factors to 0.}$$

$$y = \frac{1}{4} \quad \text{or} \quad y = -2$$

Now we replace y with x^2 in order to solve for x:

$$x^2 = \frac{1}{4} \quad \text{or} \quad x^2 = -2$$

$$x = \pm\sqrt{\frac{1}{4}} \quad \text{or} \quad x = \pm\sqrt{-2} \qquad \text{Theorem 8.1}$$

$$= \pm\frac{1}{2} \quad \text{or} \quad = \pm i\sqrt{2}$$

The solution set is $\{\frac{1}{2}, -\frac{1}{2}, i\sqrt{2}, -i\sqrt{2}\}$.

EXAMPLE 3 Solve for x: $x + \sqrt{x} - 6 = 0$

Solution To see that this equation is quadratic in form, we have to notice that $(\sqrt{x})^2 = x$. That is, the equation can be rewritten as

$$(\sqrt{x})^2 + \sqrt{x} - 6 = 0$$

Replacing $\sqrt{x}$ with y and solving as usual, we have

$$y^2 + y - 6 = 0$$

$$(y + 3)(y - 2) = 0$$

$$y + 3 = 0 \qquad \text{or} \qquad y - 2 = 0$$

$$y = -3 \qquad \text{or} \qquad y = 2$$

Again, to find x, we replace y with $\sqrt{x}$ and solve:

$$\sqrt{x} = -3 \qquad \text{or} \qquad \sqrt{x} = 2$$

$$x = 9 \qquad\qquad x = 4 \qquad \text{Square both sides of each equation.}$$

Since we squared both sides of each equation, we have the possibility of obtaining extraneous solutions. We have to check both solutions in our original equation.

When	$x = 9$	When	$x = 4$
the equation	$x + \sqrt{x} - 6 = 0$	the equation	$x + \sqrt{x} - 6 = 0$
becomes	$9 + \sqrt{9} - 6 \stackrel{?}{=} 0$	becomes	$4 + \sqrt{4} - 6 \stackrel{?}{=} 0$
	$9 + 3 - 6 \stackrel{?}{=} 0$		$4 + 2 - 6 \stackrel{?}{=} 0$
	$6 \neq 0$		$0 = 0$
	This means 9 is extraneous.		This means 4 is a solution.

The only solution to the equation $x + \sqrt{x} - 6 = 0$ is $x = 4$.

We should note here that the two possible solutions, 9 and 4, to the equation in Example 3 can be obtained by another method. Instead of substituting for $\sqrt{x}$, we can isolate it on one side of the equation and then square both sides to clear the equation of radicals.

$$x + \sqrt{x} - 6 = 0$$

$$\sqrt{x} = -x + 6 \qquad \text{Isolate } \sqrt{x}.$$

$$x = x^2 - 12x + 36 \qquad \text{Square both sides.}$$

$$0 = x^2 - 13x + 36 \qquad \text{Add } -x \text{ to both sides.}$$

$$0 = (x - 4)(x - 9) \qquad \text{Factor.}$$

$$x - 4 = 0 \qquad \text{or} \qquad x - 9 = 0$$

$$x = 4 \qquad\qquad x = 9$$

We obtain the same two possible solutions. Since we squared both sides of the equation to find them, we would have to check each one in the original equation. As was the case in Example 3, only $x = 4$ is a solution; $x = 9$ is extraneous.

EXAMPLE 4   If an object is tossed into the air with an upward velocity of 12 feet per second from the top of the building h feet high, the time it takes for the object to hit the ground below is given by the formula

$$16t^2 - 12t - h = 0$$

Solve this formula for t.

Solution   The formula is in standard form and is quadratic in t. The coefficients a, b, and c that we need to apply to the quadratic formula are $a = 16$, $b = -12$, and $c = -h$. Substituting these quantities into the quadratic formula, we have

$$t = \frac{12 \pm \sqrt{144 - 4(16)(-h)}}{2(16)}$$

$$= \frac{12 \pm \sqrt{144 + 64h}}{32}$$

We can factor the perfect square 16 from the two terms under the radical and simplify our radical somewhat:

$$t = \frac{12 \pm \sqrt{16(9 + 4h)}}{32}$$

$$= \frac{12 \pm 4\sqrt{9 + 4h}}{32}$$

Now we can reduce to lowest terms by factoring a 4 from the numerator and denominator:

$$t = \frac{\cancel{4}(3 \pm \sqrt{9 + 4h})}{\cancel{4} \cdot 8}$$

$$= \frac{3 \pm \sqrt{9 + 4h}}{8}$$

If we were given a value of h, we would find that one of the solutions to this last formula would be a negative number. Since time is always measured in positive units, we wouldn't use that solution.

MORE ABOUT THE GOLDEN RATIO

In Section 7.1 we derived the golden ratio $\dfrac{1 + \sqrt{5}}{2}$ by finding the ratio of length to width for a golden rectangle. The golden ratio was actually discovered before the golden rectangle by the Greeks who lived before Euclid. The early Greeks found the golden ratio by dividing a line segment into two parts so that the ratio of the shorter part to the longer part was the same as the ratio of the longer part to the whole segment. When they divided a line segment in this manner, they said it was divided in "extreme and mean ratio." Figure 1 illustrates a line segment divided this way.

FIGURE I

If point *B* divides segment *AC* in "extreme and mean ratio," then

$$\frac{\text{Length of shorter segment}}{\text{Length of longer segment}} = \frac{\text{length of longer segment}}{\text{length of whole segment}}$$

$$\frac{AB}{BC} = \frac{BC}{AC}$$

EXAMPLE 5   If the length of segment *AB* in Figure 1 is 1 inch, find the length of *BC* so that the whole segment *AC* is divided in "extreme and mean ratio."

Solution   Using Figure 1 as a guide, if we let $x =$ the length of segment *BC*, then the length of *AC* is $x + 1$. If *B* divides *AC* into "extreme and mean ratio," then the ratio of *AB* to *BC* must equal the ratio of *BC* to *AC*. Writing this relationship using the variable *x*, we have

$$\frac{1}{x} = \frac{x}{x + 1}$$

If we multiply both sides of this equation by the LCD $x(x + 1)$ we have

$$x + 1 = x^2$$

$$0 = x^2 - x - 1 \qquad \text{Write equation in standard form.}$$

Since this last equation is not factorable, we apply the quadratic formula.

$$x = \frac{1 \pm \sqrt{(-1)^2 - 4(1)(-1)}}{2}$$

$$= \frac{1 \pm \sqrt{5}}{2}$$

Our equation has two solutions, which we approximate using decimals:

$$\frac{1 + \sqrt{5}}{2} \approx 1.618 \qquad \frac{1 - \sqrt{5}}{2} \approx -0.618$$

Since we originally let *x* equal the length of segment *BC*, we use only the positive solution to our equation. As you can see, the positive solution is the golden ratio.

GRAPHING CALCULATORS

More About Example 1

As we mentioned before, algebraic expressions entered into a graphing calculator do not have to be simplified in order to be evaluated. This fact applies to equations as well. We can graph the equation $y = (x + 3)^2 - 2(x + 3) - 8$ to assist us in solving the equation in Example 1. The graph is shown in Figure 2. Using the Zoom and Trace features at the x-intercepts gives us $x = 1$ and $x = -5$ as the solutions to the equation $0 = (x + 3)^2 - 2(x + 3) - 8$.

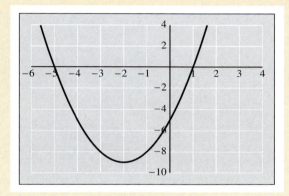

FIGURE 2

More About Example 2

Figure 3 shows the graph of $y = 4x^4 + 7x^2 - 2$. As we expect, the x-intercepts give the real number solutions to the equation $0 = 4x^4 + 7x^2 - 2$. The complex solutions do not appear on the graph.

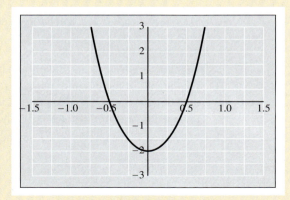

FIGURE 3

(continued)

(continued)

More About Example 3

In solving the equation in Example 3, we found that one of the possible solutions was an extraneous solution. If we solve the equation $x + \sqrt{x} - 6 = 0$ by graphing the function $y = x + \sqrt{x} - 6$, we find that the extraneous solution, 9, is not an x-intercept. Figure 4 shows that the only solution to the equation occurs at the x-intercept 4.

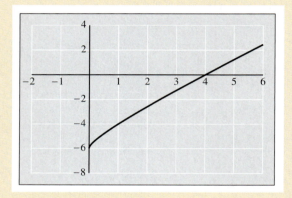

FIGURE 4

Getting Ready for Class

After reading through the preceding section, respond in your own words and in complete sentences.

A. What does it mean for an equation to be quadratic in form?

B. What are all of the circumstances in solving equations (that we have studied) in which it is necessary to check for extraneous solutions?

C. How would you start to solve the equation $x + \sqrt{x} - 6 = 0$?

D. What does it mean for a line segment to be divided in "extreme and mean ratio"?

PROBLEM SET 8.4

Solve each equation.

1. $(x - 3)^2 + 3(x - 3) + 2 = 0$

2. $(x + 4)^2 - (x + 4) - 6 = 0$

3. $2(x + 4)^2 + 5(x + 4) - 12 = 0$

4. $3(x - 5)^2 + 14(x - 5) - 5 = 0$

5. $x^4 - 6x^2 - 27 = 0$ 6. $x^4 + 2x^2 - 8 = 0$

7. $x^4 + 9x^2 = -20$ 8. $x^4 - 11x^2 = -30$

9. $(2a - 3)^2 - 9(2a - 3) = -20$

10. $(3a - 2)^2 + 2(3a - 2) = 3$

11. $2(4a + 2)^2 = 3(4a + 2) + 20$

12. $6(2a + 4)^2 = (2a + 4) + 2$

13. $6t^4 = -t^2 + 5$ 14. $3t^4 = -2t^2 + 8$

15. $9x^4 - 49 = 0$ 16. $25x^4 - 9 = 0$

Solve each of the following equations.

Remember, if you square both sides of an equation in the process of solving it, you have to check all solutions in the original equation.

17. $x - 7\sqrt{x} + 10 = 0$ 18. $x - 6\sqrt{x} + 8 = 0$

19. $t - 2\sqrt{t} - 15 = 0$ 20. $t - 3\sqrt{t} - 10 = 0$

21. $6x + 11\sqrt{x} = 35$ 22. $2x + \sqrt{x} = 15$

23. $(a - 2) - 11\sqrt{a - 2} + 30 = 0$

24. $(a - 3) - 9\sqrt{a - 3} + 20 = 0$

25. $(2x + 1) - 8\sqrt{2x + 1} + 15 = 0$

26. $(2x - 3) - 7\sqrt{2x - 3} + 12 = 0$

27. Solve the formula $16t^2 - vt - h = 0$ for t.

28. Solve the formula $16t^2 + vt + h = 0$ for t.

29. Solve the formula $kx^2 + 8x + 4 = 0$ for x.

30. Solve the formula $k^2x^2 + kx + 4 = 0$ for x.

31. Solve $x^2 + 2xy + y^2 = 0$ for x by using the quadratic formula with $a = 1$, $b = 2y$, and $c = y^2$.

32. Solve $x^2 - 2xy + y^2 = 0$ for x by using the quadratic formula, with $a = 1$, $b = -2y$, and $c = y^2$.

Applying the Concepts

For Problems 33–36, t is in seconds.

33. **Falling Object** An object is tossed into the air with an upward velocity of 8 feet per second from the top of a building h feet high. The time it takes for the object to hit the ground below is given by the formula $16t^2 - 8t - h = 0$. Solve this formula for t.

34. **Falling Object** An object is tossed into the air with an upward velocity of 6 feet per second from the top of a building h feet high. The time it takes for the object to hit the ground below is given by the formula $16t^2 - 6t - h = 0$. Solve this formula for t.

35. **Falling Object** An object is tossed into the air with an upward velocity of v feet per second from the top of a building 20 feet high. The time it takes for the object to hit the ground below is given by the formula $16t^2 - vt - 20 = 0$. Solve this formula for t.

36. **Falling Object** An object is tossed into the air with an upward velocity of v feet per second from the top of a building 40 feet high. The time it takes for the object to hit the ground below is given by the formula $16t^2 - vt - 40 = 0$. Solve this formula for t.

Use Figure 1 from this section as a guide to working Problems 37–40.

37. **Golden Ratio** If AB in Figure 1 is 4 inches, and B divides AC in "extreme and mean ratio," find BC, and then show that BC is 4 times the golden ratio.

38. **Golden Ratio** If AB in Figure 1 is $\frac{1}{2}$ inch, and B divides AC in "extreme and mean ratio," find BC, and then show that BC is half the golden ratio.

39. **Golden Ratio** If AB in Figure 1 is 2 inches, and B divides AC in "extreme and mean ratio," find BC, and then show that the ratio of BC to AB is the golden ratio.

40. **Golden Ratio** If AB in Figure 1 is $\frac{1}{2}$ inch, and B divides AC in "extreme and mean ratio," find BC, and then show that the ratio of BC to AB is the golden ratio.

41. **Saint Louis Arch** The shape of the famous "Gateway to the West" arch in Saint Louis can be modeled by a parabola. The equation for one such parabola is:

$$y = -\frac{1}{150}x^2 + \frac{21}{5}x$$

(a) Sketch the graph of the arch's equation on a coordinate axis.
(b) Approximately how far do you have to walk to get from one side of the arch to the other?

(Tom Tracy/Tony Stone Images)

42. **Area**   In the following diagram, *ABCD* is a rectangle with diagonal *AC*. Find its area.

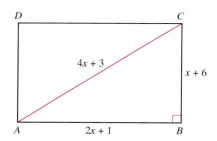

43. **Area and Perimeter**   A total of 160 yards of fencing is to be used to enclose part of a lot that borders on a river. This situation is shown in the following diagram.

(a) Write an equation that gives the relationship between the length and width and the 160 yards of fencing.
(b) The formula for the area that is enclosed by the fencing and the river is $A = lw$. Solve the equation in part (a) for *l*, and then use the result to write the area in terms of *w* only.
(c) Make a table that gives at least five possible values of *w* and associated area *A*.
(d) From the pattern in your table shown in part (c), what is the largest area that can be enclosed by the 160 yards of fencing? (Try some other table values if necessary.)

44. **Area and Perimeter**   Rework all four parts of the preceding problem if it is desired to have an opening 2 yards wide in one of the shorter sides, as shown in the following diagram.

Review Problems

The problems that follow review material we covered in Sections 7.4 and 7.5.

Combine, if possible. [7.4]

45. $5\sqrt{7} - 2\sqrt{7}$ **46.** $6\sqrt{2} - 9\sqrt{2}$

47. $\sqrt{18} - \sqrt{8} + \sqrt{32}$ **48.** $\sqrt{50} + \sqrt{72} - \sqrt{8}$

49. $9x\sqrt{20x^3y^2} + 7y\sqrt{45x^5}$

50. $5x^2\sqrt{27xy^3} - 6y\sqrt{12x^5y}$

Multiply. [7.5]

51. $(\sqrt{5} - 2)(\sqrt{5} + 8)$ **52.** $(2\sqrt{3} - 7)(2\sqrt{3} + 7)$

53. $(\sqrt{x} + 2)^2$ **54.** $(3 - \sqrt{x})(3 + \sqrt{x})$

Rationalize the denominator. [7.5]

55. $\dfrac{\sqrt{7}}{\sqrt{7} - 2}$ **56.** $\dfrac{\sqrt{5} - \sqrt{2}}{\sqrt{5} + \sqrt{2}}$

Extending the Concepts

Find the x- and y-intercepts.

57. $y = x^3 - 4x$ **58.** $y = x^4 - 10x^2 + 9$

59. $y = 3x^3 + x^2 - 27x - 9$

60. $y = 2x^3 + x^2 - 8x - 4$

61. The graph of $y = 2x^3 - 7x^2 - 5x + 4$ crosses the x-axis at $x = 4$. Where else does it cross the x-axis?

62. The graph of $y = 6x^3 + x^2 - 12x + 5$ crosses the x-axis at $x = 1$. Where else does it cross the x-axis?

8.5	**Graphing Parabolas**

The solution set to the equation

$$y = x^2 - 3$$

consists of ordered pairs. One method of graphing the solution set is to find a number of ordered pairs that satisfy the equation and to graph them. We can obtain some ordered pairs that are solutions to $y = x^2 - 3$ by use of a table as follows:

x	$y = x^2 - 3$	y	**Solutions**
-3	$y = (-3)^2 - 3 = 9 - 3 = 6$	6	$(-3, 6)$
-2	$y = (-2)^2 - 3 = 4 - 3 = 1$	1	$(-2, 1)$
-1	$y = (-1)^2 - 3 = 1 - 3 = -2$	-2	$(-1, -2)$
0	$y = 0^2 \quad\; - 3 = 0 - 3 = -3$	-3	$(0, -3)$
1	$y = 1^2 \quad\; - 3 = 1 - 3 = -2$	-2	$(1, -2)$
2	$y = 2^2 \quad\; - 3 = 4 - 3 = 1$	1	$(2, 1)$
3	$y = 3^2 \quad\; - 3 = 9 - 3 = 6$	6	$(3, 6)$

Graphing these solutions and then connecting them with a smooth curve, we have the graph of $y = x^2 - 3$. (See Figure 1.)

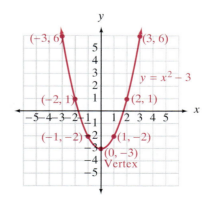

FIGURE I

This graph is an example of a *parabola*. All equations of the form $y = ax^2 + bx + c$, $a \neq 0$, have parabolas for graphs.

Although it is always possible to graph parabolas by making a table of values of x and y that satisfy the equation, there are other methods that are faster and, in some cases, more accurate.

The important points associated with the graph of a parabola are the highest (or lowest) point on the graph and the x-intercepts. The y-intercepts can also be useful.

INTERCEPTS FOR PARABOLAS

The graph of the equation $y = ax^2 + bx + c$ crosses the y-axis at $y = c$, since substituting $x = 0$ into $y = ax^2 + bx + c$ yields $y = c$.

Since the graph crosses the x-axis when $y = 0$, the x-intercepts are those values of x that are solutions to the quadratic equation $0 = ax^2 + bx + c$.

THE VERTEX OF A PARABOLA

The highest or lowest point on a parabola is called the *vertex*. The vertex for the graph of $y = ax^2 + bx + c$ will always occur when

$$x = \frac{-b}{2a}$$

To see this, we must transform the right side of $y = ax^2 + bx + c$ into an expression that contains x in just one of its terms. This is accomplished by completing the square on the first two terms. Here is what it looks like:

$$y = ax^2 + bx + c$$

$$= a\left(x^2 + \frac{b}{a}x\right) + c$$

$$= a\left[x^2 + \frac{b}{a}x + \left(\frac{b}{2a}\right)^2\right] + c - a\left(\frac{b}{2a}\right)^2$$

$$= a\left(x + \frac{b}{2a}\right)^2 + \frac{4ac - b^2}{4a}$$

It may not look like it, but this last line indicates that the vertex of the graph of $y = ax^2 + bx + c$ has an x-coordinate of $\dfrac{-b}{2a}$. Since a, b, and c are constants, the only quantity that is varying in the last expression is the x in $\left(x + \dfrac{b}{2a}\right)^2$.

Since the quantity $\left(x + \dfrac{b}{2a}\right)^2$ is the square of $x + \dfrac{b}{2a}$, the smallest it will ever be is 0, and that will happen when $x = \dfrac{-b}{2a}$.

We can use the vertex point along with the x- and y-intercepts to sketch the graph of any equation of the form $y = ax^2 + bx + c$. Here is a summary of the preceding information.

GRAPHING PARABOLAS

The graph of $y = ax^2 + bx + c$, $a \neq 0$, will have

1. A y-intercept at $y = c$
2. x-intercepts (if they exist) at

$$x = \frac{-b \pm \sqrt{b^2 - 4ac}}{2a}$$

3. A vertex when $x = \dfrac{-b}{2a}$

EXAMPLE 1   Sketch the graph of $y = x^2 - 6x + 5$.

Solution   To find the x-intercepts, we let $y = 0$ and solve for x:

$$0 = x^2 - 6x + 5$$

$$0 = (x - 5)(x - 1)$$

$$x = 5 \quad \text{or} \quad x = 1$$

To find the coordinates of the vertex, we first find

$$x = \frac{-b}{2a} = \frac{-(-6)}{2(1)} = 3$$

The x-coordinate of the vertex is 3. To find the y-coordinate, we substitute 3 for x in our original equation:

$$y = 3^2 - 6(3) + 5 = 9 - 18 + 5 = -4$$

The graph crosses the x-axis at 1 and 5 and has its vertex at $(3, -4)$. Plotting these points and connecting them with a smooth curve, we have the graph shown in Figure 2. The graph is a parabola that opens up, so we say the graph is *concave up*. The

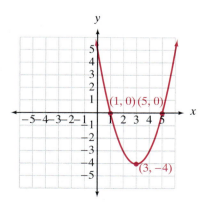

FIGURE 2

vertex is the lowest point on the graph. (Note that the graph crosses the y-axis at 5, which is the value of y we obtain when we let $x = 0$.)

FINDING THE VERTEX BY COMPLETING THE SQUARE

Another way to locate the vertex of the parabola in Example 1 is by completing the square on the first two terms on the right side of the equation $y = x^2 - 6x + 5$. In this case, we would do so by adding 9 to and subtracting 9 from the right side of the equation. This amounts to adding 0 to the equation, so we know we haven't changed its solutions. This is what it looks like:

$$y = (x^2 - 6x \quad) + 5$$
$$= (x^2 - 6x + 9) + 5 - \mathbf{9}$$
$$= (x - 3)^2 - 4$$

You may have to look at this last equation awhile to see this, but when $x = 3$, then $y = (x - 3)^2 - 4 = 0^2 - 4 = -4$ is the smallest y will ever be. And that is why the vertex is at $(3, -4)$. As a matter of fact, this is the same kind of reasoning we used when we derived the formula $x = \dfrac{-b}{2a}$ for the x-coordinate of the vertex.

EXAMPLE 2   Graph $y = -x^2 - 2x + 3$.

Solution   To find the x-intercepts, we let $y = 0$:

$$0 = -x^2 - 2x + 3$$
$$0 = x^2 + 2x - 3 \qquad \text{Multiply each side by } -1.$$
$$0 = (x + 3)(x - 1)$$
$$x = -3 \quad \text{or} \quad x = 1$$

The x-coordinate of the vertex is given by

$$x = \frac{-b}{2a} = \frac{-(-2)}{2(-1)} = \frac{2}{-2} = -1$$

To find the y-coordinate of the vertex, we substitute -1 for x in our original equation to get

$$y = -(-1)^2 - 2(-1) + 3 = -1 + 2 + 3 = 4$$

Our parabola has x-intercepts at -3 and 1, and a vertex at $(-1, 4)$. Figure 3 shows the graph. We say the graph is *concave down* since it opens downward. Again, we could have obtained the coordinates of the vertex by completing the square on the first two terms on the right side of our equation. To do so, we must first factor -1 from the first two terms. (Remember, the leading coefficient must be 1 in order to

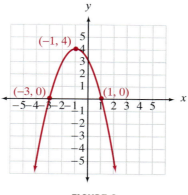

FIGURE 3

complete the square.) When we complete the square, we add 1 inside the parentheses, which actually decreases the right side of the equation by -1 since everything in the parentheses is multiplied by -1. To make up for it, we add 1 outside the parentheses.

$$y = -1(x^2 + 2x \quad) + 3$$
$$= -1(x^2 + 2x + \mathbf{1}) + 3 + \mathbf{1}$$
$$= -1(x + 1)^2 + 4$$

The last line tells us that the *largest* value of y will be 4, and that will occur when $x = -1$.

EXAMPLE 3   Graph $y = 3x^2 - 6x + 1$.

Solution   To find the x-intercepts, we let $y = 0$ and solve for x:

$$0 = 3x^2 - 6x + 1$$

Since the right side of this equation does not factor, we can look at the discriminant to see what kind of solutions are possible. The discriminant for this equation is

$$b^2 - 4ac = 36 - 4(3)(1) = 24$$

Since the discriminant is a positive number but not a perfect square, the equation will have irrational solutions. This means that the x-intercepts are irrational numbers and will have to be approximated with decimals using the quadratic formula. Rather than use the quadratic formula, we will find some other points on the graph, but first let's find the vertex.

Here are both methods of finding the vertex:

Using the formula that gives us the x-coordinate of the vertex, we have:

$$x = \frac{-b}{2a} = \frac{-(-6)}{2(3)} = 1$$

Substituting 1 for x in the equation gives us the y-coordinate of the vertex:

$$y = 3 \cdot 1^2 - 6 \cdot 1 + 1 = -2$$

To complete the square on the right side of the equation, we factor 3 from the first two terms, add 1 inside the parentheses, and add -3 outside the parentheses (this amounts to adding 0 to the right side):

$$y = 3(x^2 - 2x \qquad) + 1$$
$$= 3(x^2 - 2x + \mathbf{1}) + 1 - \mathbf{3}$$
$$= 3(x - 1)^2 - 2$$

In either case, the vertex is $(1, -2)$.

If we can find two points, one on each side of the vertex, we can sketch the graph. Let's let $x = 0$ and $x = 2$, since each of these numbers is the same distance from $x = 1$, and $x = 0$ will give us the y-intercept.

When $x = 0$ | When $x = 2$
$$y = 3(0)^2 - 6(0) + 1 \qquad\qquad y = 3(2)^2 - 6(2) + 1$$
$$= 0 - 0 + 1 \qquad\qquad\qquad = 12 - 12 + 1$$
$$= 1 \qquad\qquad\qquad\qquad\qquad = 1$$

The two points just found are $(0, 1)$ and $(2, 1)$. Plotting these two points along with the vertex $(1, -2)$, we have the graph shown in Figure 4.

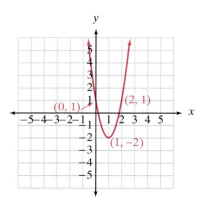

FIGURE 4

EXAMPLE 4   Graph $y = -2x^2 + 6x - 5$.

Solution   Letting $y = 0$, we have

$$0 = -2x^2 + 6x - 5$$

Again, the right side of this equation does not factor. The discriminant is $b^2 - 4ac = 36 - 4(-2)(-5) = -4$, which indicates that the solutions are complex numbers. This means that our original equation does not have x-intercepts. The graph does not cross the x-axis.

Let's find the vertex.

Using our formula for the x-coordinate of the vertex, we have

$$x = \frac{-b}{2a} = \frac{-6}{2(-2)} = \frac{6}{4} = \frac{3}{2}$$

To find the y-coordinate, we let $x = \frac{3}{2}$:

$$y = -2\left(\frac{3}{2}\right)^2 + 6\left(\frac{3}{2}\right) - 5$$

$$= \frac{-18}{4} + \frac{18}{2} - 5$$

$$= \frac{-18 + 36 - 20}{4}$$

$$= -\frac{1}{2}$$

Finding the vertex by completing the square is a more complicated matter. In order to make the coefficient of x^2 a 1, we must factor -2 from the first two terms. To complete the square inside the parentheses, we add $\frac{9}{4}$. Since each term inside the parentheses is multiplied by -2, we add $\frac{9}{2}$ outside the parentheses so that the net result is the same as adding 0 to the right side:

$$y = -2(x^2 - 3x \quad) - 5$$

$$= -2\left(x^2 - 3x + \frac{9}{4}\right) - 5 + \frac{9}{2}$$

$$= -2\left(x - \frac{3}{2}\right)^2 - \frac{1}{2}$$

The vertex is $(\frac{3}{2}, -\frac{1}{2})$. Since this is the only point we have so far, we must find two others. Let's let $x = 3$ and $x = 0$, since each point is the same distance from $x = \frac{3}{2}$ and on either side:

When $x = 3$

$$y = -2(3)^2 + 6(3) - 5$$

$$= -18 + 18 - 5$$

$$= -5$$

When $x = 0$

$$y = -2(0)^2 + 6(0) - 5$$

$$= 0 + 0 - 5$$

$$= -5$$

The two additional points on the graph are $(3, -5)$ and $(0, -5)$. Figure 5 shows the graph.

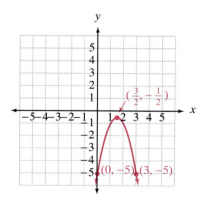

FIGURE 5

The graph is concave down. The vertex is the highest point on the graph.

By looking at the equations and graphs in Examples 1 through 4, we can conclude that the graph of $y = ax^2 + bx + c$ will be concave up when a is positive, and concave down when a is negative. Taking this even further, if $a > 0$, then the vertex is the lowest point on the graph, and if $a < 0$, the vertex is the highest point on the graph. We can use this information to solve some problems in which we are interested in finding the largest or smallest value of a variable.

EXAMPLE 5 A company selling copies of an accounting program for home computers finds that it will make a weekly profit of P dollars from selling x copies of the program, according to the equation

$$P(x) = -0.1x^2 + 27x - 500$$

How many copies of the program should it sell to make the largest possible profit, and what is the largest possible profit?

Solution Since the coefficient of x^2 is negative, we know the graph of this parabola will be concave down, meaning that the vertex is the highest point of the curve. We find the vertex by first finding its x-coordinate:

$$x = \frac{-b}{2a} = \frac{-27}{2(-0.1)} = \frac{27}{0.2} = 135$$

This represents the number of programs the company needs to sell each week in order to make a maximum profit. To find the maximum profit, we substitute 135 for x in the original equation. (A calculator is helpful for these kinds of calculations.)

$$P(135) = -0.1(135)^2 + 27(135) - 500$$

$$= -0.1(18{,}225) + 3{,}645 - 500$$

$$= -1{,}822.5 + 3{,}645 - 500$$

$$= 1{,}322.5$$

The maximum weekly profit is $1,322.50 and is obtained by selling 135 programs a week.

EXAMPLE 6 An art supply store finds that they can sell x sketch pads each week at p dollars each, according to the equation $x = 900 - 300p$. Graph the revenue equation $R = xp$. Then use the graph to find the price p that will bring in the maximum revenue. Finally, find the maximum revenue.

Solution As it stands, the revenue equation contains three variables. Since we are asked to find the value of p that gives us the maximum value of R, we rewrite the equation using just the variables R and p. Since $x = 900 - 300p$, we have

$$R = xp = (900 - 300p)p$$

The graph of this equation is shown in Figure 6. The graph appears in the first quadrant only, since R and p are both positive quantities.

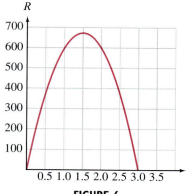

FIGURE 6

From the graph we see that the maximum value of R occurs when $p = \$1.50$. We can calculate the maximum value of R from the equation:

$$\text{When} \qquad p = 1.5$$

the equation $R = (900 - 300p)p$

becomes $R = (900 - 300 \cdot 1.5)1.5$

$$= (900 - 450)1.5$$

$$= 450 \cdot 1.5$$

$$= 675$$

The maximum revenue is \$675. It is obtained by setting the price of each sketch pad at $p = \$1.50$.

Using TECHNOLOGY

GRAPHING CALCULATORS

If you have been using a graphing calculator for some of the material in this course, you are well aware that your calculator can draw all the graphs in this section very easily. It is important, however, that you be able to recognize and sketch the graph of any parabola by hand. It is a skill that all successful intermediate algebra students should possess, even if they are proficient in the use of a graphing calculator. My suggestion is that you work the problems in this section and problem set without your calculator. Then use your calculator to check your results.

FINDING THE EQUATION FROM THE GRAPH

EXAMPLE 7   At the 1997 Washington Country Fair in Oregon, David Smith, Jr., The Bullet, was shot from a cannon. As a human cannonball, he reached a height of 70 feet before landing in a net 160 feet from the cannon. Sketch the graph of his path, and then find the equation of the graph.

Solution   We assume that the path taken by the human cannonball is a parabola. If the origin of the coordinate system is at the opening of the cannon, then the net that catches him will be at 160 on the *x*-axis. Figure 7 shows the graph:

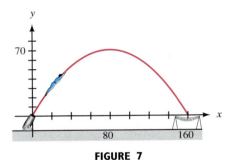

FIGURE 7

Since the curve is a parabola, we know the equation will have the form

$$y = a(x - h)^2 + k$$

Since the vertex of the parabola is at (80, 70), we can fill in two of the three constants in our equation, giving us

$$y = a(x - 80)^2 + 70$$

To find *a*, we note that the landing point will be (160, 0). Substituting the coordinates of this point into the equation, we solve for *a*:

$$0 = a(160 - 80)^2 + 70$$

$$0 = a(80)^2 + 70$$

$$0 = 6{,}400a + 70$$

$$a = -\frac{70}{6{,}400} = -\frac{7}{640}$$

The equation that describes the path of the human cannonball is

$$y = -\frac{7}{640}(x - 80)^2 + 70 \quad \text{for } 0 \le x \le 160$$

Using TECHNOLOGY

GRAPHING CALCULATORS

Graph the equation found in Example 7 by graphing it on a graphing calculator using the window shown here. (We will use this graph later in the book to find the angle between the cannon and the horizontal.)

Window: X from 0 to 180, increment 20

Y from 0 to 80, increment 10

On the TI-83, an increment of 20 for X means Xscl=20.

Getting Ready for Class

After reading through the preceding section, respond in your own words and in complete sentences.

A. What is a parabola?

B. What part of the equation of a parabola determines whether the graph is concave up or concave down?

B. How do you find the *x*-intercepts of a parabola?

C. Suppose $f(x) = ax^2 + bx + c$ is the equation of a parabola. Explain how $f(4) = 1$ relates to the graph of the parabola.

D. A line can be graphed with two points. How many points are necessary to get a reasonable sketch of a parabola? Explain.

PROBLEM SET 8.5

For each of the following equations, give the x-intercepts and the coordinates of the vertex, and sketch the graph.

1. $y = x^2 + 2x - 3$ **2.** $y = x^2 - 2x - 3$

3. $y = -x^2 - 4x + 5$ **4.** $y = x^2 + 4x - 5$

5. $y = x^2 - 1$ **6.** $y = x^2 - 4$

7. $y = -x^2 + 9$ **8.** $y = -x^2 + 1$

9. $y = 2x^2 - 4x - 6$ **10.** $y = 2x^2 + 4x - 6$

11. $y = x^2 - 2x - 4$ **12.** $y = x^2 - 2x - 2$

Find the vertex and any two convenient points to sketch the graphs of the following equations.

13. $y = x^2 - 4x - 4$ **14.** $y = x^2 - 2x + 3$

15. $y = -x^2 + 2x - 5$ **16.** $y = -x^2 + 4x - 2$

17. $y = x^2 + 1$ **18.** $y = x^2 + 4$

19. $y = -x^2 - 3$ **20.** $y = -x^2 - 2$

21. $y = 3x^2 + 4x + 1$ **22.** $y = 2x^2 + 4x + 3$

For each of the following equations, find the coordinates of the vertex, and indicate whether the vertex is the highest point on the graph or the lowest point on the graph. (Do not graph.)

23. $y = x^2 - 6x + 5$ **24.** $y = -x^2 + 6x - 5$

25. $y = -x^2 + 2x + 8$ **26.** $y = x^2 - 2x - 8$

27. $y = 12 + 4x - x^2$ **28.** $y = -12 - 4x + x^2$

29. $y = -x^2 - 8x$ **30.** $y = x^2 + 8x$

Applying the Concepts

31. Maximum Profit A company earns a weekly profit of P dollars by selling x items, according to the equation $P(x) = -0.5x^2 + 40x - 300$. Find the number of items the company must sell each week in order to obtain the largest possible profit. Then, find the largest possible profit.

32. Maximum Profit A company earns a weekly profit of P dollars by selling x items, according to the equation $P(x) = -0.5x^2 + 100x - 1,600$. Find the number of items the company must sell each week in order to obtain the largest possible profit. Then, find the largest possible profit.

33. Maximum Profit A company finds that it can make a profit of P dollars each month by selling x patterns, according to the formula $P(x) = -0.002x^2 + 3.5x - 800$. How many patterns must it sell each month in order to have a maximum profit? What is the maximum profit?

34. Maximum Profit A company selling picture frames finds that it can make a profit of P dollars each month by selling x frames, according to the formula $P(x) = -0.002x^2 + 5.5x - 1,200$. How many frames must it sell each month in order to have a maximum profit? What is the maximum profit?

35. Maximum Height Chaudra is tossing a softball into the air with an underhand motion. The distance of the ball above her hand at any time is given by the function

$$h(t) = 32t - 16t^2 \quad \text{for } 0 \le t \le 2$$

where $h(t)$ is the height of the ball (in feet) and t is the time (in seconds). Find the times at which the ball is in her hand, and the maximum height of the ball.

36. Maximum Height Hali is tossing a quarter into the air with an underhand motion. The distance of the quarter above her hand at any time is given by the function

$$h(t) = 16t - 16t^2 \quad \text{for } 0 \le t \le 1$$

where $h(t)$ is the height of the quarter (in feet) and t is the time (in seconds). Find the times at which the quarter is in her hand, and the maximum height of the quarter.

37. Maximum Height An arrow is shot straight up into the air with an initial velocity of 128 feet per second. If h represents the height (in feet) of the arrow at any time t (in seconds), then the equation that gives h in terms of t is $h(t) = 128t - 16t^2$. Find the maximum height attained by the arrow.

38. Maximum Height A ball is projected into the air with an upward velocity of 64 feet per second. The equation that gives the height h (in feet) of the ball at any time t (in seconds) is $h(t) = 64t - 16t^2$. Find the maximum height attained by the ball.

39. Maximum Area Justin wants to fence three sides of a rectangular exercise yard for his dog. The fourth side of the exercise yard will be a side of the house. He has 80 feet of fencing available. Find the dimensions of the exercise yard that will enclose the maximum area.

$80 - 2x$

40. Maximum Area Repeat Problem 39, assuming that Justin has 60 feet of fencing available.

41. Maximum Revenue A company that manufactures typewriter ribbons knows that the number of ribbons x it can sell each week is related to the price p of each ribbon by the equation $x = 1,200 - 100p$. Graph the revenue equation $R = xp$. Then use the graph to find the price p that will bring in the maximum revenue. Finally, find the maximum revenue.

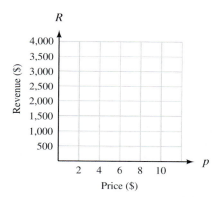

42. Maximum Revenue A company that manufactures diskettes for home computers finds that it can sell x diskettes each day at p dollars per diskette, according to the equation $x = 800 - 100p$. Graph the revenue equation $R = xp$. Then use the graph to find the price p that will bring in the maximum revenue. Finally, find the maximum revenue.

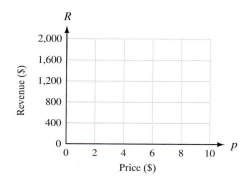

43. Maximum Revenue The relationship between the number of calculators x a company sells each day and the price p of each calculator is given by the equation $x = 1,700 - 100p$. Graph the revenue equation $R = xp$, and use the graph to find the

price p that will bring in the maximum revenue. Then find the maximum revenue.

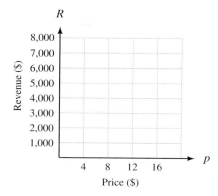

44. Maximum Revenue The relationship between the number x of pencil sharpeners a company sells each week and the price p of each sharpener is given by the equation $x = 1,800 - 100p$. Graph the revenue equation $R = xp$, and use the graph to find the price p that will bring in the maximum revenue. Then find the maximum revenue.

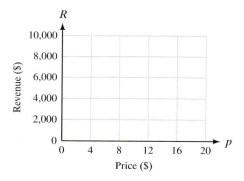

Review Problems

The problems that follow review material we covered in Section 7.7.

Perform the indicated operations.

45. $(3 - 5i) - (2 - 4i)$ **46.** $2i(5 - 6i)$

47. $(3 + 2i)(7 - 3i)$ **48.** $(4 + 5i)^2$

49. $\dfrac{i}{3 + i}$ **50.** $\dfrac{2 + 3i}{2 - 3i}$

Extending the Concepts

Finding the Equation From the Graph For each of the following problems, the graph is a parabola. In each case, find an equation in the form $y = a(x - h)^2 + k$ that describes the graph.

51.

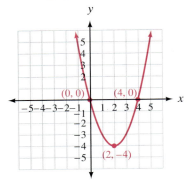

52.

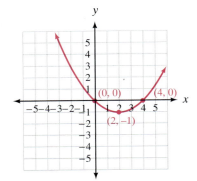

53. Human Cannonball A human cannonball is shot from a cannon at the county fair. He reaches a height of 60 feet before landing in a net 180 feet from the cannon. Sketch the graph of his path, and then find the equation of the graph.

54. Human Cannonball Referring to Problem 53, find the height reached by the human cannonball after he has traveled 30 feet horizontally, and after he has traveled 150 feet horizontally.

55. Comparing Expressions, Equations, and Functions Four problems follow. The solution to Problem 3 is shown in Figure 8. Solve the other two problems, and then explain how the solutions to the three problems are related.

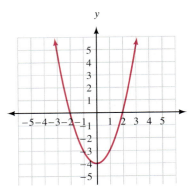

FIGURE 8

Problem 1 Factor the expression $x^2 - 4$.

Problem 2 Solve the equation $x^2 - 4 = 0$.

Problem 3 Graph the function $y = x^2 - 4$.

Problem 4 If $f(x) = x^2 - 4$, find the value of x for which $f(x) = 0$.

56. Comparing Expressions, Equations, and Functions Four problems are shown here. The solution to Problem 3 is shown in Figure 9. Solve the other two problems, and then explain how the solutions to the three problems are related.

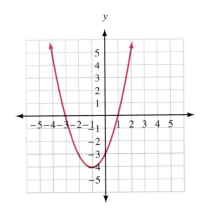

FIGURE 9

Problem 1 Factor the expression $x^2 + 2x - 3$.

Problem 2 Solve the equation $x^2 + 2x - 3 = 0$.

Problem 3 Graph the function $y = x^2 + 2x - 3$.

Problem 4 If $f(x) = x^2 + 2x - 3$, find the value of x for which $f(x) = 0$.

Quadratic inequalities in one variable are inequalities of the form

$$ax^2 + bx + c < 0 \qquad ax^2 + bx + c > 0$$
$$ax^2 + bx + c \leq 0 \qquad ax^2 + bx + c \geq 0$$

where a, b, and c are constants, with $a \neq 0$. The technique we will use to solve inequalities of this type involves graphing. Suppose, for example, we wish to find the solution set for the inequality $x^2 - x - 6 > 0$. We begin by factoring the left side to obtain

$$(x - 3)(x + 2) > 0$$

We have two real numbers $x - 3$ and $x + 2$ whose product $(x - 3)(x + 2)$ is greater than zero. That is, their product is positive. The only way the product can be positive is either if both factors, $(x - 3)$ and $(x + 2)$, are positive or if they are both negative. To help visualize where $x - 3$ is positive and where it is negative, we draw a real number line and label it accordingly:

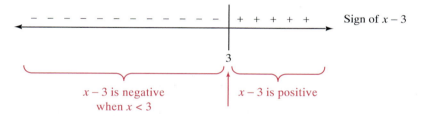

Here is a similar diagram showing where the factor $x + 2$ is positive and where it is negative:

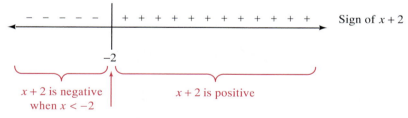

Drawing the two number lines together and eliminating the unnecessary numbers, we have

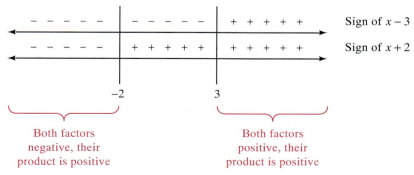

We can see from the preceding diagram that the graph of the solution to $x^2 - x - 6 > 0$ is

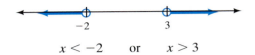

$$x < -2 \quad \text{or} \quad x > 3$$

Using TECHNOLOGY

GRAPHICAL SOLUTIONS TO QUADRATIC INEQUALITIES

We can solve the preceding problem by using a graphing calculator to visualize where the product $(x - 3)(x + 2)$ is positive. First, we graph the function $y = (x - 3)(x + 2)$ as shown in Figure 1.

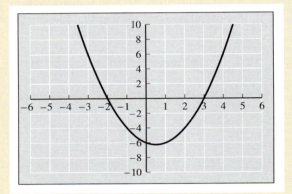

FIGURE 1

Next, we observe where the graph is above the x-axis. As you can see, the graph is above the x-axis to the right of 3 and to the left of -2, as shown in Figure 2.

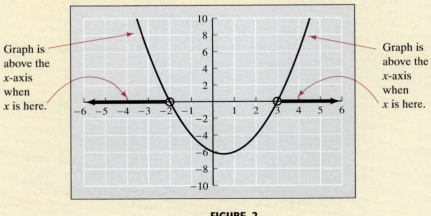

Graph is above the x-axis when x is here.

Graph is above the x-axis when x is here.

FIGURE 2

When the graph is above the x-axis, we have points whose y-coordinates are positive. Since these y-coordinates are the same as the expression $(x - 3)(x + 2)$, the values of x for which the graph of $y = (x - 3)(x + 2)$ is above the x-axis are the values of x for which the inequality $(x - 3)(x + 2) > 0$ is true. Our solution set is therefore

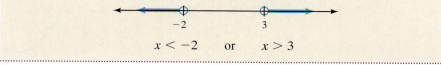

$$x < -2 \quad \text{or} \quad x > 3$$

EXAMPLE 1   Solve for x: $x^2 - 2x - 8 \leq 0$

Algebraic Solution   We begin by factoring:

$$x^2 - 2x - 8 \leq 0$$

$$(x - 4)(x + 2) \leq 0$$

The product $(x - 4)(x + 2)$ is negative or zero. The factors must have opposite signs. We draw a diagram showing where each factor is positive and where each factor is negative:

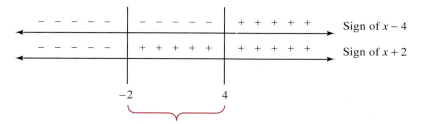

From the diagram we have the graph of the solution set:

$$-2 \leq x \leq 4$$

Graphical Solution

To solve this inequality with a graphing calculator, we graph the function $y = (x - 4)(x + 2)$ and observe where the graph is below the x-axis. These points have negative y-coordinates, which means that the product $(x - 4)(x + 2)$ is negative for these points. Figure 3 shows the graph of $y = (x - 4)(x + 2)$, along with the region on the x-axis where the graph contains points with negative y-coordinates.

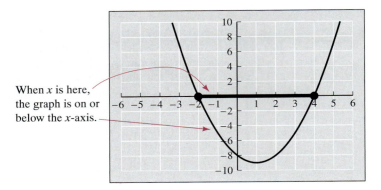

FIGURE 3

As you can see, the graph is below the x-axis when x is between -2 and 4. Since our original inequality includes the possibility that $(x - 4)(x + 2)$ is 0, we include the endpoints, -2 and 4, with our solution set.

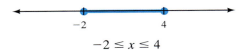

$$-2 \leq x \leq 4$$

EXAMPLE 2 Solve for x: $6x^2 - x \geq 2$

Algebraic Solution

$$6x^2 - x \geq 2$$

$$6x^2 - x - 2 \geq 0 \leftarrow \text{Standard form}$$

$$(3x - 2)(2x + 1) \geq 0$$

The product is positive, so the factors must agree in sign. Here is the diagram showing where that occurs:

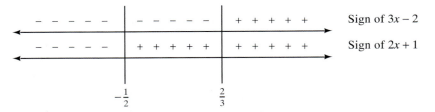

Since the factors agree in sign below $-\frac{1}{2}$ and above $\frac{2}{3}$, the graph of the solution set is

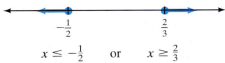

$$x \leq -\frac{1}{2} \quad \text{or} \quad x \geq \frac{2}{3}$$

Graphical Solution   To solve this inequality with a graphing calculator, we graph the function $y = (3x - 2)(2x + 1)$ and observe where the graph is above the x-axis. These are the points that have positive y-coordinates, which means that the product $(3x - 2)(2x + 1)$ is positive for these points. Figure 4 shows the graph of $y = (3x - 2)(2x + 1)$, along with the regions on the x-axis where the graph is on or above the x-axis.

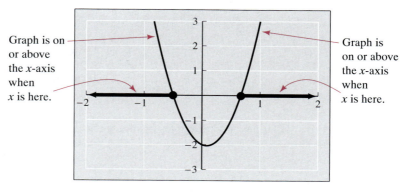

Graph is on or above the x-axis when x is here.

Graph is on or above the x-axis when x is here.

FIGURE 4

To find the points where the graph crosses the x-axis, we need to use either the Trace and Zoom features to zoom in on each point, or the calculator function that finds the intercepts automatically (on the TI-82/83 this is the root/zero function under the CALC key). Whichever method we use, we will obtain the following result:

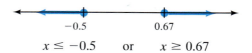

$$x \le -0.5 \qquad \text{or} \qquad x \ge 0.67$$

EXAMPLE 3   Solve $x^2 - 6x + 9 \ge 0$.

Algebraic Solution

$$x^2 - 6x + 9 \ge 0$$

$$(x - 3)^2 \ge 0$$

This is a special case in which both factors are the same. Since $(x - 3)^2$ is always positive or zero, the solution set is all real numbers. That is, any real number that is used in place of x in the original inequality will produce a true statement.

Graphical Solution   The graph of $y = (x - 3)^2$ is shown in Figure 5. Notice that it touches the x-axis at 3 and is above the x-axis everywhere else. This means that every point on the graph has a y-coordinate greater than or equal to 0, no matter what the value of x. The conclusion that we draw from the graph is that the inequality $(x - 3)^2 \ge 0$ is true for all values of x.

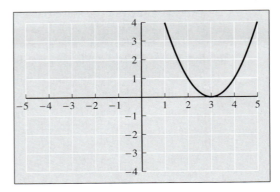

FIGURE 5

Our next two examples involve inequalities that contain rational expressions.

EXAMPLE 4   Solve: $\dfrac{x-4}{x+1} \le 0$

Solution   The inequality indicates that the quotient of $(x-4)$ and $(x+1)$ is negative or 0 (less than or equal to 0). We can use the same reasoning we used to solve the first three examples, because quotients are positive or negative under the same conditions that products are positive or negative. Here is the diagram that shows where each factor is positive and where each factor is negative:

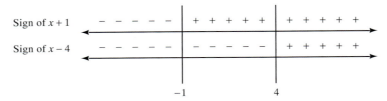

Between -1 and 4 the factors have opposite signs, making the quotient negative. Thus, the region between -1 and 4 is where the solutions lie, since the original inequality indicates the quotient $\dfrac{x-4}{x+1}$ is negative. The solution set and its graph are shown here:

$$-1 < x \le 4$$

Notice that the left endpoint is open — that is, it is not included in the solution set — because $x = -1$ would make the denominator in the original inequality 0. It is important to check all endpoints of solution sets to inequalities that involve rational expressions.

EXAMPLE 5   Solve: $\dfrac{3}{x-2} - \dfrac{2}{x-3} > 0$

Solution   We begin by adding the two rational expressions on the left side. The common denominator is $(x-2)(x-3)$:

$$\frac{3}{x-2} \cdot \frac{(x-3)}{(x-3)} - \frac{2}{x-3} \cdot \frac{(x-2)}{(x-2)} > 0$$

$$\frac{3x - 9 - 2x + 4}{(x-2)(x-3)} > 0$$

$$\frac{x-5}{(x-2)(x-3)} > 0$$

This time the quotient involves three factors. Here is the diagram that shows the signs of the three factors:

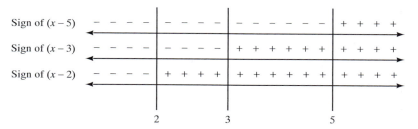

The original inequality indicates that the quotient is positive. In order for this to happen, either all three factors must be positive, or exactly two factors must be negative. Looking back at the diagram, we see the regions that satisfy these conditions are between 2 and 3 or above 5. Here is our solution set:

$$2 < x < 3 \qquad \text{or} \qquad x > 5$$

Getting Ready for Class

After reading through the preceding section, respond in your own words and in complete sentences.

A. What is the first step in solving a quadratic inequality?

B. How do you show that the endpoint of a line segment is not part of the graph of a quadratic inequality?

C. Where would you use the graph of $y = ax^2 + bx + c$ to help you find the graph of $ax^2 + bx + c < 0$?

D. Can a quadratic inequality have exactly one solution? Give an example.

PROBLEM SET 8.6

Solve each of the following inequalities, and graph the solution set.

1. $x^2 + x - 6 > 0$

2. $x^2 + x - 6 < 0$

3. $x^2 - x - 12 \leq 0$

4. $x^2 - x - 12 \geq 0$

5. $x^2 + 5x \geq -6$

6. $x^2 - 5x > 6$

7. $6x^2 < 5x - 1$

8. $4x^2 \geq -5x + 6$

9. $x^2 - 9 < 0$

10. $x^2 - 16 \geq 0$

11. $4x^2 - 9 \geq 0$

12. $9x^2 - 4 < 0$

13. $2x^2 - x - 3 < 0$

14. $3x^2 + x - 10 \geq 0$

15. $x^2 - 4x + 4 \geq 0$

16. $x^2 - 4x + 4 < 0$

17. $x^2 - 10x + 25 < 0$

18. $x^2 - 10x + 25 > 0$

19. $(x - 2)(x - 3)(x - 4) > 0$

20. $(x - 2)(x - 3)(x - 4) < 0$

21. $(x + 1)(x + 2)(x + 3) \leq 0$

22. $(x + 1)(x + 2)(x + 3) \geq 0$

23. $\dfrac{x - 1}{x + 4} \leq 0$

24. $\dfrac{x + 4}{x - 1} \leq 0$

25. $\dfrac{3x}{x + 6} - \dfrac{8}{x + 6} < 0$

26. $\dfrac{5x}{x + 1} - \dfrac{3}{x + 1} < 0$

27. $\dfrac{4}{x - 6} + 1 > 0$

28. $\dfrac{2}{x - 3} + 1 \geq 0$

29. $\dfrac{x - 2}{(x + 3)(x - 4)} < 0$

30. $\dfrac{x - 1}{(x + 2)(x - 5)} < 0$

31. $\dfrac{2}{x - 4} - \dfrac{1}{x - 3} > 0$

32. $\dfrac{4}{x + 3} - \dfrac{3}{x + 2} > 0$

33. $\dfrac{x + 7}{2x + 12} + \dfrac{6}{x^2 - 36} \leq 0$

34. $\dfrac{x + 1}{2x - 2} - \dfrac{2}{x^2 - 1} \leq 0$

35. The graph of $y = x^2 - 4$ is shown in Figure 6. Use the graph to write the solution set for each of the following:

(a) $x^2 - 4 < 0$

(b) $x^2 - 4 > 0$

(c) $x^2 - 4 = 0$

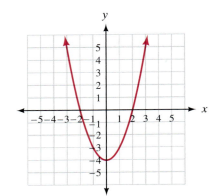

FIGURE 6

36. The graph of $y = 4 - x^2$ is shown in Figure 7. Use the graph to write the solution set for each of the following:

(a) $4 - x^2 < 0$

(b) $4 - x^2 > 0$

(c) $4 - x^2 = 0$

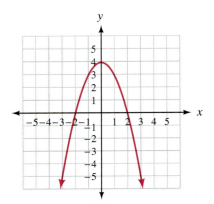

FIGURE 7

37. The graph of $y = x^2 - 3x - 10$ is shown in Figure 8. Use the graph to write the solution set for each of the following:
(a) $x^2 - 3x - 10 < 0$
(b) $x^2 - 3x - 10 > 0$
(c) $x^2 - 3x - 10 = 0$

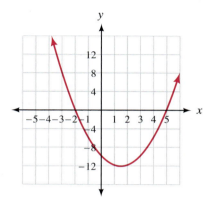

FIGURE 8

38. The graph of $y = x^2 + x - 12$ is shown in Figure 9. Use the graph to write the solution set for each of the following:
(a) $x^2 + x - 12 < 0$
(b) $x^2 + x - 12 > 0$
(c) $x^2 + x - 12 = 0$

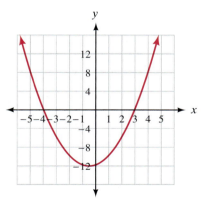

FIGURE 9

39. The graph of $y = x^3 - 3x^2 - x + 3$ is shown in Figure 10. Use the graph to write the solution set for each of the following:

(a) $x^3 - 3x^2 - x + 3 < 0$
(b) $x^3 - 3x^2 - x + 3 > 0$
(c) $x^3 - 3x^2 - x + 3 = 0$

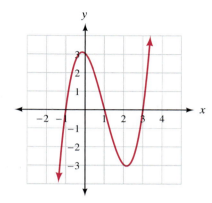

FIGURE 10

40. The graph of $y = x^3 + 4x^2 - 4x - 16$ is shown in Figure 11. Use the graph to write the solution set for each of the following:
(a) $x^3 + 4x^2 - 4x - 16 < 0$
(b) $x^3 + 4x^2 - 4x - 16 > 0$
(c) $x^3 + 4x^2 - 4x - 16 = 0$

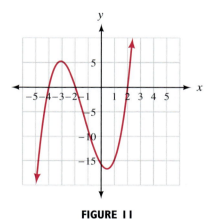

FIGURE 11

Applying the Concepts

41. Dimensions of a Rectangle   The length of a rectangle is 3 inches more than twice the width. If the area is to be at least 44 square inches, what are the possibilities for the width?

42. Dimensions of a Rectangle The length of a rectangle is 5 inches less than three times the width. If the area is to be less than 12 square inches, what are the possibilities for the width?

43. Revenue   A manufacturer of portable radios knows that the weekly revenue produced by selling x radios is given by the equation $R = 1,300p - 100p^2$, where p is the price of each radio (in dollars). What price should be charged for each radio if the weekly revenue is to be at least $4,000?

44. Revenue   A manufacturer of small calculators knows that the weekly revenue produced by selling x calculators is given by the equation $R = 1,700p - 100p^2$, where p is the price of each calculator (in dollars). What price should be charged for each calculator if the revenue is to be at least $7,000 each week?

45. Union Dues A labor union has 10,000 members. For every $10 increase in union dues, membership is decreased by 200 people. If the current dues are $100, what should be the new dues (to the nearest multiple of $10) so that income from dues is greatest, and what is that income? *Hint:* Since Income = (membership)(dues), we can let $x =$ the number of $10 increases in dues, and then this will give us income of $y = (10,000 - 200x)$ $(100 + 10x)$.

46. Bookstore Receipts The owner of a used bookstore charges $2 for quality paperbacks and usually sells 40 per day. For every 10-cent increase in the price of these paperbacks, he thinks that he will sell two fewer per day. What is the price he should charge (to the nearest 10 cents) for these books, in order to maximize his income, and what would be that income? *Hint:* Income = (price per book)

(number of books sold); let $x =$ the number of 10-cent increases in price.

47. Jiffy-Lube The owner of a quick oil-change business charges $20 per oil change and has 40 customers per day. If each increase of $2 results in 2 fewer daily customers, what price should the owner charge (to the nearest $2) for an oil change if the income from this business is to be as great as possible?

48. Computer Sales   A computer manufacturer charges $2,200 for its basic model and sells 1,500 computers per month at this price. For every $200 increase in price, it is believed that 75 fewer computers will be sold. What price should the company place on its basic model of computer (to the nearest $100) to have the greatest income?

Review Problems

Problems 49–54 review material we covered in Section 7.6.

Solve each equation.

49. $\sqrt{3t - 1} = 2$

50. $\sqrt{4t + 5} + 7 = 3$

51. $\sqrt{x + 3} = x - 3$

52. $\sqrt{x + 3} = \sqrt{x} - 3$

Graph each equation.

53. $y = \sqrt[3]{x} - 1$

54. $y = \sqrt[3]{x - 1}$

Extending the Concepts

Graph the solution set for each inequality.

55. $x^2 - 2x - 1 < 0$

56. $x^2 - 6x + 7 < 0$

57. $x^2 - 8x + 13 > 0$

58. $x^2 - 10x + 18 > 0$

CHAPTER 8   SUMMARY

Examples

1. If $\qquad (x - 3)^2 = 25$

then $\qquad x - 3 = \pm 5$

$\qquad\qquad x = 3 \pm 5$

$x = 8 \qquad$ or $\qquad x = -2$

Theorem 8.1 [8.1]

If $a^2 = b$, where b is a real number, then

$$a = \sqrt{b} \qquad \text{or} \qquad a = -\sqrt{b} \qquad \text{or} \qquad a = \pm\sqrt{b}$$

2. Solve $x^2 - 6x - 6 = 0$.

$x^2 - 6x = 6$

$x^2 - 6x + \mathbf{9} = 6 + \mathbf{9}$

$(x - 3)^2 = 15$

$x - 3 = \pm\sqrt{15}$

$x = 3 \pm \sqrt{15}$

To Solve a Quadratic Equation by Completing the Square [8.1]

Step 1: Write the equation in the form $ax^2 + bx = c$.

Step 2: If $a \neq 1$, divide through by the constant a so the coefficient of x^2 is 1.

Step 3: Complete the square on the left side by adding the square of $\frac{1}{2}$ the coefficient of x to both sides.

Step 4: Write the left side of the equation as the square of a binomial. Simplify the right side if possible.

Step 5: Apply Theorem 8.1, and solve as usual.

3. If $2x^2 + 3x - 4 = 0$, then

$$x = \frac{-3 \pm \sqrt{9 - 4(2)(-4)}}{2(2)}$$

$$= \frac{-3 \pm \sqrt{41}}{4}$$

The Quadratic Theorem [8.2]

For any quadratic equation in the form $ax^2 + bx + c = 0$, $a \neq 0$, the two solutions are

$$x = \frac{-b \pm \sqrt{b^2 - 4ac}}{2a}$$

This last expression is known as the *quadratic formula*.

4. The discriminant for

$$x^2 + 6x + 9 = 0$$

is $D = 36 - 4(1)(9) = 0$, which means the equation has one rational solution.

The Discriminant [8.3]

The expression $b^2 - 4ac$ that appears under the radical sign in the quadratic formula is known as the *discriminant*.

We can classify the solutions to $ax^2 + bx + c = 0$:

The Solutions Are	When the Discriminant Is
Two complex numbers containing i	Negative
One rational number	Zero
Two rational numbers	A positive perfect square
Two irrational numbers	A positive number, but not a perfect square

5. The equation $x^4 - x^2 - 12 = 0$ is quadratic in x^2. Letting $y = x^2$ we have

$$y^2 - y - 12 = 0$$

$$(y - 4)(y + 3) = 0$$

$$y = 4 \quad \text{or} \quad y = -3$$

Resubstituting x^2 for y, we have

$$x^2 = 4 \quad \text{or} \quad x^2 = -3$$

$$x = \pm 2 \quad \text{or} \quad x = \pm i\sqrt{3}$$

6. The graph of $y = x^2 - 4$ will be a parabola. It will cross the x-axis at 2 and -2, and the vertex will be $(0, -4)$.

7. Solve $x^2 - 2x - 8 > 0$. We factor and draw the sign diagram:

$$(x - 4)(x + 2) > 0$$

$(x-4)(x+2) > 0$

The solution is $x < -2$ or $x > 4$.

Equations Quadratic in Form [8.4]

There are a variety of equations whose form is quadratic. We solve most of them by making a substitution so that the equation becomes quadratic, and then solving that equation by factoring or the quadratic formula. For example,

The Equation	Is Quadratic in
$(2x - 3)^2 + 5(2x - 3) - 6 = 0$	$2x - 3$
$4x^4 - 7x^2 - 2 = 0$	x^2
$2x - 7\sqrt{x} + 3 = 0$	$\sqrt{x}$

Graphing Parabolas [8.5]

The graph of any equation of the form

$$y = ax^2 + bx + c \quad a \neq 0$$

is a *parabola*. The graph is *concave up* if $a > 0$, and *concave down* if $a < 0$. The highest or lowest point on the graph is called the *vertex* and always has an x-coordinate of $x = \dfrac{-b}{2a}$.

Quadratic Inequalities [8.6]

We solve quadratic inequalities by manipulating the inequality to get 0 on the right side and then factoring the left side. We then make a diagram that indicates where the factors are positive and where they are negative. From this sign diagram and the original inequality we graph the appropriate solution set.

CHAPTER 8 REVIEW

Solve each equation. [8.1]

1. $(2t - 5)^2 = 25$ **2.** $(3t - 2)^2 = 4$

3. $(3y - 4)^2 = -49$ **4.** $(2x + 6)^2 = 12$

Solve by completing the square. [8.1]

5. $2x^2 + 6x - 20 = 0$ **6.** $3x^2 + 15x = -18$

7. $a^2 + 9 = 6a$ **8.** $a^2 + 4 = 4a$

9. $2y^2 + 6y = -3$ **10.** $3y^2 + 3 = 9y$

Solve each equation. [8.2]

11. $\frac{1}{6}x^2 + \frac{1}{2}x - \frac{5}{3} = 0$ **12.** $8x^2 - 18x = 0$

13. $4t^2 - 8t + 19 = 0$

14. $100x^2 - 200x = 100$

15. $0.06a^2 + 0.05a = 0.04$

16. $9 - 6x = -x^2$

17. $(2x + 1)(x - 5) - (x + 3)(x - 2) = -17$

18. $2y^3 + 2y = 10y^2$ **19.** $5x^2 = -2x + 3$

20. $x^3 - 27 = 0$ **21.** $3 - \frac{2}{x} + \frac{1}{x^2} = 0$

22. $\dfrac{1}{x - 3} + \dfrac{1}{x + 2} = 1$

23. Profit The total cost (in dollars) for a company to produce x items per week is $C = 7x + 400$. The revenue for selling all x items is $R = 34x - 0.1x^2$. How many items must it produce and sell each week for its weekly profit to be $1,300? [8.2]

24. Profit The total cost (in dollars) for a company to produce x items per week is $C = 70x + 300$. The revenue for selling all x items is $R = 110x - 0.5x^2$. How many items must it produce and sell each week for its weekly profit to be $300? [8.2]

Use the discriminant to find the number and kind of solutions for each equation. [8.3]

25. $2x^2 - 8x = -8$ **26.** $4x^2 - 8x = -4$

27. $2x^2 + x - 3 = 0$ **28.** $5x^2 + 11x = 12$

29. $x^2 - x = 1$ **30.** $x^2 - 5x = -5$

31. $3x^2 + 5x = -4$ **32.** $4x^2 - 3x = -6$

Determine k so that each equation has exactly one real solution. [8.3]

33. $25x^2 - kx + 4 = 0$ **34.** $4x^2 + kx + 25 = 0$

35. $kx^2 + 12x + 9 = 0$ **36.** $kx^2 - 16x + 16 = 0$

37. $9x^2 + 30x + k = 0$ **38.** $4x^2 + 28x + k = 0$

For each of the following problems, find an equation that has the given solutions. [8.3]

39. $x = 3, x = 5$ **40.** $x = -2, x = 4$

41. $y = \frac{1}{2}, y = -4$ **42.** $t = 3, t = -3, t = 5$

Find all solutions. [8.4]

43. $(x - 2)^2 - 4(x - 2) - 60 = 0$

44. $6(2y + 1)^2 - (2y + 1) - 2 = 0$

45. $x^4 - x^2 = 12$ **46.** $x - \sqrt{x} - 2 = 0$

47. $2x - 11\sqrt{x} = -12$ **48.** $\sqrt{x + 5} = \sqrt{x} + 1$

49. $\sqrt{y + 21} + \sqrt{y} = 7$

50. $\sqrt{y + 9} - \sqrt{y - 6} = 3$

51. Projectile Motion An object is tossed into the air with an upward velocity of 10 feet per second from the top of a building h feet high. The time it takes for the object to hit the ground below is given by the formula $16t^2 - 10t - h = 0$. Solve this formula for t. [8.4]

52. Projectile Motion An object is tossed into the air with an upward velocity of v feet per second from the top of a 10-foot wall. The time it takes for the object to hit the ground below is given by the formula $16t^2 - vt - 10 = 0$. Solve this formula for t. [8.4]

Solve each inequality and graph the solution set. [8.6]

53. $x^2 - x - 2 < 0$ **54.** $3x^2 - 14x + 8 \le 0$

55. $2x^2 + 5x - 12 \ge 0$

56. $(x + 2)(x - 3)(x + 4) > 0$

Find the x-intercepts, if they exist, and the vertex for each parabola. Then use them to sketch the graph. [8.5]

57. $y = x^2 - 6x + 8$ **58.** $y = x^2 - 4$

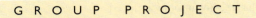

GROUP PROJECT

MAXIMUM VOLUME OF A BOX

Number of People: 4

Time Needed: 30 minutes

Equipment: Graphing calculator and five pieces of graph paper

Background: For many people, having a concrete model to work with allows them to visualize situations that they would have difficulty with if they had only a written description to work with. The purpose of this project is to rework a problem we have worked previously, but this time with a concrete model.

Procedure: You are going to make boxes of varying dimensions from rectangles that are 11 centimeters wide and 17 centimeters long.

1. Cut a rectangle from your graph paper that is 11 squares for the width and 17 squares for the length. Pretend that each small square is 1 centimeter by 1 centimeter. Do this with four pieces of paper.

2. One person tears off one square from each corner of their paper, then folds up the sides to form a box. Write down the length, width, and height of this box. Then calculate its volume.

3. The next person tears a square that is two units on a side from each corner of their paper, then folds up the sides to form a box. Write down the length, width, and height of this box. Then calculate its volume.

4. The next person follows the same procedure, tearing a still larger square from each corner of their paper. This continues until the squares that are to be torn off are larger than the original piece of paper.

5. Enter the data from each box you have created into the table below. Then graph all the points (x, V) from the table.

TABLE I	Volume of a Box			
Side of Square x (cm)	Length of Box L	Width of Box W	Height of Box H	Volume of Box V
1				
2				
3				
4				
5				

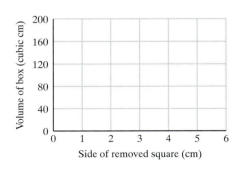

6. Using what you have learned from filling in the table, write a formula for the volume of the box that is created when a square of side x is cut from each corner of the original piece of paper.

7. Graph the equation you found in Problem 6 on a graphing calculator. Use that result to connect the points you plotted in the graph.

8. Use the graphing calculator to find the value of x that will give the maximum volume of the box. Your answer should be accurate to the nearest hundredth.

RESEARCH PROJECT

A CONTINUED FRACTION AND GOLDEN RATIO

If you successfully completed the research project in Chapter 7, then you used number sequences and inductive reasoning to conclude that the continued fraction

$$1 + \cfrac{1}{1 + \cfrac{1}{1 + \cfrac{1}{1 + \cdots}}}$$

is equal to the golden ratio.

The same conclusion can be found with the quadratic formula by using the following conditional statement

$$\text{If } x = 1 + \cfrac{1}{1 + \cfrac{1}{1 + \cfrac{1}{1 + \cdots}}}, \text{ then } x = 1 + \frac{1}{x}$$

Work with this conditional statement until you see that it is true. Then solve the equation $x = 1 + \dfrac{1}{x}$. Write an essay in which you explain in your own words why the conditional statement is true. Then show the details of the solution to the equation $x = 1 + \dfrac{1}{x}$. The message that you want to get across in your essay is that the continued fraction shown here is actually the same as the golden ratio.

CHAPTER 8 TEST

Solve each equation. [8.1, 8.2]

1. $(2x + 4)^2 = 25$ **2.** $(2x - 6)^2 = -8$

3. $y^2 - 10y + 25 = -4$

4. $(y + 1)(y - 3) = -6$

5. $8t^3 - 125 = 0$ **6.** $\dfrac{1}{a + 2} - \dfrac{1}{3} = \dfrac{1}{a}$

7. Solve the formula $64(1 + r)^2 = A$ for r. [8.1]

8. Solve $x^2 - 4x = -2$ by completing the square. [8.1]

9. **Projectile Motion** An object projected upward with an initial velocity of 32 feet per second will rise and fall according to the equation $s(t) = 32t - 16t^2$, where s is its distance above the ground at time t. At what times will the object be 12 feet above the ground? [8.2]

10. **Revenue** The total weekly cost for a company to make x ceramic coffee cups is given by the formula $C(x) = 2x + 100$. If the weekly revenue from selling all x cups is $R(x) = 25x - 0.2x^2$, how many cups must it sell a week to make a profit of $200 a week? [8.2]

11. Find k so that $kx^2 = 12x - 4$ has one rational solution. [8.3]

12. Use the discriminant to identify the number and kind of solutions to $2x^2 - 5x = 7$. [8.3]

Find equations that have the given solutions. [8.3]

13. $x = 5, x = -\dfrac{2}{3}$

14. $x = 2, x = -2, x = 7$

Solve each equation. [8.4]

15. $4x^4 - 7x^2 - 2 = 0$

16. $(2t + 1)^2 - 5(2t + 1) + 6 = 0$

17. $2t - 7\sqrt{t} + 3 = 0$

18. **Projectile Motion** An object is tossed into the air with an upward velocity of 14 feet per second from the top of a building h feet high. The time it takes for the object to hit the ground below is given by the formula $16t^2 - 14t - h = 0$. Solve this formula for t. [8.4]

Sketch the graph of each of the following equations. Give the coordinates of the vertex in each case. [8.5]

19. $y = x^2 - 2x - 3$ **20.** $y = -x^2 + 2x + 8$

Graph each of the following inequalities. [8.6]

21. $x^2 - x - 6 \leq 0$ **22.** $2x^2 + 5x > 3$

23. **Profit** Find the maximum weekly profit for a company with weekly costs of $C = 5x + 100$ and weekly revenue of $R = 25x - 0.1x^2$. [8.5]

Simplify.

1. $11 + 20 \div 5 - 3 \cdot 5$ **2.** Evaluate: $\left(-\dfrac{2}{3}\right)^3$.

3. $4(15 - 19)^2 - 3(17 - 19)^3$

4. $4 + 8x - 3(5x - 2)$

5. $3 - 5[2x - 4(x - 2)]$ **6.** $\left(\dfrac{x^{-5}y^4}{x^{-2}y^{-3}}\right)^{-1}$

7. $\sqrt[3]{32}$ **8.** $8^{-2/3} + 25^{-1/2}$

9. $\dfrac{1 - \dfrac{3}{4}}{1 + \dfrac{3}{4}}$

Reduce.

10. $\dfrac{468}{585}$

11. $\dfrac{5x^2 - 26xy - 24y^2}{5x + 4y}$ **12.** $\dfrac{x^2 - x - 6}{x + 2}$

Multiply.

13. $(3x - 2)(x^2 - 3x - 2)$

14. $(1 + i)^2$ (Remember, $i^2 = -1$.)

15. Divide $\dfrac{7 - i}{3 - 2i}$ ($i^2 = -1$)

16. Subtract $\dfrac{3}{4}$ from the product of -7 and $\dfrac{5}{28}$.

Solve.

17. $\dfrac{7}{5}a - 6 = 15$ **18.** $|a| - 6 = 3$

19. $\dfrac{a}{2} + \dfrac{3}{a - 3} = \dfrac{a}{a - 3}$ **20.** $\sqrt{y + 3} = y + 3$

21. $(3x - 4)^2 = 18$

22. $\dfrac{2}{15}x^2 + \dfrac{1}{3}x + \dfrac{1}{5} = 0$

23. $3y^3 - y = 5y^2$

24. $0.06a^2 - 0.01a = 0.02$

25. $\sqrt{x - 2} = 2 - \sqrt{x}$

26. Solve for x: $ax - 3 = bx + 5$

Solve each inequality, and graph the solution.

27. $5 \le \dfrac{1}{4}x + 3 \le 8$ **28.** $|4x - 3| \ge 5$

Solve each system.

29. $\begin{aligned} 3x - y &= 2 \\ -6x + 2y &= -4 \end{aligned}$ **30.** $\begin{aligned} 4x - 8y &= 6 \\ 6x - 12y &= 6 \end{aligned}$

31. $\begin{aligned} 3x - 2y &= 5 \\ y &= 3x - 7 \end{aligned}$ **32.** $\begin{aligned} 2x + 3y - 8z &= 2 \\ 3x - y + 2z &= 10 \\ 4x + y + 8z &= 16 \end{aligned}$

Graph on a rectangular coordinate system.

33. $2x - 3y = 12$ **34.** $y = x^2 - x - 2$

35. Find the equation of the line passing through the two points $\left(\dfrac{3}{2}, \dfrac{4}{3}\right)$, $\left(\dfrac{1}{4}, -\dfrac{1}{3}\right)$.

36. Factor completely: $x^2 + 8x + 16 - y^2$.

37. Rationalize the denominator $\dfrac{7}{\sqrt[3]{9}}$.

38. If $A = \{1, 2, 3, 4\}$ and $B = \{0, 2, 4, 6\}$, find $\{x \mid x \notin B \text{ and } x \in A\}$.

39. **Geometry** Find all three angles in a triangle if the smallest angle is one-fourth the largest angle and the remaining angle is 30° more than the smallest angle.

40. **Inverse Variation** y varies inversely with the square of x. If $y = 4$ when $x = \dfrac{5}{3}$, find y when $x = \dfrac{8}{3}$.

Exponential and Logarithmic Functions

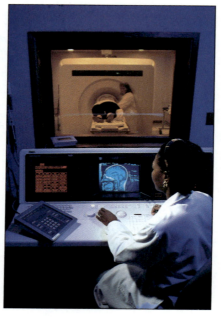

INTRODUCTION

If you have had any problems with or had testing done on your thyroid gland, then you may have come in contact with radioactive iodine-131. Like all radio-active elements, iodine-131 decays naturally. The half-life of iodine-131 is 8 days, which means that every 8 days a sample of iodine-131 will decrease to half of its original amount. The following table and graph show what happens to a 1,600-microgram sample of iodine-131 over time.

TABLE 1	Iodine-131 as a Function of Time
t (days)	**A** (micrograms)
0	1,600
8	800
16	400
24	200
32	100

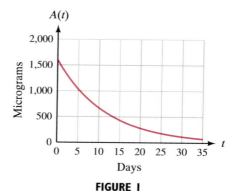

FIGURE 1

The function represented by the information in Table 1 and Figure 1 is

$$A(t) = 1{,}600 \cdot 2^{-t/8}$$

It is called an exponential function, and it is one of the types of functions we will study in this chapter.

Exponential Functions

To obtain an intuitive idea of how exponential functions behave, we can consider the heights attained by a bouncing ball. When a ball used in the game of racquetball is dropped from any height, the first bounce will reach a height that is $\frac{2}{3}$ of the original height. The second bounce will reach $\frac{2}{3}$ of the height of the first bounce, and so on, as shown in Figure 1.

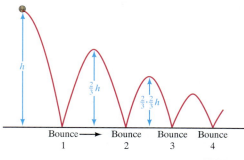
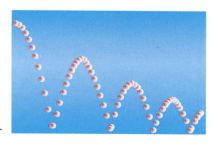

Bounce 1 → Bounce 2 Bounce 3 Bounce 4

FIGURE 1

If the ball is initially dropped from a height of 1 meter, then during the first bounce it will reach a height of $\frac{2}{3}$ meter. The height of the second bounce will reach $\frac{2}{3}$ of the height reached on the first bounce. The maximum height of any bounce is $\frac{2}{3}$ of the height of the previous bounce.

$$\text{Initial height:} \qquad h = 1$$

$$\text{Bounce 1:} \qquad h = \tfrac{2}{3}(1) = \tfrac{2}{3}$$

$$\text{Bounce 2:} \qquad h = \tfrac{2}{3}(\tfrac{2}{3}) = (\tfrac{2}{3})^2$$

$$\text{Bounce 3:} \qquad h = \tfrac{2}{3}(\tfrac{2}{3})^2 = (\tfrac{2}{3})^3$$

$$\text{Bounce 4:} \qquad h = \tfrac{2}{3}(\tfrac{2}{3})^3 = (\tfrac{2}{3})^4$$

$$\vdots \qquad\qquad\qquad \vdots$$

$$\text{Bounce } n: \qquad h = \tfrac{2}{3}(\tfrac{2}{3})^{n-1} = (\tfrac{2}{3})^n$$

This last equation is exponential in form. We classify all exponential functions together with the following definition.

DEFINITION   An **exponential function** is any function that can be written in the form

$$f(x) = b^x$$

where b is a positive real number other than 1.

Each of the following is an exponential function:

$$f(x) = 2^x \qquad y = 3^x \qquad f(x) = \left(\frac{1}{4}\right)^x$$

The first step in becoming familiar with exponential functions is to find some values for specific exponential functions.

EXAMPLE 1   If the exponential functions f and g are defined by

$$f(x) = 2^x \qquad \text{and} \qquad g(x) = 3^x$$

then

$$f(0) = 2^0 = 1 \qquad\qquad g(0) = 3^0 = 1$$
$$f(1) = 2^1 = 2 \qquad\qquad g(1) = 3^1 = 3$$
$$f(2) = 2^2 = 4 \qquad\qquad g(2) = 3^2 = 9$$
$$f(3) = 2^3 = 8 \qquad\qquad g(3) = 3^3 = 27$$
$$f(-2) = 2^{-2} = \frac{1}{2^2} = \frac{1}{4} \qquad g(-2) = 3^{-2} = \frac{1}{3^2} = \frac{1}{9}$$
$$f(-3) = 2^{-3} = \frac{1}{2^3} = \frac{1}{8} \qquad g(-3) = 3^{-3} = \frac{1}{3^3} = \frac{1}{27}$$

In the introduction to this chapter we indicated that the half-life of iodine-131 is 8 days, which means that every 8 days a sample of iodine-131 will decrease to half of its original amount. If we start with A_0 micrograms of iodine-131, then after t days the sample will contain

$$A(t) = A_0 \cdot 2^{-t/8}$$

micrograms of iodine-131.

EXAMPLE 2   A patient is administered a 1,200-microgram dose of iodine-131. How much iodine-131 will be in the patient's system after 10 days, and after 16 days?

Solution   The initial amount of iodine-131 is $A_0 = 1,200$, so the function that gives the amount left in the patient's system after t days is

$$A(t) = 1,200 \cdot 2^{-t/8}$$

After 10 days, the amount left in the patient's system is

$$A(10) = 1,200 \cdot 2^{-10/8} = 1,200 \cdot 2^{-1.25} \approx 504.5 \text{ micrograms}$$

After 16 days, the amount left in the patient's system is

$$A(16) = 1,200 \cdot 2^{-16/8} = 1,200 \cdot 2^{-2} = 300 \text{ micrograms}$$

We will now turn our attention to the graphs of exponential functions. Since the notation y is easier to use when graphing, and $y = f(x)$, for convenience we will write the exponential functions as

$$y = b^x$$

EXAMPLE 3   Sketch the graph of the exponential function $y = 2^x$.

Solution   Using the results of Example 1, we produce the following table. Graphing the ordered pairs given in the table and connecting them with a smooth curve, we have the graph of $y = 2^x$ shown in Figure 2.

x	y
-3	$\frac{1}{8}$
-2	$\frac{1}{4}$
-1	$\frac{1}{2}$
0	1
1	2
2	4
3	8

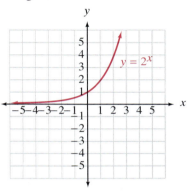

FIGURE 2

Notice that the graph does not cross the x-axis. It *approaches* the x-axis — in fact, we can get it as close to the x-axis as we want without it actually intersecting the x-axis. In order for the graph of $y = 2^x$ to intersect the x-axis, we would have to find a value of x that would make $2^x = 0$. Because no such value of x exists, the graph of $y = 2^x$ cannot intersect the x-axis.

EXAMPLE 4   Sketch the graph of $y = \left(\frac{1}{3}\right)^x$.

Solution   The table beside Figure 3 gives some ordered pairs that satisfy the equation. Using the ordered pairs from the table, we have the graph shown in Figure 3.

x	y
-3	27
-2	9
-1	3
0	1
1	$\frac{1}{3}$
2	$\frac{1}{9}$
3	$\frac{1}{27}$

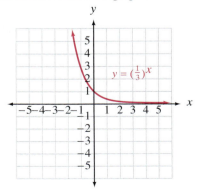

FIGURE 3

The graphs of all exponential functions have two things in common: (1) Each crosses the y-axis at (0, 1), since $b^0 = 1$; and (2) none can cross the x-axis, since $b^x = 0$ is impossible because of the restrictions on b.

Figures 4 and 5 show some families of exponential curves to help you become more familiar with them on an intuitive level.

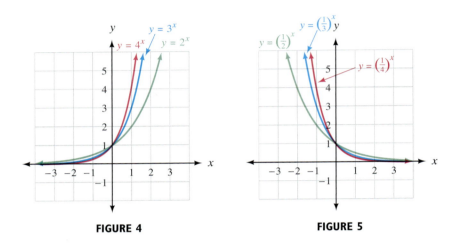

FIGURE 4   **FIGURE 5**

Among the many applications of exponential functions are the applications having to do with interest-bearing accounts. Here are the details.

Compound Interest

If P dollars are deposited in an account with annual interest rate r, compounded n times per year, then the amount of money in the account after t years is given by the formula

$$A(t) = P\left(1 + \frac{r}{n}\right)^{nt}$$

EXAMPLE 5 Suppose you deposit $500 in an account with an annual interest rate of 8% compounded quarterly. Find an equation that gives the amount of money in the account after t years. Then find

(a) The amount of money in the account after 5 years.

(b) The number of years it will take for the account to contain $1,000.

Solution First, we note that $P = 500$ and $r = 0.08$. Interest that is compounded quarterly is compounded four times a year, giving us $n = 4$. Substituting these numbers into the preceding formula, we have our function

$$A(t) = 500\left(1 + \frac{0.08}{4}\right)^{4t} = 500(1.02)^{4t}$$

(a) To find the amount after 5 years, we let $t = 5$:

$$A(5) = 500(1.02)^{4 \cdot 5} = 500(1.02)^{20} \approx \$742.97$$

Our answer is found on a calculator, and then rounded to the nearest cent.

(b) To see how long it will take for this account to total $1,000, we graph the equation $Y_1 = 500(1.02)^{4X}$ on a graphing calculator, and then look to see where it intersects the line $Y_2 = 1,000$. The two graphs are shown in Figure 6.

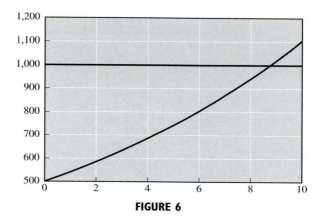

FIGURE 6

Using Zoom and Trace, or the Intersect function on the graphing calculator, we find that the two curves intersect at $X \approx 8.75$ and $Y = 1,000$. This means that our account will contain $1,000 after the money has been on deposit for 8.75 years.

THE NATURAL EXPONENTIAL FUNCTION

A very commonly occurring exponential function is based on a special number we denote with the letter e. The number e is a number like π. It is irrational and occurs in many formulas that describe the world around us. Like π, it can be approximated with a decimal number. Whereas π is approximately 3.1416, e is approximately 2.7183. (If you have a calculator with a key labeled $\boxed{e^x}$, you can use it to find e^1 to find a more accurate approximation to e.) We cannot give a more precise definition of the number e without using some of the topics taught in calculus. For the work we are going to do with the number e, we only need to know that it is an irrational number that is approximately 2.7183.

Here are a table and graph (Figure 7) for the natural exponential function

$$y = f(x) = e^x$$

x	$f(x) = e^x$
-2	$f(-2) = e^{-2} = \dfrac{1}{e^2} \approx 0.135$
-1	$f(-1) = e^{-1} = \dfrac{1}{e} \approx 0.368$
0	$f(0) = e^0 = 1$
1	$f(1) = e^1 = e \approx 2.72$
2	$f(2) = e^2 \approx 7.39$
3	$f(3) = e^3 \approx 20.09$

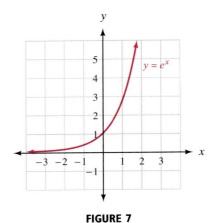

FIGURE 7

One common application of natural exponential functions is with interest-bearing accounts. In Example 5 we worked with the formula

$$A = P\left(1 + \frac{r}{n}\right)^{nt}$$

that gives the amount of money in an account if P dollars are deposited for t years at annual interest rate r, compounded n times per year. In Example 5 the number of compounding periods was four. What would happen if we let the number of compounding periods become larger and larger, so that we compounded the interest every day, then every hour, then every second, and so on? If we take this as far as it can go, we end up compounding the interest every moment. When this happens, we have an account with interest that is compounded continuously, and the amount of money in such an account depends on the number e. Here are the details.

Continuously Compounded Interest

If P dollars are deposited in an account with annual interest rate r, compounded continuously, then the amount of money in the account after t years is given by the formula

$$A(t) = Pe^{rt}$$

EXAMPLE 6 Suppose you deposit $500 in an account with an annual interest rate of 8% compounded continuously. Find an equation that gives the amount of money in the account after t years. Then find the amount of money in the account after 5 years.

Solution Since the interest is compounded continuously, we use the formula $A(t) = Pe^{rt}$. Substituting $P = 500$ and $r = 0.08$ into this formula we have

$$A(t) = 500e^{0.08t}$$

After 5 years, this account will contain

$$A(5) = 500e^{0.08 \cdot 5} = 500e^{0.4} \approx \$745.91$$

to the nearest cent. Compare this result with the answer to Example 5a.

Getting Ready for Class

After reading through the preceding section, respond in your own words and in complete sentences.

A. What is an exponential function?

B. In an exponential function, explain why the base b cannot equal 1. (What kind of function would you get if the base was equal to 1?)

C. Explain continuously compounded interest.

D. What characteristics do the graphs of $y = 2^x$ and $y = (\frac{1}{2})^x$ have in common?

PROBLEM SET 9.1

Let $f(x) = 3^x$ and $g(x) = (\frac{1}{2})^x$, and evaluate each of the following.

1. $g(0)$ **2.** $f(0)$

3. $g(-1)$ **4.** $g(-4)$

5. $f(-3)$ **6.** $f(-1)$

7. $f(2) + g(-2)$ **8.** $f(2) - g(-2)$

Graph each of the following functions.

9. $y = 4^x$ **10.** $y = 2^{-x}$

11. $y = 3^{-x}$ **12.** $y = (\frac{1}{3})^{-x}$

13. $y = 2^{x+1}$ **14.** $y = 2^{x-3}$

15. $y = e^x$ **16.** $y = e^{-x}$

Graph each of the following functions on the same co-ordinate system for positive values of x only.

17. $y = 2x, y = x^2, y = 2^x$

18. $y = 3x, y = x^3, y = 3^x$

19. On a graphing calculator, graph the family of curves $y = b^x$, $b = 2, 4, 6, 8$.

20. On a graphing calculator, graph the family of curves $y = b^x$, $b = \frac{1}{2}, \frac{1}{4}, \frac{1}{6}, \frac{1}{8}$.

Applying the Concepts

21. Bouncing Ball Suppose the ball mentioned in the introduction to this section is dropped from a height of 6 feet above the ground. Find an exponential equation that gives the height h the ball will attain during the nth bounce. How high will it bounce on the fifth bounce?

22. Bouncing Ball A golf ball is manufactured so that if it is dropped from A feet above the ground onto a hard surface, the maximum height of each bounce will be one half of the height of the previous bounce. Find an exponential equation that gives the height h the ball will attain during the nth bounce. If the ball is dropped from 10 feet above the ground onto a hard surface, how high will it bounce on the eighth bounce?

23. Exponential Decay Twinkies on the shelf of a convenience store lose their fresh tastiness over time. We say that the taste quality is 1 when the Twinkies are first put on the shelf at the store, and that the quality of tastiness declines according to the function $Q(t) = 0.85^t$. Graph this function on a graphing calculator, and determine when the taste quality will be one half of its original value.

24. Exponential Growth Automobiles built before 1993 use Freon in their air conditioners. The federal government now prohibits the manufacture of Freon. Because the supply of Freon is decreasing, the price per pound is increasing exponentially. Current estimates put the formula for the price per pound of Freon at $p(t) = 1.89(1.25)^t$, where t is the number of years since 1990. Find the price of Freon in 1995 and 1990. How much will Freon cost in the year 2000?

25. **Compound Interest** Suppose you deposit $1,200 in an account with an annual interest rate of 6% compounded quarterly.
    (a) Find an equation that gives the amount of money in the account after t years.
    (b) Find the amount of money in the account after 8 years.
    (c) How many years will it take for the account to contain $2,400?
    (d) If the interest were compounded continuously, how much money would the account contain after 8 years?

26. **Compound Interest** Suppose you deposit $500 in an account with an annual interest rate of 8% compounded monthly.
    (a) Find an equation that gives the amount of money in the account after t years.
    (b) Find the amount of money in the account after 5 years.
    (c) How many years will it take for the account to contain $1,000?
    (d) If the interest were compounded continuously, how much money would the account contain after 5 years?

Declining-Balance Depreciation The declining-balance method of depreciation is an accounting method businesses use to deduct most of the cost of new equipment during the first few years of purchase. Unlike other methods, the declining-balance formula does not consider salvage value.

27. **Value of a Crane** The function $V(t) = 450,000(1 - 0.30)^t$, where V is value and t is time in years, can be used to find the book value of a crane for the first 6 years of use.
    (a) What is the book value of the crane after 3 years and 6 months?
    (b) State the domain of this function.
    (c) Sketch the graph of this function.
    (d) State the range of this function.

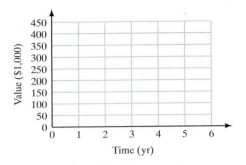

(SuperStock)

(e) After how many years will the crane be worth only $85,000?

28. **Value of a Printing Press** The function $V(t) = 375,000(1 - 0.25)^t$, where V is value and t is time in years, can be used to find the book value of a printing press during the first 7 years of use.
    (a) What is the book value of the printing press after 4 years and 9 months?
    (b) State the domain of this function.
    (c) Sketch the graph of this function.
    (d) State the range of this function.

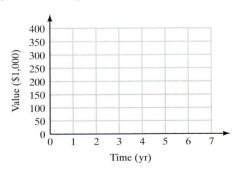

(Jim McCrary/Tony Stone Images)

(Kactus Foto, Santiago, Chile/SuperStock)

(e) After how many years will the printing press be worth only $65,000?

29. Bacteria Growth   Suppose it takes 12 hours for a certain strain of bacteria to reproduce by dividing in half. If 50 bacteria are present to begin with, then the total number present after x days will be $f(x) = 50 \cdot 4^x$. Find the total number present after 1 day, 2 days, and 3 days.

30. Bacteria Growth   Suppose it takes 1 day for a certain strain of bacteria to reproduce by dividing in half. If 100 bacteria are present to begin with, then the total number present after x days will be $f(x) = 100 \cdot 2^x$. Find the total number present after 1 day, 2 days, 3 days, and 4 days. How many days must elapse before over 100,000 bacteria are present?

31. Value of a Painting   A painting is purchased as an investment for $150. If the painting's value doubles every 3 years, then its value is given by the function

$$V(t) = 150 \cdot 2^{t/3} \quad \text{for } t \geq 0$$

where t is the number of years since it was purchased, and $V(t)$ is its value (in dollars) at that time. Graph this function.

32. Value of a Painting   A painting is purchased as an investment for $125. If the painting's value doubles every 5 years, then its value is given by the function

$$V(t) = 125 \cdot 2^{t/5} \quad \text{for } t \geq 0$$

where t is the number of years since it was pur-

chased, and $V(t)$ is its value (in dollars) at that time. Graph this function.

Review Problems

The following problems review material from Sections 3.5 and 3.6.

For each of the following relations, specify the domain and range; then indicate which are also functions. [3.5]

33. $\{(-2, 6), (-2, 8), (2, 3)\}$

34. $\{(1, 2), (3, 4), (4, 1)\}$

State the domain for each of the following functions. [3.5]

35. $y = \dfrac{-4}{x^2 + 2x - 35}$ **36.** $y = \sqrt{3x + 1}$

If $f(x) = 2x^2 - 18$ and $g(x) = 2x - 6$, find [3.6]

37. $f(0)$ **38.** $g[f(0)]$

39. $\dfrac{g(x + h) - g(x)}{h}$ **40.** $\dfrac{g}{f}(x)$

Extending the Concepts

41. Reading Graphs   The graphs of two exponential functions are given in Figures 8 and 9. Use the graphs to find the following:

(a) $f(0)$ (b) $f(-1)$

(c) $f(1)$ (d) $g(0)$

(e) $g(1)$ (f) $g(-1)$

(g) $f[g(0)]$ (h) $g[f(0)]$

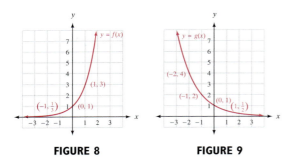

FIGURE 8   **FIGURE 9**

42. Drag Racing In Chapter 3 we mentioned the dragster equipped with a computer. Table 1 gives the speed of the dragster every second during one

TABLE I Speed of a Dragster	
Elapsed Time (sec)	**Speed (mi/hr)**
0	0.0
1	72.7
2	129.9
3	162.8
4	192.2
5	212.4
6	228.1

FIGURE 10

race at the 1993 Winternationals. Figure 10 is a line graph constructed from the data in Table 1. The graph of the function $s(t) = 250(1 - 1.5^{-t})$ contains the first point and the last point shown in Figure 10. That is, both $(0, 0)$ and $(6, 228.1)$ satisfy the function. Graph the function to see how close it comes to the other points in Figure 10.

43. Analyzing Graphs The goal of this problem is to obtain a sense of the growth rate of an exponential function. We will compare the exponential function $y = 2^x$ to the familiar function $y = x^2$.

(a) Graph $y = 2^x$ and $y = x^2$ using the window X: 0 to 3 and Y: 0 to 10. Which function appears to be taking over as x grows larger?

(b) Show algebraically that both graphs contain the points $(2, 4)$ and $(4, 16)$. You may also want to confirm this on your graph by enlarging the window.

(c) Graph $y = 2^x$ and $y = x^2$ with the window X: 0 to 15 and Y: 0 to 2,500. Which function dominates (grows the fastest) as x gets larger in the positive direction? Confirm your answer by evaluating $2^{(100)}$ and $(100)^2$.

9.2 The Inverse of a Function

The following diagram (Figure 1) shows the route Justin takes to school. He leaves his home and drives 3 miles east, and then turns left and drives 2 miles north. When he leaves school to drive home, he drives the same two segments, but in the reverse order and the opposite direction; that is, he drives 2 miles south, turns right, and drives 3 miles west. When he arrives home from school, he is right where he started. His route home "undoes" his route to school, leaving him where he began.

As you will see, the relationship between a function and its inverse function is similar to the relationship between Justin's route from home to school and his route from school to home.

Suppose the function f is given by

$$f = \{(1, 4), (2, 5), (3, 6), (4, 7)\}$$

The inverse of f is obtained by reversing the order of the coordinates in each or-

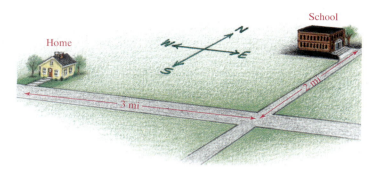

FIGURE 1

dered pair in f. The inverse of f is the relation given by

$$g = \{(4, 1), (5, 2), (6, 3), (7, 4)\}$$

It is obvious that the domain of f is now the range of g, and the range of f is now the domain of g. Every function (or relation) has an inverse that is obtained from the original function by interchanging the components of each ordered pair.

Suppose a function f is defined with an equation instead of a list of ordered pairs. We can obtain the equation of the inverse of f by interchanging the role of x and y in the equation for f.

EXAMPLE 1   If the function f is defined by $f(x) = 2x - 3$, find the equation that represents the inverse of f.

Solution   Since the inverse of f is obtained by interchanging the components of all the ordered pairs belonging to f, and each ordered pair in f satisfies the equation $y = 2x - 3$, we simply exchange x and y in the equation $y = 2x - 3$ to get the formula for the inverse of f:

$$x = 2y - 3$$

We now solve this equation for y in terms of x:

$$x + 3 = 2y$$

$$\frac{x + 3}{2} = y$$

$$y = \frac{x + 3}{2}$$

The last line gives the equation that defines the inverse of f. Let's compare the graphs of f and its inverse as given here. (See Figure 2.)

The graphs of f and its inverse have symmetry about the line $y = x$. This is a reasonable result since the one function was obtained from the other by interchang-

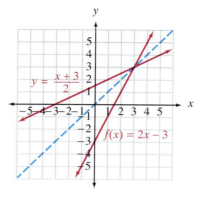

FIGURE 2

ing x and y in the equation. The ordered pairs (a, b) and (b, a) always have symmetry about the line $y = x$.

EXAMPLE 2   Graph the function $y = x^2 - 2$ and its inverse. Give the equation for the inverse.

Solution   We can obtain the graph of the inverse of $y = x^2 - 2$ by graphing $y = x^2 - 2$ by the usual methods, and then reflecting the graph about the line $y = x$. The equation that corresponds to the inverse of $y = x^2 - 2$ is obtained by interchanging x and y to get $x = y^2 - 2$. (See Figure 3.)

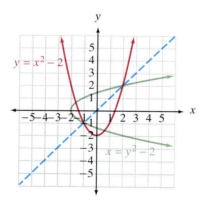

FIGURE 3

We can solve the equation $x = y^2 - 2$ for y in terms of x as follows:

$$x = y^2 - 2$$
$$x + 2 = y^2$$
$$y = \pm\sqrt{x + 2}$$

Comparing the graphs from Examples 1 and 2, we observe that the inverse of a function is not always a function. In Example 1, both f and its inverse have graphs that are nonvertical straight lines and therefore both represent functions. In Example 2, the inverse of function f is not a function, since a vertical line crosses it in more than one place.

ONE-TO-ONE FUNCTIONS

We can distinguish between those functions with inverses that are also functions and those functions with inverses that are not functions with the following definition.

> **DEFINITION**   A function is a **one-to-one function** if every element in the range comes from exactly one element in the domain.

This definition indicates that a one-to-one function will yield a set of ordered pairs in which no two different ordered pairs have the same second coordinates. For example, the function

$$f = \{(2, 3), (-1, 3), (5, 8)\}$$

is not one-to-one because the element 3 in the range comes from both 2 and -1 in the domain. On the other hand, the function

$$g = \{(5, 7), (3, -1), (4, 2)\}$$

is a one-to-one function because every element in the range comes from only one element in the domain.

HORIZONTAL LINE TEST

If we have the graph of a function, we can determine if the function is one-to-one with the following test. If a horizontal line crosses the graph of a function in more than one place, then the function is not a one-to-one function because the points at which the horizontal line crosses the graph will be points with the same y-coordinates, but different x-coordinates. Therefore, the function will have an element in the range (the y-coordinate) that comes from more than one element in the domain (the x-coordinates).

Of the functions we have covered previously, all the linear functions and exponential functions are one-to-one functions because no horizontal lines can be found that will cross their graphs in more than one place.

FUNCTIONS WHOSE INVERSES ARE ALSO FUNCTIONS

Because one-to-one functions do not repeat second coordinates, when we reverse the order of the ordered pairs in a one-to-one function, we obtain a relation in which no two ordered pairs have the same first coordinate — by definition, this relation must be a function. In other words, every one-to-one function has an inverse that is itself a function. Because of this, we can use function notation to represent that inverse.

INVERSE FUNCTION NOTATION

If $y = f(x)$ is a one-to-one function, then the inverse of f is also a function and can be denoted by $y = f^{-1}(x)$.

To illustrate, in Example 1 we found the inverse of $f(x) = 2x - 3$ was the function $y = \dfrac{x + 3}{2}$. We can write this inverse function with inverse function notation as

$$f^{-1}(x) = \frac{x + 3}{2}$$

On the other hand, the inverse of the function in Example 2 is not itself a function, so we do not use the notation $f^{-1}(x)$ to represent it.

Note: The notation f^{-1} does not represent the reciprocal of f. That is, the -1 in this notation is not an exponent. The notation f^{-1} is defined as representing the inverse function for a one-to-one function.

EXAMPLE 3   Find the inverse of $g(x) = \dfrac{x - 4}{x - 2}$.

Solution   To find the inverse for g, we begin by replacing $g(x)$ with y to obtain

$$y = \frac{x - 4}{x - 2} \qquad \text{The original function}$$

To find an equation for the inverse, we exchange x and y.

$$x = \frac{y - 4}{y - 2} \qquad \text{The inverse of the original function}$$

To solve for y, we first multiply each side by $y - 2$ to obtain

$$x(y - 2) = y - 4$$

$$xy - 2x = y - 4 \qquad \text{Distributive property}$$

$$xy - y = 2x - 4 \qquad \text{Collect all terms containing } y \text{ on the left side.}$$

$$y(x - 1) = 2x - 4 \qquad \text{Factor } y \text{ from each term on the left side.}$$

$$y = \frac{2x - 4}{x - 1} \qquad \text{Divide each side by } x - 1.$$

Figure 4 shows that the graph of this function passes the horizontal line test. Therefore, it is a one-to-one function.

Since our original function is one-to-one (see Figure 4), its inverse is also a function. Therefore, we can use inverse function notation to write

$$g^{-1}(x) = \frac{2x - 4}{x - 1}$$

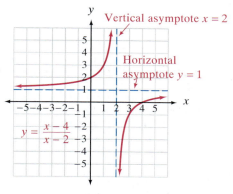

FIGURE 4

EXAMPLE 4   Graph the function $y = 2^x$ and its inverse $x = 2^y$.

Solution   We graphed $y = 2^x$ in the preceding section. We simply reflect its graph about the line $y = x$ to obtain the graph of its inverse $x = 2^y$. (See Figure 5.)

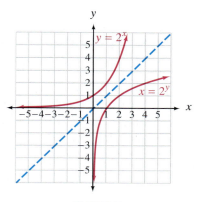

FIGURE 5

As you can see from the graph, $x = 2^y$ is a function. We do not have the mathematical tools to solve this equation for y, however. We are therefore unable to use the inverse function notation to represent this function. In the next section, we will give a definition that solves this problem. For now, we simply leave the equation as $x = 2^y$.

FUNCTIONS, RELATIONS, AND INVERSES—A SUMMARY

Here is a summary of some of the things we know about functions, relations, and their inverses:

1. Every function is a relation, but not every relation is a function.
2. Every function has an inverse, but only one-to-one functions have inverses that are also functions.
3. The domain of a function is the range of its inverse, and the range of a function is the domain of its inverse.
4. If $y = f(x)$ is a one-to-one function, then we can use the notation $y = f^{-1}(x)$ to represent its inverse function.
5. The graph of a function and its inverse have symmetry about the line $y = x$.
6. If (a, b) belongs to the function f, then the point (b, a) belongs to its inverse.

 Getting Ready for Class

After reading through the preceding section, respond in your own words and in complete sentences.

A. What is the inverse of a function?
B. What is the relationship between the graph of a function and the graph of its inverse?
C. Explain why only one-to-one functions have inverses that are also functions.
D. Describe the vertical line test, and explain the difference between the vertical line test and the horizontal line test.

PROBLEM SET 9.2

For each of the following one-to-one functions, find the equation of the inverse. Write the inverse using the notation $f^{-1}(x)$.

1. $f(x) = 3x - 1$
2. $f(x) = 2x - 5$
3. $f(x) = x^3$
4. $f(x) = x^3 - 2$
5. $f(x) = \dfrac{x - 3}{x - 1}$
6. $f(x) = \dfrac{x - 2}{x - 3}$
7. $f(x) = \dfrac{x - 3}{4}$
8. $f(x) = \dfrac{x + 7}{2}$
9. $f(x) = \dfrac{1}{2}x - 3$
10. $f(x) = \dfrac{1}{3}x + 1$
11. $f(x) = \dfrac{2x + 1}{3x + 1}$
12. $f(x) = \dfrac{3x + 2}{5x + 1}$

For each of the following relations, sketch the graph of the relation and its inverse, and write an equation for the inverse.

13. $y = 2x - 1$
14. $y = 3x + 1$
15. $y = x^2 - 3$
16. $y = x^2 + 1$
17. $y = x^2 - 2x - 3$
18. $y = x^2 + 2x - 3$
19. $y = 3^x$
20. $y = \left(\dfrac{1}{2}\right)^x$
21. $y = 4$
22. $y = -2$
23. $y = \dfrac{1}{2}x^3$
24. $y = x^3 - 2$
25. $y = \dfrac{1}{2}x + 2$
26. $y = \dfrac{1}{3}x - 1$

27. $y = \sqrt{x + 2}$ **28.** $y = \sqrt{x} + 2$

29. Determine if the following functions are one-to-one.

(a)

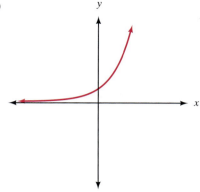

(b)

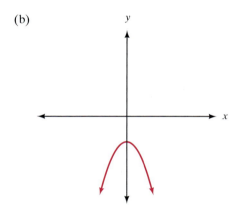

(c)

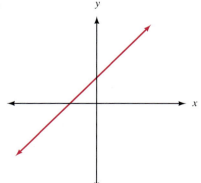

30. Could the following tables of values represent ordered pairs from one-to-one functions? Explain your answer.

(a)

x	y
-2	5
-1	4
0	3
1	4
2	5

(b)

x	y
1.5	0.1
2.0	0.2
2.5	0.3
3.0	0.4
3.5	0.5

31. If $f(x) = 3x - 2$, then $f^{-1}(x) = \dfrac{x + 2}{3}$. Use these two functions to find
(a) $f(2)$ (b) $f^{-1}(2)$
(c) $f[f^{-1}(2)]$ (d) $f^{-1}[f(2)]$

32. If $f(x) = \frac{1}{2}x + 5$, then $f^{-1}(x) = 2x - 10$. Use these two functions to find
(a) $f(-4)$ (b) $f^{-1}(-4)$
(c) $f[f^{-1}(-4)]$ (d) $f^{-1}[f(-4)]$

33. Let $f(x) = \dfrac{1}{x}$, and find $f^{-1}(x)$.

34. Let $f(x) = \dfrac{a}{x}$, and find $f^{-1}(x)$. (a is a real number constant.)

Applying the Concepts

35. Inverse Functions in Words   Inverses may also be found by *inverse reasoning.* For example, to find the inverse of $f(x) = 3x + 2$, first list, in order, the operations done to variable x:
(a) Multiply by 3.
(b) Add 2.
Then, to find the inverse, simply apply the inverse operations, in reverse order, to the variable x. That is:
(a) Subtract 2.
(b) Divide by 3.

The inverse function then becomes $f^{-1}(x) = \dfrac{x - 2}{3}$.

Use this method of "inverse reasoning" to find the inverse of the function $f(x) = \dfrac{x}{7} - 2$.

36. Inverse Functions in Words   Refer to the method of *inverse reasoning* explained in Problem 35. Use *inverse reasoning* to find the following inverses:
(a) $f(x) = 2x + 7$
(b) $f(x) = \sqrt{x} - 9$

(c) $f(x) = x^3 - 4$
(d) $f(x) = \sqrt{x^3 - 4}$

37. Reading Tables Evaluate each of the following functions using the functions defined by Tables 1 and 2.

(a) $f[g(-3)]$
(b) $g[f(-6)]$
(c) $g[f(2)]$
(d) $f[g(3)]$
(e) $f[g(-2)]$
(f) $g[f(3)]$
(g) What can you conclude about the relationship between functions f and g?

TABLE 1

x	-6	2	3	6
$f(x)$	3	-3	-2	4

TABLE 2

x	-3	-2	3	4
$g(x)$	2	3	-6	6

38. Reading Tables Use the functions defined in Tables 1 and 2 in Problem 37 to answer the following questions.
(a) What are the domain and range of f?
(b) What are the domain and range of g?
(c) How are the domain and range of f related to the domain and range of g?
(d) Is f a one-to-one function?
(e) Is g a one-to-one function?

Review Problems

The problems that follow review material we covered in Section 8.1.

Solve each equation.

39. $(2x - 1)^2 = 25$ **40.** $(3x + 5)^2 = -12$

41. What number would you add to $x^2 - 10x$ to make it a perfect square trinomial?

42. What number would you add to $x^2 - 5x$ to make it a perfect square trinomial?

Solve by completing the square.

43. $x^2 - 10x + 8 = 0$

44. $x^2 - 5x + 4 = 0$

45. $3x^2 - 6x + 6 = 0$

46. $4x^2 - 16x - 8 = 0$

Extending the Concepts

For each of the following functions, find $f^{-1}(x)$. Then show that $f[f^{-1}(x)] = x$.

47. $f(x) = 3x + 5$ **48.** $f(x) = 6 - 8x$

49. $f(x) = x^3 + 1$ **50.** $f(x) = x^3 - 8$

51. $f(x) = \dfrac{x - 4}{x - 2}$ **52.** $f(x) = \dfrac{x - 3}{x - 1}$

53. Reading Graphs The graphs of a function and its inverse are shown in Figure 6. Use the graphs to find the following:
(a) $f(0)$
(b) $f(1)$
(c) $f(2)$
(d) $f^{-1}(1)$
(e) $f^{-1}(2)$
(f) $f^{-1}(5)$
(g) $f^{-1}[f(2)]$
(h) $f[f^{-1}(5)]$

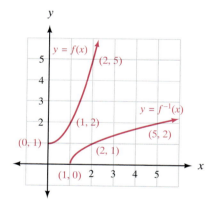

FIGURE 6

54. Domain From a Graph The function f is defined by the equation $f(x) = \sqrt{x + 4}$ for $x \geq -4$, meaning the domain is all x in the interval $[-4, \infty)$.
(a) Graph the function f.
(b) On the same set of axes, graph the line $y = x$ and the inverse function f^{-1}.
(c) State the domain for the inverse function for f^{-1}.
(d) Find the equation for $f^{-1}(x)$.

Logarithms Are Exponents

In January 1999, ABC News reported that an earthquake had occurred in Colombia, causing massive destruction. They reported the strength of the quake by indicating that it measured 6.0 on the Richter scale. For comparison, Table 1 gives the Richter magnitude of a number of other earthquakes.

TABLE 1 Earthquakes		
Year	Earthquake	Richter Magnitude
1971	Los Angeles	6.6
1985	Mexico City	8.1
1989	San Francisco	7.1
1992	Kobe, Japan	7.2
1994	Northridge	6.6
1999	Armenia, Colombia	6.0

(Reuters/José Miguel Gomez/Archives Photos)

Although the size of the numbers in the table do not seem to be very different, the intensity of the earthquakes they measure can be very different. For example, the 1989 San Francisco earthquake was more than 10 times stronger than the 1999 earthquake in Colombia. The reason behind this is that the Richter scale is a **logarithmic scale.** In this section we start our work with logarithms, which will give you an understanding of the Richter scale. Let's begin.

As you know from your work in the previous sections, equations of the form

$$y = b^x \quad b > 0, b \neq 1$$

are called exponential functions. Since the equation of the inverse of a function can be obtained by exchanging x and y in the equation of the original function, the inverse of an exponential function must have the form

$$x = b^y \quad b > 0, b \neq 1$$

Now, this last equation is actually the equation of a logarithmic function, as the following definition indicates:

DEFINITION   The expression $y = \log_b x$ is read "y is the logarithm to the base b of x" and is equivalent to the expression

$$x = b^y \quad b > 0, b \neq 1$$

In words, we say "y is the number we raise b to in order to get x."

Notation   When an expression is in the form $x = b^y$, it is said to be in exponential form. On the other hand, if an expression is in the form $y = \log_b x$, it is said to be in logarithmic form.

Here are some equivalent statements written in both forms.

Exponential Form		Logarithmic Form
$8 = 2^3$	$\Leftrightarrow$	$\log_2 8 = 3$
$25 = 5^2$	$\Leftrightarrow$	$\log_5 25 = 2$
$0.1 = 10^{-1}$	$\Leftrightarrow$	$\log_{10} 0.1 = -1$
$\dfrac{1}{8} = 2^{-3}$	$\Leftrightarrow$	$\log_2 \dfrac{1}{8} = -3$
$r = z^s$	$\Leftrightarrow$	$\log_z r = s$

EXAMPLE 1 Solve for x: $\log_3 x = -2$

Solution In exponential form the equation looks like this:

$$x = 3^{-2}$$

or

$$x = \frac{1}{9}$$

The solution is $\frac{1}{9}$.

EXAMPLE 2 Solve $\log_x 4 = 3$.

Solution Again, we use the definition of logarithms to write the expression in exponential form:

$$4 = x^3$$

Taking the cube root of both sides, we have

$$\sqrt[3]{4} = \sqrt[3]{x^3}$$
$$x = \sqrt[3]{4}$$

The solution set is $\{\sqrt[3]{4}\}$.

EXAMPLE 3 Solve $\log_8 4 = x$.

Solution We write the expression again in exponential form:

$$4 = 8^x$$

Since both 4 and 8 can be written as powers of 2, we write them in terms of powers of 2:

$$2^2 = (2^3)^x$$
$$2^2 = 2^{3x}$$

The only way the left and right sides of this last line can be equal is if the exponents are equal — that is, if

$$2 = 3x$$

or $$x = \frac{2}{3}$$

The solution is $\frac{2}{3}$. We check as follows:

$$\log_8 4 = \frac{2}{3} \Leftrightarrow 4 = 8^{2/3}$$

$$4 = (\sqrt[3]{8})^2$$

$$4 = 2^2$$

$$4 = 4$$

The solution checks when used in the original equation.

GRAPHING LOGARITHMIC FUNCTIONS

Graphing logarithmic functions can be done using the graphs of exponential functions and the fact that the graphs of inverse functions have symmetry about the line $y = x$. Here's an example to illustrate.

EXAMPLE 4   Graph the equation $y = \log_2 x$.

Solution   The equation $y = \log_2 x$ is, by definition, equivalent to the exponential equation

$$x = 2^y$$

which is the equation of the inverse of the function

$$y = 2^x$$

The graph of $y = 2^x$ was given in Figure 2 of Section 9.1. We simply reflect the graph of $y = 2^x$ about the line $y = x$ to get the graph of $x = 2^y$, which is also the graph of $y = \log_2 x$. (See Figure 1.)

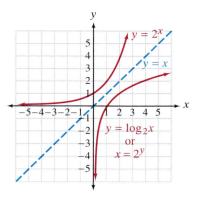

FIGURE 1

It is apparent from the graph that $y = \log_2 x$ is a function, since no vertical line will cross its graph in more than one place. The same is true for all logarithmic equations of the form $y = \log_b x$, where b is a positive number other than 1. Note also that the graph of $y = \log_b x$ will always appear to the right of the y-axis, meaning that x will always be positive in the expression $y = \log_b x$.

TWO SPECIAL IDENTITIES

If b is a positive real number other than 1, then each of the following is a consequence of the definition of a logarithm:

$$(1) \quad b^{\log_b x} = x \qquad \text{and} \qquad (2) \quad \log_b b^x = x$$

The justifications for these identities are similar. Let's consider only the first one. Consider the expression

$$y = \log_b x$$

By definition, it is equivalent to

$$x = b^y$$

Substituting $\log_b x$ for y in the last line gives us

$$x = b^{\log_b x}$$

The next examples in this section show how these two special properties can be used to simplify expressions involving logarithms.

EXAMPLE 5 Simplify $\log_2 8$.

Solution Substitute 2^3 for 8:

$$\log_2 8 = \log_2 2^3$$

$$= 3$$

EXAMPLE 6 Simplify $\log_{10} 10{,}000$.

Solution 10,000 can be written as 10^4:

$$\log_{10} 10{,}000 = \log_{10} 10^4$$

$$= 4$$

EXAMPLE 7 Simplify $\log_b b \ (b > 0, b \neq 1)$.

Solution Since $b^1 = b$, we have

$$\log_b b = \log_b b^1$$

$$= 1$$

EXAMPLE   8   Simplify $\log_b 1$ $(b > 0, b \neq 1)$.

Solution   Since $1 = b^0$, we have

$$\log_b 1 = \log_b b^0$$
$$= 0$$

EXAMPLE   9   Simplify $\log_4(\log_5 5)$.

Solution   Since $\log_5 5 = 1$,

$$\log_4(\log_5 5) = \log_4 1$$
$$= 0$$

APPLICATION

As we mentioned in the introduction to this section, one application of logarithms is in measuring the magnitude of an earthquake. If an earthquake has a shock wave T times greater than the smallest shock wave that can be measured on a seismograph, then the magnitude M of the earthquake, as measured on the Richter scale, is given by the formula

$$M = \log_{10} T$$

(When we talk about the size of a shock wave, we are talking about its amplitude. The amplitude of a wave is half the difference between its highest point and its lowest point.)

To illustrate the discussion, an earthquake that produces a shock wave that is 10,000 times greater than the smallest shock wave measurable on a seismograph will have a magnitude M on the Richter scale of

$$M = \log_{10} 10{,}000 = 4$$

EXAMPLE   I 0   If an earthquake has a magnitude of $M = 5$ on the Richter scale, what can you say about the size of its shock wave?

Solution   To answer this question, we put $M = 5$ into the formula $M = \log_{10} T$ to obtain

$$5 = \log_{10} T$$

Writing this expression in exponential form, we have

$$T = 10^5 = 100{,}000$$

We can say that an earthquake that measures 5 on the Richter scale has a shock wave 100,000 times greater than the smallest shock wave measurable on a seismograph.

From Example 10 and the discussion that preceded it, we find that an earthquake of magnitude 5 has a shock wave that is 10 times greater than an earthquake of magnitude 4, because 100,000 is 10 times 10,000.

Getting Ready for Class

After reading through the preceding section, respond in your own words and in complete sentences.

A. What is a logarithm?

B. What is the relationship between $y = 2^x$ and $y = \log_2 x$? How are their graphs related?

C. Will the graph of $y = \log_b x$ ever appear in the second or third quadrants? Explain why or why not.

D. Explain why $\log_2 0 = x$ has no solution for x.

PROBLEM SET 9.3

Write each of the following expressions in logarithmic form.

1. $2^4 = 16$ **2.** $3^2 = 9$

3. $125 = 5^3$ **4.** $16 = 4^2$

5. $0.01 = 10^{-2}$ **6.** $0.001 = 10^{-3}$

7. $2^{-5} = \dfrac{1}{32}$ **8.** $4^{-2} = \dfrac{1}{16}$

9. $\left(\dfrac{1}{2}\right)^{-3} = 8$ **10.** $\left(\dfrac{1}{3}\right)^{-2} = 9$

11. $27 = 3^3$ **12.** $81 = 3^4$

Write each of the following expressions in exponential form.

13. $\log_{10} 100 = 2$ **14.** $\log_2 8 = 3$

15. $\log_2 64 = 6$ **16.** $\log_2 32 = 5$

17. $\log_8 1 = 0$ **18.** $\log_9 9 = 1$

19. $\log_{10} 0.001 = -3$ **20.** $\log_{10} 0.0001 = -4$

21. $\log_6 36 = 2$ **22.** $\log_7 49 = 2$

23. $\log_5 \dfrac{1}{25} = -2$ **24.** $\log_3 \dfrac{1}{81} = -4$

Solve each of the following equations for x.

25. $\log_3 x = 2$ **26.** $\log_4 x = 3$

27. $\log_5 x = -3$ **28.** $\log_2 x = -4$

29. $\log_2 16 = x$ **30.** $\log_3 27 = x$

31. $\log_8 2 = x$ **32.** $\log_{25} 5 = x$

33. $\log_x 4 = 2$ **34.** $\log_x 16 = 4$

35. $\log_x 5 = 3$ **36.** $\log_x 8 = 2$

Sketch the graph of each of the following logarithmic equations.

37. $y = \log_3 x$ **38.** $y = \log_{1/2} x$

39. $y = \log_{1/3} x$ **40.** $y = \log_4 x$

41. $y = \log_5 x$ **42.** $y = \log_{1/5} x$

43. $y = \log_{10} x$ **44.** $y = \log_{1/4} x$

Each of the following graphs has an equation of the form $y = b^x$ or $y = \log_b x$. Find the equation for each graph.

45.

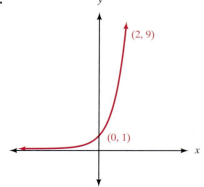

46.

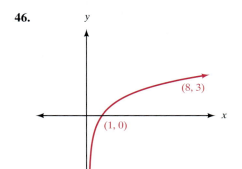

47.

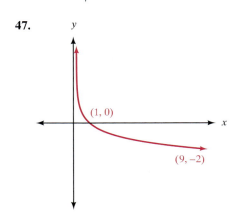

48.

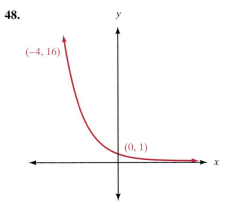

Simplify each of the following.

49. $\log_2 16$

50. $\log_3 9$

51. $\log_{25} 125$

52. $\log_9 27$

53. $\log_{10} 1{,}000$

54. $\log_{10} 10{,}000$

55. $\log_3 3$

56. $\log_4 4$

57. $\log_5 1$

58. $\log_{10} 1$

59. $\log_3(\log_6 6)$

60. $\log_5(\log_3 3)$

61. $\log_4[\log_2(\log_2 16)]$

62. $\log_4[\log_3(\log_2 8)]$

Applying the Concepts

63. Metric System The metric system uses logical and systematic prefixes for multiplication. For instance, to multiply a unit by 100, the prefix "hecto" is applied, so a hectometer is equal to 100 meters. For each of the prefixes in the following table find the logarithm, base 10, of the multiplying factor.

Prefix	Multiplying Factor	log 10 (Multiplying Factor)
Nano	0.000 000 001	
Micro	0.000 001	
Deci	0.1	
Giga	1,000,000,000	
Peta	1,000,000,000,000,000	

64. Domain and Range Use the graphs of $y = 2^x$ and $y = \log_2 x$ shown in Figure 1 of this section to find the domain and range for each function. Explain how the domain and range found for $y = 2^x$ relate to the domain and range found for $y = \log_2 x$.

65. Magnitude of an Earthquake Find the magnitude M of an earthquake with a shock wave that measures $T = 100$ on a seismograph.

66. Magnitude of an Earthquake Find the magnitude M of an earthquake with a shock wave that measures $T = 100{,}000$ on a seismograph.

67. Shock Wave If an earthquake has a magnitude of 8 on the Richter scale, how many times greater is its shock wave than the smallest shock wave measurable on a seismograph?

68. Shock Wave If the 1999 Colombia earthquake had a magnitude of 6 on the Richter scale, how many times greater was its shock wave than the smallest shock wave measurable on a seismograph?

Review Problems

The following problems review material we covered in Section 8.2.

Solve.

69. $2x^2 + 4x - 3 = 0$

70. $3x^2 + 4x - 2 = 0$

71. $(2y - 3)(2y - 1) = -4$

72. $(y - 1)(3y - 3) = 10$ **73.** $t^3 - 125 = 0$

74. $8t^3 + 1 = 0$ **75.** $4x^5 - 16x^4 = 20x^3$

76. $3x^4 + 6x^2 = 6x^3$ **77.** $\dfrac{1}{x - 3} + \dfrac{1}{x + 2} = 1$

78. $\dfrac{1}{x + 3} + \dfrac{1}{x - 2} = 1$

Extending the Concepts

79. From the graph of the exponential function $y = f(x)$, and the fact that the inverse of an exponential function is a logarithmic function, complete the following.

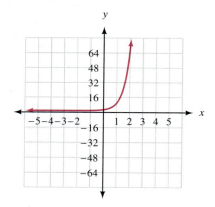

(a) Fill in the table.

x	-1	0	1	2
$f(x)$				

(b) Fill in the table.

x				
$f^{-1}(x)$	-1	0	1	2

(c) Find the equation for $f(x)$.

(d) Find the equation for $f^{-1}(x)$.

80. From the graph of the exponential function $y = f(x)$, and the fact that the inverse of an exponential function is a logarithmic function, complete the following.

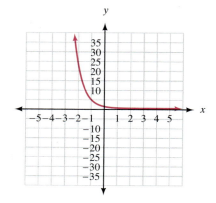

(a) Fill in the table.

x	-1	0	1	2
$f(x)$				

(b) Fill in the table.

x				
$f^{-1}(x)$	-1	0	1	2

(c) Find the equation for $f(x)$.

(d) Find the equation for $f^{-1}(x)$.

If we search for the word *decibel* in *Microsoft Bookshelf 98,* we find the following definition:

> A unit used to express relative difference in power or intensity, usually between two acoustic or electric signals, equal to ten times the common logarithm of the ratio of the two levels.

Decibels	Comparable to
10	A light whisper
20	Quiet conversation
30	Normal conversation
40	Light traffic
50	Typewriter, loud conversation
60	Noisy office
70	Normal traffic, quiet train
80	Rock music, subway
90	Heavy traffic, thunder
100	Jet plane at takeoff

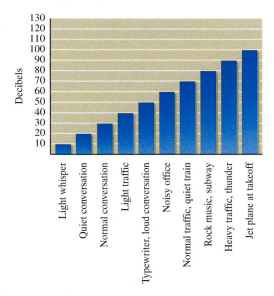

The precise definition for a *decibel* is shown below

$$D = 10 \log_{10}\left(\frac{I}{I_0}\right)$$

where I is the intensity of the sound being measured, and I_0 is the intensity of the least audible sound. (Sound intensity is related to the amplitude of the sound wave that models the sound and is given in units of watts per meter^2.) In this section we will see that the preceding formula can also be written as

$$D = 10(\log_{10} I - \log_{10} I_0)$$

The rules we use to rewrite expressions containing logarithms are called the *properties of logarithms.* There are three of them.

For the following three properties, x, y, and b are all positive real numbers, $b \neq 1$, and r is any real number.

PROPERTY I

$$\log_b(xy) = \log_b x + \log_b y$$

In words: The logarithm of a *product* is the *sum* of the logarithms.

593

PROPERTY 2

$$\log_b \left(\frac{x}{y} \right) = \log_b x - \log_b y$$

In words: The logarithm of a *quotient* is the *difference* of the logarithms.

PROPERTY 3

$$\log_b x^r = r \log_b x$$

In words: The logarithm of a number raised to a *power* is the *product* of the power and the logarithm of the number.

Proof of Property 1   To prove Property 1, we simply apply the first identity for logarithms given at the end of the preceding section:

$$b^{\log_b xy} = xy = (b^{\log_b x})(b^{\log_b y}) = b^{\log_b x + \log_b y}$$

Since the first and last expressions are equal and the bases are the same, the exponents $\log_b xy$ and $\log_b x + \log_b y$ must be equal. Therefore,

$$\log_b xy = \log_b x + \log_b y$$

The proofs of Properties 2 and 3 proceed in much the same manner, so we will omit them here. The examples that follow show how the three properties can be used.

EXAMPLE 1   Expand, using the properties of logarithms: $\log_5 \dfrac{3xy}{z}$

Solution   Applying Property 2, we can write the quotient of $3xy$ and z in terms of a difference:

$$\log_5 \frac{3xy}{z} = \log_{5,} 3xy - \log_5 z$$

Applying Property 1 to the product $3xy$, we write it in terms of addition:

$$\log_5 \frac{3xy}{z} = \log_5 3 + \log_5 x + \log_5 y - \log_5 z$$

EXAMPLE 2   Expand, using the properties of logarithms:

$$\log_2 \frac{x^4}{\sqrt{y} \cdot z^3}$$

Solution   We write $\sqrt{y}$ as $y^{1/2}$ and apply the properties:

$$\log_2 \frac{x^4}{\sqrt{y} \cdot z^3} = \log_2 \frac{x^4}{y^{1/2}z^3} \qquad\qquad \sqrt{y} = y^{1/2}$$

$$= \log_2 x^4 - \log_2(y^{1/2} \cdot z^3) \qquad \text{Property 2}$$

$$= \log_2 x^4 - (\log_2 y^{1/2} + \log_2 z^3) \qquad \text{Property 1}$$

$$= \log_2 x^4 - \log_2 y^{1/2} - \log_2 z^3 \qquad \text{Remove parentheses.}$$

$$= 4 \log_2 x - \frac{1}{2} \log_2 y - 3 \log_2 z \qquad \text{Property 3}$$

We can also use the three properties to write an expression in expanded form as just one logarithm.

EXAMPLE 3   Write as a single logarithm:

$$2 \log_{10} a + 3 \log_{10} b - \frac{1}{3} \log_{10} c$$

Solution   We begin by applying Property 3:

$$2 \log_{10} a + 3 \log_{10} b - \frac{1}{3} \log_{10} c = \log_{10} a^2 + \log_{10} b^3 - \log_{10} c^{1/3} \qquad \text{Property 3}$$

$$= \log_{10} (a^2 \cdot b^3) - \log_{10} c^{1/3} \qquad \text{Property 1}$$

$$= \log_{10} \frac{a^2 b^3}{c^{1/3}} \qquad \text{Property 2}$$

$$= \log_{10} \frac{a^2 b^3}{\sqrt[3]{c}} \qquad c^{1/3} = \sqrt[3]{c}$$

The properties of logarithms along with the definition of logarithms are useful in solving equations that involve logarithms.

EXAMPLE 4   Solve for x: $\log_2(x + 2) + \log_2 x = 3$

Solution   Applying Property 1 to the left side of the equation allows us to write it as a single logarithm:

$$\log_2(x + 2) + \log_2 x = 3$$

$$\log_2[(x + 2)(x)] = 3$$

The last line can be written in exponential form using the definition of logarithms:

$$(x + 2)(x) = 2^3$$

Solve as usual:

$$x^2 + 2x = 8$$

$$x^2 + 2x - 8 = 0$$

$$(x + 4)(x - 2) = 0$$

$$x + 4 = 0 \quad \text{or} \quad x - 2 = 0$$

$$x = -4 \quad \text{or} \quad x = 2$$

In the previous section we noted the fact that x in the expression $y = \log_b x$ cannot be a negative number. Since substitution of $x = -4$ into the original equation gives

$$\log_2(-2) + \log_2(-4) = 3$$

which contains logarithms of negative numbers, we cannot use -4 as a solution. The solution set is $\{2\}$.

Getting Ready for Class

After reading through the preceding section, respond in your own words and in complete sentences.

A. Explain why the following statement is false: "The logarithm of a product is the product of the logarithms."

B. Explain why the following statement is false: "The logarithm of a quotient is the quotient of the logarithms."

C. Explain the difference between $\log_b m + \log_b n$ and $\log_b(m + n)$. Are they equivalent?

D. Explain the difference between $\log_b(mn)$ and $(\log_b m)(\log_b n)$. Are they equivalent?

PROBLEM SET 9.4

Use the three properties of logarithms given in this section to expand each expression as much as possible.

1. $\log_3 4x$

2. $\log_2 5x$

3. $\log_6 \dfrac{5}{x}$

4. $\log_3 \dfrac{x}{5}$

5. $\log_2 y^5$

6. $\log_7 y^3$

7. $\log_9 \sqrt[3]{z}$

8. $\log_8 \sqrt{z}$

9. $\log_6 x^2 y^4$

10. $\log_{10} x^2 y^4$

11. $\log_5 \sqrt{x} \cdot y^4$

12. $\log_8 \sqrt[3]{xy^6}$

13. $\log_b \dfrac{xy}{z}$

14. $\log_b \dfrac{3x}{y}$

15. $\log_{10} \dfrac{4}{xy}$

16. $\log_{10} \dfrac{5}{4y}$

17. $\log_{10} \dfrac{x^2 y}{\sqrt{z}}$

18. $\log_{10} \dfrac{\sqrt{x} \cdot y}{z^3}$

19. $\log_{10} \dfrac{x^3 \sqrt{y}}{z^4}$

20. $\log_{10} \dfrac{x^4 \sqrt[3]{y}}{\sqrt{z}}$

21. $\log_b \sqrt[3]{\dfrac{x^2 y}{z^4}}$

22. $\log_b \sqrt[4]{\dfrac{x^4 y^3}{z^5}}$

Write each expression as a single logarithm.

23. $\log_b x + \log_b z$

24. $\log_b x - \log_b z$

25. $2 \log_3 x - 3 \log_3 y$

26. $4 \log_2 x + 5 \log_2 y$

27. $\dfrac{1}{2} \log_{10} x + \dfrac{1}{3} \log_{10} y$

28. $\dfrac{1}{3} \log_{10} x - \dfrac{1}{4} \log_{10} y$

29. $3 \log_2 x + \dfrac{1}{2} \log_2 y - \log_2 z$

30. $2 \log_3 x + 3 \log_3 y - \log_3 z$

31. $\dfrac{1}{2} \log_2 x - 3 \log_2 y - 4 \log_2 z$

32. $3 \log_{10} x - \log_{10} y - \log_{10} z$

33. $\dfrac{3}{2} \log_{10} x - \dfrac{3}{4} \log_{10} y - \dfrac{4}{5} \log_{10} z$

34. $3 \log_{10} x - \dfrac{4}{3} \log_{10} y - 5 \log_{10} z$

Solve each of the following equations.

35. $\log_2 x + \log_2 3 = 1$

36. $\log_3 x + \log_3 3 = 1$

37. $\log_3 x - \log_3 2 = 2$

38. $\log_3 x + \log_3 2 = 2$

39. $\log_3 x + \log_3(x - 2) = 1$

40. $\log_6 x + \log_6(x - 1) = 1$

41. $\log_3(x + 3) - \log_3(x - 1) = 1$

42. $\log_4(x - 2) - \log_4(x + 1) = 1$

43. $\log_2 x + \log_2(x - 2) = 3$

44. $\log_4 x + \log_4(x + 6) = 2$

45. $\log_8 x + \log_8(x - 3) = \dfrac{2}{3}$

46. $\log_{27} x + \log_{27}(x + 8) = \dfrac{2}{3}$

47. $\log_5 \sqrt{x} + \log_5 \sqrt{6x + 5} = 1$

48. $\log_2 \sqrt{x} + \log_2 \sqrt{6x + 5} = 1$

Applying the Concepts

49. Decibel Formula   Use the properties of logarithms to rewrite the decibel formula $D = 10 \log_{10}(\frac{I}{I_0})$ as $D = 10(\log_{10} I - \log_{10} I_0)$.

50. Decibel Formula   In the decibel formula $D = 10 \log_{10}(\frac{I}{I_0})$, the threshold of hearing, I_0, is

$$I_0 = 10^{-12} \text{ watts/meter}^2$$

Substitute 10^{-12} for I_0 in the decibel formula, and then show that it simplifies to

$$D = 10(\log_{10} I + 12)$$

51. Finding Logarithms   If $\log_{10} 8 = 0.903$ and $\log_{10} 5 = 0.699$, find the following without using a calculator.

(a) $\log_{10} 40$

(b) $\log_{10} 320$

(c) $\log_{10} 1{,}600$

52. Matching   Match each expression in the first column with an equivalent expression in the second column:

(a) $\log_2(ab)$ (i) b

(b) $\log_2(\frac{a}{b})$ (ii) 2

(c) $\log_5 a^b$ (iii) $\log_2 a + \log_2 b$

(d) $\log_a b^a$ (iv) $\log_2 a - \log_2 b$

(e) $\log_a a^b$ (v) $a \log_a b$

(f) $\log_3 9$ (vi) $b \log_5 a$

53. Henderson-Hasselbalch Formula   Doctors use the Henderson-Hasselbalch formula to calculate the pH of a person's blood. pH is a measure of the acidity and/or the alkalinity of a solution. We will say more about this in Section 9.5. This formula is represented as

$$\text{pH} = 6.1 + \log_{10}\left(\frac{x}{y}\right)$$

where x is the base concentration and y is the acidic concentration. Rewrite the Henderson–Hasselbalch formula so that the logarithm of a quotient is not involved.

54. Henderson-Hasselbalch Formula   Refer to the information in the preceding problem about the Henderson-Hasselbalch formula. If most people have a blood pH of 7.4, use the Henderson-Hasselbalch formula to find the ratio of x/y for an average person.

55. Food Processing The formula $M = 0.21(\log_{10} a - \log_{10} b)$ is used in the food processing industry to find the number of minutes M of heat processing a certain food should undergo at 250°F to reduce the probability of survival of *Clostridium botulinum* spores. The letter a represents the number of spores per can before heating, and b represents the number of spores per can after heating. Find M if $a = 1$ and $b = 10^{-12}$. Then find M using the same values for a and b in the formula

$$M = 0.21 \log_{10} \frac{a}{b}.$$

56. Acoustic Powers The formula $N = \log_{10} \dfrac{P_1}{P_2}$ is used in radio electronics to find the ratio of the acoustic powers of two electric circuits in terms of their electric powers. Find N if P_1 is 100 and P_2 is

1. Then use the same two values of P_1 and P_2 to find N in the formula $N = \log_{10} P_1 - \log_{10} P_2$.

Review Problems

The problems that follow review material we covered in Section 8.3.

Use the discriminant to find the number and kind of solutions to the following equations.

57. $2x^2 - 5x + 4 = 0$ **58.** $4x^2 - 12x = -9$

For each of the following problems, find an equation with the given solutions.

59. $x = -3, x = 5$

60. $x = 2, x = -2, x = 1$

61. $y = \dfrac{2}{3}, y = 3$ **62.** $y = -\dfrac{3}{5}, y = 2$

| 9.5 | **Common Logarithms and Natural Logarithms** |

Acid rain was first discovered in the 1960s by Gene Likens and his research team who studied the damage caused by acid rain to Hubbard Brook in New Hampshire. Acid rain is rain with a pH of 5.6 and below. As you will see as you work your way through this section, pH is defined in terms of common logarithms — one of the topics we present in this section. So, when you are finished with this section, you will have a more detailed knowledge of pH and acid rain.

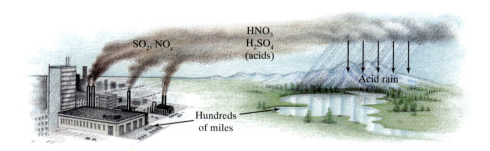

Two kinds of logarithms occur more frequently than other logarithms. Logarithms with a base of 10 are very common because our number system is a base-10 number system. For this reason, we call base-10 logarithms *common logarithms*.

> **DEFINITION**   A **common logarithm** is a logarithm with a base of 10. Since common logarithms are used so frequently, it is customary, in order to save time, to omit notating the base. That is,
>
> $$\log_{10} x = \log x$$
>
> When the base is not shown, it is assumed to be 10.

COMMON LOGARITHMS

Common logarithms of powers of 10 are very simple to evaluate. We need only recognize that $\log 10 = \log_{10} 10 = 1$ and apply the third property of logarithms: $\log_b x^r = r \log_b x$.

$$
\begin{aligned}
\log 1{,}000 &= \log 10^3 &= 3 \log 10 &= 3(1) &= 3 \\
\log 100 &= \log 10^2 &= 2 \log 10 &= 2(1) &= 2 \\
\log 10 &= \log 10^1 &= 1 \log 10 &= 1(1) &= 1 \\
\log 1 &= \log 10^0 &= 0 \log 10 &= 0(1) &= 0 \\
\log 0.1 &= \log 10^{-1} &= -1 \log 10 &= -1(1) &= -1 \\
\log 0.01 &= \log 10^{-2} &= -2 \log 10 &= -2(1) &= -2 \\
\log 0.001 &= \log 10^{-3} &= -3 \log 10 &= -3(1) &= -3
\end{aligned}
$$

To find common logarithms of numbers that are not powers of 10, we use a calculator with a $\boxed{\log}$ key or a table of logarithms. We will assume the use of a scientific calculator for the rest of this chapter.

Check the following logarithms to be sure you know how to use your calculator. (These answers have been rounded to the nearest ten-thousandth.)

$$\log 7.02 = 0.8463$$

$$\log 1.39 = 0.1430$$

$$\log 6.00 = 0.7782$$

$$\log 9.99 = 0.9996$$

EXAMPLE I   Use a calculator to find log 2,760.

Solution
$$\log 2{,}760 = 3.4409$$

To work this problem on a scientific calculator, we simply enter the number 2,760 and press the key labeled $\boxed{\log}$.

$$2{,}760 \boxed{\log}$$

The 3 in the answer is called the *characteristic,* and the decimal part of the logarithm is called the *mantissa.*

EXAMPLE 2   Find log 0.0391.

Solution $\log 0.0391 = -1.4078$

EXAMPLE 3   Find log 0.00523.

Solution $\log 0.00523 = -2.2815$

EXAMPLE 4   Find x if $\log x = 3.8774$.

Solution   We are looking for the number whose logarithm is 3.8774. On a calculator, we enter 3.8774 and press the key labeled $\boxed{10^x}$. (Sometimes it is the inverse of the $\boxed{\log}$ key.) The result is 7,540 to four significant digits.

$$\text{If} \qquad \log x = 3.8774$$
$$\text{then} \qquad x = 10^{3.8774}$$
$$= 7,540$$

The number 7,540 is called the *antilogarithm* or just *antilog* of 3.8774. That is, 7,540 is the number whose logarithm is 3.8774.

EXAMPLE 5   Find x if $\log x = -2.4179$.

Solution   We enter 2.4179, change it to a negative number with $\boxed{+/-}$, then press the $\boxed{10^x}$ key. The result is 0.00382.

$$\text{If} \qquad \log x = -2.4179$$
$$\text{then} \qquad x = 10^{-2.4179}$$
$$= 0.00382$$

The antilog of -2.4179 is 0.00382. That is, the logarithm of 0.00382 is -2.4179.

In Section 9.3 we found that the magnitude M of an earthquake that produces a shock wave T times larger than the smallest shock wave that can be measured on a seismograph is given by the formula

$$M = \log_{10} T$$

We can rewrite this formula using our shorthand notation for common logarithms as

$$M = \log T$$

EXAMPLE 6   The San Francisco earthquake of 1906 is estimated to have measured 8.3 on the Richter scale. The San Fernando earthquake of 1971 measured 6.6 on the Richter scale. Find T for each earthquake, and then give some indication of how much stronger the 1906 earthquake was than the 1971 earthquake.

Solution   For the 1906 earthquake:

$$\text{If } \log T = 8.3, \text{ then } T = 2.00 \times 10^8.$$

For the 1971 earthquake:

$$\text{If } \log T = 6.6, \text{ then } T = 3.98 \times 10^6.$$

Dividing the two values of T and rounding our answer to the nearest whole number, we have

$$\frac{2.00 \times 10^8}{3.98 \times 10^6} = 50$$

The shock wave for the 1906 earthquake was approximately 50 times larger than the shock wave for the 1971 earthquake.

8.3

In chemistry, the pH of a solution is the measure of the acidity of the solution. The definition for pH involves common logarithms. Here it is:

$$\text{pH} = -\log[\text{H}^+]$$

where $[\text{H}^+]$ is the concentration of the hydrogen ion in moles per liter. The range for pH is from 0 to 14. Pure water, a neutral solution, has a pH of 7. An acidic solution, such as vinegar, will have a pH less than 7, and an alkaline solution, such as ammonia, has a pH above 7.

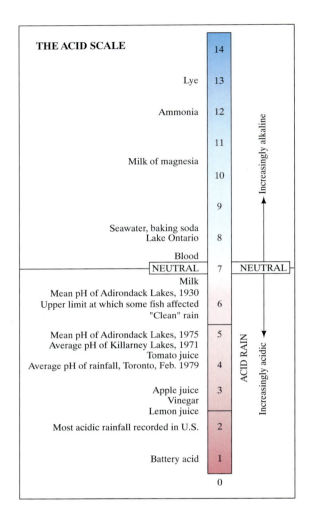

EXAMPLE 7   Normal rainwater has a pH of 5.6. What is the concentration of the hydrogen ion in normal rainwater?

Solution   Substituting 5.6 for pH in the formula $pH = -\log[H^+]$, we have

$$5.6 = -\log[H^+] \qquad \text{Substitution}$$

$$\log[H^+] = -5.6 \qquad \text{Isolate the logarithm.}$$

$$[H^+] = 10^{-5.6} \qquad \text{Write in exponential form.}$$

$$\approx 2.5 \times 10^{-6} \text{ moles per liter} \qquad \text{Answer in scientific notation.}$$

EXAMPLE 8   The concentration of the hydrogen ion in a sample of acid rain known to kill fish is 3.2×10^{-5} mole per liter. Find the pH of this acid rain to the nearest tenth.

Solution   Substituting 3.2×10^{-5} for $[H^+]$ in the formula $pH = -\log[H^+]$, we have

$$pH = -\log[3.2 \times 10^{-5}] \qquad \text{Substitution}$$
$$\approx -(-4.5) \qquad \text{Evaluate the logarithm.}$$
$$\approx 4.5 \qquad \text{Simplify.}$$

NATURAL LOGARITHMS

> **DEFINITION**   A **natural logarithm** is a logarithm with a base of e. The natural logarithm of x is denoted by $\ln x$. That is,
>
> $$\ln x = \log_e x$$

We can assume that all our properties of exponents and logarithms hold for expressions with a base of e, since e is a real number. Here are some examples intended to make you more familiar with the number e and natural logarithms.

EXAMPLE 9   Simplify each of the following expressions.

(a) $e^0 = 1$

(b) $e^1 = e$

(c) $\ln e = 1$ In exponential form, $e^1 = e$.

(d) $\ln 1 = 0$ In exponential form, $e^0 = 1$.

(e) $\ln e^3 = 3$

(f) $\ln e^{-4} = -4$

(g) $\ln e^t = t$

EXAMPLE 10   Use the properties of logarithms to expand the expression $\ln Ae^{5t}$.

Solution   Since the properties of logarithms hold for natural logarithms, we have

$$\ln Ae^{5t} = \ln A + \ln e^{5t}$$
$$= \ln A + 5t \ln e$$
$$= \ln A + 5t \qquad \text{Because } \ln e = 1$$

EXAMPLE 11   If ln 2 = 0.6931 and ln 3 = 1.0986, find

(a) ln 6   (b) ln 0.5   (c) ln 8

Solution

(a) Since 6 = 2·3, we have

$$\ln 6 = \ln 2 \cdot 3$$
$$= \ln 2 + \ln 3$$
$$= 0.6931 + 1.0986$$
$$= 1.7917$$

(b) Writing 0.5 as $\frac{1}{2}$ and applying Property 2 for logarithms gives us

$$\ln 0.5 = \ln \frac{1}{2}$$
$$= \ln 1 - \ln 2$$
$$= 0 - 0.6931$$
$$= -0.6931$$

(c) Writing 8 as 2^3 and applying Property 3 for logarithms, we have

$$\ln 8 = \ln 2^3$$
$$= 3 \ln 2$$
$$= 3(0.6931)$$
$$= 2.0793$$

Getting Ready for Class

After reading through the preceding section, respond in your own words and in complete sentences.

A. What is a common logarithm?

B. What is a natural logarithm?

C. Is *e* a rational number? Explain.

D. Find ln *e*, and explain how you arrived at your answer.

PROBLEM SET 9.5

Find the following logarithms.

1. log 378
2. log 426
3. log 37.8
4. log 42,600
5. log 3,780
6. log 0.4260
7. log 0.0378
8. log 0.0426
9. log 37,800
10. log 4,900

11. $\log 600$

12. $\log 900$

13. $\log 2{,}010$

14. $\log 10{,}200$

15. $\log 0.00971$

16. $\log 0.0312$

17. $\log 0.0314$

18. $\log 0.00052$

19. $\log 0.399$

20. $\log 0.111$

Find x in the following equations.

21. $\log x = 2.8802$

22. $\log x = 4.8802$

23. $\log x = -2.1198$

24. $\log x = -3.1198$

25. $\log x = 3.1553$

26. $\log x = 5.5911$

27. $\log x = -5.3497$

28. $\log x = -1.5670$

29. $\log x = -7.0372$

30. $\log x = -4.2000$

31. $\log x = 10$

32. $\log x = -1$

33. $\log x = -10$

34. $\log x = 1$

35. $\log x = 20$

36. $\log x = -20$

37. $\log x = -2$

38. $\log x = 4$

39. $\log x = \log_2 8$

40. $\log x = \log_3 9$

41. Solve the following for x: $\log x^2 = (\log x)^2$. (*Hint:* Use properties of logarithms and factor.)

42. Solve the following for x: $\ln x^2 = (\ln x)^2$.

Applying the Concepts

43. **Interpreting Calculator Results** Use your calculator to find $\log(-10)$. Explain the result your calculator gives you.

44. **Interpreting Calculator Results** Use your calculator to find $\ln 0$. Explain the result your calculator gives you.

45. **Atomic Bomb Tests** The Bikini Atoll in the Pacific Ocean was used as a location for atomic bomb tests by the United States government in the 1950s. One such test resulted in an earthquake measurement of 5.0 on the Richter scale. Compare the 1906 San Francisco earthquake of estimated magnitude 8.3 on the Richter scale to this atomic bomb test. Use the shock wave T for purposes of comparison.

46. **Atomic Bomb Tests** Today's nuclear weapons are 1,000 times more powerful than the atomic bombs tested in the Bikini Atoll mentioned in Problem 45. Use the shock wave T to determine the Richter scale measurement of a nuclear test today.

47. **Getting Close to** e Use a calculator to complete the following table.

x	$(1 + x)^{1/x}$
1	
0.5	
0.1	
0.01	
0.001	
0.0001	
0.00001	

What number does the expression $(1 + x)^{1/x}$ seem to approach as x gets closer and closer to zero?

48. **Getting Close to** e Use a calculator to complete the following table.

x	$(1 + \frac{1}{x})^{x}$
1	
10	
50	
100	
500	
1,000	
10,000	
1,000,000	

What number does the expression $(1 + \frac{1}{x})^{x}$ seem to approach as x gets larger and larger?

Kepler's Law Johannes Kepler (German astronomer, 1571–1630) found that for every planet in the solar system, the following relationship holds: If R is the radius of the orbit of a planet and T is its period (the time for one complete revolution around the sun), then the quotient R^3/T^2 is a constant.

49. Show that the expression $3 \log R - 2 \log T$ must also be constant if R^3/T^2 is a constant.

50. The same relationship holds for satellites orbiting planets: The quantity R^3/T^2 is a constant. Table 1 gives the orbit radius and the period for five of the satellites of Jupiter. Verify that the expression

$3 \log R - 2 \log T$ is a constant for the satellites of Jupiter.

(JPL/NASA)

TABLE I Satellites of Jupiter		
Name	**Radius of the Orbit (mi)**	**Period (h)**
V	112,000	11.96
I (Io)	262,000	42.46
II (Europa)	417,000	85.22
III (Ganymede)	666,000	171.71
IV (Callisto)	1,170,000	400.54

Use the following figure to solve Problems 51–54.

pH Scale

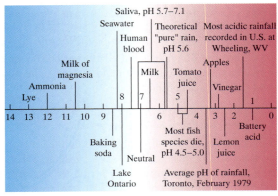

51. pH Find the pH of orange juice if the concentration of the hydrogen ion in the juice is $[H^+] = 6.50 \times 10^{-4}$.

52. pH Find the pH of milk if the concentration of the hydrogen ions in milk is $[H^+] = 1.88 \times 10^{-6}$.

53. pH Find the concentration of hydrogen ions in a glass of wine if the pH is 4.75.

54. pH Find the concentration of hydrogen ions in a bottle of vinegar if the pH is 5.75.

The Richter Scale Find the relative size T of the shock wave of earthquakes with the following magnitudes, as measured on the Richter scale.

55. 5.5   **56.** 6.6   **57.** 8.3   **58.** 8.7

59. Shock Wave How much larger is the shock wave of an earthquake that measures 6.5 on the Richter scale than one that measures 5.5 on the same scale?

60. Shock Wave How much larger is the shock wave of an earthquake that measures 8.5 on the Richter scale than one that measures 5.5 on the same scale?

Depreciation The annual rate of depreciation r on a car that is purchased for P dollars and is worth W dollars t years later can be found from the formula

$$\log(1 - r) = \frac{1}{t} \log \frac{W}{P}$$

61. Find the annual rate of depreciation on a car that is purchased for $9,000 and sold 5 years later for $4,500.

62. Find the annual rate of depreciation on a car that is purchased for $9,000 and sold 4 years later for $3,000.

Two cars depreciate in value according to the following depreciation tables. In each case, find the annual rate of depreciation.

63.

Age in Years	Value in Dollars
New	7,550
5	5,750

64.

Age in Years	Value in Dollars
New	7,550
3	5,750

Simplify each of the following expressions.

65. $\ln e$

66. $\ln 1$

67. $\ln e^5$

68. $\ln e^{-3}$

69. $\ln e^x$

70. $\ln e^y$

Use the properties of logarithms to expand each of the following expressions.

71. $\ln 10e^{3t}$

72. $\ln 10e^{4t}$

73. $\ln Ae^{-2t}$

74. $\ln Ae^{-3t}$

If $\ln 2 = 0.6931$, $\ln 3 = 1.0986$, and $\ln 5 = 1.6094$, find each of the following.

75. $\ln 15$

76. $\ln 10$

77. $\ln \dfrac{1}{3}$

78. $\ln \dfrac{1}{5}$

79. $\ln 9$

80. $\ln 25$

81. $\ln 16$

82. $\ln 81$

Review Problems

The following problems review material we covered in Section 8.4.

Solve each equation.

83. $x^4 - 2x^2 - 8 = 0$

84. $x^4 - 8x^2 - 9 = 0$

85. $x^{2/3} - 5x^{1/3} + 6 = 0$

86. $x^{2/3} - 3x^{1/3} + 2 = 0$

87. $2x - 5\sqrt{x} + 3 = 0$

88. $3x - 8\sqrt{x} + 4 = 0$

89. $(3x + 1) - 6\sqrt{3x + 1} + 8 = 0$

90. $(2x - 1) - 2\sqrt{2x - 1} - 15 = 0$

91. Solve $kx^2 + 4x - k = 0$ for x.

92. Solve $4x^2 - 4x + k = 0$ for x.

9.6 Exponential Equations and Change of Base

For items involved in exponential growth, the time it takes for a quantity to double is called the *doubling time.* For example, if you invest $5,000 in an account that pays 5% annual interest, compounded quarterly, you may want to know how long it will take for your money to double in value. You can find this doubling time if you can solve the equation

$$10,000 = 5,000(1.015)^{4t}$$

As you will see as you progress through this section, logarithms are the key to solving equations of this type.

Logarithms are very important in solving equations in which the variable appears as an exponent. The equation

$$5^x = 12$$

is an example of one such equation. Equations of this form are called *exponential equations.* Since the quantities 5^x and 12 are equal, so are their common logarithms. We begin our solution by taking the logarithm of both sides:

$$\log 5^x = \log 12$$

We now apply Property 3 for logarithms, $\log x^r = r \log x$, to turn x from an exponent into a coefficient:

$$x \log 5 = \log 12$$

Dividing both sides by $\log 5$ gives us

$$x = \frac{\log 12}{\log 5}$$

If we want a decimal approximation to the solution, we can find log 12 and log 5 on a calculator and divide:

$$x = \frac{1.0792}{0.6990}$$

$$= 1.5439$$

The complete problem looks like this:

$$5^x = 12$$

$$\log 5^x = \log 12$$

$$x \log 5 = \log 12$$

$$x = \frac{\log 12}{\log 5}$$

$$= \frac{1.0792}{0.6990}$$

$$= 1.5439$$

Here is another example of solving an exponential equation using logarithms.

EXAMPLE 1   Solve for x: $25^{2x+1} = 15$

Solution   Taking the logarithm of both sides and then writing the exponent $(2x + 1)$ as a coefficient, we proceed as follows:

$$25^{2x+1} = 15$$

$$\log 25^{2x+1} = \log 15 \qquad \text{Take the log of both sides.}$$

$$(2x + 1)\log 25 = \log 15 \qquad \text{Property 3}$$

$$2x + 1 = \frac{\log 15}{\log 25} \qquad \text{Divide by log 25.}$$

$$2x = \frac{\log 15}{\log 25} - 1 \qquad \text{Add } -1 \text{ to both sides.}$$

$$x = \frac{1}{2}\left(\frac{\log 15}{\log 25} - 1\right) \qquad \text{Multiply both sides by } \frac{1}{2}.$$

Using a calculator, we can write a decimal approximation to the answer:

$$x = \frac{1}{2}\left(\frac{1.1761}{1.3979} - 1\right)$$

$$= \frac{1}{2}(0.8413 - 1)$$

$$= \frac{1}{2}(-0.1587)$$

$$= -0.0794$$

If you invest P dollars in an account with an annual interest rate r that is compounded n times a year, then t years later the amount of money in that account will be

$$A = P\left(1 + \frac{r}{n}\right)^{nt}$$

EXAMPLE 2 How long does it take for $5,000 to double if it is deposited in an account that yields 5% interest compounded once a year?

Solution Substituting $P = 5,000$, $r = 0.05$, $n = 1$, and $A = 10,000$ into our formula, we have

$$10,000 = 5,000(1 + 0.05)^t$$

$$10,000 = 5,000(1.05)^t$$

$$2 = (1.05)^t \qquad \text{Divide by 5,000.}$$

This is an exponential equation. We solve by taking the logarithm of both sides:

$$\log 2 = \log(1.05)^t$$

$$= t \log 1.05$$

Dividing both sides by $\log 1.05$, we have

$$t = \frac{\log 2}{\log 1.05}$$

$$= 14.2 \text{ to the nearest tenth}$$

It takes a little over 14 years for $5,000 to double if it earns 5% interest per year, compounded once a year.

There is a fourth property of logarithms we have not yet considered. This last property allows us to change from one base to another and is therefore called the *change-of-base property.*

PROPERTY 4 (CHANGE OF BASE)

If a and b are both positive numbers other than 1, and if $x > 0$, then

$$\underset{\underset{\text{Base } a}{\uparrow}}{\log_a x} = \frac{\overset{}{\log_b x}}{\underset{\underset{\text{Base } b}{\uparrow}}{\log_b a}}$$

The logarithm on the left side has a base of a, and both logarithms on the right side have a base of b. This allows us to change from base a to any other base b that is a positive number other than 1. Here is a proof of Property 4 for logarithms.

Proof   We begin by writing the identity

$$a^{\log_a x} = x$$

Taking the logarithm base b of both sides and writing the exponent $\log_a x$ as a coefficient, we have

$$\log_b a^{\log_a x} = \log_b x$$

$$\log_a x \log_b a = \log_b x$$

Dividing both sides by $\log_b a$, we have the desired result:

$$\frac{\log_a x \log_b a}{\log_b a} = \frac{\log_b x}{\log_b a}$$

$$\log_a x = \frac{\log_b x}{\log_b a}$$

We can use this property to find logarithms we could not otherwise compute on our calculators — that is, logarithms with bases other than 10 or e. The next example illustrates the use of this property.

EXAMPLE 3   Find $\log_8 24$.

Solution   Since we do not have base-8 logarithms on our calculators, we can change this expression to an equivalent expression that contains only base-10 logarithms:

$$\log_8 24 = \frac{\log 24}{\log 8} \qquad \text{Property 4}$$

Don't be confused. We did not just drop the base, we changed to base 10. We could have written the last line like this:

$$\log_8 24 = \frac{\log_{10} 24}{\log_{10} 8}$$

From our calculators, we write

$$\log_8 24 = \frac{1.3802}{0.9031}$$

$$= 1.5283$$

Here is the complete calculator solution to Example 3:

$$24 \;\boxed{\log}\; \boxed{\div}\; 8 \;\boxed{\log}\; \boxed{=}$$

APPLICATION

EXAMPLE 4   Suppose that the population in a small city is 32,000 in the beginning of 1994 and that the city council assumes that the population size t

years later can be estimated by the equation

$$P = 32{,}000e^{0.05t}$$

Approximately when will the city have a population of 50,000?

Solution   We substitute 50,000 for P in the equation and solve for t:

$$50{,}000 = 32{,}000e^{0.05t}$$

$$1.56 = e^{0.05t} \qquad \frac{50{,}000}{32{,}000} \text{ is approximately 1.56.}$$

To solve this equation for t, we can take the natural logarithm of each side:

$$
\begin{aligned}
\ln 1.56 &= \ln e^{0.05t} \\
&= 0.05t \ln e && \text{Property 3 for logarithms} \\
&= 0.05t && \text{Because } \ln e = 1 \\
t &= \frac{\ln 1.56}{0.05} && \text{Divide each side by 0.05.} \\
&= \frac{0.4447}{0.05} \\
&= 8.89 \text{ years}
\end{aligned}
$$

We can estimate that the population will reach 50,000 toward the end of 2002.

 Using TECHNOLOGY

GRAPHING CALCULATOR

We can evaluate many logarithmic expressions on a graphing calculator by using the fact that logarithmic functions and exponential functions are inverses.

(continued)

(Using Technology, continued)

EXAMPLE 5   Evaluate the logarithmic expression $\log_3 7$ from the graph of an exponential function.

Solution   First, we let $\log_3 7 = x$. Next, we write this expression in exponential form as $3^x = 7$. We can solve this equation graphically by finding the intersection of the graphs $Y_1 = 3^x$ and $Y_2 = 7$, as shown in Figure 1.

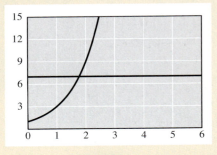

FIGURE 1

Using the calculator, we find the two graphs intersect at $(1.77, 7)$. Therefore, $\log_3 7 = 1.77$ to the nearest hundredth. We can check our work by evaluating the expression $3^{1.77}$ on our calculator with the key strokes

$$3 \;\boxed{\wedge}\; 1.77 \;\boxed{\text{ENTER}}$$

The result is 6.99 to the nearest hundredth, which seems reasonable since 1.77 is accurate to the nearest hundredth. To get a result closer to 7, we would need to find the intersection of the two graphs more accurately.

Getting Ready for Class

After reading through the preceding section, respond in your own words and in complete sentences.

A. What is an exponential equation?

B. How do logarithms help you solve exponential equations?

C. What is the change-of-base property?

D. Write an application modeled by the equation $A = 10{,}000 \left(1 + \dfrac{0.08}{2}\right)^{2 \cdot 5}$.

PROBLEM SET 9.6

Solve each exponential equation. Use a calculator to write the answer in decimal form.

1. $3^x = 5$
2. $4^x = 3$
3. $5^x = 3$
4. $3^x = 4$
5. $5^{-x} = 12$
6. $7^{-x} = 8$
7. $12^{-x} = 5$
8. $8^{-x} = 7$
9. $8^{x+1} = 4$
10. $9^{x+1} = 3$
11. $4^{x-1} = 4$
12. $3^{x-1} = 9$
13. $3^{2x+1} = 2$
14. $2^{2x+1} = 3$
15. $3^{1-2x} = 2$
16. $2^{1-2x} = 3$
17. $15^{3x-4} = 10$
18. $10^{3x-4} = 15$
19. $6^{5-2x} = 4$
20. $9^{7-3x} = 5$

Applying the Concepts

Use the change-of-base property and a calculator to find a decimal approximation to each of the following logarithms.

21. $\log_8 16$
22. $\log_9 27$
23. $\log_{16} 8$
24. $\log_{27} 9$
25. $\log_7 15$
26. $\log_3 12$
27. $\log_{15} 7$
28. $\log_{12} 3$
29. $\log_8 240$
30. $\log_6 180$
31. $\log_4 321$
32. $\log_5 462$

Find a decimal approximation to each of the following natural logarithms.

33. $\ln 345$
34. $\ln 3,450$
35. $\ln 0.345$
36. $\ln 0.0345$
37. $\ln 10$
38. $\ln 100$
39. $\ln 45,000$
40. $\ln 450,000$

41. **Compound Interest** How long will it take for $500 to double if it is invested at 6% annual interest compounded 2 times a year?

42. **Compound Interest** How long will it take for $500 to double if it is invested at 6% annual interest compounded 12 times a year?

43. **Compound Interest** How long will it take for $1,000 to triple if it is invested at 12% annual interest compounded 6 times a year?

44. **Compound Interest** How long will it take for $1,000 to become $4,000 if it is invested at 12% annual interest compounded 6 times a year?

45. **Doubling Time** How long does it take for an amount of money P to double itself if it is invested at 8% interest compounded 4 times a year?

46. **Tripling Time** How long does it take for an amount of money P to triple itself if it is invested at 8% interest compounded 4 times a year?

47. **Tripling Time** If a $25 investment is worth $75 today, how long ago must that $25 have been invested at 6% interest computed twice a year?

48. **Doubling Time** If a $25 investment is worth $50 today, how long ago must that $25 have been invested at 6% interest computed twice a year?

Recall from Section 9.1 that if P dollars are invested in an account with annual interest rate r, compounded continuously, then the amount of money in the account after t years is given by the formula

$$A(t) = Pe^{rt}$$

49. **Continuously Compounded Interest** Repeat Problem 41 if the interest is compounded continuously.

50. **Continuously Compounded Interest** Repeat Problem 44 if the interest is compounded continuously.

51. **Continuously Compounded Interest** How long will it take $500 to triple if it is invested at 6% annual interest, compounded continuously?

52. **Continuously Compounded Interest** How long will it take $500 to triple if it is invested at 12% annual interest, compounded continuously?

53. **Continuously Compounded Interest** How long will it take for $1,000 to be worth $2,500 at 8% interest, compounded continuously?

54. **Continuously Compounded Interest** How long will it take for $1,000 to be worth $5,000 at 8% interest, compounded continuously?

55. **Exponential Growth** Suppose that the population in a small city is 32,000 at the beginning of 1994 and that the city council assumes that the population size t years later can be estimated by the equation

$$P(t) = 32,000e^{0.05t}$$

Approximately when will the city have a population of 64,000?

56. **Exponential Growth** Suppose the population of a city is given by the equation

$$P(t) = 100,000e^{0.05t}$$

where t is the number of years from the present time. How large is the population now? (*Now* corresponds to a certain value of t. Once you realize what that value of t is, the problem becomes very simple.)

57. **Exponential Growth** Suppose the population of a city is given by the equation

$$P(t) = 15{,}000e^{0.04t}$$

where t is the number of years from the present time. How long will it take for the population to reach 45,000?

58. **Exponential Growth** Suppose the population of a city is given by the equation

$$P(t) = 15{,}000e^{0.08t}$$

where t is the number of years from the present time. How long will it take for the population to reach 45,000?

Review Problems

The following problems review material we covered in Section 8.5.

Find the vertex for each of the following parabolas, and then indicate if it is the highest or lowest point on the graph.

59. $y = 2x^2 + 8x - 15$ 60. $y = 3x^2 - 9x - 10$

61. $y = 12x - 4x^2$ 62. $y = 18x - 6x^2$

63. **Maximum Height** An object is projected into the air with an initial upward velocity of 64 feet per second. Its height h at any time t is given by the formula $h = 64t - 16t^2$. Find the time at which

the object reaches its maximum height. Then, find the maximum height.

64. **Maximum Height** An object is projected into the air with an initial upward velocity of 64 feet per second from the top of a building 40 feet high. If the height h of the object t seconds after it is projected into the air is $h = 40 + 64t - 16t^2$, find the time at which the object reaches its maximum height. Then, find the maximum height it attains.

Extending the Concepts

65. Solve the formula $A = Pe^{rt}$ for t.

66. Solve the formula $A = Pe^{-rt}$ for t.

67. Solve the formula $A = P2^{-kt}$ for t.

68. Solve the formula $A = P2^{kt}$ for t.

69. Solve the formula $A = P(1 - r)^t$ for t.

70. Solve the formula $A = P(1 + r)^t$ for t.

Use Example 5 as a model to evaluate the following logarithmic expressions by using the graph of an exponential function.

71. $\log_6 23$ 72. $\log_4 14$

73. $\log_7 29$ 74. $\log_5 34$

75. In Problem 1 you solved $3^x = 5$ algebraically. Now graph $y = 3^x$ and $y = 5$, and approximate their points of intersection. How do these points of intersection compare to your answer in Problem 1?

76. Now try approximating a solution(s) to $x^2 = 2^x$, which cannot be solved with traditional algebraic techniques (other than "guess and check"). Graph $y = x^2$ and $y = 2^x$, and approximate the points of intersection. What would be the estimate of your solution(s) to $x^2 = 2^x$?

CHAPTER 9   SUMMARY

Examples

1. For the exponential function
$f(x) = 2^x$,

$$f(0) = 2^0 = 1$$

$$f(1) = 2^1 = 2$$

$$f(2) = 2^2 = 4$$

$$f(3) = 2^3 = 8$$

Exponential Functions [9.1]

Any function of the form

$$f(x) = b^x$$

where $b > 0$ and $b \neq 1$, is an **exponential function.**

2. The function $f(x) = x^2$ is not one-to-one because 9, which is in the range, comes from both 3 and -3 in the domain.

One-to-One Functions [9.2]

A function is a **one-to-one function** if every element in the range comes from exactly one element in the domain.

3. The inverse of $f(x) = 2x - 3$ is

$$f^{-1}(x) = \frac{x + 3}{2}$$

Inverse Functions [9.2]

The **inverse** of a function is obtained by reversing the order of the coordinates of the ordered pairs belonging to the function. Only one-to-one functions have inverses that are also functions.

4. The definition allows us to write expressions like

$$y = \log_3 27$$

equivalently in exponential form as

$$3^y = 27$$

which makes it apparent that y is 3.

Definition of Logarithms [9.3]

If b is a positive number not equal to 1, then the expression

$$y = \log_b x$$

is equivalent to $x = b^y$. That is, in the expression $y = \log_b x$, y is the number to which we raise b in order to get x. Expressions written in the form $y = \log_b x$ are said to be in *logarithmic form.* Expressions like $x = b^y$ are in *exponential form.*

5. Examples of the two special identities are

$$5^{\log_5 12} = 12$$

and

$$\log_8 8^3 = 3$$

Two Special Identities [9.3]

For $b > 0$, $b \neq 1$, the following two expressions hold for all positive real numbers x:

$$(1)\ b^{\log_b x} = x$$

$$(2)\ \log_b b^x = x$$

6. We can rewrite the expression

$$\log_{10} \frac{45^6}{273}$$

using the properties of logarithms, as

$$6 \log_{10} 45 - \log_{10} 273$$

7.
$$\log_{10} 10,000 = \log 10,000$$
$$= \log 10^4$$
$$= 4$$

8.
$$\ln e = 1$$
$$\ln 1 = 0$$

9.
$$\log_6 475 = \frac{\log 475}{\log 6}$$
$$= \frac{2.6767}{0.7782}$$
$$= 3.44$$

Properties of Logarithms [9.4]

If x, y, and b are positive real numbers, $b \neq 1$, and r is any real number, then

1. $\log_b(xy) = \log_b x + \log_b y$

2. $\log_b \left(\dfrac{x}{y} \right) = \log_b x - \log_b y$

3. $\log_b x^r = r \log_b x$

Common Logarithms [9.5]

Common logarithms are logarithms with a base of 10. To save time in writing, we omit the base when working with common logarithms. That is,

$$\log x = \log_{10} x$$

Natural Logarithms [9.5]

Natural logarithms, written **ln x**, are logarithms with a base of e, where the number e is an irrational number (like the number π). A decimal approximation for e is 2.7183. All the properties of exponents and logarithms hold when the base is e.

Change of Base [9.6]

If x, a, and b are positive real numbers, $a \neq 1$ and $b \neq 1$, then

$$\log_a x = \frac{\log_b x}{\log_b a}$$

COMMON MISTAKES

The most common mistakes that occur with logarithms come from trying to apply the three properties of logarithms to situations in which they don't apply. For example, a very common mistake looks like this:

$$\frac{\log 3}{\log 2} = \log 3 - \log 2 \qquad \text{Mistake}$$

This is not a property of logarithms. In order to write the expression $\log 3 - \log 2$, we would have to start with

$$\log \frac{3}{2} \qquad NOT \qquad \frac{\log 3}{\log 2}$$

There is a difference.

CHAPTER 9 REVIEW

Let $f(x) = 2^x$ and $g(x) = \left(\frac{1}{3}\right)^x$, and find the following. [9.1]

1. $f(4)$
2. $f(-1)$
3. $g(2)$
4. $f(2) - g(-2)$
5. $f(-1) + g(1)$
6. $g(-1) + f(2)$
7. The graph of $y = f(x)$
8. The graph of $y = g(x)$

For each relation that follows, sketch the graph of the relation and its inverse, and write an equation for the inverse. [9.2]

9. $y = 2x + 1$
10. $y = x^2 - 4$

For each of the following functions, find the equation of the inverse. Write the inverse using the notation $f^{-1}(x)$ if the inverse is itself a function. [9.2]

11. $f(x) = 2x + 3$
12. $f(x) = x^2 - 1$
13. $f(x) = \frac{1}{2}x + 2$
14. $f(x) = 4 - 2x^2$

Write each expression in logarithmic form. [9.3]

15. $3^4 = 81$
16. $7^2 = 49$
17. $0.01 = 10^{-2}$
18. $2^{-3} = \frac{1}{8}$

Write each expression in exponential form. [9.3]

19. $\log_2 8 = 3$
20. $\log_3 9 = 2$
21. $\log_4 2 = \frac{1}{2}$
22. $\log_4 4 = 1$

Solve for x. [9.3]

23. $\log_5 x = 2$
24. $\log_{16} 8 = x$
25. $\log_x 0.01 = -2$

Graph each equation. [9.3]

26. $y = \log_2 x$
27. $y = \log_{1/2} x$

Simplify each expression. [9.3]

28. $\log_4 16$
29. $\log_{27} 9$
30. $\log_4(\log_3 3)$

Use the properties of logarithms to expand each expression. [9.4]

31. $\log_2 5x$
32. $\log_{10} \dfrac{2x}{y}$

33. $\log_a \dfrac{y^3\sqrt{x}}{z}$
34. $\log_{10} \dfrac{x^2}{y^3 z^4}$

Write each expression as a single logarithm. [9.4]

35. $\log_2 x + \log_2 y$
36. $\log_3 x - \log_3 4$
37. $2\log_a 5 - \frac{1}{2}\log_a 9$
38. $3\log_2 x + 2\log_2 y - 4\log_2 z$

Solve each equation. [9.4]

39. $\log_2 x + \log_2 4 = 3$
40. $\log_2 x - \log_2 3 = 1$
41. $\log_3 x + \log_3(x - 2) = 1$
42. $\log_4(x + 1) - \log_4(x - 2) = 1$
43. $\log_6(x - 1) + \log_6 x = 1$
44. $\log_4(x - 3) + \log_4 x = 1$

Evaluate each expression. [9.5]

45. $\log 346$
46. $\log 0.713$

Find x. [9.5]

47. $\log x = 3.9652$
48. $\log x = -1.6003$

Simplify. [9.5]

49. $\ln e$
50. $\ln 1$
51. $\ln e^2$
52. $\ln e^{-4}$

Use the formula $pH = -\log[H^+]$ to find the pH of a solution with the given hydrogen ion concentration. [9.5]

53. $[H^+] = 7.9 \times 10^{-3}$
54. $[H^+] = 8.1 \times 10^{-6}$

Find $[H^+]$ for a solution with the given pH. [9.5]

55. $pH = 2.7$
56. $pH = 7.5$

Solve each equation. [9.6]

57. $4^x = 8$
58. $4^{3x+2} = 5$

Use the change-of-base property and a calculator to evaluate each expression. Round your answers to the nearest hundredth. [9.6]

59. $\log_{16} 8$
60. $\log_{12} 421$

Use the formula $A = P\left(1 + \dfrac{r}{n}\right)^{nt}$ to solve each of the following problems. [9.6]

61. Investing   How long does it take $5,000 to double if it is deposited in an account that pays 16% annual interest compounded once a year?

62. Investing   How long does it take $10,000 to triple if it is deposited in an account that pays 12% annual interest compounded 6 times a year?

If the current price of an item is P_0 dollars and the annual inflation rate is r, then in t years the price of the item will be $P(t) = P_0(1 + r)^t$ dollars. [9.1]

63. Inflation   If the price of a new home is currently $100,000, what will it sell for in 8 years if the annual inflation rate is 4%?

64. Inflation   Suppose tuition at a college in Ohio is currently $1,980 per year. If the inflation rate is a constant 3% per year, what will tuition at that school cost in 10 years?

CHAPTER 9   PROJECTS

EXPONENTIAL AND LOGARITHMIC FUNCTIONS

GROUP PROJECT — TWO DEPRECIATION MODELS

Number of People: 3

Time Needed: 20 minutes

Equipment: Paper, pencils, and graphing calculator

Background: Recently, one of the consumer magazines contained an article on leasing a computer. The original price of the computer was $2,500. The term of the lease was 24 months. At the end of the lease, the computer could be purchased for its residual value, $188. This is enough information to find an equation that will give us the value of the computer t months after it has been purchased.

Procedure: We will find models for two types of depreciation: linear depreciation and exponential depreciation. Here are the general equations:

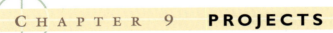

$$\textit{Linear Depreciation} \qquad \textit{Exponential Depreciation}$$
$$V = mt + b \qquad\qquad V = V_0 e^{-kt}$$

1. Let t represent time in months and V represent the value of the computer at time t; then find two ordered pairs (t, V) that corresponds to the initial price of the computer, and another ordered pair that corresponds to the residual value of $188 when the computer is 2 years old.

2. Use the two ordered pairs to find m and b in the linear depreciation model; then write the equation that gives us linear depreciation.

3. Use the two ordered pairs to find V_0 and k in the exponential depreciation model; then write the equation that gives us exponential depreciation.

4. Graph each of the equations on your graphing calculator; then sketch the graphs on the following templates.

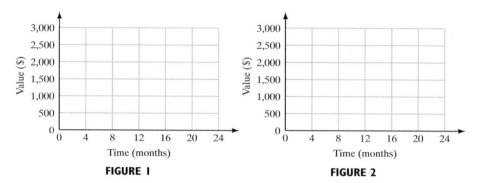

FIGURE 1   FIGURE 2

5. Find the value of the computer after 1 year, using both models.

6. Which of the two models do you think best describes the depreciation of a computer?

RESEARCH PROJECT

DRAG RACING

(The Kobal Collection/20th Century Fox)

The movie *Heart Like a Wheel* is based on the racing career of drag racer Shirley Muldowney. The movie includes a number of races. Choose four races as the basis for your report. For each race you choose, give a written description of the events leading to the race, along with the details of the race itself. Then draw a graph similar to the one for Problem 42 in Section 9.1. (For each graph, label the horizontal axis from 0 to 12 seconds, and the vertical axis from 0 to 200 miles per hour.) Follow each graph with a description of any significant events that happen during the race (such as a motor malfunction or a crash) and their correlation to the graph.

CHAPTER 9   TEST

Graph each exponential function. [9.1]

1. $f(x) = 2^x$

2. $g(x) = 3^{-x}$

Sketch the graph of each function and its inverse. Find $f^{-1}(x)$ for Problem 3. [9.2]

3. $f(x) = 2x - 3$

4. $f(x) = x^2 - 4$

Solve for x. [9.3]

5. $\log_4 x = 3$

6. $\log_x 5 = 2$

Graph each of the following. [9.3]

7. $y = \log_2 x$

8. $y = \log_{1/2} x$

Evaluate each of the following. [9.3, 9.4, 9.5]

9. $\log_8 4$

10. $\log_7 21$

11. $\log 23{,}400$

12. $\log 0.0123$

13. $\ln 46.2$

14. $\ln 0.0462$

Use the properties of logarithms to expand each expression. [9.4]

15. $\log_2 \dfrac{8x^2}{y}$

16. $\log \dfrac{\sqrt{x}}{(y^4)\sqrt[5]{z}}$

Write each expression as a single logarithm. [9.4]

17. $2 \log_3 x - \frac{1}{2} \log_3 y$

18. $\frac{1}{3} \log x - \log y - 2 \log z$

Use a calculator to find x. [9.5]

19. $\log x = 4.8476$

20. $\log x = -2.6478$

Solve for x. [9.4, 9.6]

21. $5 = 3^x$

22. $4^{2x-1} = 8$

23. $\log_5 x - \log_5 3 = 1$

24. $\log_2 x + \log_2(x - 7) = 3$

25. pH   Find the pH of a solution in which $[H^+] = 6.6 \times 10^{-7}$. [9.5]

26. Compound Interest   If $400 is deposited in an account that earns 10% annual interest compounded twice a year, how much money will be in the account after 5 years? [9.1]

27. Compound Interest   How long will it take $600 to become $1,800 if the $600 is deposited in an account that earns 8% annual interest compounded 4 times a year? [9.6]

28. Depreciation   If a car depreciates in value 20% per year for the first 5 years after it is purchased for P_0 dollars, then its value after t years will be $V(t) = P_0(1 - r)^t$ for $0 \le t \le 5$. To the nearest dollar, find the value of a car 4 years after it is purchased for $18,000. [9.1]

Simplify.

1. $-8 + 2[5 - 3(-2 - 3)]$

2. $6(2x - 3) + 4(3x - 2)$

3. $\left(\dfrac{3}{5}\right)^{-2} - \left(\dfrac{3}{13}\right)^{-2}$ 4. $\dfrac{3}{4} - \dfrac{1}{8} + \dfrac{3}{2}$

5. $\sqrt[3]{27x^4y^3}$

6. $3\sqrt{48} - 3\sqrt{75} + 2\sqrt{27}$

7. $[(6 + 2i) - (3 - 4i)] - (5 - i)$

8. $\log_5[\log_2(\log_3 9)]$ 9. $1 + \dfrac{x}{1 + \dfrac{1}{x}}$

10. Reduce to lowest terms: $\dfrac{452}{791}$

Multiply.

11. $\left(3t^2 + \dfrac{1}{4}\right)\left(4t^2 - \dfrac{1}{3}\right)$

12. $(\sqrt{6} + 3\sqrt{2})(2\sqrt{6} + \sqrt{2})$

13. Divide: $\dfrac{9x^2 + 9x - 18}{3x - 4}$

14. Rationalize the denominator: $\dfrac{5\sqrt{6}}{2\sqrt{6} + 7}$

15. Subtract: $\dfrac{7}{4x^2 - x - 3} - \dfrac{1}{4x^2 - 7x + 3}$

Solve.

16. $6 - 3(2x - 4) = 2$

17. $\dfrac{2}{3}(6x - 5) + \dfrac{1}{3} = 13$

18. $28x^2 = 3x + 1$ 19. $|3x - 5| + 6 = 2$

20. $\dfrac{2}{x + 1} = \dfrac{4}{5}$

21. $\dfrac{1}{x + 3} + \dfrac{1}{x - 2} = 1$

22. $\dfrac{1}{x^2 + 3x - 4} + \dfrac{3}{x^2 - 1} = \dfrac{-1}{x^2 + 5x + 4}$

23. $2x - 1 = x^2$

24. $3(4y - 1)^2 + (4y - 1) - 10 = 0$

25. $\sqrt{7x - 4} = -2$

26. $\sqrt{y - 3} - \sqrt{y} = -1$ 27. $x - 3\sqrt{x} + 2 = 0$

28. $\begin{vmatrix} 2x & 2x \\ -5 & 4x \end{vmatrix} = 3$ 29. $\log_3 x = 3$

30. $\log_x 0.1 = -1$

31. $\log_3(x - 3) - \log_3(x + 2) = 1$

Solve each system.

32. $\begin{aligned} -9x + 3y &= 1 \\ 5x - 2y &= -2 \end{aligned}$

33. $\begin{aligned} 4x + 7y &= -3 \\ x &= -2y - 2 \end{aligned}$ 34. $\begin{aligned} x + y &= 4 \\ x + z &= 1 \\ y - 2z &= 5 \end{aligned}$

Solve each inequality and graph the solution.

35. $3y - 6 \geq 3$ or $3y - 6 \leq -3$

36. $|4x + 3| - 6 > 5$

37. $x^3 + 2x^2 - 9x - 18 < 0$

38. Graph the line: $5x + 4y = 20$

39. Write the inequality of the line in slope-intercept form if the slope is $-\dfrac{5}{3}$ and $(3, -3)$ is a point on the line.

40. Write in symbols: The difference of $9a$ and $4b$ is less than their sum.

41. Write 0.0000972 in scientific notation.

42. Factor completely: $50a^4 + 10a^2 - 4$

43. If $f(x) = \dfrac{1}{2}x + 3$, find $f^{-1}(x)$.

44. Evaluate: $\begin{vmatrix} 3 & 4 & -1 \\ 0 & 0 & 2 \\ -5 & 1 & 2 \end{vmatrix}$

45. Specify the domain and range for the relation $\{(2, -3), (2, -1), (-3, 3)\}$. Is this relation also a function?

46. State the domain: $y = \sqrt{3 - x}$

47. Give the next number of the sequence, and state whether arithmetic or geometric: 2, -6, 18, . . .

48. **Direct Variation** w varies directly with the square root of c. If w is 8 when c is 16, find w when c is 9.

49. **Mixture** How many gallons of 25% alcohol solution and 50% alcohol solution should be mixed to get 20 gallons of 42.5% alcohol solution?

50. **Compound Interest** A $10,000 Treasury bill earns 6% compounded twice a year. How much is the Treasury bill worth after 4 years?

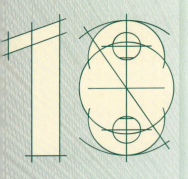

Sequences and Series

(FPG International/Telegraph Colour Library)

INTRODUCTION

Suppose you run up a balance of $1,000 on a credit card that charges 1.65% interest each month (i.e., an annual rate of 19.8%). If you stop using the card and make the minimum payment of $20 each month, how long will it take you to pay off the balance on the card? The answer can be found by using the formula

$$U_n = (1.0165)U_{n-1} - 20$$

where U_n stands for the current unpaid balance on the card, and U_{n-1} is the previous month's balance. Table 1 and Figure 1 were created from this formula and a graphing calculator. As you can see from the table, the balance on the credit card decreases very little each month.

TABLE I Monthly Credit Card Balances

Previous Balance $U_{(n-1)}$	Monthly Interest Rate	Payment Number n	Monthly Payment	New Balance $U_{(n)}$
$1,000.00	1.65%	1	$20	$996.58
$996.50	1.65%	2	$20	$993.11
$992.94	1.65%	3	$20	$989.58
$989.32	1.65%	4	$20	$985.99
$985.64	1.65%	5	$20	$982.34
⋮	⋮	⋮	⋮	⋮

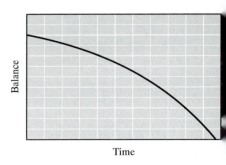

FIGURE I

In the group project at the end of the chapter, you will use a graphing calculator to continue this table and in so doing find out just how many months it will take to pay off this credit card balance.

Many of the sequences in this chapter will be familiar to you on an intuitive level because you have worked with them for some time now. Here are some of those sequences:

The sequence of odd numbers

$$1, 3, 5, 7, \ldots$$

The sequence of even numbers

$$2, 4, 6, 8, \ldots$$

The sequence of squares

$$1^2, 2^2, 3^2, 4^2, \ldots = 1, 4, 9, 16, \ldots$$

The numbers in each of these sequences can be found from the formulas that define functions. For example, the sequence of even numbers can be found from the function

$$f(x) = 2x$$

by finding $f(1)$, $f(2)$, $f(3)$, $f(4)$, and so forth. This gives us justification for the formal definition of a sequence.

> **DEFINITION** A **sequence** is a function whose domain is the set of positive integers $\{1, 2, 3, 4, \ldots\}$.

As you can see, sequences are simply functions with a specific domain. If we want to form a sequence from the function $f(x) = 3x + 5$, we simply find $f(1)$, $f(2)$, $f(3)$, and so on. Doing so gives us the sequence

$$8, 11, 14, 17, \ldots$$

because $f(1) = 3(1) + 5 = 8$, $f(2) = 3(2) + 5 = 11$, $f(3) = 3(3) + 5 = 14$, and $f(4) = 3(4) + 5 = 17$.

Notation   Since the domain for a sequence is always the set $\{1, 2, 3, \ldots\}$, we can simplify the notation we use to represent the terms of a sequence. Using the letter a instead of f, and subscripts instead of numbers enclosed by parentheses, we can represent the sequence from the previous discussion as follows:

$$a_n = 3n + 5$$

Instead of $f(1)$ we write a_1 for the *first term* of the sequence.

Instead of $f(2)$ we write a_2 for the *second term* of the sequence.

Instead of $f(3)$ we write a_3 for the *third term* of the sequence.

Instead of $f(4)$ we write a_4 for the *fourth term* of the sequence.

Instead of $f(n)$ we write a_n for the *nth term* of the sequence.

The *n*th term is also called the **general term** of the sequence. The general term is used to define the other terms of the sequence. That is, if we are given the formula for the general term a_n, we can find any other term in the sequence. The following examples illustrate.

EXAMPLE 1   Find the first four terms of the sequence whose general term is given by $a_n = 2n - 1$.

Solution   The subscript notation a_n works the same way function notation works. To find the first, second, third, and fourth terms of this sequence, we simply substitute 1, 2, 3, and 4 for *n* in the formula $2n - 1$:

$$\text{If} \qquad \text{the general term is } a_n = 2n - 1$$

$$\text{then} \qquad \text{the first term is } a_1 = 2(1) - 1 = 1$$

$$\text{the second term is } a_2 = 2(2) - 1 = 3$$

$$\text{the third term is } a_3 = 2(3) - 1 = 5$$

$$\text{the fourth term is } a_4 = 2(4) - 1 = 7$$

The first four terms of this sequence are the odd numbers 1, 3, 5, and 7. The whole sequence can be written as

$$1, 3, 5, \ldots, 2n - 1, \ldots$$

Since each term in this sequence is larger than the preceding term, we say the sequence is an **increasing sequence.**

EXAMPLE 2   Write the first four terms of the sequence defined by

$$a_n = \frac{1}{n + 1}$$

Solution   Replacing *n* with 1, 2, 3, and 4, we have, respectively, the first four terms:

$$\text{First term} = a_1 = \frac{1}{1 + 1} = \frac{1}{2}$$

$$\text{Second term} = a_2 = \frac{1}{2 + 1} = \frac{1}{3}$$

$$\text{Third term} = a_3 = \frac{1}{3 + 1} = \frac{1}{4}$$

$$\text{Fourth term} = a_4 = \frac{1}{4 + 1} = \frac{1}{5}$$

The sequence defined by

$$a_n = \frac{1}{n + 1}$$

can be written as

$$\frac{1}{2}, \frac{1}{3}, \frac{1}{4}, \; \cdots \; , \frac{1}{n + 1}, \; \cdots$$

Since each term in the sequence is smaller than the term preceding it, the sequence is said to be a **decreasing sequence.**

E X A M P L E   3   Find the fifth and sixth terms of the sequence whose general term is given by $a_n = \dfrac{(-1)^n}{n^2}$.

Solution   For the fifth term, we replace n with 5. For the sixth term, we replace n with 6:

$$\text{Fifth term} = a_5 = \frac{(-1)^5}{5^2} = \frac{-1}{25}$$

$$\text{Sixth term} = a_6 = \frac{(-1)^6}{6^2} = \frac{1}{36}$$

The sequence in Example 3 can be written as

$$-1, \frac{1}{4}, -\frac{1}{9}, \frac{1}{16}, \; \cdots \; , \frac{(-1)^n}{n^2}, \; \cdots$$

Since the terms alternate in sign — if one term is positive, then the next term is negative — we call this an **alternating sequence.** The first three examples all illustrate how we work with a sequence in which we are given a formula for the general term. The next example gives us another way to write the general term.

Using TECHNOLOGY

FINDING SEQUENCES ON A GRAPHING CALCULATOR

Method 1: Using a Table

We can use the table function on a graphing calculator to view the terms of a sequence. To view the terms of the sequence $a_n = 3n + 5$, we set $Y_1 = 3X + 5$. Then we use the table setup feature on the calculator to set the table minimum to 1, and the table increment to 1 also. Here is the setup and result for a TI-83.

Table Setup	Y Variables Setup	Resulting Table	
Table minimum = 1 Table increment = 1 Independent variable: Auto Dependent variable: Auto	$Y_1 = 3X + 5$	**X**	**Y**
		1	8
		2	11
		3	14
		4	17
		5	20

To find any particular term of a sequence, we change the dependent variable setting to Ask, and then input the number of the term of the sequence we want to find. For example, if we want term a_{100}, we input 100 for the independent variable, and the table gives us the value of 305 for that term.

Method 2: Using the Built-in seq(Function

Using this method, first find the seq(function. On a TI-83 it is found in the LIST OPS menu. To find terms a_1 through a_7 for $a_n = 3n + 5$, we first bring up the seq(function on our calculator, then we input the following four items, in order, separated by commas: 3X+5, X, 1, 7. Then we close the parentheses. Our screen will look like this:

$$\text{seq}(3X+5, X, 1, 7)$$

Pressing $\boxed{\text{ENTER}}$ displays the first five terms of the sequence. Pressing the right arrow key repeatedly brings the remaining members of the sequence into view.

Method 3: Using the Built-in Seq Mode

Press the $\boxed{\text{MODE}}$ key on your TI-83 and then select Seq (it's next to Func Par and Pol). Go to the Y variables list set $n\text{Min} = 1$ and $u(n) = 3n+5$. Then go to the TBLSET key to set up your table like the one shown in Method 1. Pressing $\boxed{\text{TABLE}}$ will display the sequence you have defined.

RECURSION FORMULAS

Let's go back to one of the first sequences we looked at in this section:

$$8, 11, 14, 17, \ldots$$

Each term in the sequence can be found by simply substituting positive integers for n in the formula $a_n = 3n + 5$. Another way to look at this sequence, however, is to notice that each term can be found by adding 3 to the preceding term; so, we could give all the terms of this sequence by simply saying

Start with 8, and then add 3 to each term to get the next term.

The same idea, expressed in symbols, looks like this:

$$a_1 = 8 \quad \text{and} \quad a_n = a_{n-1} + 3 \quad \text{for } n > 1$$

This formula is called a **recursion formula** because each term is written *recursively* in terms of the term or terms that precede it.

E X A M P L E   4   Write the first four terms of the sequence given recursively by

$$a_1 = 4 \quad \text{and} \quad a_n = 5a_{n-1} \quad \text{for } n > 1$$

Solution   The formula tells us to start the sequence with the number 4, and then multiply each term by 5 to get the next term. Therefore,

$$a_1 = 4$$
$$a_2 = 5a_1 = 5(4) = 20$$
$$a_3 = 5a_2 = 5(20) = 100$$
$$a_4 = 5a_3 = 5(100) = 500$$

The sequence is 4, 20, 100, 500, ...

Using TECHNOLOGY

RECURSION FORMULAS ON A GRAPHING CALCULATOR

We can use a TI-83 graphing calculator to view the sequence defined recursively as

$$a_1 = 8, a_n = a_{n-1} + 3$$

First, put your TI-83 calculator in sequence mode by pressing the $\boxed{\text{MODE}}$ key, and then selecting Seq (it's next to Func Par and Pol). Go to the Y variables list set $n\text{Min} = 1$ and $u(n) = u(n-1)+3$. (The u is above the 7, and the n is on the $\boxed{\text{X, T, }\theta, n}$ and is automatically displayed if that key is pressed when the calculator is in the Seq mode. Pressing $\boxed{\text{TABLE}}$ will display the sequence you have defined.

FINDING THE GENERAL TERM

In the first four examples, we found some terms of a sequence after being given the general term. In the next two examples, we will do the reverse. That is, given some terms of a sequence, we will find the formula for the general term.

EXAMPLE 5   Find a formula for the nth term of the sequence 2, 8, 18, 32, ...

Solution   Solving a problem like this involves some guessing. Looking over the first four terms, we see each is twice a perfect square:

$$2 = 2(1)$$

$$8 = 2(4)$$

$$18 = 2(9)$$

$$32 = 2(16)$$

If we write each square with an exponent of 2, the formula for the nth term becomes obvious:

$$a_1 = 2 \ = 2(1)^2$$

$$a_2 = 8 \ = 2(2)^2$$

$$a_3 = 18 = 2(3)^2$$

$$a_4 = 32 = 2(4)^2$$

$$\vdots$$

$$a_n = \qquad 2(n)^2 = 2n^2$$

The general term of the sequence 2, 8, 18, 32, . . . is $a_n = 2n^2$.

EXAMPLE 6   Find the general term for the sequence $2, \frac{3}{8}, \frac{4}{27}, \frac{5}{64}, \ldots$.

Solution   The first term can be written as $\frac{2}{1}$. The denominators are all perfect cubes. The numerators are all 1 more than the base of the cubes in the denominators:

$$a_1 = \frac{2}{1} = \frac{1+1}{1^3}$$

$$a_2 = \frac{3}{8} = \frac{2+1}{2^3}$$

$$a_3 = \frac{4}{27} = \frac{3+1}{3^3}$$

$$a_4 = \frac{5}{64} = \frac{4+1}{4^3}$$

Observing this pattern, we recognize the general term to be

$$a_n = \frac{n + 1}{n^3}$$

Note: Finding the nth term of a sequence from the first few terms is not always automatic. That is, it sometimes takes awhile to recognize the pattern. Don't be afraid to guess at the formula for the general term. Many times an incorrect guess leads to the correct formula.

Getting Ready for Class

After reading through the preceding section, respond in your own words and in complete sentences.

A. How are subscripts used to denote the terms of a sequence?

B. What is the relationship between the subscripts used to denote the terms of a sequence and function notation?

C. What is a decreasing sequence?

D. What is meant by a recursion formula for a sequence?

PROBLEM SET 10.1

Write the first five terms of the sequences with the following general terms.

1. $a_n = 3n + 1$

2. $a_n = 2n + 3$

3. $a_n = 4n - 1$

4. $a_n = n + 4$

5. $a_n = n$

6. $a_n = -n$

7. $a_n = n^2 + 3$

8. $a_n = n^3 + 1$

9. $a_n = \dfrac{n}{n + 3}$

10. $a_n = \dfrac{n}{n + 2}$

11. $a_n = \dfrac{n + 1}{n + 2}$

12. $a_n = \dfrac{n + 3}{n + 4}$

13. $a_n = \dfrac{1}{n^2}$

14. $a_n = \dfrac{1}{n^3}$

15. $a_n = 2^n$

16. $a_n = 3^n$

17. $a_n = 3^{-n}$

18. $a_n = 2^{-n}$

19. $a_n = 1 + \dfrac{1}{n}$

20. $a_n = 1 - \dfrac{1}{n}$

21. $a_n = n - \dfrac{1}{n}$

22. $a_n = n + \dfrac{1}{n}$

23. $a_n = (-2)^n$

24. $a_n = (-3)^n$

Write the first five terms of the sequences defined by the following recursion formulas.

25. $a_1 = 3$ $a_n = -3a_{n-1}$ $n > 1$

26. $a_1 = -3$ $a_n = 3a_{n-1}$ $n > 1$

27. $a_1 = 3$ $a_n = a_{n-1} - 3$ $n > 1$

28. $a_1 = -3$ $a_n = a_{n-1} - 3$ $n > 1$

29. $a_1 = 1$ $a_n = 2a_{n-1} + 3$ $n > 1$

30. $a_1 = 1$ $a_n = 3a_{n-1} + 2$ $n > 1$

31. $a_1 = 1$ $a_n = a_{n-1} + n$ $n > 1$

32. $a_1 = 2$ $a_n = a_{n-1} - n$ $n > 1$

Determine the general term for each of the following sequences.

33. 2, 3, 4, 5, . . .

34. 3, 6, 9, 12, . . .

35. 4, 8, 12, 16, 20, . . .

36. 3, 4, 5, 6, . . .

37. 7, 10, 13, 16, . . .

38. 4, 9, 14, 19, . . .

39. 1, 4, 9, 16, . . .

40. 1, 8, 27, 64, . . .

41. 3, 12, 27, 48, . . .

42. 2, 16, 54, 128, . . .

43. 4, 8, 16, 32, . . .

44. 3, 9, 27, 81, . . .

45. $-2, 4, -8, 16, \ldots$

46. $-3, 9, -27, 81, \ldots$

47. $\frac{1}{4}, \frac{1}{8}, \frac{1}{16}, \frac{1}{32}, \cdots$

48. $\frac{1}{3}, \frac{1}{9}, \frac{1}{27}, \frac{1}{81}, \cdots$

49. $\frac{1}{4}, \frac{2}{9}, \frac{3}{16}, \frac{4}{25}, \cdots$

50. $\frac{1}{4}, \frac{2}{10}, \frac{3}{28}, \frac{4}{82}, \cdots$

Review Problems

The problems that follow review material we covered in Section 9.3.

51. Salary Increase The entry level salary for a teacher is $28,000 with 4% increases after every year of service.

(a) Write a sequence for this teacher's salary for the first 5 years.

(b) Find the general term of the sequence in part (a).

52. Holiday Account To save money for holiday presents, a person deposits $5 in a savings account on January 1, and then deposits an additional $5 every week thereafter until Christmas.

(a) Write a sequence for the money in that savings account for the first 10 weeks of the year.

(b) Write the general term of the sequence in part (a).

(c) If there are 50 weeks from January 1 to Christmas, how much money will be available for spending on Christmas presents?

53. Saving for College To save money for his son's college education, a person deposits $100 in a savings account on the first day of each month after his son was born.

(a) Write a sequence to show the money in this college education account at the end of the first, second, third, fourth, and fifth years.

(b) Write the general term for this sequence in terms of m, if m represents the number of months.

(c) Write the general term for this sequence in terms of y, if y represents the number of years.

54. Saving for College To save money for his daughter's college education, a person deposits $100 in a savings account on the first day of each month after his daughter is born. After every year, this monthly deposit is increased by $10 — that is, in the second year $110 is deposited each month, in the third year $120 is deposited each month, and so on.

(a) Write a sequence to show the money in this college education account at the end of the first, second, third, fourth, and fifth years.

(b) Refer to the preceding exercise and determine what percentage more money has been saved for the daughter than for the son.

Find x in each of the following.

55. $\log_9 x = \frac{3}{2}$

56. $\log_x \frac{1}{4} = -2$

Simplify each expression.

57. $\log_2 32$

58. $\log_{10} 10,000$

59. $\log_3 [\log_2 8]$

60. $\log_5 [\log_6 6]$

Extending the Concepts

61. As n increases, the terms in the sequence

$$a_n = \left(1 + \frac{1}{n}\right)^n$$

get closer and closer to the number e (that's the same e we used in defining natural logarithms). It takes some fairly large values of n, however, before we can see this happening. Use a calculator to find $a_{100}, a_{1,000}, a_{10,000}$, and $a_{100,000}$, and compare them to the decimal approximation we gave for the number e.

62. The sequence

$$a_n = \left(1 + \frac{1}{n}\right)^{-n}$$

gets close to the number $1/e$ as n becomes large. Use a calculator to find approximations for a_{100} and $a_{1,000}$, and then compare them to $\dfrac{1}{2.7183}$.

63. Write the first ten terms of the sequence defined by the recursion formula

$$a_1 = 1, a_2 = 1, a_n = a_{n-1} + a_{n-2} \qquad n > 2$$

64. Write the first ten terms of the sequence defined by the recursion formula

$$a_1 = 2, a_2 = 2, a_n = a_{n-1} + a_{n-2} \qquad n > 2$$

65. Simplify each complex fraction in the following sequence, and then compare this sequence with the sequence you wrote in Problem 63.

$$1 + \frac{1}{1+1}, 1 + \frac{1}{1 + \frac{1}{1+1}}, 1 + \frac{1}{1 + \frac{1}{1 + \frac{1}{1+1}}}, \ldots$$

66. Write the first five terms of the sequence given by the following recursion formula.

$$a_1 = \frac{3}{2}, a_n = 1 + \frac{1}{a_{n-1}} \qquad n > 1$$

10.2 Series

There is an interesting relationship between the sequence of odd numbers and the sequence of squares that is found by adding the terms in the sequence of odd numbers.

$$
\begin{aligned}
1 &= 1 \\
1 + 3 &= 4 \\
1 + 3 + 5 &= 9 \\
1 + 3 + 5 + 7 &= 16
\end{aligned}
$$

When we add the terms of a sequence the result is called a series.

> **DEFINITION** The sum of a number of terms in a sequence is called a series.

A sequence can be finite or infinite depending on whether or not the sequence ends at the nth term. For example,

$$1, 3, 5, 7, 9$$

is a finite sequence, but

$$1, 3, 5, \ldots$$

is an infinite sequence. Associated with each of the preceding sequences is a series found by adding the terms of the sequence:

$$1 + 3 + 5 + 7 + 9 \qquad \text{Finite series}$$

$$1 + 3 + 5 + \ldots \qquad \text{Infinite series}$$

In this section we will consider only finite series. We can introduce a new kind of notation here that is a compact way of indicating a finite series. The notation is called **summation notation,** or **sigma notation** since it is written using the Greek letter sigma. The expression

$$\sum_{i=1}^{4} (8i - 10)$$

is an example of an expression that uses summation notation. The summation notation in this expression is used to indicate the sum of all the expressions $8i - 10$ from $i = 1$ up to and including $i = 4$. That is,

$$\sum_{i=1}^{4} (8i - 10) = (8 \cdot 1 - 10) + (8 \cdot 2 - 10) + (8 \cdot 3 - 10) + (8 \cdot 4 - 10)$$

$$= -2 + 6 + 14 + 22$$

$$= 40$$

The letter i as used here is called the **index of summation,** or just **index** for short.

Here are some examples illustrating the use of summation notation.

EXAMPLE 1 Expand and simplify $\displaystyle\sum_{i=1}^{5} (i^2 - 1)$.

Solution We replace i in the expression $i^2 - 1$ with all consecutive integers from 1 up to 5, including 1 and 5:

$$\sum_{i=1}^{5} (i^2 - 1) = (1^2 - 1) + (2^2 - 1) + (3^2 - 1) + (4^2 - 1) + (5^2 - 1)$$

$$= 0 + 3 + 8 + 15 + 24$$

$$= 50$$

EXAMPLE 2 Expand and simplify $\displaystyle\sum_{i=3}^{6} (-2)^i$.

Solution We replace i in the expression $(-2)^i$ with the consecutive integers beginning at 3 and ending at 6:

$$\sum_{i=3}^{6} (-2)^i = (-2)^3 + (-2)^4 + (-2)^5 + (-2)^6$$

$$= -8 + 16 + (-32) + 64$$

$$= 40$$

Using TECHNOLOGY

SUMMING SERIES ON A GRAPHING CALCULATOR

A TI-83 graphing calculator has a built-in sum(function that, when used with the seq(function, allows us to add the terms of a series. Let's repeat Example 1 using our graphing calculator. First, we go to LIST and select MATH. The fifth option in that list is sum(, which we select. Then we go to LIST again and select OPS. From that list we select seq(. Next we enter X^2−1, X, 1, 5, and then we close both sets of parentheses. Our screen shows the following:

$$\text{sum(seq(X\textasciicircum2}-1, X, 1, 5)) \qquad \text{which will give us } \sum_{i=1}^{5} (i^2 - 1)$$

When we press ENTER the calculator displays 50, which is the same result we obtained in Example 1.

EXAMPLE 3   Expand $\displaystyle\sum_{i=2}^{5} (x^i - 3)$.

Solution   We must be careful not to confuse the letter x with i. The index i is the quantity we replace by the consecutive integers from 2 to 5, not x:

$$\sum_{i=2}^{5} (x^i - 3) = (x^2 - 3) + (x^3 - 3) + (x^4 - 3) + (x^5 - 3)$$

In the first three examples, we were given an expression with summation notation and asked to expand it. The next examples in this section illustrate how we can write an expression in expanded form as an expression involving summation notation.

EXAMPLE 4   Write with summation notation $1 + 3 + 5 + 7 + 9$.

Solution   A formula that gives us the terms of this sum is

$$a_i = 2i - 1$$

where i ranges from 1 up to and including 5. Notice we are using the subscript i in exactly the same way we used the subscript n in the previous section — to indicate the general term. Writing the sum

$$1 + 3 + 5 + 7 + 9$$

with summation notation looks like this:

$$\sum_{i=1}^{5} (2i - 1)$$

EXAMPLE 5   Write with summation notation $3 + 12 + 27 + 48$.

Solution   We need a formula, in terms of i, that will give each term in the sum. Writing the sum as

$$3 \cdot 1^2 + 3 \cdot 2^2 + 3 \cdot 3^2 + 3 \cdot 4^2$$

we see the formula

$$a_i = 3 \cdot i^2$$

where i ranges from 1 up to and including 4. Using this formula and summation notation, we can represent the sum

$$3 + 12 + 27 + 48$$

as

$$\sum_{i=1}^{4} 3i^2$$

EXAMPLE 6   Write with summation notation

$$\frac{x + 3}{x^3} + \frac{x + 4}{x^4} + \frac{x + 5}{x^5} + \frac{x + 6}{x^6}$$

Solution   A formula that gives each of these terms is

$$a_i = \frac{x + i}{x^i}$$

where i assumes all integer values between 3 and 6, including 3 and 6. The sum can be written as

$$\sum_{i=3}^{6} \frac{x + i}{x^i}$$

 # Getting Ready for Class

After reading through the preceding section, respond in your own words and in complete sentences.

A. What is the difference between a sequence and a series?

B. Explain the summation notation $\sum_{i=1}^{4}$ in the series $\sum_{i=1}^{4} (2i + 1)$.

C. When will a finite series result in a numerical value versus an algebraic expression?

D. Determine for what values of n the series $\sum_{i=1}^{n} (-1)^i$ will be equal to 0. Explain your answer.

PROBLEM SET 10.2

Expand and simplify each of the following.

1. $\sum_{i=1}^{4} (2i + 4)$

2. $\sum_{i=1}^{5} (3i - 1)$

3. $\sum_{i=1}^{3} (2i - 1)$

4. $\sum_{i=1}^{4} (2i - 1)$

5. $\sum_{i=2}^{3} (i^2 - 1)$

6. $\sum_{i=3}^{6} (i^2 + 1)$

7. $\sum_{i=1}^{4} \frac{i}{1 + i}$

8. $\sum_{i=1}^{4} \frac{i^2}{1 + i}$

9. $\sum_{i=1}^{3} \frac{i^2}{2i - 1}$

10. $\sum_{i=3}^{5} (i^3 + 4)$

11. $\sum_{i=1}^{4} (-3)^i$

12. $\sum_{i=1}^{4} \left(-\frac{1}{3}\right)^i$

13. $\sum_{i=3}^{6} (-2)^i$

14. $\sum_{i=4}^{6} \left(-\frac{1}{2}\right)^i$

Expand the following.

15. $\sum_{i=1}^{5} (x + i)$

16. $\sum_{i=3}^{6} (x - i)$

17. $\sum_{i=2}^{7} (x + 1)^i$

18. $\sum_{i=1}^{4} (x + 3)^i$

19. $\sum_{i=1}^{5} \frac{x + i}{x - 1}$

20. $\sum_{i=1}^{6} \frac{x - 3i}{x + 3i}$

21. $\sum_{i=3}^{8} (x + i)^i$

22. $\sum_{i=4}^{7} (x - 2i)^i$

23. $\sum_{i=1}^{5} (x + i)^{i+1}$

24. $\sum_{i=2}^{6} (x + i)^{i-1}$

Write each of the following sums with summation notation.

25. $2 + 4 + 8 + 16$

26. $3 + 5 + 7 + 9 + 11$

27. $4 + 8 + 16 + 32 + 64$

28. $1 + 3 + 5$

29. $5 + 10 + 17 + 26 + 37$

30. $3 + 8 + 15 + 24$

31. $\frac{3}{4} + \frac{4}{5} + \frac{5}{6} + \frac{6}{7} + \frac{7}{8}$ **32.** $\frac{1}{2} + \frac{2}{3} + \frac{3}{4} + \frac{4}{5}$

33. $\frac{1}{3} + \frac{2}{5} + \frac{3}{7} + \frac{4}{9}$ **34.** $\frac{3}{1} + \frac{5}{3} + \frac{7}{5} + \frac{9}{7}$

35. $(x - 3) + (x - 4) + (x - 5) + (x - 6)$

36. $x^2 + x^3 + x^4 + x^5 + x^6$

37. $\frac{x}{x + 3} + \frac{x}{x + 4} + \frac{x}{x + 5}$

38. $\frac{x - 3}{x^3} + \frac{x - 4}{x^4} + \frac{x - 5}{x^5} + \frac{x - 6}{x^6}$

39. $x^2(x + 2) + x^3(x + 3) + x^4(x + 4)$

40. $x(x + 2)^2 + x(x + 3)^3 + x(x + 4)^4$

41. Repeating Decimals Any repeating, nonterminating decimal may be viewed as a series. For instance, $\frac{2}{3} = 0.6 + 0.06 + 0.006 + 0.0006 + \cdots$. Write the following fractions as series.
 (a) $\frac{1}{3}$
 (b) $\frac{2}{9}$
 (c) $\frac{3}{11}$

42. Repeating Decimals Refer to the previous exercise, and express the following repeating decimals as fractions.
 (a) $0.55555 \cdots$
 (b) $1.33333 \cdots$
 (c) $0.29292929 \cdots$

Applying the Concepts

43. Skydiving A skydiver jumps from a plane and falls 16 feet the first second, 48 feet the second second, and 80 feet the third second. If he continues to fall in the same manner, how far will he fall the seventh second? What is the distance he falls in 7 seconds?

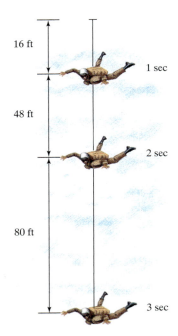

16 ft

48 ft

80 ft

1 sec

2 sec

3 sec

44. Bacterial Growth After 1 day, a colony of 50 bacteria reproduces to become 200 bacteria. After 2 days, they reproduce to become 800 bacteria. If they continue to reproduce at this rate, how many bacteria will be present after 4 days?

45. Converging Series Given the series

$$\frac{1}{1\cdot3} + \frac{1}{3\cdot5} + \frac{1}{5\cdot7} + \cdots$$

(a) Write the first six terms of this series.

(b) Use a graphing calculator to find the sum of the first six terms of this series.

(c) What do you think is the exact sum of this series? (You may need to compute some more terms.)

46. Converging Series Given the series

$$\frac{1}{1\cdot3} + \frac{1}{2\cdot4} + \frac{1}{3\cdot5} + \cdots$$

(a) Write the first six terms of this series.

(b) Use a graphing calculator to find the sum of the first six terms of this series.

(c) What do you think is the exact sum of this series? (You may need to compute some more terms.)

Review Problems

The following problems review material we covered in Section 9.4.

Use the properties of logarithms to expand each of the following expressions.

47. $\log_2 x^3 y$

48. $\log_7 \dfrac{x^2}{y^4}$

49. $\log_{10} \dfrac{\sqrt[3]{x}}{y^2}$

50. $\log_{10} \sqrt[3]{\dfrac{x}{y^2}}$

Write each expression as a single logarithm.

51. $\log_{10} x - \log_{10} y^2$

52. $\log_{10} x^2 + \log_{10} y^2$

53. $2 \log_3 x - 3 \log_3 y - 4 \log_3 z$

54. $\frac{1}{2} \log_6 x + \frac{1}{3} \log_6 y + \frac{1}{4} \log_6 z$

Solve each equation.

55. $\log_4 x - \log_4 5 = 2$

56. $\log_3 6 + \log_3 x = 4$

57. $\log_2 x + \log_2(x - 7) = 3$

58. $\log_5(x + 1) + \log_5(x - 3) = 1$

Extending the Concepts

59. Solve for x by first expanding:

$$\sum_{i=1}^{3} (2^i \cdot x - 4) = 20$$

In this and the following section, we will review and extend two major types of sequences, which we have worked with previously — arithmetic sequences and geometric sequences.

> **DEFINITION**   An **arithmetic sequence** is a sequence of numbers in which each term is obtained from the preceding term by adding the same amount each time. An arithmetic sequence is also called an **arithmetic progression.**

The sequence

$$2, 6, 10, 14, \ldots$$

is an example of an arithmetic sequence, since each term is obtained from the preceding term by adding 4 each time. The amount we add each time — in this case, 4 — is called the **common difference,** since it can be obtained by subtracting any two consecutive terms. (The term with the larger subscript must be written first.) The common difference is denoted by d.

EXAMPLE 1   Give the common difference d for the arithmetic sequence 4, 10, 16, 22, ...

Solution   Since each term can be obtained from the preceding term by adding 6, the common difference is 6. That is, $d = 6$.

EXAMPLE 2   Give the common difference for 100, 93, 86, 79, ...

Solution   The common difference in this case is $d = -7$, since adding -7 to any term always produces the next consecutive term.

EXAMPLE 3   Give the common difference for $\frac{1}{2}, 1, \frac{3}{2}, 2, \ldots$.

Solution   The common difference is $d = \frac{1}{2}$.

THE GENERAL TERM

The general term a_n of an arithmetic progression can always be written in terms of the first term a_1 and the common difference d. Consider the sequence from Example 1:

$$4, 10, 16, 22, \ldots$$

We can write each term in terms of the first term 4 and the common difference 6:

$$4, \qquad 4 + (1 \cdot 6), \qquad 4 + (2 \cdot 6), \qquad 4 + (3 \cdot 6), \; . \, . \, .$$

$$a_1, \qquad\quad a_2, \qquad\qquad a_3, \qquad\qquad a_4, \qquad . \, . \, .$$

Observing the relationship between the subscript on the terms in the second line and the coefficients of the 6's in the first line, we write the general term for the sequence as

$$a_n = 4 + (n - 1)6$$

We generalize this result to include the general term of any arithmetic sequence.

ARITHMETIC SEQUENCES

The **general term** of an arithmetic progression with the first term a_1 and common difference d is given by

$$a_n = a_1 + (n - 1)d$$

EXAMPLE 4   Find the general term for the sequence

$$7, 10, 13, 16, \; . \, . \, .$$

Solution   The first term is $a_1 = 7$, and the common difference is $d = 3$. Substituting these numbers into the formula given earlier, we have

$$a_n = 7 + (n - 1)3$$

which we can simplify, if we choose, to

$$a_n = 7 + 3n - 3$$
$$= 3n + 4$$

EXAMPLE 5   Find the general term of the arithmetic progression whose third term a_3 is 7 and whose eighth term a_8 is 17.

Solution   According to the formula for the general term, the third term can be written as $a_3 = a_1 + 2d$, and the eighth term can be written as $a_8 = a_1 + 7d$. Since these terms are also equal to 7 and 17, respectively, we can write

$$a_3 = a_1 + 2d = 7$$

$$a_8 = a_1 + 7d = 17$$

To find a_1 and d, we simply solve the system:

$$a_1 + 2d = 7$$

$$a_1 + 7d = 17$$

We add the opposite of the top equation to the bottom equation. The result is

$$5d = 10$$

$$d = 2$$

To find a_1, we simply substitute 2 for d in either of the original equations and get

$$a_1 = 3$$

The general term for this progression is

$$a_n = 3 + (n - 1)2$$

which we can simplify to

$$a_n = 2n + 1$$

The sum of the first n terms of an arithmetic sequence is denoted by S_n. The following theorem gives the formula for finding S_n, which is sometimes called the **nth partial sum.**

THEOREM 10.1

The sum of the first n terms of an arithmetic sequence whose first term is a_1 and whose nth term is a_n is given by

$$S_n = \frac{n}{2}(a_1 + a_n)$$

Proof   We can write S_n in expanded form as

$$S_n = a_1 + [a_1 + d] + [a_1 + 2d] + \cdots + [a_1 + (n - 1)d]$$

We can arrive at this same series by starting with the last term a_n and subtracting d each time. Writing S_n this way, we have

$$S_n = a_n + [a_n - d] + [a_n - 2d] + \cdots + [a_n - (n - 1)d]$$

If we add the preceding two expressions term by term, we have

$$2S_n = (a_1 + a_n) + (a_1 + a_n) + (a_1 + a_n) + \cdots + (a_1 + a_n)$$

$$2S_n = n(a_1 + a_n)$$

$$S_n = \frac{n}{2}(a_1 + a_n)$$

E X A M P L E   6   Find the sum of the first ten terms of the arithmetic progression 2, 10, 18, 26, ...

Solution   The first term is 2, and the common difference is 8. The tenth term is

$$a_{10} = 2 + 9(8)$$
$$= 2 + 72$$
$$= 74$$

Substituting $n = 10$, $a_1 = 2$, and $a_{10} = 74$ into the formula

$$S_n = \frac{n}{2}(a_1 + a_n)$$

we have

$$S_{10} = \frac{10}{2}(2 + 74)$$
$$= 5(76)$$
$$= 380$$

The sum of the first 10 terms is 380.

Getting Ready for Class

After reading through the preceding section, respond in your own words and in complete sentences.

A. Explain how to determine if a sequence is arithmetic.

B. What is a common difference?

C. Suppose the value of a_5 is given. What other possible pieces of information could be given in order to have enough information to obtain the first 10 terms of the sequence?

D. Explain the formula $a_n = a_1 + (n - 1)d$ in words so that someone who wanted to find the nth term of an arithmetic sequence could do so from your description.

PROBLEM SET 10.3

Determine which of the following sequences are arithmetic progressions. For those that are arithmetic progressions, identify the common difference d.

1. 1, 2, 3, 4, . . .

2. 4, 6, 8, 10, . . .

3. 1, 2, 4, 7, . . .

4. 1, 2, 4, 8, . . .

5. 50, 45, 40, . . .

6. $1, \frac{1}{2}, \frac{1}{4}, \frac{1}{8}$, . . .

7. 1, 4, 9, 16, . . .

8. 5, 7, 9, 11, . . .

9. $\frac{1}{3}, 1, \frac{5}{3}, \frac{7}{3}$, . . .

10. 5, 11, 17, . . .

Each of the following problems refers to arithmetic sequences.

11. If $a_1 = 3$ and $d = 4$, find a_n and a_{24}.

12. If $a_1 = 5$ and $d = 10$, find a_n and a_{100}.

13. If $a_1 = 6$ and $d = -2$, find a_{10} and S_{10}.

14. If $a_1 = 7$ and $d = -1$, find a_{24} and S_{24}.

15. If $a_6 = 17$ and $a_{12} = 29$, find the term a_1, the common difference d, and then find a_{30}.

16. If $a_5 = 23$ and $a_{10} = 48$, find the first term a_1, the common difference d, and then find a_{40}.

17. If the third term is 16 and the eighth term is 26, find the first term, the common difference, and then find a_{20} and S_{20}.

18. If the third term is 16 and the eighth term is 51, find the first term, the common difference, and then find a_{50} and S_{50}.

19. Find the sum of the first 100 terms of the sequence 5, 9, 13, 17, ...

20. Find the sum of the first 50 terms of the sequence 8, 11, 14, 17, ...

21. Find a_{35} for the sequence 12, 7, 2, -3, ...

22. Find a_{45} for the sequence 25, 20, 15, 10, ...

23. Find the tenth term and the sum of the first ten terms of the sequence $\frac{1}{2}$, 1, $\frac{3}{2}$, 2, ...

24. Find the 15th term and the sum of the first 15 terms of the sequence $-\frac{1}{3}$, 0, $\frac{1}{3}$, $\frac{2}{3}$, ...

Straight-Line Depreciation   Recall from Section 3.6 that straight-line depreciation is an accounting method used to help spread the cost of new equipment over a number of years. The value at any time during the life of the machine can be found with a linear equation in two variables. For income tax purposes, however, it is the value at the end of the year that is most important, and for this reason sequences can be used.

25. Value of a Copy Machine   A large copy machine sells for $18,000 when it is new. Its value decreases $3,300 each year after that. We can use an arithmetic sequence to find the value of the machine at the end of each year. If we let a_0 represent the value when it is purchased, then a_1 is the value after 1 year, a_2 is the value after 2 years, and so on.
  (a) Write the first 5 terms of the sequence.
  (b) What is the common difference?
  (c) Construct a line graph for the first 5 terms of the sequence.
  (d) Use the line graph to estimate the value of the copy machine 2.5 years after it is purchased.
  (e) Write the sequence from part (a) using a recursive formula.

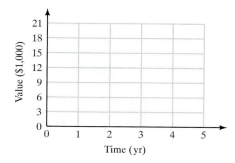

(John Turner / Tony Stone Images)

26. Value of a Forklift   An electric forklift sells for $125,000 when new (see top of p. 642). Each year after that, it decreases $16,500 in value.
  (a) Write an arithmetic sequence that gives the value of the forklift at the end of each of the first 5 years after it is purchased.
  (b) What is the common difference for this sequence?
  (c) Construct a line graph for this sequence.
  (d) Use the line graph to estimate the value of the forklift 3.5 years after it is purchased.
  (e) Write the sequence from part (a) using a recursive formula

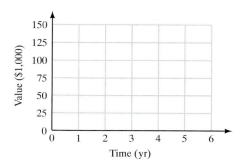

(FPG International LLC/Telegraph Colour Library)

27. Distance A rocket travels vertically 1,500 feet in its first second of flight, and then about 40 feet less each succeeding second. Use these estimates to answer the following questions.
(a) Write a sequence of the vertical distance traveled by a rocket in each of its first 6 seconds.
(b) Is the sequence in part (a) an arithmetic sequence? Explain why or why not.
(c) What is the general term of the sequence in part (a)?

28. Depreciation Suppose an automobile sells for N dollars new, and then depreciated 40% each year.
(a) Write a sequence for the value of this automobile (in terms of N) for each year.
(b) What is the general term of the sequence in part (a)?
(c) Is the sequence in part (a) an arithmetic sequence? Explain why it is or is not.

29. Triangular Numbers The first four triangular numbers are $\{1, 3, 6, 10, \ldots\}$, and are illustrated in the following diagram.

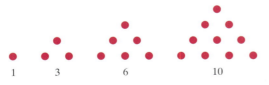

(a) Write a sequence of the first 15 triangular numbers.
(b) Write the recursive general term for the sequence of triangular numbers.

(c) Is the sequence of triangular numbers an arithmetic sequence? Explain why it is or is not.

30. Arithmetic Means Three (or more) arithmetic means between two numbers may be found by forming an arithmetic sequence using the original two numbers and the arithmetic means. For example, three arithmetic means between 10 and 34 may be found by examining the sequence $\{10, a, b, c, 34\}$. For the sequence to be arithmetic, the common difference must be 6; therefore, $a = 16$, $b = 22$, and $c = 28$. Use this idea to answer the following questions.
(a) Find four arithmetic means between 10 and 35.
(b) Find three arithmetic means between 2 and 62.
(c) Find five arithmetic means between 4 and 28.

Applying the Concepts

31. Increasing Salary Suppose a woman earns $28,000 the first year she works and then gets a raise of $850 every year after that. Write a sequence that gives her salary for each of the first 5 years she works. What is the general term of this sequence? At this rate, how much will she be making the tenth year she works?

32. Increasing Salary Suppose a schoolteacher makes $26,500 the first year he works and then gets a $900 raise every year after that. Write a sequence that gives his salary for the first 5 years he works. What is the general term of this sequence? How much will he be making the 20th year he works?

Review Problems

The problems that follow review material we covered in Section 9.5.

Find the following logarithms.

33. log 576

34. log 57,600

35. log 0.0576

36. log 0.000576

Find x.

37. $\log x = 2.6484$

38. $\log x = 7.9832$

39. $\log x = -7.3516$

40. $\log x = -2.0168$

Geometric Sequences

This section is concerned with the second major classification of sequences, called geometric sequences. The problems in this section are very similar to the problems in the preceding section.

DEFINITION A sequence of numbers in which each term is obtained from the previous term by multiplying by the same amount each time is called a **geometric sequence**. Geometric sequences are also called **geometric progressions**.

The sequence

$$3, 6, 12, 24, \ldots$$

is an example of a geometric progression. Each term is obtained from the previous term by multiplying by 2. The amount by which we multiply each time — in this case, 2 — is called the **common ratio.** The common ratio is denoted by r and can be found by taking the ratio of any two consecutive terms. (The term with the larger subscript must be in the numerator.)

EXAMPLE 1 Find the common ratio for the geometric progression.

$$\frac{1}{2}, \frac{1}{4}, \frac{1}{8}, \frac{1}{16}, \ldots$$

Solution Since each term can be obtained from the term before it by multiplying by $\frac{1}{2}$, the common ratio is $\frac{1}{2}$. That is, $r = \frac{1}{2}$.

EXAMPLE 2 Find the common ratio for $\sqrt{3}, 3, 3\sqrt{3}, 9, \ldots$

Solution If we take the ratio of the third term to the second term, we have

$$\frac{3\sqrt{3}}{3} = \sqrt{3}$$

The common ratio is $r = \sqrt{3}$.

GEOMETRIC SEQUENCES

The **general term** a_n of a geometric sequence with the first term a_1 and common ratio r is given by

$$a_n = a_1 r^{n-1}$$

To see how we arrive at this formula, consider the following geometric progression whose common ratio is 3:

$$2, 6, 18, 54, \ldots$$

We can write each term of the sequence in terms of the first term 2 and the common ratio 3:

$$2 \cdot 3^0, \quad 2 \cdot 3^1, \quad 2 \cdot 3^2, \quad 2 \cdot 3^3, \ldots$$
$$a_1, \quad\quad a_2, \quad\quad a_3, \quad\quad a_4, \quad \ldots$$

Observing the relationship between the two preceding lines, we find we can write the general term of this progression as

$$a_n = 2 \cdot 3^{n-1}$$

Since the first term can be designated by a_1 and the common ratio by r, the formula

$$a_n = 2 \cdot 3^{n-1}$$

coincides with the formula

$$a_n = a_1 r^{n-1}$$

EXAMPLE 3   Find the general term for the geometric progression

$$5, 10, 20, \ldots$$

Solution   The first term is $a_1 = 5$, and the common ratio is $r = 2$. Using these values in the formula

$$a_n = a_1 r^{n-1}$$

we have

$$a_n = 5 \cdot 2^{n-1}$$

EXAMPLE 4   Find the tenth term of the sequence $3, \frac{3}{2}, \frac{3}{4}, \frac{3}{8}, \ldots$.

Solution   The sequence is a geometric progression with first term $a_1 = 3$ and common ratio $r = \frac{1}{2}$. The tenth term is

$$a_{10} = 3\left(\frac{1}{2}\right)^9 = \frac{3}{512}$$

EXAMPLE 5   Find the general term for the geometric progression whose fourth term is 16 and whose seventh term is 128.

Solution   The fourth term can be written as $a_4 = a_1 r^3$, and the seventh term can be written as $a_7 = a_1 r^6$.

$$a_4 = a_1 r^3 = 16$$
$$a_7 = a_1 r^6 = 128$$

We can solve for r by using the ratio a_7/a_4.

$$\frac{a_7}{a_4} = \frac{a_1 r^6}{a_1 r^3} = \frac{128}{16}$$

$$r^3 = 8$$

$$r = 2$$

The common ratio is 2. To find the first term, we substitute $r = 2$ into either of the original two equations. The result is

$$a_1 = 2$$

The general term for this progression is

$$a_n = 2 \cdot 2^{n-1}$$

which we can simplify by adding exponents, since the bases are equal:

$$a_n = 2^n$$

As was the case in the preceding section, the sum of the first n terms of a geometric progression is denoted by S_n, which is called the **nth partial sum** of the progression.

THEOREM 10.2

The sum of the first n terms of a geometric progression with first term a_1 and common ratio r is given by the formula

$$S_n = \frac{a_1(r^n - 1)}{r - 1}$$

Proof   We can write the sum of the first n terms in expanded form:

$$S_n = a_1 + a_1 r + a_1 r^2 + \cdots + a_1 r^{n-1} \qquad (1)$$

Then multiplying both sides by r, we have

$$r S_n = a_1 r + a_1 r^2 + a_1 r^3 + \cdots + a_1 r^n \qquad (2)$$

If we subtract the left side of equation (1) from the left side of equation (2) and do the same for the right sides, we end up with

$$r S_n - S_n = a_1 r^n - a_1$$

We factor S_n from both terms on the left side and a_1 from both terms on the right side of this equation:

$$S_n(r - 1) = a_1(r^n - 1)$$

Dividing both sides by $r - 1$ gives the desired result:

$$S_n = \frac{a_1(r^n - 1)}{r - 1}$$

EXAMPLE 6   Find the sum of the first ten terms of the geometric progression 5, 15, 45, 135, ...

Solution   The first term is $a_1 = 5$, and the common ratio is $r = 3$. Substituting these values into the formula for S_{10}, we have the sum of the first ten terms of the sequence:

$$S_{10} = \frac{5(3^{10} - 1)}{3 - 1}$$

$$= \frac{5(3^{10} - 1)}{2}$$

The answer can be left in this form. A calculator will give the result as 147,620.

INFINITE GEOMETRIC SERIES

Suppose the common ratio for a geometric sequence is a number whose absolute value is less than 1 — for instance, $\frac{1}{2}$. The sum of the first n terms is given by the formula

$$S_n = \frac{a_1\left[\left(\frac{1}{2}\right)^n - 1\right]}{\frac{1}{2} - 1}$$

As n becomes larger and larger, the term $(\frac{1}{2})^n$ will become closer and closer to 0. That is, for $n = 10$, 20, and 30, we have the following approximations:

$$\left(\frac{1}{2}\right)^{10} \approx 0.001$$

$$\left(\frac{1}{2}\right)^{20} \approx 0.000001$$

$$\left(\frac{1}{2}\right)^{30} \approx 0.000000001$$

so that for large values of n, there is very little difference between the expression

$$\frac{a_1(r^n - 1)}{r - 1}$$

and the expression

$$\frac{a_1(0 - 1)}{r - 1} = \frac{-a_1}{r - 1} = \frac{a_1}{1 - r} \qquad \text{if} \qquad |r| < 1$$

In fact, the sum of the terms of a geometric sequence in which $|r| < 1$ actually becomes the expression

$$\frac{a_1}{1 - r}$$

as n approaches infinity. To summarize, we have the following:

THE SUM OF AN INFINITE GEOMETRIC SERIES

If a geometric sequence has first term a_1 and common ratio r such that $|r| < 1$, then the following is called an **infinite geometric series:**

$$S = \sum_{i=0}^{\infty} a_1 r^i = a_1 + a_1 r + a_1 r^2 + a_1 r^3 + \cdots$$

Its sum is given by the formula

$$S = \frac{a_1}{1 - r}$$

EXAMPLE 7   Find the sum of the infinite geometric series

$$\frac{1}{5} + \frac{1}{10} + \frac{1}{20} + \frac{1}{40} + \cdots$$

Solution   The first term is $a_1 = \frac{1}{5}$, and the common ratio is $r = \frac{1}{2}$, which has an absolute value less than 1. Therefore, the sum of this series is

$$S = \frac{a_1}{1 - r} = \frac{\frac{1}{5}}{1 - \frac{1}{2}} = \frac{\frac{1}{5}}{\frac{1}{2}} = \frac{2}{5}$$

EXAMPLE 8   Show that $0.999 \ldots$ is equal to 1.

Solution   We begin by writing $0.999 \ldots$ as an infinite geometric series:

$$0.999 \ldots = 0.9 + 0.09 + 0.009 + 0.0009 + \cdots$$

$$= \frac{9}{10} + \frac{9}{100} + \frac{9}{1,000} + \frac{9}{10,000} + \cdots$$

$$= \frac{9}{10} + \frac{9}{10}\left(\frac{1}{10}\right) + \frac{9}{10}\left(\frac{1}{10}\right)^2 + \frac{9}{10}\left(\frac{1}{10}\right)^3 + \cdots$$

As the last line indicates, we have an infinite geometric series with $a_1 = \frac{9}{10}$ and $r = \frac{1}{10}$. The sum of this series is given by

$$S = \frac{a_1}{1-r} = \frac{\frac{9}{10}}{1-\frac{1}{10}} = \frac{\frac{9}{10}}{\frac{9}{10}} = 1$$

 Getting Ready for Class

After reading through the preceding section, respond in your own words and in complete sentences.

A. What is a common ratio?

B. Explain the formula $a_n = a_1 r^{n-1}$ in words so that someone who wanted to find the nth term of a geometric sequence could do so from your description.

C. When is the sum of an infinite geometric series a finite number?

D. Explain how a repeating decimal can be represented as an infinite geometric series.

PROBLEM SET 10.4

Identify those sequences that are geometric progressions. For those that are geometric, give the common ratio r.

1. 1, 5, 25, 125, . . .

2. 6, 12, 24, 48, . . .

3. $\frac{1}{2}, \frac{1}{6}, \frac{1}{18}, \frac{1}{54}, \ldots$

4. 5, 10, 15, 20, . . .

5. 4, 9, 16, 25, . . .

6. $-1, \frac{1}{3}, -\frac{1}{9}, \frac{1}{27}, \ldots$

7. $-2, 4, -8, 16, \ldots$

8. 1, 8, 27, 64, . . .

9. 4, 6, 8, 10, . . .

10. 1, -3, 9, -27, . . .

Each of the following problems gives some information about a specific geometric progression.

11. If $a_1 = 4$ and $r = 3$, find a_n.

12. If $a_1 = 5$ and $r = 2$, find a_n.

13. If $a_1 = -2$ and $r = -\frac{1}{2}$, find a_6.

14. If $a_1 = 25$ and $r = -\frac{1}{5}$, find a_6.

15. If $a_1 = 3$ and $r = -1$, find a_{20}.

16. If $a_1 = -3$ and $r = -1$, find a_{20}.

17. If $a_1 = 10$ and $r = 2$, find S_{10}.

18. If $a_1 = 8$ and $r = 3$, find S_5.

19. If $a_1 = 1$ and $r = -1$, find S_{20}.

20. If $a_1 = 1$ and $r = -1$, find S_{21}.

21. Find a_8 for $\frac{1}{5}, \frac{1}{10}, \frac{1}{20}, \ldots$

22. Find a_8 for $\frac{1}{2}, \frac{1}{10}, \frac{1}{50}, \ldots$

23. Find S_5 for $-\frac{1}{2}, -\frac{1}{4}, -\frac{1}{8}, \ldots$

24. Find S_6 for $-\frac{1}{2}, 1, -2, \ldots$

25. Find a_{10} and S_{10} for $\sqrt{2}, 2, 2\sqrt{2}, \ldots$

26. Find a_8 and S_8 for $\sqrt{3}, 3, 3\sqrt{3}, \ldots$

27. Find a_6 and S_6 for 100, 10, 1, ...

28. Find a_6 and S_6 for 100, -10, 1, ...

29. If $a_4 = 40$ and $a_6 = 160$, find r.

30. If $a_5 = \frac{1}{8}$ and $a_8 = \frac{1}{64}$, find r.

Find the sum of each geometric series.

31. $\frac{1}{2} + \frac{1}{4} + \frac{1}{8} + \cdots$

32. $\frac{1}{3} + \frac{1}{9} + \frac{1}{27} + \cdots$

33. $4 + 2 + 1 + \cdots$

34. $8 + 4 + 2 + \cdots$

35. $\frac{2}{5} + \frac{4}{25} + \frac{8}{125} + \cdots$ **36.** $\frac{3}{4} + \frac{9}{16} + \frac{27}{64} + \cdots$

37. $\frac{3}{4} + \frac{1}{4} + \frac{1}{12} + \cdots$ **38.** $\frac{5}{3} + \frac{1}{3} + \frac{1}{15} + \cdots$

39. Show that $0.444 \ldots$ is the same as $\frac{4}{9}$.

40. Show that $0.333 \ldots$ is the same as $\frac{1}{3}$.

41. Show that $0.272727 \ldots$ is the same as $\frac{3}{11}$.

42. Show that $0.545454 \ldots$ is the same as $\frac{6}{11}$.

Applying the Concepts

Declining-Balance Depreciation The declining-balance method of depreciation is an accounting method businesses use to deduct most of the cost of new equipment during the first few years of purchase. The value at any time during the life of the machine can be found with a linear equation in two variables. For income tax purposes, however, it is the value at the end of the year that is most important, and for this reason sequences can be used.

43. Value of a Crane A construction crane sells for $450,000 if purchased new. After that, the value decreases by 30% each year. We can use a geometric sequence to find the value of the crane at the end of each year. If we let a_0 represent the value when it is purchased, then a_1 is the value after 1 year, a_2 is the value after 2 years, and so on.
   (a) Write the first 5 terms of the sequence.
   (b) What is the common ratio?
   (c) Construct a line graph for the first 5 terms of the sequence.
   (d) Use the line graph to estimate the value of the crane 4.5 years after it is purchased.
   (e) Write the sequence from part (a) using a recursive formula.

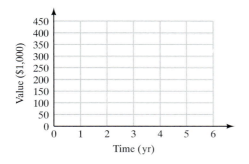

(SuperStock)

44. Value of a Printing Press A large printing press sells for $375,000 when it is new. After that, its value decreases 25% each year.
   (a) Write a geometric sequence that gives the value of the press at the end of each of the first 5 years after it is purchased.
   (b) What is the common ratio for this sequence?
   (c) Construct a line graph for this sequence.
   (d) Use the line graph to estimate the value of the printing press 1.5 years after it is purchased.
   (e) Write the sequence from part (a) using a recursive formula.

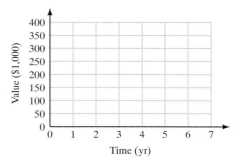

(Jim McCrary / Tony Stone Images)

45. Adding Terms Given the geometric series $\frac{1}{3} + \frac{1}{9} + \frac{1}{27} + \cdots$,
(a) Find the sum of all the terms.
(b) Find the sum of the first 6 terms.
(c) Find the sum of all but the first 6 terms.

46. Perimeter Triangle ABC has a perimeter of 40 inches. A new triangle XYZ is formed by connecting the midpoints of the sides of the first triangle, as shown in the following figure. Since midpoints are joined, the perimeter of triangle XYZ will be one-half the perimeter of triangle ABC, or 20 inches.

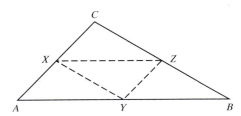

Midpoints of the sides of triangle XYZ are used to form a new triangle RST. If this pattern of using midpoints to draw triangles is continued seven more times, so that there is a total of 10 triangles drawn, what will be the sum of the perimeters of these ten triangles?

47. Bouncing Ball A ball is dropped from a height of 20 feet. Each time it bounces it returns to $\frac{7}{8}$ of the height it fell from. If the ball is allowed to bounce an infinite number of times, find the total vertical distance that the ball travels.

48. Stacking Paper Assume that a thin sheet of paper is 0.002 inch thick. The paper is torn in half, and the two halves placed together.
(a) How thick is the pile of torn paper?
(b) The pile of paper is torn in half again, and then the two halves placed together and torn in half again. The paper is large enough so that this process may be performed a total of 5 times. How thick is the pile of torn paper?
(c) Refer to the tearing and piling process described in part (b). Assuming that somehow the original paper is large enough, how thick is the pile of torn paper if 25 tears are made?

Review Problems

The following problems review material we covered in Section 5.3. Reviewing these problems will help you understand the next section.

Expand and multiply.

49. $(x + 5)^2$

50. $(x + y)^2$

51. $(x + y)^3$

52. $(x - 2)^3$

53. $(x + y)^4$

54. $(x - 1)^4$

Extending the Concepts

Use a calculator to find the given term or partial sum.

55. Find a_{20} if $a_1 = 100$ and $r = 2$.

56. Find a_{20} if $a_1 = 81$ and $r = \frac{1}{3}$.

57. Find S_{18} if $a_1 = 100$ and $r = \frac{1}{2}$.

58. Find S_{22} if $a_1 = 64$ and $r = -\frac{1}{2}$.

59. Find the sum for $a + \frac{a}{2} + \frac{a}{4} + \cdots$ if $a > 1$.

60. Find the sum for $\frac{1}{a} + \frac{1}{a^2} + \frac{1}{a^3} + \cdots$ if $a > 1$.

61. Find the sum for $\frac{a}{b} + \frac{a^2}{b^2} + \frac{a^3}{b^3} + \cdots$ if $\left| \frac{a}{b} \right| < 1$.

62. Sierpinski Triangle In the sequence that follows, the figures are moving toward what is known as the Sierpinski triangle. To construct the figure in stage 2, we remove the triangle formed from the midpoints of the sides of the shaded region in stage 1. Likewise, the figure in stage 3 is found by removing the triangles formed by connecting the midpoints of the sides of the shaded regions in stage 2. If we repeat this process infinitely many times, we arrive at the Sierpinski triangle.
(a) If the shaded region in stage 1 has an area of 1, find the area of the shaded regions in stages 2 through 4.
(b) Do the areas you found in part (a) form an arithmetic sequence or a geometric sequence?

Stage 1   Stage 2   Stage 3   Stage 4

(c) The Sierpinski triangle is the triangle that is formed after the process of forming the stages shown in the figure is repeated infinitely many times. What do you think the area of the shaded region of the Sierpinski triangle will be?

(d) Suppose the perimeter of the shaded region of the triangle in stage 1 is 1. If we were to find the perimeters of the shaded regions in the other stages, would we have an increasing sequence or a decreasing sequence?

10.5 The Binomial Expansion

The purpose of this section is to write and apply the formula for the expansion of expressions of the form $(x + y)^n$, where n is any positive integer. In order to write the formula, we must generalize the information in the following chart:

$$(x + y)^0 = \qquad\qquad\qquad\qquad 1$$
$$(x + y)^1 = \qquad\qquad\qquad x \;+\; y$$
$$(x + y)^2 = \qquad\qquad x^2 \;+\; 2xy \;+\; y^2$$
$$(x + y)^3 = \qquad x^3 \;+\; 3x^2y \;+\; 3xy^2 \;+\; y^3$$
$$(x + y)^4 = \; x^4 \;+\; 4x^3y \;+\; 6x^2y^2 \;+\; 4xy^3 \;+\; y^4$$
$$(x + y)^5 = x^5 + 5x^4y \;+\; 10x^3y^2 \;+\; 10x^2y^3 \;+\; 5xy^4 + y^5$$

Note: The polynomials to the right have been found by expanding the binomials on the left — we just haven't shown the work.

There are a number of similarities to notice among the polynomials on the right. Here is a list:

1. In each polynomial, the sequence of exponents on the variable x decreases to 0 from the exponent on the binomial at the left. (The exponent 0 is not shown, since $x^0 = 1$.)

2. In each polynomial, the exponents on the variable y increase from 0 to the exponent on the binomial at the left. (Since $y^0 = 1$, it is not shown in the first term.)

3. The sum of the exponents on the variables in any single term is equal to the exponent on the binomial at the left.

The pattern in the coefficients of the polynomials on the right can best be seen by writing the right side again without the variables. It looks like this:

row 0						1					
row 1					1		1				
row 2				1		2		1			
row 3			1		3		3		1		
row 4		1		4		6		4		1	
row 5	1		5		10		10		5		1

This triangle-shaped array of coefficients is called **Pascal's triangle.** Each entry in the triangular array is obtained by adding the two numbers above it. Each row begins and ends with the number 1. If we were to continue Pascal's triangle, the next two rows would be

row 6		1	6	15	20	15	6	1	
row 7	1	7	21	35	35	21	7	1	

The coefficients for the terms in the expansion of $(x + y)^n$ are given in the nth row of Pascal's triangle.

There is an alternative method of finding these coefficients that does not involve Pascal's triangle. The alternative method involves **factorial notation.**

DEFINITION The expression $n!$ is read "n factorial" and is the product of all the consecutive integers from n down to 1. For example,

$$1! = 1$$
$$2! = 2 \cdot 1 = 2$$
$$3! = 3 \cdot 2 \cdot 1 = 6$$
$$4! = 4 \cdot 3 \cdot 2 \cdot 1 = 24$$
$$5! = 5 \cdot 4 \cdot 3 \cdot 2 \cdot 1 = 120$$

The expression $0!$ is defined to be 1. We use factorial notation to define binomial coefficients as follows.

DEFINITION The expression $\binom{n}{r}$ is called a **binomial coefficient** and is defined by

$$\binom{n}{r} = \frac{n!}{r!(n - r)!}$$

EXAMPLE 1 Calculate the following binomial coefficients:

$$\binom{7}{5}, \binom{6}{2}, \binom{3}{0}$$

Solution We simply apply the definition for binomial coefficients:

$$\binom{7}{5} = \frac{7!}{5!(7 - 5)!}$$
$$= \frac{7!}{5! \cdot 2!}$$

$$= \frac{7 \cdot 6 \cdot \cancel{5} \cdot \cancel{4} \cdot \cancel{3} \cdot \cancel{2} \cdot \cancel{1}}{(\cancel{5} \cdot \cancel{4} \cdot \cancel{3} \cdot \cancel{2} \cdot \cancel{1})(2 \cdot 1)}$$

$$= \frac{42}{2}$$

$$= 21$$

$$\binom{6}{2} = \frac{6!}{2!(6-2)!}$$

$$= \frac{6!}{2! \cdot 4!}$$

$$= \frac{6 \cdot 5 \cdot \cancel{4} \cdot \cancel{3} \cdot \cancel{2} \cdot \cancel{1}}{(2 \cdot 1)(\cancel{4} \cdot \cancel{3} \cdot \cancel{2} \cdot \cancel{1})}$$

$$= \frac{30}{2}$$

$$= 15$$

$$\binom{3}{0} = \frac{3!}{0!(3-0)!}$$

$$= \frac{3!}{0! \cdot 3!}$$

$$= \frac{\cancel{3} \cdot \cancel{2} \cdot \cancel{1}}{(1)(\cancel{3} \cdot \cancel{2} \cdot \cancel{1})}$$

$$= 1$$

If we were to calculate all the binomial coefficients in the following array, we would find they match exactly with the numbers in Pascal's triangle. That is why they are called binomial coefficients — because they are the coefficients of the expansion of $(x + y)^n$.

$$\binom{0}{0}$$

$$\binom{1}{0} \qquad \binom{1}{1}$$

$$\binom{2}{0} \qquad \binom{2}{1} \qquad \binom{2}{2}$$

$$\binom{3}{0} \qquad \binom{3}{1} \qquad \binom{3}{2} \qquad \binom{3}{3}$$

$$\binom{4}{0} \qquad \binom{4}{1} \qquad \binom{4}{2} \qquad \binom{4}{3} \qquad \binom{4}{4}$$

$$\binom{5}{0} \qquad \binom{5}{1} \qquad \binom{5}{2} \qquad \binom{5}{3} \qquad \binom{5}{4} \qquad \binom{5}{5}$$

Using the new notation to represent the entries in Pascal's triangle, we can summarize everything we have noticed about the expansion of binomial powers of the form $(x + y)^n$.

THE BINOMIAL EXPANSION

If x and y represent real numbers and n is a positive integer, then the following formula is known as the **binomial expansion** or **binomial formula:**

$$(x + y)^n = \binom{n}{0} x^n y^0 + \binom{n}{1} x^{n-1} y^1 + \binom{n}{2} x^{n-2} y^2 + \cdots + \binom{n}{n} x^0 y^n$$

It does not make any difference, when expanding binomial powers of the form $(x + y)^n$, whether we use Pascal's triangle or the formula

$$\binom{n}{r} = \frac{n!}{r!(n - r)!}$$

to calculate the coefficients. We will show examples of both methods.

EXAMPLE 2   Expand $(x - 2)^3$.

Solution   Applying the binomial formula, we have

$$(x - 2)^3 = \binom{3}{0} x^3(-2)^0 + \binom{3}{1} x^2(-2)^1 + \binom{3}{2} x^1(-2)^2 + \binom{3}{3} x^0(-2)^3$$

The coefficients

$$\binom{3}{0}, \binom{3}{1}, \binom{3}{2}, \text{ and } \binom{3}{3}$$

can be found in the third row of Pascal's triangle. They are 1, 3, 3, and 1:

$$(x - 2)^3 = 1x^3(-2)^0 + 3x^2(-2)^1 + 3x^1(-2)^2 + 1x^0(-2)^3$$
$$= x^3 - 6x^2 + 12x - 8$$

EXAMPLE 3   Expand $(3x + 2y)^4$.

Solution   The coefficients can be found in the fourth row of Pascal's triangle.

$$1, 4, 6, 4, 1$$

Here is the expansion of $(3x + 2y)^4$:

$$(3x + 2y)^4 = 1(3x)^4 + 4(3x)^3(2y) + 6(3x)^2(2y)^2 + 4(3x)(2y)^3 + 1(2y)^4$$
$$= 81x^4 + 216x^3y + 216x^2y^2 + 96xy^3 + 16y^4$$

EXAMPLE 4   Write the first three terms in the expansion of $(x + 5)^9$.

Solution   The coefficients of the first three terms are

$$\binom{9}{0}, \binom{9}{1}, \text{ and } \binom{9}{2}$$

which we calculate as follows:

$$\binom{9}{0} = \frac{9!}{0! \cdot 9!} = \frac{9 \cdot 8 \cdot 7 \cdot 6 \cdot 5 \cdot 4 \cdot 3 \cdot 2 \cdot 1}{(1)(9 \cdot 8 \cdot 7 \cdot 6 \cdot 5 \cdot 4 \cdot 3 \cdot 2 \cdot 1)} = \frac{1}{1} = 1$$

$$\binom{9}{1} = \frac{9!}{1! \cdot 8!} = \frac{9 \cdot 8 \cdot 7 \cdot 6 \cdot 5 \cdot 4 \cdot 3 \cdot 2 \cdot 1}{(1)(8 \cdot 7 \cdot 6 \cdot 5 \cdot 4 \cdot 3 \cdot 2 \cdot 1)} = \frac{9}{1} = 9$$

$$\binom{9}{2} = \frac{9!}{2! \cdot 7!} = \frac{9 \cdot 8 \cdot 7 \cdot 6 \cdot 5 \cdot 4 \cdot 3 \cdot 2 \cdot 1}{(2 \cdot 1)(7 \cdot 6 \cdot 5 \cdot 4 \cdot 3 \cdot 2 \cdot 1)} = \frac{72}{2} = 36$$

From the binomial formula, we write the first three terms:

$$(x + 5)^9 = 1 \cdot x^9 + 9 \cdot x^8(5) + 36x^7(5)^2 + \cdots$$
$$= x^9 + 45x^8 + 900x^7 + \cdots$$

THE KTH TERM OF A BINOMIAL EXPANSION

If we look at each term in the expansion of $(x + y)^n$ as a term in a sequence, a_1, a_2, a_3, . . . , we can write

$$a_1 = \binom{n}{0} x^n y^0$$

$$a_2 = \binom{n}{1} x^{n-1} y^1$$

$$a_3 = \binom{n}{2} x^{n-2} y^2$$

$$a_4 = \binom{n}{3} x^{n-3} y^3 \qquad \text{and so on}$$

To write the formula for the general term, we simply notice that the exponent on y and the number below n in the coefficient are both 1 less than the term number. This observation allows us to write the following:

THE GENERAL TERM OF A BINOMIAL EXPANSION

The kth term in the expansion of $(x + y)^n$ is

$$a_k = \binom{n}{k-1} x^{n-(k-1)} y^{k-1}$$

EXAMPLE 5   Find the fifth term in the expansion of $(2x + 3y)^{12}$.

Solution   Applying the preceding formula, we have

$$a_5 = \binom{12}{4} (2x)^8 (3y)^4$$

$$= \frac{12!}{4! \cdot 8!} (2x)^8 (3y)^4$$

Notice that once we have one of the exponents, the other exponent and the denominator of the coefficient are determined: The two exponents add to 12 and match the denominator of the coefficient.

Making the calculations from the preceding formula, we have

$$a_5 = 495(256x^8)(81y^4)$$

$$= 10{,}264{,}320x^8y^4$$

Getting Ready for Class

After reading through the preceding section, respond in your own words and in complete sentences.

A. What is Pascal's triangle?

B. Why is $\binom{n}{0} = 1$ for any natural number?

C. State the binomial formula.

D. When is the binomial formula more efficient than multiplying to expand a binomial raised to a whole-number exponent?

PROBLEM SET 10.5

Use the binomial formula to expand each of the following.

1. $(x + 2)^4$

2. $(x - 2)^5$

3. $(x + y)^6$

4. $(x - 1)^6$

5. $(2x + 1)^5$

6. $(2x - 1)^4$

7. $(x - 2y)^5$

8. $(2x + y)^5$

9. $(3x - 2)^4$

10. $(2x - 3)^4$

11. $(4x - 3y)^3$

12. $(3x - 4y)^3$

13. $(x^2 + 2)^4$

14. $(x^2 - 3)^3$

15. $(x^2 + y^2)^3$

16. $(x^2 - 3y)^4$

17. $\left(\dfrac{x}{2} - 4\right)^3$

18. $\left(\dfrac{x}{3} + 6\right)^3$

19. $\left(\dfrac{x}{3} + \dfrac{y}{2}\right)^4$

20. $\left(\dfrac{x}{2} - \dfrac{y}{3}\right)^4$

Write the first four terms in the expansion of the following.

21. $(x + 2)^9$

22. $(x - 2)^9$

23. $(x - y)^{10}$ **24.** $(x + y)^{10}$

25. $(x + 2y)^{10}$ **26.** $(x - 2y)^{10}$

Write the first three terms in the expansion of each of the following.

27. $(x + 1)^{15}$ **28.** $(x - 1)^{15}$

29. $(x - y)^{12}$ **30.** $(x + y)^{12}$

31. $(x + 2)^{20}$ **32.** $(x - 2)^{20}$

Write the first two terms in the expansion of each of the following.

33. $(x + 2)^{100}$

34. $(x - 2)^{50}$

35. $(x + y)^{50}$

36. $(x - y)^{100}$

37. Find the ninth term in the expansion of $(2x + 3y)^{12}$.

38. Find the sixth term in the expansion of $(2x + 3y)^{12}$.

39. Find the fifth term of $(x - 2)^{10}$.

40. Find the fifth term of $(2x - 1)^{10}$.

41. Find the fourth term of $(x + 3)^9$.

42. Find the fifth term of $(x + 3)^9$.

43. Write the formula for the 12th term of $(2x + 5y)^{20}$. Do not simplify.

44. Write the formula for the eighth term of $(2x + 5y)^{20}$. Do not simplify.

Applying the Concepts

45. Probability   The third term in the expansion of $(\frac{1}{2} + \frac{1}{2})^7$ will give the probability that in a family with 7 children, 5 will be boys and 2 will be girls. Find the third term.

46. Probability   The fourth term in the expansion of $(\frac{1}{2} + \frac{1}{2})^8$ will give the probability that in a family with 8 children, 3 will be boys and 5 will be girls. Find the fourth term.

47. Multiplication   Without using a calculator, evaluate 1.01^4 by using the binomial formula to expand $(1 + 0.01)^4$.

48. Multiplication   Without using a calculator, evaluate 1.02^5 by using the binomial formula.

Review Problems

The problems that follow review material we covered in Section 9.6.

Solve each equation. Write your answers to the nearest hundredth.

49. $5^x = 7$ **50.** $10^x = 15$

51. $8^{2x+1} = 16$ **52.** $9^{3x-1} = 27$

53. Compound Interest   How long will it take $400 to double if it is invested in an account with an annual interest rate of 10% compounded four times a year?

54. Compound Interest   How long will it take $200 to become $800 if it is invested in an account with an annual interest rate of 8% compounded four times a year?

Find each of the following to the nearest hundredth.

55. $\log_4 20$ **56.** $\log_7 21$

57. $\ln 576$ **58.** $\ln 5{,}760$

59. Solve the formula $A = 10e^{5t}$ for t.

60. Solve the formula $A = P2^{-5t}$ for t.

Extending the Concepts

61. Calculate both $\binom{8}{5}$ and $\binom{8}{3}$ to show that they are equal.

62. Calculate both $\binom{10}{8}$ and $\binom{10}{2}$ to show that they are equal.

63. Simplify $\binom{20}{12}$ and $\binom{20}{8}$.

64. Simplify $\binom{15}{10}$ and $\binom{15}{5}$.

65. Show that $\binom{n}{r}$ and $\binom{n}{n-r}$ are equal.

66. Use the results of Problem 63 to find an easy way to compute $\binom{24}{23}$.

67. Choose three values of n, say $n = 3$, $n = 5$, and $n = 7$, and then verify the following formula for each value of n.

$$\binom{n}{0} + \binom{n}{1} + \binom{n}{2} + \cdots + \binom{n}{n} = 2^n$$

68. Pascal's Triangle Copy the first eight rows of Pascal's triangle into the eight rows of the following triangular array. (Each number in Pascal's triangle will go into one of the hexagons in the array.) Next, color in each hexagon that contains an odd number. What pattern begins to emerge from this coloring process?

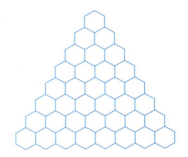

CHAPTER 10   SUMMARY

Examples

1. In the sequence $1, 3, 5, \ldots,$ $2n - 1, \ldots, a_1 = 1, a_2 = 3,$ $a_3 = 5$, and $a_n = 2n - 1$.

Sequences [10.1]

A *sequence* is a function whose domain is the set of positive integers. The terms of a sequence are denoted by

$$a_1, a_2, a_3, \ldots, a_n, \ldots$$

where a_1 (read "a sub 1") is the first term, a_2 the second term, and a_n the nth or *general term*.

2. $\displaystyle\sum_{i=3}^{6} (-2)^i$

$\quad = (-2)^3 + (-2)^4 + (-2)^5$
$\quad\quad + (-2)^6$

$\quad = -8 + 16 + (-32) + 64$

$\quad = 40$

Summation Notation [10.2]

The notation

$$\sum_{i=1}^{n} a_i = a_1 + a_2 + a_3 + \cdots + a_n$$

is called *summation notation* or *sigma notation*. The letter i as used here is called the *index of summation* or just *index*.

3. For the sequence 3, 7, 11, $15, \ldots, a_1 = 3$ and $d = 4$. The general term is

$$a_n = 3 + (n - 1)4$$
$$\quad = 4n - 1$$

Using this formula to find the tenth term, we have

$$a_{10} = 4(10) - 1 = 39$$

The sum of the first ten terms is

$$S_{10} = \frac{10}{2}(3 + 39) = 210$$

Arithmetic Sequences [10.3]

An *arithmetic sequence* is a sequence in which each term comes from the preceding term by adding a constant amount each time. If the first term of an arithmetic sequence is a_1 and the amount we add each time (called the *common difference*) is d, then the nth term of the progression is given by

$$a_n = a_1 + (n - 1)d$$

The sum of the first n terms of an arithmetic sequence is

$$S_n = \frac{n}{2}(a_1 + a_n)$$

S_n is called the nth *partial sum*.

4. For the geometric progression 3, $6, 12, 24, \ldots, a_1 = 3$ and $r = 2$. The general term is

$$a_n = 3 \cdot 2^{n-1}$$

The sum of the first ten terms is

$$S_{10} = \frac{3(2^{10} - 1)}{2 - 1} = 3{,}069$$

Geometric Sequences [10.4]

A *geometric sequence* is a sequence of numbers in which each term comes from the previous term by multiplying by a constant amount each time. The constant by which we multiply each term to get the next term is called the *common ratio*. If the first term of a geometric sequence is a_1 and the common ratio is r, then the formula that gives the general term a_n is

$$a_n = a_1 r^{n-1}$$

The sum of the first n terms of a geometric sequence is given by the formula

$$S_n = \frac{a_1(r^n - 1)}{r - 1}$$

5. The sum of the series

$$\frac{1}{3} + \frac{1}{6} + \frac{1}{12} + \cdots$$

is

$$S = \frac{\frac{1}{3}}{1 - \frac{1}{2}} = \frac{\frac{1}{3}}{\frac{1}{2}} = \frac{2}{3}$$

The Sum of an Infinite Geometric Series [10.4]

If a geometric sequence has first term a_1 and common ratio r such that $|r| < 1$, then the following is called an *infinite geometric series:*

$$S = \sum_{i=0}^{\infty} a_1 r^i = a_1 + a_1 r + a_1 r^2 + a_1 r^3 + \cdots$$

Its sum is given by the formula

$$S = \frac{a_1}{1 - r}$$

Factorials [10.5]

The notation $n!$ is called *n factorial* and is defined to be the product of each consecutive integer from n down to 1. That is,

$$0! = 1 \qquad \text{(By definition)}$$
$$1! = 1$$
$$2! = 2 \cdot 1$$
$$3! = 3 \cdot 2 \cdot 1$$
$$4! = 4 \cdot 3 \cdot 2 \cdot 1$$

and so on.

6. $\displaystyle \binom{7}{3} = \frac{7!}{3!(7-3)!}$

$$= \frac{7!}{3! \cdot 4!}$$

$$= \frac{7 \cdot 6 \cdot 5 \cdot 4 \cdot 3 \cdot 2 \cdot 1}{(3 \cdot 2 \cdot 1)(4 \cdot 3 \cdot 2 \cdot 1)}$$

$$= 35$$

Binomial Coefficients [10.5]

The notation $\displaystyle \binom{n}{r}$ is called a *binomial coefficient* and is defined by

$$\binom{n}{r} = \frac{n!}{r!(n-r)!}$$

Binomial coefficients can be found by using the formula above or by *Pascal's triangle,* which is

```
          1
        1   1
      1   2   1
    1   3   3   1
  1   4   6   4   1
1   5   10   10   5   1
```

and so on.

7. $(x + 2)^4$

$= x^4 + 4x^3 \cdot 2 + 6x^2 \cdot 2^2$

$+ 4x \cdot 2^3 + 2^4$

$= x^4 + 8x^3 + 24x^2 + 32x + 16$

Binomial Expansion [10.5]

If n is a positive integer, then the formula for expanding $(x + y)^n$ is given by

$$(x + y)^n = \binom{n}{0} x^n y^0 + \binom{n}{1} x^{n-1} y^1 + \binom{n}{2} x^{n-2} y^2 + \cdots + \binom{n}{n} x^0 y^n$$

CHAPTER 10   REVIEW

Write the first four terms of the sequence with the following general terms. [10.1]

1. $a_n = 2n + 5$

2. $a_n = 3n - 2$

3. $a_n = n^2 - 1$

4. $a_n = \dfrac{n + 3}{n + 2}$

5. $a_1 = 4, a_n = 4a_{n-1}, n > 1$

6. $a_1 = \frac{1}{4}, a_n = \frac{1}{4}a_{n-1}, n > 1$

Determine the general term for each of the following sequences. [10.1]

7. $2, 5, 8, 11, \ldots$

8. $-3, -1, 1, 3, 5, \ldots$

9. $1, 16, 81, 256, \ldots$

10. $2, 5, 10, 17, \ldots$

11. $\frac{1}{2}, \frac{1}{4}, \frac{1}{8}, \frac{1}{16}, \ldots$

12. $2, \frac{3}{4}, \frac{4}{9}, \frac{5}{16}, \frac{6}{25}, \ldots$

Expand and simplify each of the following. [10.2]

13. $\displaystyle\sum_{i=1}^{4} (2i + 3)$

14. $\displaystyle\sum_{i=1}^{3} (2i^2 - 1)$

15. $\displaystyle\sum_{i=2}^{3} \frac{i^2}{i + 2}$

16. $\displaystyle\sum_{i=1}^{4} (-2)^{i-1}$

17. $\displaystyle\sum_{i=3}^{5} (4i + i^2)$

18. $\displaystyle\sum_{i=4}^{6} \frac{i + 2}{i}$

Write each of the following sums with summation notation. [10.2]

19. $3 + 6 + 9 + 12$

20. $3 + 7 + 11 + 15$

21. $5 + 7 + 9 + 11 + 13$

22. $4 + 9 + 16$

23. $\frac{1}{3} + \frac{1}{4} + \frac{1}{5} + \frac{1}{6}$

24. $\frac{1}{3} + \frac{2}{9} + \frac{3}{27} + \frac{4}{81} + \frac{5}{243}$

25. $(x - 2) + (x - 4) + (x - 6)$

26. $\dfrac{x}{x + 1} + \dfrac{x}{x + 2} + \dfrac{x}{x + 3} + \dfrac{x}{x + 4}$

Determine which of the following sequences are arithmetic progressions, geometric progressions, or neither. [10.3, 10.4]

27. $1, -3, 9, -27, \ldots$

28. $7, 9, 11, 13, \ldots$

29. $5, 11, 17, 23, \ldots$

30. $\frac{1}{2}, \frac{1}{3}, \frac{1}{4}, \frac{1}{5}, \ldots$

31. $4, 8, 16, 32, \ldots$

32. $\frac{1}{2}, \frac{1}{4}, \frac{1}{8}, \frac{1}{16}, \ldots$

33. $12, 9, 6, 3, \ldots$

34. $2, 5, 9, 14, \ldots$

Each of the following problems refers to arithmetic progressions. [10.3]

35. If $a_1 = 2$ and $d = 3$, find a_n and a_{20}.

36. If $a_1 = 5$ and $d = -3$, find a_n and a_{16}.

37. If $a_1 = -2$ and $d = 4$, find a_{10} and S_{10}.

38. If $a_1 = 3$ and $d = 5$, find a_{16} and S_{16}.

39. If $a_5 = 21$ and $a_8 = 33$, find the first term a_1, the common difference d, and then find a_{10}.

40. If $a_3 = 14$ and $a_7 = 26$, find the first term a_1, the common difference d, and then find a_9 and S_9.

41. If $a_4 = -10$ and $a_8 = -18$, find the first term a_1, the common difference d, and then find a_{20} and S_{20}.

42. Find the sum of the first 100 terms of the sequence 3, 7, 11, 15, 19, ...

43. Find a_{40} for the sequence 100, 95, 90, 85, 80, ...

Each of the following problems refers to infinite geometric progressions. [10.4]

44. If $a_1 = 3$ and $r = 2$, find a_n and a_{20}.

45. If $a_1 = 5$ and $r = -2$, find a_n and a_{16}.

46. If $a_1 = 4$ and $r = \frac{1}{2}$, find a_n and a_{10}.

47. If $a_1 = -2$ and $r = \frac{1}{3}$, find the sum.

48. If $a_1 = 4$ and $r = \frac{1}{2}$, find the sum.

49. If $a_3 = 12$ and $a_4 = 24$, find the first term a_1, the common ratio r, and then find a_6.

50. Find the tenth term of the sequence 3, $3\sqrt{3}$, 9, $9\sqrt{3}$, ...

Evaluate each of the following. [10.5]

51. $\begin{pmatrix} 8 \\ 2 \end{pmatrix}$

52. $\begin{pmatrix} 7 \\ 4 \end{pmatrix}$

53. $\begin{pmatrix} 6 \\ 3 \end{pmatrix}$

54. $\begin{pmatrix} 9 \\ 2 \end{pmatrix}$

55. $\begin{pmatrix} 10 \\ 8 \end{pmatrix}$

56. $\begin{pmatrix} 100 \\ 3 \end{pmatrix}$

Use the binomial formula to expand each of the following. [10.5]

57. $(x - 2)^4$

58. $(2x + 3)^4$

59. $(3x + 2y)^3$

60. $(x^2 - 2)^5$

61. $\left(\frac{x}{2} + 3\right)^4$

62. $\left(\frac{x}{3} - \frac{y}{2}\right)^3$

Use the binomial formula to write the first three terms in the expansion of the following. [10.5]

63. $(x + 3y)^{10}$

64. $(x - 3y)^9$

65. $(x + y)^{11}$

66. $(x - 2y)^{12}$

Use the binomial formula to write the first two terms in the expansion of the following. [10.5]

67. $(x - 2y)^{16}$

68. $(x + 2y)^{32}$

69. $(x - 1)^{50}$

70. $(x + y)^{150}$

71. Find the sixth term in $(x - 3)^{10}$.

72. Find the fourth term in $(2x + 1)^9$.

CHAPTER 10 **PROJECTS**

SEQUENCES AND SERIES

GROUP PROJECT

CREDIT CARD PAYMENTS

Number of People: 2

Time Needed: 20–30 minutes

Equipment: Paper, pencil, and graphing calculator

Background: In the beginning of this chapter, you were given a recursive function that you can use to find how long it will take to pay off a credit card. A graphing calculator can be used to model each payment by using the recall function on the graphing calculator. To set up this problem do the following:

(1)   1000   ENTER

(2)   Round (1.0165ANS−20,2)   ENTER

Note: The *Round* function is under the MATH key in the NUM list.

Procedure: Enter the preceding commands into your calculator. While one person in the group hits the ENTER key, another person counts, by keeping a tally of the "payments" made. The credit card is paid off when the calculator displays a negative number.

1. How many months did it take to pay off the credit card?
2. What was the amount of the last payment?
3. What was the total interest paid to the credit card company?
4. How much would you save if you paid $25 per month instead of $20?
5. If the credit card company raises the interest rate to 21.5% annual interest, how long will it take to pay off the balance? How much more would it cost in interest?
6. On larger balances, many credit card companies require a minimum payment of 2% of the outstanding balance. What is the recursion formula for this? How much would this save or cost you in interest?

Detach here and return with check or money order.

Summary of Corporate Card Account
Retain this portion for your files

Corporate Cardmember Name	Account Number	Closing Date
Leonardo Fibonacci	00000-1000-001	08-01-99

New Balance	Other Debits	Interest Rate	Minimim Payment
$1,000.00	$.00	19.8%	$20.00

R E S E A R C H P R O J E C T

BUILDING SQUARES FROM ODD NUMBERS

Leonardo Fibonacci
(Corbis/Bettmann)

A relationship exists between the sequence of squares and the sequence of odd numbers. In *The Book of Squares,* written in 1225, Leonardo Fibonacci has this to say about that relationship:

> I thought about the origin of all square numbers and discovered that they arise out of the increasing sequence of odd numbers.

Work with the sequence of odd numbers until you discover how it can be used to produce the sequence of squares. Then write an essay in which you give a written description of the relationship between the two sequences, along with a diagram that illustrates the relationship. Then see if you can use summation notation to write an equation that summarizes the whole relationship. Your essay should be clear and concise and written so that any of your classmates can read it and understand the relationship you are describing.

CHAPTER 10   TEST

Write the first five terms of the sequences with the following general terms [10.1]

1. $a_n = 3n - 5$

2. $a_1 = 3, a_n = a_{n-1} + 4, n > 1$

3. $a_n = n^2 + 1$ **4.** $a_n = 2n^3$

5. $a_n = \dfrac{n + 1}{n^2}$

6. $a_1 = 4, a_n = -2a_{n-1}, n > 1$

Give the general term for each sequence. [10.1]

7. $6, 10, 14, 18, \ldots$ **8.** $1, 2, 4, 8, \ldots$

9. $\frac{1}{2}, \frac{1}{4}, \frac{1}{8}, \frac{1}{16}, \ldots$ **10.** $-3, 9, -27, 81, \ldots$

11. Expand and simplify each of the following. [10.2]

(a) $\displaystyle\sum_{i=1}^{5} (5i + 3)$ (b) $\displaystyle\sum_{i=3}^{5} (2^i - 1)$

(c) $\displaystyle\sum_{i=2}^{6} (i^2 + 2i)$

12. Find the first term of an arithmetic progression if $a_5 = 11$ and $a_9 = 19$. [10.3]

13. Find the second term of a geometric progression for which $a_3 = 18$ and $a_5 = 162$. [10.4]

Find the sum of the first 10 terms of the following arithmetic progressions. [10.3]

14. $5, 11, 17, \ldots$ **15.** $25, 20, 15, \ldots$

16. Write a formula for the sum of the first 50 terms of the geometric progression $3, 6, 12, \ldots$ [10.4]

17. Find the sum of $\frac{1}{2} + \frac{1}{6} + \frac{1}{18} + \frac{1}{54} + \cdots$. [10.4]

Use the binomial formula to expand each of the following. [10.5]

18. $(x - 3)^4$ **19.** $(2x - 1)^5$

20. Find the first 3 terms in the expansion of $(x - 1)^{20}$. [10.5]

21. Find the sixth term in $(2x - 3y)^8$. [10.5]

Simplify.

1. $\dfrac{5(-6) + 3(-2)}{4(-3) + 3}$

2. $9 + 5(4y + 8) + 10y$

3. $\dfrac{18a^7b^{-4}}{36a^2b^{-8}}$

4. $\dfrac{y^2 - y - 6}{y^2 - 4}$

5. $8^{-2/3}$

6. $\log_3 27$

Factor completely.

7. $ab^3 + b^3 + 6a + 6$

8. $8x^2 - 5x - 3$

Solve.

9. $6 - 2(5x - 1) + 4x = 20$

10. $|4x - 3| + 2 = 3$

11. $(x + 1)(x + 2) = 12$

12. $1 - \dfrac{2}{x} = \dfrac{8}{x^2}$

13. $t - 6 = \sqrt{t - 4}$

14. $(4x - 3)^2 = -50$

15. $8t^3 - 27 = 0$

16. $6x^4 - 13x^2 = 5$

Solve and graph the solution on the number line.

17. $-3y - 2 < 7$

18. $|2x + 5| - 2 < 9$

Graph on a rectangular coordinate system.

19. $2x - 3y = 6$

20. $y < \frac{1}{2}x + 3$

21. $y = x^2 - 2x - 3$

22. $y = \left(\frac{1}{2}\right)^x$

Solve each system.

23. $\begin{aligned} 5x - 3y &= -4 \\ x + 2y &= 7 \end{aligned}$

24. $\begin{aligned} x + 2y &= 0 \\ 3y + z &= -3 \\ 2x - z &= 5 \end{aligned}$

Multiply.

25. $\dfrac{3y^2 - 3y}{3y - 12} \cdot \dfrac{y^2 - 2y - 8}{y^2 + 3y + 2}$

26. $(x^{3/5} + 2)(x^{3/5} - 2)$

27. $(2 + 3i)(1 - 4i)$

28. Add: $\dfrac{-3}{x^2 - 2x - 8} + \dfrac{4}{x^2 - 16}$

29. Combine: $4\sqrt{50} + 3\sqrt{8}$

30. Rationalize the denominator:

$$\dfrac{3}{\sqrt{7} - \sqrt{3}}$$

31. If $f(x) = 4x + 1$, find $f^{-1}(x)$.

32. Find x if $\log x = 3.9786$.

33. Find $\log_6 14$ to the nearest hundredth.

Find the general term for each sequence.

34. $3, 14, 25, \ldots$

35. $16, 8, 4, \ldots$

36. Solve $S = 2x^2 + 4xy$ for y.

37. Expand and simplify.

$$\sum_{i=1}^{5} (2i - 1)$$

38. Find the slope of the line through $(2, -3)$ and $(-1, -3)$.

39. Find the equation of the line through $(2, 5)$ and $(6, -3)$. Write your answer in slope-intercept form.

40. If $C(t) = 80(\frac{1}{2})^{t/5}$, find $C(5)$ and $C(10)$.

41. Find an equation with solutions $t = -\frac{3}{4}$ and $t = \frac{1}{5}$.

42. Evaluate.

$$\begin{vmatrix} 2 & 1 \\ 4 & 3 \end{vmatrix}$$

43. Use Cramer's rule to solve.

$$\begin{aligned} 3x - 5y &= 2 \\ 2x + 4y &= 1 \end{aligned}$$

44. Find the first term in the expansion of $(2x - y)^5$.

45. Add -8 to the difference of -7 and 4.

46. If the conditional statement "If $a = 3$, then $a^2 = 9$" is true, what other conditional statement must also be true?

47. Geometry Two supplementary angles are such that one is 5 times as large as the other. Find the two angles.

48. Variation y varies directly with x. If y is 24 when x is 8, find y when x is 2.

49. Number Problem One number is 3 times another. The sum of their reciprocals is $\frac{4}{3}$. Find the numbers.

50. Mixture How much 30% alcohol solution and 70% alcohol solution must be mixed to get 16 gallons of 60% alcohol solution?

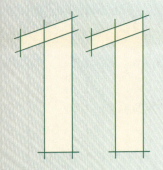

Conic Sections

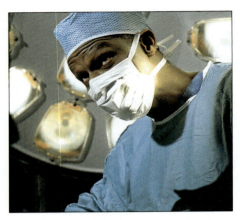

(FPG International/Telegraph Colour Library)

INTRODUCTION

One of the curves we will study in this chapter has interesting reflective properties. Figure 1(a) shows how you can draw one of these curves (an ellipse) using thumbtacks, string, pencil, and paper. Elliptical surfaces will reflect sound waves that originate at one focus through the other focus. This property of ellipses allows doctors to treat patients with kidney stones using a procedure called lithotripsy. A lithotripter is an elliptical device that creates sound waves that crush the kidney stone into small pieces, without surgery. The sound wave originates at one focus of the lithotripter. The energy from it reflects off the surface of the lithotripter and converges at the other focus, where the kidney stone is positioned. Figure 1(b) shows a cross section of a lithotripter, with a patient positioned so that the kidney stone is at the other focus.

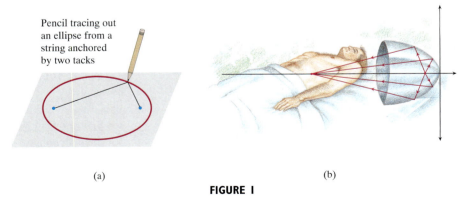

Pencil tracing out an ellipse from a string anchored by two tacks

(a)

(b)

FIGURE 1

By studying the conic sections in this chapter, you will be better equipped to understand some of the more technical equipment that exists in the world outside of class.

Conic sections include ellipses, circles, hyperbolas, and parabolas. They are called conic sections because each can be found by slicing a cone with a plane as shown in Figure 1. We begin our work with conic sections by studying circles. Before we find the general equation of a circle, we must first derive what is known as the *distance formula.*

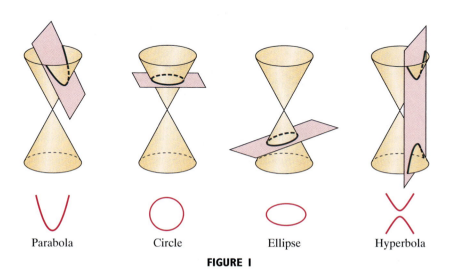

| Parabola | Circle | Ellipse | Hyperbola |

FIGURE 1

Suppose (x_1, y_1) and (x_2, y_2) are any two points in the first quadrant. (Actually, we could choose the two points to be anywhere on the coordinate plane. It is just more convenient to have them in the first quadrant.) We can name the points P_1 and P_2, respectively, and draw the diagram shown in Figure 2.

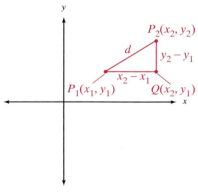

FIGURE 2

Notice the coordinates of point Q. The x-coordinate is x_2 since Q is directly below point P_2. The y-coordinate of Q is y_1 since Q is directly across from point P_1. It is evident from the diagram that the length of P_2Q is $y_2 - y_1$ and the length of P_1Q is $x_2 - x_1$. Using the Pythagorean theorem, we have

$$(P_1P_2)^2 = (P_1Q)^2 + (P_2Q)^2$$

or

$$d^2 = (x_2 - x_1)^2 + (y_2 - y_1)^2$$

Taking the square root of both sides, we have

$$d = \sqrt{(x_2 - x_1)^2 + (y_2 - y_1)^2}$$

We know this is the positive square root, since d is the distance from P_1 to P_2 and must therefore be positive. This formula is called the *distance formula.*

EXAMPLE 1   Find the distance between $(3, 5)$ and $(2, -1)$.

Solution   If we let $(3, 5)$ be (x_1, y_1) and $(2, -1)$ be (x_2, y_2) and apply the distance formula, we have

$$d = \sqrt{(2 - 3)^2 + (-1 - 5)^2}$$
$$= \sqrt{(-1)^2 + (-6)^2}$$
$$= \sqrt{1 + 36}$$
$$= \sqrt{37}$$

EXAMPLE 2   Find x if the distance from $(x, 5)$ to $(3, 4)$ is $\sqrt{2}$.

Solution   Using the distance formula, we have

$$\sqrt{2} = \sqrt{(x - 3)^2 + (5 - 4)^2}$$
$$2 = (x - 3)^2 + 1^2$$
$$2 = x^2 - 6x + 9 + 1$$
$$0 = x^2 - 6x + 8$$
$$0 = (x - 4)(x - 2)$$
$$x = 4 \qquad \text{or} \qquad x = 2$$

The two solutions are 4 and 2, which indicates that two points, $(4, 5)$ and $(2, 5)$, are $\sqrt{2}$ units from $(3, 4)$.

We can use the distance formula to derive the equation of a circle.

> ### THEOREM 11.1
>
> The equation of the circle with center at (a, b) and radius r is given by
> $$(x - a)^2 + (y - b)^2 = r^2$$

Proof   By definition, all points on the circle are a distance r from the center (a, b). If we let (x, y) represent any point on the circle, then (x, y) is r units from (a, b). Applying the distance formula, we have

$$r = \sqrt{(x - a)^2 + (y - b)^2}$$

Squaring both sides of this equation gives the equation of the circle:

$$(x - a)^2 + (y - b)^2 = r^2$$

We can use Theorem 11.1 to find the equation of a circle, given its center and radius, or to find its center and radius, given the equation.

EXAMPLE 3   Find the equation of the circle with center at $(-3, 2)$ having a radius of 5.

Solution   We have $(a, b) = (-3, 2)$ and $r = 5$. Applying Theorem 11.1 yields

$$[x - (-3)]^2 + (y - 2)^2 = 5^2$$
$$(x + 3)^2 + (y - 2)^2 = 25$$

EXAMPLE 4   Give the equation of the circle with radius 3 whose center is at the origin.

Solution   The coordinates of the center are $(0, 0)$, and the radius is 3. The equation must be

$$(x - 0)^2 + (y - 0)^2 = 3^2$$
$$x^2 + y^2 = 9$$

We can see from Example 4 that the equation of any circle with its center at the origin and radius r will be

$$x^2 + y^2 = r^2$$

EXAMPLE 5   Find the center and radius, and sketch the graph, of the circle whose equation is

$$(x - 1)^2 + (y + 3)^2 = 4$$

Solution   Writing the equation in the form

$$(x - a)^2 + (y - b)^2 = r^2$$

we have

$$(x - 1)^2 + [y - (-3)]^2 = 2^2$$

The center is at $(1, -3)$, and the radius is 2. (See Figure 3.)

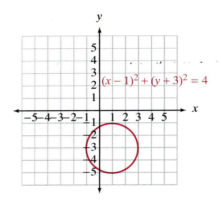

FIGURE 3

EXAMPLE 6   Sketch the graph of $x^2 + y^2 = 9$.

Solution   Since the equation can be written in the form

$$(x - 0)^2 + (y - 0)^2 = 3^2$$

it must have its center at $(0, 0)$ and a radius of 3. (See Figure 4.)

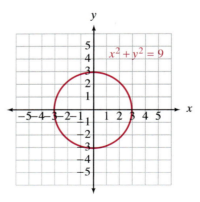

FIGURE 4

EXAMPLE 7   Sketch the graph of $x^2 + y^2 + 6x - 4y - 12 = 0$.

Solution   To sketch the graph, we must find the center and radius. The center and radius can be identified if the equation has the form

$$(x - a)^2 + (y - b)^2 = r^2$$

The original equation can be written in this form by completing the squares on x and y:

$$x^2 + y^2 + 6x - 4y - 12 = 0$$

$$x^2 + 6x \quad\;\; + y^2 - 4y \quad\;\; = 12$$

$$x^2 + 6x + \mathbf{9} + y^2 - 4y + \mathbf{4} = 12 + \mathbf{9} + \mathbf{4}$$

$$(x + 3)^2 + (y - 2)^2 = 25$$

$$(x + 3)^2 + (y - 2)^2 = 5^2$$

From the last line it is apparent that the center is at $(-3, 2)$ and the radius is 5. (See Figure 5.)

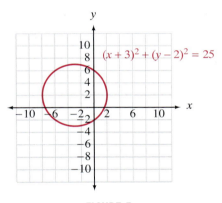

FIGURE 5

Getting Ready for Class

After reading through the preceding section, respond in your own words and in complete sentences.

A. Describe the distance formula in words, as if you were explaining to someone how they should go about finding the distance between two points.

B. What is the mathematical definition of a circle?

C. How are the distance formula and the equation of a circle related?

D. When graphing a circle from its equation, why is completing the square sometimes useful?

PROBLEM SET 11.1

Find the distance between the following points.

1. (3, 7) and (6, 3)　　　　**2.** (4, 7) and (8, 1)

3. (0, 9) and (5, 0)　　　　**4.** (−3, 0) and (0, 4)

5. (3, −5) and (−2, 1)　　　**6.** (−8, 9) and (−3, −2)

7. (−1, −2) and (−10, 5)　**8.** (−3, −8) and (−1, 6)

9. Find x so that the distance between $(x, 2)$ and $(1, 5)$ is $\sqrt{13}$.

10. Find x so that the distance between $(−2, 3)$ and $(x, 1)$ is 3.

11. Find y so that the distance between $(7, y)$ and $(8, 3)$ is 1.

12. Find y so that the distance between $(3, −5)$ and $(3, y)$ is 9.

Write the equation of the circle with the given center and radius.

13. Center (2, 3); $r = 4$

14. Center (3, −1); $r = 5$

15. Center (3, −2); $r = 3$

16. Center (−2, 4); $r = 1$

17. Center (−5, −1); $r = \sqrt{5}$

18. Center (−7, −6); $r = \sqrt{3}$

19. Center (0, −5); $r = 1$

20. Center (0, −1); $r = 7$

21. Center (0, 0); $r = 2$

22. Center (0, 0); $r = 5$

Give the center and radius, and sketch the graph of each of the following circles.

23. $x^2 + y^2 = 4$　　　　**24.** $x^2 + y^2 = 16$

25. $(x − 1)^2 + (y − 3)^2 = 25$

26. $(x − 4)^2 + (y − 1)^2 = 36$

27. $(x + 2)^2 + (y − 4)^2 = 8$

28. $(x − 3)^2 + (y + 1)^2 = 12$

29. $(x + 1)^2 + (y + 1)^2 = 1$

30. $(x + 3)^2 + (y + 2)^2 = 9$

31. $x^2 + y^2 − 6y = 7$　　　**32.** $x^2 + y^2 − 4y = 5$

33. $x^2 + y^2 + 2x = 1$　　　**34.** $x^2 + y^2 + 10x = 0$

35. $x^2 + y^2 − 4x − 6y = −4$

36. $x^2 + y^2 − 4x + 2y = 4$

37. $x^2 + y^2 + 2x + y = \dfrac{11}{4}$

38. $x^2 + y^2 − 6x − y = −\dfrac{1}{4}$

Each of the following circles passes through the origin. In each case, find the equation.

39.

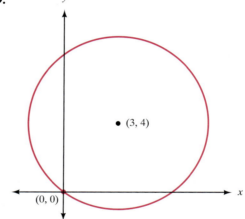

40.

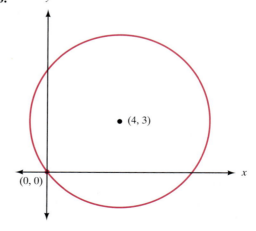

41. Find the equations of circles *A, B,* and *C* in the following diagram. The three points are the centers of the three circles.

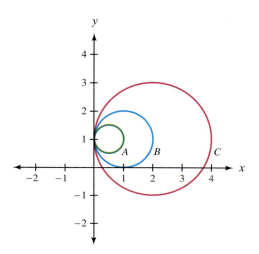

42. Each of the following circles passes through the origin. The centers are as shown. Find the equation of each circle.

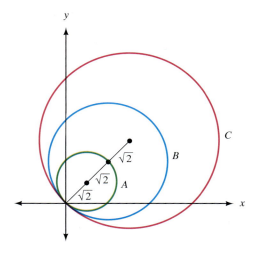

43. Find the equation of the circle with center at the origin that contains the point (3, 4).

44. Find the equation of the circle with center at the origin that contains the point (−5, 12).

45. Find the equation of the circle with center at the origin and *x*-intercepts 3 and −3.

46. Find the equation of the circle with *y*-intercepts 4 and −4 and center at the origin.

47. A circle with center at (−1, 3) passes through the point (4, 3). Find the equation.

48. A circle with center at (2, 5) passes through the point (−1, 4). Find the equation.

Review Problems

The problems that follow are a review of material we covered in Sections 10.1 and 10.2.

Find the general term of each sequence. [10.1]

49. 5, 9, 13, 17, . . .

50. 3, 8, 15, 24, . . .

Expand and simplify each series. [10.2]

51. $\displaystyle\sum_{i=2}^{5} \left(\frac{1}{2}\right)^i$ **52.** $\displaystyle\sum_{i=3}^{6} (i^2 - 5)$

Write using summation notation. [10.2]

53. $1 + 3 + 5 + 7 + 9$

54. $\frac{2}{3} + \frac{3}{4} + \frac{4}{5} + \frac{5}{6}$

Extending the Concepts

A circle is *tangent to* a line if it touches, but does not cross, the line.

55. Find the equation of the circle with center at (2, 3) if the circle is tangent to the *y*-axis.

56. Find the equation of the circle with center at (3, 2) if the circle is tangent to the *x*-axis.

57. Find the equation of the circle with center at (2, 3) if the circle is tangent to the vertical line *x* = 4.

58. Find the equation of the circle with center at (3, 2) if the circle is tangent to the horizontal line *y* = 6.

Find the distance from the origin to the center of each of the following circles.

59. $x^2 + y^2 - 6x + 8y = 144$

60. $x^2 + y^2 - 8x + 6y = 144$

61. $x^2 + y^2 - 6x - 8y = 144$

62. $x^2 + y^2 + 8x + 6y = 144$

Ellipses and Hyperbolas

This section is concerned with the graphs of ellipses and hyperbolas. To simplify matters somewhat, we will consider only those graphs that are centered about the origin.

Suppose we want to graph the equation

$$\frac{x^2}{25} + \frac{y^2}{9} = 1$$

We can find the y-intercepts by letting $x = 0$, and the x-intercepts by letting $y = 0$:

When $x = 0$

$$\frac{0^2}{25} + \frac{y^2}{9} = 1$$

$$y^2 = 9$$

$$y = \pm 3$$

When $y = 0$

$$\frac{x^2}{25} + \frac{0^2}{9} = 1$$

$$x^2 = 25$$

$$x = \pm 5$$

The graph crosses the y-axis at $(0, 3)$ and $(0, -3)$ and the x-axis at $(5, 0)$ and $(-5, 0)$. Graphing these points and then connecting them with a smooth curve gives the graph shown in Figure 1.

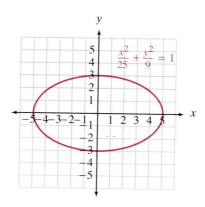

FIGURE 1

We can find other ordered pairs on the graph by substituting in values for x (or y) and then solving for y (or x). For example, if we let $x = 3$, then

$$\frac{3^2}{25} + \frac{y^2}{9} = 1$$

$$\frac{9}{25} + \frac{y^2}{9} = 1$$

$$0.36 + \frac{y^2}{9} = 1$$

$$\frac{y^2}{9} = 0.64$$

$$y^2 = 5.76$$

$$y = \pm 2.4$$

This would give us the two ordered pairs $(3, -2.4)$ and $(3, 2.4)$.

A graph of the type shown in Figure 1 is called an *ellipse*. If we were to find some other ordered pairs that satisfy our original equation, we would find that their graphs lie on the ellipse. Also, the coordinates of any point on the ellipse will satisfy the equation. We can generalize these results as follows.

THE ELLIPSE

The graph of any equation of the form

$$\frac{x^2}{a^2} + \frac{y^2}{b^2} = 1 \qquad \text{Standard form}$$

will be an **ellipse** centered at the origin. The ellipse will cross the x-axis at $(a, 0)$ and $(-a, 0)$. It will cross the y-axis at $(0, b)$ and $(0, -b)$. When a and b are equal, the ellipse will be a circle. Each of the points $(a, 0)$, $(-a, 0)$, $(0, b)$, and $(0, -b)$ is a **vertex** (intercept) of the graph.

The most convenient way to graph an ellipse is to locate the intercepts (vertices).

EXAMPLE 1 Sketch the graph of $4x^2 + 9y^2 = 36$.

Solution To write the equation in the form

$$\frac{x^2}{a^2} + \frac{y^2}{b^2} = 1$$

we must divide both sides by 36:

$$\frac{4x^2}{36} + \frac{9y^2}{36} = \frac{36}{36}$$

$$\frac{x^2}{9} + \frac{y^2}{4} = 1$$

The graph crosses the x-axis at $(3, 0)$, $(-3, 0)$ and the y-axis at $(0, 2)$, $(0, -2)$. (See Figure 2.)

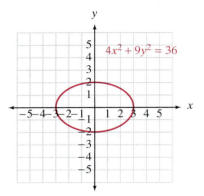

FIGURE 2

Consider the equation

$$\frac{x^2}{9} - \frac{y^2}{4} = 1$$

If we were to find a number of ordered pairs that are solutions to the equation and connect their graphs with a smooth curve, we would have Figure 3.

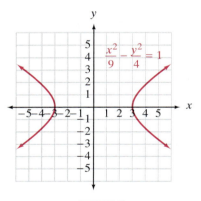

FIGURE 3

This graph is an example of a *hyperbola*. Notice that the graph has x-intercepts at $(3, 0)$ and $(-3, 0)$. The graph has no y-intercepts and hence does not cross the y-axis, since substituting $x = 0$ into the equation yields

$$\frac{0^2}{9} - \frac{y^2}{4} = 1$$

$$-y^2 = 4$$

$$y^2 = -4$$

for which there is no real solution. We can, however, use the number below y^2 to help sketch the graph. If we draw a rectangle that has its sides parallel to the x- and y-axes and that passes through the x-intercepts and the points on the y-axis corresponding to the square roots of the number below y^2, $+2$ and -2, it looks like the rectangle in Figure 4. The lines that connect opposite corners of the rectangle are called *asymptotes*. The graph of the hyperbola

$$\frac{x^2}{9} - \frac{y^2}{4} = 1$$

will approach these lines. Figure 4 is the graph.

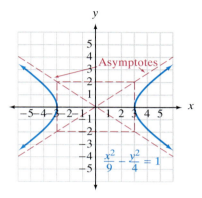

FIGURE 4

EXAMPLE 2 Graph the equation $\dfrac{y^2}{9} - \dfrac{x^2}{16} = 1$.

Solution   In this case the y-intercepts are 3 and -3, and the x-intercepts do not exist. We can use the square roots of the number below x^2, however, to find the asymptotes associated with the graph. The sides of the rectangle used to draw the asymptotes must pass through 3 and -3 on the y-axis, and 4 and -4 on the x-axis. (See Figure 5.)

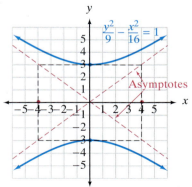

FIGURE 5

Here is a summary of what we have for hyperbolas.

HYPERBOLAS CENTERED AT THE ORIGIN

The graph of the equation

$$\frac{x^2}{a^2} - \frac{y^2}{b^2} = 1$$

will be a **hyperbola centered at the origin.** The graph will have **x-intercepts (vertices)** at $-a$ and a.

The graph of the equation

$$\frac{y^2}{a^2} - \frac{x^2}{b^2} = 1$$

will be a **hyperbola centered at the origin.** The graph will have **y-intercepts (vertices)** at $-a$ and a.

As an aid in sketching either of these equations, the asymptotes can be found by drawing lines through opposite corners of the rectangle whose sides pass through $-a$, a, $-b$, and b on the axes.

ELLIPSES AND HYPERBOLAS NOT CENTERED AT THE ORIGIN

The following equation is that of an ellipse with its center at the point (4, 1):

$$\frac{(x-4)^2}{9} + \frac{(y-1)^2}{4} = 1$$

To see why the center is at (4, 1) we substitute x' (read "x prime") for $x - 4$ and y' for $y - 1$ in the equation. That is:

If $x' = x - 4$

and $y' = y - 1$

the equation $\dfrac{(x-4)^2}{9} + \dfrac{(y-1)^2}{4} = 1$

becomes $\dfrac{(x')^2}{9} + \dfrac{(y')^2}{4} = 1$

This is the equation of an ellipse in a coordinate system with an x'-axis and a y'-axis. We call this new coordinate system the $x'y'$**-coordinate system.** The center of our ellipse is at the origin in the $x'y'$-coordinate system. The question is this: What are the coordinates of the center of this ellipse in the original xy-coordinate system? To answer this question we go back to our original substitutions:

$$x' = x - 4$$
$$y' = y - 1$$

In the $x'y'$-coordinate system, the center of our ellipse is at $x' = 0$, $y' = 0$ (the origin of the $x'y'$ system). Substituting these numbers for x' and y', we have

$$0 = x - 4$$
$$0 = y - 1$$

Solving these equations for x and y will give us the coordinates of the center of our ellipse in the xy-coordinate system. As you can see, the solutions are $x = 4$ and

$y = 1$. Therefore, in the xy-coordinate system, the center of our ellipse is at the point (4, 1). Figure 6 shows the graph.

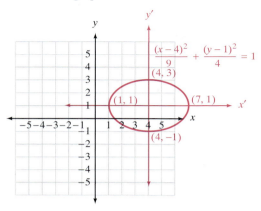

FIGURE 6

The coordinates of all points labeled in Figure 6 are given with respect to the xy-coordinate system. The x'- and y'-axes are shown simply for reference in our discussion. Note that the horizontal distance from the center to the vertices is 3 — the square root of the denominator of the $(x - 4)^2$ term. Likewise, the vertical distance from the center to the other vertices is 2 — the square root of the denominator of the $(y - 1)^2$ term.

We summarize the information above with the following:

AN ELLIPSE WITH CENTER AT (h, k)

The graph of the equation

$$\frac{(x - h)^2}{a^2} + \frac{(y - k)^2}{b^2} = 1$$

will be an **ellipse with center at (h, k)**. The vertices of the ellipse will be at the points $(h + a, k)$, $(h - a, k)$, $(k, k + b)$, and $(h, k - b)$.

EXAMPLE 3   Graph the ellipse: $x^2 + 9y^2 + 4x - 54y + 76 = 0$

Solution   To identify the coordinates of the center, we must complete the square on x and also on y. To begin, we rearrange the terms so that those containing x are together, those containing y are together, and the constant term is on the other side of the equal sign. Doing so gives us the following equation:

$$x^2 + 4x + 9y^2 - 54y = -76$$

Before we can complete the square on y, we must factor 9 from each term containing y:

$$x^2 + 4x + 9(y^2 - 6y) = -76$$

To complete the square on x, we add 4 to each side of the equation. To complete the square on y, we add 9 inside the parentheses. This increases the left side of the equation by 81 since each term within the parentheses is multiplied by 9. There-fore, we must add 81 to the right side of the equation also.

$$x^2 + 4x + 4 + 9(y^2 - 6y + 9) = -76 + 4 + 81$$
$$(x + 2)^2 + 9(y - 3)^2 = 9$$

To identify the distances to the vertices, we divide each term on both sides by 9:

$$\frac{(x + 2)^2}{9} + \frac{9(y - 3)^2}{9} = \frac{9}{9}$$

$$\frac{(x + 2)^2}{9} + \frac{(y - 3)^2}{1} = 1$$

The graph is an ellipse with center at $(-2, 3)$, as shown in Figure 7.

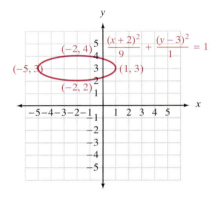

FIGURE 7

The ideas associated with graphing hyperbolas whose centers are not at the ori-gin parallel the ideas just presented about graphing ellipses whose centers have been moved off the origin. Without showing the justification for doing so, we state the following guidelines for graphing hyperbolas:

HYPERBOLAS WITH CENTERS AT (h, k)

The graphs of the equations

$$\frac{(x - h)^2}{a^2} - \frac{(y - k)^2}{b^2} = 1 \quad \text{and} \quad \frac{(y - k)^2}{b^2} - \frac{(x - h)^2}{a^2} = 1$$

will be hyperbolas with their centers at (h, k). The vertices of the graph of the first equation will be at the points $(h + a, k)$ and $(h - a, k)$, and the vertices for the graph of the second equation will be at $(h, k + b)$ and $(h, k - b)$. In either case, the asymptotes can be found by connecting opposite corners of the rect-angle that contains the four points $(h + a, k)$, $(h - a, k)$, $(h, k + b)$, and $(h, k - b)$.

EXAMPLE 4   Graph the hyperbola: $4x^2 - y^2 + 4y - 20 = 0$

Solution   To identify the coordinates of the center of the hyperbola, we need to complete the square on y. (Since there is no linear term in x, we do not need to complete the square on x. The x-coordinate of the center will be $x = 0$.)

$$4x^2 - y^2 + 4y - 20 = 0$$

$$4x^2 - y^2 + 4y = 20 \qquad \text{Add 20 to each side.}$$

$$4x^2 - 1(y^2 - 4y) = 20 \qquad \text{Factor } -1 \text{ from each term containing } y.$$

To complete the square on y, we add 4 to the terms inside the parentheses. Doing so adds -4 to the left side of the equation since everything inside the parentheses is multiplied by -1. To keep from changing the equation we must add -4 to the right side also.

$$4x^2 - 1(y^2 - 4y + \mathbf{4}) = 20 - \mathbf{4}$$

$$4x^2 - 1(y - 2)^2 = 16$$

$$\frac{4x^2}{16} - \frac{(y - 2)^2}{16} = \frac{16}{16}$$

$$\frac{x^2}{4} - \frac{(y - 2)^2}{16} = 1$$

This is the equation of a hyperbola with center at $(0, 2)$. The graph opens to the right and left as shown in Figure 8.

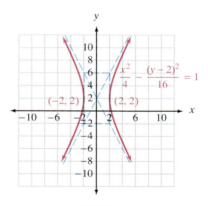

FIGURE 8

Getting Ready for Class

After reading through the preceding section, respond in your own words and in complete sentences.

A. How do we find the x-intercepts of a graph from the equation?

B. What is an ellipse?

C. How can you tell by looking at an equation, if its graph will be an ellipse or a hyperbola?

D. Are the points on the asymptotes of a hyperbola in the solution set of the equation of the hyperbola? Explain. (That is, are the asymptotes actually part of the graph?)

PROBLEM SET 11.2

Graph each of the following. Be sure to label both the x- and y-intercepts.

1. $\dfrac{x^2}{9} + \dfrac{y^2}{16} = 1$

2. $\dfrac{x^2}{25} + \dfrac{y^2}{4} = 1$

3. $\dfrac{x^2}{16} + \dfrac{y^2}{9} = 1$

4. $\dfrac{x^2}{4} + \dfrac{y^2}{25} = 1$

5. $\dfrac{x^2}{3} + \dfrac{y^2}{4} = 1$

6. $\dfrac{x^2}{4} + \dfrac{y^2}{3} = 1$

7. $4x^2 + 25y^2 = 100$

8. $4x^2 + 9y^2 = 36$

9. $x^2 + 8y^2 = 16$

10. $12x^2 + y^2 = 36$

Graph each of the following. Show the intercepts and the asymptotes in each case.

11. $\dfrac{x^2}{9} - \dfrac{y^2}{16} = 1$

12. $\dfrac{x^2}{25} - \dfrac{y^2}{4} = 1$

13. $\dfrac{x^2}{16} - \dfrac{y^2}{9} = 1$

14. $\dfrac{x^2}{4} - \dfrac{y^2}{25} = 1$

15. $\dfrac{y^2}{9} - \dfrac{x^2}{16} = 1$

16. $\dfrac{y^2}{25} - \dfrac{x^2}{4} = 1$

17. $\dfrac{y^2}{36} - \dfrac{x^2}{4} = 1$

18. $\dfrac{y^2}{4} - \dfrac{x^2}{36} = 1$

19. $x^2 - 4y^2 = 4$

20. $y^2 - 4x^2 = 4$

21. $16y^2 - 9x^2 = 144$

22. $4y^2 - 25x^2 = 100$

Find the x- and y-intercepts, if they exist, for each of the following. Do not graph.

23. $0.4x^2 + 0.9y^2 = 3.6$

24. $1.6x^2 + 0.9y^2 = 14.4$

25. $\dfrac{x^2}{0.04} - \dfrac{y^2}{0.09} = 1$

26. $\dfrac{y^2}{0.16} - \dfrac{x^2}{0.25} = 1$

27. $\dfrac{25x^2}{9} + \dfrac{25y^2}{4} = 1$

28. $\dfrac{16x^2}{9} + \dfrac{16y^2}{25} = 1$

Graph each of the following ellipses. In each case, label the coordinates of the center and the vertices.

29. $\dfrac{(x-4)^2}{4} + \dfrac{(y-2)^2}{9} = 1$

30. $\dfrac{(x-2)^2}{4} + \dfrac{(y-4)^2}{9} = 1$

31. $4x^2 + y^2 - 4y - 12 = 0$

32. $4x^2 + y^2 - 24x - 4y + 36 = 0$

33. $x^2 + 9y^2 + 4x - 54y + 76 = 0$

34. $4x^2 + y^2 - 16x + 2y + 13 = 0$

Graph each of the following hyperbolas. In each case, label the coordinates of the center and the vertices and show the asymptotes.

35. $\dfrac{(x-2)^2}{16} - \dfrac{y^2}{4} = 1$

36. $\dfrac{(y-2)^2}{16} - \dfrac{x^2}{4} = 1$

37. $9y^2 - x^2 - 4x + 54y + 68 = 0$

38. $4x^2 - y^2 - 24x + 4y + 28 = 0$

39. $4y^2 - 9x^2 - 16y + 72x - 164 = 0$

40. $4x^2 - y^2 - 16x - 2y + 11 = 0$

Find the equation for the following ellipses and hyperbolas.

41.

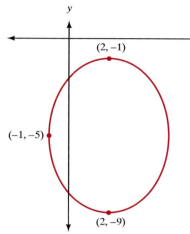

42.

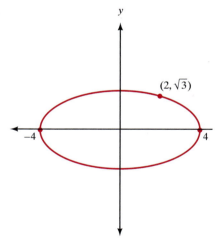

43.

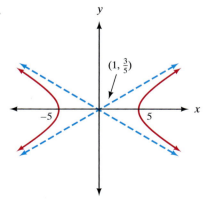

44.

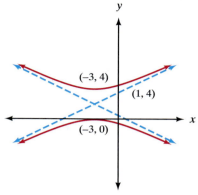

45. Give the equations of the two asymptotes in the graph you found in Problem 15.

46. Give the equations of the two asymptotes in the graph you found in Problem 16.

47. The longer line segment connecting opposite vertices of an ellipse is called the **major axis** of the ellipse. Give the length of the major axis of the ellipse you graphed in Problem 3.

48. The shorter line segment connecting opposite vertices of an ellipse is called the **minor axis** of the ellipse. Give the length of the minor axis of the ellipse you graphed in Problem 3.

Review Problems

The following problems review material we covered in Sections 10.1, 10.3, and 10.4.

Find the general term of each sequence.

49. 5, 11, 17, 23, . . .

50. $-3, 9, -27, 81, . . .$

51. An arithmetic sequence has a first term of $a_1 = 4$ and a common difference of $d = 5$. Find the sum of the first 20 terms, S_{20}.

52. An arithmetic sequence is such that $a_4 = 23$ and $a_9 = 48$. Find a_{40}.

53. A geometric sequence has a first term of $a_1 = 8$ and a common ratio of $r = \frac{1}{2}$. Find the sum of the first 6 terms.

54. Find the sum: $1 + \frac{1}{2} + \frac{1}{4} + \frac{1}{8} + \cdots$

In Section 3.4 we graphed linear inequalities by first graphing the boundary and then choosing a test point not on the boundary to indicate the region used for the solution set. The problems in this section are very similar. We will use the same general methods for graphing the inequalities in this section that we used in Section 3.4.

EXAMPLE 1 Graph $x^2 + y^2 < 16$.

Solution The boundary is $x^2 + y^2 = 16$, which is a circle with center at the origin and a radius of 4. Since the inequality sign is $<$, the boundary is not included in the solution set and must therefore be represented with a broken line. The graph of the boundary is shown in Figure 1.

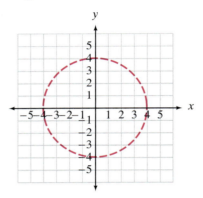

FIGURE 1

The solution set for $x^2 + y^2 < 16$ is either the region inside the circle or the region outside the circle. To see which region represents the solution set, we choose a convenient point not on the boundary and test it in the original inequality. The origin $(0, 0)$ is a convenient point. Since the origin satisfies the inequality $x^2 + y^2 < 16$, all points in the same region will also satisfy the inequality. The graph of the solution set is shown in Figure 2.

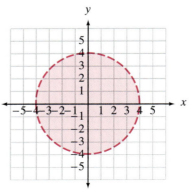

FIGURE 2

EXAMPLE   2   Graph the inequality $y \leq x^2 - 2$.

Solution   The parabola $y = x^2 - 2$ is the boundary and is included in the solution set. Using $(0, 0)$ as the test point, we see that $0 \leq 0^2 - 2$ is a false statement, which means that the region containing $(0, 0)$ is not in the solution set. (See Figure 3.)

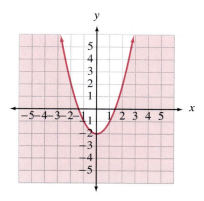

FIGURE 3

EXAMPLE   3   Graph $4y^2 - 9x^2 < 36$.

Solution   The boundary is the hyperbola $4y^2 - 9x^2 = 36$ and is not included in the solution set. Testing $(0, 0)$ in the original inequality yields a true statement, which means that the region containing the origin is the solution set. (See Figure 4.)

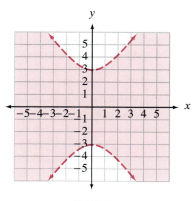

FIGURE 4

EXAMPLE   4   Solve the system.

$$x^2 + y^2 = 4$$
$$x - 2y = 4$$

Solution   In this case the substitution method is the most convenient. Solving the second equation for x in terms of y, we have

$$x - 2y = 4$$
$$x = 2y + 4$$

We now substitute $2y + 4$ for x in the first equation in our original system and proceed to solve for y:

$$(2y + 4)^2 + y^2 = 4$$
$$4y^2 + 16y + 16 + y^2 = 4$$
$$5y^2 + 16y + 12 = 0$$
$$(5y + 6)(y + 2) = 0$$
$$5y + 6 = 0 \quad \text{or} \quad y + 2 = 0$$
$$y = -\frac{6}{5} \quad \text{or} \quad y = -2$$

These are the y-coordinates of the two solutions to the system. Substituting $y = -\frac{6}{5}$ into $x - 2y = 4$ and solving for x gives us $x = \frac{8}{5}$. Using $y = -2$ in the same equation yields $x = 0$. The two solutions to our system are $\left(\frac{8}{5}, -\frac{6}{5}\right)$ and $(0, -2)$. Although graphing the system is not necessary, it does help us visualize the situation. (See Figure 5.)

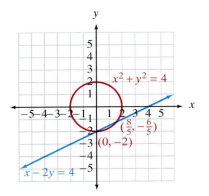

FIGURE 5

EXAMPLE 5   Solve the system.

$$16x^2 - 4y^2 = 64$$
$$x^2 + y^2 = 9$$

Solution   Since each equation is of the second degree in both x and y, it is easier to solve this system by eliminating one of the variables by addition. To eliminate y, we multiply the bottom equation by 4 and add the result to the top equation:

$$16x^2 - 4y^2 = 64$$
$$\underline{4x^2 + 4y^2 = 36}$$
$$20x^2 = 100$$
$$x^2 = 5$$
$$x = \pm\sqrt{5}$$

The x-coordinates of the points of intersection are $\sqrt{5}$ and $-\sqrt{5}$. We substitute each back into the second equation in the original system and solve for y:

When $\hspace{8em} x = \sqrt{5}$

$$(\sqrt{5})^2 + y^2 = 9$$
$$5 + y^2 = 9$$
$$y^2 = 4$$
$$y = \pm 2$$

When $\hspace{8em} x = -\sqrt{5}$

$$(-\sqrt{5})^2 + y^2 = 9$$
$$5 + y^2 = 9$$
$$y^2 = 4$$
$$y = \pm 2$$

The four points of intersection are $(\sqrt{5}, 2)$, $(\sqrt{5}, -2)$, $(-\sqrt{5}, 2)$, and $(-\sqrt{5}, -2)$. Graphically the situation is as shown in Figure 6.

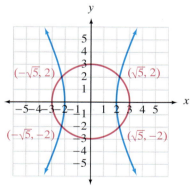

FIGURE 6

EXAMPLE 6   Solve the system.

$$x^2 - 2y = 2$$
$$y = x^2 - 3$$

Solution We can solve this system using the substitution method. Replacing y in the first equation with $x^2 - 3$ from the second equation, we have

$$x^2 - 2(x^2 - 3) = 2$$
$$-x^2 + 6 = 2$$
$$x^2 = 4$$
$$x = \pm 2$$

Using either $+2$ or -2 in the equation $y = x^2 - 3$ gives us $y = 1$. The system has two solutions: $(2, 1)$ and $(-2, 1)$.

EXAMPLE 7 The sum of the squares of two numbers is 34. The difference of their squares is 16. Find the two numbers.

Solution Let x and y be the two numbers. The sum of their squares is $x^2 + y^2$, and the difference of their squares is $x^2 - y^2$. (We can assume here that x^2 is the larger number.) The system of equations that describes the situation is

$$x^2 + y^2 = 34$$
$$x^2 - y^2 = 16$$

We can eliminate y by simply adding the two equations. The result of doing so is

$$2x^2 = 50$$
$$x^2 = 25$$
$$x = \pm 5$$

Substituting $x = 5$ into either equation in the system gives $y = \pm 3$. Using $x = -5$ gives the same results, $y = \pm 3$. The four pairs of numbers that are solutions to the original problem are

$$(5, 3) \qquad (-5, 3) \qquad (5, -3) \qquad (-5, -3)$$

We now turn our attention to systems of inequalities. To solve a system of inequalities by graphing, we simply graph each inequality on the same set of axes. The solution set for the system is the region common to both graphs — the intersection of the individual solution sets.

EXAMPLE 8 Graph the solution set for the system

$$x^2 + y^2 \leq 9$$
$$\frac{x^2}{4} + \frac{y^2}{25} \geq 1$$

Solution The boundary for the top inequality is a circle with center at the origin and a radius of 3. The solution set lies inside the boundary. The boundary for the second inequality is an ellipse. In this case the solution set lies outside the boundary. (See Figure 7.)

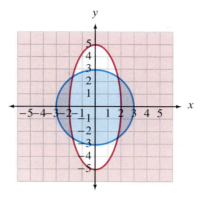

FIGURE 7

The solution set is the intersection of the two individual solution sets.

EXAMPLE 9   Graph the solution set for the following system.

$$x - 2y \leq 4$$

$$x + \ y \leq 4$$

$$x \geq -1$$

Solution   We have three linear inequalities, representing three sections of the coordinate plane. The graph of the solution set for this system will be the intersection of these three sections. The graph of $x - 2y \leq 4$ is the section above and including the boundary $x - 2y = 4$. The graph of $x + y \leq 4$ is the section below and including the boundary line $x + y = 4$. The graph of $x \geq -1$ is all the points to the right of, and including, the vertical line $x = -1$. The intersection of these three graphs is shown in Figure 8.

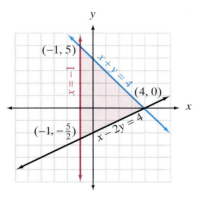

FIGURE 8

Getting Ready for Class

After reading through the preceding section, respond in your own words and in complete sentences.

A. What is the significance of a broken line when graphing inequalities?

B. Describe, in words, the set of points described by $(x - 3)^2 + (y - 2)^2 < 9$.

C. When solving the nonlinear systems whose graphs are a line and a circle, how many possible solutions can you expect?

D. When solving the nonlinear systems whose graphs are both circles, how many possible solutions can you expect?

PROBLEM SET 11.3

Graph each of the following inequalities.

1. $x^2 + y^2 \leq 49$ **2.** $x^2 + y^2 < 49$

3. $(x - 2)^2 + (y + 3)^2 < 16$

4. $(x + 3)^2 + (y - 2)^2 \geq 25$

5. $y < x^2 - 6x + 7$ **6.** $y \geq x^2 + 2x - 8$

7. $\dfrac{x^2}{25} - \dfrac{y^2}{9} \geq 1$ **8.** $\dfrac{x^2}{25} - \dfrac{y^2}{9} \leq 1$

9. $4x^2 + 25y^2 \leq 100$ **10.** $25x^2 - 4y^2 > 100$

Graph the solution sets to the following systems.

11. $x^2 + y^2 < 9$ **12.** $x^2 + y^2 \leq 16$
$\quad\ y \geq x^2 - 1$ $\quad\ y < x^2 + 2$

13. $\dfrac{x^2}{9} + \dfrac{y^2}{25} \leq 1$ **14.** $\dfrac{x^2}{4} + \dfrac{y^2}{16} \geq 1$

$\quad\ \dfrac{x^2}{4} - \dfrac{y^2}{9} > 1$ $\quad\ \dfrac{x^2}{9} - \dfrac{y^2}{25} < 1$

15. $4x^2 + 9y^2 \leq 36$ **16.** $9x^2 + 4y^2 \geq 36$
$\quad\ y > x^2 + 2$ $\quad\ y < x^2 + 1$

17. $x + \ \ y \leq \ \ 3$ **18.** $x - \ \ y \leq \ \ 4$
$\quad x - 3y \leq \ \ 3$ $\quad x + 2y \leq \ \ 4$
$\qquad\ x \geq -2$ $\qquad\ x \geq -1$

19. $\ \ x + y \leq \ \ 2$ **20.** $\ \ \ x - y \leq \ \ 3$
$\quad -x + y \leq \ \ 2$ $\quad -x - y \leq \ \ 3$
$\qquad\quad y \geq -2$ $\qquad\quad y \leq -1$

21. $x + y \leq 4$ **22.** $x - y \leq 2$
$\qquad x \geq 0$ $\qquad x \geq 0$
$\qquad y \geq 0$ $\qquad y \leq 0$

Solve each of the following systems of equations.

23. $\ \ x^2 + y^2 = 9$ **24.** $x^2 + \ \ y^2 = 9$
$\quad\ 2x + \ \ y = 3$ $\quad\ x + 2y = 3$

25. $x^2 + \ \ y^2 = 16$ **26.** $x^2 + \ \ y^2 = 16$
$\quad\ x + 2y = \ \ 8$ $\quad\ x - 2y = \ \ 8$

27. $x^2 + y^2 = 25$ **28.** $\ x^2 + y^2 = 4$
$\quad x^2 - y^2 = 25$ $\quad 2x^2 - y^2 = 5$

29. $x^2 + y^2 = 9$ **30.** $x^2 + y^2 = 4$
$\qquad\quad y = x^2 - 3$ $\qquad\quad y = x^2 - 2$

31. $x^2 + y^2 = 16$ **32.** $x^2 + y^2 = 1$
$\qquad\quad y = x^2 - 4$ $\qquad\quad y = x^2 - 1$

33. $3x + 2y = 10$ **34.** $4x + 2y = 10$
$\qquad\quad y = x^2 - 5$ $\qquad\quad y = x^2 - 10$

35. $y = x^2 + 2x - 3$ **36.** $y = -x^2 - 2x + 3$
$\quad y = \qquad\ -x + 1$ $\quad y = \qquad\qquad x - 1$

37. $y = x^2 - 6x + 5$ **38.** $y = x^2 - 2x - 4$
$\quad y = \qquad\qquad x - 5$ $\quad y = \qquad\qquad x - 4$

39. $4x^2 - 9y^2 = 36$ **40.** $4x^2 + 25y^2 = 100$
$\quad 4x^2 + 9y^2 = 36$ $\quad 4x^2 - 25y^2 = 100$

41. $x \ - y = \ \ 4$ **42.** $x \ + y = 2$
$\quad x^2 + y^2 = 16$ $\quad x^2 - y^2 = 4$

Applying the Concepts

43. Number Problem   The sum of the squares of two numbers is 89. The difference of their squares is 39. Find the numbers.

44. Number Problem   The difference of the squares of two numbers is 35. The sum of their squares is 37. Find the numbers.

45. Number Problem   One number is 3 less than the square of another. Their sum is 9. Find the numbers.

46. Number Problem   The square of one number is 2 less than twice the square of another. The sum of the squares of the two numbers is 25. Find the numbers.

Review Problems

The following problems review material we covered in Section 10.5.

Expand and simplify.

47. $(x + 2)^4$

48. $(x - 2)^4$

49. $(2x + y)^3$

50. $(x - 2y)^3$

51. Find the first two terms in the expansion of $(x + 3)^{50}$.

52. Find the first two terms in the expansion of $(x - y)^{75}$.

Examples

1. The distance between (5, 2) and (−1, 1) is

$$d = \sqrt{(5 + 1)^2 + (2 - 1)^2}$$
$$= \sqrt{37}$$

Distance Formula [11.1]

The distance between the two points (x_1, y_1) and (x_2, y_2) is given by the formula

$$d = \sqrt{(x_2 - x_1)^2 + (y_2 - y_1)^2}$$

2. The graph of the circle

$$(x - 3)^2 + (y + 2)^2 = 25$$

has its center at (3, −2) and the radius is 5.

The Circle [11.1]

The graph of any equation of the form

$$(x - a)^2 + (y - b)^2 = r^2$$

is a circle having its center at (a, b) and a radius of r.

3.

An Ellipse With Center at (*h, k*) [11.2]

The graph of the equation

$$\frac{(x - h)^2}{a^2} + \frac{(y - k)^2}{b^2} = 1$$

is an ellipse with center at (h, k). The vertices of the ellipse are at the points (h + a, k), (h − a, k), (h, k + b), and (h, k − b).

4.

Hyperbolas With Centers at (*h, k*) [11.2]

The graphs of the equations

$$\frac{(x - h)^2}{a^2} - \frac{(y - k)^2}{b^2} = 1 \quad \text{and} \quad \frac{(y - k)^2}{b^2} - \frac{(x - h)^2}{a^2} = 1$$

are hyperbolas with their centers at (h, k). The vertices of the graph of the first equation are at the points (h + a, k) and (h − a, k), and the vertices for the graph of the second equation are at (h, k + b) and (h, k − b). In either case, the asymptotes can be found by connecting opposite corners of the rectangle that contains the points (h + a, k), (h − a, k), (h, k + b), and (h, k − b).

5. The graph of the inequality

$$x^2 + y^2 < 9$$

is all points inside the circle with center at the origin and radius 3. The circle itself is not part of the solution and is therefore shown with a broken curve.

6. We can solve the system

$$x^2 + y^2 = 4$$

$$x = 2y + 4$$

by substituting $2y + 4$ from the second equation for x in the first equation:

$$(2y + 4)^2 + y^2 = 4$$

$$4y^2 + 16y + 16 + y^2 = 4$$

$$5y^2 + 16y + 12 = 0$$

$$(5y + 6)(y + 2) = 0$$

$$y = -\frac{6}{5} \quad \text{or} \quad y = -2$$

Substituting these values of y into the second equation in our system gives $x = \frac{8}{5}$ and $x = 0$. The solutions are $(\frac{8}{5}, -\frac{6}{5})$ and $(0, -2)$.

Second-Degree Inequalities in Two Variables [11.3]
We graph second-degree inequalities in two variables in much the same way that we graphed linear inequalities. That is, we begin by graphing the boundary, using a solid curve if the boundary is included in the solution (this happens when the inequality symbol is $\geq$ or $\leq$), or a broken curve if the boundary is not included in the solution (when the inequality symbol is $>$ or $<$). After we have graphed the boundary, we choose a test point that is not on the boundary and try it in the original inequality. A true statement indicates we are in the region of the solution. A false statement indicates we are not in the region of the solution.

Systems of Nonlinear Equations [11.3]
A system of nonlinear equations is two equations, at least one of which is not linear, considered at the same time. The solution set for the system consists of all ordered pairs that satisfy both equations. In most cases we use the substitution method to solve these systems; however, the addition method can be used if like variables are raised to the same power in both equations. It is sometimes helpful to graph each equation in the system on the same set of axes in order to anticipate the number and approximate positions of the solutions.

CHAPTER 11 REVIEW

Find the distance between the following points. [11.1]

1. $(2, 6), (-1, 5)$

2. $(3, -4), (1, -1)$

3. $(0, 3), (-4, 0)$

4. $(-3, 7), (-3, -2)$

5. Find x so that the distance between $(x, -1)$ and $(2, -4)$ is 5. [11.1]

6. Find y so that the distance between $(3, -4)$ and $(-3, y)$ is 10. [11.1]

Write the equation of the circle with the given center and radius. [11.1]

7. Center $(3, 1), r = 2$

8. Center $(3, -1)$, $r = 4$

9. Center $(-5, 0)$, $r = 3$

10. Center $(-3, 4)$, $r = 3\sqrt{2}$

Find the equation of each circle. [11.1]

11. Center at the origin, x-intercepts ± 5

12. Center at the origin, y-intercepts ± 3

13. Center at $(-2, 3)$ and passing through the point $(2, 0)$

14. Center at $(-6, 8)$ and passing through the origin

Give the center and radius of each circle, and then sketch the graph. [11.1]

15. $x^2 + y^2 = 4$

16. $(x - 3)^2 + (y + 1)^2 = 16$

17. $x^2 + y^2 - 6x + 4y = -4$

18. $x^2 + y^2 + 4x - 2y = 4$

Graph each of the following. Label the x- and y-intercepts. [11.2]

19. $\dfrac{x^2}{4} + \dfrac{y^2}{9} = 1$ 20. $4x^2 + y^2 = 16$

Graph the following. Show the asymptotes. [11.2]

21. $\dfrac{x^2}{4} - \dfrac{y^2}{9} = 1$ 22. $4x^2 - y^2 = 16$

Graph each equation. [11.2]

23. $\dfrac{(x + 2)^2}{9} + \dfrac{(y - 3)^2}{1} = 1$

24. $\dfrac{(x - 2)^2}{16} - \dfrac{y^2}{4} = 1$

25. $9y^2 - x^2 - 4x + 54y + 68 = 0$

26. $9x^2 + 4y^2 - 72x - 16y + 124 = 0$

Graph each of the following inequalities. [11.3]

27. $x^2 + y^2 < 9$

28. $(x + 2)^2 + (y - 1)^2 \le 4$

29. $y \ge x^2 - 1$ 30. $9x^2 + 4y^2 \le 36$

Graph the solution set for each system. [11.3]

31. $x^2 + y^2 < 16$ 32. $\quad x + y \le 2$
    $\quad\; y > x^2 - 4$ $\quad -x + y \le 2$
                                 $\quad\quad\; y \ge -2$

Solve each system of equations. [11.3]

33. $x^2 + y^2 = 16$ 34. $x^2 + y^2 = 4$
    $\;2x + y = 4$ $\quad\; y = x^2 - 2$

35. $9x^2 - 4y^2 = 36$ 36. $2x^2 - 4y^2 = 8$
    $9x^2 + 4y^2 = 36$ $\; x^2 + 2y^2 = 10$

CHAPTER 11 PROJECTS

CONIC SECTIONS

CONSTRUCTING ELLIPSES

Number of People: 4

Time Needed: 20 minutes

Equipment: Graph paper, pencils, string, and thumbtacks

Background: The geometric definition for an ellipse is the set of points the sum of whose distances from two fixed points (called foci) is a constant. We can use this definition to draw an ellipse using thumbtacks, string, and a pencil.

Procedure:

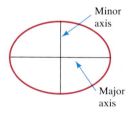

FIGURE 1

1. Start with a piece of string 7 inches long. Place thumbtacks through the string $\frac{1}{2}$ inch from each end, then tack the string to a pad of graph paper so that the tacks are 4 inches apart. Pull the string tight with the tip of a pencil, then trace all the way around the two tacks. (See Figure 1.) The resulting diagram will be an ellipse.

FIGURE 2

2. The line segment that passes through the tacks (these are the foci) and connects opposite ends of the ellipse is called the major axis. The line segment perpendicular to the major axis that passes through the center of the ellipse and connects opposites ends of the ellipse is called the minor axis. (See Figure 2.) Measure the length of the major axis and the length of the minor axis. Record your results in Table 1.

3. Explain how drawing the ellipse as you have in step 2 shows that the geometric definition of an ellipse given at the beginning of this project is, in fact, correct.

4. Next, move the tacks so that they are 3 inches apart. Trace out that ellipse. Measure the length of the major axis and the length of the minor axis, and record your results in Table 1.

5. Repeat step 4 with the tacks 2 inches apart.

TABLE 1 Ellipses (All Lengths Are Inches)			
Length of String	Distance Between Foci	Length of Major Axis	Length of Minor Axis
6	4		
6	3		
6	2		

6. If the length of the string between the tacks stays at 6 inches, and the tacks were placed 6 inches apart, then the resulting ellipse would be a _____. If the tacks were placed 0 inches apart, then the resulting ellipse would be a _____.

RESEARCH PROJECT

HYPATIA OF ALEXANDRIA

The first woman mentioned in the history of mathematics is Hypatia of Alexandria. Research the life of Hypatia, and then write an essay that begins with a description of the time and place in which she lived and then goes on to give an indication of the type of person she was, her accomplishments in areas other than mathematics, and how she was viewed by her contemporaries.

CHAPTER 11   TEST

1. Find x so that $(x, 2)$ is $2\sqrt{5}$ units from $(-1, 4)$. [11.1]

2. Give the equation of the circle with center at $(-2, 4)$ and radius 3. [11.1]

3. Give the equation of the circle with center at the origin that contains the point $(-3, -4)$. [11.1]

4. Find the center and radius of the circle. [11.1]

$$x^2 + y^2 - 10x + 6y = 5$$

Graph each of the following. [11.2, 11.3]

5. $4x^2 - y^2 = 16$

6. $\dfrac{x^2}{25} + \dfrac{y^2}{4} = 1$

7. $(x - 2)^2 + (y + 1)^2 \leq 9$

8. $9x^2 + 4y^2 - 72x - 16y + 124 = 0$

Solve the following systems. [11.3]

9. $x^2 + y^2 = 25$
   $2x + y = 5$

10. $x^2 + y^2 = 16$
    $y = x^2 - 4$

Simplify.

1. $2^3 + 3(2 + 20 \div 4)$ **2.** $-|-5|$

3. $-5(2x + 3) + 8x$

4. $4 - 2[3x - 4(x + 2)]$

5. $(3y + 2)^2 - (3y - 2)^2$

6. $\dfrac{\frac{1}{5} - \frac{1}{4}}{\frac{1}{2} + \frac{3}{4}}$ **7.** $x^{2/3} \cdot x^{1/5}$

8. $\sqrt{48x^5y^3}$ (Assume x and y are positive.)

Solve.

9. $5y - 2 = -3y + 6$

10. $3 - 2(3x - 4) = -1$

11. $|3x - 1| - 2 = 6$ **12.** $2x^2 = 5x + 3$

13. $x^3 - 3x^2 - 4x + 12 = 0$

14. $\dfrac{6}{a + 2} = \dfrac{5}{a - 3}$ **15.** $x - 2 = \sqrt{3x + 4}$

16. $(x - 3)^2 = -3$ **17.** $4x^2 + 6x = -5$

18. $\log_2 x = 3$ **19.** $\log_2 x + \log_2 5 = 1$

20. $8^{x+3} = 4$ **21.** $4x + 2y = 4$
$y = -3x + 1$

22. $2x - 5y = 3$ **23.** $x + 2y - z = 4$
$3x + 2y = -5$ $2x - y - 3z = -1$
$-x + 2y + 2z = 3$

Multiply.

24. $\dfrac{x^2 - 16}{x^2 + 5x + 6} \cdot \dfrac{x^2 + 6x + 9}{x^3 + 4x^2}$

25. $(x^{1/5} + 3)(x^{1/5} - 3)$

Divide.

26. $\dfrac{12x^2y^3 - 16x^2y + 8xy^3}{4xy}$

27. $\dfrac{3 - 2i}{1 + 2i}$

Graph.

28. $2x - 3y = 12$ **29.** $3x - y < -2$

30. $y = 2^x$ **31.** $y = \log_2 x$

32. $x^2 + 4x + y^2 - 6y = 12$

33. $9x^2 - 4y^2 = 36$

34. $x^2 + y^2 < 4$ **35.** $y = (x - 2)^2 - 3$

Solve.

36. Find the next number in the sequence: $1, -4, 9, \ldots$

37. Identify the hypothesis and conclusion: If $x = -2$, then $x^3 = 8$.

38. Solve $mx + 2 = nx - 3$ for x.

39. If $f(x) = -\frac{3}{2}x + 1$, find $f(4)$.

40. Find the slope and y-intercept for $2x - 3y = 12$.

41. Find the equation of the line through $(-6, -1)$ and $(-3, -5)$.

42. y varies inversely with the square of x. If y is 3 when x is 3, find y when x is 6.

43. Find the distance between $(-3, 1)$ and $(4, 5)$.

44. **Mixture** How many gallons of 20% alcohol solution and 60% alcohol solution must be mixed to get 16 gallons of 30% alcohol solution? Be sure to show the system of equations used to solve the problem.

45. Solve for x.

$$\begin{vmatrix} 2x & -4 \\ 4 & x \end{vmatrix} = 18x$$

46. Find the second term in the expansion of $(x - 2y)^5$.

47. **Projectile Motion** An object projected upward with an initial velocity of 48 feet per second will rise and fall according to the equation $s = 48t - 16t^2$, where s is its distance above the ground at the time t. At what times will the object be 20 feet above the ground?

48. Find an equation that has solutions $x = 1$ and $x = \frac{2}{3}$.

49. Graph the solution set for $x^2 + x - 6 > 0$.

50. What is the domain for $f(x) = \dfrac{x + 3}{x - 2}$?

51. Find the general term of the sequence $5, 15, 45, 135, \ldots$.

52. Expand and simplify.

$$\sum_{i=1}^{5} (2i + 1)$$

Appendix A

Synthetic Division

S *ynthetic division* is a short form of long division with polynomials. We will consider synthetic division only for those cases in which the divisor is of the form $x + k$, where k is a constant.

Let's begin by looking over an example of long division with polynomials as done in Section 6.2:

$$
\begin{array}{r}
3x^2 - 2x + 4 \\
x + 3 \overline{)\, 3x^3 + 7x^2 - 2x - 4} \\
\underline{3x^3 + 9x^2} \\
-2x^2 - 2x \\
\underline{-2x^2 - 6x} \\
4x - 4 \\
\underline{4x + 12} \\
-16
\end{array}
$$

We can rewrite the problem without showing the variable, since the variable is written in descending powers and similar terms are in alignment. It looks like this:

$$
\begin{array}{r}
3 \quad -2 \quad +4 \\
1 + 3 \overline{)\, 3 \quad\;\; 7 \quad -2 \quad -4} \\
\underline{(3) \quad +9} \\
-2 \;\; (-2) \\
\underline{(-2) \;\; -6} \\
4 \;\; (-4) \\
\underline{(4) \quad 12} \\
-16
\end{array}
$$

We have used parentheses to enclose the numbers that are repetitions of the numbers above them. We can compress the problem by eliminating all repetitions, except the first one:

$$
\begin{array}{r}
3 \quad -2 \quad\;\; 4 \\
1 + 3 \overline{)\, 3 \quad\;\; 7 \quad -2 \quad\;\; -4} \\
\underline{9 \quad -6 \quad\;\; 12} \\
3 \quad -2 \quad\;\; 4 \quad -16
\end{array}
$$

The top line is the same as the first three terms of the bottom line, so we eliminate the top line. Also, the 1 that was the coefficient of x in the original problem can be eliminated, since we will consider only division problems where the divisor is of the form $x + k$. The following is the most compact form of the original division problem:

$$+3\overline{)3 \quad 7 \quad -2 \quad -4}$$
$$\underline{ \quad 9 \quad -6 \quad 12}$$
$$3 \quad -2 \quad 4 \quad -16$$

If we check over the problem, we find that the first term in the bottom row is exactly the same as the first term in the top row — and it always will be in problems of this type. Also, the last three terms in the bottom row come from multiplication by $+3$ and then subtraction. We can get an equivalent result by multiplying by -3 and adding. The problem would then look like this:

$$\begin{array}{r|rrrr} -3 & 3 & 7 & -2 & -4 \\ & \downarrow & -9 & 6 & -12 \\ \hline & 3 & -2 & 4 & \boxed{-16} \end{array}$$

We have used the brackets $\rfloor\lfloor$ to separate the divisor and the remainder. This last expression is synthetic division. It is an easy process to remember. Simply change the sign of the constant term in the divisor, then bring down the first term of the dividend. The process is then just a series of multiplications and additions, as indicated in the following diagram by the arrows:

$$\begin{array}{r|rrrr} -3 & 3 & 7 & -2 & -4 \\ & \downarrow & 9 & 6 & 12 \\ \hline & 3 & -2 & 4 & \boxed{-16} \end{array}$$

The last term of the bottom row is always the remainder.

Here are some additional examples of synthetic division with polynomials.

EXAMPLE I   Divide $x^4 - 2x^3 + 4x^2 - 6x + 2$ by $x - 2$.

Solution   We change the sign of the constant term in the divisor to get $+2$ and then complete the procedure:

$$\begin{array}{r|rrrrr} +2 & 1 & -2 & 4 & -6 & 2 \\ & \downarrow & 2 & 0 & 8 & 4 \\ \hline & 1 & 0 & 4 & 2 & \boxed{6} \end{array}$$

From the last line we have the answer:

$$1x^3 + 0x^2 + 4x + 2 + \frac{6}{x - 2}$$

EXAMPLE 2   Divide $\dfrac{3x^3 - 4x + 5}{x + 4}$.

Solution   Since we cannot skip any powers of the variable in the polynomial $3x^3 - 4x + 5$, we rewrite it as $3x^3 + 0x^2 - 4x + 5$ and proceed as we did in Example 1:

$$
\begin{array}{r|rrrr}
-4 & 3 & 0 & -4 & 5 \\
 & \downarrow & -12 & 48 & -176 \\
\hline
 & 3 & -12 & 44 & \boxed{-171}
\end{array}
$$

From the synthetic division, we have

$$\frac{3x^3 - 4x + 5}{x + 4} = 3x^2 - 12x + 44 - \frac{171}{x + 4}$$

EXAMPLE 3   Divide $\dfrac{x^3 - 1}{x - 1}$.

Solution   Writing the numerator as $x^3 + 0x^2 + 0x - 1$ and using synthetic division, we have

$$
\begin{array}{r|rrrr}
+1 & 1 & 0 & 0 & -1 \\
 & \downarrow & 1 & 1 & 1 \\
\hline
 & 1 & 1 & 1 & \boxed{0}
\end{array}
$$

which indicates

$$\frac{x^3 - 1}{x - 1} = x^2 + x + 1$$

PROBLEM SET A

Use synthetic division to find the following quotients.

1. $\dfrac{x^2 - 5x + 6}{x + 2}$

2. $\dfrac{x^2 + 8x - 12}{x - 3}$

3. $\dfrac{3x^2 - 4x + 1}{x - 1}$

4. $\dfrac{4x^2 - 2x - 6}{x + 1}$

5. $\dfrac{x^3 + 2x^2 + 3x + 4}{x - 2}$

6. $\dfrac{x^3 - 2x^2 - 3x - 4}{x - 2}$

7. $\dfrac{3x^3 - x^2 + 2x + 5}{x - 3}$

8. $\dfrac{2x^3 - 5x^2 + x + 2}{x - 2}$

9. $\dfrac{2x^3 + x - 3}{x - 1}$

10. $\dfrac{3x^3 - 2x + 1}{x - 5}$

11. $\dfrac{x^4 + 2x^2 + 1}{x + 4}$

12. $\dfrac{x^4 - 3x^2 + 1}{x - 4}$

13. $\dfrac{x^5 - 2x^4 + x^3 - 3x^2 - x + 1}{x - 2}$

14. $\dfrac{2x^5 - 3x^4 + x^3 - x^2 + 2x + 1}{x + 2}$

15. $\dfrac{x^2 + x + 1}{x - 1}$

16. $\dfrac{x^2 + x + 1}{x + 1}$

17. $\dfrac{x^4 - 1}{x + 1}$

18. $\dfrac{x^4 + 1}{x - 1}$

19. $\dfrac{x^3 - 1}{x - 1}$

20. $\dfrac{x^3 - 1}{x + 1}$

Appendix B

Matrix Solutions to Linear Systems

In mathematics, a **matrix** is a rectangular array of elements considered as a whole. We can use matrices to represent systems of linear equations. To do so, we write the coefficients of the variables and the constant terms in the same position in the matrix as they occur in the system of equations. To show where the coefficients end and the constant terms begin, we use vertical lines instead of equal signs. For example, the system

$$2x + 5y = -4$$
$$x - 3y = 9$$

can be represented by the matrix

$$\left[\begin{array}{rr|r} 2 & 5 & -4 \\ 1 & -3 & 9 \end{array}\right]$$

which is called an **augmented matrix** because it includes both the coefficients of the variables and the constant terms.

To solve a system of linear equations by using the augmented matrix for that system, we need the following row operations as the tools of that solution process. The row operations tell us what we can do to an augmented matrix that may change the numbers in the matrix, but will always produce a matrix that represents a system of equations with the same solution as that of our original system.

ROW OPERATIONS

1. We can interchange any two rows of a matrix.

2. We can multiply any row by a nonzero constant.

3. We can add to any row a constant multiple of another row.

The three row operations are simply a list of the properties we use to solve systems of linear equations, translated to fit an augmented matrix. For instance, the second operation in our list is actually just another way to state the multiplication property of equality.

We solve a system of linear equations by transforming the augmented matrix into a matrix that has 1's down the diagonal of the coefficient matrix, and 0's below it. For instance, we will have solved the system

$$2x + 5y = -4$$
$$x - 3y = 9$$

When the matrix

$$\left[\begin{array}{cc|c} 2 & 5 & -4 \\ 1 & -3 & 9 \end{array}\right]$$

has been transformed, using the row operations listed earlier, we get a matrix of the form

$$\left[\begin{array}{cc|c} 1 & - & - \\ 0 & 1 & - \end{array}\right]$$

To accomplish this, we begin with the first column and try to produce a 1 in the first position and a 0 below it. Interchanging rows 1 and 2 gives us a 1 in the top position of the first column:

$\downarrow$ Interchange rows 1 and 2.

$$\left[\begin{array}{cc|c} 1 & -3 & 9 \\ 2 & 5 & -4 \end{array}\right]$$

Multiplying row 1 by -2 and adding the result to row 2 gives us a 0 where we want it.

$\downarrow$ Multiply row 1 by -2 and add the result to row 2.

$$\left[\begin{array}{cc|c} 1 & -3 & 9 \\ 0 & 11 & -22 \end{array}\right]$$

$\downarrow$ Multiply row 2 by $\frac{1}{11}$.

$$\left[\begin{array}{cc|c} 1 & -3 & 9 \\ 0 & 1 & -2 \end{array}\right]$$

Taking this last matrix and writing the system of equations it represents, we have

$$x - 3y = 9$$
$$y = -2$$

Substituting -2 for y in the top equation gives us

$$x = 3$$

The solution to our system is $(3, -2)$.

EXAMPLE I   Solve the following system using an augmented matrix:

$$x + y - z = 2$$
$$2x + 3y - z = 7$$
$$3x - 2y + z = 9$$

Solution   We begin by writing the system in terms of an augmented matrix:

$$\begin{vmatrix} 1 & 1 & -1 & | & 2 \\ 2 & 3 & -1 & | & 7 \\ 3 & -2 & 1 & | & 9 \end{vmatrix}$$

Next, we want to produce 0's in the second two positions of column 1:

↓ Multiply row 1 by -2 and add the result to row 2.

$$\begin{bmatrix} 1 & 1 & -1 & | & 2 \\ 0 & 1 & 1 & | & 3 \\ 3 & -2 & 1 & | & 9 \end{bmatrix}$$

↓ Multiply row 1 by -3 and add the result to row 3.

$$\begin{bmatrix} 1 & 1 & -1 & | & 2 \\ 0 & 1 & 1 & | & 3 \\ 0 & -5 & 4 & | & 3 \end{bmatrix}$$

Note that we could have done these two steps in one single step. As you become more familiar with this method of solving systems of equations, you will do just that.

↓ Multiply row 2 by 5 and add the result to row 3.

$$\begin{bmatrix} 1 & 1 & -1 & | & 2 \\ 0 & 1 & 1 & | & 3 \\ 0 & 0 & 9 & | & 18 \end{bmatrix}$$

↓ Multiply row 3 by $\frac{1}{9}$.

$$\begin{bmatrix} 1 & 1 & -1 & | & 2 \\ 0 & 1 & 1 & | & 3 \\ 0 & 0 & 1 & | & 2 \end{bmatrix}$$

Converting back to a system of equations, we have

$$x + y - z = 2$$

$$y + z = 3$$

$$z = 2$$

This system is equivalent to our first one, but much easier to solve.

Substituting $z = 2$ into the second equation, we have

$$y = 1$$

Substituting $z = 2$ and $y = 1$ into the first equation, we have

$$x = 3$$

The solution to our original system is (3, 1, 2). It satisfies each of our original equations. You can check this, if you like.

PROBLEM SET B

Solve the following systems of equations by using matrices.

1. $\quad x + y = 5$
$\quad 3x - y = 3$

2. $\quad x + y = -2$
$\quad 2x - y = -10$

3. $\quad 3x - 5y = 7$
$\quad -x + y = -1$

4. $2x - y = 4$
$\quad x + 3y = 9$

5. $2x - 8y = 6$
$\quad 3x - 8y = 13$

6. $\quad 3x - 6y = 3$
$\quad -2x + 3y = -4$

7. $x + y + z = 4$
$\quad x - y + 2z = 1$
$\quad x - y - z = -2$

8. $x - y - 2z = -1$
$\quad x + y + z = 6$
$\quad x + y - z = 4$

9. $\quad x + 2y + z = 3$
$\quad 2x - y + 2z = 6$
$\quad 3x + y - z = 5$

10. $\quad x - 3y + 4z = -4$
$\quad 2x + y - 3z = 14$
$\quad 3x + 2y + z = 10$

11. $\quad x + 2y = 3$
$\quad y + z = 3$
$\quad 4x - z = 2$

12. $\quad x + y = 2$
$\quad 3y - 2z = -8$
$\quad x + z = 5$

13. $\quad x + 3y = 7$
$\quad 3x - 4z = -8$
$\quad 5y - 2z = -5$

14. $\quad x + 4y = 13$
$\quad 2x - 5z = -3$
$\quad 4y - 3z = 9$

Solve each system using matrices. Remember, multiplying a row by a nonzero constant will not change the solution to the system.

15. $\frac{1}{3}x + \frac{1}{5}y = 2$

$\quad \frac{1}{3}x - \frac{1}{2}y = -\frac{1}{3}$

16. $\frac{1}{2}x + \frac{1}{3}y = 13$

$\quad \frac{1}{5}x + \frac{1}{8}y = 5$

The systems that follow are inconsistent systems. In both cases, the lines are parallel. Try solving each system using matrices and see what happens.

17. $2x - 3y = 4$
$\quad 4x - 6y = 4$

18. $\quad 10x - 15y = 5$
$\quad -4x + 6y = -4$

The systems that follow are dependent systems. In each case, the lines coincide. Try solving each system using matrices and see what happens.

19. $-6x + 4y = 8$
$\quad -3x + 2y = 4$

20. $\quad x + 2y = 5$
$\quad -x - 2y = -5$

Appendix C

Algebra With Functions

A company produces and sells copies of an accounting program for home computers. The price they charge for the program is related to the number of copies sold by the demand function

$$p(x) = 35 - 0.1x$$

We find the revenue for this business by multiplying the number of items sold by the price per item. When we do so, we are forming a new function by combining two existing functions. That is, if $n(x) = x$ is the number of items sold and $p(x) = 35 - 0.1x$ is the price per item, then revenue is

$$R(x) = n(x) \cdot p(x) = x(35 - 0.1x) = 35x - 0.1x^2$$

In this case, the revenue function is the product of two functions. When we combine functions in this manner, we are applying our rules for algebra to functions.

To carry this situation further, we know the profit function is the difference between two functions. If the cost function for producing x copies of the accounting program is $C(x) = 8x + 500$, then the profit function is

$$P(x) = R(x) - C(x) = (35x - 0.1x^2) - (8x + 500) = -500 + 27x - 0.1x^2$$

The relationship between these last three functions is shown visually in Figure 1.

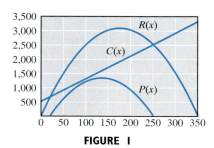

FIGURE 1

Again, when we combine functions in the manner shown we are applying our rules for algebra to functions. To begin this section we take a formal look at addition, subtraction, multiplication, and division with functions.

If we are given two functions f and g with a common domain, we can define four other functions as follows.

DEFINITION

$(f + g)(x) = f(x) + g(x)$ The function $f + g$ is the sum of the functions f and g.

$(f - g)(x) = f(x) - g(x)$ The function $f - g$ is the difference of the functions f and g.

$(fg)(x) = f(x)g(x)$ The function fg is the product of the functions f and g.

$\dfrac{f}{g}(x) = \dfrac{f(x)}{g(x)}$ The function f/g is the quotient of the functions f and g, where $g(x) \neq 0$.

EXAMPLE 1   If $f(x) = 4x^2 + 3x + 2$ and $g(x) = 2x^2 - 5x - 6$, write the formula for the functions $f + g$, $f - g$, fg, and f/g.

Solution   The function $f + g$ is defined by

$$(f + g)(x) = f(x) + g(x)$$
$$= (4x^2 + 3x + 2) + (2x^2 - 5x - 6)$$
$$= 6x^2 - 2x - 4$$

The function $f - g$ is defined by

$$(f - g)(x) = f(x) - g(x)$$
$$= (4x^2 + 3x + 2) - (2x^2 - 5x - 6)$$
$$= 4x^2 + 3x + 2 - 2x^2 + 5x + 6$$
$$= 2x^2 + 8x + 8$$

The function fg is defined by

$$(fg)(x) = f(x)g(x)$$
$$= (4x^2 + 3x + 2)(2x^2 - 5x - 6)$$
$$= 8x^4 - 20x^3 - 24x^2 + 6x^3 - 15x^2 - 18x + 4x^2 - 10x - 12$$
$$= 8x^4 - 14x^3 - 35x^2 - 28x - 12$$

The function f/g is defined by

$$\left(\frac{f}{g}\right)(x) = \frac{f(x)}{g(x)}$$
$$= \frac{4x^2 + 3x + 2}{2x^2 - 5x - 6}$$

EXAMPLE 2   Let $f(x) = 4x - 3$, $g(x) = 4x^2 - 7x + 3$, and $h(x) = x - 1$. Find $f + g$, fh, fg, and g/f.

Solution   The function $f + g$, the sum of functions f and g, is defined by

$$(f + g)(x) = f(x) + g(x)$$
$$= (4x - 3) + (4x^2 - 7x + 3)$$
$$= 4x^2 - 3x$$

The function fh, the product of functions f and h, is defined by

$$(fh)(x) = f(x)h(x)$$
$$= (4x - 3)(x - 1)$$
$$= 4x^2 - 7x + 3$$
$$= g(x)$$

The product of the functions f and g, fg, is given by

$$(fg)(x) = f(x)g(x)$$
$$= (4x - 3)(4x^3 - 7x + 3)$$
$$= 16x^2 - 28x^2 + 12x - 12x^2 + 21x - 9$$
$$= 16x^3 - 40x^2 + 33x - 9$$

The quotient of the functions g and f, g/f, is defined as

$$\frac{g}{f}(x) = \frac{g(x)}{f(x)}$$
$$= \frac{4x^2 - 7x + 3}{4x - 3}$$

Factoring the numerator, we can reduce to lowest terms:

$$\frac{g}{f}(x) = \frac{(4x - 3)(x - 1)}{4x - 3}$$
$$= x - 1$$
$$= h(x)$$

EXAMPLE 3   If f, g, and h are the same functions defined in Example 2, evaluate $(f + g)(2)$, $(fh)(-1)$, $(fg)(0)$, and $(g/f)(5)$.

Solution   We use the formulas for $f + g$, fh, fg and g/f found in Example 2:

$$(f + g)(2) = 4(2)^2 - 3(2)$$
$$= 16 - 6$$
$$= 10$$

$$(fh)(-1) = 4(-1)^2 - 7(-1) + 3$$
$$= 4 + 7 + 3$$
$$= 14$$
$$(fg)(0) = 16(0)^3 - 40(0)^2 + 33(0) - 9$$
$$= 0 - 0 + 0 - 9$$
$$= -9$$
$$\frac{g}{f}(5) = 5 - 1$$
$$= 4$$

COMPOSITION OF FUNCTIONS

In addition to the four operations used to combine functions shown so far in this section, there is a fifth way to combine two functions to obtain a new function. It is called **composition of functions.** To illustrate the concept, recall from Chapter 2 the definition of training heart rate: training heart rate, in beats per minute, is resting heart rate plus 60% of the difference between maximum heart rate and resting heart rate. If your resting heart rate is 70 beats per minute, then your training heart rate is a function of your maximum heart rate M

$$T(M) = 70 + 0.6(M - 70) = 70 + 0.6M - 42 = 28 + 0.6M$$

But your maximum heart rate is found by subtracting your age in years from 220. So, if x represents your age in years, then your maximum heart rate is

$$M(x) = 220 - x$$

Therefore, if your resting heart rate is 70 beats per minute and your age in years is x, then your training heart rate can be written as a function of x.

$$T(x) = 28 + 0.6(220 - x)$$

This last line is the composition of functions T and M. We input x into function M, which outputs $M(x)$. Then, we input $M(x)$ into function T, which outputs $T(M(x))$, which is the training heart rate as a function of age x (Fig. 2). Here is a diagram of the situation, which is called a function map:

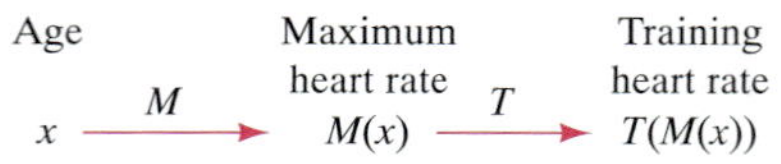

FIGURE 2

Now let's generalize the preceding ideas into a formal development of composition of functions. To find the composition of two functions f and g, we first require that the range of g have numbers in common with the domain of f. Then the composition of f with g, is defined this way:

$$(f \circ g)(x) = f(g(x))$$

To understand this new function, we begin with a number x, and we operate on it with g, giving us $g(x)$. Then we take $g(x)$ and operate on it with f, giving us $f(g(x))$. The only numbers we can use for the domain of the composition of f with g are numbers x in the domain of g, for which $g(x)$ is in the domain of f. The diagrams in Figure 3 illustrate the composition of f with g.

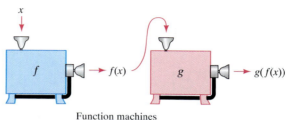

Function machines

$$x \xrightarrow{\ f\ } f(x) \xrightarrow{\ g\ } g(f(x))$$

FIGURE 3 A function machine

Composition of functions is not commutative. The composition of f with g, $f \circ g$, may therefore be different from the composition of g with f, $g \circ f$.

$$(g \circ f)(x) = g(f(x))$$

Again, the only numbers we can use for the domain of the composition of g with f are numbers in the domain of f, for which $f(x)$ is in the domain of g. The diagrams in Figure 4 illustrate the composition of g with f.

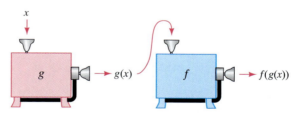

Function machines

$$x \xrightarrow{\ g\ } g(x) \xrightarrow{\ f\ } f(g(x))$$

FIGURE 4

EXAMPLE 4   If $f(x) = x + 5$ and $g(x) = x^2 - 2x$, find $(f \circ g)(x)$ and $(g \circ f)(x)$.

Solution   The composition of f with g is

$$\begin{aligned}
(f \circ g)(x) &= f(g(x)) \\
&= f(x^2 - 2x) \\
&= (x^2 - 2x) + 5 \\
&= x^2 - 2x + 5
\end{aligned}$$

The composition of g with f is

$$(g \circ f)(x) = g(f(x))$$
$$= g(x + 5)$$
$$= (x + 5)^2 - 2(x + 5)$$
$$= (x^2 + 10x + 25) - 2(x + 5)$$
$$= x^2 + 8x + 15$$

PROBLEM SET C

Let $f(x) = 4x - 3$ and $g(x) = 2x + 5$. Write a formula for each of the following functions.

1. $f + g$

2. $f - g$

3. $g - f$

4. $g + f$

5. fg

6. f/g

7. g/f

8. ff

If the functions f, g, and h are defined by $f(x) = 3x - 5$, $g(x) = x - 2$, and $h(x) = 3x^2 - 11x + 10$, write a formula for each of the following functions.

9. $g + f$

10. $f + h$

11. $g + h$

12. $f - g$

13. $g - f$

14. $h - g$

15. fg

16. gf

17. fh

18. gh

19. h/f

20. h/g

21. f/h

22. g/h

23. $f + g + h$

24. $h - g + f$

25. $h + fg$

26. $h - fg$

Let $f(x) = 2x + 1$, $g(x) = 4x + 2$, and $h(x) = 4x^2 + 4x + 1$, and find the following.

27. $(f + g)(2)$

28. $(f - g)(-1)$

29. $(fg)(3)$

30. $(f/g)(-3)$

31. $(h/g)(1)$

32. $(hg)(1)$

33. $(fh)(0)$

34. $(h - g)(-4)$

35. $(f + g + h)(2)$

36. $(h - f + g)(0)$

37. $(h + fg)(3)$

38. $(h - fg)(5)$

39. Let $f(x) = x^2$ and $g(x) = x + 4$, and find
  (a) $(f \circ g)(5)$
  (b) $(g \circ f)(5)$
  (c) $(f \circ g)(x)$
  (d) $(g \circ f)(x)$

40. Let $f(x) = 3 - x$ and $g(x) = x^3 - 1$, and find
  (a) $(f \circ g)(0)$
  (b) $(g \circ f)(0)$
  (c) $(f \circ g)(x)$
  (d) $(g \circ f)(x)$

41. Let $f(x) = x^2 + 3x$ and $g(x) = 4x - 1$, and find
  (a) $(f \circ g)(0)$
  (b) $(g \circ f)(0)$
  (c) $(f \circ g)(x)$
  (d) $(g \circ f)(x)$

42. Let $f(x) = (x - 2)^2$ and $g(x) = x + 1$, and find the following
  (a) $(f \circ g)(-1)$
  (b) $(g \circ f)(-1)$
  (c) $(f \circ g)(x)$
  (d) $(g \circ f)(x)$

For each of the following pairs of functions f and g, show that $(f \circ g)(x) = (g \circ f)(x) = x$.

43. $f(x) = 5x - 4$ and $g(x) = \dfrac{x + 4}{5}$

44. $f(x) = \dfrac{x}{6} - 2$ and $g(x) = 6x + 12$

Applying the Concepts

45. Profit, Revenue, and Cost   A company manufactures and sells prerecorded videotapes. Here are the equations they use in connection with their business.

Number of tapes sold each day: $n(x) = x$

Selling price for each tape: $p(x) = 11.5 - 0.05x$

Daily fixed costs: $f(x) = 200$
Daily variable costs: $v(x) = 2x$

Find the following functions.

(a) Revenue = $R(x)$ = the product of the number of tapes sold each day and the selling price of each tape.

(b) Cost = $C(x)$ = the sum of the fixed costs and the variable costs.

(c) Profit = $P(x)$ = the difference between revenue and cost.

(d) Average cost = $\overline{C}(x)$ = the quotient of cost and the number of tapes sold each day.

46. **Training Heart Rate**   Find an equation for the training heart rate for a person with a resting heart rate of 60 beats per minute. What is the training heart rate for a 24-year-old person?

Appendix D

Answers to Chapter Reviews, Cumulative Reviews, Chapter Tests, and Odd-Numbered Problems

CHAPTER 1

PROBLEM SET 1.1

1. $x + 5 = 2$ **3.** $6 - x = y$ **5.** $2t < y$ **7.** $x + y < x - y$ **9.** $3(x - 5) > y$ **11.** 36   **13.** 100
15. 8   **17.** 16   **19.** 10,000   **21.** 121   **23.** 19   **25.** 27   **27.** 42   **29.** 50   **31.** 16   **33.** 12
35. 18   **37.** 33   **39.** 33   **41.** 23   **43.** 41   **45.** 65   **47.** 39   **49.** 7   **51.** 5   **53.** 10   **55.** 1
57. 5,431   **59.** 32   **61.** 24   **63.** 41   **65.** 95   **67.** 138   **69.** 78   **71.** 152   **73.** 48
75. $\{0, 1, 2, 3, 4, 5, 6\}$ **77.** $\varnothing$ **79.** $\{0, 1, 2, 3, 4, 5, 6\}$ **81.** $\{0, 2\}$ **83.** $\{0, 6\}$
85. $\{0, 1, 2, 3, 4, 5, 6, 7\}$ **87.** \$7,371   **89.** \$2,638   **91.** \$6,609   **93.** \$4,086   **95.** \$6,500
97. $12 + \frac{1}{4}(12) = 15$ **99.** (a) $16 \div 4 - 8 \div 2 = 0$ (b) $16 \div 4 \div 8 \cdot 2 = 1$ or $16 \div (4 \cdot 8 \div 2) = 1$ or
$(16 \div 4) \div (8 \div 2) = 1$ (c) $16 - 4 - 8 - 2 = 2$ (d) $(16 - 4) \div (8 \div 2) = 3$
(e) $(16 + 4) \div (8 \div 2) = 5$ (f) $16 - 4 - 8 \div 2 = 8$ or $(16 \div 4) + (8 \div 2) = 8$
(g) $16 \div 4 + 8 - 2 = 10$ or $16 - [(4 + 8) \div 2] = 10$ (h) $16 \div 4 \cdot 8 \cdot 2 = 64$ **101.** 9   **103.** 13

PROBLEM SET 1.2

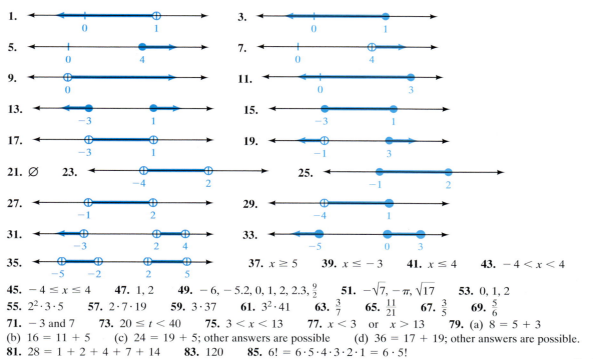

37. $x \geq 5$ **39.** $x \leq -3$ **41.** $x \leq 4$ **43.** $-4 < x < 4$
45. $-4 \leq x \leq 4$ **47.** 1, 2   **49.** $-6, -5.2, 0, 1, 2, 2.3, \frac{9}{2}$ **51.** $-\sqrt{7}, -\pi, \sqrt{17}$ **53.** 0, 1, 2
55. $2^2 \cdot 3 \cdot 5$ **57.** $2 \cdot 7 \cdot 19$ **59.** $3 \cdot 37$ **61.** $3^2 \cdot 41$ **63.** $\frac{3}{7}$ **65.** $\frac{11}{21}$ **67.** $\frac{3}{5}$ **69.** $\frac{5}{6}$
71. -3 and 7   **73.** $20 \leq t < 40$ **75.** $3 < x < 13$ **77.** $x < 3$ or $x > 13$ **79.** (a) $8 = 5 + 3$
(b) $16 = 11 + 5$ (c) $24 = 19 + 5$; other answers are possible   (d) $36 = 17 + 19$; other answers are possible.
81. $28 = 1 + 2 + 4 + 7 + 14$ **83.** 120   **85.** $6! = 6 \cdot 5 \cdot 4 \cdot 3 \cdot 2 \cdot 1 = 6 \cdot 5!$

PROBLEM SET 1.3

1. $4, -4, \frac{1}{4}$ **3.** $-\frac{1}{2}, \frac{1}{2}, -2$ **5.** $5, -5, \frac{1}{5}$ **7.** $\frac{3}{8}, -\frac{3}{8}, \frac{8}{3}$ **9.** $-\frac{1}{6}, \frac{1}{6}, -6$ **11.** $3, -3, \frac{1}{3}$ **13.** $-1, 1$

15. 0 **17.** 2 **19.** $\frac{3}{4}$ **21.** π **23.** -4 **25.** -2 **27.** $-\frac{3}{4}$ **29.** $\frac{21}{40}$ **31.** 2 **33.** $\frac{8}{27}$

35. $\frac{1}{10,000}$ **37.** $\frac{72}{385}$ **39.** 1 **41.** $6 + x$ **43.** $a + 8$ **45.** $15y$ **47.** x **49.** a **51.** x

53. $3x + 18$ **55.** $12x + 8$ **57.** $15a + 10b$ **59.** $\frac{4}{3}x + 2$ **61.** $2 + y$ **63.** $40t + 8$ **65.** $15x + 10$

67. $8y + 32$ **69.** $15t + 9$ **71.** $28x + 11$ **73.** $\frac{7}{15}$ **75.** $\frac{29}{35}$ **77.** $\frac{35}{144}$ **79.** $\frac{949}{1,260}$ **81.** $14a + 7$

83. $6y + 6$ **85.** $12x + 2$ **87.** $8y + 11$ **89.** $24a + 15$ **91.** $11x + 20$

93. Commutative property of addition **95.** Commutative property of multiplication **97.** Additive inverse

99. Commutative property of addition **101.** Associative and commutative properties of multiplication

103. Commutative and associative properties of addition **105.** Distributive

107. $7(x + 2) = 7x + 14$; $7x + 7(2) = 7x + 14$ **109.** $x(y + 4) = xy + 4x$; $xy + 4x$

111. Total = 1 million pounds $\cdot (1.212 + 3.795 + 103.543)$

= 1 million pounds $\cdot (108.55)$

$= \left(\dfrac{1 \text{ million pounds}}{2000 \text{ pounds}} \right)$ tons $\cdot (108.55)$

= 500 tons $\cdot (108.55)$

= 54,275 tons

113. (a) $6 + 7 = 13 = 1$ hour past 12 or 1.

(b) $3 + 11 + 8 = 22$

= 10 hours past 12 or 10.

(c) $3 + 11 + 8 + 12 = 34$

= 10 hours past 12 twice, or 10.

PROBLEM SET 1.4

1. 4 **3.** -4 **5.** -10 **7.** -4 **9.** $\frac{19}{12}$ **11.** $-\frac{32}{105}$ **13.** -8 **15.** -12 **17.** $-7x$ **19.** 13

21. -14 **23.** $6a$ **25.** -15 **27.** 15 **29.** -24 **31.** $-10x$ **33.** x **35.** y **37.** $-8x + 6$

39. $-3a + 4$ **41.** -14 **43.** 18 **45.** 16 **47.** -19 **49.** 50 **51.** 20 **53.** -2 **55.** 1

57. -30 **59.** 18 **61.** 277 **63.** -73 **65.** $14x + 12$ **67.** $7m - 15$ **69.** $-2x + 9$ **71.** $7y + 10$

73. $-20x + 5$ **75.** $-11x + 10$ **77.** -2 **79.** Undefined **81.** 0 **83.** $-\frac{2}{3}$ **85.** 32 **87.** 64

89. $-\frac{1}{18}$ **91.** $\frac{5}{3}$ **93.** 11 **95.** 12 **97.** -3 **99.** $2x$

101.

Pitcher, Team	Rolaids Points
Trevor Hoffman, San Diego	161
Rod Beck, Chicago	137
Jeff Shaw, Los Angeles	116
Robb Nen, San Francisco	110
Ugueth Urbina, Montreal	100
Kerry Ligtenberg, Atlanta	84

103. Your watch will say 9:10 P.M. in Santa Fe, but adding 1 hour for the time zone change, clocks in Santa Fe will say 10:10 P.M.

Your watch will say 12:30 A.M. in Detroit, but adding 3 hours for three time zone changes, clocks in Detroit will say 3:30 A.M.

105. (a) About 4.5 times deeper (b) 64,868 feet

PROBLEM SET 1.5

1. Hypothesis: You argue for your limitations.
Conclusion: They are yours.

3. Hypothesis: x is an even number.
Conclusion: x is divisible by 2.

5. Hypothesis: A triangle is equilaterial.
   Conclusion: All of its angles are equal.

7. Hypothesis: $x + 5 = -2$
   Conclusion: $x = -7$

9. Converse: If $a^2 = 64$, then $a = 8$.
   Inverse: If $a \neq 8$, then $a^2 \neq 64$.
   Contrapositive: If $a^2 \neq 64$, then $a \neq 8$.

11. Converse: If $a = b$, then $\frac{a}{b} = 1$.
   Inverse: If $\frac{a}{b} \neq 1$, then $a \neq b$.
   Contrapositive: If $a \neq b$, then $\frac{a}{b} \neq 1$.

13. Converse: If it is a rectangle, then it is a square.
   Inverse: If it is not a square, then it is not a rectangle.
   Contrapositive: If it is not a rectangle, then it is
      not a square.

15. Converse: If good is not enough, then better is possible.
   Inverse: If better is not possible, then good is enough.
   Contrapositive: If good is enough, then better
      is not possible.

17. If E, then F **19.** If it is misery, then it loves company.   **21.** If it is the squeaky wheel, then it gets the grease.

23. (c)   **25.** (a)   **27.** (c)   **29.** (b)

31. The contrapositive of the statement is "If your eyes are not closed, then you are not sleeping," which has the contrapositive: "If you are sleeping, then your eyes are closed," which is the original statement.

33.

	Statement	Inverse	Converse	Contrapositive
i)	True	True	True	True
ii)	False	True	True	False
iii)	True	False	False	True
iv)	False	True	True	False
v)		False	True	

Impossible, both must be the same.

35. *If Amy does not stay out late, then I extended her curfew. Tell her yes.

PROBLEM SET 1.6

1. 5   **3.** 10   **5.** 25   **7.** 29   **9.** 125   **11.** △   **13.** ⊙   **15.** 17, 21   **17.** $-2, -3$
19. $-4, -7$ **21.** $-\frac{1}{2}, -\frac{3}{4}$ **23.** $\frac{5}{2}, 3$ **25.** 27   **27.** -270 **29.** $\frac{1}{8}$ **31.** $\frac{5}{2}$ **33.** -625

35. $-\frac{1}{125}$ **37.** (a) 12 (b) 16   **39.** 144   **41.** 2, 3, 5, among others   **43.** 2, 8, 34

45. 0, 3, 6, 4, 7, 10, 8, 11, 14, 12   **47.** 41, 37.5, 34, 30.5, 27, 23.5; yes   **49.** 41, 45.5, 50, 54.5, 59, 63.5; yes

51. The patient on antidepressant 1 misses his morning dose; less than half of the antidepressant will remain in the body. The patient on antidepressant 2 still has most of the medication in his body even after missing a dose because of the relatively long (5-day) half-life.

53.

Hours Since Discontinuing	Concentration (ng/ml)
0	60
4	30
8	15
12	7.5
16	3.75

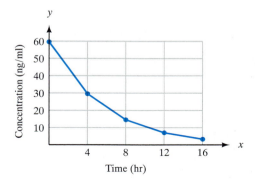

55.

Elevation (feet)	−2,000	−1,000	0	1,000	2,000	3,000
Boiling Point °F	215.6	213.8	212.0	210.2	208.4	206.6

57.

Year	1997	2032	2067	2102	2137
Population (billions)	5.852	11.704	23.408	46.816	93.632

59. (a) The patient taking the medication   (b) The patient stops taking the medication   (c) 50 ng/ml
(d) 4, 8, and 12 hours

61.

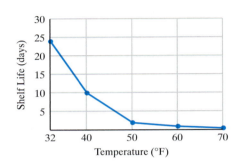

Shelf-Life of Milk	
Temperature (Fahrenheit)	**Shelf-Life (days)**
32°	24
40°	10
50°	2
60°	1
70°	$\frac{1}{2}$

63. 2, 4, 8, 16, 32 . . .   **65.** 1,024

CHAPTER 1 REVIEW

1. $x + 2$ **2.** $x - 2$ **3.** $\frac{x}{2}$ **4.** $2x$ **5.** $2(x + y)$ **6.** $2x + y$ **7.** 27   **8.** 125   **9.** 64   **10.** 1
11. 32   **12.** 81   **13.** 17   **14.** 4   **15.** 13   **16.** 7   **17.** 9   **18.** 32   **19.** 30   **20.** 43
21. {1, 2, 3, 4, 5, 6}   **22.** {1, 3}   **23.** {5}   **24.** {6}
25.

26.   **27.** $-2, \frac{1}{2}$ **28.** $\frac{2}{5}, -\frac{5}{2}$

29. 3   **30.** −5   **31.** 4   **32.** 6   **33.** −7, 0, 5   **34.** $-7, -4.2, 0, \frac{3}{4}, 5$ **35.** $-\sqrt{3}, \pi$ **36.** $2^2 \cdot 3^2 \cdot 11^2$

37. $\frac{11}{13}$ **38.** 1   **39.** $\frac{27}{64}$ **40.** 2
41.   **42.**

43.   **44.**   **45.** $x \geq 4$ **46.** $x \leq 5$

47. $0 < x < 8$ **48.** $0 \leq x \leq 8$ **49.** $2y$ **50.** $3x$ **51.** $8x + 5$ **52.** $y + 2$ **53.** a **54.** c **55.** a
56. b, d **57.** a, c **58.** f **59.** e **60.** g **61.** 2   **62.** −2   **63.** 1   **64.** 17   **65.** 3   **66.** −6
67. 2   **68.** 66   **69.** $-\frac{5}{6}$ **70.** $-\frac{5}{6}$ **71.** −42   **72.** 30   **73.** $21x$ **74.** $-6x$ **75.** $-6x + 10$

76. $-6x + 21$ **77.** $-x + 3$ **78.** $-15x + 3$ **79.** $-\frac{5}{6}$ **80.** −36   **81.** $\frac{1}{10}$ **82.** $-\frac{2}{7}$ **83.** −13

84. -17 **85.** 0 **86.** -1 **87.** -36 **88.** -34 **89.** 16 **90.** -24 **91.** 2 **92.** 39 **93.** 0
94. $-2y + 9$ **95.** $-18x - 14$ **96.** $5a - 22$
97. Converse: If $|x| = 7$, then $x = -7$.
 Inverse: If $x \neq -7$, then $|x| \neq 7$.
 Contrapositive: If $|x| \neq 7$, then $x \neq -7$.
98. Converse: If Therese lives in Texas, then she lives in Amarillo.
 Inverse: If Therese does not live in Amarillo, then she does not live in Texas.
 Contrapositive: If Therese does not live in Texas, then she does not live in Amarillo.
99. If $x^4 \neq 81$, then $x \neq 3$.   **100.** If Maria goes to school, then she does not go to the beach.   **101.** -1, arithmetic
102. 810, geometric   **103.** 8 **104.** 625 **105.** -1, arithmetic   **106.** $\frac{1}{16}$, geometric

CHAPTER 1 TEST

1. $2(3x + 4y)$ **2.** $2a - 3b < 2a + 3b$ **3.** 57 **4.** 10 **5.** 16 **6.** 0 **7.** $\{2, 4\}$ **8.** $\varnothing$
9. $-5, 0, 1, 4$ **10.** $-5, -4.1, -3.75, -\frac{5}{6}, 0, 1, 1.8, 4$ **11.** $-\sqrt{2}, \sqrt{3}$ **12.** $3^2 \cdot 5 \cdot 13$ **13.** $3, -\frac{1}{3}$ **14.** $-\frac{4}{3}, \frac{3}{4}$
15. 3 **16.** -2 **17.** ⟵━━━●━━━⊕━━━⟶
                    -1 5
18. ⟵━━━━━━━━━━━━⟶
      -2 4
19. Commutative property of addition   **20.** Multiplicative identity property
21. Commutative and associative properties of multiplication   **22.** Associative and commutative properties of addition
23. -19 **24.** 14 **25.** -149 **26.** 213 **27.** 2 **28.** 0 **29.** $\frac{59}{72}$ **30.** $-4x$ **31.** $-5x - 8$
32. $4y - 10$ **33.** $3x - 17$ **34.** $11a - 10$ **35.** $-\frac{7}{3}$ **36.** $-\frac{5}{2}$
37. Converse: If Emily does not study, then she goes out at night.
 Inverse: If Emily does not go out at night, then she studies.
 Contrapositive: If Emily studies, then she does not go out at night.
38. If $|x| \neq 3$, then $x \neq -3$.   **39.** -320, geometric   **40.** -2, arithmetic   **41.** 17 **42.** $\frac{1}{125}$, geometric

CHAPTER 2

PROBLEM SET 2.1

1. 8 **3.** 5 **5.** 2 **7.** -7 **9.** $-\frac{9}{2}$ **11.** -4 **13.** $-\frac{4}{3}$ **15.** -4 **17.** $\frac{7}{2}$ **19.** $-\frac{11}{5}$ **21.** 12
23. -10 **25.** 7 **27.** -4 **29.** -2 **31.** $\frac{3}{4}$ **33.** 3 **35.** $\frac{4}{5}$ **37.** 2 **39.** 1 **41.** 2 **43.** 0
45. -3 **47.** 4 **49.** 0 **51.** 17 **53.** 2 **55.** -3 **57.** 6 **59.** $-\frac{4}{3}$ **61.** 3 **63.** $-\frac{3}{2}$
65. $\frac{5}{3}$ **67.** 1 **69.** $6{,}000$ **71.** $5{,}000$ **73.** $\frac{3}{2}$ or 1.5 **75.** 1
77. Any method of solution results in a false statement.
79. Every attempt at solving the equation results in a true statement.   **81.** No solution
83. All real numbers are solutions.   **85.** No solution   **87.** (a) $\$6.60 = \$0.4n + \$1.80$ (b) 12 miles
89. (a) $1{,}025 = \dfrac{3{,}522{,}037}{A}$ (b) $A = 3{,}436$ square miles   **91.** Commutative   **93.** Associative
95. Commutative and associative   **97.** Multiplicative identity   **99.** Commutative   **101.** Additive identity

PROBLEM SET 2.2

1. -3 **3.** 0 **5.** $\frac{3}{2}$ **7.** 4 **9.** $l = \dfrac{A}{w}$ **11.** $t = \dfrac{I}{pr}$ **13.** $T = \dfrac{PV}{nR}$ **15.** $x = \dfrac{y - b}{m}$
17. $F = \frac{9}{5}C + 32$ **19.** $v = \dfrac{h - 16t^2}{t}$ **21.** $d = \dfrac{A - a}{n - 1}$ **23.** $y = -\frac{2}{3}x + 2$ **25.** $y = \frac{3}{5}x + 3$
27. $y = \frac{1}{3}x + 2$ **29.** $x = \dfrac{5}{a - b}$ **31.** $P = \dfrac{A}{1 + rt}$ **33.** $y = -\frac{1}{4}x + 2$ **35.** $y = \frac{3}{5}x - 3$ **37.** 20.52
39. 25% **41.** 925 **43.** $4, 7, 10, 13, 16$ **45.** $4, 7, 12, 19, 28$ **47.** $\frac{1}{4}, \frac{2}{5}, \frac{1}{2}, \frac{4}{7}, \frac{5}{8}$

49. $\dfrac{1}{1}, \dfrac{1}{4}, \dfrac{1}{9}, \dfrac{1}{16}, \dfrac{1}{25} \rightarrow 1, \dfrac{1}{4}, \dfrac{1}{9}, \dfrac{1}{16}, \dfrac{1}{25}$ **51.** 2, 4, 8, 16, 32 **53.** $2, \dfrac{3}{2}, \dfrac{4}{3}, \dfrac{5}{4}, \dfrac{6}{5}$ **55.** $5.00 **57.** $10.00

59. 2 centimeters **61.** 2 inches **63.** 5 feet **65.** National League

Pitcher, Team	Rolaids Points
John Franco, New York	82
Billy Wagner, Houston	82
Gregg Olson, Arizona	80
Bob Wickman, Milwaukee	55

67. 1.8 tons **69.** Ordinary interest = $67.67 **71.** 13,330 kilobytes
Exact interest = $66.74

73. $673.68 **75.** $674.92 **77.** $760.05 **79.** $18,878,664 **81.** $2(x + 3)$ **83.** $2(x + 3) = 16$

85. $5(x - 3)$ **87.** $3x + 2 = x - 4$ **89.** $x = -\dfrac{a}{b}y + a$ **91.** $a = \dfrac{bc}{b - c}$ **93.** $R = \dfrac{abc}{bc + ac + ab}$

95. Shar's training heart rate is 128.4 beats per minute.
Sara's training heart rate is 140.4 beats per minute.

PROBLEM SET 2.3

Along with the answers to the odd-numbered problems in this problem set, we are including some of the equations used to solve each problem. Be sure that you try the problems on your own before looking here to see what the correct equations are.
1. The width is x, the length is $2x$; $2x + 4x = 60$; 10 feet by 20 feet **3.** The length of a side is x; $4x = 28$; 7 feet
5. The shortest side is x, the medium side is $x + 3$, the longest side is $2x$; $x + (x + 3) + 2x = 23$; 5 inches
7. The width is x, the length is $2x - 3$; $2(2x - 3) + 2x = 18$; 4 meters **9.** $92.00 **11.** $9,339.00
13. 860 items **15.** 41,667 workers **17.** $3,260.66 per month **19.** $99.6 million **21.** (a) 24.4% increase
(b) $14,300 (c) $7,400 **23.** 8.3 grams of fat **25.** 20°, 160° **27.** (a) 20.4°, 69.6° (b) 38.4°, 141.6°
29. 27°, 72°, 81° **31.** 102°, 44°, 34° **33.** 43°, 43°, 94° **35.** $6,000 at 8%, $3,000 at 9%
37. $5,000 at 12%, $10,000 at 10% **39.** $4,000 at 8%, $2,000 at 9%
41. Melissa spent $700 on Shell products and $2,150 on products from other companies.
43. Chad spent $4,200 at Company A and $2,480 at other companies. **45.** 30 fathers, 45 sons **47.** $54
49. 44 minutes **51.**

Width (ft)	Length (ft)	Area (sq ft)
2	22	44
4	20	80
6	18	108
8	16	128
10	14	140
12	12	144

53.

Time (sec)	Height (ft)
1	112
2	192
3	240
4	256
5	240
6	192

55.

Age (years)	Maximum Heart Rate (beats per minute)
18	202
19	201
20	200
21	199
22	198
23	197

57. For a 20- year-old person

Resting Heart Rate (beats per minute)	Training Heart Rate (beats per minute)
60	144
62	145
64	146
68	147
70	148
72	149

59.

Width	$L = 6 - W$	Length	Area
0	$L = 6 - 0$	6	0
0.5	$L = 6 - 0.5$	5.5	2.75
1	$L = 6 - 1$	5	5
1.5	$L = 6 - 1.5$	4.5	6.75
2	$L = 6 - 2$	4	8
2.5	$L = 6 - 2.5$	3.5	8.75
3	$L = 6 - 3$	3	9
3.5	$L = 6 - 3.5$	2.5	8.75
4	$L = 6 - 4$	2	8
4.5	$L = 6 - 4.5$	1.5	6.75
5	$L = 6 - 5$	1	5
5.5	$L = 6 - 5.5$	0.5	2.75
6	$L = 6 - 6$	0	0

9 square inches

61.

Time in Months	Amount in Account
0	$100.00
1	$100.50
2	$101.00
3	$101.51
4	$102.02
5	$102.53
6	$103.04
7	$103.55
8	$104.07
9	$104.59
10	$105.11
11	$105.64
12	$106.17
13	$106.70
14	$107.23
15	$107.77
16	$108.31
17	$108.85
18	$109.39
19	$109.94
20	$110.49
21	$111.04
22	$111.60
23	$112.16
24	$112.72

63.

65.

67.

69.

PROBLEM SET 2.4

1. $x \leq \frac{3}{2}$

3. $x > 4$

5. $x \geq -5$

7. $x < 4$

9. $x \geq -6$

11. $x \geq 4$

13. $x < -3$

15. $m \geq -1$

17. $x \geq -3$

19. $y \leq \frac{7}{2}$

21. $x < 6$

23. $y \geq -52$

25. $(-\infty, -2]$ **27.** $[1, \infty)$ **29.** $(-\infty, 3)$ **31.** $(-\infty, -1]$ **33.** $[-17, \infty)$ **35.** $(-\infty, -5)$

37. $[3, 7]$

39. $(-4, 2)$

41. $[4, 6]$

43. $(-4, 2)$

45. $(-3, 3)$

47. $(-\infty, -7] \cup [-3, \infty)$

49. $(-\infty, -1] \cup [\frac{3}{5}, \infty)$

51. $(-\infty, -10) \cup (6, \infty)$

53. $p \leq 2$; set the price at $2.00 or less per pad **55.** $p > 1.25$; charge more than $1.25 per pad
57. 35° to 45° Celsius; $35° \leq C \leq 45°$ **59.** $-25°$ to $-10°$ Celsius; $-25° \leq C \leq -10°$
61. (a) George Ryan: $46\% \leq R \leq 52\%$ (b) 9% (c) No, the pollsters are not 100% certain of the outcome.
  Glen Poshard: $31\% \leq P \leq 37\%$
63. (a) $15 \leq D < 20$ at the low end (b) $85 \leq H < 452$ at the low end
  $20,000 < D \leq 50,000$ at the high end $1,080 < H \leq 1100$ at the high end
  (c) A dog can hear high-frequency sounds that a human cannot. **65.** $x \leq 160$ **67.** $x = 3, y = -1, z = 0$
69. -3 **71.** The distance between x and 0 on the number line **73.** $x < \dfrac{c - b}{a}$ **75.** $\dfrac{-c - b}{a} < x < \dfrac{c - b}{a}$

PROBLEM SET 2.5

1. $-4, 4$ **3.** $-2, 2$ **5.** $\varnothing$ **7.** $-1, 1$ **9.** $\varnothing$ **11.** $-6, 6$ **13.** $-3, 7$ **15.** $\frac{17}{3}, \frac{7}{3}$ **17.** $2, 4$
19. $-\frac{5}{2}, \frac{5}{6}$ **21.** $-1, 5$ **23.** $\varnothing$ **25.** $-4, 20$ **27.** $-4, 8$ **29.** $-\frac{10}{3}, \frac{2}{3}$ **31.** $\varnothing$ **33.** $-1, \frac{3}{2}$
35. $5, 25$ **37.** $-30, 26$ **39.** $-12, 28$ **41.** $-5, \frac{3}{5}$ **43.** $1, \frac{1}{9}$ **45.** $-\frac{1}{2}$ **47.** 0 **49.** $-\frac{1}{2}$

51. $-\frac{1}{6}, -\frac{7}{4}$ **53.** All real numbers   **55.** All real numbers

57. Answers will vary depending upon the choice of numbers.

59.

n	a_n
1	2
2	1
3	0
4	1
5	2

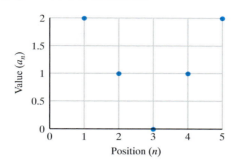

61.

n	a_n
1	4
2	2
3	0
4	2
5	4

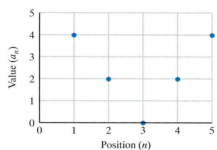

63.    **65.**

67. $t \le -\frac{3}{2}$ **69.** $x < -5$ **71.** $\frac{2}{3} < a < \frac{4}{5}$ **73.** $x = a - b$ or $x = a + b$

75. $x = \dfrac{-b - c}{a}$ or $x = \dfrac{-b + c}{a}$ **77.** $x = -\dfrac{a}{b}y - a$ or $x = -\dfrac{a}{b}y + a$

PROBLEM SET 2.6

1. $-3 < x < 3$

3. $x \le -2$ or $x \ge 2$

5. $-3 < x < 3$

7. $t < -7$ or $t > 7$

9. $\varnothing$ **11.** All real numbers   **13.** $-4 < x < 10$

15. $a \le -9$ or $a \ge -1$ **17.** $\varnothing$

19. $-1 < x < 5$

21. $y \le -5$ or $y \ge -1$

23. $k \le -5$ or $k \ge 2$

25. $-1 < x < 7$

27. $a \le -2$ or $a \ge 1$

29. $-6 < x < \frac{8}{3}$

31. $x < 2$ or $x > 8$

33. $x \le -3$ or $x \ge 12$

35. $x < 2$ or $x > 6$

37. $0.99 < x < 1.01$ **39.** $x \le -\frac{3}{5}$ or $x \ge -\frac{2}{5}$ **41.** $-\frac{1}{6} \le x \le \frac{3}{2}$ **43.** $-0.05 < x < 0.25$
45. $|x| \le 4$ **47.** $|x - 5| \le 1$ **49.** The minimum number of channels is 12, and the maximum is 36.
51. $|65 - s| \le 20$ **53.** -6 **55.** 66 **57.** -13 **59.** 11 **61.** -4 **63.** $a - b < x < a + b$
65. $x < \dfrac{b - c}{a}$ or $x > \dfrac{b + c}{a}$

CHAPTER 2 REVIEW

1. 10 **2.** 2 **3.** 2 **4.** -3 **5.** -3 **6.** 0 **7.** $\frac{2}{3}$ **8.** $\frac{10}{13}$ **9.** $\frac{5}{2}$ **10.** 1 **11.** $-\frac{5}{11}$ **12.** $-\frac{2}{9}$
13. $h = 17$ **14.** $t = 20$ **15.** $p = \dfrac{I}{rt}$ **16.** $x = \dfrac{y - b}{m}$ **17.** $y = \frac{4}{3}x - 4$ **18.** $v = \dfrac{d - 16t^2}{t}$
19. 3, 6, 9, 12; increasing **20.** 4, 5, 6, 7; increasing **21.** $-2, 4, -8, 16$; alternating **22.** $1, \frac{1}{2}, \frac{1}{3}, \frac{1}{4}$; decreasing
23. 4 feet by 12 feet **24.** 3 meters, 4 meters, 5 meters **25.** (a) 3,780 bricks (b) 625 feet long
26. \$24,875.24 **27.** $(-\infty, \frac{1}{2})$ **28.** $(-\infty, 8]$ **29.** $(-\infty, 12]$ **30.** $(-1, \infty)$ **31.** $[2, 6]$ **32.** $[0, 1]$
33. $(-\infty, -\frac{3}{2}] \cup [3, \infty)$ **34.** $(-\infty, 1) \cup [4, \infty)$ **35.** $-2, 2$ **36.** $-4, 4$ **37.** 2, 4 **38.** $-1, 4$
39. $-\frac{3}{2}, 3$ **40.** 5, 9 **41.** $-1, 1$ **42.** 0 **43.**

44.

45. $\varnothing$ **46.**

47. $\varnothing$ **48.** $\varnothing$ **49.** $\varnothing$ **50.** All real numbers except 0 **51.** $\varnothing$ **52.** All real numbers

CHAPTER 2 TEST

1. 12 **2.** $-\frac{4}{3}$ **3.** 28 **4.** -3 **5.** $-\frac{7}{4}$ **6.** 2 **7.** $w = \dfrac{P - 2l}{2}$ **8.** $B = \dfrac{2A}{h} - b$
9. $-1, 1, 3, 5$; increasing **10.** $-3, 9, -27, 81$; alternating **11.** 6 inches, 12 inches **12.** 84.6% **13.** \$56.25
14. $55°, 125°$ **15.** $t \ge -6$

$[-6, \infty)$

16. $x < 4$

$(-\infty, 4)$

17. $x < 6$

$(-\infty, 6)$

18. $y \ge -52$

$[-52, \infty)$ **19.** 6, 2

20. $3, -15$ **21.** $\varnothing$ **22.** $-5, 1$ **23.** $x < -1$ or $x > \frac{4}{3}$

24. $-\frac{2}{3} \leq x \leq 4$

25. All real numbers   **26.** $\varnothing$

CHAPTERS 1 & 2 CUMULATIVE REVIEW

1. 49   **2.** 3   **3.** -2 **4.** 5   **5.** -11 **6.** -32 **7.** 25   **8.** 36   **9.** 3   **10.** 123   **11.** 3

12. x **13.** $10x + 26$ **14.** $\frac{103}{448}$ **15.** $5x + 11$ **16.** -9 **17.** -34 **18.** 6   **19.** $\{2, 3, 4, 6, 8, 9\}$

20. $\{6\}$ **21.** $\{4, 5\}$ **22.** $0, 2$ **23.** $-13, -6.7, 0, \frac{1}{2}, 2, \frac{5}{2}$ **24.** $-13, 0, 2$ **25.** $-\sqrt{5}, \pi, \sqrt{13}$ **26.** $\frac{3}{8}$

27. -54 **28.** If $x^2 = 9$, then $x = -3$.   **29.** 10   **30.** $-\frac{1}{2}$ **31.** $\frac{14}{13}$ **32.** 1   **33.** $\frac{21}{20}$ **34.** $t = 10$

35. $n = 5$ **36.** $x = \frac{3}{4}y + 3$ **37.** $C = \frac{5}{9}(F - 32)$ **38.** $\frac{1}{3}, \frac{2}{3}, 1, \frac{4}{3}$; increasing   **39.** $1, \frac{1}{4}, \frac{1}{9}, \frac{1}{16}$; decreasing

40. 12.6 cubic inches   **41.** $25°, 65°$ **42.** $(-1, \infty)$ **43.** $[-3, 2]$ **44.** $(-\infty, \frac{4}{5}] \cup [2, \infty)$ **45.** $-5, 5$

46. $-\frac{5}{3}, 3$ **47.** $-\frac{5}{3}, \frac{7}{3}$ **48.**

49.   **50.**

CHAPTER 3

PROBLEM SET 3.1

1.

3.

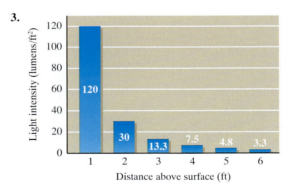

5–15. (odd)   **17.** $\left(-\frac{5}{2}, \frac{9}{2}\right)$ **19.** $\left(-3, \frac{5}{2}\right)$ **21.** $(-2, 0)$ **23.** $(-3, -2)$

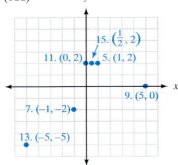

25. $(-3, -3)$ **27.** $(3, -4)$ **29.**

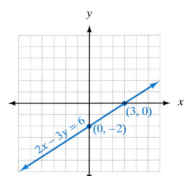

31.

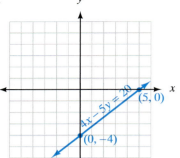

33.

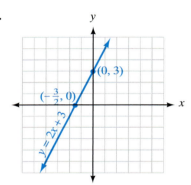

35. Table (b)   **37.**

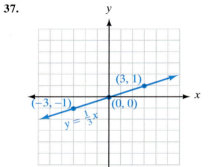

39.

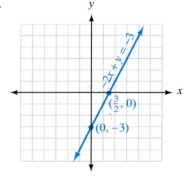

41.

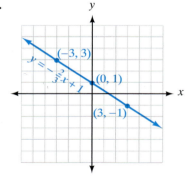

43.

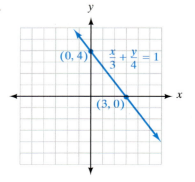

45. Equation (b)   **47.**

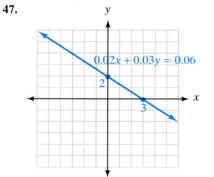

49.

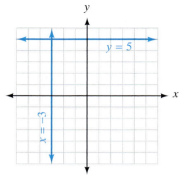

51.

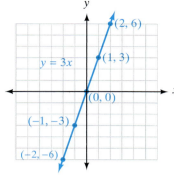

53. (a) $x > 0, y > 0$
(b) $x < 0, y > 0$
(c) $x < 0, y < 0$
(d) $x > 0, y < 0$

55.

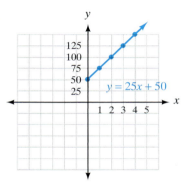

Note that the graph appears in quadrant I only since x and y represent positive numbers.

57.

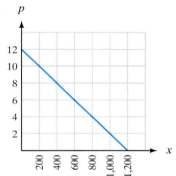

59.

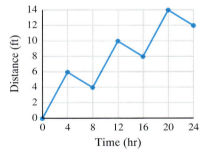

Time (hr)	Distance (ft)
0	0
4	6
8	4
12	10
16	8
20	14
24	12

61. (a) 159   (b) 19,000   (c) 11,845
(d) No, because for the year 2050, D = −80, and there cannot be a negative number of accident-related deaths.

63. (a)

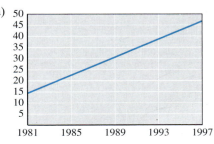

(b)

x	y
1982	16.60
1985	22.67
1989	30.76
1993	38.85
1997	46.93

65. 2

67. -2 **69.** x-intercept $= \frac{c}{a}$, y-intercept $= \frac{c}{b}$ **71.** x-intercept $= a$, y-intercept $= b$

73.

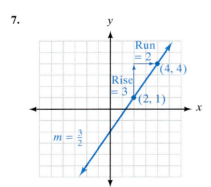

75.

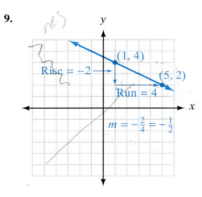

PROBLEM SET 3.2

1. $\frac{3}{2}$ **3.** No slope   **5.** $\frac{2}{3}$

7.

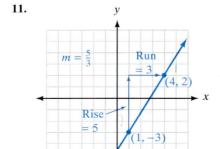

9.

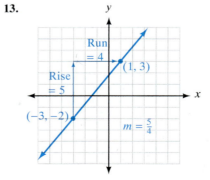

11.

13.

15.

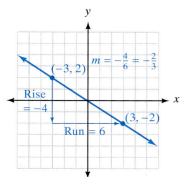

17.

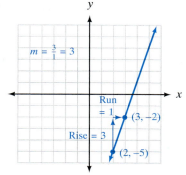

19.

x	y
0	2
3	0

Slope $= -\frac{2}{3}$

21.

x	y
0	-5
3	-3

Slope $= \frac{2}{3}$

23.

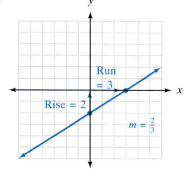

25.

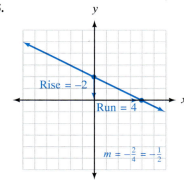

27. $\frac{1}{5}$ **29.** 0   **31.** (a) Yes   (b) No

33. 24 feet   **35.** 10 minutes   **37.** 20; °C per minute

39. 1st minute   **41.** Slope $= -1{,}250$; dollars per year

43. 2 to 3 years   **45.** (a) 15,800 feet   (b) $-\frac{7}{100}$

47. $1,875; $310,000   **49.** 0

51. $y = -\frac{3}{2}x + 6$

53. $t = \dfrac{A - P}{Pr}$

55.

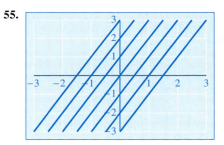

57.

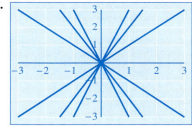

59.

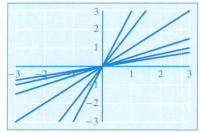

PROBLEM SET 3.3

1. $y = 2x + 3$ **3.** $y = x - 5$ **5.** $y = \frac{1}{2}x + \frac{3}{2}$ **7.** $y = 4$

9. Slope $= 3$,
   y-intercept $= -2$,
   perpendicular slope $= -\frac{1}{3}$

11. Slope $= \frac{2}{3}$,
   y-intercept $= -4$,
   perpendicular slope $= -\frac{3}{2}$

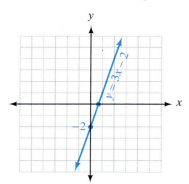

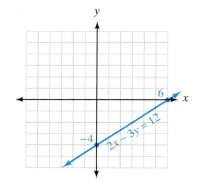

13. Slope $= -\frac{4}{5}$,
   y-intercept $= 4$,
   perpendicular slope $= \frac{5}{4}$

15. Slope $= \frac{1}{2}$, y-intercept $= -4$, $y = \frac{1}{2}x - 4$

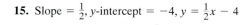

17. Slope $= -\frac{2}{3}$, y-intercept $= 3$, $y = -\frac{2}{3}x + 3$ **19.** $y = 2x - 1$ **21.** $y = -\frac{1}{2}x - 1$ **23.** $y = -3x + 1$

25. $x - y = 2$ **27.** $2x - y = 3$ **29.** $6x - 5y = 3$ **31.** $(0, -4), (2, 0); y = 2x - 4$

33. $(-2, 0), (0, 4); y = 2x + 4$ **35.** Slope $= 0$, y-intercept $= -2$ **37.** $y = 3x + 7$ **39.** $y = -\frac{5}{2}x - 13$

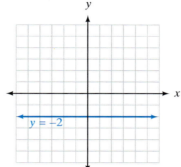

41. $y = \frac{1}{4}x + \frac{1}{4}$ **43.** $y = -\frac{2}{3}x + 2$ **45.** (b) $86°$ **47.** (a) $(y - 1) = 0.00877(x - 98)$ (b) $40 \leq x \leq 220$

49. (a) $(y - 3,000) = 4,250(x - 1,984)$ or $y = 4,250x - 8,429,000$ (b) $11,500$ cases

51. 5 inches, 23 inches   **53.** \$46.50   **55.**

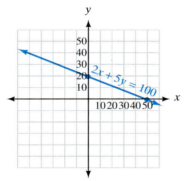

57.

59. $y = -\frac{3}{2}x + 3$; slope $= -\frac{3}{2}$, y-intercept $= 3$, x-intercept $= 2$

61. $y = \frac{3}{2}x + 3$; slope $= \frac{3}{2}$, y-intercept $= 3$, x-intercept $= -2$ **63.** Slope $= -\frac{b}{a}$, x-intercept $= a$, y-intercept $= b$

PROBLEM SET 3.4

1.

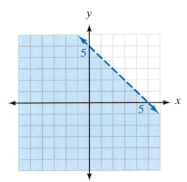

3.

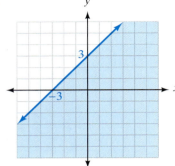

5.

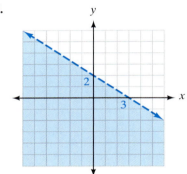

7.

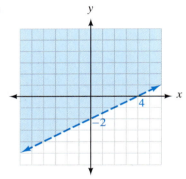

9.

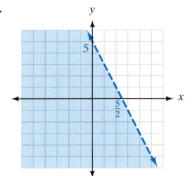

11.

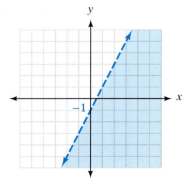

13. $x + y > 4$ **15.** $-x + 2y \le 4$ **17.**

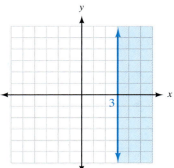

19.

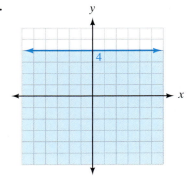

21.

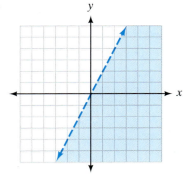

23.

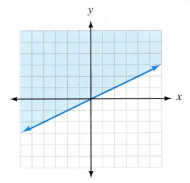

25.

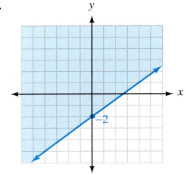

27.

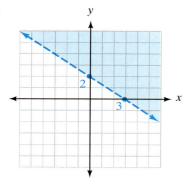

29.

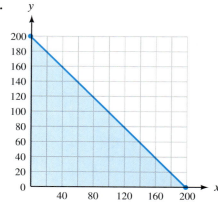

31.

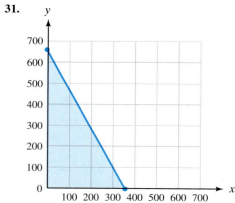

33.

35. $y \le 7$ **37.** $t > -2$ **39.** $-1 < t < 2$ **41.**

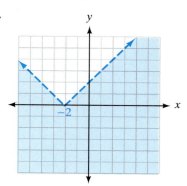

43.

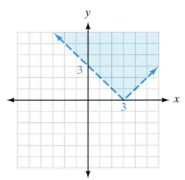

45.

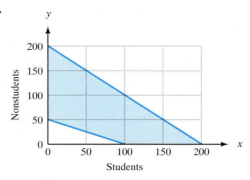

PROBLEM SET 3.5

1. (a) $y = 8.5x$ for $10 \leq x \leq 40$

(b)

Hours Worked	Function Rule	Gross Pay ($)
x	$y = 8.5x$	y
10	$y = 8.5(10) = 85$	85
20	$y = 8.5(20) = 170$	170
30	$y = 8.5(30) = 255$	255
40	$y = 8.5(40) = 340$	340

(c)

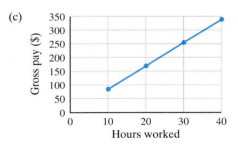

(d) Domain = $\{x \mid 10 \leq x \leq 40\}$; Range = $\{y \mid 85 \leq y \leq 340\}$ (e) Minimum = \$85; Maximum = \$340

3. Domain = $\{1, 2, 4\}$; Range = $\{3, 5, 1\}$; a function   **5.** Domain = $\{-1, 1, 2\}$; Range = $\{3, -5\}$; a function

7. Domain = $\{7, 3\}$; Range = $\{-1, 4\}$; not a function   **9.** Yes   **11.** No   **13.** No   **15.** Yes   **17.** Yes

19. (a)

Time (sec)	Function Rule	Distance (ft)
t	$h = 16t - 16t^2$	h
0	$h = 16(0) - 16(0)^2$	0
0.1	$h = 16(0.1) - 16(0.1)^2$	1.44
0.2	$h = 16(0.2) - 16(0.2)^2$	2.56
0.3	$h = 16(0.3) - 16(0.3)^2$	3.36
0.4	$h = 16(0.4) - 16(0.4)^2$	3.84
0.5	$h = 16(0.5) - 16(0.5)^2$	4
0.6	$h = 16(0.6) - 16(0.6)^2$	3.84
0.7	$h = 16(0.7) - 16(0.7)^2$	3.36
0.8	$h = 16(0.8) - 16(0.8)^2$	2.56
0.9	$h = 16(0.9) - 16(0.9)^2$	1.44
1	$h = 16(1) - 16(1)^2$	0

(b) Domain = $\{t \mid 0 \leq t \leq 1\}$;
Range = $\{h \mid 0 \leq h \leq 4\}$

(c)

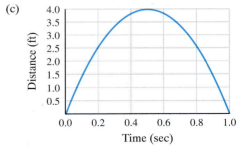

21. Domain = $\{x \mid -5 \leq x \leq 5\}$
Range = $\{y \mid 0 \leq y \leq 5\}$

23. Domain = $\{x \mid -5 \leq x \leq 3\}$
Range = $\{y \mid y = 3\}$

25. Domain = All real numbers;
Range = $\{y \mid y \geq -1\}$; a function

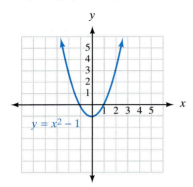

27. Domain = All real numbers;
Range = $\{y \mid y \geq 4\}$; a function

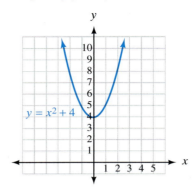

29. Domain = $\{x \mid x \geq -1\}$;
Range = All real numbers; not a function

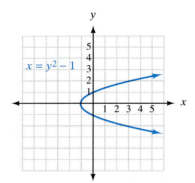

31. Domain = $\{x \mid x \geq 4\}$;
Range = All real numbers; not a function

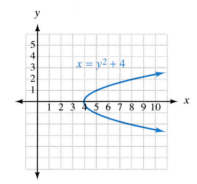

33. Domain = All real numbers;
Range = $\{y \mid y \geq 0\}$; a function

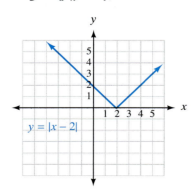

35.

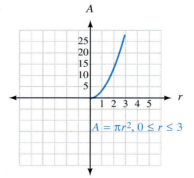

37. x = the width, so $x + 2$ = the length; $P = 2x + 2(x + 2) = 4x + 4$. The variable x must be positive.
39. $A = x(x + 2)$, where $x > 0$ **41.** (a) Yes   (b) Domain = $\{t \mid 0 \leq t \leq 6\}$; Range = $\{h \mid 0 \leq h \leq 60\}$
(c) $t = 3$ (d) $h = 60$ (e) $t = 6$ **43.** (a) F11   (b) F12   (c) F10   (d) F9   **45.** 10   **47.** -14
49. 1   **51.** -3

PROBLEM SET 3.6

1. −1   **3.** −11   **5.** 2   **7.** 4   **9.** 35   **11.** −13   **13.** 1   **15.** −9   **17.** 8   **19.** 19   **21.** 16

23. 0   **25.** $3a^2 - 4a + 1$ **27.** 4   **29.** 0   **31.** 2   **33.** −8   **35.** −1   **37.** $2a^2 - 8$ **39.** $2b^2 - 8$

41. 0   **43.** −2   **45.** −3   **47.**   **49.** $x = 4$

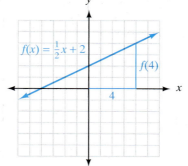

51.

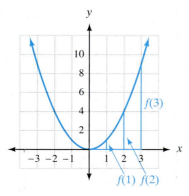

53. $V(3) = 300$, the painting is worth $300 in 3 years; $V(6) = 600$, the painting is worth $600 in 6 years.

55. $P(x) = 2x + 2(2x + 3) = 6x + 6$, where $x > 0$ **57.** $A(2) = 3.14(4) = 12.56$; $A(5) = 3.14(25) = 78.5$; $A(10) = 3.14(100) = 314$

59. (a) $2.49   (b) $1.53 for a 6-minute call   (c) 5 minutes   **61.** (a) $5,625   (b) $1,500

(c) Domain = $\{t \mid 0 \le t \le 5\}$ (d)

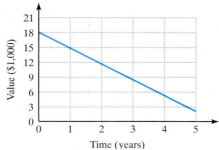

(e) Range = $\{V(t) \mid 1,500 \le V(t) \le 18,000\}$ (f) 2.42 years

63. (a)

Weight (oz)	0.6	1.0	1.1	2.5	3.0	4.8	5.0	5.3
Cost (cents)	32	32	55	78	78	124	124	147

(b) In words: Over 2 ounces, but not over 3 ounces. Inequality: $2 < x \le 3$ (c) Domain $= \{x \mid 0 < x \le 6\}$

(d) Range $= \{C(x) \mid C(x) = 32, 55, 78, 101, 124, 147\}$ **65.** $-\frac{2}{3}, 4$ **67.** $-2, 1$ **69.** $\emptyset$

71. (a) 2   (b) 0   (c) 1   (d) 4   (e) 1   (f) 3   (g) 0   (h) 2

PROBLEM SET 3.7

1. 30   **3.** 5   **5.** -6 **7.** $\frac{1}{2}$ **9.** 40   **11.** 225   **13.** $\frac{81}{5}$ **15.** 40.5   **17.** 64   **19.** 8

21. $\frac{50}{3}$ pounds   **23.** (a) $T = 4P$ (b)

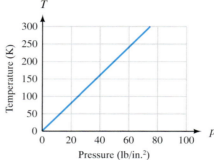

(c) 70 pounds per square inch   **25.** 12 pounds per square inch   **27.** (a) $f = \frac{80}{d}$

(b)   (c) An f-stop of 8

29. $\frac{1,504}{15}$ square inches   **31.** 1.5 ohms

33. (a) $P = 0.21\sqrt{l}$ (b)   (c) 3.15 seconds

35. 1.28 meters   **37.** $F = \dfrac{G(m_1 \cdot m_2)}{d^2}$ **39.** $x \le -9$ or $x \ge -1$

41. All real numbers   **43.** $-6 < y < \dfrac{8}{3}$

45. (a)

Height Above Surface (ft)	Illumination (foot-candles)
2	900
4	225
6	100
8	56.25
10	36

Height Above Surface (ft)	Area of Illumination Region (ft^2)
2	$\pi \approx 3.1$
4	$4\pi \approx 12.6$
6	$9\pi \approx 28.3$
8	$16\pi \approx 50.3$
10	$25\pi \approx 78.5$

(b)

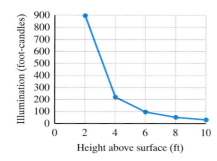

(c) Illumination F is inversely proportional to the square of distance h.

Area A is directly proportional to the square of distance h.

(d) $A = \dfrac{\pi}{4}h^2$

$F = \dfrac{3{,}600}{h^2}$

CHAPTER 3 REVIEW

1.

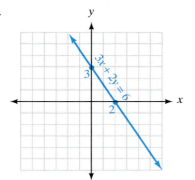

2.

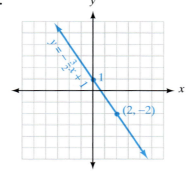

3.

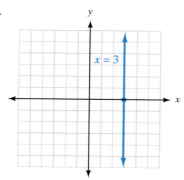

4. -2 **5.** 0 **6.** 3 **7.** 5 **8.** -5 **9.** 6 **10.** $y = 3x + 5$

11. $y = -2x$ **12.** $m = 3, b = -6$ **13.** $m = \frac{2}{3}, b = -3$ **14.** $y = 2x$ **15.** $y = -\frac{1}{3}x$ **16.** $y = 2x + 1$

17. $y = 7$ **18.** $y = -\frac{3}{2}x - \frac{17}{2}$ **19.** $y = 2x - 7$ **20.** $y = \frac{1}{3}x - \frac{2}{3}$

21.

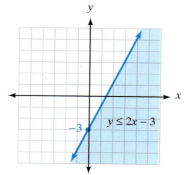

22.

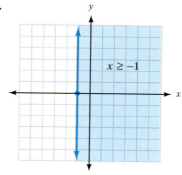

23. Domain $= \{2, 3, 4\}$;
Range $= \{4, 3, 2\}$;
a function

24. Domain $= \{6, -4, -2\}$; Range $= \{3, 0\}$; a function   **25.** 0 **26.** 1 **27.** 1 **28.** $3a + 2$ **29.** 1
30. 31 **31.** 24 **32.** 6 **33.** 4 **34.** 25 **35.** 84 pounds   **36.** 16 foot-candles

CHAPTER 3 TEST

1. x-intercept $= 3$,
y-intercept $= 6$,
slope $= -2$

2. x-intercept $= -\frac{3}{2}$,
y-intercept $= -3$,
slope $= -2$

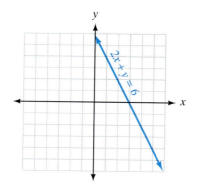

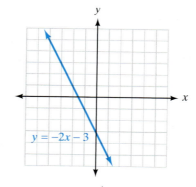

3. x-intercept $= -\frac{8}{3}$,
   y-intercept $= 4$,
   slope $= \frac{3}{2}$

4. x-intercept $= -2$,
   no y-intercept,
   no slope

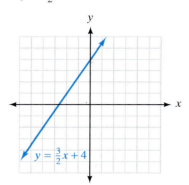

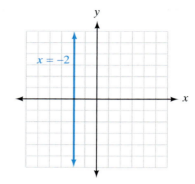

5. $y = 2x + 5$ **6.** $y = -\frac{3}{7}x + \frac{5}{7}$ **7.** $y = \frac{2}{5}x - 5$ **8.** $y = -\frac{1}{3}x - \frac{7}{3}$ **9.** $x = 4$

10.

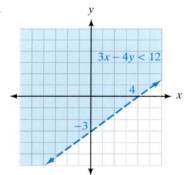

11.

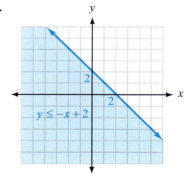

12. Domain $= \{-2, -3\}$; Range $= \{0, 1\}$; not a function
13. Domain $=$ All real numbers; Range $= \{y \mid y \geq -9\}$; a function   **14.** 11   **15.** -4 **16.** 8   **17.** 4
18. $0 < x < 4$ **19.** $V(x) = x(8 - 2x)^2$
20. $V(2) = 32$ cubic inches is the volume of the box if a square with 2-inch sides is cut from each corner.
21. 18   **22.** $\frac{81}{4}$ **23.** $\frac{2{,}000}{3}$ pounds

CHAPTERS 1-3 CUMULATIVE REVIEW

1. 12   **2.** $20x + 19$ **3.** 6   **4.** $-6x - 15$ **5.** 96   **6.** $\frac{1}{12}$ **7.** 15   **8.** -1 **9.** $\frac{5}{9}$ **10.** $\frac{7}{12}$

11. $\frac{7}{5}$ **12.** $\varnothing$ **13.** $-\frac{5}{2}, \frac{11}{2}$ **14.** 1   **15.** $[-\frac{2}{3}, \frac{4}{3}]$ **16.** $x = \dfrac{13}{b - a}$

17.

18.

19.

20.

21.

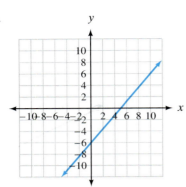

22.

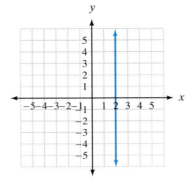

23.

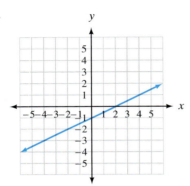

24.

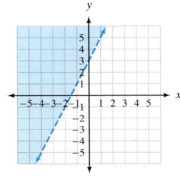

25. $m = -\frac{5}{2}$

26. $y = 6$

27. $m = 0$

28. $y = -\frac{3}{4}x + 2$

29. $m = \frac{4}{5}, b = -4$

30. $y = -\frac{1}{2}x - 1$ **31.** $y = -\frac{5}{2}x - 12$ **32.** $y = \frac{7}{3}x - 6$ **33.** -1 **34.** 7   **35.** 18   **36.** -2

37. $-\frac{8}{5}$ **38.** The commutative and associative properties of addition   **39.** $\frac{7}{9}$ **40.** $3a + 4b < 3a - 4b$

41. $A \cup B = \{0, 1, 2, 3, 6, 7\}$ **42.** Opposite: $\frac{3}{4}$, Reciprocal: $-\frac{4}{3}$

43. Hypothesis: A quadrangle is a square. Conclusion: All of its sides are equal.   **44.** 5; arithmetic

45. 300   **46.** $-3, 9, -27, 81, -243$ **47.** Domain: $\{-1, 2, 3\}$; Range: $\{3, -1\}$; Yes, the relation is a function.

48. $y = \frac{16}{9}$ **49.** $l = 17$ ft, $w = 7$ ft   **50.** $24°, 36°, 120°$

CHAPTER 4

PROBLEM SET 4.1

1. $(4, 3)$

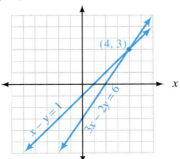

3. $(-5, -6)$

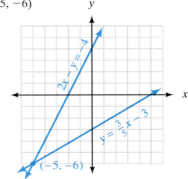

5. $(4, 2)$ **7.** Lines are parallel; there is no solution.

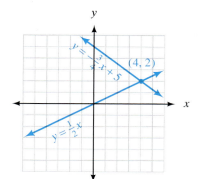

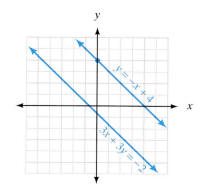

9. Lines coincide; any solution to one of the equations is a solution to the other.   **11.** $(2, 3)$ **13.** $(1, 1)$

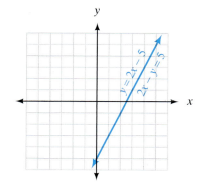

15. Lines coincide: $\{(x, y) \mid 3x - 2y = 6\}$ **17.** $(1, -\frac{1}{2})$ **19.** $(\frac{1}{2}, -3)$ **21.** $(-\frac{8}{3}, 5)$ **23.** $(2, 2)$

25. Parallel lines; $\varnothing$ **27.** $(12, 30)$ **29.** $(10, 24)$ **31.** $(4, \frac{10}{3})$ **33.** $(3, -3)$ **35.** $(\frac{4}{3}, -2)$ **37.** $(6, 2)$

39. $(2, 4)$ **41.** Lines coincide: $\{(x, y) \mid 2x - y = 5\}$ **43.** Lines coincide: $\{(x, y) \mid x = \frac{3}{2}y\}$ **45.** $(-\frac{15}{43}, -\frac{27}{43})$

47. $(\frac{60}{43}, \frac{46}{43})$ **49.** $(\frac{9}{41}, -\frac{11}{41})$ **51.** $(6{,}000, 4{,}000)$ **53.** 2   **55.** (a) $3.29   (b) $y = 30x + 45$

(c) 2 minutes   **57.** (a) $0 \le x \le 12$ (b)

Percentage vs *Year*

(c) $(9, 50)$; in year 9, or 1989, both inpatient and outpatient surgery were 50%.

59. (a)

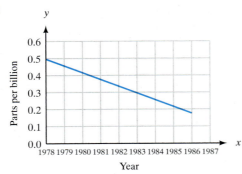

(b) $y = -0.04x + 79.62$
(c) 0.42 ppb
(d) 0.02 ppb

61. 3   **63.** $m = \frac{2}{3}, b = -2$ **65.** $y = \frac{2}{3}x + 6$ **67.** $y = \frac{2}{3}x - 2$

PROBLEM SET 4.2

1. $(1, 2, 1)$ **3.** $(2, 1, 3)$ **5.** $(2, 0, 1)$ **7.** $(\frac{1}{2}, \frac{2}{3}, -\frac{1}{2})$ **9.** No solution, inconsistent system   **11.** $(4, -3, -5)$
13. No unique solution   **15.** $(4, -5, -3)$ **17.** No unique solution   **19.** $(\frac{1}{2}, 1, 2)$ **21.** $(\frac{1}{2}, \frac{1}{3}, \frac{1}{4})$
23. $(1, 3, 1)$ **25.** $(-1, 2, -2)$ **27.** 4 amp, 3 amp, 1 amp   **29.** (a) Answers will vary   (b) Infinitely many

31.

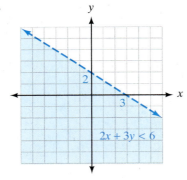

33.

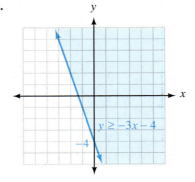

35.

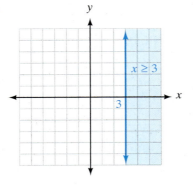

37. (a)

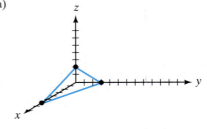

(b)

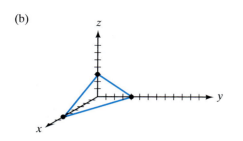

(c)

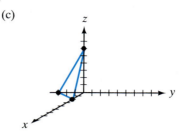

39. (1, 2, 3, 4)

PROBLEM SET 4.3

1. 3 **3.** 5 **5.** −1 **7.** 0 **9.** 10 **11.** 2 **13.** −3 **15.** −2 **17.** −2, 5 **19.** 3 **21.** 0

23. 3 **25.** 8 **27.** 6 **29.** −228 **31.** $\begin{vmatrix} y & x \\ m & 1 \end{vmatrix} = y - mx = b; y = mx + b$

33. (a) $y = 0.3x + 3.4$ (b) $y = 0.3(2) + 3.4$ **35.** Domain {1, 3, 4}; range {2, 4}; function
$\qquad\qquad\qquad\qquad = 4$ billion dollars

37. Domain {1, 2, 3}; range {1, 2, 3}; function   **39.** Function   **41.** Not a function   **43.** 4   **45.** 4

PROBLEM SET 4.4

1. (3, 1) **3.** Lines are parallel; $\varnothing$ **5.** $\left(-\frac{15}{43}, -\frac{27}{43}\right)$ **7.** $\left(\frac{60}{43}, \frac{46}{43}\right)$ **9.** (3, −1, 2) **11.** $\left(\frac{1}{2}, \frac{5}{2}, 1\right)$

13. No unique solution **15.** $\left(-\frac{10}{91}, -\frac{9}{13}, \frac{107}{91}\right)$ **17.** $\left(\frac{71}{13}, -\frac{12}{13}, \frac{24}{13}\right)$ **19.** (3, 1, 2) **21.** $x = 50$ items

23. 1986 **25.** 3 **27.** 0 **29.** 1 **31.** 3

33. $x = -\dfrac{1}{a-b} = \dfrac{1}{b-a}, y = \dfrac{1}{a-b}$ **35.** $x = \dfrac{1}{a^2+ab+b^2}, y = \dfrac{a+b}{a^2+ab+b^2}$ **37.** $\begin{array}{l} x + 2y = 1 \\ 3x + 4y = 0 \end{array}$

PROBLEM SET 4.5

1. $y = 2x + 3, x + y = 18$; the two numbers are 5 and 13. **3.** 10, 16 **5.** 1, 3, 4

7. Let $x =$ the number of adult tickets and $y =$ the number of children's tickets.
$\quad x + y = \quad 925 \qquad 225$ adult and 700 children's tickets
$\quad 2x + y = 1{,}150$

9. Let $x =$ the amount invested at 6% and $y =$ the amount invested at 7%.
$\qquad\quad x + y = 20{,}000 \qquad$ He has \$12,000 at 6% and \$8,000 at 7%.
$\quad 0.06x + 0.07y = 1{,}280$

11. \$4,000 at 6%, \$8,000 at 7.5% **13.** \$200 at 6%, \$1,400 at 8%, \$600 at 9%

15. 3 gallons of 50%, 6 gallons of 20% **17.** 5 gallons of 20%, 10 gallons of 14%

19. Let $x =$ the speed of the boat and $y =$ the speed of the current.
$\quad 3(x - y) = 18 \qquad$ The speed of the boat is 9 miles per hour.
$\quad 2(x + y) = 24 \qquad$ The speed of the current is 3 miles per hour.

21. 270 miles per hour airplane, 30 miles per hour wind **23.** 12 nickels, 8 dimes **25.** 3 of each

27. $x = -200p + 700$, when $p =$ \$3, $x = 100$ items **29.** $C = 0.75x + 21.85$; \$30.85

31. $h = -16t^2 + 64t + 80$, maximum height is 144 feet when $t = 2$ seconds

33. $\quad a + b = 50$
$\quad 0.4a + 0.6b = 0.55(50); a = 12.5$ pounds of 40%, $b = 37.5$ pounds of 60% **35.** 147 **37.** 50 **39.** 100

41.

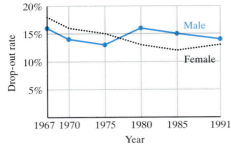

43. (a) $M = 0.6x - 1172$ (b) $F = -0.4x + 805$
(c) 1977

CHAPTER 4 REVIEW

1. $(6, -2)$ **2.** $(2, -4)$ **3.** Parallel lines   **4.** $(1, -1)$ **5.** $(0, 1)$ **6.** Lines coincide   **7.** $(3, 1)$

8. $(\frac{3}{2}, 0)$ **9.** $(3, 5)$ **10.** $(4, 8)$ **11.** $(3, 12)$ **12.** $(3, -2)$ **13.** $(\frac{3}{2}, \frac{1}{2})$ **14.** $(4, 1)$ **15.** $(6, -2)$

16. $(7, -4)$ **17.** $(-5, 3)$ **18.** Parallel lines   **19.** $(3, -1, 4)$ **20.** $(2, \frac{1}{2}, -3)$ **21.** $(-1, \frac{1}{2}, \frac{3}{2})$

22. $(2, \frac{1}{3}, \frac{2}{3})$ **23.** Dependent system   **24.** Dependent system   **25.** $(2, -1, 4)$ **26.** Dependent system

27. 23   **28.** -3 **29.** -3 **30.** -16 **31.** 36   **32.** -2 **33.** $\frac{4}{7}$ **34.** $-\frac{3}{2}, \frac{1}{2}$ **35.** $(\frac{7}{29}, -\frac{19}{29})$

36. $(\frac{34}{41}, -\frac{18}{41})$ **37.** Lines coincide   **38.** $(\frac{17}{27}, -\frac{11}{81})$ **39.** $(-\frac{50}{33}, -\frac{23}{33})$ **40.** $(\frac{11}{14}, -\frac{1}{14}, \frac{11}{14})$

41. $(\frac{113}{46}, \frac{59}{23}, -\frac{7}{23})$ **42.** Inconsistent   **43.** 47 adults, 80 children   **44.** 12 dimes, 8 quarters

45. \$5,000 at 12%, \$7,000 at 15%   **46.** 12 miles per hour boat, 2 miles per hour river current

CHAPTER 4 TEST

1. $(1, 2)$ **2.** $(3, 2)$ **3.** $(15, 12)$ **4.** $(-\frac{54}{13}, -\frac{58}{13})$ **5.** $(1, 2)$ **6.** $(3, -2, 1)$ **7.** -14 **8.** -26

9. $(-\frac{14}{3}, -\frac{19}{3})$ **10.** Lines coincide $\{(x, y) \mid 2x + 4y = 3\}$ **11.** $(\frac{5}{11}, -\frac{15}{11}, -\frac{1}{11})$ **12.** 5, 9

13. \$4,000 at 5%, \$8,000 at 6%   **14.** 340 adults, 410 children   **15.** 4 gallons 30%; 12 gallons 70%

16. Boat 8 miles per hour; current 2 miles per hour   **17.** 11 nickels, 3 dimes, 1 quarter

18. 3 ounces of cereal I; 1 ounce of cereal II

CHAPTERS 1–4 CUMULATIVE REVIEW

1. 28   **2.** $28x - 29$ **3.** 9   **4.** $3x - 18$ **5.** $\frac{16}{3}$ **6.** $\frac{11}{9}$ **7.** -21 **8.** $\frac{3}{4}$ **9.** 5   **10.** $-10, 10$

11. 1   **12.** $(-\infty, -\frac{5}{2}] \cup [3, \infty)$

13. $x \geq -3$

14. $x < -5$ or $x > -1$

15.

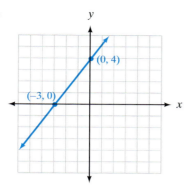

16.

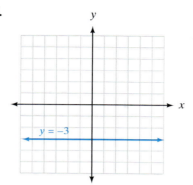

17. 0   **18.** $2 \cdot 3^3 \cdot 11$

19. $\left(\frac{2}{3}, 0\right)$ **20.** Parallel lines   **21.** Parallel lines   **22.** $(3, 2)$ **23.** $(2, 1)$ **24.** $\left(3, 0, \frac{1}{2}\right)$ **25.** 30

26. -3 **27.** 0   **28.** $\frac{5}{22}$ **29.** $\left(\frac{2}{31}, -\frac{5}{31}\right)$ **30.** $\left(\frac{59}{37}, \frac{92}{37}, \frac{54}{37}\right)$ **31.** $(-2, \infty)$ **32.** 3   **33.** 7, arithmetic

34. $-7 \le x \le 7$ **35.** $-\sqrt{2}, \sqrt{3}$ **36.** Commutative, associative   **37.** $-\frac{7}{3}, \frac{3}{7}$ **38.** 640   **39.** Function

40. 12   **41.** 4   **42.** $-\frac{5}{2}$ **43.** 1   **44.** $\frac{2}{3}, -4$ **45.** $y = -\frac{3}{4}x + 2$ **46.** $y = -\frac{1}{9}x - \frac{26}{9}$ **47.** $x = 1$

48. $P = 270$ **49.** 7.5 ounces of 30% HCl, and 7.5 ounces of 70% HCl   **50.** $120°, 36°, 24°$

CHAPTER 5

PROBLEM SET 5.1

1. 16   **3.** -16 **5.** -0.027 **7.** 32   **9.** $\frac{1}{8}$ **11.** $\frac{25}{36}$ **13.** x^9 **15.** 64   **17.** $-\frac{8}{27}x^6$ **19.** $-6a^6$

21. $\frac{1}{9}$ **23.** $-\frac{1}{32}$ **25.** $\frac{16}{9}$ **27.** 17   **29.** x^3 **31.** $\frac{a^6}{b^{15}}$ **33.** $\frac{8}{125y^{18}}$ **35.** $\frac{1}{5}$ **37.** $\frac{24a^{12}c^6}{b^3}$ **39.** $\frac{8x^{22}}{81y^{23}}$

41. $\frac{1}{x^{10}}$ **43.** a^{10} **45.** $\frac{1}{t^6}$ **47.** x^{12} **49.** x^{18} **51.** $\frac{1}{x^{22}}$ **53.** $\frac{a^3b^7}{4}$ **55.** $\frac{y^{38}}{x^{16}}$ **57.** $\frac{16y^{16}}{x^8}$

59. x^4y^6 **61.** $\frac{b^3}{a^4c^3}$ **63.** 3.78×10^5 **65.** 4.9×10^3 **67.** 3.7×10^{-4} **69.** 4.95×10^{-3}

71. 5,340   **73.** 7,800,000   **75.** 0.00344   **77.** 0.49   **79.** 8×10^4 **81.** 2×10^9 **83.** 2.5×10^{-6}

85. 1.8×10^{-7} **87.** 2.37×10^6 **89.** 6.3×10^8 **91.** 22   **93.** 1.003×10^{19} **95.** 3.43×10^{10}

97.

Planet	Minimum Distance from Earth (in miles)
Mercury	$48,000,000 = 4.8 \times 10^7$
Venus	$24,000,000 = 2.4 \times 10^7$
Mars	$34,000,000 = 3.4 \times 10^7$
Jupiter	$366,000,000 = 3.66 \times 10^8$
Saturn	$743,000,000 = 7.43 \times 10^8$
Uranus	$1,604,000,000 = 1.604 \times 10^9$
Neptune	$2,676,000,000 = 2.676 \times 10^9$
Pluto	$2,668,000,000 = 2.668 \times 10^9$

99. (a)

Planet	Approximate Volume (in cubic miles)
Mercury	1.459×10^{10}
Venus	2.227×10^{11}
Earth	2.598×10^{11}
Mars	3.911×10^{10}
Jupiter	3.432×10^{14}
Saturn	1.983×10^{14}
Uranus	1.639×10^{13}
Neptune	1.500×10^{13}
Pluto	1.480×10^9

(b) Because the planets are not perfectly round.

101. $(1, 2)$ **103.** $\left(\frac{5}{2}, -1\right)$ **105.** $(0, 3)$ **107.** $(3, 4)$ **109.** $\frac{1}{x^3}$ **111.** y^3 **113.** x^5

PROBLEM SET 5.2

1. Trinomial, 2, 5 **3.** Binomial, 1, 3 **5.** Trinomial, 2, 8 **7.** Polynomial, 3, 4 **9.** Monomial, 0, $-\frac{3}{4}$

11. Trinomial, 3, 6 **13.** $7x + 1$ **15.** $2x^2 + 7x - 15$ **17.** $12a^2 - 7ab - 10b^2$ **19.** $x^2 - 13x + 3$

21. $\frac{1}{4}x^2 - \frac{7}{12}x - \frac{1}{4}$ **23.** $-y^3 - y^2 - 4y + 7$ **25.** $2x^3 + x^2 - 3x - 17$ **27.** $\frac{1}{14}x^2 + \frac{1}{7}xy + \frac{5}{7}y^2$

29. $-3a^3 + 6a^2b - 5ab^2$ **31.** $-3x$ **33.** $3x^2 - 12xy$ **35.** $17x^5 - 12$ **37.** $14a^2 - 2ab + 8b^2$

39. $2 - x$ **41.** $10x - 5$ **43.** $9x - 35$ **45.** $9y - 4x$ **47.** $9a + 2$ **49.** -2 **51.** 208

53. -15 **55.** $(3 + 4)^2 = 49; 3^2 + 16 = 25; 3^2 + 8 \cdot 3 + 16 = 49$

57.

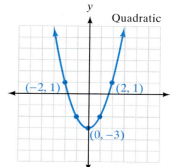

59.

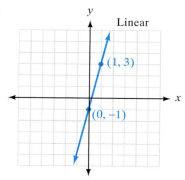

61.

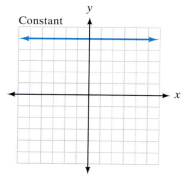

63.

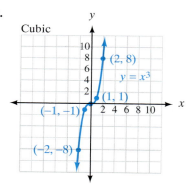

65.

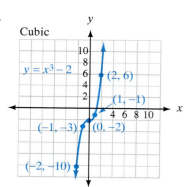

67. 240 feet for both

69. $P(x) = -300 + 40x - 0.5x^2$; $P(60) = \$300$ **71.** $P(x) = -800 + 3.5x - 0.002x^2$; $P(1,000) = \$700$
73. $20x^5$ **75.** $2a^3b^3$ **77.** (a) 5   (b) -4 (c) -3 (d) 3   (e) 4   (f) 0   (g) -4 (h) 4

PROBLEM SET 5.3

1. $12x^3 - 10x^2 + 8x$ **3.** $-3a^5 + 18a^4 - 21a^2$ **5.** $2a^5b - 2a^3b^2 + 2a^2b^4$ **7.** $x^2 - 2x - 15$
9. $6x^4 - 19x^2 + 15$ **11.** $x^3 + 9x^2 + 23x + 15$ **13.** $a^3 - b^3$ **15.** $8x^3 + y^3$
17. $2a^3 - a^2b - ab^2 - 3b^3$ **19.** $x^2 + x - 6$ **21.** $6a^2 + 13a + 6$ **23.** $20 - 2t - 6t^2$
25. $x^6 - 2x^3 - 15$ **27.** $20x^2 - 9xy - 18y^2$ **29.** $18t^2 - \frac{2}{9}$ **31.** $25x^2 + 20xy + 4y^2$
33. $25 - 30t^3 + 9t^6$ **35.** $4a^2 - 9b^2$ **37.** $9r^4 - 49s^2$ **39.** $\frac{1}{9}x^2 - \frac{4}{25}$ **41.** $x^3 - 6x^2 + 12x - 8$
43. $x^3 - \frac{3}{2}x^2 + \frac{3}{4}x - \frac{1}{8}$ **45.** $3x^3 - 18x^2 + 33x - 18$ **47.** $a^2b^2 + b^2 + 8a^2 + 8$
49. $3x^2 + 12x + 14$ **51.** $24x$ **53.** $(x + y)^2 + (x + y) - 20 = x^2 + 2xy + y^2 + x + y - 20$
55. $2^4 - 3^4 = -65$; $(2 - 3)^4 = 1$; $(2^2 + 3^2)(2 + 3)(2 - 3) = -65$
57. $R(p) = 900p - 300p^2$; $R(x) = 3x - \frac{x^2}{300}$; $\$672$ **59.** $R(p) = 350p - 10p^2$; $R(x) = 35x - \frac{x^2}{10}$; $\$1,852.50$
61. $P(x) = -500 + 30x - \frac{x^2}{10}$; $P(60) = \$940$ **63.** $A = 100 + 400r + 600r^2 + 400r^3 + 100r^4$

65. $(1, 2, 3)$ **67.** $(1, 3, 1)$ **69.**

71.

PROBLEM SET 5.4

1. $5x^2(2x - 3)$ **3.** $9y^3(y^3 + 2)$ **5.** $3ab(3a - 2b)$ **7.** $7xy^2(3y^2 + x)$ **9.** $3(a^2 - 7a + 10)$
11. $4x(x^2 - 4x - 5)$ **13.** $10x^2y^2(x^2 + 2xy - 3y^2)$ **15.** $xy(-x + y - xy)$ **17.** $2xy^2z(2x^2 - 4xz + 3z^2)$
19. $5abc(4abc - 6b + 5ac)$ **21.** $(a - 2b)(5x - 3y)$ **23.** $3(x + y)^2(x^2 - 2y^2)$ **25.** $(x + 5)(2x^2 + 7x + 6)$
27. $(x + 1)(3y + 2a)$ **29.** $(xy + 1)(x + 3)$ **31.** $(x - 2)(3y^2 + 4)$ **33.** $(x - a)(x - b)$
35. $(b + 5)(a - 1)$ **37.** $(b^2 + 1)(a^4 - 5)$ **39.** $(x + 3)(x^2 - 4)$ **41.** $(x + 2)(x^2 - 25)$
43. $(2x + 3)(x^2 - 4)$ **45.** $(x + 3)(4x^2 - 9)$ **47.** 6

49. $P(1 + r) + P(1 + r)r = (1 + r)(P + Pr) = (1 + r)P(1 + r) = P(1 + r)^2$ **51.** $p = 11.5 - 0.05x$; $5.25
53. $28.50   **55.** $x^2 + 5x + 6$ **57.** $6y^2 + y - 35$ **59.** $20 - 19a + 3a^2$ **61.** 36   **63.** 0
65. $x^n(x^2 + x + 1)$ **67.** $x^{n+3}(x^2 + x + 1)$

PROBLEM SET 5.5

1. $(x + 3)(x + 4)$ **3.** $(x + 3)(x - 4)$ **5.** $(y + 3)(y - 2)$ **7.** $(2 - x)(8 + x)$ **9.** $(2 + x)(6 + x)$
11. $3(a - 2)(a - 5)$ **13.** $4x(x - 5)(x + 1)$ **15.** $(x + 2y)(x + y)$ **17.** $(a + 6b)(a - 3b)$
19. $(x - 8a)(x + 6a)$ **21.** $(x - 6b)^2$ **23.** $3(x - 3y)(x + y)$ **25.** $2a^3(a^2 + 2ab + 2b^2)$
27. $10x^2y^2(x + 3y)(x - y)$ **29.** $(2x - 3)(x + 5)$ **31.** $(2x - 5)(x + 3)$ **33.** $(2x - 3)(x - 5)$
35. $(2x - 5)(x - 3)$ **37.** Prime   **39.** $(2 + 3a)(1 + 2a)$ **41.** $15(4y + 3)(y - 1)$ **43.** $x^2(3x - 2)(2x + 1)$
45. $10r(2r - 3)^2$ **47.** $(4x + y)(x - 3y)$ **49.** $(2x - 3a)(5x + 6a)$ **51.** $(3a + 4b)(6a - 7b)$
53. $200(1 + 2t)(3 - 2t)$ **55.** $y^2(3y - 2)(3y + 5)$ **57.** $2a^2(3 + 2a)(4 - 3a)$ **59.** $2x^2y^2(4x + 3y)(x - y)$
61. $100(3x^2 + 1)(x^2 + 3)$ **63.** $(5a^2 + 3)(4a^2 + 5)$ **65.** $3(3 + 4r^2)(1 - r^2)$ **67.** $(x + 5)(2x + 3)(x + 2)$
69. $(2x + 3)(x + 5)(x + 2)$ **71.** $9x^2 - 25y^2$ **73.** $a + 250$
75. $y = 2(2x - 1)(x + 5)$; $y = 0$ when $x = \frac{1}{2}$ or $x = -5$, $y = 42$ when $x = 2$
77. $h = 16(6 - t)(1 + t)$; $h = 0$ when $t = 6$, $h = 192$ when $t = 3$ **79.** $4x^2 - 9$ **81.** $4x^2 - 12x + 9$
83. $8x^3 - 27$ **85.** $(2x^3 + 5y^2)(4x^3 + 3y^2)$ **87.** $(3x - 5)(x + 100)$ **89.** $(\frac{1}{4}x + 1)(\frac{1}{2}x + 2)$
91. $(2x + 0.5)(x + 0.5)$ **93.**

	x	1
x	x^2	x
1	x	1
1	x	1
1	x	1
1	x	1

PROBLEM SET 5.6

1. $(x - 3)^2$ **3.** $(a - 6)^2$ **5.** $(5 - t)^2$ **7.** $(2y^2 - 3)^2$ **9.** $(4a + 5b)^2$ **11.** $(\frac{1}{5} + \frac{1}{4}t^2)^2$
13. $(x + 2 + 3)^2 = (x + 5)^2$ **15.** $(7x - 8y)(7x + 8y)$ **17.** $(2a - \frac{1}{2})(2a + \frac{1}{2})$ **19.** $(x - \frac{3}{5})(x + \frac{3}{5})$
21. $(5 - t)(5 + t)$ **23.** $(4a^2 + 9)(2a - 3)(2a + 3)$ **25.** $(x - 5 + y)(x - 5 - y)$
27. $(a + 4 + b)(a + 4 - b)$ **29.** $(x + 2)(x + 5)(x - 5)$ **31.** $(2x + 3)(x + 2)(x - 2)$
33. $(x + 3)(2x + 3)(2x - 3)$ **35.** $(x - y)(x^2 + xy + y^2)$ **37.** $(a + 2)(a^2 - 2a + 4)$
39. $(y - 1)(y^2 + y + 1)$ **41.** $10(r - 5)(r^2 + 5r + 25)$ **43.** $(4 + 3a)(16 - 12a + 9a^2)$
45. $(t + \frac{1}{3})(t^2 - \frac{1}{3}t + \frac{1}{9})$ **47.** $(x + 9)(x - 9)$ **49.** $(x - 3)(x + 5)$ **51.** $(x^2 + 2)(y^2 + 1)$
53. $2ab(a^2 + 3a + 1)$ **55.** Does not factor   **57.** $3(2a + 5)(2a - 5)$ **59.** $(5 - t)^2$ **61.** $4x(x^2 + 4y^2)$
63. $(x + 5)(x + 3)(x - 3)$ **65.** Does not factor   **67.** $(x - 3)(x - 7)^2$ **69.** $(2 - 5x)(4 + 3x)$
71. $(r + \frac{1}{5})(r - \frac{1}{5})$ **73.** Does not factor   **75.** $100(x - 3)(x + 2)$ **77.** $(3x^2 + 1)(x^2 - 5)$
79. $3a^2b(2a - 1)(4a^2 + 2a + 1)$ **81.** $(4 - r)(16 + 4r + r^2)$ **83.** $5x^2(2x + 3)(2x - 3)$

85. $2x^3(4x - 5)(2x - 3)$ **87.** $(y + 1)(y - 1)(y^2 - y + 1)(y^2 + y + 1)$ **89.** $2(5 + a)(5 - a)$
91. $(x - 2 + y)(x - 2 - y)$ **93.** 30 and -30 **95.** \$112.36   **97.** $p^3 - r^3 = (p - r)(p^2 + pr + r^2)$
99.   **101.** $\left(-\frac{15}{43}, -\frac{27}{43}\right)$ **103.** $(1, 3, 1)$

PROBLEM SET 5.7

1. $-1, 6$ **3.** $0, 2, 3$ **5.** $-4, \frac{1}{3}$ **7.** $\frac{2}{3}, \frac{3}{2}$ **9.** $-5, 5$ **11.** $0, -3, 7$ **13.** $-4, \frac{5}{2}$ **15.** $0, \frac{4}{3}$
17. $-\frac{1}{5}, \frac{1}{3}$ **19.** $0, \frac{4}{3}, -\frac{4}{3}$ **21.** $-10, 0$ **23.** $-5, 1$ **25.** $1, 2$ **27.** $-2, 3$ **29.** $-2, \frac{1}{4}$ **31.** $-3, -2, 2$
33. $-5, -2, 5$ **35.** $-\frac{3}{2}, -2, 2$ **37.** $-3, -\frac{3}{2}, \frac{3}{2}$ **39.** $-5, -3$ or $3, 5$ **41.** $-5, -4$ or $4, 5$ **43.** $\sqrt{527}$ ft
45. $6, 8, 10$ **47.** 2 ft $\times$ 8 ft   **49.** Base is 18 inches; height is 4 inches.   **51.** At 0 sec and 2 sec
53. At 1 sec and 2 sec   **55.** At 0 sec and $\frac{3}{2}$ sec   **57.** At 2 sec and 3 sec   **59.** At \$4 or \$8   **61.** At \$7 or \$10
63. 5 ft   **65.** 60 geese; 48 ducks   **67.** 150 oranges; 144 apples

CHAPTER 5 REVIEW

1. x^{10} **2.** $25x^6$ **3.** $-32x^{18}y^8$ **4.** $\frac{1}{8}$ **5.** $\frac{9}{4}$ **6.** $\frac{1}{2}$ **7.** 3.45×10^7 **8.** 3.57×10^{-3} **9.** 44,500
10. 0.000445   **11.** $\frac{1}{a^9}$ **12.** $\frac{x^{12}}{4}$ **13.** x^2 **14.** 8×10^{-2} **15.** 4×10^{-10} **16.** $2x^2 - 5x + 7$
17. $2x^3 - 2x^2 - 2x - 4$ **18.** $x^2 - 2x - 3$ **19.** $30x + 12$ **20.** 15   **21.** $12x^3 - 6x^2 + 3x$
22. $2a^4b^3 + 4a^3b^4 + 2a^2b^5$ **23.** $18 - 9y + y^2$ **24.** $6x^4 + 5x^2 - 4$ **25.** $2t^3 - 4t^2 - 6t$ **26.** $x^3 + 27$
27. $8x^3 - 27$ **28.** $a^4 - 4a^2 + 4$ **29.** $9x^2 + 30x + 25$ **30.** $16x^2 - 24xy + 9y^2$ **31.** $x^2 - \frac{1}{9}$
32. $4a^2 - b^2$ **33.** $x^3 - 3x^2 + 3x - 1$ **34.** $x^{2m} - 4$ **35.** $3xy(2x^3 - 3y^3 + 6x^2y^2)$
36. $4(x + y)^2(x^2 - 2y^2)$ **37.** $(4x^2 + 5)(2 - y)$ **38.** $(1 - y)(1 + y)(x^3 + 8b^2)$ **39.** $(x - 2)(x - 3)$
40. $2x(x + 5)(x - 3)$ **41.** $(5a - 4b)(4a - 5b)$ **42.** $x^2(3x + 2)(2x - 5)$ **43.** $3y(4x + 5)(2x - 3)$
44. $(x^2 + 4)(x + 2)(x - 2)$ **45.** $3(a^2 + 3)^2$ **46.** $(a - 2)(a^2 + 2a + 4)$ **47.** $5x(x + 3y)^2$
48. $3ab(a - 3b)(a + 3b)$ **49.** $(x - 5 + y)(x - 5 - y)$ **50.** $(6 - 5a)(6 + 5a)$ **51.** $(x + 3)(x - 3)(x + 4)$
52. $-3, -2$ **53.** $-\frac{1}{2}, \frac{4}{5}$ **54.** $-\frac{5}{3}, \frac{5}{3}$ **55.** $0, -2$ **56.** $-3, 6$ **57.** $-4, 3, -3$ **58.** $-10, -8$ or $8, 10$
59. $-5, -4$ or $4, 5$ **60.** $3, 4, 5$ **61.** $6, 8, 10$

CHAPTER 5 TEST

1. x^8 **2.** $\frac{1}{32}$ **3.** $\frac{16}{9}$ **4.** $32x^{12}y^{11}$ **5.** a^2 **6.** x^6 **7.** $\frac{2a^{12}}{b^{15}}$ **8.** 6.53×10^6 **9.** 8.7×10^{-4}
10. 8.7×10^7 **11.** 3×10^8 **12.** $\frac{3}{4}x^3 - \frac{5}{4}x^2 - 2x - 1$ **13.** $4x + 75$ **14.** $6y^2 + y - 35$
15. $2x^3 + 3x^2 - 26x + 15$ **16.** $64 - 48t^3 + 9t^6$ **17.** $1 - 36y^2$ **18.** $4x^3 - 2x^2 - 30x$ **19.** $10t^4 - \frac{1}{10}$
20. $(x + 4)(x - 3)$ **21.** $2(3x^2 - 1)(2x^2 + 5)$ **22.** $(4a^2 + 9y^2)(2a + 3y)(2a - 3y)$
23. $(7a - b^2)(x^2 - 2y)$ **24.** $\left(t + \frac{1}{2}\right)\left(t^2 - \frac{1}{2}t + \frac{1}{4}\right)$ **25.** $4a^3b(a - 8b)(a + 2b)$

26. $(x - 5 + b)(x - 5 - b)$ **27.** $(9 + x^2)(3 + x)(3 - x)$ **28.** $-\frac{1}{3}, 2$ **29.** $0, 5$ **30.** $-5, 2$
31. $-2, -4, 4$ **32.** $-5, -2$ or $2, 5$ **33.** $6, 8, 10$ inches **34.** $R = 25x - 0.2x^2$
35. $P = -100 + 23x - 0.2x^2$ **36.** \$500 **37.** \$300 **38.** \$200

CHAPTERS 1–5 CUMULATIVE REVIEW

1. 6 **2.** -93 **3.** $\frac{25}{4}$ **4.** $18x + 7$ **5.** $6x - 19$ **6.** $24y$ **7.** $2x^3 - 5x^2 - 8x + 6$ **8.** -1
9. 0 **10.** $5, \frac{2}{3}$ **11.** $\varnothing$ **12.** $(4, 3)$ **13.** Parallel lines **14.** Lines coincide **15.** $(4, 0, 9)$

16. $(-4, 4)$

17. $[2, 5]$

18. $(-\infty, -2] \cup [9, \infty)$

19. $(-\infty, -4]$ **20.** $\dfrac{13}{a - b}$ **21.** $\frac{1}{2}, \frac{4}{5}, 1, \frac{8}{7}, \frac{5}{4}$ **22.** $7, -\frac{1}{7}$ **23.** -1 **24.** $-5 < x < 5$
25. $-1, 0, 2.35, 4$ **26.** $\{0, 1, 2, 3, 5, 6, 9\}$ **27.** Hypothesis: x is a multiple of 12; Conclusion: x is divisible by 3.
28. 36 **29.** $\dfrac{30y^6}{x^8 z^5}$ **30.** 4.69×10^{-4} **31.** 4.9×10^2 **32.** $D = \{-1, 2, 3\}, R = \{-10, 3\}$, yes
33. $\{x \mid x \le 5\}$ **34.**

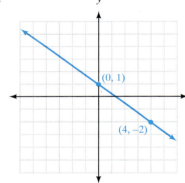

35.

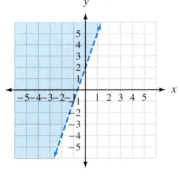

36. 2 **37.** $\frac{3}{5}, -3$ **38.** $y = -\frac{2}{3}x - 3$ **39.** $y = \frac{7}{3}x - 6$ **40.** $y = 3x + 1$ **41.** $(4y + \frac{1}{4})^2$
42. $(2x + y^2)(3a^2 + 1)$ **43.** $(x - 2)(x^2 + 2x + 4)$ **44.** $(4a^2 + 9b^2)(2a - 3b)(2a - 3b)$ **45.** $(-\frac{1}{22}, \frac{17}{66})$
46. $(\frac{1}{7}, -\frac{18}{77}, \frac{45}{77})$ **47.** 23 **48.** -14 **49.** 8 mph; 4 mph **50.** $\pi(a - b)(a + b)$

CHAPTER 6

PROBLEM SET 6.1

1. $-\frac{1}{3}$ **3.** $\frac{b^3}{2}$ **5.** $-\frac{3y^3}{2x}$ **7.** $\frac{18c^2}{7a^2}$ **9.** $\frac{x - 4}{6}$ **11.** $\frac{4x - 3y}{x(x + y)}$ **13.** $(a^2 + 9)(a + 3)$ **15.** $\frac{a - 6}{a + 6}$

17. $\dfrac{2y + 3}{y + 1}$ **19.** $\dfrac{5x - 2}{x + 2}$ **21.** $\dfrac{2 - x}{1 - x}$ or $\dfrac{x - 2}{x - 1}$ **23.** $\dfrac{x - 3}{x + 2}$ **25.** $\dfrac{a^2 - ab + b^2}{a - b}$ **27.** $\dfrac{2(x - 1)}{x}$

29. $\dfrac{2x + 3y}{2x + y}$ **31.** $\dfrac{x + 3}{y - 4}$ **33.** $\dfrac{x + b}{x - 2b}$ **35.** $x + 2$ **37.** $\dfrac{1}{x - 3}$ **39.** $\dfrac{2x^2 - 5}{3x - 2}$ **41.** -1

43. $-(y + 6)$ **45.** $\dfrac{-(3a + 1)}{3a - 1}$ or $-\dfrac{3a + 1}{3a - 1}$ **47.** 0   **49.** $4x$

51. $g(0) = -3$, $g(-3) = 0$, $g(3) = 3$, $g(-1) = -1$, $g(1)$ is undefined

53. $h(0) = -3$, $h(-3) = 3$, $h(3) = 0$, $h(-1)$ is undefined, $h(1) = -1$ **55.** $\{x \mid x \neq 1\}$ **57.** $\{x \mid x \neq 2\}$

59. $\{t \mid t \neq 4, t \neq -4\}$ **61.** 2; 2   **63.** Undefined; 4   **65.** 1; 1   **67.** 3; 3

69. The two graphs differ by the point (2, 4).

71.

Weeks	Weight (lb)
x	$W(x)$
0	200
1	194
4	184
12	173
24	168

73. 0.1 mile per minute   **75.** 10.8 miles per gallon   **77.** 6.8 feet per second

79. 3,768 inches per minute; 2,826 inches per minute   **81.** (a) Domain $= \{t \mid 20 \leq t \leq 50\}$

(b)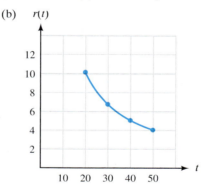

83. (a) Domain $= \{d \mid 1 \leq d \leq 6\}$

(b)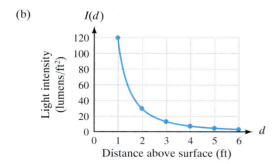

85. (a) $10:25$ (b) $20:25$ **87.** 3,080 vehicles   **89.** $3x^2 - 7x + 4$ **91.** 9   **93.** $8x^2$
95. (a) 2   (b) -4 (c) Undefined   (d) 2   (e) 1   (f) -6 (g) 4   (h) -3

PROBLEM SET 6.2

1. $2x^2 - 4x + 3$ **3.** $-2x^2 - 3x + 4$ **5.** $2y^2 + \dfrac{5}{2} - \dfrac{3}{2y^2}$ **7.** $-\dfrac{5}{2}x + 4 + \dfrac{3}{x}$ **9.** $4ab^3 + 6a^2b$

11. $-xy + 2y^2 + 3xy^2$ **13.** $x + 2$ **15.** $a - 3$ **17.** $5x + 6y$ **19.** $x^2 + xy + y^2$ **21.** $(y^2 + 4)(y + 2)$

23. $(x + 2)(x + 5)$ **25.** $(2x + 3)(2x - 3)$ **27.** $x - 7 + \dfrac{7}{x + 2}$ **29.** $2x + 5 + \dfrac{2}{3x - 4}$

31. $2x^2 - 5x + 1 + \dfrac{4}{x + 1}$ **33.** $y^2 - 3y - 13$ **35.** $x - 3$ **37.** $3y^2 + 6y + 8 + \dfrac{37}{2y - 4}$

39. $a^3 + 2a^2 + 4a + 6 + \dfrac{17}{a - 2}$ **41.** $y^3 + 2y^2 + 4y + 8$ **43.** $x^2 - 2x + 1$

45. $(x + 3)(x + 2)(x + 1)$ **47.** $(x + 3)(x + 4)(x - 2)$ **49.** Yes **51.** 7

53. (a) $x^3 - 3x^2 + 5x - 6 = (x - 2)(x^2 - x + 3)$; $P(2) = 2^3 - 3 \cdot 2^2 + 5 \cdot 2 - 6 = 0$
(b) $x^4 - 5x^3 - x^2 + 6x - 5 = (x - 5)(x^3 - x + 1)$; $P(5) = 5^4 - 5 \cdot 5^3 - 5^2 + 6 \cdot 5 - 5 = 0$

55. (a)

x	1	5	10	15	20
$C(x)$	2.15	2.75	3.50	4.25	5.00

(b) $\overline{C}(x) = \dfrac{C(x)}{x}$

(c)

x	1	5	10	15	20
$\overline{C}(x)$	2.15	0.55	0.35	0.28	0.25

(d) It decreases. (e)

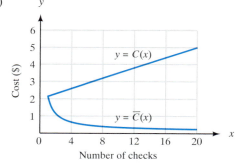

57. $\dfrac{21}{10}$ **59.** $\dfrac{11}{8}$ **61.** $\dfrac{1}{18}$ **63.** 32 **65.** 17 **67.** $\dfrac{x^{16}}{y^{22}}$ **69.** $4x^3 - x^2 + 3$ **71.** $0.5x^2 - 0.4x + 0.3$

73. $\dfrac{3}{2}x - \dfrac{5}{2}$ **75.** $\dfrac{2}{3}x + \dfrac{1}{3} + \dfrac{2}{3x - 1}$

PROBLEM SET 6.3

1. $\dfrac{1}{6}$ **3.** $\dfrac{9}{4}$ **5.** $\dfrac{1}{2}$ **7.** $\dfrac{15y}{x^2}$ **9.** $\dfrac{b}{a}$ **11.** $\dfrac{2y^5}{z^3}$ **13.** $\dfrac{x + 3}{x + 2}$ **15.** $y + 1$ **17.** $\dfrac{3(x + 4)}{x - 2}$

19. 1 **21.** $\dfrac{(a - 2)(a + 2)}{a - 5}$ **23.** $\dfrac{9t^2 - 6t + 4}{4t^2 - 2t + 1}$ **25.** $\dfrac{x + 3}{x + 4}$ **27.** $\dfrac{5a - b}{9a^2 + 15ab + 25b^2}$ **29.** 2

31. $\dfrac{x(x - 1)}{x^2 + 1}$ **33.** $\dfrac{(a + 4b)(a - 3b)}{(a - 4b)(a + 5b)}$ **35.** $\dfrac{2y - 1}{2y - 3}$ **37.** $\dfrac{(y - 2)(y + 1)}{(y + 2)(y - 1)}$ **39.** $\dfrac{x - 1}{x + 1}$ **41.** $\dfrac{x - 2}{x + 3}$

43. $3x$ **45.** $2(x + 5)$ **47.** $x - 2$ **49.** $-(y - 4)$ **51.** $(a - 5)(a + 1)$

53.

Number of Copies	Price per Copy ($)
1	20.33
10	9.33
20	6.40
50	4.00
100	3.05

55. $305.00

57. (a) $\dfrac{a^2b}{a^3} \cdot \dfrac{ab^2}{b^3} \div \dfrac{a^4}{b} = \dfrac{b}{a^4}$ (b) $\dfrac{a^2b}{a^3} \cdot \dfrac{ab^2}{b^3} \cdot \dfrac{a^4}{b} = \dfrac{a^4}{b}$ (c) $\dfrac{a^2b}{a^3} \div \dfrac{ab^2}{b^3} \cdot \dfrac{a^4}{b} = a^2b$ **59.** $(r^2 + rh) = 3$

61. $10x^5 + 8x^3 - 6x^2$ **63.** $12a^2 + 11a - 5$ **65.** $12xy - 6x + 28y - 14$ **67.** $9 - 6t^2 + t^4$

69. $3x^3 + 18x^2 + 33x + 18$ **71.** $\dfrac{47}{105}$ **73.** $\dfrac{1}{35}$

PROBLEM SET 6.4

1. $\dfrac{5}{4}$ **3.** $\dfrac{1}{3}$ **5.** $\dfrac{41}{24}$ **7.** $\dfrac{19}{144}$ **9.** $\dfrac{31}{24}$ **11.** 1 **13.** -1 **15.** $\dfrac{1}{x+y}$ **17.** 1 **19.** $\dfrac{a^2 + 2a - 3}{a^3}$

21. 1 **23.** $\dfrac{4-3t}{2t^2}$ **25.** $\dfrac{1}{2}$ **27.** $\dfrac{1}{5}$ **29.** $\dfrac{x+3}{2(x+1)}$ **31.** $\dfrac{a-b}{a^2 + ab + b^2}$ **33.** $\dfrac{2y-3}{4y^2 + 6y + 9}$

35. $\dfrac{2(2x-3)}{(x-3)(x-2)}$ **37.** $\dfrac{1}{2t-7}$ **39.** $\dfrac{4}{(a-3)(a+1)}$ **41.** $\dfrac{-4x^2}{(2x+1)(2x-1)(4x^2 + 2x + 1)}$

43. $\dfrac{2}{(2x+3)(4x+3)}$ **45.** $\dfrac{a}{(a+4)(a+5)}$ **47.** $\dfrac{x+1}{(x-2)(x+3)}$ **49.** $\dfrac{x-1}{(x+1)(x+2)}$ **51.** $\dfrac{1}{(x+2)(x+1)}$

53. $\dfrac{1}{(x+2)(x+3)}$ **55.** $\dfrac{4x+5}{2x+1}$ **57.** $\dfrac{22-5t}{4-t}$ **59.** $\dfrac{2x^2 + 3x - 4}{2x+3}$ **61.** $\dfrac{2x-3}{2x}$ **63.** $\dfrac{1}{2}$

65. $\dfrac{51}{10} = 5.1$ **67.** $(3+4)^{-1} = 7^{-1} = \dfrac{1}{7}$; $3^{-1} + 4^{-1} = \dfrac{1}{3} + \dfrac{1}{4} = \dfrac{7}{12}$ **69.** (a) $T = 120$ months

(b) If $t_A = t_B$, then $\dfrac{1}{T} = 0$. Since no value of T satisfies this equation, this implies that the two objects will never meet.

71. $x + \dfrac{4}{x} = \dfrac{x^2 + 4}{x}$ **73.** $\dfrac{1}{x} + \dfrac{1}{x+1} = \dfrac{2x+1}{x(x+1)}$ **75.** 5.4×10^4 **77.** 3.4×10^{-4} **79.** $6{,}440$

81. 0.00644 **83.** 1.2×10^4 **85.** $\dfrac{x-1}{x+3}$

PROBLEM SET 6.5

1. $\dfrac{9}{8}$ **3.** $\dfrac{2}{15}$ **5.** $\dfrac{119}{20}$ **7.** $\dfrac{1}{x+1}$ **9.** $\dfrac{a+1}{a-1}$ **11.** $\dfrac{y-x}{y+x}$ **13.** $\dfrac{1}{(x+5)(x-2)}$ **15.** $\dfrac{1}{a^2 - a + 1}$

17. $\dfrac{x+3}{x+2}$ **19.** $\dfrac{a+3}{a-2}$ **21.** $\dfrac{x-3}{x}$ **23.** $\dfrac{x+4}{x+2}$ **25.** $\dfrac{x-3}{x+3}$ **27.** $\dfrac{a-1}{a+1}$ **29.** $-\dfrac{x}{3}$ **31.** $\dfrac{y^2 + 1}{2y}$

33. $\dfrac{-x^2 + x - 1}{x-1}$ **35.** $\dfrac{5}{3}$ **37.** $\dfrac{2x-1}{2x+3}$ **39.** $(a^{-1} + b^{-1})^{-1} = \left(\dfrac{1}{a} + \dfrac{1}{b}\right)^{-1} = \left(\dfrac{a+b}{ab}\right)^{-1} = \dfrac{ab}{a+b}$

41. (a) As v approaches 0, the denominator approaches 1.   (b) $v = \dfrac{fs}{h} - s$ **43.** (a) $\dfrac{ab}{a+b}$ (b) $\dfrac{12}{7}$ hours

45. 1 **47.** $-3, 4$ **49.** $2, 3$

PROBLEM SET 6.6

1. $-\frac{35}{3}$ **3.** $-\frac{18}{5}$ **5.** $\frac{36}{11}$ **7.** 2   **9.** 5   **11.** 2   **13.** $-3, 4$ **15.** $-\frac{4}{3}, 1$

17. Possible solution -1, which does not check; $\varnothing$ **19.** 5   **21.** $-\frac{1}{2}, \frac{5}{3}$ **23.** $\frac{2}{3}$ **25.** 18

27. Possible solution 4, which does not check; $\varnothing$ **29.** Possible solutions 3 and -4; only -4 checks; -4

31. -6 **33.** -5 **35.** $\frac{53}{17}$ **37.** Possible solutions 1 and 2; only 2 checks; 2

39. Possible solution 3, which does not check; $\varnothing$ **41.** $\frac{22}{3}$ **43.** 2   **45.** 1, 5   **47.** $x = \dfrac{ab}{a - b}$

49. $R = \dfrac{R_1 R_2}{R_1 + R_2}$ **51.** $y = \dfrac{x - 3}{x - 1}$ **53.** $y = \dfrac{1 - x}{3x - 2}$

55.

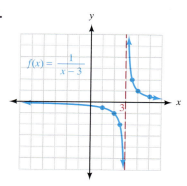

57.

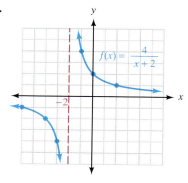

59.

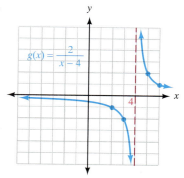

61.

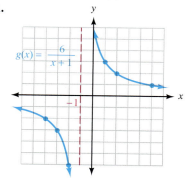

63. $f(0) = -\frac{1}{3}; f(6) = \frac{1}{3}$ **65.** $f(1) = -\frac{1}{2}; f(5) = \frac{1}{2}$ **67.** Domain $= \{x \mid x \neq 3\}$ **69.** $f(0) = 2; f(-4) = -2$

71. $f(2) = 1; f(-6) = -1$ **73.** Domain $= \{x \mid x \neq -2\}$ **75.** 5   **77.** 8 meters; 13 meters

79. $-6, -5$ or $5, 6$ **81.** 3, 4, 5   **83.** $\frac{24}{5}$ ft

85. $\dfrac{2}{x - y} - \dfrac{1}{y - x} = \dfrac{2y - 2x - x + y}{(x - y)(y - x)} = \dfrac{3y - 3x}{xy - x^2 - y^2 + xy} = \dfrac{+3(y - x)}{-1(x - y)^2} = \dfrac{3}{x - y}$

87. (a)

Time t (seconds)	Speed of Kayak Relative to the Water v (meters/second)	Current of the River c (meters/second)
240	4	1
300	4	2
514	4	3
338	3	1
540	3	2
NA*	3	3

* Since the current and the kayak are traveling at the same speed, the kayak will not be able to travel back up the river.

(b) Equal to the time in the river because they are racing both upstream and downstream.

(c) He will not go anywhere. The denominator of $\dfrac{450}{v - c}$ will be 0.

89. $-\dfrac{5}{2}, -\dfrac{2}{3}, \dfrac{5}{2}$ **91.** $-2, 2, 3$

PROBLEM SET 6.7

1. $\dfrac{1}{x} + \dfrac{1}{3x} = \dfrac{20}{3}; \dfrac{1}{5}$ and $\dfrac{3}{5}$ **3.** $x + \dfrac{1}{x} = \dfrac{10}{3}; 3$ or $\dfrac{1}{3}$ **5.** $\dfrac{1}{x} + \dfrac{1}{x + 1} = \dfrac{7}{12}; 3, 4$ **7.** $\dfrac{7 + x}{9 + x} = \dfrac{5}{6}; 3$

9. Let x = speed of current; $\dfrac{1.5}{5 - x} = \dfrac{3}{5 + x}; \dfrac{5}{3}$ miles per hour   **11.** $\dfrac{8}{x + 2} + \dfrac{8}{x - 2} = 3; 6$ miles per hour

13. Train A: 75 miles per hour, train B: 60 miles per hour; let x = speed of B; $\dfrac{150}{x + 15} = \dfrac{120}{x}$

15. 540 miles per hour; let x = speed of 747; $\dfrac{810}{270} - 1\dfrac{1}{2} = \dfrac{810}{x}$

17. 54 miles per hour; let x = usual speed; $\dfrac{270}{x + 6} + \dfrac{1}{2} = \dfrac{270}{x}$

19. Let x = time to fill the tank with both open; $\dfrac{1}{8} - \dfrac{1}{16} = \dfrac{1}{x}; 16$ hours   **21.** $\dfrac{1}{10} - \dfrac{1}{15} = \dfrac{1}{2x}; 15$ hours

23. Let x = time to fill with hot water; $\dfrac{1}{x} + \dfrac{1}{3.5} = \dfrac{1}{2.1}; 5\dfrac{1}{4}$ minutes   **25.** 51.1 acres   **27.** 5.9 miles per hour

29. 20.7 miles per hour   **31.** 4.6 miles per hour   **33.** 3.6 miles per hour

35. 3,241,440 minutes is a little more than 6 years.   **37.** $\dfrac{1}{3}[(x + \dfrac{2}{3}x) + \dfrac{1}{3}(x + \dfrac{2}{3}x)] = 10$ **39.** (a) 30 grams

$x = \dfrac{27}{2}$ (b) 3.25 moles

41.

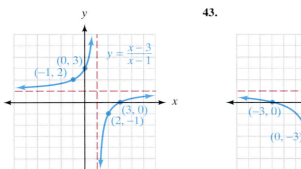

43.

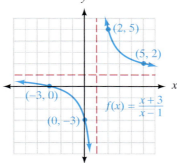

45.

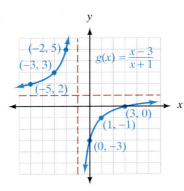

47. $\dfrac{2}{3a}$ **49.** $(x-3)(x+2)$ **51.** 1   **53.** $\dfrac{3-x}{3+x}$

55. Possible solution $x = 3$, which does not check; $\varnothing$

CHAPTER 6 REVIEW

1. $\dfrac{25x^2}{7y^3}$ **2.** $\dfrac{a(a-b)}{4}$ **3.** $\dfrac{x-5}{x+5}$ **4.** $\dfrac{a+1}{a-1}$ **5.** $3x + 2 + \dfrac{4}{x}$ **6.** $-9b + 5a - 7a^2b^2$ **7.** $x^{3n} - x^{2n}$

8. $x + 2$ **9.** $5x + 6y$ **10.** $y^3 + 2y^2 + 4y + 8$ **11.** $4x + 1 - \dfrac{2}{2x-7}$ **12.** $y^2 - 3y - 13$

13. $\dfrac{9}{5}$ **14.** $\dfrac{3x}{4y^2}$ **15.** $\dfrac{x-1}{x^2+1}$ **16.** 1   **17.** $\dfrac{x+2}{x-2}$ **18.** $(2x-3)(x+3)$ **19.** $\dfrac{31}{30}$ **20.** -1

21. $\dfrac{x^2+x+1}{x^3}$ **22.** $\dfrac{1}{(y+4)(y+3)}$ **23.** $\dfrac{x-1}{2(x+1)(x+2)}$ **24.** $\dfrac{15x-2}{5x-2}$ **25.** 5   **26.** $\dfrac{1}{a^2-a+1}$

27. $\dfrac{x^2+x+1}{x^2+1}$ **28.** $\dfrac{x+3}{x+2}$ **29.** 6   **30.** 1   **31.** -6

32. Possible solution -3, which does not check; $\varnothing$ **33.** $\dfrac{22}{3}$

34. Possible solutions 4 and -5; only 4 checks

35.

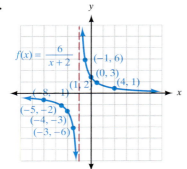

36.

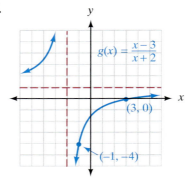

37. Car: 30 miles per hour; truck: 20 miles per hour   **38.** 7.5 miles per hour   **39.** 742 miles per hour

CHAPTER 6 TEST

1. $x + y$ **2.** $\dfrac{x-1}{x+1}$ **3.** $6x^2 + 3xy - 4y^2$ **4.** $x^2 - 4x - 2 + \dfrac{8}{2x-1}$ **5.** $2(a+4)$ **6.** $4(a+3)$

7. $x + 3$ **8.** $\dfrac{38}{105}$ **9.** $\dfrac{7}{8}$ **10.** $\dfrac{1}{a-3}$ **11.** $\dfrac{3(x-1)}{x(x-3)}$ **12.** $\dfrac{x}{(x+4)(x+5)}$ **13.** $\dfrac{x+4}{(x+1)(x+2)}$

14. $\dfrac{3a+8}{3a+10}$ **15.** $\dfrac{x-3}{x-2}$ **16.** $-\dfrac{3}{5}$ **17.** Possible solution 3, which does not check; $\varnothing$ **18.** $\dfrac{3}{13}$ **19.** $-2, 3$

20.

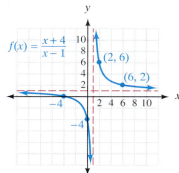

$f(x) = \dfrac{x+4}{x-1}$

(2, 6)

(6, 2)

21. $\dfrac{10}{23 - x} = \dfrac{1}{3}; -7$ **22.** $\dfrac{8}{x + 2} - \dfrac{8}{x - 2} = 3$; 6 miles per hour

23. $\dfrac{1}{10} - \dfrac{1}{15} = \dfrac{1}{2x}$; 15 hours   **24.** 2.7 miles

25. 1,102 miles per hour

CHAPTERS 1–6: CUMULATIVE REVIEW

1. 18   **2.** $\dfrac{36}{25}$ **3.** x^3 **4.** $\dfrac{x^3}{y^7}$ **5.** $-3x - 12$ **6.** $\dfrac{1}{a^2 + a + 1}$ **7.** $12x$ **8.** 2   **9.** -2

10. $5a - 7b > 5a + 7b$ **11.** $\{1, 7\}$ **12.** 12, arithmetic   **13.** Commutative, associative   **14.** $\dfrac{1}{(y + 3)(y + 2)}$

15. $-\dfrac{1}{x + y}$ **16.** $12t^4 - \dfrac{1}{12}$ **17.** $x + 2$ **18.** $2x^{2n} - 3x^{3n}$ **19.** $a^3 - a^2 + 2a - 4 + \dfrac{7}{a + 2}$

20. -20 **21.** 3   **22.** 1   **23.** $-12, 12$ **24.** $-5, 3, 5$ **25.** 5   **26.** $\dfrac{3}{2}, 4$ **27.** $(-3, -7)$

28. $(0, -\frac{1}{2})$ **29.** Lines are parallel.   **30.** $(0, \frac{6}{5}, \frac{-14}{5})$ **31.** $(\frac{31}{50}, \frac{13}{50})$ **32.** $(\frac{51}{31}, \frac{10}{31}, \frac{19}{31})$

33. $x \geq -1$ **34.** $-4 < x < 1$

35. -1 **36.**

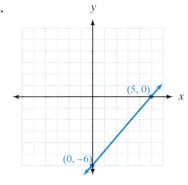

(5, 0)

(0, −6)

37.

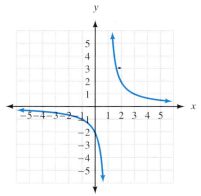

38.

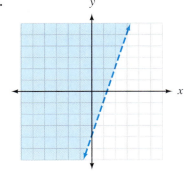

39. $2^3 \times 3 \times 7$ **40.** $(x - 10)(x + 7)$

41. $(x + 5 - y)(x + 5 + y)$ **42.** $x + 3$ **43.** 9.27×10^6 **44.** 225   **45.** 0   **46.** $\frac{3}{4}$ **47.** $x = -2$
48. -108 **49.** $b = 10$ feet, $h = 15$ feet   **50.** $120°, 40°, 20°$

CHAPTER 7

PROBLEM SET 7.1

1. 12   **3.** Not a real number   **5.** -7 **7.** -3 **9.** 2   **11.** Not a real number   **13.** 0.2   **15.** 0.2
17. $6a^4$ **19.** $3a^4$ **21.** xy^2 **23.** $2x^2y$ **25.** $2a^3b^5$ **27.** 6   **29.** -3 **31.** 2   **33.** -2 **35.** 2
37. $\frac{9}{5}$ **39.** $\frac{4}{5}$ **41.** 9   **43.** 125   **45.** 8   **47.** $\frac{1}{3}$ **49.** $\frac{1}{27}$ **51.** $\frac{6}{5}$ **53.** $\frac{8}{27}$ **55.** 7   **57.** $\frac{3}{4}$
59. $x^{4/5}$ **61.** a **63.** $\frac{1}{x^{2/5}}$ **65.** $x^{1/6}$ **67.** $x^{9/25}y^{1/2}z^{1/5}$ **69.** $\frac{b^{7/4}}{a^{1/8}}$ **71.** $y^{3/10}$ **73.** $\frac{1}{a^2b^4}$ **75.** $\frac{s^{1/2}}{r^{20}}$
77. $10b^3$ **79.** $(9^{1/2} + 4^{1/2})^2 = (3 + 2)^2 = 5^2 = 25 \neq 9 + 4$ **81.** $\sqrt{\sqrt{a}} = (a^{1/2})^{1/2} = a^{1/4} = \sqrt[4]{a}$ **83.** 25 mph
85. 1.618   **87.** $\frac{13}{8}$, numerator and denominator are consecutive members of the Fibonacci sequence.
89. (a) 420 pm   (b) 594 pm   (c) 5.94×10^{-10} m   **91.** (a) $\sqrt{2}$ (b) $\sqrt{3}$
93. (a) B   (b) A   (c) C   (d) $(0, 0)$ and $(1, 1)$ **95.** $x^6 - x^3$ **97.** $x^2 + 2x - 15$
99. $x^4 - 10x^2 + 25$ **101.** $x^3 - 27$ **103.** When $x = 2, y = 1.7$ **105.** When $x = 10, y = 5.6$
107. Graphs intersect at $x = 1, y = 1$ and $x = 0, y = 0$
109. (a) 1.62 micrograms   (b) 0.87 microgram   (c) 0.00293 microgram   (d) 0.0000029 microgram

PROBLEM SET 7.2

1. $x + x^2$ **3.** $a^2 - a$ **5.** $6x^3 - 8x^2 + 10x$ **7.** $12x^2 - 36y^2$ **9.** $x^{4/3} - 2x^{2/3} - 8$
11. $a - 10a^{1/2} + 21$ **13.** $20y^{2/3} - 7y^{1/3} - 6$ **15.** $10x^{4/3} + 21x^{2/3}y^{1/2} + 9y$ **17.** $t + 10t^{1/2} + 25$
19. $x^3 + 8x^{3/2} + 16$ **21.** $a - 2a^{1/2}b^{1/2} + b$ **23.** $4x - 12x^{1/2}y^{1/2} + 9y$ **25.** $a - 3$ **27.** $x^3 - y^3$
29. $t - 8$ **31.** $4x^3 - 3$ **33.** $x + y$ **35.** $a - 8$ **37.** $8x + 1$ **39.** $t - 1$ **41.** $2x^{1/2} + 3$
43. $3x^{1/3} - 4y^{1/3}$ **45.** $3a - 2b$ **47.** $3(x - 2)^{1/2}(4x - 11)$ **49.** $5(x - 3)^{7/5}(x - 6)$
51. $3(x + 1)^{1/2}(3x^2 + 3x + 2)$ **53.** $(x^{1/3} - 2)(x^{1/3} - 3)$ **55.** $(a^{1/5} - 4)(a^{1/5} + 2)$ **57.** $(2y^{1/3} + 1)(y^{1/3} - 3)$
59. $(3t^{1/5} + 5)(3t^{1/5} - 5)$ **61.** $(2x^{1/7} + 5)^2$ **63.** $\frac{3 + x}{x^{1/2}}$ **65.** $\frac{x + 5}{x^{1/3}}$ **67.** $\frac{x^3 + 3x^2 + 1}{(x^3 + 1)^{1/2}}$
69. $\frac{-4}{(x^2 + 4)^{1/2}}$ **71.** 2   **73.** 27   **75.** 0.871   **77.** 15.8%   **79.** 5.9%   **81.** $\frac{1}{x^2 + 9}$ **83.** $3x - 4x^3y$
85. $5x - 4$ **87.** $x^2 + 5x + 25$

PROBLEM SET 7.3

1. $2\sqrt{2}$ **3.** $7\sqrt{2}$ **5.** $12\sqrt{2}$ **7.** $4\sqrt{5}$ **9.** $4\sqrt{3}$ **11.** $15\sqrt{3}$ **13.** $3\sqrt[3]{2}$ **15.** $4\sqrt[3]{2}$ **17.** $6\sqrt[3]{2}$
19. $2\sqrt[5]{2}$ **21.** $3x\sqrt{2x}$ **23.** $2y\sqrt[4]{2y^3}$ **25.** $2xy^2\sqrt[3]{5xy}$ **27.** $4abc^2\sqrt{3b}$ **29.** $2bc\sqrt[3]{6a^2c}$ **31.** $2xy^2\sqrt[5]{2x^3y^2}$
33. $3xy^2z\sqrt[5]{x^2}$ **35.** $2\sqrt{3}$ **37.** $\sqrt{-20}$, which is not a real number   **39.** $\frac{\sqrt{11}}{2}$ **41.** $\frac{2\sqrt{3}}{3}$ **43.** $\frac{5\sqrt{6}}{6}$
45. $\frac{\sqrt{2}}{2}$ **47.** $\frac{\sqrt{5}}{5}$ **49.** $2\sqrt[3]{4}$ **51.** $\frac{2\sqrt[3]{3}}{3}$ **53.** $\frac{\sqrt[4]{24x^2}}{2x}$ **55.** $\frac{\sqrt[4]{8y^3}}{y}$ **57.** $\frac{\sqrt[3]{36xy^2}}{3y}$ **59.** $\frac{\sqrt[3]{6xy^2}}{3y}$
61. $\frac{\sqrt[4]{2x}}{2x}$ **63.** $\frac{3x\sqrt{15xy}}{5y}$ **65.** $\frac{5xy\sqrt{6xz}}{2z}$ **67.** $\frac{2ab\sqrt[3]{6ac^2}}{3c}$ **69.** $\frac{2xy^2\sqrt[3]{3z^2}}{3z}$ **71.** $5|x|$ **73.** $3|xy|\sqrt{3x}$
75. $|x - 5|$ **77.** $|2x + 3|$ **79.** $2|a(a + 2)|$ **81.** $2|x|\sqrt{x - 2}$
83. $\sqrt{9 + 16} = \sqrt{25} = 5; \sqrt{9} + \sqrt{16} = 3 + 4 = 7$ **85.** $5\sqrt{13}$ feet
89. $\sqrt{2}, \sqrt{3}, \sqrt{4}, \sqrt{5}, \sqrt{6}, \sqrt{7}, \ldots$; 10th term $= \sqrt{11}$; 100th term $= \sqrt{101}$ **91.** $\frac{y^3}{x^2}$ **93.** 1   **95.** $\frac{4x^2 - 6x + 9}{9x^2 - 3x + 1}$

97. $12\sqrt[3]{5}$ **99.** $6\sqrt[3]{49}$ **101.** $\dfrac{\sqrt[10]{a^7}}{a}$ **103.** $\dfrac{\sqrt[20]{a^9}}{a}$ **105.**

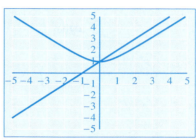

107. About $\dfrac{3}{4}$ **109.** $x = 0$

111. (a) $s = \dfrac{a + b + c}{2}$ (b) $A \approx 14.70$

PROBLEM SET 7.4

1. $7\sqrt{5}$ **3.** $-x\sqrt{7}$ **5.** $\sqrt[3]{10}$ **7.** $9\sqrt[5]{6}$ **9.** 0 **11.** $\sqrt{5}$ **13.** $-32\sqrt{2}$ **15.** $-3x\sqrt{2}$ **17.** $-2\sqrt[3]{2}$

19. $8x\sqrt[3]{xy^2}$ **21.** $3a^2b\sqrt{3ab}$ **23.** $11ab\sqrt[3]{3a^2b}$ **25.** $10xy\sqrt[4]{3y}$ **27.** $\sqrt{2}$ **29.** $\dfrac{8\sqrt{5}}{15}$ **31.** $\dfrac{(x-1)\sqrt{x}}{x}$

33. $\dfrac{3\sqrt{2}}{2}$ **35.** $\dfrac{5\sqrt{6}}{6}$ **37.** $\dfrac{8\sqrt[3]{25}}{5}$ **39.** $\sqrt{12} \approx 3.464;\ 2\sqrt{3} = 2(1.732) \approx 3.464$

41. $\sqrt{8} + \sqrt{18} \approx 2.828 + 4.243 = 7.071;\ \sqrt{50} \approx 7.071;\ \sqrt{26} \approx 5.099$ **43.** $8\sqrt{2x}$ **45.** 5

53.   The ratio of the hypotenuse to a leg is $\sqrt{2}:1$, or approximately $1.414:1$.

55. (a) $\sqrt{2}:1 \approx 1.414:1$ (b) $5:\sqrt{2}$ (c) $5:4$ **57.** 1 **59.** $\dfrac{13 - 3t}{3 - t}$ **61.** $\dfrac{6}{4x + 3}$ **63.** $\dfrac{x - y}{x^2 + xy + y^2}$

PROBLEM SET 7.5

1. $3\sqrt{2}$ **3.** $10\sqrt{21}$ **5.** 720 **7.** 54 **9.** $\sqrt{6} - 9$ **11.** $24 + 6\sqrt[3]{4}$ **13.** $7 + 2\sqrt{6}$ **15.** $x + 2\sqrt{x} - 15$
17. $34 + 20\sqrt{3}$ **19.** $19 + 8\sqrt{3}$ **21.** $x - 6\sqrt{x} + 9$ **23.** $4a - 12\sqrt{ab} + 9b$ **25.** $x + 4\sqrt{x} - 4$

27. $x - 6\sqrt{x - 5} + 4$ **29.** 1 **31.** $a - 49$ **33.** $25 - x$ **35.** $x - 8$ **37.** $10 + 6\sqrt{3}$ **39.** $\dfrac{\sqrt{3} + 1}{2}$

41. $\dfrac{5 - \sqrt{5}}{4}$ **43.** $\dfrac{x + 3\sqrt{x}}{x - 9}$ **45.** $\dfrac{10 + 3\sqrt{5}}{11}$ **47.** $\dfrac{3\sqrt{x} + 3\sqrt{y}}{x - y}$ **49.** $2 + \sqrt{3}$ **51.** $\dfrac{11 - 4\sqrt{7}}{3}$

53. $\dfrac{a + 2\sqrt{ab} + b}{a - b}$ **55.** $\dfrac{x + 4\sqrt{x} + 4}{x - 4}$ **57.** $\dfrac{5 - \sqrt{21}}{4}$ **59.** $\dfrac{\sqrt{x} - 3x + 2}{1 - x}$ **63.** $10\sqrt{3}$

65. $x + 6\sqrt{x} + 9$ **67.** 75 **69.** $\dfrac{5\sqrt{2}}{4}$ second; $\dfrac{5}{2}$ second   **75.** $-\dfrac{1}{8}$ **77.** $\dfrac{y - 2}{y + 2}$ **79.** $\dfrac{2x + 1}{2x - 1}$

PROBLEM SET 7.6

1. 4 **3.** $\varnothing$ **5.** 5 **7.** $\varnothing$ **9.** $\dfrac{39}{2}$ **11.** $\varnothing$ **13.** 5 **15.** 3 **17.** $-\dfrac{32}{3}$ **19.** $3, 4$ **21.** $-1, -2$

23. -1 **25.** $\varnothing$ **27.** 7 **29.** $0, 3$ **31.** -4 **33.** 8 **35.** 0 **37.** 9 **39.** 0 **41.** 8
43. Possible solution 9, which does not check; $\varnothing$ **45.** Possible solutions 0 and 32; only 0 checks; 0
47. Possible solutions -2 and 6; only 6 checks; 6   **49.** $h = 100 - 16t^2$ **51.** $\dfrac{392}{121} \approx 3.24$ feet

53. $w = \dfrac{10}{3}(\sqrt{x} + 1)$ **55.** 5 meters   **57.** 2,500 meters   **59.** $0 \le y \le 100$

61.

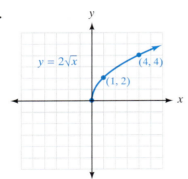

63.

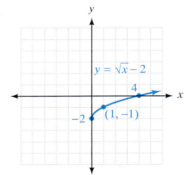

65.

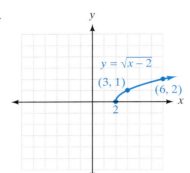

67.

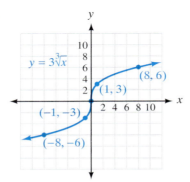

69.

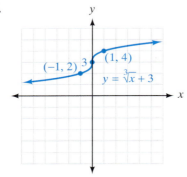

71.

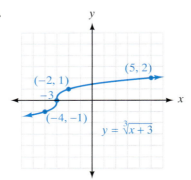

73. $\sqrt{6} - 2$ **75.** $x + 10\sqrt{x} + 25$ **77.** $\dfrac{x - 3\sqrt{x}}{x - 9}$ **79.** Possible solutions 2 and 6; only 6 checks; 6

81. $1, -1, -3$ **83.** $y = \frac{1}{8}x^2$ **85.**

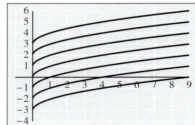

87. The value of b shifts the curve b units along the y-axis.

89.

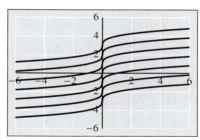

91. The value of b shifts the curve b units along the y-axis.

93.

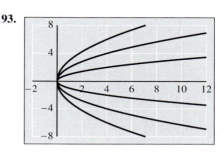

95. The smaller the absolute value of a, the more slowly the graph rises or falls. If a is negative, the graph lies below the x-axis.

PROBLEM SET 7.7

1. $6i$ **3.** $-5i$ **5.** $6i\sqrt{2}$ **7.** $-2i\sqrt{3}$ **9.** 1 **11.** -1 **13.** $-i$ **15.** $x = 3, y = -1$
17. $x = -2, y = -\frac{1}{2}$ **19.** $x = -8, y = -5$ **21.** $x = 7, y = \frac{1}{2}$ **23.** $x = \frac{3}{7}, y = \frac{2}{5}$ **25.** $5 + 9i$
27. $5 - i$ **29.** $2 - 4i$ **31.** $1 - 6i$ **33.** $2 + 2i$ **35.** $-1 - 7i$ **37.** $6 + 8i$ **39.** $2 - 24i$
41. $-15 + 12i$ **43.** $18 + 24i$ **45.** $10 + 11i$ **47.** $21 + 23i$ **49.** $-2 + 2i$ **51.** $2 - 11i$
53. $-21 + 20i$ **55.** $-2i$ **57.** $-7 - 24i$ **59.** 5 **61.** 40 **63.** 13 **65.** 164 **67.** $-3 - 2i$
69. $-2 + 5i$ **71.** $\frac{8}{13} + \frac{12}{13}i$ **73.** $-\frac{18}{13} - \frac{12}{13}i$ **75.** $-\frac{5}{13} + \frac{12}{13}i$ **77.** $\frac{13}{15} - \frac{2}{5}i$
79. $R = -11 - 7i$ ohms **81.** $-\frac{3}{2}$ **83.** $-3, \frac{1}{2}$ **85.** $\frac{5}{4}$ or $\frac{4}{5}$

CHAPTER 7 REVIEW

1. 7 **2.** -3 **3.** 2 **4.** 27 **5.** $2x^3y^2$ **6.** $\frac{1}{16}$ **7.** x^2 **8.** a^2b^4 **9.** $a^{7/20}$ **10.** $a^{5/12}b^{8/3}$
11. $12x + 11x^{1/2}y^{1/2} - 15y$ **12.** $a^{2/3} - 10a^{1/3} + 25$ **13.** $4x^{1/2} + 2x^{5/6}$ **14.** $2(x - 3)^{1/4}(4x - 13)$
15. $\dfrac{x + 5}{x^{1/4}}$ **16.** $2\sqrt{3}$ **17.** $5\sqrt{2}$ **18.** $2\sqrt[3]{2}$ **19.** $3x\sqrt{2}$ **20.** $4ab^2c\sqrt{5a}$ **21.** $2abc\sqrt[4]{2bc^2}$ **22.** $\dfrac{3\sqrt{2}}{2}$
23. $3\sqrt[3]{4}$ **24.** $\dfrac{4x\sqrt{21xy}}{7y}$ **25.** $\dfrac{2y\sqrt[3]{45x^2z^2}}{3z}$ **26.** $-2x\sqrt{6}$ **27.** $3\sqrt{3}$ **28.** $\dfrac{8\sqrt{5}}{5}$ **29.** $7\sqrt{2}$
30. $11a^2b\sqrt{3ab}$ **31.** $-4xy\sqrt[3]{xz^2}$ **32.** $\sqrt{6} - 4$ **33.** $x - 5\sqrt{x} + 6$ **34.** $3\sqrt{5} + 6$ **35.** $6 + \sqrt{35}$
36. $\dfrac{63 + 12\sqrt{7}}{47}$ **37.** 0 **38.** 3 **39.** 5 **40.** $\varnothing$

41.

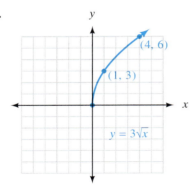

42.

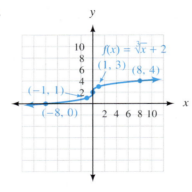

43. 1 **44.** $-i$ **45.** $x = -\frac{3}{2}, y = -\frac{1}{2}$ **46.** $x = -2, y = -4$ **47.** $9 + 3i$ **48.** $-7 + 4i$

49. $-6 + 12i$ **50.** $5 + 14i$ **51.** $12 + 16i$ **52.** 25 **53.** $1 - 3i$ **54.** $-\frac{6}{5} + \frac{3}{5}i$

55. 22.5 ft **56.** 65 yd

CHAPTER 7 TEST

1. $\frac{1}{9}$ **2.** $\frac{7}{5}$ **3.** $a^{5/12}$ **4.** $\frac{x^{13/12}}{y}$ **5.** $7x^4y^5$ **6.** $2x^2y^4$ **7.** $2a$ **8.** $x^{n^2-n}y^{1-n^3}$ **9.** $6a^2 - 10a$

10. $16a^3 - 40a^{3/2} + 25$ **11.** $(3x^{1/3} - 1)(x^{1/3} + 2)$ **12.** $(3x^{1/3} - 7)(3x^{1/3} + 7)$ **13.** $\frac{x + 4}{x^{1/2}}$

14. $\frac{3}{(x^2 - 3)^{1/2}}$ **15.** $5xy^2\sqrt{5xy}$ **16.** $2x^2y^2\sqrt[3]{5xy^2}$ **17.** $\frac{\sqrt{6}}{3}$ **18.** $\frac{2a^2b\sqrt{15bc}}{5c}$ **19.** $-6\sqrt{3}$ **20.** $-ab\sqrt[3]{3}$

21. $x + 3\sqrt{x} - 28$ **22.** $21 - 6\sqrt{6}$ **23.** $\frac{5 + 5\sqrt{3}}{2}$ **24.** $\frac{x - 2\sqrt{2x} + 2}{x - 2}$

25. Possible solutions 1 and 8; only 8 checks; 8 **26.** -4 **27.** -3

28.

29.

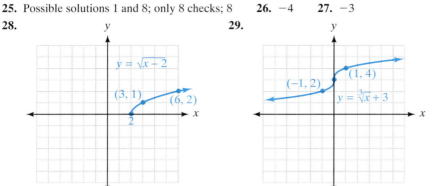

30. $x = \frac{1}{2}, y = 7$ **31.** $6i$

32. $17 - 6i$ **33.** $9 - 40i$ **34.** $-\frac{5}{13} - \frac{12}{13}i$ **35.** $i^{38} = (i^2)^{19} = (-1)^{19} = -1$

CHAPTERS 1–7 CUMULATIVE REVIEW

1. 23 **2.** 1 **3.** 59 **4.** $22x + 11$ **5.** $\frac{8}{y^3}$ **6.** 4 **7.** $6y^2\sqrt{2y}$ **8.** $\frac{4\sqrt{15}}{5}$ **9.** $-12, \frac{1}{12}$

10. $D = \{-1, 3\}, R = \{2, -1, 0\}$; no **11.** 4.15×10^4 **12.** 1 **13.** $(5a - 2b)(5a + 2b)(25a^2 + 4b^2)$

14. $2(4a^2 + 3)(3a^2 - 1)$ **15.** $\frac{2}{7}$ **16.** $\frac{2z}{x}$ **17.** $\frac{x-5}{x-3}$ **18.** $\frac{18}{7}$ **19.** $6x^2 - 19xy + 10y^2$

20. $3x^2 + 20x + 25$ **21.** $x - 4\sqrt{x} + 4$ **22.** $-2a^2 + 1 - 3b^2$ **23.** $\frac{6}{y^2}$ **24.** $\frac{x + 2\sqrt{xy} + y}{x - y}$

25. $\frac{2}{3}x^3 - \frac{1}{12}x^2 + \frac{1}{3}x + \frac{5}{12}$ **26.** $\frac{1}{(x+3)(x+5)}$ **27.** 2 **28.** $\frac{5}{6}$ **29.** $-\frac{7}{3}, 3$ **30.** $-4, -2, 4$

31. Possible solution -1, which does not check; $\varnothing$ **32.** $x = 3$ **33.** $-10, 7$ **34.** 2 **35.** 6

36. $y \le -2$ or $y \ge 3$ **37.** $x \le -\frac{2}{5}$ or $x \ge 2$ **38.** $(-9, -7)$

39. $(5, -2)$ **40.** $(3, -5, 0)$ **41.** $(9, -11, 1)$ **42.** -37

43.

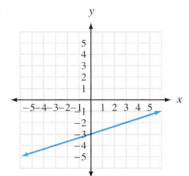

44.

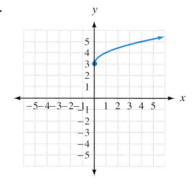

45.

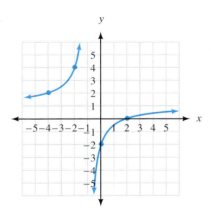

46. 6 **47.** $\frac{4}{5}, -3$ **48.** $y = -\frac{2}{5}x + 5$ **49.** $y = \frac{2}{3}x - 3$

50. $L = 24$ ft, $W = 10$ ft

CHAPTER 8

PROBLEM SET 8.1

1. ± 5 **3.** $\pm 3i$ **5.** $\pm\frac{\sqrt{3}}{2}$ **7.** $\pm 2i\sqrt{3}$ **9.** $\pm\frac{3\sqrt{5}}{2}$ **11.** $-2, 3$ **13.** $\frac{-3 \pm 3i}{2}$ **15.** $\frac{-2 \pm 2i\sqrt{2}}{5}$

17. $-4 \pm 3i\sqrt{3}$ **19.** $\frac{3 \pm 2i}{2}$ **21.** ± 1 **23.** ± 1 **25.** ± 1 **27.** $x^2 + 12x + 36 = (x + 6)^2$

29. $x^2 - 4x + 4 = (x - 2)^2$ **31.** $a^2 - 10a + 25 = (a - 5)^2$ **33.** $x^2 + 5x + \frac{25}{4} = (x + \frac{5}{2})^2$

35. $y^2 - 7y + \frac{49}{4} = (y - \frac{7}{2})^2$ **37.** $-6, 2$ **39.** $-3, -9$ **41.** $1 \pm 2i$ **43.** $4 \pm \sqrt{15}$ **45.** $\frac{5 \pm \sqrt{37}}{2}$

47. $1 \pm \sqrt{5}$ **49.** $\frac{4 \pm \sqrt{13}}{3}$ **51.** $\frac{3 \pm i\sqrt{71}}{8}$ **53.** $\frac{\sqrt{3}}{2}$ inch, 1 inch   **55.** $x\sqrt{3}, 2x$ **57.** $\sqrt{2}$ inches

59. $\frac{\sqrt{2}}{2}$ inch   **61.** $x\sqrt{2}$ **63.** 781 feet   **65.** $\frac{1,170}{5,630} = 0.21$ to the nearest hundredth   **67.** $20\sqrt{2} \approx 28$ feet

69. 7.3% to the nearest tenth   **71.** 19.6 to the nearest tenth

73. $3\sqrt{5}$ **75.** $3y^2\sqrt{3y}$ **77.** $3x^2y\sqrt[3]{2y^2}$ **79.** 13   **81.** $\frac{3\sqrt{2}}{2}$ **83.** $\sqrt[3]{2}$ **85.** $x = \pm 2a$ **87.** $x = -a$

89. $x = 0, -2a$ **91.** $x = \frac{-p \pm \sqrt{p^2 - 4q}}{2}$

PROBLEM SET 8.2

1. $-2, -3$ **3.** $2 \pm \sqrt{3}$ **5.** $1, 2$ **7.** $\frac{2 \pm i\sqrt{14}}{3}$ **9.** $0, 5$ **11.** $0, -\frac{4}{3}$ **13.** $\frac{3 \pm \sqrt{5}}{4}$ **15.** $-3 \pm \sqrt{17}$

17. $\frac{-1 \pm i\sqrt{5}}{2}$ **19.** 1   **21.** $\frac{1 \pm i\sqrt{47}}{6}$ **23.** $4 \pm \sqrt{2}$ **25.** $\frac{1}{2}, 1$ **27.** $-\frac{1}{2}, 3$ **29.** $\frac{-1 \pm i\sqrt{7}}{2}$

31. $1 \pm \sqrt{2}$ **33.** $\frac{-3 \pm \sqrt{5}}{2}$ **35.** $3, -5$ **37.** $2, -1 \pm i\sqrt{3}$ **39.** $-\frac{3}{2}, \frac{3 \pm 3i\sqrt{3}}{4}$ **41.** $\frac{1}{5}, \frac{-1 \pm i\sqrt{3}}{10}$

43. $0, \frac{-1 \pm i\sqrt{5}}{2}$ **45.** $0, 1 \pm i$ **47.** $0, \frac{-1 \pm i\sqrt{2}}{3}$ **49.** $\frac{-3 - 2i}{5}$ **51.** 2 seconds

53. $\frac{1}{4}$ second and 1 second   **55.** $40 \pm 20 = 20$ or 60 items   **57.** $\frac{3.5 \pm 0.5}{0.004} = 750$ or 1,000 patterns

59. $(10.5 - 2x)(8.2 - 2x) = 0.8(10.5)(8.2); 4x^2 - 37.4x + 17.22 = 0; x \approx 8.86$ centimeters (impossible) or 0.49 centimeter

61. (a) $21 + 2w = 20$, or $l + w = 10; l \cdot w = 15$
 (b)   8.16 or 1.84 to the nearest hundredth
 (c)   Two answers are possible because either dimension (long or short) may be considered the length.

63. $4y + 1 + \frac{-2}{2y - 7}$ **65.** $x^2 + 7x + 12$ **67.** 5   **69.** $\frac{27}{125}$ **71.** $\frac{1}{4}$ **73.** $21x^3y$ **75.** $-2\sqrt{3}, \sqrt{3}$

77. $\frac{-1 \pm \sqrt{3}}{\sqrt{2}} = \frac{-\sqrt{2} \pm \sqrt{6}}{2}$ **79.** $-2i, i$ **81.** $-i, 4i$

PROBLEM SET 8.3

1. $D = 16$; two rational   **3.** $D = 0$; one rational   **5.** $D = 5$; two irrational   **7.** $D = 17$; two irrational

9. $D = 36$; two rational   **11.** $D = 116$; two irrational   **13.** ± 10 **15.** ± 12 **17.** 9   **19.** -16

21. $\pm 2\sqrt{6}$ **23.** $x^2 - 7x + 10 = 0$ **25.** $t^2 - 3t - 18 = 0$ **27.** $y^3 - 4y^2 - 4y + 16 = 0$

29. $2x^2 - 7x + 3 = 0$ **31.** $4t^2 - 9t - 9 = 0$ **33.** $6x^3 - 5x^2 - 54x + 45 = 0$ **35.** $10a^2 - a - 3 = 0$

37. $9x^3 - 9x^2 - 4x + 4 = 0$ **39.** $x^4 - 13x^2 + 36 = 0$ **41.** $(x - 3)(x + 5)^2 = 0$ or $x^3 + 7x^2 - 5x - 75 = 0$

43. $(x - 3)^2(x + 3)^2 = 0$ or $x^4 - 18x^2 + 81 = 0$ **45.** $-3, -2, -1$ **47.** $-3, -4, 2$ **49.** $1 \pm i$

51. $5, 4 \pm 3i$ **53.** $a^{11/2} - a^{9/2}$ **55.** $x^3 - 6x^{3/2} + 9$ **57.** $6x^{1/2} - 5x$ **59.** $5(x - 3)^{1/2}(2x - 9)$

61. $(2x^{1/3} - 3)(x^{1/3} - 4)$ **63.** $-2, 1, i, -i$ **65.** $x = -a$ is a solution of multiplicity 3.   **67.** $-1, 5$

69. $1 + \sqrt{2} \approx 2.41, 1 - \sqrt{2} \approx -0.41$ **71.** $-1, \frac{1}{2}, 1$ **73.** $\frac{1}{2}, \sqrt{2} \approx 1.41, -\sqrt{2} \approx -1.41$ **75.** $\frac{2}{3}, 1 + i, 1 - i$

PROBLEM SET 8.4

1. $1, 2$ **3.** $-8, -\frac{5}{2}$ **5.** $\pm 3, \pm i\sqrt{3}$ **7.** $\pm 2i, \pm i\sqrt{5}$ **9.** $\frac{7}{2}, 4$ **11.** $-\frac{9}{8}, \frac{1}{2}$ **13.** $\pm \frac{\sqrt{30}}{6}, \pm i$

15. $\pm \frac{\sqrt{21}}{3}, \pm \frac{i\sqrt{21}}{3}$ **17.** $4, 25$ **19.** Possible solutions 25 and 9; only 25 checks; 25

21. Possible solutions $\frac{25}{9}$ and $\frac{49}{4}$; only $\frac{25}{9}$ checks; $\frac{25}{9}$ **23.** $27, 38$ **25.** $4, 12$ **27.** $t = \dfrac{v \pm \sqrt{v^2 + 64h}}{32}$

29. $x = \dfrac{-4 \pm 2\sqrt{4 - k}}{k}$ **31.** $x = -y$ **33.** $t = \dfrac{1 \pm \sqrt{1 + h}}{4}$ **35.** $t = \dfrac{v \pm \sqrt{v^2 + 1{,}280}}{32}$

37. $2 + 2\sqrt{5}; 4\left(\dfrac{1 + \sqrt{5}}{2}\right) = 2(1 + \sqrt{5}) = 2 + 2\sqrt{5}$ **39.** $1 + \sqrt{5}; \dfrac{BC}{AB} = \dfrac{1 + \sqrt{5}}{2}$

41. (a)

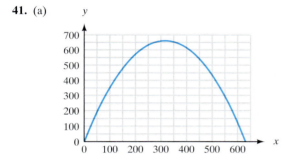

(b) The legs are 630 feet apart.

43. (a) $l + 2w = 160$
(b) $A = (160 - 2w) \cdot w$ or $A = -2w^2 + 160w$
(c) Some possible values are shown below:

w	$A = -2w^2 + 160w$
20	2,400
25	2,750
30	3,000
35	3,150
40	3,200
45	3,150
50	3,000
55	2,750

(d) 3,200 square yards

45. $3\sqrt{7}$ **47.** $5\sqrt{2}$ **49.** $39x^2y\sqrt{5x}$ **51.** $-11 + 6\sqrt{5}$ **53.** $x + 4\sqrt{x} + 4$ **55.** $\dfrac{7 + 2\sqrt{7}}{3}$

57. x-intercepts $= -2, 0, 2$; y-intercept $= 0$ **59.** x-intercepts $= -3, -\frac{1}{3}, 3$; y-intercept $= -9$

61. $\frac{1}{2}$ and -1

PROBLEM SET 8.5

1. x-intercepts $= -3, 1$;
Vertex $= (-1, -4)$

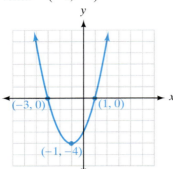

3. x-intercepts $= -5, 1$;
Vertex $= (-2, 9)$

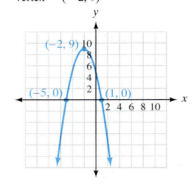

5. x-intercepts $= -1, 1$;
Vertex $= (0, -1)$

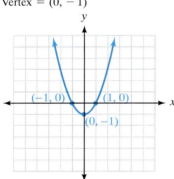

7. x-intercepts $= 3, -3$;
Vertex $= (0, 9)$

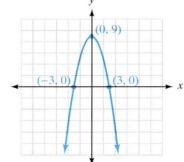

9. x-intercepts $= -1, 3$;
Vertex $= (1, -8)$

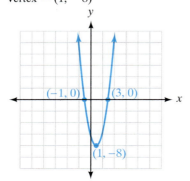

11. x-intercepts $= 1 + \sqrt{5}, 1 - \sqrt{5}$;
Vertex $= (1, -5)$

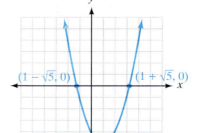

13. Vertex $= (2, -8)$

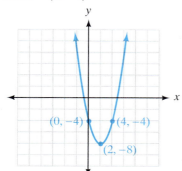

15. Vertex $= (1, -4)$

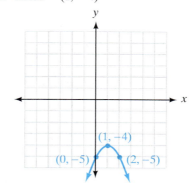

17. Vertex $= (0, 1)$

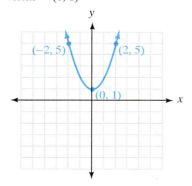

19. Vertex $= (0, -3)$

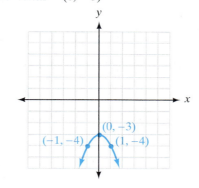

21. Vertex $= \left(-\frac{2}{3}, -\frac{1}{3}\right)$

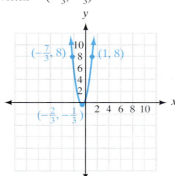

23. $(3, -4)$ lowest

25. $(1, 9)$ highest

27. $(2, 16)$ highest

29. $(-4, 16)$ highest

31. 40 items; maximum profit $500

33. 875 patterns; maximum profit $731.25

35. The ball is in her hand when $h(t) = 0$, which means $t = 0$ or $t = 2$ seconds. Maximum height is $h(1) = 16$ feet.

37. 256 feet   **39.** 40 feet by 20 feet

41. Maximum $R = \$3,600$
when $p = \$6.00$

43. Maximum $R = \$7,225$
when $p = \$8.50$

45. $1 - i$

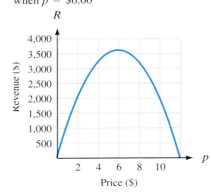

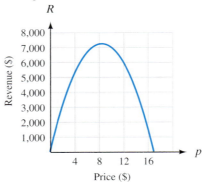

47. $27 + 5i$ **49.** $\frac{1}{10} + \frac{3}{10}i$ **51.** $y = (x - 2)^2 - 4$ **53.** $y = -\frac{1}{135}(x - 90)^2 + 60$

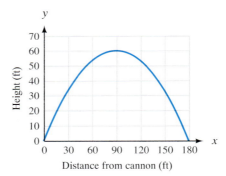

55. Problem 1: $(x - 2)(x + 2)$
Problem 2: $(x - 2)(x + 2) = 0; x = 2, -2$
Problem 3:

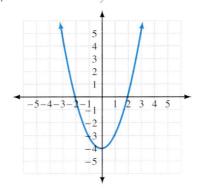

Relationship: If $x - 2$ is a factor of polynomial $P(x)$,
then $x = 2$ is a solution to the equation $P(x) = 0$, and
the graph of $y = P(x)$ has an x-intercept at 2.

Problem 4: $0 = x^2 - 4; (x - 2)(x + 2) = 0; x = -2, 2$

PROBLEM SET 8.6

1.

3.

5.

7.

9.

11.

13.

15. All real numbers   **17.** No solution; ∅

19.

21.

23.

25.

27.

29.

31.

33.

35. (a) $-2 < x < 2$ (b) $x < -2$ or $x > 2$ (c) $x = -2$ or $x = 2$

37. (a) $-2 < x < 5$ (b) $x < -2$ or $x > 5$ (c) $x = -2$ or $x = 5$

39. (a) $x < -1$ or $1 < x < 3$ (b) $-1 < x < 1$ or $x > 3$ (c) $x = -1$ or $x = 1$ or $x = 3$

41. $x \geq 4$; the width is at least 4 inches.   **43.** $5 \leq p \leq 8$; charge at least $5 but no more than $8 for each radio.

45. Let $x =$ the number of $10 increases in dues. Then,
Income $= (10{,}000 - 200x)(100 + 10x) = -2{,}000x^2 + 80{,}000x + 1{,}000{,}000.$ The vertex of this parabola is at
$x = -b/(2a) = (-80{,}000)/[2(-2{,}000)] = 20.$ Thus, there should be 20 increases of $10, or an increase of $200, making
the new maximum income $= (10{,}000 - 200 \cdot 20)(100 + 10 \cdot 20) = \$1{,}800{,}000.$

47. Let $x =$ the number of $2 increases. Then,
Income $= (20 + 2x)(40 - 2x) = -4x^2 + 40x + 800.$ The vertex of this parabola is at $x = -b/(2a) =$
$(-40)/[2(-4)] = 5.$ This implies there should be five increases of $2 for maximum income; that is, $\$20 + 5 \cdot \$2 = \$30$
for an oil change.

49. $\frac{5}{3}$ **51.** Possible solutions 1 and 6; only 6 checks; 6

53.

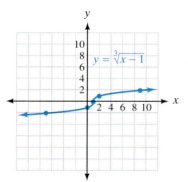

55.

57.

CHAPTER 8 REVIEW

1. $0, 5$ **2.** $0, \frac{4}{3}$ **3.** $\dfrac{4 \pm 7i}{3}$ **4.** $-3 \pm \sqrt{3}$ **5.** $-5, 2$ **6.** $-3, -2$ **7.** 3 **8.** 2 **9.** $\dfrac{-3 \pm \sqrt{3}}{2}$

10. $\dfrac{3 \pm \sqrt{5}}{2}$ **11.** $-5, 2$ **12.** $0, \frac{9}{4}$ **13.** $\dfrac{2 \pm i\sqrt{15}}{2}$ **14.** $1 \pm \sqrt{2}$ **15.** $-\frac{4}{3}, \frac{1}{2}$ **16.** 3 **17.** $5 \pm \sqrt{7}$

18. $0, \dfrac{5 \pm \sqrt{21}}{2}$ **19.** $-1, \frac{3}{5}$ **20.** $3, \dfrac{-3 \pm 3i\sqrt{3}}{2}$ **21.** $\dfrac{1 \pm i\sqrt{2}}{3}$ **22.** $\dfrac{3 \pm \sqrt{29}}{2}$ **23.** 100 or 170 items

24. 20 or 60 items   **25.** $D = 0$; one rational   **26.** $D = 0$; one rational   **27.** $D = 25$; two rational

28. $D = 361$; two rational   **29.** $D = 5$; two irrational   **30.** $D = 5$; two irrational   **31.** $D = -23$; two complex

32. $D = -87$; two complex   **33.** ± 20 **34.** ± 20 **35.** 4   **36.** 4   **37.** 25   **38.** 49

39. $x^2 - 8x + 15 = 0$ **40.** $x^2 - 2x - 8 = 0$ **41.** $2y^2 + 7y - 4 = 0$ **42.** $t^3 - 5t^2 - 9t + 45 = 0$

43. $-4, 12$ **44.** $-\frac{3}{4}, -\frac{1}{6}$ **45.** $\pm 2, \pm i\sqrt{3}$ **46.** Possible solutions 4 and 1; only 4 checks; 4   **47.** $\frac{9}{4}, 16$

48. 4   **49.** 4   **50.** 7   **51.** $t = \dfrac{5 \pm \sqrt{25 + 16h}}{16}$ **52.** $t = \dfrac{v \pm \sqrt{v^2 + 640}}{32}$

53.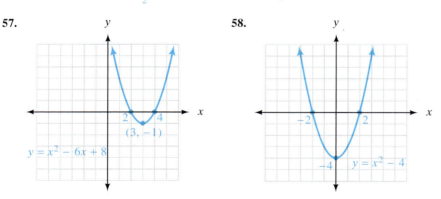

57.

$y = x^2 - 6x + 8$

58.

$y = x^2 - 4$

CHAPTER 8 TEST

1. $-\frac{9}{2}, \frac{1}{2}$ **2.** $3 \pm i\sqrt{2}$ **3.** $5 \pm 2i$ **4.** $1 \pm i\sqrt{2}$ **5.** $\frac{5}{2}, \dfrac{-5 \pm 5i\sqrt{3}}{4}$ **6.** $-1 \pm i\sqrt{5}$ **7.** $r = \pm\dfrac{\sqrt{A}}{8} - 1$

8. $2 \pm \sqrt{2}$ **9.** $\frac{1}{2}$ and $\frac{3}{2}$ seconds   **10.** 15 or 100 cups   **11.** 9   **12.** $D = 81$; two rational

13. $3x^2 - 13x - 10 = 0$ **14.** $x^3 - 7x^2 - 4x + 28 = 0$ **15.** $\pm\sqrt{2}, \pm\frac{1}{2}i$ **16.** $1, \frac{1}{2}$ **17.** $\frac{1}{4}, 9$

18. $t = \dfrac{7 \pm \sqrt{49 + 16h}}{16}$

19.

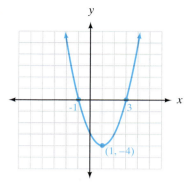

20.

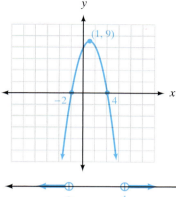

21.

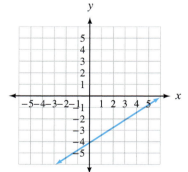

22.

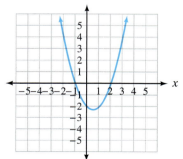

23. Maximum profit = \$900 by selling 100 items per week

CHAPTERS 1–8: CUMULATIVE REVIEW

1. 0   **2.** $-\frac{8}{27}$ **3.** 88   **4.** $-7x + 10$ **5.** $10x - 37$ **6.** $\dfrac{x^3}{y^7}$ **7.** $2\sqrt[3]{4}$ **8.** $\frac{9}{20}$ **9.** $\frac{1}{7}$ **10.** $\frac{4}{5}$

11. $x - 6y$ **12.** $x - 3$ **13.** $3x^3 - 11x^2 + 4$ **14.** $2i$ **15.** $\dfrac{23}{13} + \dfrac{11}{13}i$ **16.** -2 **17.** 15

18. $-9, 9$ **19.** Possible solutions 3 and 2; only 2 checks.   **20.** $-3, -2$ **21.** $\frac{4}{3} - \sqrt{2}$ and $\frac{4}{3} + \sqrt{2}$

22. $-\frac{3}{2}, -1$ **23.** $0, \dfrac{5 + \sqrt{37}}{6}, \dfrac{5 - \sqrt{37}}{6}$ **24.** $-\frac{1}{2}, \frac{2}{3}$ **25.** $\frac{9}{4}$ **26.** $\dfrac{8}{a - b}$

27. $8 \leq x \leq 20$

28. $x \leq -\frac{1}{2}$ or $x \geq 2$ **29.** Lines coincide.

30. Lines are parallel.   **31.** $(3, 2)$ **32.** $(3, 0, \frac{1}{2})$

33.

34.

35. $4x - 3y = 2$ **36.** $(x + 4 + y)(x + 4 - y)$ **37.** $\dfrac{7\sqrt[3]{3}}{3}$ **38.** $\{1, 3\}$ **39.** $100°, 55°, 25°$ **40.** $\frac{25}{16}$

CHAPTER 9

PROBLEM SET 9.1

1. 1   **3.** 2   **5.** $\frac{1}{27}$ **7.** 13

9.

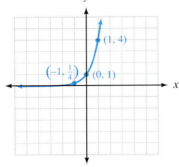

11.

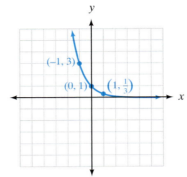

13.

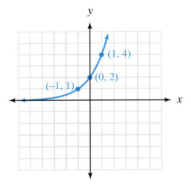

15.

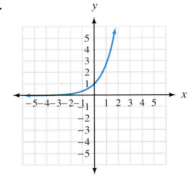

17.

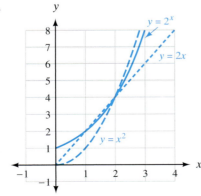

19.

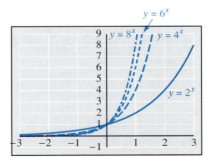

21. $h = 6 \cdot \left(\frac{2}{3}\right)^n$; fifth bounce: $6\left(\frac{2}{3}\right)^5 \approx 0.79$ feet   **23.**

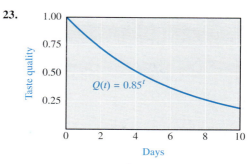

$Q(t) = \frac{1}{2}$ when $t \approx 4.3$ days

$Q(t) = \frac{1}{2}$ when $t \approx 4.3$ days

25. (a) $A(t) = 1,200\left(1 + \dfrac{0.06}{4}\right)^{4t}$ (b) \$1,932.39   (c) About 11.64 years   (d) \$1,939.29

27. (a) \$129,138.48   (b) Domain: $\{t \mid 0 \le t \le 6\}$ (c)
(d) Range: $\{V(t) \mid 52,942.05 < V(t) \le 450,000\}$
(e) After approximately 4 years and 8 months

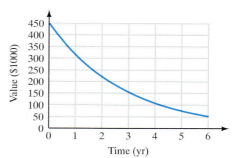

29. $f(1) = 200, f(2) = 800, f(3) = 3,200$

31.

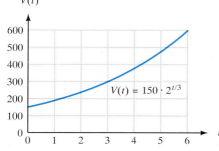

33. Domain $= \{-2, 2\}$, range $= \{3, 6, 8\}$, not a function

35. All real numbers except -7 and 5 **37.** -18 **39.** 2

41. (a) 1 (b) $\frac{1}{3}$ (c) 3 (d) 1 (e) $\frac{1}{2}$ (f) 2 (g) 3 (h) $\frac{1}{2}$

43. (a)

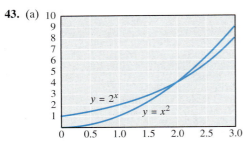

(b) $2^2 = 4$; $2^4 = 16$ and $4^2 = 16$

(c)

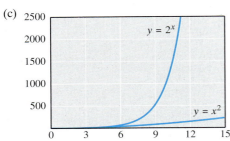

From the graph, $y = 2^x$ grows faster. This is confirmed because 2^{100} is much larger than 100^2, which is only 10,000.

PROBLEM SET 9.2

1. $f^{-1}(x) = \dfrac{x + 1}{3}$ **3.** $f^{-1}(x) = \sqrt[3]{x}$ **5.** $f^{-1}(x) = \dfrac{x - 3}{x - 1}$ **7.** $f^{-1}(x) = 4x + 3$

9. $f^{-1}(x) = 2(x + 3) = 2x + 6$ **11.** $f^{-1}(x) = \dfrac{1 - x}{3x - 2}$ **13.**

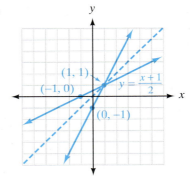

15.

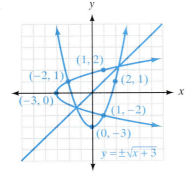

17.

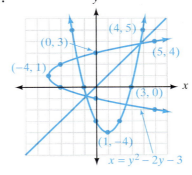

19.

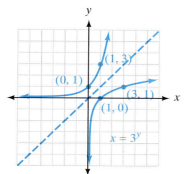

21.

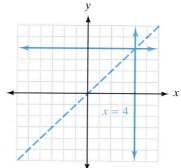

23.

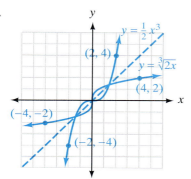

25.

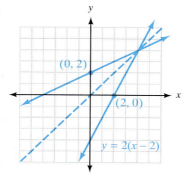

27.

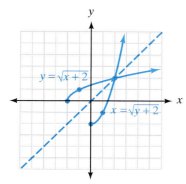

29. (a) Yes (b) No (c) Yes   **31.** (a) 4 (b) $\frac{4}{3}$ (c) 2 (d) 2

33. $f^{-1}(x) = \frac{1}{x}$ **35.** $f^{-1}(x) = 7(x + 2)$

37. (a) -3 (b) -6 (c) 2 (d) 3 (e) -2 (f) 3 (g) Each is the inverse of the other.   **39.** $-2, 3$ **41.** 25

43. $5 \pm \sqrt{17}$ **45.** $1 \pm i$ **47.** $f^{-1}(x) = \dfrac{x - 5}{3}$ **49.** $f^{-1}(x) = \sqrt[3]{x - 1}$ **51.** $f^{-1}(x) = \dfrac{2x - 4}{x - 1}$

53. (a) 1 (b) 2 (c) 5 (d) 0 (e) 1 (f) 2 (g) 2 (h) 5

PROBLEM SET 9.3

1. $\log_2 16 = 4$ **3.** $\log_5 125 = 3$ **5.** $\log_{10} 0.01 = -2$ **7.** $\log_2 \frac{1}{32} = -5$ **9.** $\log_{1/2} 8 = -3$

11. $\log_3 27 = 3$ **13.** $10^2 = 100$ **15.** $2^6 = 64$ **17.** $8^0 = 1$ **19.** $10^{-3} = 0.001$ **21.** $6^2 = 36$

23. $5^{-2} = \frac{1}{25}$ **25.** 9   **27.** $\frac{1}{125}$ **29.** 4   **31.** $\frac{1}{3}$ **33.** 2   **35.** $\sqrt[3]{5}$

37.

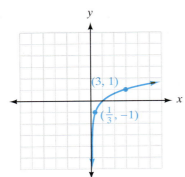

39.

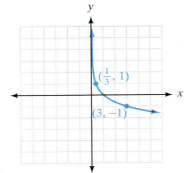

41.

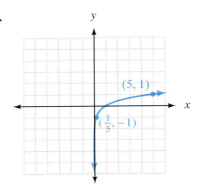

43.

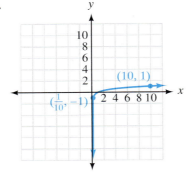

45. $y = 3^x$ **47.** $y = \log_{1/3} x$ **49.** 4   **51.** $\frac{3}{2}$ **53.** 3   **55.** 1   **57.** 0   **59.** 0   **61.** $\frac{1}{2}$

63.

Prefix	Multiplying Factor	$\log_{10}$ (Multiplying Factor)
Nano	0.000 000 001	−9
Micro	0.000 001	−6
Deci	0.1	−1
Giga	1,000,000,000	9
Peta	1,000,000,000,000,000	15

65. 2   **67.** 10^8 times as large   **69.** $\dfrac{-2 \pm \sqrt{10}}{2}$ **71.** $\dfrac{2 \pm i\sqrt{3}}{2}$ **73.** 5, $\dfrac{-5 \pm 5i\sqrt{3}}{2}$ **75.** 0, 5, −1

77. $\dfrac{3 \pm \sqrt{29}}{2}$

79. (a)

x	−1	0	1	2
$f(x)$	$\frac{1}{8}$	1	8	64

(b)

x	$\frac{1}{8}$	1	8	64
$f^{-1}(x)$	−1	0	1	2

(c) $f(x) = 8^x$

(d) $f^{-1}(x) = \log_8 x$

PROBLEM SET 9.4

1. $\log_3 4 + \log_3 x$ **3.** $\log_6 5 - \log_6 x$ **5.** $5 \log_2 y$ **7.** $\frac{1}{3} \log_9 z$ **9.** $2 \log_6 x + 4 \log_6 y$

11. $\frac{1}{2}\log_5 x + 4\log_5 y$ **13.** $\log_b x + \log_b y - \log_b z$ **15.** $\log_{10} 4 - \log_{10} x - \log_{10} y$

17. $2\log_{10} x + \log_{10} y - \frac{1}{2}\log_{10} z$ **19.** $3\log_{10} x + \frac{1}{2}\log_{10} y - 4\log_{10} z$ **21.** $\frac{2}{3}\log_b x + \frac{1}{3}\log_b y - \frac{4}{3}\log_b z$

23. $\log_b xz$ **25.** $\log_3 \dfrac{x^2}{y^3}$ **27.** $\log_{10} \sqrt{x}\,\sqrt[3]{y}$ **29.** $\log_2 \dfrac{x^3\sqrt{y}}{z}$ **31.** $\log_2 \dfrac{\sqrt{x}}{y^3 z^4}$ **33.** $\log_{10} \dfrac{x^{3/2}}{y^{3/4}z^{4/5}}$ **35.** $\frac{2}{3}$

37. 18   **39.** Possible solutions -1 and 3; only 3 checks; 3   **41.** 3

43. Possible solutions -2 and 4; only 4 checks; 4   **45.** Possible solutions -1 and 4; only 4 checks; 4

47. Possible solutions $-\frac{5}{2}$ and $\frac{5}{3}$; only $\frac{5}{3}$ checks; $\frac{5}{3}$ **51.** (a) 1.602 (b) 2.505 (c) 3.204

53. $\text{pH} = 6.1 + \log_{10} x - \log_{10} y$ **55.** 2.52   **57.** $D = -7$; two complex   **59.** $x^2 - 2x - 15 = 0$

61. $3y^2 - 11y + 6 = 0$

PROBLEM SET 9.5

1. 2.5775   **3.** 1.5775   **5.** 3.5775   **7.** -1.4225 **9.** 4.5775   **11.** 2.7782   **13.** 3.3032   **15.** -2.0128

17. -1.5031 **19.** -0.3990 **21.** 759   **23.** 0.00759   **25.** 1,430   **27.** 0.00000447   **29.** 0.0000000918

31. 10^{10} **33.** 10^{-10} **35.** 10^{20} **37.** $\frac{1}{100}$ **39.** 1,000

41. Possible solutions 1 and 100; both check

43. The calculator gives a "domain error," indicating that -10 is not in the domain of $f(x) = \log x$.

45. The San Francisco earthquake was approximately 2,000 times greater.

47.

x	$(1 + x)^{(1/x)}$
1	2
0.5	2.25
0.1	2.5937
0.01	2.7048
0.001	2.7169
0.0001	2.7181
0.00001	2.7183

$(1 + x)^{(1/x)}$ appears to approach e.

49. $3\log R - 2\log T = \log R^3 - \log T^2$

$$= \log \frac{R^3}{T^2}$$

So if $\dfrac{R^3}{T^2}$ is constant, then its logarithm $\left(\log \dfrac{R^3}{T^2}\right)$ will also be constant.

51. Approximately 3.19   **53.** 1.78×10^{-5} **55.** 3.16×10^5 **57.** 2.00×10^8 **59.** 10 times as large

61. 12.9%   **63.** 5.3%   **65.** 1   **67.** 5   **69.** x **71.** $\ln 10 + 3t$ **73.** $\ln A - 2t$ **75.** 2.7080

77. -1.0986 **79.** 2.1972   **81.** 2.7724   **83.** $\pm 2, \pm i\sqrt{2}$ **85.** 8, 27   **87.** $1, \frac{9}{4}$ **89.** 5, 1

91. $\dfrac{-2 \pm \sqrt{4 + k^2}}{k}$

PROBLEM SET 9.6

1. 1.4650   **3.** 0.6826   **5.** -1.5440 **7.** -0.6477 **9.** -0.3333 **11.** 2.0000   **13.** -0.1845

15. 0.1845   **17.** 1.6168   **19.** 2.1131   **21.** 1.333   **23.** 0.7500   **25.** 1.3917   **27.** 0.7186

29. 2.6356   **31.** 4.1632   **33.** 5.8435   **35.** -1.0642 **37.** 2.3026   **39.** 10.7144   **41.** 11.72 years

43. 9.25 years   **45.** 8.75 years   **47.** 18.58 years   **49.** 11.55 years   **51.** 18.31 years   **53.** 11.45 years

55. 13.9 years later or toward the end of 2007   **57.** 27.5 years   **59.** $(-2, -23)$; lowest   **61.** $(\frac{3}{2}, 9)$; highest

63. 2 seconds, 64 feet   **65.** $t = \dfrac{1}{r}\ln\dfrac{A}{P}$ **67.** $t = \dfrac{1}{k}\dfrac{\log P - \log A}{\log 2}$ **69.** $t = \dfrac{\log A - \log P}{\log(1 - r)}$ **71.** 1.75

73. 1.73   **75.** $(1.46, 5)$; they are nearly the same point

CHAPTER 9 REVIEW

1. 16   **2.** $\frac{1}{2}$ **3.** $\frac{1}{9}$ **4.** -5 **5.** $\frac{5}{6}$ **6.** 7   **7.**

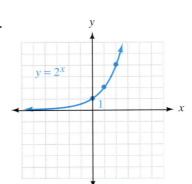

8.

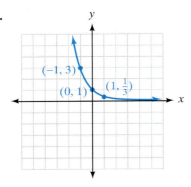

9.

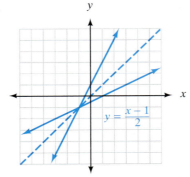

10.

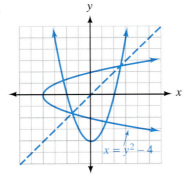

11. $f^{-1}(x) = \dfrac{x - 3}{2}$ **12.** $y = \pm\sqrt{x + 1}$ **13.** $f^{-1}(x) = 2x - 4$

14. $y = \pm\sqrt{\dfrac{4 - x}{2}}$ **15.** $\log_3 81 = 4$ **16.** $\log_7 49 = 2$ **17.** $\log_{10} 0.01 = -2$ **18.** $\log_2 \frac{1}{8} = -3$

19. $2^3 = 8$ **20.** $3^2 = 9$ **21.** $4^{1/2} = 2$ **22.** $4^1 = 4$ **23.** 25   **24.** $\frac{3}{4}$ **25.** 10

26.

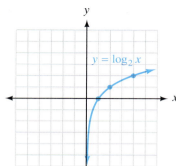

27.

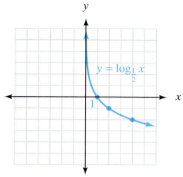

28. 2   **29.** $\frac{2}{3}$ **30.** 0   **31.** $\log_2 5 + \log_2 x$ **32.** $\log_{10} 2 + \log_{10} x - \log_{10} y$

33. $3 \log_a y + \frac{1}{2} \log_a x - \log_a z$ **34.** $2 \log_{10} x - 3 \log_{10} y - 4 \log_{10} z$ **35.** $\log_2 xy$ **36.** $\log_3 \frac{x}{4}$ **37.** $\log_a \frac{25}{3}$

38. $\log_2 \frac{x^3 y^2}{z^4}$ **39.** 2   **40.** 6   **41.** Possible solutions -1 and 3; only 3 checks; 3   **42.** 3

43. Possible solutions -2 and 3; only 3 checks; 3   **44.** Possible solutions -1 and 4; only 4 checks; 4   **45.** 2.5391

46. -0.1469 **47.** 9,230   **48.** 0.0251   **49.** 1   **50.** 0   **51.** 2   **52.** -4 **53.** 2.1   **54.** 5.1

55. 2.0×10^{-3} **56.** 3.2×10^{-8} **57.** $\frac{3}{2}$ **58.** $x = \frac{1}{3}\left(\frac{\log 5}{\log 4} - 2\right) = -0.28$ **59.** 0.75   **60.** 2.43

61. About 4.67 years   **62.** About 9.25 years   **63.** \$136,856.91   **64.** \$2,660.95 per year

CHAPTER 9 TEST

1.

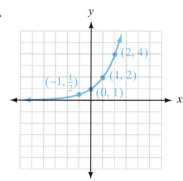

2.

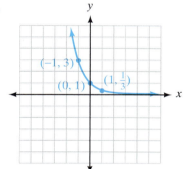

3.

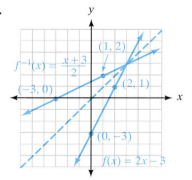

4.

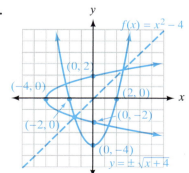

5. 64   **6.** $\sqrt{5}$

7.

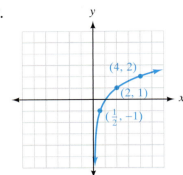

8.

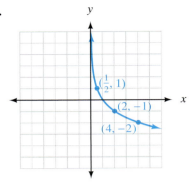

9. $\frac{2}{3}$ **10.** 1.5646 **11.** 4.3692 **12.** -1.9101 **13.** 3.8330 **14.** -3.0748 **15.** $3 + 2\log_2 x - \log_2 y$

16. $\frac{1}{2}\log x - 4\log y - \frac{1}{5}\log z$ **17.** $\log_3 \dfrac{x^2}{\sqrt{y}}$ **18.** $\log \dfrac{\sqrt[3]{x}}{yz^2}$ **19.** 7.04×10^4 **20.** 2.25×10^{-3}

21. 1.46 **22.** $\frac{5}{4}$ or 1.25 **23.** 15 **24.** Possible solutions -1 and 8; only 8 checks; 8

25. 6.2 **26.** \$651.56 **27.** About 13.9 years **28.** \$7,372.80

CHAPTERS 1–9: CUMULATIVE REVIEW

1. 32 **2.** $24x - 26$ **3.** -16 **4.** $\frac{17}{8}$ **5.** $3xy\sqrt[3]{x}$ **6.** $3\sqrt{3}$ **7.** $-2 + 7i$ **8.** 0 **9.** $\dfrac{x^2 + x + 1}{x + 1}$

10. $\frac{4}{7}$ **11.** $12t^4 - \frac{1}{12}$ **12.** $18 + 14\sqrt{3}$ **13.** $3x + 7 + \dfrac{10}{3x - 4}$ **14.** $\dfrac{-12 + 7\sqrt{6}}{5}$ **15.** $\dfrac{24}{(4x + 3)(4x - 3)}$

16. $\frac{8}{3}$ **17.** 4 **18.** $\frac{1}{4}, -\frac{1}{7}$ **19.** $\varnothing$ **20.** $\frac{3}{2}$ **21.** $\dfrac{1 \pm \sqrt{29}}{2}$ **22.** $-\frac{12}{5}$ **23.** 1 **24.** $-\frac{1}{4}, \frac{2}{3}$ **25.** $\varnothing$

26. 4 **27.** 1, 4 **28.** $-\frac{3}{2}, \frac{1}{4}$ **29.** 27 **30.** 10 **31.** Possible solution $-\frac{9}{2}$; does not check; $\varnothing$ **32.** $\left(\frac{4}{3}, \frac{13}{3}\right)$

33. $(8, -5)$ **34.** $(3, 1, -2)$ **35.** $y \le 1$ or $y \ge 3$

36. $x < -\frac{7}{2}$ or $x > 2$

37. $x < -3$ or $-2 < x < 3$

38.

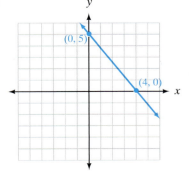

39. $y = -\frac{5}{3}x + 2$ **40.** $9a - 4b < 9a + 4b$ **41.** 9.72×10^{-5} **42.** $2(5a^2 + 2)(5a^2 - 1)$
43. $f^{-1}(x) = 2x - 6$ **44.** -46 **45.** $D = \{2, -3\}, R = \{-3, -1, 3\}$, no   **46.** $\{x \mid x \le 3\}$
47. -54, geometric   **48.** 6   **49.** 6 gallons 25%; 14 gallons 50%   **50.** \$12,667.70

CHAPTER 10

PROBLEM SET 10.1

1. 4, 7, 10, 13, 16   **3.** 3, 7, 11, 15, 19   **5.** 1, 2, 3, 4, 5   **7.** 4, 7, 12, 19, 28   **9.** $\frac{1}{4}, \frac{2}{5}, \frac{3}{6}, \frac{4}{7}, \frac{5}{8}$ **11.** $\frac{2}{3}, \frac{3}{4}, \frac{4}{5}, \frac{5}{6}, \frac{6}{7}$
13. $1, \frac{1}{4}, \frac{1}{9}, \frac{1}{16}, \frac{1}{25}$ **15.** 2, 4, 8, 16, 32   **17.** $\frac{1}{3}, \frac{1}{9}, \frac{1}{27}, \frac{1}{81}, \frac{1}{243}$ **19.** $2, \frac{3}{2}, \frac{4}{3}, \frac{5}{4}, \frac{6}{5}$ **21.** $0, \frac{3}{2}, \frac{8}{3}, \frac{15}{4}, \frac{24}{5}$
23. $-2, 4, -8, 16, -32$ **25.** $3, -9, 27, -81, 243$ **27.** $3, 0, -3, -6, -9$ **29.** 1, 5, 13, 29, 61
31. 1, 3, 6, 10, 15   **33.** $a_n = n + 1$ **35.** $a_n = 4n$ **37.** $a_n = 3n + 4$ or recursively as $a_1 = 7, a_n = a_{n-1} + 3$
39. $a_n = n^2$ **41.** $a_n = 3n^2$ **43.** $a_n = 2^{n+1}$ or recursively as $a_1 = 4, a_n = 2a_{n-1}$
45. $a_n = (-2)^n$ or recursively as $a_1 = -2, a_n = -2a_{n-1}$ **47.** $a_n = \frac{1}{2^{n+1}}$ or recursively as $a_1 = \frac{1}{4}, a_n = \frac{1}{2}a_{n-1}$
49. $a_n = \frac{n}{(n + 1)^2}$ **51.** (a) \$28,000, \$29,120, \$30,284.80, \$31,496.19, \$32,756.04   (b) $a_n = 28,000(1.04)^{n-1}$
53. (a) \$1,200, \$2,400, \$3,600, \$4,800, \$6,000   (b) $a_m = 100m$ (c) $a_y = 100(12y)$ **55.** 27   **57.** 5   **59.** 1
61. $a_{100} \approx 2.7048; a_{1,000} \approx 2.7169; a_{10,000} \approx 2.7181; a_{100,000} \approx 2.7183$ **63.** 1, 1, 2, 3, 5, 8, 13, 21, 34, 55
65. $\frac{3}{2}, \frac{5}{3}, \frac{8}{5}$

PROBLEM SET 10.2

1. 36   **3.** 9   **5.** 11   **7.** $\frac{163}{60}$ **9.** $\frac{62}{15}$ **11.** 60   **13.** 40   **15.** $5x + 15$ when simplified
17. $(x + 1)^2 + (x + 1)^3 + (x + 1)^4 + (x + 1)^5 + (x + 1)^6 + (x + 1)^7$
19. $\frac{x + 1}{x - 1} + \frac{x + 2}{x - 1} + \frac{x + 3}{x - 1} + \frac{x + 4}{x - 1} + \frac{x + 5}{x - 1}$
21. $(x + 3)^3 + (x + 4)^4 + (x + 5)^5 + (x + 6)^6 + (x + 7)^7 + (x + 8)^8$
23. $(x + 1)^2 + (x + 2)^3 + (x + 3)^4 + (x + 4)^5 + (x + 5)^6$ **25.** $\sum_{i=1}^{4} 2^i$ **27.** $\sum_{i=2}^{6} 2^i$ **29.** $\sum_{i=2}^{6} (i^2 + 1)$
31. $\sum_{i=3}^{7} \frac{i}{i + 1}$ **33.** $\sum_{i=1}^{4} \frac{i}{2i + 1}$ **35.** $\sum_{i=3}^{6} (x - i)$ **37.** $\sum_{i=3}^{5} \frac{x}{x + i}$ **39.** $\sum_{i=2}^{4} x^i(x + i)$
41. (a) $\frac{1}{3} = 0.3 + 0.03 + 0.003 + 0.0003 + \cdots$ (b) $\frac{2}{9} = 0.2 + 0.02 + 0.002 + 0.0002 + \cdots$
(c) $\frac{3}{11} = 0.27 + 0.0027 + 0.000027 + \cdots$ **43.** 208 feet, 784 feet   **45.** (a) $\frac{1}{3}, \frac{1}{15}, \frac{1}{35}, \frac{1}{63}, \frac{1}{99}, \frac{1}{143}$ (b) $\frac{6}{13}$ (c) $\frac{1}{2}$
47. $3 \log_2 x + \log_2 y$ **49.** $\frac{1}{3} \log_{10} x - 2 \log_{10} y$ **51.** $\log_{10} \frac{x}{y^2}$ **53.** $\log_3 \frac{x^2}{y^3 z^4}$
55. 80   **57.** Possible solutions -1 and 8, only 8 checks; 8   **59.** $\frac{16}{7}$

PROBLEM SET 10.3

1. 1   **3.** Not an arithmetic progression   **5.** -5 **7.** Not an arithmetic progression   **9.** $\frac{2}{3}$
11. $a_n = 4n - 1; a_{24} = 95$ **13.** $a_{10} = -12; S_{10} = -30$ **15.** $a_1 = 7; d = 2; a_{30} = 65$
17. $a_1 = 12; d = 2; a_{20} = 50; S_{20} = 620$ **19.** 20,300   **21.** -158 **23.** $a_{10} = 5; S_{10} = \frac{55}{2}$
25. (a) \$18,000, \$14,700, \$11,400, \$8,100, \$4,800   (b) $-3,300$

(c)

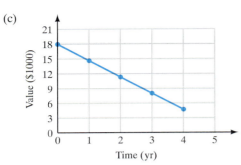

(d) $9,750 (e) $a_n = 18,000 - (n - 1)3,300$

27. (a) 1,500, 1,460, 1,420, 1,380, 1,340, 1,300
(b) It is arithmetic because the same amount is subtracted from each succeeding term. (c) $a_n = 1,500 - (n - 1)40$
29. (a) 1, 3, 6, 10, 15, 21, 28, 36, 45, 55, 66, 78, 91, 105, 120 (b) $a_n = n + a_{n-1}$
(c) No, it is not arithmetic because the same amount is not added to each term.
31. $28,000, $28,850, $29,700, $30,550, $31,400; $a_n = 27,150 + 850n$; $a_{10} = $35,650 **33.** 2.7604
35. -1.2396 **37.** 445 **39.** 4.45×10^{-8}

PROBLEM SET 10.4

1. 5 **3.** $\frac{1}{3}$ **5.** Not geometric **7.** -2 **9.** Not geometric **11.** $a_n = 4 \cdot 3^{n-1}$

13. $a_6 = -2\left(-\frac{1}{2}\right)^5 = \frac{1}{16}$ **15.** $a_{20} = 3(-1)^{19} = -3$ **17.** $S_{10} = \frac{10(2^{10} - 1)}{2 - 1} = 10,230$

19. $S_{20} = \frac{1[(-1)^{20} - 1]}{-1 - 1} = 0$ **21.** $a_8 = \frac{1}{5}\left(\frac{1}{2}\right)^7 = \frac{1}{640}$ **23.** $S_5 = \frac{-\frac{1}{2}\left[\left(\frac{1}{2}\right)^5 - 1\right]}{\frac{1}{2} - 1} = -\frac{31}{32}$

25. $a_{10} = \sqrt{2}(\sqrt{2})^9 = (\sqrt{2})^{10} = 32; S_{10} = \frac{\sqrt{2}[(\sqrt{2})^{10} - 1]}{\sqrt{2} - 1} = \frac{31\sqrt{2}}{\sqrt{2} - 1} = 62 + 31\sqrt{2}$

27. $a_6 = 100\left(\frac{1}{10}\right)^5 = \frac{1}{10^3} = \frac{1}{1,000}; S_6 = \frac{100\left[\left(\frac{1}{10}\right)^6 - 1\right]}{\frac{1}{10} - 1} = \frac{100(-0.999999)}{-0.9} = 111.111$ **29.** $r = \pm 2$

31. $S = \frac{\frac{1}{2}}{1 - \frac{1}{2}} = 1$ **33.** $S = \frac{4}{1 - \frac{1}{2}} = 8$ **35.** $S = \frac{\frac{2}{5}}{1 - \frac{2}{5}} = \frac{2}{3}$ **37.** $S = \frac{\frac{3}{4}}{1 - \frac{1}{3}} = \frac{9}{8}$

43. (a) $450,000, $315,000, $220,500, $154,350, $108,045 (b) 0.7
(c) (d) $90,000 (e) $a_n = 0.7a_{n-1}$

45. (a) $\frac{1}{2}$ (b) $\frac{364}{729}$ (c) $\frac{1}{1,458}$ **47.** 300 feet   **49.** $x^2 + 10x + 25$ **51.** $x^3 + 3x^2y + 3xy^2 + y^3$
53. $x^4 + 4x^3y + 6x^2y^2 + 4xy^3 + y^4$ **55.** 52,428,800   **57.** 199.99924   **59.** $S = 2a$ **61.** $S = \dfrac{a}{b - a}$

PROBLEM SET 10.5

1. $x^4 + 8x^3 + 24x^2 + 32x + 16$ **3.** $x^6 + 6x^5y + 15x^4y^2 + 20x^3y^3 + 15x^2y^4 + 6xy^5 + y^6$
5. $32x^5 + 80x^4 + 80x^3 + 40x^2 + 10x + 1$ **7.** $x^5 - 10x^4y + 40x^3y^2 - 80x^2y^3 + 80xy^4 - 32y^5$
9. $81x^4 - 216x^3 + 216x^2 - 96x + 16$ **11.** $64x^3 - 144x^2y + 108xy^2 - 27y^3$
13. $x^8 + 8x^6 + 24x^4 + 32x^2 + 16$ **15.** $x^6 + 3x^4y^2 + 3x^2y^4 + y^6$ **17.** $\dfrac{x^3}{8} - 3x^2 + 24x - 64$

19. $\dfrac{x^4}{81} + \dfrac{2x^3y}{27} + \dfrac{x^2y^2}{6} + \dfrac{xy^3}{6} + \dfrac{y^4}{16}$ **21.** $x^9 + 18x^8 + 144x^7 + 672x^6$ **23.** $x^{10} - 10x^9y + 45x^8y^2 - 120x^7y^3$
25. $x^{10} + 20x^9y + 180x^8y^2 + 960x^7y^3$ **27.** $x^{15} + 15x^{14} + 105x^{13}$ **29.** $x^{12} - 12x^{11}y + 66x^{10}y^2$
31. $x^{20} + 40x^{19} + 760x^{18}$ **33.** $x^{100} + 200x^{99}$ **35.** $x^{50} + 50x^{49}y$

37. $a_9 = \dfrac{12!}{8!4!}(2x)^4(3y)^8 = 495(16x^4)(6,561y^8) = 51,963,120x^4y^8$ **39.** $a_5 = \dfrac{10!}{4!6!}x^6(-2)^4 = 210x^6(16) = 3,360x^6$

41. $a_4 = \dfrac{9!}{3!6!}x^6(3)^3 = 84x^6(27) = 2,268x^6$ **43.** $a_{12} = \dfrac{20!}{11!9!}(2x)^9(5y)^{11}$ **45.** $\dfrac{21}{128}$ **47.** 1.04060401

49. $x = \dfrac{\log 7}{\log 5} \approx 1.21$ **51.** $\frac{1}{6}$ or 0.17   **53.** Approximately 7 years   **55.** 2.16   **57.** 6.36   **59.** $t = \dfrac{1}{5}\ln\dfrac{A}{10}$
61. 56   **63.** 125,970
67. As one example, if $n = 5$, $\binom{5}{0} + \binom{5}{1} + \binom{5}{2} + \binom{5}{3} + \binom{5}{4} + \binom{5}{5} = 1 + 5 + 10 + 10 + 5 + 1 = 32 = 2^5$.

CHAPTER 10 REVIEW

1. 7, 9, 11, 13   **2.** 1, 4, 7, 10   **3.** 0, 3, 8, 15   **4.** $\frac{4}{3}, \frac{5}{4}, \frac{6}{5}, \frac{7}{6}$ **5.** 4, 16, 64, 256   **6.** $\frac{1}{4}, \frac{1}{16}, \frac{1}{64}, \frac{1}{256}$

7. $3n - 1$ **8.** $2n - 5$ **9.** n^4 **10.** $n^2 + 1$ **11.** 2^{-n} **12.** $\dfrac{n + 1}{n^2}$ **13.** 32   **14.** 25   **15.** $\frac{14}{5}$

16. -5 **17.** 98   **18.** $\frac{127}{30}$ **19.** $\displaystyle\sum_{i=1}^{4} 3i$ **20.** $\displaystyle\sum_{i=1}^{4}(4i - 1)$ **21.** $\displaystyle\sum_{i=1}^{5}(2i + 3)$ **22.** $\displaystyle\sum_{i=2}^{4} i^2$ **23.** $\displaystyle\sum_{i=1}^{4}\dfrac{1}{i + 2}$

24. $\displaystyle\sum_{i=1}^{5}\dfrac{i}{3^i}$ **25.** $\displaystyle\sum_{i=1}^{3}(x - 2i)$ **26.** $\displaystyle\sum_{i=1}^{4}\dfrac{x}{x + i}$ **27.** Geometric   **28.** Arithmetic   **29.** Arithmetic
30. Neither   **31.** Geometric   **32.** Geometric   **33.** Arithmetic   **34.** Neither   **35.** $a_n = 3n - 1$: 59
36. $a_n = 8 - 3n$: -40 **37.** 34: 160   **38.** 78: 648   **39.** $a_1 = 5, d = 4, a_{10} = 41$
40. $a_1 = 8, d = 3, a_9 = 32, S_9 = 180$ **41.** $a_1 = -4, d = -2, a_{20} = -42, S_{20} = -460$ **42.** 20,100
43. -95 **44.** $a_n = 3(2)^{n-1}, a_{20} = 3(2)^{19}$ **45.** $a_n = 5(-2)^{n-1}, a_{16} = 5(-2)^{15}$
46. $a_n = 4(\frac{1}{2})^{n-1}, a_{10} = 4(\frac{1}{2})^9 = \frac{1}{128}$ **47.** -3 **48.** 8   **49.** $a_1 = 3, r = 2, a_6 = 96$ **50.** $243\sqrt{3}$
51. 28   **52.** 35   **53.** 20   **54.** 36   **55.** 45   **56.** 161,700   **57.** $x^4 - 8x^3 + 24x^2 - 32x + 16$
58. $16x^4 + 96x^3 + 216x^2 + 216x + 81$ **59.** $27x^3 + 54x^2y + 36xy^2 + 8y^3$
60. $x^{10} - 10x^8 + 40x^6 - 80x^4 + 80x^2 - 32$ **61.** $\frac{1}{16}x^4 + \frac{3}{2}x^3 + \frac{27}{2}x^2 + 54x + 81$
62. $\frac{1}{27}x^3 - \frac{1}{6}x^2y + \frac{1}{4}xy^2 - \frac{1}{8}y^3$ **63.** $x^{10} + 30x^9y + 405x^8y^2$ **64.** $x^9 - 27x^8y + 324x^7y^2$
65. $x^{11} + 11x^{10}y + 55x^9y^2$ **66.** $x^{12} - 24x^{11}y + 264x^{10}y^2$ **67.** $x^{16} - 32x^{15}y$ **68.** $x^{32} + 64x^{31}y$

69. $x^{50} - 50x^{49}$ **70.** $x^{150} + 150x^{149}y$ **71.** $\dfrac{10!}{5!5!}x^5(-3)^5 = 252x^5(-243) = -61,236x^5$

72. $\dfrac{9!}{3!6!}(2x)^6 = 84(64x^6) = 5,376x^6$

CHAPTER 10 TEST

1. $-2, 1, 4, 7, 10$ **2.** $3, 7, 11, 15, 19$ **3.** $2, 5, 10, 17, 26$ **4.** $2, 16, 54, 128, 250$ **5.** $2, \frac{3}{4}, \frac{4}{9}, \frac{5}{16}, \frac{6}{25}$

6. $4, -8, 16, -32, 64$ **7.** $a_n = 4n + 2$ **8.** $a_n = 2^{n-1}$ **9.** $a_n = (\frac{1}{2})^n = 1/2^n$ **10.** $a_n = (-3)^n$

11. (a) 90   (b) 53   (c) 130   **12.** 3   **13.** ± 6 **14.** 320   **15.** 25   **16.** $S_{50} = 3(2^{50} - 1)$ **17.** $\frac{3}{4}$

18. $x^4 - 12x^3 + 54x^2 - 108x + 81$ **19.** $32x^5 - 80x^4 + 80x^3 - 40x^2 + 10x - 1$ **20.** $x^{20} - 20x^{19} + 190x^{18}$

21. $\dfrac{8!}{5!3!}(2x)^3(-3y)^5 = 56(8x^3)(-243y^5) = -108{,}864x^3y^5$

CHAPTERS 1–10: CUMULATIVE REVIEW

1. 4   **2.** $30y + 49$ **3.** $\dfrac{a^5b^4}{2}$ **4.** $\dfrac{y-3}{y-2}$ **5.** $\dfrac{1}{4}$ **6.** 3   **7.** $(a + 1)(b^3 + 6)$ **8.** $(x - 1)(8x + 3)$

9. -2 **10.** $1, \frac{1}{2}$ **11.** $-5, 2$ **12.** $-2, 4$ **13.** 8   **14.** $\dfrac{3 \pm 5i\sqrt{2}}{4}$ **15.** $\dfrac{3}{2}, \dfrac{-3 \pm 3i\sqrt{3}}{4}$

16. $\pm\dfrac{\sqrt{10}}{2}$ and $\pm\dfrac{\sqrt{3}}{3}i$ **17.**    **18.**

19.   **20.**

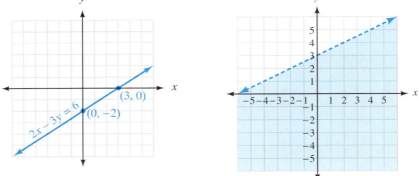

21.   **22.**

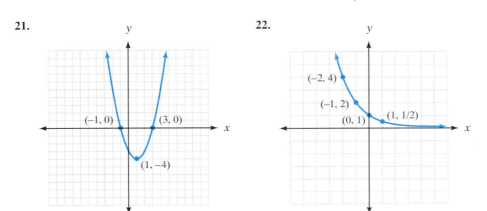

23. $(1, 3)$ **24.** $(4, -2, 3)$ **25.** $\dfrac{y(y-1)}{y+1}$ **26.** $x^{6/5} - 4$ **27.** $14 - 5i$ **28.** $\dfrac{1}{(x+4)(x+2)}$

29. $26\sqrt{2}$ **30.** $\dfrac{3\sqrt{7} + 3\sqrt{3}}{4}$ **31.** $f^{-1}(x) = \dfrac{x-1}{4}$ **32.** $9{,}519$ **33.** 1.47 **34.** $a_n = a_{n-1} + 11$

35. $a_n = \frac{1}{2}(a_{n-1})$ **36.** $y = \dfrac{S - 2x^2}{4x}$ **37.** 25 **38.** 0 **39.** $y = -2x + 9$ **40.** $C(5) = 40, C(10) = 20$

41. $20t^2 + 11t - 3$ **42.** 2 **43.** $\left(\frac{13}{22}, -\frac{1}{22}\right)$ **44.** $32x^5$ **45.** -19 **46.** If $a^2 \neq 9$, then $a \neq 3$.

47. $30°, 150°$ **48.** 6 **49.** $1, 3$ **50.** 4 gal. of 30%, 12 gal. of 70%

CHAPTER 11

PROBLEM SET 11.1

1. 5 **3.** $\sqrt{106}$ **5.** $\sqrt{61}$ **7.** $\sqrt{130}$ **9.** 3 or -1 **11.** 3 **13.** $(x-2)^2 + (y-3)^2 = 16$

15. $(x-3)^2 + (y+2)^2 = 9$ **17.** $(x+5)^2 + (y+1)^2 = 5$ **19.** $x^2 + (y+5)^2 = 1$ **21.** $x^2 + y^2 = 4$

23. Center $= (0, 0)$ **25.** Center $= (1, 3)$
Radius $= 2$ Radius $= 5$

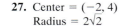

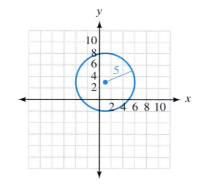

27. Center $= (-2, 4)$ **29.** Center $= (-1, -1)$
Radius $= 2\sqrt{2}$ Radius $= 1$

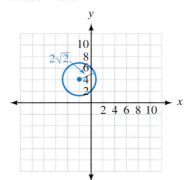

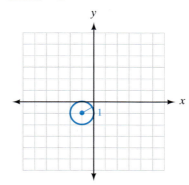

31. Center = $(0, 3)$
Radius = 4

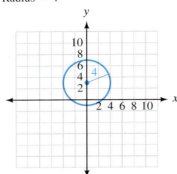

33. Center = $(-1, 0)$
Radius = $\sqrt{2}$

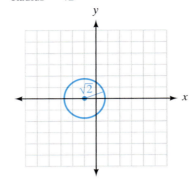

35. Center = $(2, 3)$,
Radius = 3

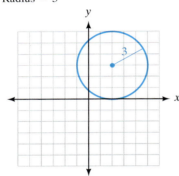

37. Center = $(-1, -\frac{1}{2})$,
Radius = 2

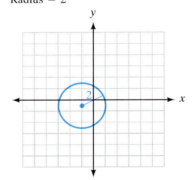

39. $(x - 3)^2 + (y - 4)^2 = 25$

41. (a) $(x - \frac{1}{2})^2 + (y - 1)^2 = \frac{1}{4}$ (b) $(x - 1)^2 + (y - 1)^2 = 1$ (c) $(x - 2)^2 + (y - 1)^2 = 4$ **43.** $x^2 + y^2 = 25$

45. $x^2 + y^2 = 9$ **47.** $(x + 1)^2 + (y - 3)^2 = 25$ **49.** $a_n = 4n + 1$ **51.** $\frac{15}{32}$ **53.** $\sum\limits_{i=1}^{5} (2i - 1)$

55. $(x - 2)^2 + (y - 3)^2 = 4$ **57.** $(x - 2)^2 + (y - 3)^2 = 4$ **59.** 5   **61.** 5

PROBLEM SET 11.2

1.

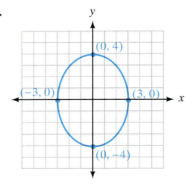

3.

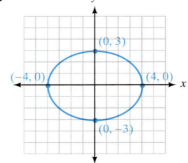

5.

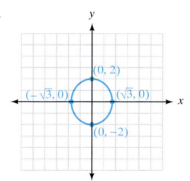

7.

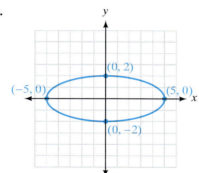

9.

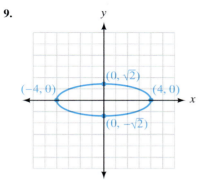

11.

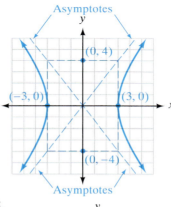

13.

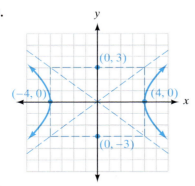

15.

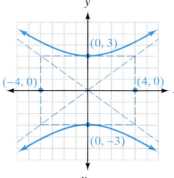

17.

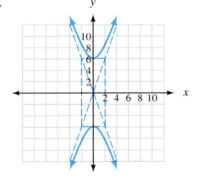

19.

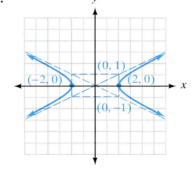

21.

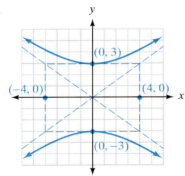

23. x-intercepts $= \pm 3$, y-intercepts $= \pm 2$

25. x-intercepts $= \pm 0.2$, no y-intercepts

27. x-intercepts $= \pm \frac{3}{5}$, y-intercepts $= \pm \frac{2}{5}$

29.

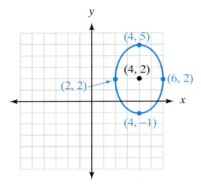

31.

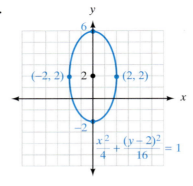

33.

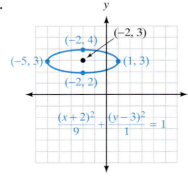

35.

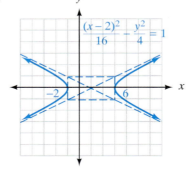

37.

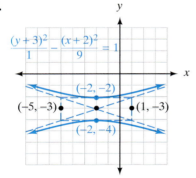

39.

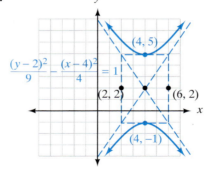

41. $\dfrac{(x-2)^2}{9} + \dfrac{(y+5)^2}{16} = 1$ **43.** $\dfrac{x^2}{25} - \dfrac{y^2}{9} = 1$ **45.** $y = \dfrac{3}{4}x,\ y = -\dfrac{3}{4}x$ **47.** 8   **49.** $a_n = 6n - 1$

51. 1,030   **53.** $\dfrac{63}{4}$

PROBLEM SET 11.3

1.

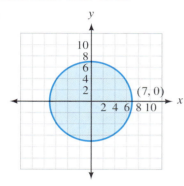

3.

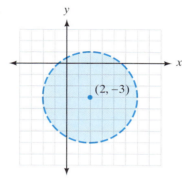

5.

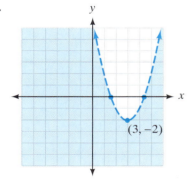

7.

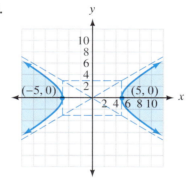

9.

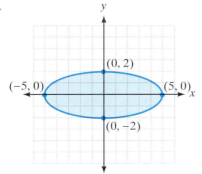

11.

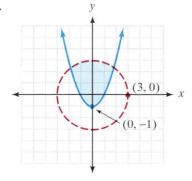

13.

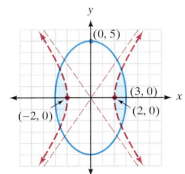

15. No intersection

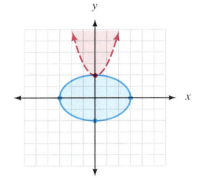

17.

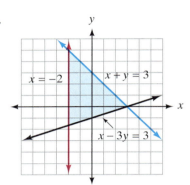

19.

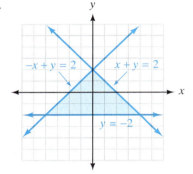

21.

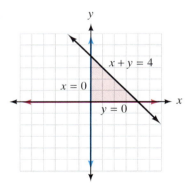

23. $(0, 3), (\frac{12}{5}, -\frac{9}{5})$ **25.** $(0, 4), (\frac{16}{5}, \frac{12}{5})$ **27.** $(5, 0), (-5, 0)$ **29.** $(0, -3), (\sqrt{5}, 2), (-\sqrt{5}, 2)$

31. $(0, -4), (\sqrt{7}, 3), (-\sqrt{7}, 3)$ **33.** $(-4, 11), (\frac{5}{2}, \frac{5}{4})$ **35.** $(-4, 5), (1, 0)$ **37.** $(2, -3), (5, 0)$

39. $(3, 0), (-3, 0)$ **41.** $(4, 0), (0, -4)$ **43.** $8, 5$ or $-8, -5$ or $8, -5$ or $-8, 5$ **45.** $6, 3$ or $13, -4$

47. $x^4 + 8x^3 + 24x^2 + 32x + 16$ **49.** $8x^3 + 12x^2y + 6xy^2 + y^3$ **51.** $x^{50} + 150x^{49}$

CHAPTER 11 REVIEW

1. $\sqrt{10}$ **2.** $\sqrt{13}$ **3.** 5 **4.** 9 **5.** $-2, 6$ **6.** $-12, 4$ **7.** $(x - 3)^2 + (y - 1)^2 = 4$
8. $(x - 3)^2 + (y + 1)^2 = 16$ **9.** $(x + 5)^2 + y^2 = 9$ **10.** $(x + 3)^2 + (y - 4)^2 = 18$ **11.** $x^2 + y^2 = 25$

12. $x^2 + y^2 = 9$ **13.** $(x + 2)^2 + (y - 3)^2 = 25$ **14.** $(x + 6)^2 + (y - 8)^2 = 100$

15. $(0, 0); r = 2$ **16.** $(3, -1); r = 4$

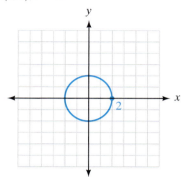

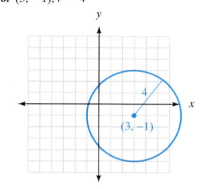

17. $(3, -2); r = 3$ **18.** $(-2, 1); r = 3$

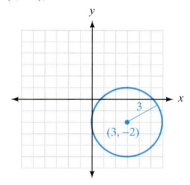

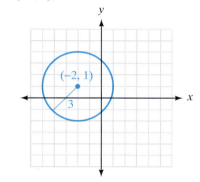

19.

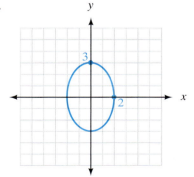

20.

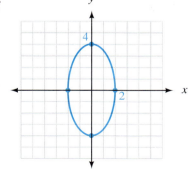

21.

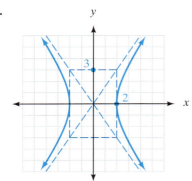

22.

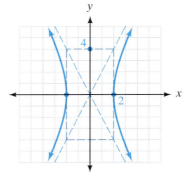

23.

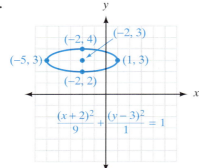

$$\frac{(x+2)^2}{9} + \frac{(y-3)^2}{1} = 1$$

24.

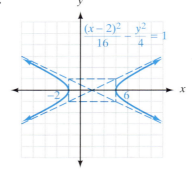

$$\frac{(x-2)^2}{16} - \frac{y^2}{4} = 1$$

25.

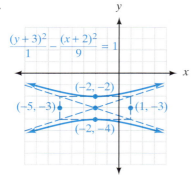

$$\frac{(y+3)^2}{1} - \frac{(x+2)^2}{9} = 1$$

26.

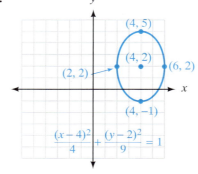

$$\frac{(x-4)^2}{4} + \frac{(y-2)^2}{9} = 1$$

27.

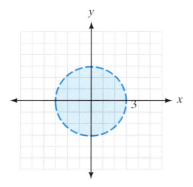

28.

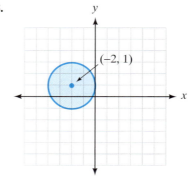

29.

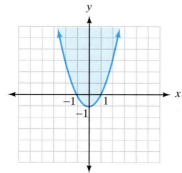

30.

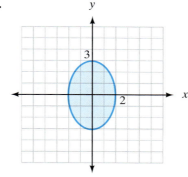

31.

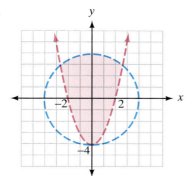

32.

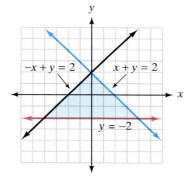

33. $(0, 4), (\frac{16}{5}, -\frac{12}{5})$ **34.** $(0, -2), (\sqrt{3}, 1), (-\sqrt{3}, 1)$ **35.** $(-2, 0), (2, 0)$

36. $\left(-\sqrt{7}, -\frac{\sqrt{6}}{2}\right), \left(-\sqrt{7}, \frac{\sqrt{6}}{2}\right), \left(\sqrt{7}, -\frac{\sqrt{6}}{2}\right), \left(\sqrt{7}, \frac{\sqrt{6}}{2}\right)$

CHAPTER 11 TEST

1. -5 and 3   **2.** $(x + 2)^2 + (y - 4)^2 = 9$ **3.** $x^2 + y^2 = 25$ **4.** Center $= (5, -3)$, Radius $= \sqrt{39}$

5.

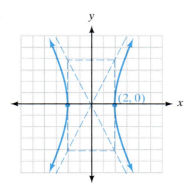

6.

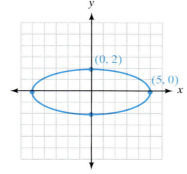

7.

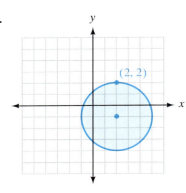

8. $\dfrac{(x-4)^2}{4} + \dfrac{(y-2)^2}{9} = 1$

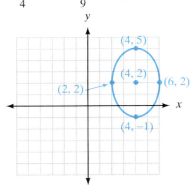

9. $(0, 5), (4, -3)$ **10.** $(0, -4), (\sqrt{7}, 3), (-\sqrt{7}, 3)$

CHAPTERS 1–11: CUMULATIVE REVIEW

1. 29   **2.** -5 **3.** $-2x - 15$ **4.** $2x + 20$ **5.** $24y$ **6.** $-\dfrac{1}{25}$ **7.** $x^{13/15}$ **8.** $4x^2 y \sqrt{3xy}$ **9.** 1

10. 2   **11.** $-\dfrac{7}{3}, 3$ **12.** $-\dfrac{1}{2}, 3$ **13.** $-2, 2, 3$ **14.** 28   **15.** Possible solutions 0 and 7; only 7 checks.

16. $3 \pm i\sqrt{3}$ **17.** $-\dfrac{3}{4} \pm \dfrac{i\sqrt{11}}{4}$ **18.** 8   **19.** $\dfrac{2}{5}$ **20.** $-\dfrac{7}{3}$ **21.** $(-1, 4)$ **22.** $(-1, -1)$

23. $(-1, 2, -1)$ **24.** $\dfrac{x^2 - x - 12}{x^3 + 2x^2}$ **25.** $x^{2/5} - 9$ **26.** $3xy^2 - 4x + 2y^2$ **27.** $-\dfrac{1}{5} - \dfrac{8}{5}i$

28.

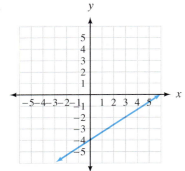

29.

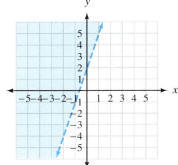

30.

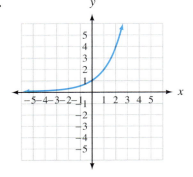

31.

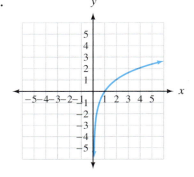

32.

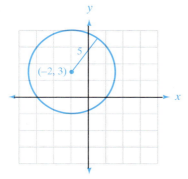

33.

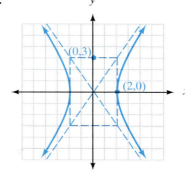

34.

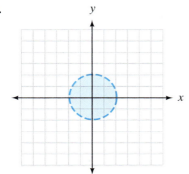

35.

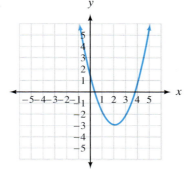

36. -16 **37.** Hypothesis: $x = -2$, Conclusion: $x^3 = 8$ **38.** $-\dfrac{5}{m - n}$ **39.** -5

40. Slope $= \frac{2}{3}$; y-intercept $= -4$ **41.** $y = -\frac{4}{3}x - 9$ **42.** $\frac{3}{4}$ **43.** $\sqrt{65}$

44. 4 gallons of 60%, 12 gallons of 20%; the equations are $x + y = 16$ and $0.2x + 0.6y = 0.3(16)$ **45.** 1, 8

46. $-10x^4y$ **47.** 0.5 second and 2.5 seconds   **48.** $3x^2 - 5x + 2 = 0$

49. $x < -3$ or $x > 2$

50. Domain $= \{x \mid x \neq 2\}$ **51.** $a_n = 5 \cdot 3^{n-1}$ **52.** 35

Index